Geophysical Monograph Series

Including

IUGG Volumes

Maurice Ewing Volumes

Mineral Physics Volumes

GEOPHYSICAL MONOGRAPH SERIES

Geophysical Monograph Volumes

1. Antarctica in the International Geophysical Year *A. P. Crary, L. M. Gould, E. O. Hulburt, Hugh Odishaw, and Waldo E. Smith (Eds.)*
2. Geophysics and the IGY *Hugh Odishaw and Stanley Ruttenberg (Eds.)*
3. Atmospheric Chemistry of Chlorine and Sulfur Compounds *James P. Lodge, Jr. (Ed.)*
4. Contemporary Geodesy *Charles A. Whitten and Kenneth H. Drummond (Eds.)*
5. Physics of Precipitation *Helmut Weickmann (Ed.)*
6. The Crust of the Pacific Basin *Gordon A. Macdonald and Hisashi Kuno (Eds.)*
7. Antarctic Research: The Matthew Fontaine Maury Memorial Symposium *H. Wexler, M. J. Rubin, and J. E. Caskey, Jr. (Eds.)*
8. Terrestrial Heat Flow *William H. K. Lee (Ed.)*
9. Gravity Anomalies: Unsurveyed Areas *Hyman Orlin (Ed.)*
10. The Earth Beneath the Continents: A Volume of Geophysical Studies in Honor of Merle A. Tuve *John S. Steinhart and T. Jefferson Smith (Eds.)*
11. Isotope Techniques in the Hydrologic Cycle *Glenn E. Stout (Ed.)*
12. The Crust and Upper Mantle of the Pacific Area *Leon Knopoff, Charles L. Drake, and Pembroke J. Hart (Eds.)*
13. The Earth's Crust and Upper Mantle *Pembroke J. Hart (Ed.)*
14. The Structure and Physical Properties of the Earth's Crust *John G. Heacock (Ed.)*
15. The Use of Artificial Satellites for Geodesy *Soren W. Henricksen, Armando Mancini, and Bernard H. Chovitz (Eds.)*
16. Flow and Fracture of Rocks *H. C. Heard, I. Y. Borg, N. L. Carter, and C. B. Raleigh (Eds.)*
17. Man-Made Lakes: Their Problems and Environmental Effects *William C. Ackermann, Gilbert F. White, and E. B. Worthington (Eds.)*
18. The Upper Atmosphere in Motion: A Selection of Papers With Annotation *C. O. Hines and Colleagues*
19. The Geophysics of the Pacific Ocean Basin and Its Margin: A Volume in Honor of George P. Woollard *George H. Sutton, Murli H. Manghnani, and Ralph Moberly (Eds.)*
20. The Earth's Crust: Its Nature and Physical Properties *John C. Heacock (Ed.)*
21. Quantitative Modeling of Magnetospheric Processes *W. P. Olson (Ed.)*
22. Derivation, Meaning, and Use of Geomagnetic Indices *P. N. Mayaud*
23. The Tectonic and Geologic Evolution of Southeast Asian Seas and Islands *Dennis E. Hayes (Ed.)*
24. Mechanical Behavior of Crustal Rocks: The Handin Volume *N. L. Carter, M. Friedman, J. M. Logan, and D. W. Stearns (Eds.)*
25. Physics of Auroral Arc Formation *S.-I. Akasofu and J. R. Kan (Eds.)*
26. Heterogeneous Atmospheric Chemistry *David R. Schryer (Ed.)*
27. The Tectonic and Geologic Evolution of Southeast Asian Seas and Islands: Part 2 *Dennis E. Hayes (Ed.)*
28. Magnetospheric Currents *Thomas A. Potemra (Ed.)*
29. Climate Processes and Climate Sensitivity (Maurice Ewing Volume 5) *James E. Hansen and Taro Takahashi (Eds.)*
30. Magnetic Reconnection in Space and Laboratory Plasmas *Edward W. Hones, Jr. (Ed.)*
31. Point Defects in Minerals (Mineral Physics Volume 1) *Robert N. Schock (Ed.)*
32. The Carbon Cycle and Atmospheric CO_2: Natural Variations Archean to Present *E. T. Sundquist and W. S. Broecker (Eds.)*
33. Greenland Ice Core: Geophysics, Geochemistry, and the Environment *C. C. Langway, Jr., H. Oeschger, and W. Dansgaard (Eds.)*
34. Collisionless Shocks in the Heliosphere: A Tutorial Review *Robert G. Stone and Bruce T. Tsurutani (Eds.)*
35. Collisionless Shocks in the Heliosphere: Reviews of Current Research *Bruce T. Tsurutani and Robert G. Stone (Eds.)*
36. Mineral and Rock Deformation: Laboratory Studies —The Paterson Volume *B. E. Hobbs and H. C. Heard (Eds.)*
37. Earthquake Source Mechanics (Maurice Ewing Volume 6) *Shamita Das, John Boatwright, and Christopher H. Scholz (Eds.)*
38. Ion Acceleration in the Magnetosphere and Ionosphere *Tom Chang (Ed.)*
39. High Pressure Research in Mineral Physics (Mineral Physics Volume 2) *Murli H. Manghnani and Yasuhiko Syono (Eds.)*
40. Gondwana Six: Structure, Tectonics, and Geophysics *Gary D. McKenzie (Ed.)*
41. Gondwana Six: Stratigraphy, Sedimentology, and Paleontology *Garry D. McKenzie (Ed.)*
42. Flow and Transport Through Unsaturated Fractured Rock *Daniel D. Evans and Thomas J. Nicholson (Eds.)*
43. Seamounts, Islands, and Atolls *Barbara H. Keating, Patricia Fryer, Rodey Batiza, and George W. Boehlert (Eds.)*

44 Modeling Magnetospheric Plasma   *T. E. Moore and J. H. Waite, Jr. (Eds.)*

45 Perovskite: A Structure of Great Interest to Geophysics and Materials Science   *Alexandra Navrotsky and Donald J. Weidner (Eds.)*

46 Structure and Dynamics of Earth's Deep Interior (IUGG Volume 1)   *D. E. Smylie and Raymond Hide (Eds.)*

47 Hydrological Regimes and Their Subsurface Thermal Effects (IUGG Volume 2)   *Alan E. Beck, Grant Garven, and Lajos Stegena (Eds.)*

48 Origin and Evolution of Sedimentary Basins and Their Energy and Mineral Resources (IUGG Volume 3)   *Raymond A. Price (Ed.)*

49 Slow Deformation and Transmission of Stress in the Earth (IUGG Volume 4)   *Steven C. Cohen and Petr Vaníček (Eds.)*

50 Deep Structure and Past Kinematics of Accreted Terranes (IUGG Volume 5)   *John W. Hillhouse (Ed.)*

51 Properties and Processes of Earth's Lower Crust (IUGG Volume 6)   *Robert F. Mereu, Stephan Mueller, and David M. Fountain (Eds.)*

52 Understanding Climate Change (IUGG Volume 7)   *Andre L. Berger, Robert E. Dickinson, and J. Kidson (Eds.)*

53 Plasma Waves and Instabilities at Comets and in Magnetospheres   *Bruce T. Tsurutani and Hiroshi Oya (Eds.)*

54 Solar System Plasma Physics   *J. H. Waite, Jr., J. L. Burch, and R. L. Moore (Eds.)*

55 Aspects of Climate Variability in the Pacific and Western Americas   *David H. Peterson (Ed.)*

56 The Brittle-Ductile Transition in Rocks   *A. G. Duba, W. B. Durham, J. W. Handin, and H. F. Wang (Eds.)*

57 Evolution of Mid Ocean Ridges (IUGG Volume 8)   *John M. Sinton (Ed.)*

58 Physics of Magnetic Flux Ropes   *C. T. Russell, E. R. Priest, and L. C. Lee (Eds.)*

59 Variations in Earth Rotation (IUGG Volume 9)   *Dennis D. McCarthy and Williams E. Carter (Eds.)*

60 Quo Vadimus Geophysics for the Next Generation (IUGG Volume 10)   *George D. Garland and John R. Apel (Eds.)*

61 Cometary Plasma Processes   *Alan D. Johnstone (Ed.)*

62 Modeling Magnetospheric Plasma Processes   *Gordon R. Wilson (Ed.)*

63 Marine Particles: Analysis and Characterization   *David C. Hurd and Derek W. Spencer (Eds.)*

64 Magnetospheric Substorms   *Joseph R. Kan, Thomas A. Potemra, Susumu Kokubun, and Takesi Iijima (Eds.)*

65 Explosion Source Phenomenology   *Steven R. Taylor, Howard J. Patton, and Paul G. Richards (Eds.)*

66 Venus and Mars: Atmospheres, Ionospheres, and Solar Wind Interactions   *Janet G. Luhmann, Mariella Tatrallyay, and Robert O. Pepin (Eds.)*

67 High-Pressure Research: Application to Earth and Planetary Sciences (Mineral Physics Volume 3)   *Yasuhiko Syono and Murli H. Manghnani (Eds.)*

68 Microwave Remote Sensing of Sea Ice   *Frank Carsey, Roger Barry, Josefino Comiso, D. Andrew Rothrock, Robert Shuchman, W. Terry Tucker, Wilford Weeks, and Dale Winebrenner*

69 Sea Level Changes: Determination and Effects (IUGG Volume 11)   *P. L. Woodworth, D. T. Pugh, J. G. DeRonde, R. G. Warrick, and J. Hannah*

70 Synthesis of Results from Scientific Drilling in the Indian Ocean   *Robert A. Duncan, David K. Rea, Robert B. Kidd, Ulrich von Rad, and Jeffrey K. Weissel (Eds.)*

71 Mantle Flow and Melt Generation at Mid-Ocean Ridges   *Jason Phipps Morgan, Donna K. Blackman, and John M. Sinton (Eds.)*

72 Dynamics of Earth's Deep Interior and Earth Rotation (IUGG Volume 12)   *Jean-Louis Le Mouël, D.E. Smylie, and Thomas Herring (Eds.)*

73 Environmental Effects on Spacecraft Positioning and Trajectories (IUGG Volume 13)   *A. Vallance Jones (Ed.)*

74 Evolution of the Earth and Planets (IUGG Volume 14)   *E. Takahashi, Raymond Jeanloz, and David Rubie (Eds.)*

75 Interactions Between Global Climate Subsystems: The Legacy of Hann (IUGG Volume 15)   *G. A. McBean and M. Hantel (Eds.)*

76 Relating Geophysical Structures and Processes: The Jeffreys Volume (IUGG Volume 16)   *K. Aki and R. Dmowska (Eds.)*

77 The Mesozoic Pacific: Geology, Tectonics, and Volcanism—A Volume in Memory of Sy Schlanger   *Malcolm S. Pringle, William W. Sager, William V. Sliter, and Seth Stein (Eds.)*

78 Climate Change in Continental Isotopic Records   *P. K. Swart, K. C. Lohmann, J. McKenzie, and S. Savin (Eds.)*

79 The Tornado: Its Structure, Dynamics, Prediction, and Hazards   *C. Church, D. Burgess, C. Doswell, R. Davies-Jones (Eds.)*

80 Auroral Plasma Dynamics   *R. L. Lysak (Ed.)*

81 Solar Wind Sources of Magnetospheric Ultra-Low Frequency Waves   *M. J. Engebretson, K. Takahashi, and M. Scholer (Eds.)*

82 **Gravimetry and Space Techniques Applied to Geodynamics and Ocean Dynamics** *Bob E. Schutz, Allen Anderson, Claude Froidevaux, and Michael Parke (Eds.)*

83 **Nonlinear Dynamics and Predictability of Geophysical Phenomena** *William I. Newman, Andrei Gabrielov, and Donald L. Turcotte (Eds.)*

84 **Solar System Plasmas in Space and Time** *J. Burch, J. H. Waite, Jr. (Eds.)*

85 **The Polar Oceans and Their Role in Shaping the Global Environment** *O. M. Johannessen, R. D. Muench, and J. E. Overland (Eds.)*

86 **Space Plasmas: Coupling Between Small and Medium Scale Processes** *Maha Ashour-Abdalla, Tom Chang, and Paul Dusenbery (Eds.)*

Maurice Ewing Volumes

1 **Island Arcs, Deep Sea Trenches, and Back-Arc Basins** *Manik Talwani and Walter C. Pitman III (Eds.)*

2 **Deep Drilling Results in the Atlantic Ocean: Ocean Crust** *Manik Talwani, Christopher G. Harrison, and Dennis E. Hayes (Eds.)*

3 **Deep Drilling Results in the Atlantic Ocean: Continental Margins and Paleoenvironment** *Manik Talwani, William Hay, and William B. F. Ryan (Eds.)*

4 **Earthquake Prediction—An International Review** *David W. Simpson and Paul G. Richards (Eds.)*

5 **Climate Processes and Climate Sensitivity** *James E. Hansen and Taro Takahashi (Eds.)*

6 **Earthquake Source Mechanics** *Shamita Das, John Boatwright, and Christopher H. Scholz (Eds.)*

IUGG Volumes

1 **Structure and Dynamics of Earth's Deep Interior** *D. E. Smylie and Raymond Hide (Eds.)*

2 **Hydrological Regimes and Their Subsurface Thermal Effects** *Alan E. Beck, Grant Garven, and Lajos Stegena (Eds.)*

3 **Origin and Evolution of Sedimentary Basins and Their Energy and Mineral Resources** *Raymond A. Price (Ed.)*

4 **Slow Deformation and Transmission of Stress in the Earth** *Steven C. Cohen and Petr Vaníček (Eds.)*

5 **Deep Structure and Past Kinematics of Accreted Terrances** *John W. Hillhouse (Ed.)*

6 **Properties and Processes of Earth's Lower Crust** *Robert F. Mereu, Stephan Mueller, and David M. Fountain (Eds.)*

7 **Understanding Climate Change** *Andre L. Berger, Robert E. Dickinson, and J. Kidson (Eds.)*

8 **Evolution of Mid Ocean Ridges** *John M. Sinton (Ed.)*

9 **Variations in Earth Rotation** *Dennis D. McCarthy and William E. Carter (Eds.)*

10 **Quo Vadimus Geophysics for the Next Generation** *George D. Garland and John R. Apel (Eds.)*

11 **Sea Level Changes: Determinations and Effects** *Philip L. Woodworth, David T. Pugh, John G. DeRonde, Richard G. Warrick, and John Hannah (Eds.)*

12 **Dynamics of Earth's Deep Interior and Earth Rotation** *Jean-Louis Le Mouël, D.E. Smylie, and Thomas Herring (Eds.)*

13 **Environmental Effects on Spacecraft Positioning and Trajectories** *A. Vallance Jones (Ed.)*

14 **Evolution of the Earth and Planets** *E. Takahashi, Raymond Jeanloz, and David Rubie (Eds.)*

15 **Interactions Between Global Climate Subsystems: The Legacy of Hann** *G. A. McBean and M. Hantel (Eds.)*

16 **Relating Geophysical Structures and Processes: The Jeffreys Volume** *K. Aki and R. Dmowska (Eds.)*

17 **Gravimetry and Space Techniques Applied to Geodynamics and Ocean Dynamics** *Bob E. Schutz, Allen Anderson, Claude Froidevaux, and Michael Parke (Eds.)*

18 **Nonlinear Dynamics and Predictability of Geophysical Phenomena** *William I. Newman, Andrei Gabrielov, and Donald L. Turcotte (Eds.)*

Mineral Physics Volumes

1 **Point Defects in Minerals** *Robert N. Schock (Ed.)*

2 **High Pressure Research in Mineral Physics** *Murli H. Manghnani and Yasuhiko Syona (Eds.)*

3 **High Pressure Research: Application to Earth and Planetary Sciences** *Yasuhiko Syono and Murli H. Manghnani (Eds.)*

Geophysical Monograph 87

The Upper Mesosphere and Lower Thermosphere: A Review of Experiment and Theory

R. M. Johnson
T. L. Killeen
Editors

American Geophysical Union

Published under the aegis of the AGU Books Board.

Library of Congress Cataloging-in-Publication Data

The upper mesosphere and lower thermosphere : a review of experiment
  and theory / R. M. Johnson, T. L. Killeen, editors.
      p.   cm. — (Geophysical monograph : 87)
   Includes bibliographical references.
   ISBN 0-87590-044-5
   1. Mesosphere—Research. 2. Thermosphere—Research.
 I. Johnson. R. M. (Robert M.), 1958—   . II. Killeen, T. L.   III. Series.
 QC881.2.M3U67   1995   .
 551.5'14—dc20   95-9945
                                                                              CIP

ISSN 0065-8448

ISBN 0-87590-044-5

This book is printed on acid-free paper. ∞

Copyright 1995 by the American Geophysical Union, 2000 Florida Avenue, NW, Washington, DC 20009, USA

Figures, tables, and short excerpts may be reprinted in scientific books and journals if the source is properly cited.

   Authorization to photocopy items for internal or personal use, or the internal or personal use of specific clients, is granted by the American Geophysical Union for libraries and other users registered with the Copyright Clearance Center (CCC) Transactional Reporting Service, provided that the base fee of $1.00 per copy plus $0.10 per page is paid directly to CCC, 222 Rosewood Dr., Danvers, MA 01923. 0065-8448/95/$01.+.10.
   This consent does not extend to other kinds of copying, such as copying for creating new collective works or for resale. The reproduction of multiple copies and the use of full articles or the use of extracts, including figures and tables, for commercial purposes requires permission from AGU.

Printed in the United States of America.

CONTENTS

Preface
R. M. Johnson and T. L. Killeen   xi

Energetics, Dynamics, and Electrodynamics of the Mesosphere and Lower Thermosphere

Energetics of the Mesosphere and Thermosphere
R. G. Roble   1

The Dynamics of the Lower Thermosphere
T. J. Fuller-Rowell   23

The Lower Ionosphere at High Latitudes
R. W. Schunk and J. J. Sojka   37

The Ionospheric Wind Dynamo: Effects of Its Coupling with Different Atmospheric Regions
A. D. Richmond   49

Planetary Waves, Tides, and Gravity Waves in the Upper Mesosphere and Lower Thermosphere

Tidal and Planetary Waves
J. M. Forbes   67

Gravity Wave Forcing and Effects in the Mesosphere and Lower Thermosphere
D. C. Fritts   89

Observations of the Meridional Quasi Two-Day Wave in the Mesosphere and Lower Thermosphere at Christmas Island
S. E. Palo and S. K. Avery   101

Analysis of Wave Signatures in the Equatorial Ionosphere
H. F. Parish, J. M. Forbes, and F. Kamalabadi   111

Gravity Wave-Tidal Interactions in the Middle Atmosphere: Observations and Theory
D. C. Fritts   121

Gravity Wave Mean State Interactions in the Upper Mesosphere and Lower Thermosphere
R. L. Walterscheid   133

Modulation of Gravity Wave Drag in the Mesosphere and Lower Thermosphere by Tides and the Effects on the Time Mean Flow-Model Results
C. McLandress and W. E. Ward   145

Scale-Independent Diffusive Filtering Theory of Gravity Wave Spectra in the Atmosphere
C. S. Gardner   153

An Investigation of Thunderstorms as a Source of Short Period Mesospheric Gravity Waves
M. J. Taylor, V. Taylor, and R. Edwards   177

CONTENTS

Mesopause and Lower Thermosphere/Ionosphere Phenomena

Climatology of Polar Mesospheric Clouds: Interannual Variability and Implications for Long-Term Trends
G. E. Thomas   185

Charge Balance at the Summer Polar Mesopause: Ice Particles, Electrons, and PMSE
G. C. Reid   201

Ionic Nucleation of Ice Particles in Noctilucent Clouds
T. Sugiyama   209

Changes in the Concentration of Mesospheric O_3 and OH During a Highly Relativistic Electron Precipitation Event
R. A. Goldberg, C. H. Jackman, D. N. Baker, and F. A. Herrero   215

Nitric Oxide in the Lower Thermosphere
C. Barth   225

The Role of Fast $N(^4S)$ Atoms and Energetic Photoelectrons on the Distribution of NO in the Thermosphere
J. C. Gérard, V. I. Shematovich, and D. V. Bisikalo   235

The State of O_2 in the Mesopause Region
D. P. Murtagh   243

Measurements of Electron Density Profiles in the Ionosphere Using Artificial Periodic Inhomogeneties
V. Belikovich, E. A. Benediktov, and A. V. Tolmacheva   251

Advances in Models of the Upper Mesosphere and Lower Thermosphere

Aspects of Mesospheric Simulation in a Comprehensive General Circulation Model
K. Hamilton   255

A Numerical Spectral Model for the Mean Zonal Circulation and the Tides in the Middle and Upper Atmosphere
K. L. Chan, H. G. Mayr, J. G. Mengel, and I. Harris   265

Energy Spectrum of the NCAR-TIGCM Lower Thermosphere
R. G. Raskin, A. G. Burns, T. L. Killeen, and R. G. Roble   279

SHARC, a Model for Calculating Atmospheric Infrared Radiation Under Non-Equilibrium Conditions
R. L. Sundberg, J. W. Duff, J. H. Gruninger, L. S. Bernstein, M. W. Matthew, S. M. Adler-Golden, D. C. Robertson, R. D. Sharma, J. H. Brown, and R. J. Healey   287

CONTENTS

New Space-Borne Results

WINDII on UARS—Status Report and Preliminary Results
G. G. Shepherd, G. Thuillier, B. H. Solheim, Y. J. Rochon, J. Criswick, W. A. Gault, R. N. Peterson, R. H. Wiens, and S.-P. Zhang   297

Preliminary Results from the Imaging Spectrometric Observatory Flown on ATLAS 1
D. G. Torr, M. R. Torr, M. F. Morgan, T. Chang, J. K. Owens, J. A. Fennelly, P. G. Richards, and T. W. Baldridge   305

Spatial Variability in $O(^1S)$ and $O_2\ (b^1\Sigma_g^+)$ Emissions as Observed with the Wind Imaging Interferometer (WINDII) on UARS
W. E. Ward, E. J. Llewellyn, Y. Rochon, C. C. Tai, W. S. C. Brooks, B. H. Solheim, and G. G. Shepherd   323

A Sequential Estimation Technique for Recovering Atmospheric Data from Orbiting Satellites
D. A. Ortland, P. B. Hays, W. R. Skinner, M. D. Burrage, A. R. Marshall, and D. A. Gell   329

Satellite Observations of Neutral Density Cells in the Lower Thermosphere at High Latitudes
G. Crowley, J. Schoendorf, R. G. Roble, and F. A. Marcos   339

Recommended Drag Coefficients for Aeronomic Satellites
M. M. Moe, S. D. Wallace, and K. Moe   349

PREFACE

This volume provides a review of progress made in recent years in experimental and theoretical investigation of the upper mesosphere and lower thermosphere and coupling between these regions and the ionosphere. Detailed study of the mesosphere/lower thermosphere/ionosphere (MLTI) region has historically been difficult because of its relative inaccessibility to direct measurement techniques and the complex and highly coupled processes which occur there. Although we have still not successfully unraveled all these complex interactions, we have made significant recent progress toward a fuller understanding of the basic state of the MLTI and of the dominant wave and coupling processes. This monograph includes a set of tutorial papers, which review our current understanding of aspects of the MLTI. These tutorials are interspersed with a selection of papers describing research progress on various topics of current interest in this region. The book should therefore be useful both to the newcomer, as an introduction to this field of research, and to the more experienced researcher, providing an overview of research in progress as well as a convenient reference collection of papers describing our current understanding.

In contrast to the lower atmosphere, the upper mesosphere, lower thermosphere and ionosphere are strongly driven by a multitude of external sources. These include forcing by wave activity penetrating upward from the lower atmosphere, solar EUV and UV radiation, auroral and energetic particle precipitation, and magnetospheric plasma convection. Heating, dissociation, and ionization result from the action of these sources on the coupled system, which in turn drive the global wind system, lead to increases in electrical conductivity, and induce chemical changes in neutral and ionized species. Magnetospheric convection drives ionospheric plasma into motion and, through collisions with neutrals, provides momentum and energy to the system via ion drag and Joule heating. At low and middle latitudes, ion drag acts as a break to the dynamo wind field, generates polarization electric fields, and drives horizontal and field-aligned currents.

Tides, planetary waves, and gravity waves, forced either in situ or at lower levels, all participate in and frequently dominate the dynamics of the region. These waves can dissipate within the region, where they deposit momentum and energy, and mix atmospheric species or nonlinearly interact with each other or with the mean flow, generating additional wave activity. The wind and wave fields also transport chemically and radiatively active species, such as NO, O, CO_2, O_3 and water vapor. These species affect the energy budget either by radiating to space or to adjacent atmospheric layers, or by absorbing solar radiation. In addition to these complex interacting processes, the possibility of anthropogenic effects on the mesosphere has been raised to explain the increasing frequency of occurrence of "noctilucent" clouds at the summer polar mesopause - the coldest spot in the Earth's atmosphere.

The first section of the book includes tutorials which provide an overview of the energetics and dynamics of the mesosphere, lower thermosphere, and ionosphere and the electrodynamic coupling between these regions. The second section focuses on wave phenomena occurring within the mesosphere and lower thermosphere. Tutorials review our understanding of tides and planetary waves, gravity waves and their effects, and coupling of gravity waves with tides and the mean state. Research papers present observational and theoretical results on mesospheric and lower thermospheric wave phenomena, including studies of the "2-day" wave, gravity wave coupling and filtering, and possible sources of mesospheric gravity waves. Some of the more interesting aspects of the MLTI transition region are the topic of the third section of the book. Tutorial papers in this section describe the climatology of polar mesospheric clouds, charge balance at the mesopause, and the roles of nitric oxide and molecular oxygen in the lower thermosphere and upper mesosphere. Research papers include studies of the formation of noctilucent clouds, the effects of highly relativistic electron precipitation on mesospheric ozone and hydroxyl, and ionospheric measurements using artificial periodic inhomogeneities. The fourth section presents recent advances in models of the upper mesosphere and lower thermosphere. Tutorial papers describe models of the upper and middle atmosphere, and research papers present an infrared radiation model and an analysis of the energy spectrum of the NCAR-TIGCM. The final section of the book focuses on satellite observations, including results for UARS and ATLAS-1 and satellite data analysis techniques.

R. M. Johnson and T. L. Killeen
The University of Michigan
Ann Arbor

ACKNOWLEDGMENTS

The papers published here are based on presentations made at the AGU Chapman conference on the Upper Mesosphere and Lower Thermosphere held on November 16-20, 1992, at Asilomar, California. This conference and book would not have been possible without the support and cooperation of many scientists. The conveners of the conference were assisted by a Program Committee consisting of S. Avery and J. M. Forbes of the University of Colorado; A. Manson of the University of Saskatchewan; I. McDade of York University; J. Meriwether of Clemson University; D. Rees of University College London; R. G. Roble of the NCAR High Altitude Observatory; and R. Walterscheid of the Aerospace Corporation. We gratefully acknowledge the funds provided by the National Science Foundation and the National Aeronautics and Space Agency in support of this conference. Finally, we would like to acknowledge and thank the following scientists, among others, who served as referees for the papers submitted to this volume.

D. Baker
B. Balsley
A. Burns
M. Burrage
J. Cho
T. Cravens
F. Djuth
R. F. Donnelly
S. R. Drayson
C. Fesen
J. Forbes
D. Fritts
T. J. Fuller-Rowell
R. Garcia
C. Gardner
M. Geller
K. Hamilton
P. Hays
J. H. Hecht
A. Hedin
R. Heelis

G. Hernandez
F. Herrero
C. Hines
M. Kelley
M. Keskinen
S. Kirkwood
M. Larsen
J. Liu
A. Manson
F. Marcos
H. Mayr
I. McDade
J. Meriwether
I. Mikkelsen
M. Mlynczak
R. Niciejewski
J. J. Olivero
H. Parish
R. Pfaff
D. Rees
M. Rees

G. Reid
A. Richmond
R. Roble
C. Rodgers
J. Salah
D. Siskind
G. G. Sivjee
T. Slanger
R. Smith
S. Solomon
S. C. Solomon
M. F. Storz
G. Swensen
E. Szuszczewicz
J. Thayer
G. E. Thomas
T. E. VanZandt
F. Vial
U. von Zahn
R. L. Walterscheid
S. Zasadil

R. M. Johnson and T. L. Killeen
The University of Michigan
Ann Arbor

Energetics of the Mesosphere and Thermosphere

Raymond G. Roble

High Altitude Observatory, National Center for Atmospheric Research, Boulder, Colorado

A brief review of the main physical and chemical processes that affect the energy balance of the thermosphere and mesosphere is presented first. These processes are then assimilated into a globally averaged model of the coupled mesosphere, thermosphere, and ionosphere to quantitatively evaluate our understanding of the aeronomy of these regions and the ability of the model to calculate a thermal and compositional structure consistent with observations. Starting from arbitrary initial conditions and specified boundary conditions at 30 km altitude, the model calculates the globally averaged temperature and compositional structure of the upper atmosphere between 30 and 500 km altitude. The model solves for the neutral, electron and ion temperature profiles as well as for the compositional profiles of $O_x = (O + O_3)$, O_2, and N_2 coupled through major species diffusion equations. It also includes as minor species with transport and appropriate photochemistry $N(^4S)$, $NO_x = (NO + NO_2)$, H_2O, H_2, CH_4, CO, CO_2, and $HO_x = (H + OH + HO_2)$, and in photochemical equilibrium with the above (including O_x, NO_x, and HO_x family partitioning), O, O_3, NO, NO_2, $N(^2D)$, $O(^1D)$, $O_2(^1\Sigma_g)$, $O_2(^1\Delta_g)$, OH, H, HO_2, and H_2O_2. The distribution of the passive tracers He and Ar are also calculated. The model also includes appropriate F-, E-, D-region and upper stratospheric ion chemistry.

Using a measured solar spectrum for solar minimum conditions, the model calculated temperature and compositional profiles for O, O_2, N_2, H, He, Ar, and $N(^4S)$ were adjusted so that reasonable agreement with similar profiles from the MSIS-90 empirical model were obtained. This was achieved by crudely adjusting eddy thermal and compositional diffusion vertical profiles. The results of the calculation suggest that the global mean atmosphere above 30 km can be reasonably modeled considering our present day knowledge of aeronomic processes. The calculated component heating and cooling rates contributing to the total rates are presented throughout the altitude range illustrating the changing physical and chemical processes responsible for the maintenance of the global mean energy and compositional balance.

1. INTRODUCTION

The global mean temperature structure of the mesosphere and thermosphere is governed primarily by the absorption of solar UV and EUV radiation, auroral heating by particles and Joule dissipation of ionospheric currents, release of chemical energy and dissipation of wave disturbances propagating upward from the lower atmosphere balanced by molecular and eddy heat conduction, and radiative cooling mainly from CO_2, NO, O_3, and O infrared emissions. Previous studies of the energetics of this atmospheric region have concentrated either on the thermospheric structure [*Roble and Dickinson*, 1973; *Gordiets et al.*, 1982; *Roble and Emery*, 1983; *Roble et al.*, 1987] or on mesospheric structure [*Park and London*, 1974; *London*, 1980; *Appruzese et al.*, 1984; *Fomichev and Shved*, 1988; *Kiehl and Solomon*, 1986; *Fomichev et al.*, 1993]. Lower boundaries for the thermosphere models and upper boundaries for the mesosphere models have usually been specified in the lower thermosphere near 80–120 km. This can be an awkward region for the specification of boundary conditions on chemical and energy balance models because constituent and heat transport occur readily

The Upper Mesosphere and Lower Thermosphere:
A Review of Experiment and Theory
Geophysical Monograph 87
Copyright 1995 by the American Geophysical Union

between the thermosphere and mesosphere.

There are a number of uncertainties in the various physical, chemical, radiative and transport processes that occur in the mesopause region: (1) chemical models appear to underestimate the ozone number density in the mesosphere by about a factor of two [e.g., *Solomon et al.*, 1983; *Allen et al.*, 1984; *Rusch and Eckman*, 1985; *Clancy et al.*, 1987; *Bates*, 1988]; (2) microwave measurements of water vapor and carbon monoxide seem to imply low eddy diffusion transport coefficients in the upper mesosphere [*Strobel et al.*, 1987]; (3) vertical eddy diffusion coefficients derived from momentum stress, constituent transport and heat transport suggest a large value (~ 10) of the eddy Prandtl number in the mesosphere, defined as the ratio of the momentum diffusion coefficient divided by the thermal diffusion coefficient [*Strobel et al.*, 1985, 1987]; (4) CO_2 cooling in the upper mesosphere is complicated by both local thermodynamic equilibrium (LTE) and non-LTE effects, importance of hot bands and isotopic bands, complex line shapes, radiative transfer processes and an uncertainty in the magnitude of the energy exchange rate between O and $CO_2(\nu_2 = 1)$ [*Dickinson*, 1984; *Dickinson et al.*, 1987; *Rogers et al.*, 1992; *Fomichev et al.*, 1994; *Lopez-Puertas et al.*, 1993]; (5) the magnitude of the solar EUV and UV radiation and its long and short term variations [e.g., *Lean*, 1987; *Donnelly et al.*, 1986; *Torr and Torr*, 1985]; and (6) complex circulations, wave-mean flow interactions and turbulence may introduce complex transport processes and non-linear couplings [*Garcia and Solomon*, 1985; *Brasseur and Offerman*, 1986; *Akmaev*, 1992; *Ward and Fomichev*, 1993].

A brief review of the energetics of the coupled mesosphere-thermosphere system is presented first. The important energetics of this region are then incorporated into a self-consistent global mean model of coupled mesosphere, thermosphere, and ionosphere system and the model run to steady state to determine the magnitude of the component processes responsible for the maintenance of the global mean thermal and compositional structure. The model is an extension of the global mean model of the thermosphere developed by *Roble et al.* [1987] and the global mean model of the mesosphere-thermosphere (60–500 km) developed by *Roble and Dickinson* [1989]. The model in this paper has been extended further, downward to 10 mb (30 km) to better resolve processes in the vicinity of the stratopause. The 10 mb pressure surface was selected so as to move the boundary sufficiently far from the stratopause region to avoid possible boundary layers that could affect calculation of the stratopause temperature. The motivation behind this model is not only to examine the energetics of the global mean structure, but also to use the model as a testbed for the development of a self-consistent aeronomic scheme that can then be incorporated into a new version of the NCAR thermosphere-ionosphere-electrodynamics general circulation model (TIE-GCM) [*Richmond et al.*, 1992] that extends downward to 10 mb (30 km altitude).

2. ENERGETICS OF THE MESOSPHERE AND THERMOSPHERE

At wavelengths of less than 200 nm, the absorption cross-section of atmospheric gases is so large that, even given the low densities of species in the upper mesosphere and thermosphere, all of the incident solar flux below 200 nm wavelength is completely absorbed above 80 km altitude. Just how the absorbed solar energy is then channeled through the myriad of atomic and molecular chemical and physical processes that can occur has been a subject of considerable interest for many years. It is necessary to understand the processes by which the absorbed solar energy is redistributed into radiated airglow, stored chemical energy, and direct heating of the neutral gas in order to understand the energy balance of the upper atmosphere and the global distribution of heating that drives the atmospheric circulation.

Virtually all solar photons at wavelengths less that 102.5 nm are absorbed by the major constituents of the thermosphere, N_2, O_2, and O, leading to ionization of the absorbing species. The photoionization event splits the energy of the absorbed photon approximately equally into two channels shown schematically in Figure 1; the kinetic energy of the ejected fast photoelectron and the chemical energy of the ion product. The latter can be regained in recombination with an electron. The ejected fast photoelectron slows down by Coulomb collisions with the ambient thermal electrons of the ionosphere and by elastic and inelastic collisions with the neutral particles. Fast-photoelectron/thermal-electron collisions result in a local heating of the background ionospheric electron gas, which in turn is directly transferred to the neutral gas by collisions, thus providing the thermal energy to the neutral gas. Also the fast photoelectrons may be sufficiently energetic that on collisions with neutrals, they can cause further ionization, creating more electron-ion pairs, or they may excite internal atomic levels of an atom or molecule. These excited levels can then be either deactivated by collisions with neutral particles, the excess energy appearing as local

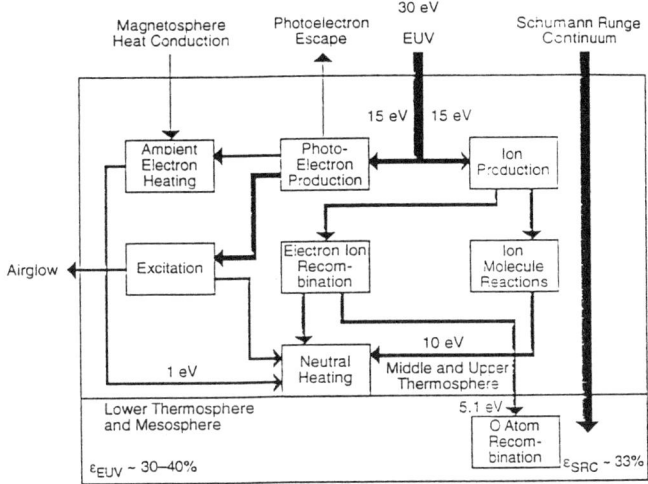

Fig. 1. Energy flow diagram of the processes leading to the conversion of absorbed solar EUV radiation energy into local thermal energy of the neutral gas in the thermosphere. The width of the arrow illustrates the relative energy in each process.

thermal energy of the particles, or, if the optical transition probability is sufficiently high radiated as airglow. The airglow can be either lost by radiation to space or reabsorbed – possibly causing dissociation or additional ionization, depending on the wavelength and the optical depth.

The chemical energy of the ion, on the other hand, is shuffled from one ion to another by a series of charge transfer or ion-molecule interchange reactions, eventually ending in a dissociative recombination yielding two atoms of oxygen or nitrogen [*Stolarski*, 1976]. The N atoms participate in local neutral chemistry that leads to the production of O atoms by breaking the molecular oxygen bond, with the N atoms ending up as nitric oxide, NO, or molecular nitrogen, N_2. The O atoms do not recombine locally to O_2 because the reaction for atomic oxygen recombination is a three-body reaction that requires a high neutral particle density to proceed rapidly. Atomic oxygen is transferred by molecular and eddy diffusion downward to about 90 km. As a result, the energy required to dissociate O_2 is lost from the local region of EUV absorption, as it is transported to lower altitudes and released there through recombination, reappearing as thermal energy below 90 km.

Transport of the photoelectron or ion species can further complicate the heating process because the created products may move away from the region of solar energy absorption and participate in either a collisional process or a chemical reaction at some other altitude.

Although the details of the neutral gas heating are complex, a rather simple argument can be made to show that the direct neutral gas heating efficiency (defined as the heat directly deposited at a given location through fast atomic processes to the energy of the solar radiation absorbed at that location) is 30–40% of the absorbed solar radiation. The average solar EUV photon energy is approximately 30 eV. The initial photoionization event splits this energy into two channels – about 50% or 15 eV each, for the ion formation and the fast photoelectron. Of the 15 eV potential energy of recombination in the ion formation, 5 eV – the binding energy of O_2 – is transported by atomic oxygen downward to the lower thermosphere, where the recombination to O_2 occurs and the binding energy is released. The remaining 10 eV provides direct neutral heating near the altitude where the photon was absorbed. Of the 15 eV that goes into the photoelectron channel, only about 1 eV heats the ambient electron gas, and this eventually appears as neutral heating because of collisional processes. The rest of the energy goes into atomic and molecular excitation that radiates as airglow and, depending upon the radiation optical depth, may be lost as far as local heating is concerned. Combining the two channels, an overall local neutral gas heating efficiency of 36.6% is obtained. Allowing for uncertainties in the processes, a local neutral gas heating efficiency of between 30 and 40% is expected. Approximately 20% of the absorbed solar energy is transported by atomic oxygen to the lower thermosphere below 90 km, and 40 to 50% appears as airglow radiation that is either lost to space or transferred out of the thermosphere and reabsorbed in the lower atmosphere.

The above narrative gives a brief overview of the processes involved in determining the heating rate of the thermosphere and the details of the individual processes are quite complex. These details are described in papers by *Stolarski et al.* [1975] and *Stolarski* [1976]. In addition there are numerous other aeronomic studies that have contributed to our overall understanding of the flow of energy through the various physical and chemical processes that operate in the thermosphere and these are discussed in detail in the books by *Rees* [1989] and *Banks and Kockarts* [1973].

Neutral gas heating by solar EUV, as described above, dominates in the 150–300 km altitude range. Above 300 km, the same processes occur, but the neutral gas heating there is also affected by collisional processes involving electrons and ions. The electrons and ions are heated by energy flowing as thermal conduction in the plasma down along geomagnetic field lines from the magnetosphere. The neutral gas heating is thus controlled by complex magnetospheric-ionospheric coupling interactions and because the heating is deposited high in the thermosphere it may be important for establishing the magnitude of the exospheric temperature. It is relatively unimportant in controlling thermospheric dynamics because the heating that is important for dynamics occurs mainly in the 150–300 km altitude region.

Below 150 km, molecular oxygen absorption of solar ultraviolet radiation in the Schumann-Runge continuum region (130–175 nm) is the dominant heat source with a direct heating efficiency of about 33%. The absorbed energy breaks the O_2 bond, producing two atomic oxygen atoms that are transported to the lower thermosphere below 90 km, where they recombine. The average solar energy per photon in the continuum is about 7.5 eV (at 160 nm), and the excess of this energy over the O_2 dissociation energy (5 eV) results in neutral gas heating with a direct efficiency of 33%.

At lower altitudes, below about 100 km it is also important to examine the various energy channels of solar energy partitioning as discussed by *Mlynczak and Solomon* [1991, 1993]. Upon photolysis of O_3 or O_2 significant amounts of chemical potential energy and atomic and molecular internal energy are generated and *Mlynczak and Solomon* [1993] have provided a detailed review of the processes involved in determining the heating efficiency throughout the middle atmosphere. Results of their analysis indicate that the reaction of atomic hydrogen and ozone is potentially the largest single source of heat in the vicinity of the mesopause and that airglow and chemiluminescent emission significantly reduce the amount of energy available for heat throughout the mesosphere and lower thermosphere. Between 65 and 85 km, failure to account for the combined airglow and chemiluminescent energy loss would result in an overestimation of the heating rate by 5 to 20%. Below about 65 km quenching of these emissions indicate that all of the absorbed solar energy would be available for heating at unit efficiency.

The main radiative cooling processes for the thermosphere have been discussed by *Gordiets et al.* [1982] and *Roble et al.* [1987] and for the middle atmosphere by *London* [1980]. In the upper thermosphere 63 μm IR radiation from the fine structure of atomic oxygen is the dominant radiative loss process [*Bates*, 1951]. In the middle thermosphere (120–200 km) 5.3 μm non-LTE IR radiation from NO becomes important [*Kocharts*, 1980]. Below 120 km, 15 μm LTE and non-LTE CO_2 radiative loss processes dominate the cooling of the atmosphere [*Dickinson*, 1984] with a contribution from 9.6 μm IR radiation from O_3 [*London*, 1980]. There have been numerous studies of these radiative loss processes that have been summarized by *Dickinson* [1984] for the upper middle atmosphere and by *Kiehl and Solomon* [1986] for the lower middle atmosphere. The largest uncertainty in the radiative loss processes in the thermosphere and upper middle atmosphere are associated with the uncertainties in the $CO_2 - O$ collisional energy transfer rate coefficient that is discussed further in a later section.

3. SOLAR SPECTRAL IRRADIANCE

Solar EUV (5 nm to 103 nm) spectral irradiance measurements from the Atmosphere Explorer satellites have been used by *Hinteregger* [1981] to construct an empirical model of the solar radiative output for a variety of solar conditions. These spectral irradiance values have been parameterized further by *Torr et al.* [1979] into 37 wavelength intervals and solar emission lines. They also presented the absorption and ionization cross-sections of the major atmospheric constituents at the same wavelength intervals. Furthermore, *Torr and Torr* [1985] have parameterized the solar spectral irradiance variations in the Schumann-Runge continuum region (135 nm to 175 nm) into 8 wavelength intervals. These parameterizations have been used by aeronomers to calculate solar heating, dissociation and ionization rates, and photoelectron fluxes, as well as other aeronomic parameters in the thermosphere and ionosphere. The parameterized solar spectral irradiance for two levels of solar $F_{10.7}$ values representing solar minimum and solar maximum conditions are shown in Figure 2. There are 59 specified wavelength intervals and lines used to represent the solar spectra between (0.2 nm to 200 nm). The largest variation in solar flux occurs at the shortest wavelengths where there is as much as a factor of 100 variation between solar minimum and maximum conditions, decreasing to a factor 2.0–2.5 near 100 nm and then to about 10% near 200 nm. This empirical model has been adjusted to take into consideration many of the uncertainties in the measurements that have been discussed in the literature over the past decade. An overall review of the solar EUV variations, their correction factors, and their respective uncertainties has been given by *Lean* [1987]. The model of solar EUV and UV spectral irradiance shown in Figure 2 is based on recent work by S. C. Solomon (private communication, 1992). It requires, as inputs, solar $F_{10.7}$ and $\overline{F_{10.7}}$ daily values. This parameterized solar flux model is currently being used in both the global mean model of the upper atmosphere, discussed in this paper, and also in the NCAR TIGCM and TIE-GCM for radiative inputs into the upper atmosphere.

There has also been considerable uncertainty in the magnitude of the solar spectral irradiance at longer UV wavelengths as well as the spectral variations over a solar cycle and 27 day solar rotation period. The situation, however, has become much clearer in recent years because of the solar spectral irradiance measurements made by the Solar Mesosphere Explorer (SME) spacecraft [*Rottman*, 1981]. These recent satellite measurements have been assimilated into an empirical model that describes the solar spectral irradiance from Lyman-α (121.6 nm) to 800 nm and its variability in terms of solar $F_{10.7}$ and $\overline{F_{10.7}}$ [*Rottman et al.*, 1986]. This empirical model, along with the ozone panel recommended O_3 and O_2 absorption cross-sections, are used in both the global average model and in the new thermosphere-ionosphere-mesosphere-electrodynamics general circulation model (TIME-GCM) that extends between 30 and 500 km altitude [*Roble and Ridley*, 1994].

4. GLOBAL MEAN MODEL

The model of the global mean structure of the upper stratosphere, mesosphere, thermosphere, and ionosphere is an extension of the model described by *Roble et al.* [1987] for the global mean structure of the thermosphere and ionosphere above 97 km, and the model described by *Roble and Dickinson* [1989] that included both the thermosphere and ionosphere and a portion of the mesosphere down to 60 km. This new global mean model includes the same features described in the above two papers but it also includes a further extension, down to 10 mb (30 km). The global mean model now extends between 30 and 500 km and includes the appropriate thermosphere, ionosphere, and middle atmosphere physics, chemistry and one-dimensional transport. We use the same formulation as that given in the paper by *Roble et al.* [1987], hereafter referred to as Paper I, and describe only the modifications necessary to extend our independent vertical coordinate, $Z = \log_e (P_o/P)$ from -7 to -17 (approximately 97 km to 30 km). Here P_o is a reference pressure ($5 \times 10^{-4} \mu$b or 50 μPa) and P is pressure. The extension to the physics, chemistry, radiation and transport to that discussed in Paper I include the following:

4.1. Ion Chemistry

The ion chemical model presented in Paper I was designed for the NCAR thermosphere/ionosphere gen-

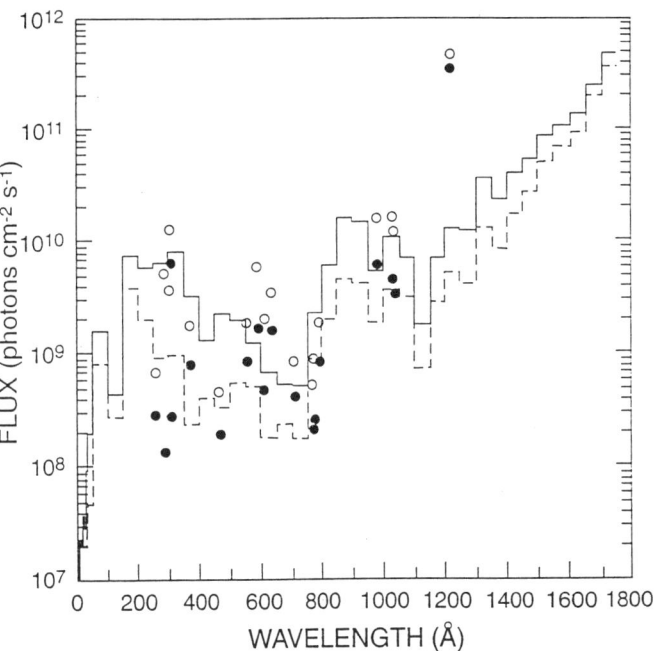

Fig. 2. Solar spectral irradiance parameterized into 59 spectral regions and emission lines for two levels of solar activity, (a) solar cycle maximum ($F_{10.7} = 243$, $\overline{F_{10.7}} = 215$) solid line and open circles, and (b) solar cycle minimum ($F_{10.7} = 67$, $\overline{F_{10.7}} = 72$) dashed line and closed circles, empirical model developed by S. C. Solomon (private communication, 1989).

eral circulation model (TIGCM) discussed by *Roble et al.* [1988]. It was a limited set of the ionospheric chemical reactions that captured the essential features of the ionospheric structure, but yet was simple enough for inclusion in a global general circulation model. This simplified chemical scheme has since been upgraded to include the chemistry of the metastable ions and additional ion chemical reactions to better describe the ionospheric structure. At the present time, the ion chemical scheme is the same as that presented in Table 1 of *Torr et al.* [1990]. The ion chemistry is now described by about 30 ion-chemical reactions that are summarized in Table 1 and this new chemical scheme has been embedded into the global mean model, the TIGCM, the TIE-GCM, and TIME-GCM that has been described by *Roble et al.* [1987], *Richmond et al.* [1992], and *Roble and Ridley* [1994], respectively. All models use the mathematical scheme presented by *Roble and Ridley* [1987] to solve for the various electron and ion distributions.

The global mean model also includes the ion chemistry code of *Reid* [1977, 1986] to calculate the distribution of ion species in the stratosphere and the D-region

TABLE 1. Chemical Reactions, Reaction Rates, and Exothermic Reaction Energy
Used in the Global Mean Model of the Thermosphere and Mesosphere

A. Nitrogen Chemistry

Reaction				Reaction Rate
(1)	$N(^4S) + O_2$	$\xrightarrow{\beta_1}$	$NO + O + 1.4 eV$	$\beta_1 = 1.5 \times 10^{-11} \exp(-3600./T_n)$
(2)	$N(^2D) + O_2$	$\xrightarrow{\beta_2}$	$NO + O(^1D) + 1.84 eV$	$\beta_2 = 5.0 \times 10^{-12}$
(3)	$N(^4S) + NO$	$\xrightarrow{\beta_3}$	$N_2 + O + 2.68 eV$	$\beta_3 = 1.6 \times 10^{-10} \exp(-460./T_n)$
(4)	$N(^2D) + O$	$\xrightarrow{\beta_4}$	$N(^4S) + O + 2.38 eV$	$\beta_4 = 4.5 \times 10^{-13}$
(5)	$N(^2D) + e$	$\xrightarrow{\beta_5}$	$N(^4S) + e + 2.38 eV$	$\beta_5 = 3.6 \times 10^{-10} (T_e/300)^{1/2}$
(6)	$N(^2D) + NO$	$\xrightarrow{\beta_6}$	$N_2 + O + 5.63 eV$	$\beta_6 = 7.0 \times 10^{-11}$
(7)	$N(^2D)$	$\xrightarrow{\beta_7}$	$N(^4S) + h\nu$	$\beta_7 = 1.06 \times 10^{-5}$
(8)	$NO + h\nu$	$\xrightarrow{\beta_8}$	$N(^4S) + O$	$\beta_8 = 4.5 \times 10^{-6}$ $\exp(-10^{-8}[N(O_2)]^{0.38})$ $\exp(-5 \times 10^{-19} N(O_3))$
(9)	$NO + h\nu/_{Ly-\alpha}$	$\xrightarrow{\beta_9}$	$NO^+ + e$	$\beta_9 = 5.88 \times 10^{-7}(1. + 0.2 (F_{10.7} - 65)/100)$ $\cdot \exp(-2.115 \times 10^{-18} \cdot N(O_2)^{0.8855})$
(10)	$N(^4S) + OH$	$\xrightarrow{\beta_{10}}$	$NO + H + 2.1 eV$	$\beta_{10} = 5 \times 10^{-11}$
(11)	$NO + O_3$	$\xrightarrow{\beta_{11}}$	$NO_2 + O_2 + 2.08 eV$	$\beta_{11} = 1.8 \times 10^{-12} \exp(-1370/T_n)$
(12)	$NO + HO_2$	$\xrightarrow{\beta_{12}}$	$NO_2 + OH + 0.35 eV$	$\beta_{12} = 3.5 \times 10^{-12} \exp(250/T_n)$
(13)	$NO_2 + O$	$\xrightarrow{\beta_{13}}$	$NO + O_2 + 1.98 eV$	$\beta_{13} = 9.3 \times 10^{-12}$
(14)	$NO_2 + h\nu$	$\xrightarrow{\beta_{14}}$	$NO + O$	$\beta_{14} = 10^{-2}$
(15)	$NO_2 + O_3$	$\xrightarrow{\beta_{15}}$	$NO_3 + O_2 + 1.08 eV$	$\beta_{15} = 1.2 \times 10^{-13} \exp(-2450/T_n)$
(16)	$NO_3 + h\nu$	$\xrightarrow{\beta_{16}}$	$NO_2 + O_2$	$\beta_{16} = 0.18$

B. Oxygen and Hydrogen Chemistry ([Allen et al., 1984], but Rate Coefficients According to JPL-92)

Reaction				Reaction Rate
(1)	$O_2 + h\nu$	$\xrightarrow{J_1}$	$O + O$	Global mean J values are calculated as described in the text by eq. 7 of paper 1
(2)	$O_2 + h\nu$	$\xrightarrow{J_2}$	$O + O(^1D)$	The derived values and components are consistent with Brasseur and Solomon (1986)
(3)	$O_3 + h\nu$	$\xrightarrow{J_3}$	$O_2 + O$	
(4)	$O_3 + h\nu$	$\xrightarrow{J_4}$	$O_2 + O(^1D)$	
(5)	$H_2O + h\nu$	$\xrightarrow{J_5}$	$H + OH$	
(6)	$H_2O + h\nu$	$\xrightarrow{J_6}$	$H_2 + O(^1D)$	
(7)	$H_2O_2 + h\nu$	$\xrightarrow{J_7}$	$2OH$	
(8)	$O(^1D) + O_2$	$\xrightarrow{k_8}$	$O + O_2 + 1.96 eV$	$k_8 = 3.2 \times 10^{-11} \exp(67/T_n)$
(9)	$O(^1D) + N_2$	$\xrightarrow{k_9}$	$O + N_2 + 1.96 eV$	$k_9 = 1.8 \times 10^{-11} \exp(107/T_n)$
(10)	$O(^1D) + H_2O$	$\xrightarrow{k_{10}}$	$OH + OH + 1.23 eV$	$k_{10} = 2.2 \times 10^{-10}$
(11)	$O(^1D) + H_2$	$\xrightarrow{k_{11}}$	$H + OH + 1.88 eV$	$k_{11} = 1.0 \times 10^{-10}$
(12)	$O + O + M$	$\xrightarrow{k_{12}}$	$O_2 + M + 5.12 eV$	$k_{12} = 9.59 \times 10^{-34} \exp(480/T_n)$
(13)	$O + O_2 + O$	$\xrightarrow{k_{13}}$	$O_3 + O + 1.10 eV$	$k_{13} = 6.0 \times 10^{-34} (T_n/300)^{-2.8}$
(14)	$O + O_2 + O_2$	$\xrightarrow{k_{14}}$	$O_3 + O_2 + 1.10 eV$	$k_{14} = 6.0 \times 10^{-34} (T_n/300)^{-2.8}$

TABLE 1. Chemical Reactions, Reaction Rates, and Exothermic Reaction Energy
Used in the Global Mean Model of the Thermosphere and Mesosphere

A. Nitrogen Chemistry

Reaction	Reaction Rate
(1) $N(^4S) + O_2 \xrightarrow{\beta_1} NO + O + 1.4 eV$	$\beta_1 = 1.5 \times 10^{-11} \exp(-3600./T_n)$
(2) $N(^2D) + O_2 \xrightarrow{\beta_2} NO + O(^1D) + 1.84 eV$	$\beta_2 = 5.0 \times 10^{-12}$
(3) $N(^4S) + NO \xrightarrow{\beta_3} N_2 + O + 2.68 eV$	$\beta_3 = 1.6 \times 10^{-10} \exp(-460./T_n)$
(4) $N(^2D) + O \xrightarrow{\beta_4} N(^4S) + O + 2.38 eV$	$\beta_4 = 4.5 \times 10^{-13}$
(5) $N(^2D) + e \xrightarrow{\beta_5} N(^4S) + e + 2.38 eV$	$\beta_5 = 3.6 \times 10^{-10} (T_e/300)^{1/2}$
(6) $N(^2D) + NO \xrightarrow{\beta_6} N_2 + O + 5.63 eV$	$\beta_6 = 7.0 \times 10^{-11}$
(7) $N(^2D) \xrightarrow{\beta_7} N(^4S) + h\nu$	$\beta_7 = 1.06 \times 10^{-5}$
(8) $NO + h\nu \xrightarrow{\beta_8} N(^4S) + O$	$\beta_8 = 4.5 \times 10^{-6}$ $\exp(-10^{-8}[N(O_2)]^{0.38})$ $\exp(-5 \times 10^{-19} N(O_3))$
(9) $NO + h\nu/_{Ly-\alpha} \xrightarrow{\beta_9} NO^+ + e$	$\beta_9 = 5.88 \times 10^{-7}(1. + 0.2$ $(F_{10.7} - 65)/100)$ $\cdot \exp(-2.115 \times 10^{-18} \cdot N(O_2)^{0.8855})$
(10) $N(^4S) + OH \xrightarrow{\beta_{10}} NO + H + 2.1 eV$	$\beta_{10} = 5 \times 10^{-11}$
(11) $NO + O_3 \xrightarrow{\beta_{11}} NO_2 + O_2 + 2.08 eV$	$\beta_{11} = 1.8 \times 10^{-12} \exp(-1370/T_n)$
(12) $NO + HO_2 \xrightarrow{\beta_{12}} NO_2 + OH + 0.35 eV$	$\beta_{12} = 3.5 \times 10^{-12} \exp(250/T_n)$
(13) $NO_2 + O \xrightarrow{\beta_{13}} NO + O_2 + 1.98 eV$	$\beta_{13} = 9.3 \times 10^{-12}$
(14) $NO_2 + h\nu \xrightarrow{\beta_{14}} NO + O$	$\beta_{14} = 10^{-2}$
(15) $NO_2 + O_3 \xrightarrow{\beta_{15}} NO_3 + O_2 + 1.08 eV$	$\beta_{15} = 1.2 \times 10^{-13} \exp(-2450/T_n)$
(16) $NO_3 + h\nu \xrightarrow{\beta_{16}} NO_2 + O_2$	$\beta_{16} = 0.18$

B. Oxygen and Hydrogen Chemistry ([*Allen et al.*, 1984], but Rate Coefficients According to JPL-92)

Reaction	Reaction Rate
(1) $O_2 + h\nu \xrightarrow{J_1} O + O$	Global mean J values are calculated as described in the text by eq. 7 of paper 1
(2) $O_2 + h\nu \xrightarrow{J_2} O + O(^1D)$	The derived values and components are consistent with Brasseur and Solomon (1986)
(3) $O_3 + h\nu \xrightarrow{J_3} O_2 + O$	
(4) $O_3 + h\nu \xrightarrow{J_4} O_2 + O(^1D)$	
(5) $H_2O + h\nu \xrightarrow{J_5} H + OH$	
(6) $H_2O + h\nu \xrightarrow{J_6} H_2 + O(^1D)$	
(7) $H_2O_2 + h\nu \xrightarrow{J_7} 2OH$	
(8) $O(^1D) + O_2 \xrightarrow{k_8} O + O_2 + 1.96 eV$	$k_8 = 3.2 \times 10^{-11} \exp(67/T_n)$
(9) $O(^1D) + N_2 \xrightarrow{k_9} O + N_2 + 1.96 eV$	$k_9 = 1.8 \times 10^{-11} \exp(107/T_n)$
(10) $O(^1D) + H_2O \xrightarrow{k_{10}} OH + OH + 1.23 eV$	$k_{10} = 2.2 \times 10^{-10}$
(11) $O(^1D) + H_2 \xrightarrow{k_{11}} H + OH + 1.88 eV$	$k_{11} = 1.0 \times 10^{-10}$
(12) $O + O + M \xrightarrow{k_{12}} O_2 + M + 5.12 eV$	$k_{12} = 9.59 \times 10^{-34} \exp(480/T_n)$

TABLE 1. (continued)

Reaction			Reaction Rate
(13) $O + O_2 + O$	$\xrightarrow{k_{13}}$	$O_3 + O + 1.10 eV$	$k_{13} = 6.0 \times 10^{-34} (T_n/300)^{-2.8}$
(14) $O + O_2 + O_2$	$\xrightarrow{k_{14}}$	$O_3 + O_2 + 1.10 eV$	$k_{14} = 6.0 \times 10^{-34} (T_n/300)^{-2.8}$
(15) $O + O_2 + N_2$	$\xrightarrow{k_{15}}$	$O_3 + N_2 + 1.10 eV$	$k_{15} = 6.0 \times 10^{-34} (T_n/300)^{-2.8}$
(16) $O + O_3$	$\xrightarrow{k_{16}}$	$O_2 + O + 4.06 eV$	$k_{16} = 8.0 \times 10^{-12} \exp(-2060/T_n)$
(17) $O + OH$	$\xrightarrow{k_{17}}$	$O_2 + H + 0.72 eV$	$k_{17} = 2.2 \times 10^{-11} \exp(117/T_n)$
(18) $O + HO_2$	$\xrightarrow{k_{18}}$	$OH + O_2 + 2.33 eV$	$k_{18} = 3.0 \times 10^{-11} \exp(200/T_n)$
(19) $O + H_2O_2$	$\xrightarrow{k_{19}}$	$OH + HO_2 + 3.44 eV$	$k_{19} = 1.4 \times 10^{-12} \exp(-2000/T_n)$
(20) $O + H_2$	$\xrightarrow{k_{20}}$	$OH + H + 0.08 eV$	$k_{20} = 1.6 \times 10^{-11} \exp(-4570/T_n)$
(21) $OH + O_3$	$\xrightarrow{k_{21}}$	$HO_2 + O_2 + 1.73 eV$	$k_{21} = 1.6 \times 10^{-12} \exp(-940/T_n)$
(22) $OH + OH$	$\xrightarrow{k_{22}}$	$H_2O + O + 0.73 eV$	$k_{22} = 4.2 \times 10^{-12} \exp(-242/T_n)$
(23) $OH + HO_2$	$\xrightarrow{k_{23}}$	$H_2O + O_2 + 3.06 eV$	$k_{23} = 4.8 \times 10^{-11} \exp(215/T_n)$
(24) $OH + H_2O_2$	$\xrightarrow{k_{24}}$	$H_2O + HO_2 + 1.35 eV$	$k_{24} = 3.1 \times 10^{-12} \exp(-187/T_n)$
(25) $OH + H_2$	$\xrightarrow{k_{25}}$	$H_2O + H + 0.65 eV$	$k_{25} = 7.7 \times 10^{-12} \exp(-2100/T_n)$
(26) $HO_2 + O_3$	$\xrightarrow{k_{26}}$	$OH + O_2 + O_2 + 1.23 eV$	$k_{26} = 1.4 \times 10^{-14} \exp(-580/T_n)$
(27) $HO_2 + HO_2$	$\xrightarrow{k_{27}}$	$H_2O_2 + O_2 + 1.71 eV$	$k_{27} = 2.3 \times 10^{-13} \exp(590/T_n)$
(28) $H + O_2 + M$	$\xrightarrow{k_{28}}$	$HO_2 + M + 2.11 eV$	$k_{28} = 5.5 \times 10^{-32} (T_n/300)^{-1.6}$
(29) $H + O_3$	$\xrightarrow{k_{29}}$	$OH + O_2 + 3.34 eV$	$k_{29} = 1.4 \times 10^{-10} \exp(-470/T_n)$
(30) $H + HO_2$	$\xrightarrow{k_{30}}$	$H_2 + O_2 + 2.41 eV$	$k_{30} = 4.2 \times 10^{-11} \exp(-350/T_n)$
(31) $H + HO_2$	$\xrightarrow{k_{31}}$	$OH + OH + 1.61 eV$	$k_{31} = 4.2 \times 10^{-10} \exp(-950/T_n)$
(32) $H + HO_2$	$\xrightarrow{k_{32}}$	$H_2O + O + 2.34 eV$	$k_{32} = 8.3 \times 10^{-11} \exp(-500/T_n)$
(33) $H + H + M$	$\xrightarrow{k_{33}}$	$H_2 + M + 4.52 eV$	$k_{33} = 5.7 \times 10^{-32} (300/T_n)^{1.6}$
(34) $CH_4 + OH$	$\xrightarrow{k_{34}}$	$CO + OH + 2H_2O + 5.38 eV$	$k_{34} = 2.4 \times 10^{-12} \exp(-1710/T_n)$
(35) $CH_4 + O$	$\xrightarrow{k_{35}}$	$CO + 2OH + H_2O + 4.65 eV$	$k_{35} = 3.5 \times 10^{-11} \exp(-4550/T_n)$
(36) $CH_4 + O(^1D)$	$\xrightarrow{k_{36}}$	$CO + 2OH + H_2O + 6.61 eV$	$k_{36} = 1.4 \times 10^{-10}$

C. Carbon Dioxide and Monoxide

Reaction			Reaction Rate
(37) $CO + O + M$	$\xrightarrow{k_{37}}$	$CO_2 + M + 5.51 eV$	$k_{37} = 6.6 \times 10^{-33} \exp(-1103/T_n)$
(38) $CO + OH$	$\xrightarrow{k_{38}}$	$CO_2 + H + 1.07 eV$	$k_{41} = 1.47 \times 10^{-13}(1 + 0.59 P_{atm})$

The chemical reactions and reaction rates are standard for mesospheric models and they are taken from the papers by *Allen et al.* [1984], *De More et al.* [1990], *Garcia and Solomon* [1983], *Brasseur and Solomon* [1984], *Solomon et al.* [1985], *Clancy et al.* [1987], and *Roble et al.* [1987]. Appropriate reference to laboratory measurements are given in these papers.

D. $O_2(^1\Sigma_g)$ and $O_2(^1\Delta_g)$

Reaction			Reaction Rate
(40) $O_3 + h\nu(210 < \lambda < 310 nm)$	$\xrightarrow{\epsilon J_{1\Delta}}$	$O_2(^1\Delta g) + O(^1D)$	$\epsilon J_{1\Delta} = 0.9 \times 8 \times 10^{-3} s^{-1}$

TABLE 1. (continued)

(41) $O_2(^1\Sigma g) + M \xrightarrow{K_{41}} O_2(^1\Delta g) + M + 0.65 eV$ $K_{41} = 2.2 \times 10^{-15}$

(42) $O_2(^1\Delta g) + O_2 \xrightarrow{K_{42}} O_2 + O_2 + 0.98 eV$ $K = 2.22 \times 10^{-18} \cdot (T_n/300)^{0.78}$

(43) $O_2(^1\Delta g) \xrightarrow{A_{DL}} O_2 + h\nu/_{1.27\mu m}$ $A_{DL} = 2.58 \times 10^{-4} s^{-1}$

(44) $O_2(^1\Sigma g) \xrightarrow{A_{SG}} O_2 + h\nu/_{762 nm}$ $A_{SG} = 0.085 s^{-1}$

E. F- and E-Region Ion Chemistry

(1) $O^+(^4S) + O_2 \xrightarrow{K_1} O_2^+ + O + 1.556 eV$

$k_1 = 2.82 \times 10^{-11} - 7.74 \times 10^{-12}(T_1/300.) + 1.073 \times 10^{-12}(T_1/300)^2$
$- 5.17 \times 10^{-14}(T_1/300.)^3 + 9.65 \times 10^{-16}(T_1/300)^4$

where $T_1 = 0.667 T_i + 0.333 T_n$; $300 \leq T_1 \leq 6000 K$

(2) $O^+(^4S) + N_2 \xrightarrow{k_2} NO^+ + N(^4S) + 1.0888 eV$

$k_2 = 1.5333 \times 10^{-12} - 5.92 \times 10^{-13}(T_2/300.) + 8.6 \times 10^{-14}(T_2/300.)^2$

where $T_2 = 0.6363 T_i + 0.3637 T_n$; $300 \leq T_2 \leq 1700 K$

(3) $N_2^+ + O \xrightarrow{k_3} NO^+ + N(^2D) + 0.70 eV$

$k_3 = 1.4 \times 10^{-10}(300./T_R)^{0.44}$ $T_R < 1500 K$
$= 5.2 \times 10^{-11}(T_R/300.)^{0.2}$ $T_R \geq 1500 K$

where $T_R = (T_i + T_n)/2$.

Reaction			Reaction Rate
(4) $O_2^+ + N(^4S)$	$\xrightarrow{k_4}$	$NO^+ + O + 4.21 eV$	$k_4 = 1.0 \times 10^{-10}$
(5) $O_2^+ + NO$	$\xrightarrow{k_5}$	$NO^+ + O_2 + 2.813 eV$	$k_5 = 4.4 \times 10^{-10}$
(6) $N^+ + O_2$	$\xrightarrow{k_6}$	$O_2^+ + N(^4S) + 2.486 eV$	$k_6 = 4.0 \times 10^{-10}$
(7) $N^+ + O_2$	$\xrightarrow{k_7}$	$NO^+ + O + 6.699 eV$	$k_7 = 2.0 \times 10^{-10}$
(8) $N^+ + O$	$\xrightarrow{k_8}$	$O^+ + N + 0.98 eV$	$k_8 = 1.0 \times 10^{-12}$
(9) $N_2^+ + O_2$	$\xrightarrow{k_9}$	$O_2^+ + N_2 + 3.52 eV$	$k_9 = 6.0 \times 10^{-11}$
(10) $O^+(^4S) + N(^2D)$	$\xrightarrow{k_{10}}$	$N^+ + O(^3P) + 1.45 eV$	$k_{10} = 1.3 \times 10^{-10}$
(11) $H^+ + O$	$\xrightarrow{k_{11}}$	$O^+ + H$	$k_{11} = k_{12} \cdot \frac{8}{9} \left[\frac{T_i + T_n/16.}{T_n + T_i/16.}\right]^{1/2}$
(12)	$\xleftarrow{k_{12}}$		
(13) $O^+ + CO_2$	$\xrightarrow{k_{13}}$	$O_2^+ + CO$	$k_{13} = 9.4 \times 10^{-10}$
(14) $O^+ + H_2$	$\xrightarrow{k_{14}}$	$OH^+ + H$	$k_{14} = 2.0 \times 10^{-9}$
(15) $O^+ + H_2O$	$\xrightarrow{k_{15}}$	$H_2O^+ + O$	$k_{15} = 2.4 \times 10^{-9}$
(16) $O^+(^2P) + N_2$	$\xrightarrow{k_{16}}$	$N_2^+ + O + 3.02 eV$	$k_{16} = 4.8 \times 10^{-10}$
(17) $O^+(^2P) + N_2$	$\xrightarrow{k_{17}}$	$N^+ + NO + 0.70 eV$	$k_{17} = 1.0 \times 10^{-10}$
(18) $O^+(^2P) + O$	$\xrightarrow{k_{18}}$	$O^+(^4S) + O + 5.2 eV$	$k_{18} = 5.0 \times 10^{-11}$
(19) $O^+(^2P) + e$	$\xrightarrow{k_{19}}$	$O^+(^4S) + e + 5.0 eV$	$k_{19} = 4.0 \times 10^{-8} (300/T_e)^{1/2}$
(20) $O^+(^2P) + e$	$\xrightarrow{k_{20}}$	$O^+(^2O) + e + 1.69 eV$	$k_{20} = 1.5 \times 10^{-7} (300./T_e)^{1/2}$

TABLE 1. (continued)

(21)	$O^+(^2P)$	$\xrightarrow{k_{21}}$	$O^+(^4S) + h\nu\ (2470\ \text{Å})$	$k_{21} = 0.047 s^{-1}$
(22)	$O^+(^2P)$	$\xrightarrow{k_{22}}$	$O^+(^2D) + h\nu\ (7320\ \text{Å})$	$k_{22} = 0.171 s^{-1}$
(23)	$O^+(^2D)$	$\xrightarrow{k_{23}}$	$N_2^+O + 1.33 eV$	$k_{23} = 8 \times 10^{-10}$
(24)	$O^+(^2D) + O$	$\xrightarrow{k_{24}}$	$O^+(^4S) + O + 3.31 eV$	$k_{24} = 1.0 \times 10^{-11}$
(25)	$O^+(^2D) + e$	$\xrightarrow{k_{25}}$	$O^+(^4S) + e + 3.31 eV$	$k_{25} = 6.6 \times 10^{-8}\ (300/T_e)^{1/2}$
(26)	$O^+(^2D) + O_2$	$\xrightarrow{k_{26}}$	$O_2^+O + 4.865 eV$	$k_{26} = 7 \times 10^{-10}$

Recombination

(1) $NO^+ e \xrightarrow{\alpha_1} N(^4S) + O + 2.75 eV$ (20%)

$N(^2D) + O + 0.38 eV$ (80%)

$\alpha_1 = 4.2 \times 10^{-7} (300./T_e)^{0.85}$

(2) $O_2^+ e \xrightarrow{\alpha_2} O(^3P) + O(^3P) + 6.95 eV$ (15%)

$O(^3P) + O(^1D) + 4.98 eV$ (85%)

$\alpha_2 = 1.6 \times 10^{-7}(300./T_e)^{0.55}$ for $T_e \geq 1200$ K

$= 2.7 \times 10^{-7}(300./T_e)^{0.7}$ for $T_e < 1200$ K

(3) $N_2^+ e \xrightarrow{\alpha_3} N(^4S) + N(^4S) + 5.82 eV$ (10%)

$N(^4S) + N(^2D) + 3.44 eV$ (90%)

$\alpha_3 = 1.8 \times 10^{-7}(300./T_e)^{0.39}$

of the ionosphere.

4.2. Neutral Gas Energy Equation

The globally averaged energy equation used in the model is given by equation (1) in Paper I. For the globally averaged structure, the time rate of change of temperature is determined by a balance between neutral gas heating and cooling rates and downward molecular and eddy heat transport. In Paper I, the globally averaged neutral gas heating rate for the thermosphere above 97 km consisted of the following ten component processes: (1) absorption of solar UV radiation in the O_2 Schumann-Runge continuum region; (2) likewise in the Schumann-Runge bands; (3) heating by exothermic neutral-neutral chemical reactions primarily the thermospheric odd nitrogen system; (4) heating by exothermic ion-neutral chemical reactions, (5) heating by elastic and inelastic collisions between ambient electrons, ions and neutrals; (6) quenching of metastable species, such as $O(^1D)$ by N_2 and O_2; (7) atomic oxygen recombination; (8) absorption of solar Lyman-α radiation; (9) heating by fast photoelectron and auroral electrons; and (10) Joule heating from the dissipation of ionospheric currents. All of these processes have been included in the global mean model as well as in the TIGCM and TIE-GCM and these processes have been extended downward to the Z = -17 (10 mb) constant pressure surface near 30 km even though their influence on mesospheric and upper stratospheric structure is small.

In addition to the above, the following middle atmospheric heating processes are included: (11) absorption of solar UV radiation in the O_2 Hertzberg continuum; (12) absorption of solar UV radiation in the O_3 Hartley, Huggins, and Chappius bands; (13) heating by additional mesospheric neutral-neutral chemical reactions as given in Table 1; (14) heating from exothermic ionospheric D-region chemical reactions; and (15) an assumed gravity wave heat source using the parameterization developed by *Gavrilov and Roble* [1994] for the global mean model.

The main cooling processes for the thermosphere have also been discussed in detail in Paper I. These include: (1) molecular heat conduction; (2) eddy heat conduction; (3) non-local thermodynamic equilibrium (non-LTE) NO 5.3-μm cooling; (4) 63-μm cooling from the fine structure of atomic oxygen; and (5) LTE and non-LTE CO_2 15-μm cooling in the lower thermosphere. For the mesosphere, all of the cooling processes (1) through (4) have been extended downward and (5) has been modified to include the complete CO_2 radiation parameterization of *Fomichev et al.* [1993] for both LTE and non-LTE 15-μm radiational cooling. Also included in the model is (6) the parameterized O_3 9.6-μm radiational cooling of *Fomichev and Shved* [1985]. Above about 80 km, radiative cooling rates from CO_2 are strongly dependent upon atomic oxygen concentrations and on the rate of energy exchange between atomic oxygen and CO_2. We use a rate coefficient of 4×10^{-12} cm s^{-2} which is a compromise between the laboratory rates measured by *Shved et al.* [1991] and *Pollack et* al. [1993], and the rate derived from satellite measurements by *Sharma and Wintersteiner* [1990]. This value is also consistent with the recent analysis of satellite measurements made by *Rogers et al.* [1992] and *Lopez-Puertas et al.* [1993] and planetary atmosphere studies by *Bougher and Roble* [1991].

The upper boundary for the thermodynamic equation is the same as in Paper I, that assumes the absence of any neutral gas heat source in the exosphere. At the lower boundary, the temperature is specified as 225 K, as determined by globally averaging the 10 mb temperature distribution obtained from the MSIS-90 empirical model of *Hedin* [1991].

4.3. Major Constituent Composition Equation

The globally averaged composition equation for the major neutral gas constituents O, O_2, and N_2 is given by equation (2) of Paper I. This equation is a globally averaged version of the major composition equation used in the NCAR TGCMs that has been derived and discussed in detail by *Dickinson et al.* [1984]. We use the same equation for the extension downward into the upper stratosphere and mesosphere. The lower boundary of the model is now at the $Z = -17$ constant pressure surface (10 mb) near 30 km. The equation has also been modified to solve for $O_x = (O + O_3)$ instead of O throughout the altitude grid. Once O_x is calculated, O and O_3 are determined using a photochemical partitioning of the appropriate chemistry that is given in Table 1.

The eddy diffusion term in equation (2) of Paper I is also modified to be

$$e^Z \frac{\partial}{\partial Z} \left\{ e^{-Z} K_E(Z) \left[\frac{\partial}{\partial Z} + \frac{1}{\overline{m}} \frac{\partial \overline{m}}{\partial Z} \right] \Psi \right\} \quad (1)$$

where Z is our log pressure coordinate, $K_E(Z)$ is the eddy diffusion coefficient in units of s^{-1} (to obtain the eddy diffusion coefficient in more conventional units (cm^2/s) multiply K_E by the scale height H squared, i.e., $k_E = K_E H^2$), $\overline{m}$ is the mean mass and $\Psi = (\Psi_{O_2}, \Psi_{O_x})$ the mass mixing ratio vector for O_2 and O_x, respectively. In Paper I, the term $1/\overline{m} \, \partial \overline{m}/\partial Z$ was inadvertently omitted.

The sources and sinks for O_x are expanded to include photodissociation in the Hertzberg continuum of O_2 and the Hartley, Huggins, and Chappius bands of O_3 and additional chemical production and loss terms as derived from the chemical reactions given in Table 1. The lower boundary mass mixing ratio of O_x at 10 mb is specified at the lower boundary and diffusive equilibrium is assumed at the upper boundary.

4.4. Minor Neutral Constituent Equations

The minor neutral constituents in the global mean model include $N(^2D)$, $N(^4S)$, NO, NO_2, H_2O, H_2, CH_4, H, OH, HO_2, H_2O_2, CO, CO_2, O_3, $O(^1D)$, $O_2(^1\Sigma_g)$, $O_2(^1\Delta_g)$, He, and Ar. The chemical reactions, reaction rates and exothermic reaction energy for these species are given in Table 1. The odd nitrogen chemistry for the thermosphere is the same as that used in Paper I. For the mesosphere, we also consider the additional reactions (10–14), [*Garcia and Solomon*, 1983]. Although the lower boundary of the model extends down to 10 mb, odd nitrogen species such as, NO_3, N_2O_5, HNO_3, and N_2O are neglected but their effects are approximately included by specifying globally averaged NO_x mixing ratio at the lower boundary from a stratospheric model that considers these species [*Brasseur et al.*, 1990]. We also include the same oxygen and hydrogen chemistry as presented by *Allen et al.* [1984] but modify the chemical reaction rates to be consistent with the JPL-90 listing [*De More et al.*, 1990]. The chemistry for CO and CO_2 is the same as that used by *Solomon et al.* [1985].

Based upon the respective chemical and diffusive lifetimes of each species [e.g., *Brasseur and Solomon*, 1984], we consider $O(^1D)$, $O_2(^1\Sigma_g)$, $O_2(^1\Delta_g)$, $N(^2D)$, NO_2, OH, HO_2, and H_2O_2 to be in photochemical equilibrium and solve a minor constituent transport equation for each of the remaining species. Only $HO_x = (H + OH + OH_2)$ and $NO_x = (NO + NO_2)$ are transported as a minor chemical family and then partitioned according to photochemical considerations [*Brasseur and Solomon*, 1984].

The globally averaged minor constituent equation, transformed into the NCAR thermospheric general circulation model log pressure coordinate system is,

$$\frac{\partial \Psi_m}{\partial t} = -e^{-Z} \frac{\partial}{\partial Z} A_m \left[\frac{\partial}{\partial Z} - E_m \right] \Psi_m + S_m - R_m$$

$$+ e^Z \frac{\partial}{\partial Z} e^{-Z} K_E(Z) \left(\frac{\partial}{\partial Z} + \frac{1}{\overline{m}} \frac{\partial \overline{m}}{\partial Z} \right) \Psi_m \quad (2)$$

where $E_m = \left[1 - \frac{m_m}{\overline{m}} - \frac{1}{\overline{m}} \frac{\partial \overline{m}}{\partial Z} \right] - \alpha_m \frac{1}{T} \frac{\partial T}{\partial Z} + \underline{F} \, \Psi_m$

The three terms in E_m represent gravitational force, thermal diffusion and the frictional interaction with major species, Ψ_m is the mass mixing ratio of the minor species, t is the time, A_m is the molecular diffusion coefficient of the minor species, m_m is the mass of the minor species and $\overline{m}$ is the average mass of the major species, α_m is the thermal diffusion coefficient, T is temperature and $\underline{F}$ is a matrix operator that represents the friction interaction of the minor species with the three major species O, O_2, and N_2 that is important in the thermosphere and S_m and R_m represent mass sources and sinks of minor species m respectively.

At the lower boundary of the model, number density mixing ratios are specified as follows: 0.67 ppm for CH_4, 0.5 ppm for H_2, 350 ppm for CO_2, 11 ppb for CO, 4.4 ppm for H_2O, 5.24 ppm for He, 9340 ppm for Ar and 8 ppb for NO. Photochemical equilibrium is assumed for the remainder of the species at the lower boundary.

Diffusive equilibrium is the upper boundary condition at the $Z = +5$ constant pressure surface (~ 400 km for solar minimum and ~ 600 km for solar maximum conditions) for all species except HO_x, that is $(\partial/\partial Z - E_m)\Psi_m = 0$. For HO_x, we consider exospheric escape of atomic hydrogen as specified by *Liu and Donahue* [1974] and *Hunten and Donahue* [1976]. The three exospheric escape processes are: (1) Jean's thermal escape flux; (2) $H - H^+$ charge exchange; and (3) the polar wind. The total escape flux is determined using equation (4) of *Liu and Donahue* [1974].

Helium and argon are passive tracers and their distribution is governed solely by transport; eddy diffusion below the turbopause and molecular diffusion above. Their distributions are important for evaluating the eddy transport that is also used for chemically active species.

4.5. Component Coupling Processes

An overall schematic of the physical and chemical processes incorporated into the global mean model of the upper atmosphere is given in Figure 3. The model includes external inputs from solar EUV and UV radiation and specified auroral particle precipitation and Joule heating from the dissipation of ionospheric currents. The model solves for the vertical distribution of electron and ion temperature, electron and ion densities, and the distributions of the major and minor neutral gas species that were described in the previous sections. The upper portion of the figure includes the same coupled physical and chemical processes that have been described by *Roble et al.* [1987] for the thermosphere and ionosphere above 97 km and the lower portion of the figure specifies the additional processes included in the extension of the model downward to 30 km. All of these physical and chemical processes are coupled and calculated at each model time step. One of the purposes of this model is to try to calculate the global mean structure of the upper atmosphere above 30 km altitude using our overall knowledge of the aeronomic processes that appear to be operating in the atmosphere. There are no empirical models used to provide background properties. The model assumes mixing ratios of various long-lived constituents and a lower boundary temperature and from these specifications the atmospheric structure evolves to a steady state based only on the self-consistent internal workings of the physical and chemical processes. As discussed in the next section the major tunable parameter is the specified vertical profile of eddy diffusion.

5. RESULTS FOR SOLAR CYCLE MINIMUM

The global mean model was run initially for solar cycle minimum conditions considering solar EUV and UV heating. From arbitrary initial conditions, the self-consistent coupled mesosphere, thermosphere, and ionosphere evolve to a steady state in a few hundred days. The solar heating rates, IR cooling rates, chemical reactions, etc., are all updated at each model time step to allow interactive processes to occur. The time-scale for the approach to steady state is controlled primarily by the value of the derived eddy diffusion in the upper stratosphere near the lower boundary ($K_e = 5. \times 10^{-7} s^{-1}$). As in Paper I, the calculated exospheric temperature when only solar EUV and UV heating is considered was about 120 K too small compared with predictions by the MSIS-90 empirical model. To achieve agreement with MSIS-90 it is necessary to consider additional Joule and aurora particle heat sources of the same magnitude as described in Paper I. The total global Joule heat input for quiet geomagnetic conditions ($A_p \sim 4$) is 7×10^{10} W and the global auroral particle input is 1.3×10^{10} W.

The theoretical model is "tuned" to give reasonable agreement with respect to temperature and compositional (O_2, N_2, O, He, Ar, H, $N(^4S)$) profiles obtained from MSIS-90 above 85 km, blended with the U.S. Standard Atmosphere 1976 profiles below 85 km [*Fels*, 1986]. This "tuning" was accomplished in a trial and error process, by adjusting the eddy diffusion profile and Prandtl number in the mesosphere until reasonable agreement with the empirical model was achieved. The form of the eddy diffusion profile used in the final analysis is a simple exponential interpolation with key values as

z (km)	Z	K_E (s^{-1})	k_E (cm^{2}s^{-1})
30	-17	5.0×10^{-7}	2.24×10^5
86	-9	8.5×10^{-7}	3.17×10^5
100	-6.5	2.0×10^{-6}	5.91×10^5
386	+5	2.0×10^{-11}	3.45×10^2

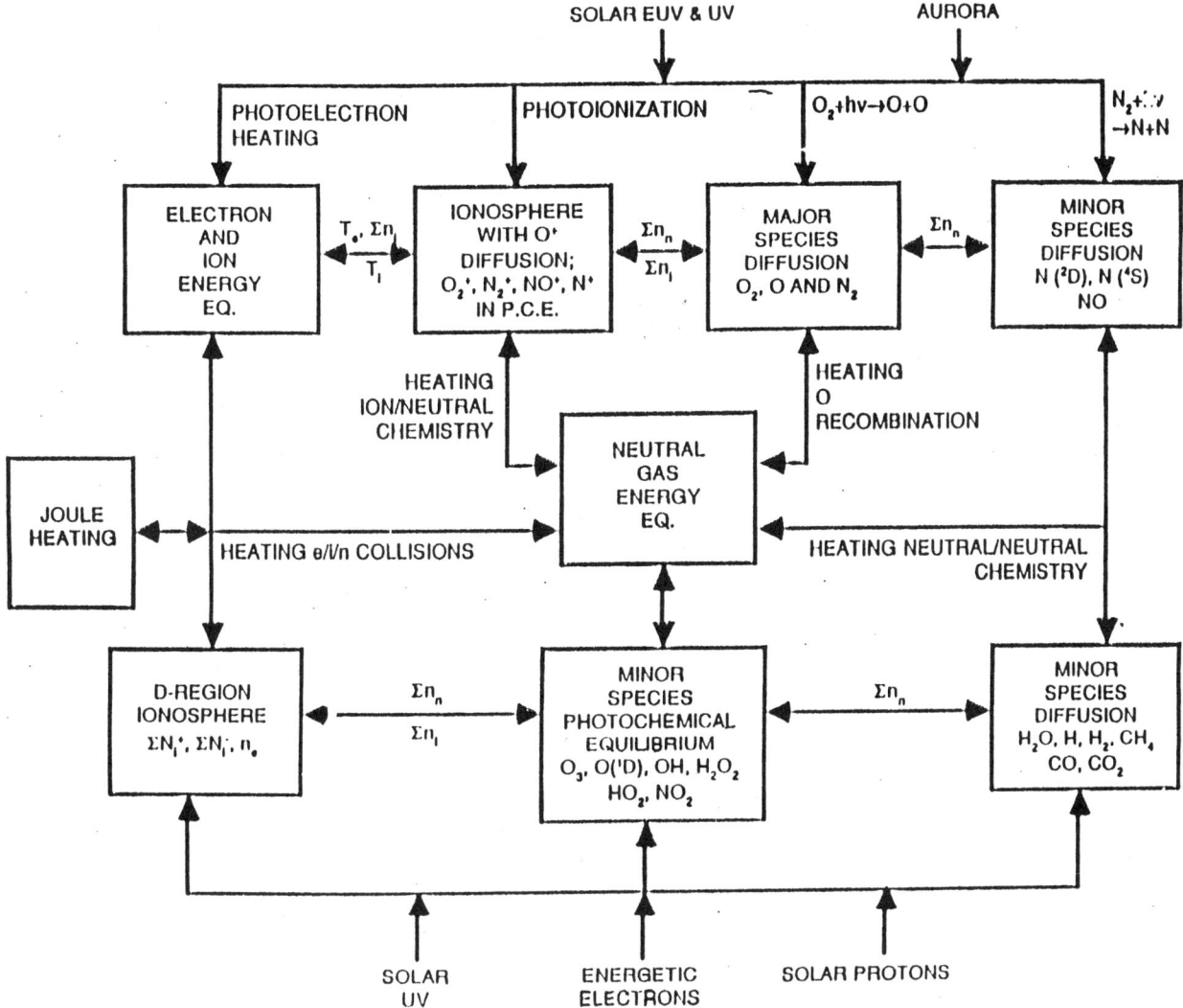

Fig. 3. Schematic of the major processes within the mesosphere, thermosphere, and ionosphere that establish their global mean structure.

The eddy diffusion profile given above is similar to model D of *Strobel* [1989] that was derived to obtain reasonable agreement of calculated H_2O, CO, O, O_3, and Ar vertical profiles with observations of the vertical profiles of these species. A Prandtl number of three is also assumed that is consistent with the analysis of Strobel. Since the main purpose of this study is primarily to illustrate the main energetic processes operating in the upper atmosphere, no attempt was made to adjust the eddy diffusion coefficients for individual chemical species in one-dimensional models as discussed by *Holton* [1986] and *Stordal and Garcia* [1987]. Three-dimensional transport will be addressed with the TIME-GCM that has been described by *Roble and Ridley* [1994].

The calculated neutral gas temperature profile, using the eddy diffusion profile given above, is shown in Figures 4a and 4b. In Figure 4a, the calculated electron, ion and neutral gas temperatures and the global mean temperature profile obtained from the empirical model are plotted over the entire altitude range 30 km to 500 km. In Figure 4b, only the lower portion of the neutral gas temperature comparisons are shown between 30 km and 120 km. In the upper stratosphere and lower mesosphere the calculated global average temperature is in reasonable agreement with the empirical model with a maximum deviation of 5 K near 40 km. There is a good agreement at the stratopause where the calculated value of 266 K is within a few degrees of the value derived from the empirical model, 263 K. Between 80 km and

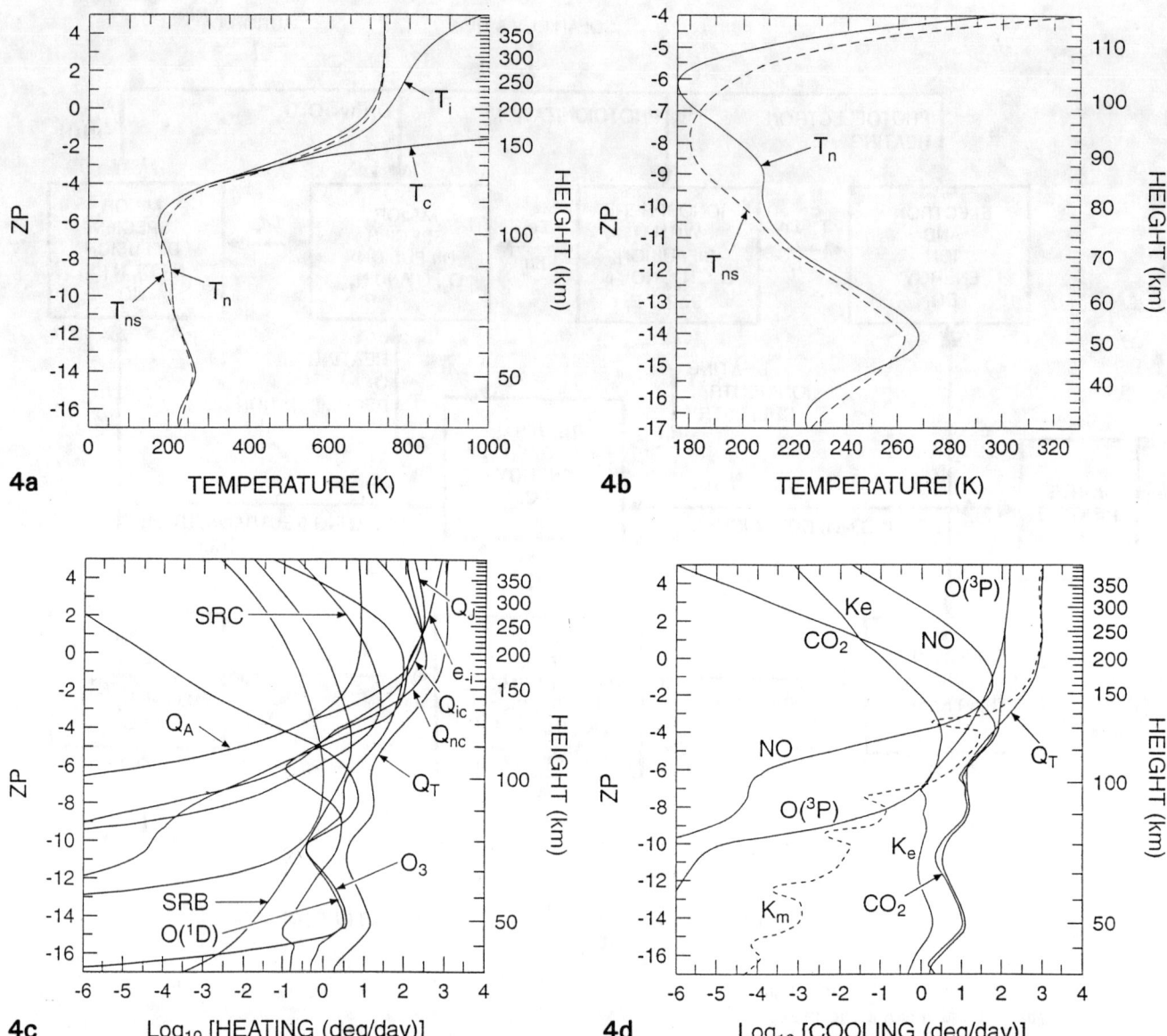

Fig. 4. (a) Calculated global mean neutral, T_n, electron, T_e, and ion T_i, temperature profiles and the global mean neutral temperature profile obtained from MSIS-90 T_{ns}, (b) neutral gas temperature from the model, T_n, and MSIS-90, T_{ns}, in the 30 to 120 km height range, (c) $\log_{10}$ total and component neutral gas heating profiles (K day^{-1}), Q_T is the total, $e - i$ is heating by collisions between thermal electrons, ions, and neutrals; Q_{ic} is heating from exothermic ion-neutral chemistry; Q_{nc} is heating from exothermic neutral-neutral chemistry; Q_J is Joule heating from a superimposed electric field; Q_A is heating from auroral particle precipitation, $O(^1D)$ is heating from quenching of $O(^1D)$, SRC and SRB are heating from O_2 absorption in the Schumann-Runge continuum and bands, respectively; O_R is heating from atomic oxygen recombinations; O_3 is heating from absorption of solar radiation in the Hartley, Huggins, and Chappins bands of ozone; (d) $\log_{10}$ total neutral gas heating rate profile and component neutral gas cooling rate profiles (K day^{-1}), Q_T is the total neutral gas heating rate, K_m is the cooling rate by downward molecular thermal conduction; and K_E is the cooling rate by eddy thermal conduction; NO is radiative cooling from the 5.3 μm emission from nitric oxide, CO_2 is total radiative cooling from carbon dioxide, and $O(^3P)$ is cooling from the fine structure of atomic oxygen; IR is the sum of radiative cooling terms from O, NO, and CO_2.

110 km, however, there is considerable difference between the magnitude and shapes of the calculated and empirical model vertical temperature profiles. The empirical temperature profile is relatively smooth with a mesopause temperature of 180K near 90 km. The calculated temperature profile shows a minor peak in the temperature near 90 km caused by exothermic chemical heating, primarily the HO_x system as discussed by *Mlynczak and Solomon* [1993]. The minimum temperature of 177 K occurs near 100 km and it is caused by enhanced CO_2 cooling by collisions with atomic oxygen.

Above 110 km, the calculated global mean temperature is in reasonable agreement with the empirical model with exospheric temperatures being 720 K and 733 K respectively. The electron temperature departs from the neutral gas temperature above about 150 km altitude and has a profile shape typical of solar minimum conditions [*Roble*, 1976]. The ion temperature follows the neutral gas temperature profile in the lower thermosphere and tends toward the electron temperature in the upper ionosphere.

The calculated total and component neutral gas heating rates are plotted in Figure 4c. The details of the thermospheric heating rates have been discussed in Paper I. Above about 250 km, the dominant neutral gas heating rate is due to collisions between thermal electrons, ions and neutrals. Exothermic ion-neutral and neutral-neutral chemical reaction heating dominates between 150 km and 250 km with a significant amount of Joule heating contributing to the total heating rate. From about 110 km to 150 km heating due to absorption of solar radiative energy in the Schumann-Runge continuum dominates. Below 100 km, atomic oxygen recombination heating, ozone heating and exothermic neutral-neutral chemical reaction heating each contribute significantly to the total heating rate. Here atomic oxygen recombination heating is given separately and subtracted from the rest of the total exothermic neutral-neutral chemical heating. The total heating rate in degrees per day varies from 12 K/day near the stratopause to 3 K/day near 70 km to 15 K/day near 100 km to 900 K/day in the upper thermosphere. Although there are separate dominant heating rates within various altitude intervals, the component heating rates all contribute, in a complex fashion, to make up the total heating rate that governs the global mean temperature distribution of the atmosphere above 30 km.

The total and component cooling rates that balance the total heating rates are shown in Figure 4d. In the upper thermosphere, downward molecular heat conduction is the dominant cooling mechanism with infrared cooling mainly from 63 μm cooling from O having a minor role. Infrared cooling from CO_2 dominates throughout the rest of the atmospheric region from 140 km downward to the lower boundary near 30 km. There are minor contributions to the total cooling rate from the 5.3 μm radiation from NO in the middle thermosphere and 9.6 μm radiation from O_3 in the stratopause region. Eddy thermal conduction cooling is small throughout the region and there is an approximate balance between radiative heating and radiative cooling rates suggesting that the global mean upper atmosphere may be in approximate radiative equilibrium consistent with the study of *Fomichev and Shved* [1988]. It is well known, however, that there are significant local departures from radiative equilibrium in the upper mesosphere caused by the dissipation of upward propagating gravity waves, but on a global average, radiative equilibrium appears to be achievable in these calculations. The heating and cooling rates from gravity waves used within the model has been discussed in detail by *Gavrilov and Roble* [1994].

Number density profiles of O, O_2, and N_2 calculated by the model are shown in Figure 5a along with global mean profiles obtained from MSIS-90, and the U.S. standard atmosphere 1976 below 85 km. There is reasonable agreement between the calculated and empirical model profiles throughout the mesosphere and thermosphere. The calculated peak O density is $3.9 \times 10^{11} \text{cm}^{-3}$ at 97 km compared with a globally averaged peak O density from MSIS-90 of $3.8 \times 10^{11} \text{cm}^{-3}$ at 97 km. The peak O density is sensitive to the assumed eddy diffusion coefficient profile in the lower thermosphere, and it can locally have a wide range of values, as discussed by *Brasseur and Offerman* [1986], but the derived profile used in the model is for global average conditions.

Both helium and argon are passive minor neutral constituents in the atmosphere with their distribution in the global mean being governed eddy and molecular diffusion. The calculated globally averaged number density distributions for these two species are shown in Figure 5b along with globally averaged values obtained from MSIS-90 and the U.S. standard atmospheres. The species distributions are governed by eddy diffusion below the turbopause, specified to occur near 105 km, and both maintain a constant mixing ratio throughout this region. Above 105 km molecular diffusion dominates and both species tend to diffusive equilibrium at high altitudes. The calculated distribution of argon is in good agreement with MSIS-90. Helium, however, is calculated to be somewhat larger than MSIS-90 in the

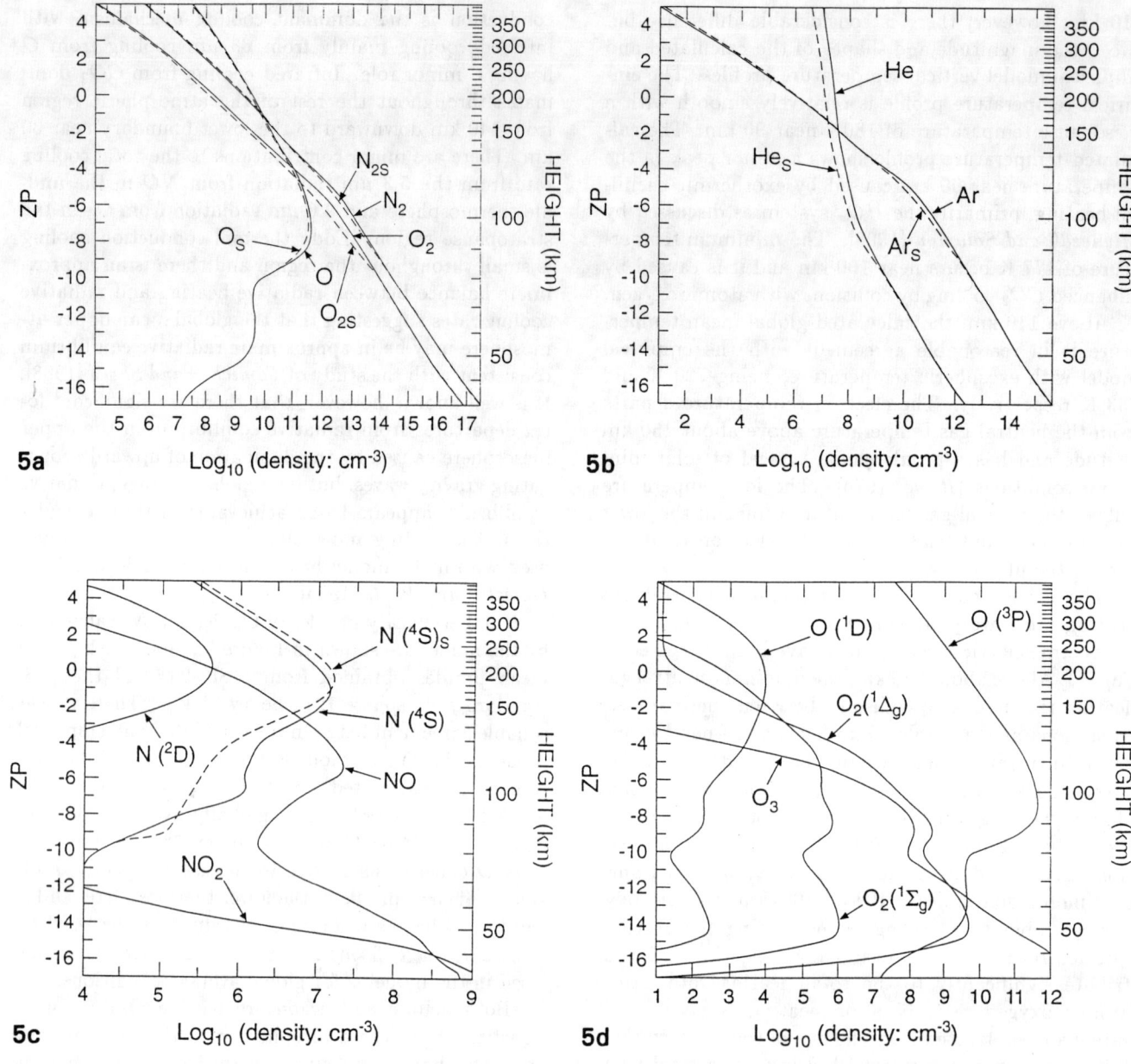

Fig. 5. (a) Calculated $\log_{10} O$, O_2, and N_2 number density profiles (cm^{-3}) and similar profiles from MSIS-86, $\log_{10} O_s$, O_{2s}, and N_{2s} number density profiles (cm^{-3}), (b) same as (a) except for He and Ar, (c) calculated $\log_{10} N(^2D)$, $N(^4S)$, NO, and NO_2 (cm^{-3}) global mean profiles and the $N(^4S)_s$ profile obtained from MSIS-86 (d) calculated $\log_{10} O(^1D)$, $O(^3P)$, $O_2(^1\Sigma_g)$, $O_2(^1\Delta_g)$, and O_3 (cm^{-3}) all for solar minimum, quiet geomagnetic conditions.

upper thermosphere, perhaps indicating that some escape may be necessary. The agreement shown gives confidence to the derived eddy diffusion profile that is used in the transport equation for all chemical species.

The calculated global average distributions for $N(^2D)$, $N(^4S)$, NO, and NO_2 are shown in Figure 5c, along with the globally averaged $N(^4S)$ profile obtained from MSIS-90. $N(^2D)$ is assumed to be in photochemical equilibrium throughout the mesosphere and thermosphere and has a peak value of $3.8 \times 10^5 \text{cm}^{-3}$ occurring near 200 km. $N(^4S)$ has a peak value of $2.0 \times 10^7 \text{cm}^{-3}$ occurring near 160 km compared with $1.6 \times 10^7 \text{cm}^{-3}$ at 180 km obtained from MSIS-90. The calculated peak NO density in the thermosphere is $9 \times 10^6 \text{cm}^{-3}$ occurring at 105 km and it is in reasonable agreement with low solar activity measurements made by the Solar Mesosphere Explorer (SME) spacecraft at low latitudes during the daytime [Barth et al., 1988]. Photodissociation results in an upper mesospheric minimum of 10^6 at 86 km. NO densities increase with decreasing height in the lower mesosphere and significant NO_2 densities occur only in the lower mesosphere.

The odd-oxygen species $O(^1D)$, $O(^3P)$, $O_2(^1\Sigma_g)$, $O_2(^1\Delta_g)$, and O_3 are shown in Figure 5d. $O(^1D)$ is in photochemical equilibrium with a peak value occurring near the ionospheric F-region peak and with a profile that decreases slowly with altitude below the peak. Ozone has its largest value at the lower boundary and decreases with increasing altitude in the mesosphere. There is a bulge in the density near 86 km with a peak of $1.5 \times 10^8 \text{cm}^{-3}$ with the density decreasing steadily above that altitude. The calculated value of $1.0 \times 10^7 \text{cm}^{-3}$ near 100 km is in reasonable agreement with the observations reported by Allen [1986].

The calculated profiles of various hydrogen species are shown in Figure 6a. H_2O and H_2 decrease steadily throughout the mesosphere with molecular diffusion affecting the H_2 profile above about 200 km. Below that altitude both species are subjected to various chemical reactions that are listed in Table 1. The number densities for both HO_2 and OH are relatively constant up to about 80 km and above that altitude decrease rapidly with increasing altitude. Atomic hydrogen is in photochemical equilibrium in the lower mesosphere with a peak value of $1.9 \times 10^8 \text{cm}^{-3}$ at 85 km. Above that altitude atomic hydrogen decreases with altitude with molecular diffusion and exospheric escape controlling the structure of the profile in the thermosphere. The calculated total escape flux is $2.3 \times 10^8 \text{cm}^{-2}\text{s}^{-1}$ with Jean's escape, charge exchange and polar wind contributing 4%, 95%, and 1% of the total flux respectively.

A globally averaged atomic hydrogen profile obtained from MSIS-90 is also shown. It has a peak density of $2.2 \times 10^8 \text{cm}^{-3}$ at 86 km. The calculated exospheric or upper boundary H density is $1.9 \times 10^5 \text{cm}^{-3}$ versus $2.8 \times 10^5 \text{cm}^{-3}$ obtained from MSIS-90 at 400 km.

The calculated globally averaged number density profiles for CH_4, CO_2 and CO are shown in Figure 6b. CO_2 is fully mixed in the mesosphere up to the turbopause and slowly decreases to a mixing ratio of 300 ppm at 100 km. Above the turbopause, CO_2 adapts a near diffusive equilibrium profile modified by dissociation from solar EUV processes above about 140 km. The CO volume mixing ratio peaks at 25 ppm near 123 km. The number density of CO exceeds that of CO_2 above 130 km. The methane number density profile decreases with altitude throughout the mesosphere and thermosphere.

Finally, the calculated electron and ion number density profiles in the mesosphere and thermosphere and electrical conductivities are shown in Figure 6c and 6d, respectively. In the upper thermosphere O^+ dominates with a peak density equal to the electron density of $3.0 \times 10^5 \text{cm}^{-3}$ occurring near 254 km. NO^+ becomes the dominant ion below 175 km and O_2^+ dominates in the E-region between 120 km and 90 km. Below 90 km NO^+ and positive ion clusters balance the electron and negative ion density structure.

Below about 75 km the electrical conductivity is isotropic. Above that altitude the increasing importance of the Earth's magnetic field in controlling electron and ion motions causes the electrical conductivity to become anisotropic with the largest electrical conductivity occurring along the geomagnetic field line. The Pedersen conductivity has a minor peak in the D-region near 80 km from the electron contribution and a peak near 140 km from the ion contribution to the total Pedersen electrical conductivity. The Hall conductivity has a peak near 105 km. These conductivities are an important components for electrodynamics in the TIE-GCM.

6. SUMMARY AND CONCLUSIONS

The self-consistent global mean model of the thermosphere, ionosphere, and mesosphere developed by Roble et al. [1987] and Roble and Dickinson [1989] has been extended downward into the upper stratosphere near 30 km (10 mb). The model is a consolidation of many of the physical and chemical processes that are known to be operating within the thermosphere, ionosphere, mesosphere, and upper stratosphere. It is used to examine the basic global mean structure of the atmosphere above 30 km for solar minimum geomag-

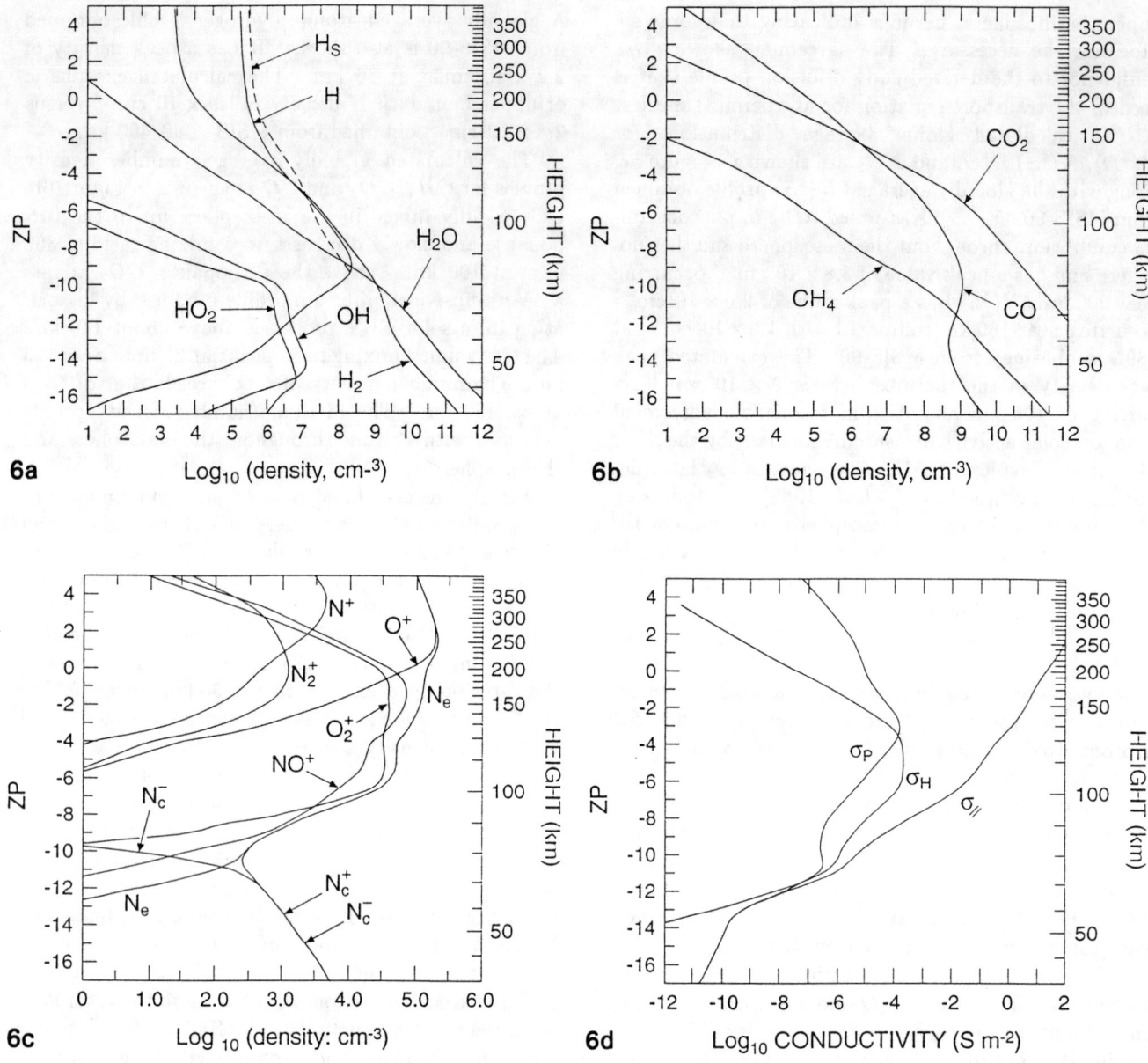

Fig. 6. Calculated (a) $\log_{10}$ number density profiles (cm^{-3}) for OH, HO_2, H_2O, H_2, and H, also shown is atomic hydrogen, H_s obtained from MSIS-86, (b) $\log_{10}$ number density profiles (cm^{-3}) for CO_2, CO, and CH_4, and (c) $\log_{10}$ number density profiles (cm^{-3}) for n_e, O^+, NO^+, O_2^+, N^+, N_2^+, and water cluster ions (being the difference between n_e and NO^+ in the D-region (d) $\log_{10}$ profiles of electrical conductivity (Sm^{-2}), σ_H, σ_P, and $\sigma_\parallel$ are the Hall, Pedersen and Parallel conductivities respectively.

netic quiet conditions and to determine the component processes that contribute to an overall energy balance. The only adjustable parameters used in the calculation to achieve reasonable agreement between model calculated profiles of neutral temperature and O, O_2, N_2, H, $N(^4S)$, He, and Ar number densities with similar global mean profiles obtained from MSIS-90 were the assumed eddy diffusion profile and Prandtl number. Reasonable agreement of the calculated temperature profile with the empirical model was obtained at the stratopause, in the lower mesosphere and in the thermosphere. There are considerable differences, however, in the 80 to 120 km region that are difficult to resolve with the assumption of a simplified eddy diffusion profile. The empirical model shows a general temperature decrease with altitude in the upper mesosphere reaching a minimum temperature near 90 km of 185 K before increasing in the thermosphere. The calculated profile differs because chemical heating [*Mlynczak and Solomon*, 1993] produces a temperature bulge near 85 km and the strong CO_2 cooling from collisions with O at a rate of 4×10^{-12} cm s^{-2} produce a temperature minimum of 177 K near 100 km. There is a growing body of observational evidence [e.g., *She et al.*, 1993] that suggest a more complex mesopause temperature profile than current empirical models indicate.

In spite of these mesopause difficulties, the study suggests that a reasonable global mean structure of the upper atmosphere, above the stratopause, can be calculated from our present understanding of the aeronomic processes that are operating in the Earth's upper atmosphere. However, there are many details that require further analysis using two-dimensional models and a three-dimensional time-dependent GCM.

REFERENCES

Akmaev, R. A., V. I. Fomichev, V. I. Gavrilov, and G. M. Shved, Simulation of the zonal mean climatology of the middle atmosphere with a three-dimensional spectral model for solstice and equinox conditions, *J. Atmos. Terr. Phys.*, *54*, 119-128, 1992.

Allen, M., A new source of ozone in the terrestrial upper atmosphere?, *J. Geophys. Res.*, *91*, 2844-2848, 1986.

Allen, M., J. I. Lunine, and Y. L. Yung, The vertical distribution of ozone in the mesosphere and lower thermosphere, *J. Geophys. Res.*, *89*, 4841-4872, 1984.

Appruzese, J. P., D. F. Strobel, and M. R. Schoeberl, Parameterization of IR cooling in a middle atmosphere dynamics model, 2, Non-LTE radiative transfer and the globally averaged temperature of the mesosphere and lower thermosphere, *J. Geophys. Res.*, *89*, 4917-4926, 1984.

Banks, P. M., and G. Kockarts, *Aeronomy*, Academic Press, New York, 1973.

Barth, C. A., K. Tobiska, D. E. Siskind, and D. D. Cleary, Solar-terrestrial coupling: Low-latitude thermospheric nitric oxide, *Geophys. Res. Lett.*, *15*, 92-94, 1988.

Bates, D. R., The temperature of the upper atmosphere, *Proc. Phys. Soc. London, Sect. B*, *64*, 805-820, 1951.

Bates, D. R., Deficiency in model ozone in the lower thermosphere and excitation of 557.7 nm in nightglow, *Planet. Space Sci.*, *36*, 1077-1084, 1988.

Bougher, S. W., and R. G. Roble, Comparative terrestrial planet thermospheres, 1, Solar cycle variation of global mean temperatures, *J. Geophys. Res.*, *96*, 11045-11055, 1991.

Brasseur, G., and D. Offerman, Recombination of atomic oxygen near the mesopause: Interpretation of rocket data, *J. Geophys. Res.*, *91*, 10,818-10,824, 1986.

Brasseur, G., and S. Solomon, *Aeronomy of the Middle Atmosphere*, D. Reidel Publishing Company, Dordrecht, 441 pp., 1984.

Brasseur, G., M. H. Hutchman, S. Waters, M. Dymek, E. Falsie, and M. Pine, An interactive chemical dynamical radiative two-dimensional model of the middle atmosphere, *J. Geophys. Res.*, *95*, 5639-5655, 1990.

Clancy, R. T., D. W. Rusch, and R. J. Thomas, Model ozone photochemistry on the basis of Solar Mesosphere Explorer mesospheric observations, *J. Geophys. Res.*, *92*, 3067-3080, 1987.

De More, W. B., et al., Chemical kinetics and photchemical data for use in stratospheric modeling, Evaluation Number 9, JPL-Publication 90-1, NASA-JPL, Pasadena, CA, 1990.

Dickinson, R. E., Infrared radiative cooling in the mesosphere and lower thermosphere, *J. Atmos. Terr. Phys.*, *46*, 995-1008, 1984.

Dickinson, R. E., E. C. Ridley, and R. G. Roble, General circulation with coupled dynamics and composition, *J. Atmos. Sci.*, *41*, 205-219, 1984.

Dickinson, R. E., R. G. Roble, and S. W. Bougher, Radiative cooling in the NLTE region of the mesosphere and lower thermosphere - global energy balance, *Adv. Space. Res.*, *7*, (10)5-(10)15, 1987.

Donnelly, R. F., H. E. Hinteregger, and D. F. Heath, Temporal variations of solar EUV, UV, and 10,830-Å radiations, *J. Geophys. Res.*, *91*, 5567-5578, 1986.

Fomichev, V. I., and G. M. Shved, Parameterization of the radiative flux divergences in the 9.6 μm O_3 band, *J. Atmos. Terr. Phys.*, *47*, 1037-1049, 1985.

Fomichev, V. I., and G. M. Shved, Net radiative heating in the middle atmosphere, *J. Atmos. Terr. Phys.*, *50*, 671-688, 1988.

Fomichev, V. I., W. E. Ward, and C. Mc Landress, The effect of variations in the oxygen mixing ratio in current climatological models on the 15 μm CO_2 cooling and associated dynamic structure in the mesosphere and lower thermosphere, *J. Atmos. Terr. Phys.*, in press, 1994.

Fomichev, V. I., A. A. Kutepov, R. A. Akmaev, and G. M. Shved, Parameterization of the 15 μm CO_2 band cooling in the middle atmosphere (15-115 km), *J. Atmos. Terr. Phys.*, *55*, 7-18, 1993.

Garcia, R. R., and S. Solomon, A numerical model of the zonally averaged dynamical and chemical structure of the middle atmosphere, *J. Geophys. Res.*, *88*, 1379-1400, 1983.

Garcia, R. R., and S. Solomon, The effect of breaking gravity waves on the dynamics and chemical composition of the mesosphere and lower thermosphere, *J. Geophys. Res.*, *90*, 3850-3868, 1985.

Gavrilov, N. M., and R. G. Roble, The effect of gravity waves on the global mean temperature and compositional structure of the upper atmosphere, *J. Geophys. Res.*, submitted, 1994.

Gordiets, B. F., Yu. N. Kulikov, M. N. Markov, and M. Ya. Marov, Numerical modelling of the thermospheric heat budget, *J. Geophys. Res.*, *87*, 4504-4514, 1982.

Hedin, A. E., Extension of the MSIS thermospheric model into the middle and lower atmosphere, *J. Geophys. Res.*, *96*, 1159-1172, 1991.

Hinteregger, H. E., Representation of solar EUV fluxes for aeronomical applications, *Adv. Space. Res.*, *1*, 39-42, 1981.

Holton, J. R., A dynamically based transport parameterization for one-dimnsional photochemical models of the stratosphere, *J. Geophys. Res.*, *91*, 2681-2686, 1986.

Hunten, D. M., and T. M. Donahue, Hydrogen loss from the terrestrial planets, *Ann. Rev. Earth and Planet. Sci.*, *4*, 265-292, 1976.

Kiehl, J. T., and S. Solomon, On the radiative balance of the stratosphere, *J. Atmos. Sci.*, *43*, 1525-1534, 1986.

Kocharts, G., Nitric oxide cooling in the terrestrial thermosphere, *Gephys. Res. Lett.*, *7*, 137-140, 1980.

Lean, J., Solar ultraviolet irradiance variations: A review, *J. Geophys. Res.*, *92*, 839-868, 1987.

Liu, S. C., and T. M. Donahue, Mesospheric hydrogen related to exospheric escape mechanisms, *J. Atmos. Sci.*, *31*, 1466-1470, 1974.

London, J., Radiative energy sources and sinks in the stratosphere and mesosphere, *Proc. NATO Adv. Study Inst.*, U.S. Dept. of Transportation, FAA, Washington, D.C., 1980.

Lopez-Puertas, M., M. A. Lopez-Valverde, C. P. Rinsland, and M. R. Gunson, Analysis of the upper atmosphere CO_2 (u2) vibrational temperatures retrieved from ATMOS/Spacelab 3 observations, *J. Geophys. Res.*, in press, 1993.

Mlynczak, M. G., and S. Solomon, On the efficiency of solar heating in the middle atmosphere, *Geophys. Res. Lett.*, *18*, 1201-1204, 1991.

Mlynczak, M. G., and S. Solomon, A detailed evaluation of the heating efficiency in the middle atmosphere, *J. Geophys. Res.*, in press, 1993.

Park, J. H., and J. London, Ozone photochemistry and radiative heating of the middle atmosphere, *J. Atmos. Sci.*, *31*, 1898-1916, 1974.

Pollock, D. S., G. B. I. Scott, and L. F. Phillips, Rate constant for quenching of CO_2 (010) by atomic oxygen, *Geophys. Res. Lett.*, *20*, 727-730, 1993.

Rees, M. H., *Physics and Chemistry of the Upper Atmosphere*, Cambridge University Press, 289 pp., 1989.

Reid, G. C., The production of water-cluster positive ions in the quiet daytime D region, *Planet. Space Sci.*, *25*, 275-290, 1977.

Reid, G. C., Ion chemistry in the D-region, *Adv. At. Mol. Phys.*, *12*, 375-411, 1986.

Richmond, A. D., E. C. Ridley, and R. G. Roble, A thermosphere/ionosphere general circulation model with coupled electrodynamics, *Geophys. Res. Lett.*, *19*, 601-604, 1992.

Roble, R. G., Solar EUV flux variation during a solar cycle as derived from ionospheric modeling considerations, *J. Geophys. Res.*, *81*, 265-270, 1976.

Roble, R. G., and R. E. Dickinson, Is there enough solar extreme ultra-violet radiation to maintain the global mean thermospheric temperature? *J. Geophys. Res.*, *78*, 249-257, 1973.

Roble, R. G., and R. E. Dickinson, How will changes in carbon dioxide and methane modify the mean structure of the mesosphere and thermosphere, *Geophys. Res. Lett.*, *16*, 1441-1444, 1989.

Roble, R. G., and B. A. Emery, On the global mean temperature of the thermosphere, *Planet. Space Sci.*, *31*, 597-614, 1983.

Roble, R. G., and E. C. Ridley, An auroral model for the NCAR thermospheric general circulation model (TGCM), *Annales. Geophysicae*, *5*, 369-382, 1987.

Roble, R. G., and E. C. Ridley, A thermosphere-ionosphere-mesosphere-electrodynamics general circulation model (TIME-GCM): Equniox solar minimum simulation, 30-500 km, *Geophys. Res. Lett.*, submitted, 1994.

Roble, R. G., E. C. Ridley, and R. E. Dickinson, On the global mean structure of the thermosphere, *J. Geophys. Res.*, *92*, 8745-8758, 1987.

Roble, R. G., E. C. Ridley, A. D. Richmond, and R. E. Dickinson, A coupled thermosphere/ionosphere general circulation model, *Geophys. Res. Lett.*, *15*, 1325-1328, 1988.

Rogers, C. D., F. W. Taylor, A. H. Muggeridge, M. Lopez-Puertas, and M. A. Lopez-Valverde, Local thermodynamic equilibrium of carbon dioxide in the upper atmosphere, *Geophys. Res. Lett.*, *19*, 589-592, 1992.

Rottman, G. J., Rocket measurements of the solar spectral irradiance during solar minimum 1972-1977, *J. Geophys. Res.*, *86*, 6697-6705, 1981.

Rottman, G. J., *et al.*, WMO, atmospheric ozone, assessment of our understanding of processes controlling its present distribution and change, report No. 16, *World Meteorological Organization Global Ozone Research and Monitoring Project*, Volume 1, Washington, D.C., 1986.

Rusch, D. W., and R. S. Eckman, Implications of the comparison of ozone abundance measured by the Solar-

Mesosphere Explorer to model calculations, *J. Geophys. Res.*, *90*, 12,991-12,998, 1985.

Sharma, R. D., and P. P. Wintersteiner, Role of carbon dioxide in cooling planetary atmospheres, *Geophys. Res. Lett.*, *17*, 2201-2204, 1990.

She, C. Y., J. R. Yu, and H. Chen, Observed thermal structure of a midlatitude mesopause, *Geophys. Res. Lett.*, *20*, 567-570, 1993.

Shved, G. M., L. E. Khvorostovskaya, I. Yu. Potekhim, A. I. Demyanikov, A. A. Kutepov, and V. I. Fomichev, Measurement of the quenching rate constant for collisions $CO_2(01^10) - O$: The importance of the rate constant magnitude for the thermal regime and radiation of the lower thermosphere, *Atmos. and Oceanic Phys.*, *27*, 431-437, 1991.

Solomon, S., D. W. Rusch, R. J. Thomas, and R. S. Eckman, Comparison of mesospheric ozone abundances measured by the solar mesosphere explorer and model calculations, *Geophys. Res. Lett.*, *10*, 249-252, 1983.

Solomon, S., R. R. Garcia, J. J. Olivero, R. M. Bevilacqua, P. R. Schwartz, R. T. Clancy, and D. O. Muhleman, Photochemistry and transport of carbon monoxide in the middle atmosphere, *J. Atmos. Sci.*, *42*, 1072-1083, 1985.

Stolarski, R. S., Energetics of the midlatitude thermosphere, *J. Atmos. Terr. Phys*, *38*, 863-868, 1976.

Stolarski, R. S., P. B. Hays, and R. G. Roble, Atmospheric heating by solar EUV radiation, *J. Geophys. Res.*, *80*, 2266-2276, 1975.

Stordal, F., and R. R. Garcia, Sensitivity studies and a simple ozone perturbation experiment with a truncated two-dimensional model of the stratosphere, *J. Geophys. Res.*, *92*, 11,909-11,918, 1987.

Strobel, D. F., Constraints on gravity wave induced diffusion in the middle atmosphere, *PAGEOPH*, *130*, 533-546, 1989.

Strobel, D. F., J. P. Appruzese, and M. R. Schoeberl, Energy balance constraints on gravity wave induced eddy diffusion in the mesosphere and lower thermosphere, *J. Geophys. Res.*, *90*, 13,067-13,072, 1985.

Strobel, D. F., M. E. Summers, R. M. Bevilacqua, M. T. DeLand, and M. Allen, Vertical constituent transport in the mesosphere, *J. Geophys. Res.*, *92*, 6691-6698, 1987.

Torr, D. G., and M. R. Torr, Ionization frequencies for solar cycle 21: Revised, *J. Geophys. Res.*, *90*, 6675-6678, 1985.

Torr, M. R., D. G. Torr, R. A. Ong, and H. E. Hinteregger, Ionization frequencies for major thermospheric constituents as a function of solar cycle 21, *Geophys. Res. Lett.*, *6*, 771-774, 1979.

Torr, M. R., D. G. Torr, P. G. Richards, and S. P. Yung, Mid- and low-latitude model of thermospheric emissions, 1, $O^+(^2P)$ 7320 Å and $N_2(\alpha P)$ 3371 Å, *J. Geophys. Res.*, *95*, 21,147-21,168, 1990.

Ward, W. E., and V. I. Fomichev, On the role of atomic oxygen in the dynamics and energy budget of the mesosphere and lower thermosphere, *Geophys. Res. Lett.*, in press, 1993.

R. G. Roble, High Altitude Observatory, National Center for Atmospheric Research, P.O. Box 3000, Boulder, CO 80307-3000.

The Dynamics of the Lower Thermosphere

T.J. Fuller-Rowell

CIRES, University of Colorado, and NOAA Space Environment Laboratory, Boulder, Colorado

The dynamics of the lower thermosphere is strongly forced by a number of external processes. On a global average, the dominant heat source from 100 to 150 km altitude is O_2 absorption in the Schumann-Runge continuum. This heat input creates global scale pressure gradients, setting the lower thermosphere into motion. Earth's rotation acts on the wind field to try to establish geostrophic balance; this is one of the most fundamental relationships in atmospheric dynamics. Solar heating drives a weak semi-diurnal wind field in the lower thermosphere, but is overshadowed by waves propagating upward from sources in the lower atmosphere. One set of waves, the diurnal and semi-diurnal propagating tides, increases the amplitude of wind and temperature oscillations significantly. The impact of planetary and gravity waves on lower thermosphere dynamics is just beginning to be addressed. At high latitudes, another driving force originating from the magnetosphere can dominate the dynamics. Magnetospheric convection drives ionospheric plasma, and, through collisions with the neutrals, provides momentum and energy to the system. The ion-drag momentum source can drive winds of hundreds of meters per second. In these circumstances non-linear effects become dominant, and inertial resonance can influence the motion. Strong, divergent neutral-wind flow, driven by ion drag, generates cold cells; and heat from ohmic dissipation disturbs the local and global-scale temperature and pressure fields.

1. INTRODUCTION

Observation of the wind field in the lower thermosphere began with studies of chemical trails from rocket releases [e.g. *Rees*, 1971; *Pereira et al.*, 1980; *Heppner and Miller*, 1982]. It was clear that the vertical structure was complex and that easy interpretation of the driving mechanisms of the wind field was not possible from the early data. Since then several other techniques have been utilized, including incoherent scatter radar [*Johnson et al.*, 1987; *Salah et al.*, 1991] and remote sensing of optical emissions such as the oxygen line at 557.7 nm from satellite platforms [*Killeen et al.*, 1992]. Other radar techniques such as meteor wind and medium frequency are unable to penetrate significantly into the lower thermosphere [*Manson et al.*, 1991], but they can be used to define the forcing of the lower thermosphere. The complexity and variability of this atmospheric region has been confirmed by all subsequent observations.

A notable characteristic of the lower thermosphere is the transition from turbulent mixing to molecular diffusion at about 110 km. The height of the turbopause can be variable since it arises from the action of breaking gravity waves, of the most variable of processes. From a number of different sources in the lower atmosphere a spectrum of waves propagate upward and interact in the mesosphere and lower thermosphere.

Gravity (or buoyancy) waves interact with the atmosphere in at least three distinct ways. As the waves propagate in altitude the negative density gradient causes their amplitudes to grow. At some point the amplitude of a wave will force local temperature gradients to exceed the lapse rate; at this point the wave becomes unstable and is expected to break [*Lindzen*, 1981]. This process is also referred to as saturation. The "dissipation" of the wave in this circumstance causes turbulent mixing of the medium and a small deposition of heat. It was suggested by *Hines* [1965] that the waves would generate sufficient turbulence to prevent further growth. Vertical mixing of gas parcels transport potential temperature between layers and pushes the temperature profile closer to the adiabatic lapse rate to apparently cool the atmosphere [*Schoeberl et al.*, 1983; *Fuller-Rowell and Rees*, 1992]. Mixing also transfers mo-

The Upper Mesosphere and Lower Thermosphere:
A Review of Experiment and Theory
Geophysical Monograph 87
Copyright 1995 by the American Geophysical Union

mentum between layers, and, if the wave has a phase speed that differs from the mean flow, the breaking wave can provide a momentum source that accelerates or retards the background motion field [*Fritts*, 1984].

A second type of interaction of gravity waves with the background medium arises when the phase speed of the wave matches the background wind field and critical absorption occurs. This can be considered a special case of saturation, and is accompanied by aforementioned turbulent mixing and momentum deposition.

If the wave does not encounter a critical layer, and breaking does not occur, a third mechanism takes over where the wave is dissipated by ion drag or molecular diffusion in the thermosphere. The altitude and type of dissipation mechanism will depend on the amplitude and speed of the wave, and its fate will depend critically on the intervening medium through which it propagates.

Gravity waves are therefore an important driver of dynamical change, particularly in the mesosphere but also in the lower thermosphere. Sources of gravity waves include airflow over topography, convective storms, and other tropospheric features such as frontal activity. The nature of the source and the characteristics of the various dissipation processes renders the outcome highly variable. Strong seasonal, and probably latitudinal, variations are likely, and the altitude of the turbopause can vary. In the mesosphere the closing of the zonal jets, the generation of meridional flow, and the high-latitude seasonal mesopause temperature anomaly are all thought to be a consequence of gravity wave momentum dissipation.

At low latitude the breaking of the propagating diurnal (1,1) tide can produce a similar effect. The diurnal tide is generated in the troposphere by absorption of solar radiation by water vapor [*Groves and Forbes*, 1984]. *Lindzen and Blake* [1971] estimated that this mode would break between 80 and 90 km, generating turbulence up to 108 km, and that its growth would be inhibited by ion drag and molecular diffusion at greater altitudes. The dynamical effect of the diurnal tide has been estimated by *Forbes et al.* [1993] in more detail. They show acceleration of the zonal flow in the lower thermosphere due to molecular and eddy dissipation, and generation of a multi-cell structure in the meridional wind within 30° of the equator.

There are, however, many large-scale propagating tidal waves that do not break but penetrate deep into the thermosphere. Tides are global-scale harmonic variations, with periods of a whole or fraction of a day, and can be thought of as a particular type of gravity wave. The oscillations we are concerned with here are westward-traveling waves that follow the Sun, rather than being fixed relative to geographic locations on Earth. With the exception of the propagating diurnal tide discussed above the important modes that drive the lower thermosphere are thermally generated in the stratosphere by the absorption of solar radiation by ozone [*Forbes*, 1994; this issue].

Propagating semi-diurnal tides tend to dominate this spectrum because the propagating diurnal tides are suppressed. Most of the diurnal modes have short vertical wavelengths [*Forbes*, 1994; this issue] so cancellation occurs over the source height. The (1,1) diurnal mode is the exception, with a 30 km vertical wavelength that enables it to reach the lower thermosphere. In general, there is a preference for exciting global tidal modes where the vertical and horizontal structures match the distribution of the forcing [*Forbes*, 1994; this issue].

Tides are typically classified by Hough modes, which describe their latitude structure and their period. The dominant oscillations that are of interest to the dynamics of the lower thermosphere are the semi-diurnal modes (2,2), (2,3), (2,4), (2,5) and (2,6). The effect that some of these modes have on the wind and temperature fields will be illustrated later.

Planetary waves are also of global scale but are not necessarily tied to the location of the Sun. Planetary waves can be forced by features at Earth's surface, such as mountains or large land masses, or by meteorological patterns. The waves propagate vertically into the stratosphere and mesosphere, or they can be generated in-situ. Planetary waves with periods of 2, 5, 10 and 16 days have been documented in the MLT region for many years [see review by *Salby*, 1984, and papers by *Vincent*, 1990; *Manson et al.*, 1981, 1982; and *Clark*, 1983]. Recent observation by the Upper Atmosphere Research Satellite (UARS) have revealed the strong influence of planetary waves on the upper mesosphere [*Hays*, 1993]; their effect on the lower thermosphere is unknown.

Gravity waves, tides, and planetary waves are all sources of energy that propagate into the lower thermosphere from below. In-situ absorption of solar radiation and magnetospheric sources are two other inputs important to the dynamics of the lower thermosphere.

On a global average the dominant heat source in the lower thermosphere is O_2 absorption in the Schumann-Runge continuum [*Roble et al.*, 1987]. This heat, deposited primarily between 100 and 150 km, creates global-scale pressure gradients that set the atmosphere into motion. This in-situ solar forcing, although driving a semi-diurnal wind field, is overshadowed by the tidal forces propagating from below.

The thermosphere can be considered as a weakly ionized

plasma pervaded by a magnetic field. If no external electric field is present, the ions are constrained by the magnetic field. Collisions between the ions and neutral are sufficiently frequent that they affect each other. Neutral winds driven into motion by pressure gradients will experience a resistance, known as ion-drag, which slows the wind. Collisions of the neutral atmosphere with electrons are not significant in the thermosphere, but ions are partially carried by the action of the neutral atmosphere so that currents flow. Divergence of the current leads to the build up of charges that creates polarization electric fields [*Richmond et al.*, 1992]. Although *Rishbeth* [1971] has shown that F-region polarization is also important, this process is often referred to as the E-region dynamo.

At high latitudes the magnetosphere imposes an important electrodynamic source. The driver can be separated into auroral precipitation and convective electric field. Empirical models of the aurora, as a function of magnetospheric activity, have been developed by *Spiro et al.* [1982], *Hardy et al.* [1985], and *Fuller-Rowell and Evans* [1987]. They all show the expansion and intensification of the auroral oval as the magnetosphere becomes more disturbed as it is driven more strongly by the solar wind. Auroral ionization enhances conductivity, particularly in the lower thermosphere, and heat is deposited.

The more important magnetospheric source for neutral dynamics is the electric field [*Heppner and Maynard*, 1987; *Foster et al.*, 1986], which is mapped by Earth's magnetic field into the high-latitude ionosphere and drives the plasma into motion. Collisions between the ions and neutrals can accelerate the neutral atmosphere to several hundreds of meters per second, even in the relatively dense lower thermosphere. At these velocities the fluid motion becomes highly non-linear. Frictional heating associated with the collisions between ions and neutrals moving at different velocities gives rise to Joule heating, which is typically four to five times more intense than heating by auroral precipitation [*Evans et al.*, 1988].

2. EQUATIONS OF MOTION

The lower thermosphere, although a tenuous gas, is still collision dominated and isotropic. The consequence is that the basic equations of fluid dynamics, i.e. the Navier-Stokes expressions, can be applied to the system [*Conrad and Schunk*, 1979]. The nature of the atmospheric response to these many sources can be interpreted by these equations. The equation of motion describing the balance of forces acting on a parcel of neutral gas and the resultant time rate of change of velocity is given by

$$\frac{D}{Dt}V = -\frac{1}{\varrho}\nabla P - 2\Omega \wedge V - v_{ni}(V - U) + \frac{1}{\varrho}\nabla(\mu\nabla V), \quad (1)$$
$$\text{pressure} \quad \text{Coriolis} \quad \text{ion drag} \quad \text{viscosity}$$
$$\text{gradient}$$

where V is the neutral wind velocity, P the gas pressure, Ω the angular rotation rate of Earth, U the ion drift velocity, v_{ni} the neutral-ion collision frequency, ϱ gas density, and μ the sum of the molecular and turbulent viscosity coefficient. The four main forces acting on a parcel of gas are pressure gradient, Coriolis, ion drag and viscosity. Pressure gradients are produced by heating from solar radiation or Joule dissipation, or are generated by tidal fields propagating from the lower atmosphere. As the atmosphere is forced into motion the Coriolis force begins to act; this tends to move the gas in a clockwise sense in the northern hemisphere and anti-clockwise in the south. The ion drag term represents the collisional interaction between the neutral and plasma components. This term can either inhibit the neutral motion or, if an electric field is present to accelerate the ions, can provide a strong source of acceleration, driving winds to high velocity. The final term is viscosity, which acts to smooth gradients in the wind field, particularly over the short vertical distances. In the lower thermosphere there is a region of cross-over from turbulent to molecular diffusion, where both are small. Such a region enables steep vertical shear to persist and is partly the cause of the high degree of vertical structure in the lower thermosphere.

The above equation is appropriate for the forces acting on a parcel of gas as it moves with respect to Earth, the so-called Lagrangian frame of reference. To transform to Earth's frame, or the so-called Eulerian frame, where latitude and longitude are the independent variables, the following transformation is required:

$$\frac{D}{Dt}X = \frac{\partial}{\partial t}X + (V \cdot \nabla)X, \quad (2)$$

where X is any property such as temperature or velocity of the fluid.

The partial derivative with respect to time represents the rate of change of X at a fixed point on Earth, and is related to the total derivative by the advection term. The process of advection simply represents the change of a property of the fluid as a result of transport by the wind field past a fixed location.

We can now express the change in velocity of the two wind components, southward (V_θ) and eastward (V_ϕ), by the following equations:

$$\frac{\partial}{\partial t}V_\theta = \frac{V_\theta}{r}\frac{\partial}{\partial \theta}V_\theta - \frac{V_\phi}{r\sin\theta}\frac{\partial}{\partial \phi}V_\theta - V_z\frac{\partial}{\partial z}V_\theta \quad (3)$$

$$+ \left(2\Omega + \frac{V_\phi}{r\sin\theta}\right)V_\phi\cos\theta + \frac{1}{\varrho}\frac{\partial}{\partial z}\left(\mu\frac{\partial}{\partial z}V_\theta\right)$$

$$- \nu_{ni}(V_\theta - U_\theta) - \frac{1}{\varrho r}\frac{\partial}{\partial \theta}P,$$

and

$$\frac{\partial}{\partial t}V_\phi = \frac{V_\theta}{r}\frac{\partial}{\partial \theta}V_\phi - \frac{V_\phi}{r\sin\theta}\frac{\partial}{\partial \phi}V_\phi - V_z\frac{\partial}{\partial z}V_\phi \quad (4)$$

$$- \left(2\Omega + \frac{V_\phi}{r\sin\theta}\right)V_\theta\cos\theta + \frac{1}{\varrho}\frac{\partial}{\partial z}\left(\mu\frac{\partial}{\partial z}V_\phi\right)$$

$$- \nu_{ni}(V_\phi - U_\phi) - \frac{1}{\varrho r}\frac{\partial}{\partial \phi}P,$$

where r is Earth's radius, θ is co-latitude, ϕ is longitude, and z is altitude.

3. DYNAMICAL BALANCES

One of the most fundamental dynamical relationships in the atmosphere of a rotating planet is the meteorological concept of geostrophic balance (Figure 1a), where wind vectors follow isobars. Although never exact, an approximate balance between the pressure gradient and the Coriolis force applies in many regions of the atmosphere, particularly in the upper troposphere and stratosphere. Nowhere in the upper atmosphere is geostrophic balance a good approximation. The midlatitude lower thermosphere, between 120 and 150 km altitude, is the region of the upper atmosphere where the closest approach to this state might be expected. This altitude range is above the region where gravity waves drag controls the dynamics and below where ion drag and molecular viscosity dominate. A requirement for such an approximation is that the wind velocities are small and the Rossby number, the ratio of acceleration to Coriolis, is much less than one. Forcing by planetary wave and tidal forcing upsets this balance. A semi-diurnal tidal oscillation of amplitude 50 m/s implies a change of wind of 100 m/s over a six hour period corresponding to an acceleration of 4.6×10^{-3} m/s^2. The Coriolis acceleration at midlatitude on a 50 m/s wind is approximately 5.2×10^{-3} m/s^2. Under these circumstances the Rossby number approaches one, which implies that quasi-geostrophic balance is not valid even in this most favorable region of the upper atmosphere.

If drag is introduced, such as that produced by gravity wave breaking or by ion drag, the wind vector reduces in magnitude and rotates toward the direction of the pressure gradient (Figure 1b). Then the Coriolis force associated with the reduced vector can no longer balance the pressure

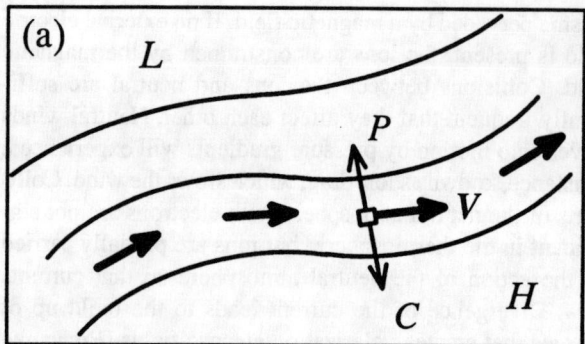
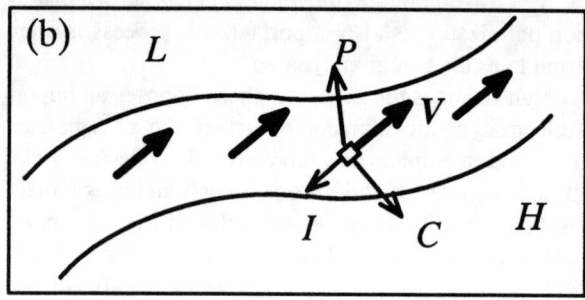
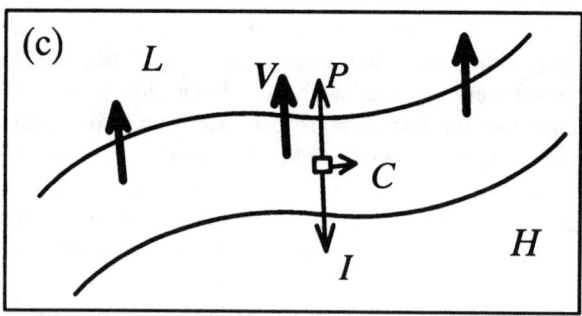

Figure 1. An illustration of dynamical balances in the atmosphere; (a) geostrophic balance: where pressure and Coriolis are the main forces, both perpendicular to the wind flow along the isobars; (b) the introduction of a small drag or friction force retards the wind, which now has a component across the isobars; and (c) where the drag and pressure are large, compared with Coriolis, and the wind field is nearly parallel to the pressure gradient. V represents the neutral wind flow; of the three forces acting on the gas, P represents pressure, C is Coriolis, and I is drag (such as ion drag). L represents a region of low pressure, and H is a region of high pressure.

force. If this drag force is large in comparison to Coriolis, the wind vector is nearly parallel to the pressure gradient (Figure 1c). This situation exists in the upper thermosphere, where pressure and ion drag are the primary forces [Killeen and Roble, 1984].

At high latitude, ion drag is no longer a sink of momentum. The magnetospheric electric field mapped into the ionosphere drives ions at several hundreds of meters per second. Collisions with the neutral atmosphere accelerate the fluid to a point where advection becomes important; the fluid motion becomes non-linear. In this case the Rossby number can approach or exceed one. If the velocity is high, another quasi dynamical balance can evolve, in which the acceleration, or curvature term, and Coriolis are the main forces; this state is close to inertial resonance (Figure 2). Such a resonance, or oscillation, has been observed in Earth's oceans. Once excited, the inertial resonance vortex can persist for many hours in the absence of dissipation. Inertial motion is the simplest form of dynamical circulation on a rotating planet where parcels of gas follow their natural motion: a clockwise vortex in the north and anticlockwise in the south.

In the equatorial region, Coriolis forces tends to zero and a new regime unfolds. The breaking of the propagating diurnal tide is important, and strong zonal winds are possible.

4. ZONAL AVERAGE CIRCULATION

Figure 3 shows the climatological annual average meridional and zonal winds from Hedin et al. (1993). In the atmosphere, equinox is a time of transition between the solstice-type circulations; perfect symmetry is not expected due to the time lag for reaching equilibrium. For this reason the annual average of the empirical model is used as representative of equinox condition. The region from the south geographic pole to the north pole is depicted from 80 to 200 km altitude. The meridional wind is positive northward, shown by the solid contours with 2 m/s spacing; the zonal wind is positive east with a 5 m/s interval.

The pattern of meridional wind is symmetric about the equator and has four features that alternate the direction of flow between low and high altitudes. At high altitudes the circulation is poleward below 40° latitude and equatorward at higher latitudes. Between 120 and 150 km the flow reverses at low latitudes to be consistently equatorward. Below 120 km the circulation reverses again, returning to poleward and interrupted only by a small region of equatorward flow that emerges at low latitudes below 110 km.

Zonal wind climatology is less symmetric. Eastward superrotation dominates the equatorial region above 140 km; at high latitude westward subrotation takes over. The flow is westward between 115 and 140 km at all latitudes. Eastward flow returns below 110 km, with peaks at the equator and at high latitudes.

One of the prominent features of the circulation over much of the region above 120 km, but particularly at mid and high latitudes, is the equatorward wind, driven by the high-latitude magnetospheric heat source. The pole-to-

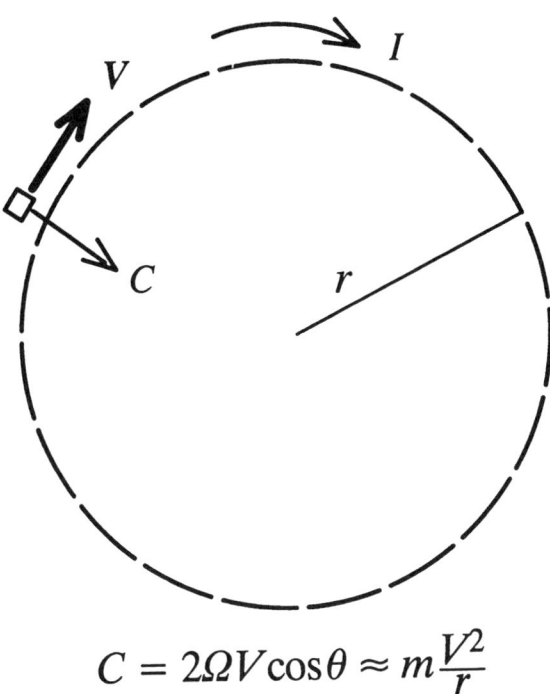

Figure 2. An illustration of the inertial oscillation where the acceleration or curvature term balances Coriolis: the natural motion of particles. V represents the neutral wind flow, and the two main forces are Coriolis (represented by C) and ion drag (represented by I).

$$C = 2\Omega V \cos\theta \approx m\frac{V^2}{r}$$

equator pressure gradient induces a westward flow by the action of Coriolis torque. The reversal of both wind components above 150 km within 45 of the equator is in response to solar heating, which generates an equator-to-midlatitude pressure gradient. The zonal superrotation here may be a consequence of the interaction of the neutral atmosphere with the ionospheric equatorial anomaly [Richmond et al., 1992]. The driver of the poleward flows between 95 and 120 km is uncertain; the magnitude of the flows exceeds that required to satisfy continuity of the circulation above, but it may be a consequence of gravity waves. The peak poleward velocity at 110 km is close to the altitude associated with the turbopause. The feature at low latitudes below 110 km may arise from the action of the propagating diurnal tide.

The equivalent zonally averaged circulation at solstice from Hedin et al. [1993] is presented in Figure 4. The December solstice is depicted, but the June solstice is similar with regard to the summer/winter pattern. At the lowest altitude the closing of the summer-westward and winter-eastward jets can be seen. These jets are closed by gravity wave momentum dissipation, which eventually reverses the circulation between 90 and 115 km. The summer-to-winter meridional flow below 100 km and the thermal structure at

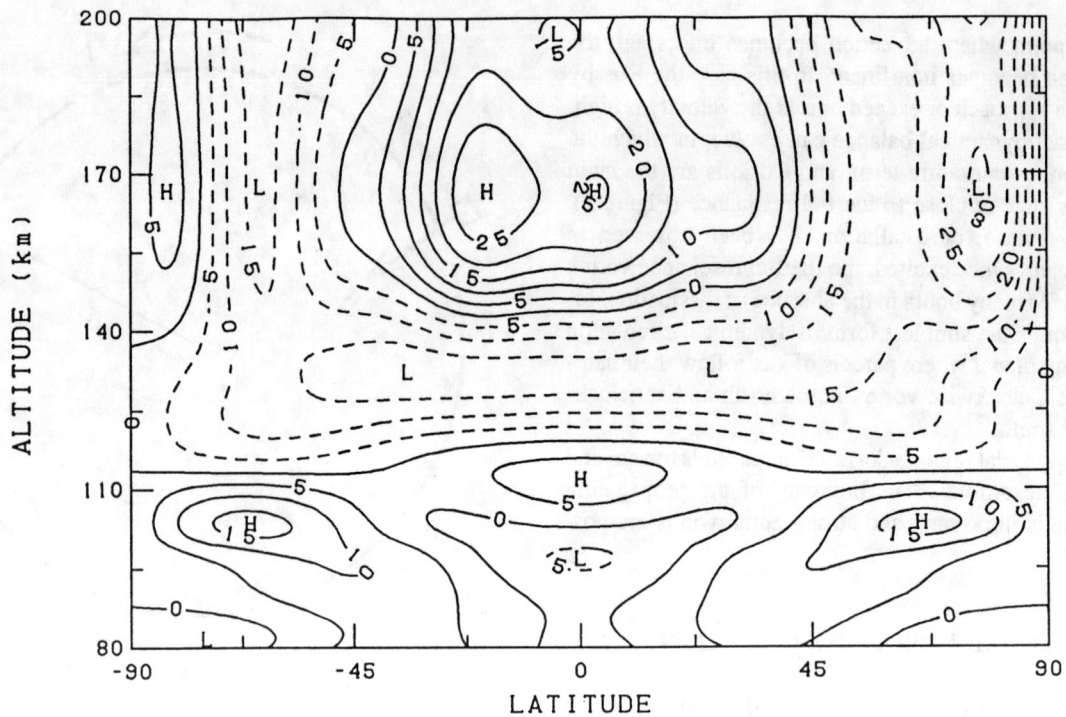

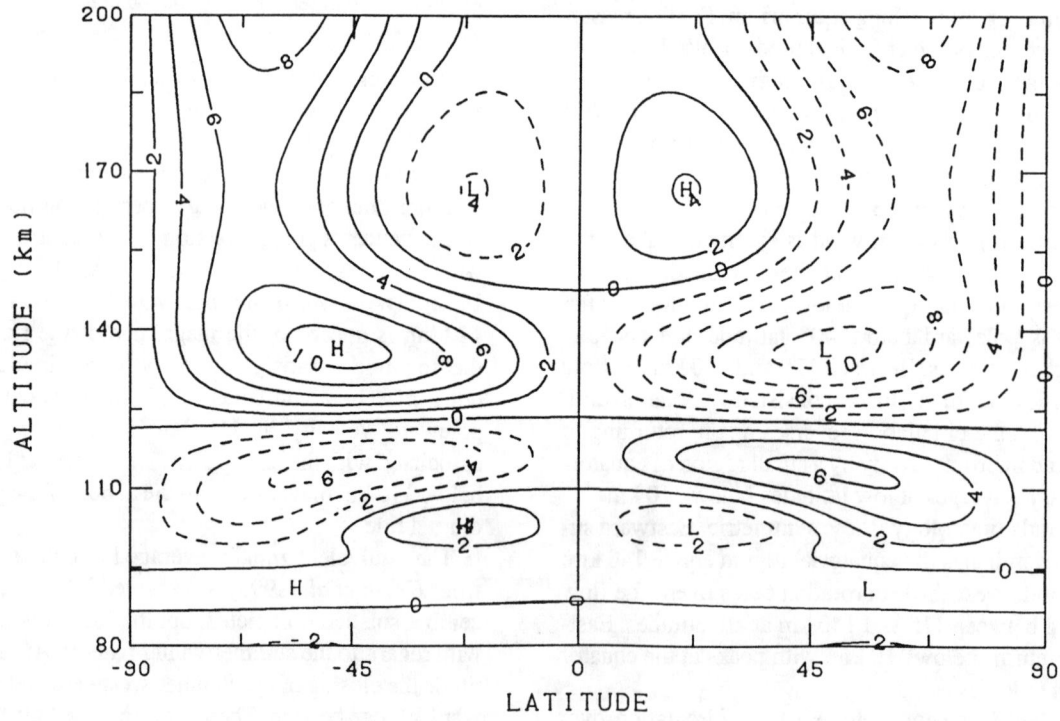

Figure 3. Observed annual average climatology of the northward and eastward wind as a function of latitude and height from the empirical model of *Hedin et al.* [1993]. Positive winds are northward and eastward; contour interval for northward wind is 2 m/s and for eastward wind is 5 m/s.

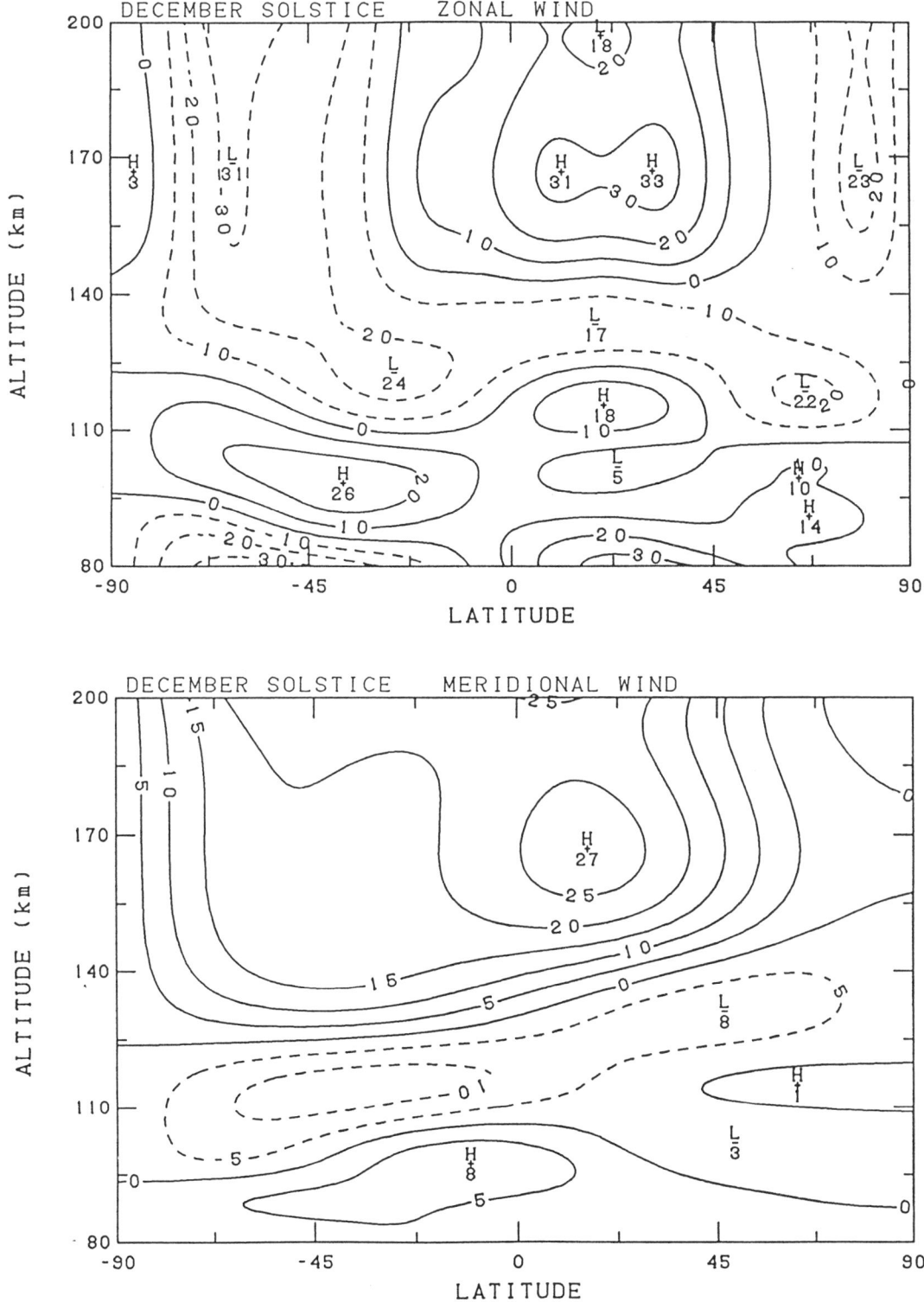

Figure 4. Observed December climatology of the northward and eastward winds as a function of latitude and height from the empirical model of *Hedin et al.* [1993]. Positive winds are northward and eastward; contour interval for northward wind is 2 m/s and for eastward wind is 5 m/s.

high latitudes is a direct consequence of the reduction in the zonal wind by gravity wave drag. Between 100 and 130 km the meridional wind reverses, flowing from winter to summer, and is accompanied by a complex structure of predominantly eastward zonal flow. Above 130 km the flow is from summer to winter, as expected for the thermosphere, with predominantly eastward flow in the winter hemisphere and westward flow in summer. In all cases, at high latitudes the higher elevations are forced by ion drag. The characteristics of this forcing are not well represented by the zonal average presentation.

5. RESPONSE TO SEMI-DIURNAL TIDAL FORCING

The dynamics of the thermosphere from 100 to 200 km is observed to have strong semi-diurnal oscillations [e.g. *Johnson et al.*, 1987; *Salah et al.*, 1991] that are inconsistent with solely in-situ forcing. Tides, driven in the middle atmosphere by absorption of solar radiation by ozone, propagate into the lower thermosphere and increase the magnitude of 12-hour waves. The important Hough functions include the symmetric (2,2), (2,4), and (2,6) modes and the asymmetric (2,3) and (2,5) modes. The magnitude of the observed variations is also quite variable, presumably due to interaction of the tides with planetary waves and/or gravity waves [*Fritts*, 1994; this issue].

Within thermospheric numerical codes [*Fuller-Rowell et al.*, 1987; *Roble et al.*, 1988], tidal forcing can be imposed at or near the lower boundary by introducing a variation in the height of the geopotential surface together with the appropriate wind and temperature fields [*Fesen et al.*, 1986; *Parish et al.*, 1990]. Boundary forcing propagates through the modeled regime and mimics the forcing from the middle atmosphere.

At equinox the symmetric Hough modes (2,2) and (2,4) are the dominant modes. To illustrate the dynamic response of the thermosphere to these modes, the local time variation of the temperature, meridional, and zonal wind fields has been extracted from the 0° longitude sector of a global thermosphere-ionosphere model [*Fuller-Rowell et al.*, 1991]. The height/latitude structure of the amplitude of the semi-diurnal oscillation has been extracted from model simulations, where either the (2,2) or the (2,4) Hough modes provides the tidal forcing. The simulations were at low solar activity, September equinox, and average geomagnetic activity (K_p=2).

Figure 5 illustrates the semi-diurnal amplitude of temperature and wind field from the computation where the (2,2) Hough mode represented the tidal forcing. The mode is imposed at 97 km with an amplitude of 400 m, which corresponds to deviation in the height of the pressure surface.

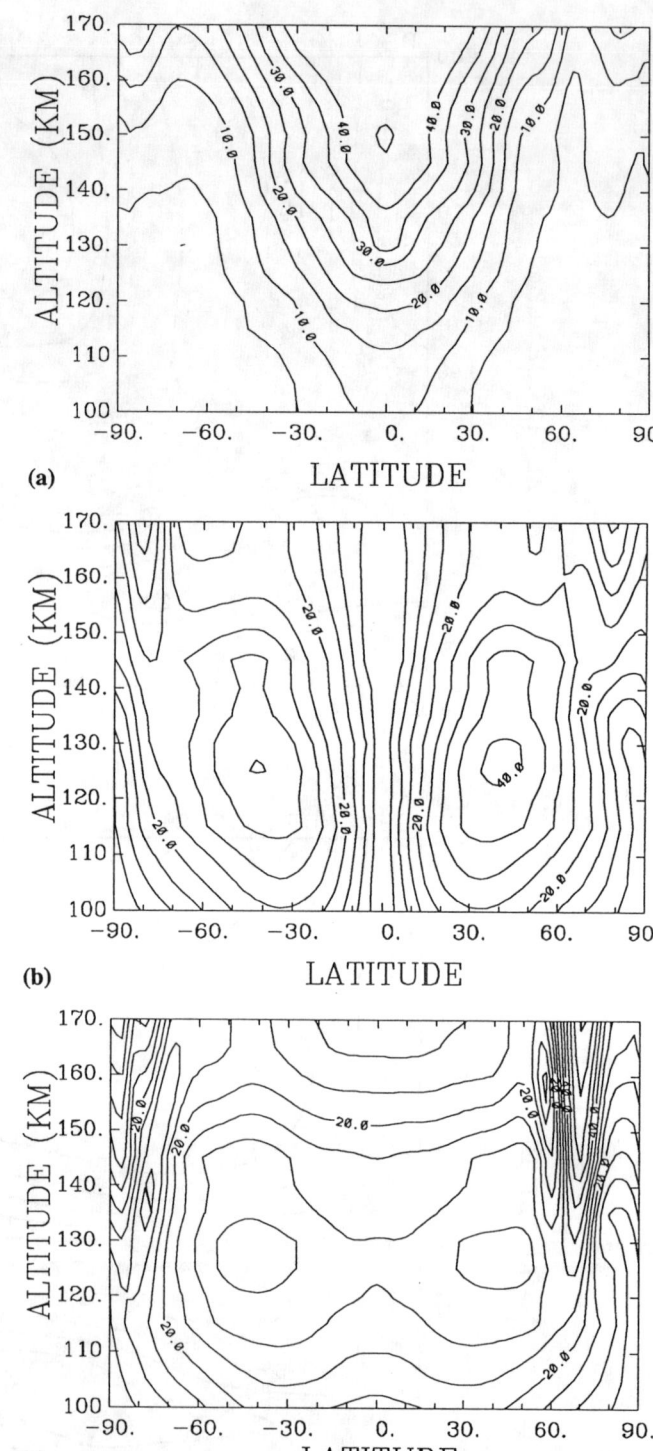

Figure 5. Amplitude of the semi-diurnal variation of temperature (a), meridional wind (b), and zonal wind (c), from a simulation driven by the (2,2) Hough mode tidal forcing of 400 m amplitude at 97 km and for average geomagnetic activity (K_p=2). This figure shows a height/latitude grid from the north to south geographic poles and from 100 to 170 km altitude.

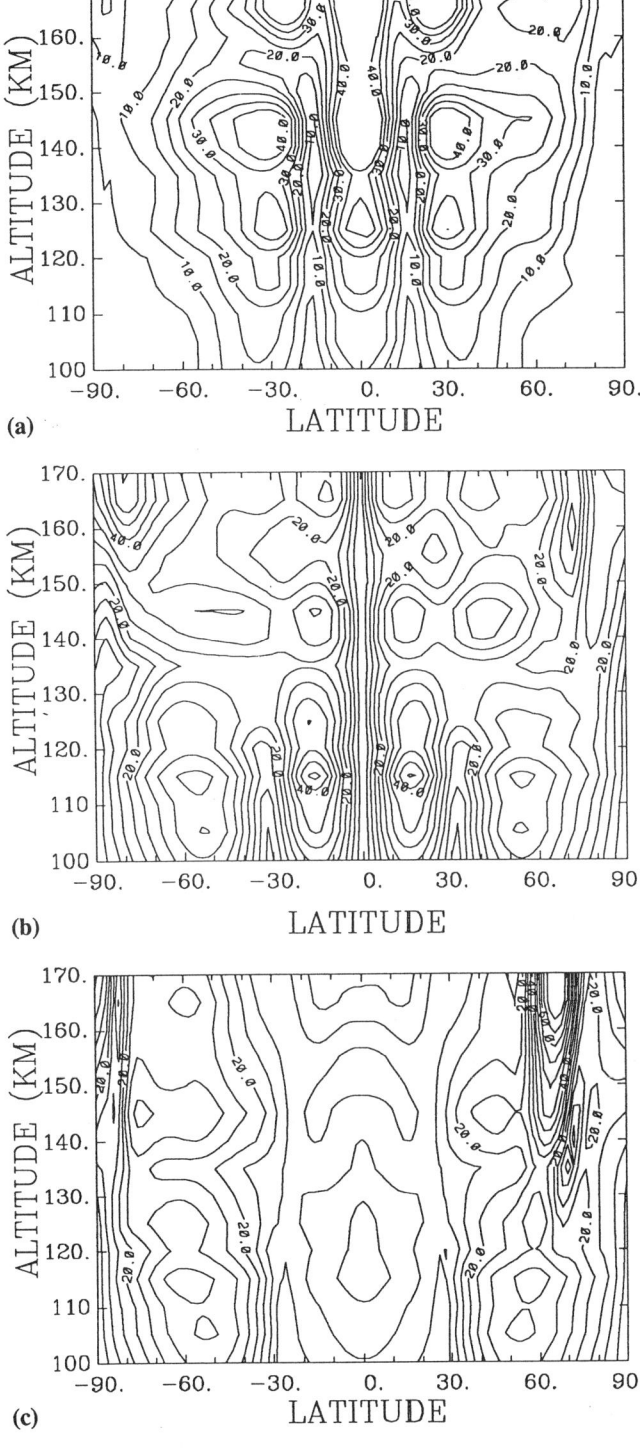

Figure 6. Amplitude of the semi-diurnal variation of temperature (a), meridional wind (b), and zonal wind (c), from a simulation driven by the (2,4) Hough mode tidal forcing of 400 m amplitude at 97 km and for average geomagnetic activity (K_p=2). This figure shows a height/latitude grid from the north to south geographic poles and from 100 to 170 km altitude.

The region from the south to the north geographic pole is depicted from 100 to 170 km altitude. The latitude structure of the mode [*Chapman and Lindzen*, 1970] is clearly depicted by the temperature response, with the peak occurring in the equatorial zone. The long vertical wavelength enables the mode to penetrate very effectively to high altitudes [*Forbes*, 1982], and only above 160 km are there signs that the growth of the wave is limited by dissipation. The meridional wind response occurs at midlatitudes; there is a peak at 120 km in spite of the apparent penetration to higher altitudes indicated by the temperature structure. The distribution of the zonal wind follows a similar shape. The latitude structure in the lower thermosphere is consistent with thermospheric extensions of the classical expansion functions for semi-diurnal tides [*Forbes et al.*, 1982]. The additional effects of high-latitude magnetospheric sources are clearly depicted in both wind components.

Figures such as these can be used to place observing sites into context. For example, the Arecibo incoherent scatter radar facility, at 18°N lies firmly within the main temperature peak; in contrast Millstone Hill, at 43°N, lies nearer the edge of the response. Millstone Hill would be expected to observe a strong signal in the wind field to the (2,2) mode, but Arecibo is toward the equatorward side of the midlatitude peak.

Figure 6 depicts the same information but from the model forced by the (2,4) Hough mode. The temperature structure illustrates the more complex latitude shape of the (2,4) mode, with its three peaks: at the equator and in northern and southern midlatitudes. The shorter vertical wavelength increases dissipation by viscosity and ion drag, causing the temperature to peak sharply at 140 km altitude. The meridional wind response is quite complex. Peaks in wind occur on either side of the temperature maxima. At high midlatitudes, tidal forcing merges with the high latitude features driven by magnetospheric input. The zonal wind responds weakly to the (2,4) mode at low latitude, the main feature occurring poleward of 40°. The latitude structure in the lower thermosphere generated by the (2,4) mode is again consistent with *Forbes et al.* [1982]. The simulation indicates that Arecibo is close to the null point in the latitude temperature structure for this mode and that Millstone lies firmly within the northern midlatitude peak.

When a combination of modes is used to force the model, the structure becomes increasingly complex. The different vertical wavelengths of the Hough modes will induce constructive and destructive interference, depending on their phases and relative amplitudes, and deconvolving observations into component tidal modes becomes increasing difficult.

Figure 7 shows a global snapshot of the meridional cir-

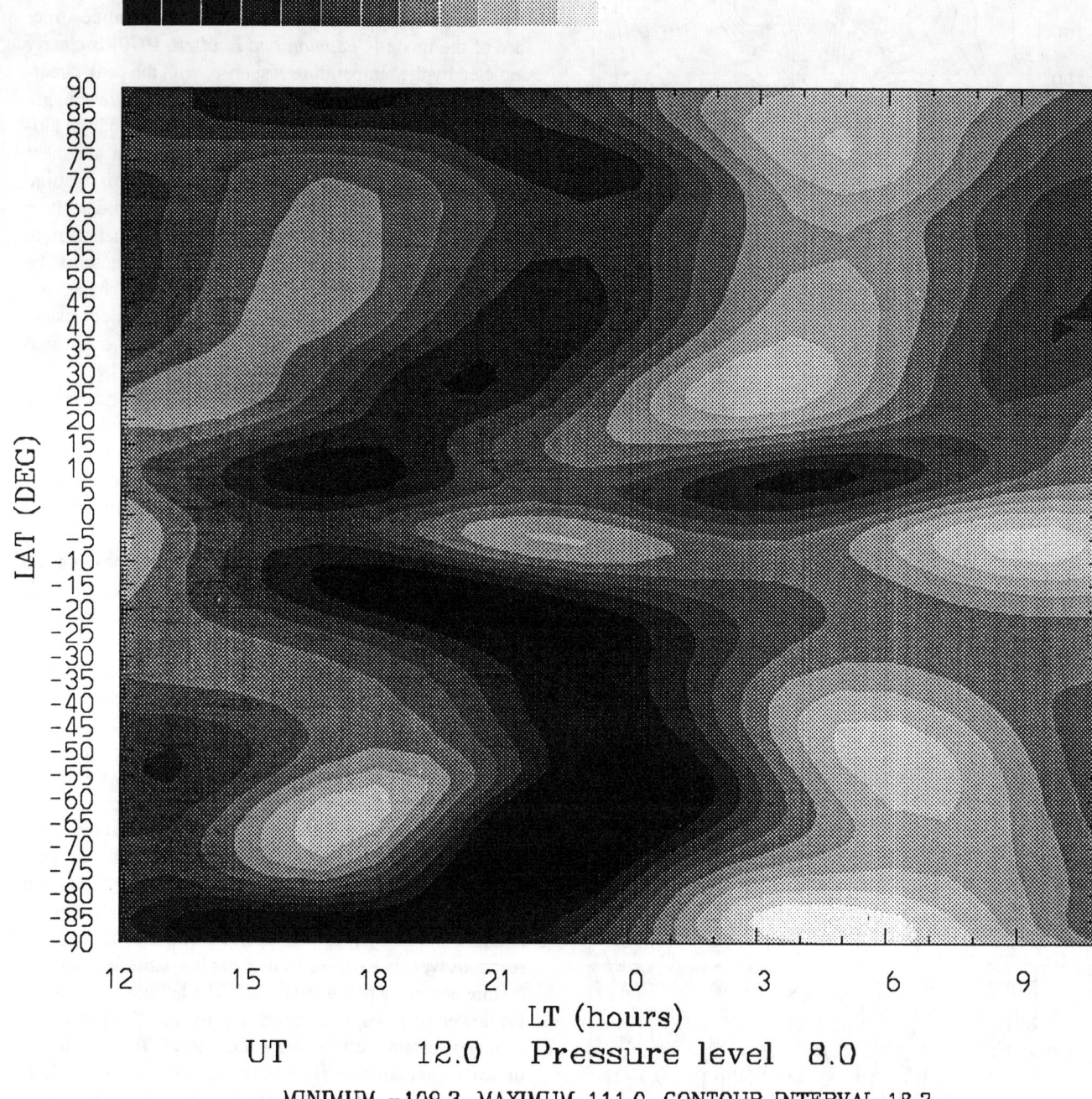

Figure 7. Global structure of meridional wind at 135 km altitude as simulated by a coupled thermosphere-ionosphere global model under the influence of tidal forcing. This figure is an attempt to simulate the December 1988 LTCS period.

culation at 135 km under the influence of the diurnal (1,1) mode and four semi-diurnal modes as computed by a coupled thermosphere-ionosphere model. The figure shows the global structure of meridional wind as a function of latitude and local time at 12 UT. The simulation is an attempt to reproduce a period in December 1988 during the second Lower Thermosphere Coupling Study campaign. The tidal amplitudes used to force the model at 97 km are given in Table 1 [*Forbes*, 1992; private communication]. As the modes propagate in altitude they interfere and eventually dissipate by viscosity or ion drag. The pattern that develops at a particular altitude can be quite complex, as can be seen in Figure 7.

TABLE 1. Amplitude and Phases of the Tidal Forcing

Mode	Amplitude	Phase (h)
(1,1)	100 m	12.0
(2,2)	304 m	9.6
(2,3)	250 m	4.6
(2,4)	67 m	1.2
(2,5)	67 m	2.2

6. RESPONSE TO MAGNETOSPHERIC FORCING

Poleward of 50° latitude the lower thermosphere is forced by magnetospheric processes; the dominance of the source depends on the level of geomagnetic activity. Evidence of ion drag driving the neutral wind emerged from observations of chemical trails released from rockets, but the first indication of the large-scale flow pattern came from a synthesis of satellite data. *Killeen et al.* [1992] combined data from multiple orbital passes to construct a map of the averaged vector wind field at 120 km altitude. Typical magnitudes of 300 m/s exceed those expected from solar heating or tidal-driven winds. The observations were, however, in reasonable agreement with model predictions, assuming the presence of a significant magnetospheric source.

Figure 8 illustrates the development of the high-latitude wind and temperature field, in transition from quiet to disturbed geomagnetic conditions, as simulated by a global thermosphere-ionosphere model [*Fuller-Rowell et al.*, 1994]. The simulation is for equinox at moderate solar activity, and the information is extracted from the model at approximately 135 km altitude. The six panels depict the region poleward of 40°N latitude. The wind and the temperature field are shown during various phases of a geomagnetic storm simulation. The storm is defined by a substantial increase in the magnetospheric source for a period of 12 hours from 12 to 24 UT.

The first panel illustrates the conditions at 18 UT, 18 hours before the start of the storm simulation. Toward midlatitudes the semi-diurnal tidal forcing controls both the wind and temperature fields. The influence of quiet geomagnetic forcing induces a clockwise vortex at high latitudes, although both clockwise and anti clockwise ion drift cell are driving the atmosphere. The preference for clockwise motion is an example of inertial resonance, which has also been shown to be a characteristic of the upper thermosphere [*Fuller-Rowell et al.*, 1984; *Hays et al.*, 1984]. The temperature scale is from 500 to 580 K, and the maximum wind is about 150 m/s.

As storm forcing is imposed, neutral winds are accelerated rapidly by ion drag. The storm source is characterized by an increase in the magnetospheric convection electric field [*Foster et al.*, 1986]; the cross-polar cap potential increases from 40 to 130 kV, and there is an appropriate increase in auroral precipitation [*Fuller-Rowell and Evans*, 1987]. Three hours into the storm (panel (b)) neutral winds of over 400 m/s develop. Again the dominant vortex is clockwise, due to inertial motion, but forcing is now strong enough to excite the opposite circulation in the dawn sector. The maximum temperature has increased by 60 K, and a cold region has developed close to the center of the strong, clockwise wind cell.

As seen in panel (c), 3 hours later, at 18 UT, the cold cell has deepened by more than 100 K, a result of the divergent nature of the wind field. Peak winds are close to 500 m/s at this time and, within the inertial oscillation characterization, are bound to be divergent given their radius of curvature. Back pressure induced by the divergent wind field does not provide a complete balance. Air rises within the core, to maintain continuity, and cools adiabatically to produce the low temperatures. This cooling takes place in spite of the strong Joule heating that must accompany the ion drag. The range of temperature of 300 K is nearly four times that at the commencement of the storm. The meridional wind divergence transports zonal momentum to midlatitude and dominates the wind field poleward of 40°N in the dusk sector.

By the end of the storm at 24 UT (panel (d)), peak temperatures have risen to above 800 K but the wind field has not grown. The cold core has warmed slightly as the divergence weakens. Zonal winds in excess of 250 m/s now penetrate to 40°N in the dusk sector due to advective transport by the meridional wind. The mechanism for transport of the wind field is quite different from that which drives

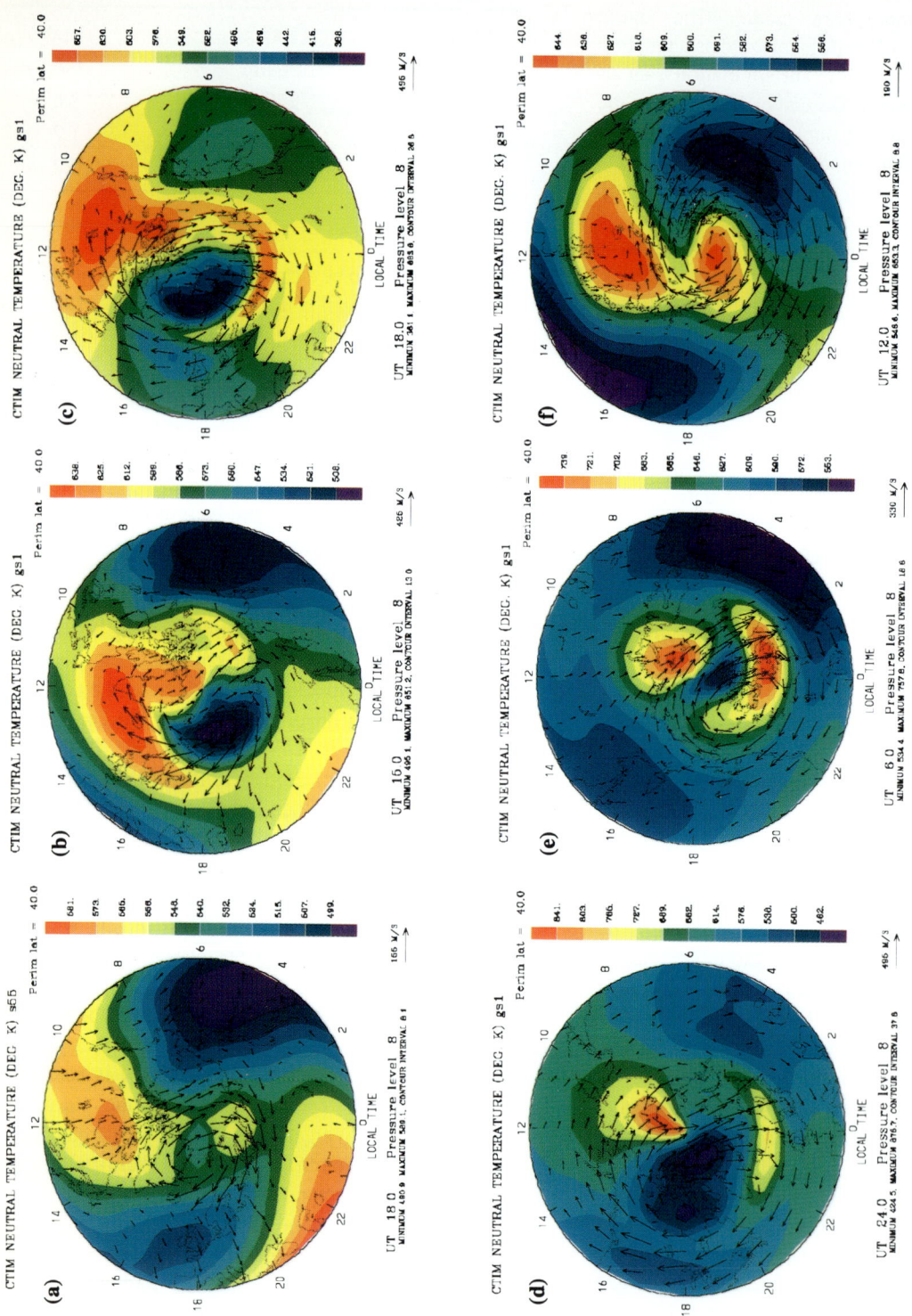

Figure 8. An illustration of the development of the high-latitude wind and temperature structure in the lower thermosphere in response to a 12 h geomagnetic storm. The six panels show the parameters at 135 km from 40°N to the pole before the storm (panel a) and at storm times of 3, 6, 12, 18 and 24 h.

equatorward wind surges at high altitudes. In the latter case the wind surges are due to gravity waves, which can be explained by linearized equations of motion [*Hines*, 1960]. In the lower thermosphere, during extreme events such as are modeled here, the expansion is not purely a pressure wave but is augmented by the non-linear advection of momentum.

The recovery phase of the storm at 6 and 12 UT is depicted in panels (e) and (f) respectively. The clockwise vortex is dynamically stable [*Fuller-Rowell and Rees*, 1984] and persists for many hours after cessation of the storm input. The wind magnitude gradually decays, but the time scale is slow due to weak molecular viscosity. The clockwise vortex is characteristic of a planetary wave feature that rotates with Earth, remaining quasi-stationary over a longitude sector; "flywheel" vortices such as these contribute to the E-region dynamo [*Deng et al.*, 1991].

7. SUMMARY

The lower thermosphere is a complex dynamical regime driven by a multitude of sources. Tides propagating from the lower atmosphere are clearly a major driver at middle and low latitudes, but the variability of observed oscillations in the wind field has yet to be explained in the thermosphere. Interaction of wave fields such as tides, planetary waves and gravity waves is beginning to be addressed (see accompanying papers in this issue); these coupling studies will explore some of the causes of high daily variability of the wind field in the lower thermosphere.

At high latitude, magnetospheric sources exert a strong influence. Observations provide a first indication of the pattern of large-scale circulation, and numerical simulations augment this picture. The paucity of observations makes a conclusive picture impossible at this time.

Many other questions regarding the dynamics of the lower thermosphere remain. Are there circumstances in which gravity waves can penetrate more deeply into the lower thermosphere, and what is the impact of spatial variability in gravity wave forcing? How dominant is magnetospheric control at high latitudes, and to what altitude does this source influence the dynamics? Are the long-lived vortices a reality or merely an artifact of the numerical simulations? These are some examples of the many questions that remain unanswered.

Acknowledgments. The author is grateful to Alan Hedin for Figures 3 and 4, and to Lawrence Puga for helpful comments on the manuscript. The constructive comments from two referees are also appreciated. This material was presented as an invited paper at the Chapman Conference on the Upper Mesosphere and Lower Thermosphere, Asilomar, California, in November 1992.

REFERENCES

Chapman, S. and R. S. Lindzen, *Atmospheric Tides*, D. Reidel, Hingham, Mass., 1970.

Clark, R. R., Upper atmosphere wind observations of waves and tides with the UNH meteor wind radar system at Durham 43°N (1977, 1978 and 1979). *J. Atmos. Terr. Phys.*, 45, 621–627, 1983.

Conrad, J. R. and R. W. Schunk, On the validity of the Navier-Stokes equations for thermospheric dynamics calculations. *J. Geophys. Res.*, 84, 5355–5360, 1979.

Deng, W., T. L. Killeen, A. G. Burns, and R. G. Roble, The flywheel effect: Ionospheric currents after a geomagnetic storm. *Geophys. Res. Lett.*, 18, 1845–1848, 1991.

Evans, D. S., T. J. Fuller-Rowell, S. Maeda, and J. Foster, Specification of the heat input to the thermosphere from magnetospheric processes using TIROS/NOAA auroral particle observations, *Adv. Astron. Sci.*, 65, 1649–1667, 1988.

Fesen, C. G., R. E. Dickinson, and R. G. Roble, Simulation of thermospheric tides at equinox with the NCAR thermospheric general circulation model. *J. Geophys. Res.*, 91, 4471–4489, 1986.

Forbes, J. M., Atmospheric Tides 2. The solar and lunar semidiurnal components. *J. Geophys. Res.*, 87, 5241–5252, 1982.

Forbes, J. M., M. E. Hagan, E. Dicesare, and D. F. Gillette, *A Compendium of Theoretical Atmospheric Tidal Structures, Part II: Therospheric Extensions of the Classical Expansions Functions for Semidiurnal Tides.* Air Force Geophysics Laboratory report number AFGL-TR-82-0173(II), Hanscom AFB, Mass 01731, 1982.

Forbes, J. M., R. G. Roble, and C. G. Fesen, Acceleration, heating, and compositional mixing of the thermosphere due to upward propagating tides. *J. Geophys. Res.*, 98, 311–322, 1993.

Foster, J. C., J. M. Holt, R. G. Musgrove and D. S. Evans, Ionospheric convection associated with discrete levels of particle precipitation. *Geophys. Res. Lett.*, 13, 656–659, 1986.

Fritts, D. C., Gravity wave saturation on the middle atmosphere, a review of theory and observations. *Rev. Geophys.*, 22, 275–308, 1984.

Fuller-Rowell, T.J. and D. Rees, Interpretation of an anticipated long-lived vortex in the lower thermosphere following simulation of an isolated substorm. *Planet. Space Sci.*, 32, 69–85, 1984.

Fuller-Rowell, T. J., and D. Rees, Turbulent diffusion variability and implications for the upper thermosphere. *Adv. Space Res.*, 12, (10), 45–56, 1992.

Fuller-Rowell, T. J. and D. S. Evans, Height-integrated Pedersen and Hall conductivity patterns inferred from TIROS-NOAA satellite data. *J. Geophys. Res.*, 92, 7606–7618, 1987.

Fuller-Rowell, T. J., S. Quegan, D. Rees, R. J. Moffett, and G. J. Bailey, The effect of realistic conductivities on the high-latitude neutral thermospheric circulation. *Planet. Space Sci.*, 32, 469–480, 1984.

Fuller-Rowell, T. J., D. Rees, S. Quegan, R. J. Moffett, and G. J. Bailey, Interactions between neutral thermospheric composition and the polar thermosphere using a coupled global model. *J. Geophys. Res.*, 92, 7744–7748, 1987.

Fuller-Rowell, T. J., D. Rees, H. F. Parish, T. S. Virdi, P. J. S. Williams, and R. G. Johnson, Lower Thermosphere Coupling Study: Comparison of observations with predictions of the UCL-Sheffield thermosphere-ionosphere model. *J. Geophys.*

Res., 96, 1181–1202, 1991.

Fuller-Rowell, T. J., M. V. Codrescu, R. J. Moffett, and S. Quegan, Response of the thermosphere and ionosphere to geomagnetic storms. *J. Geophys. Res.*, 99, 3893–3914, 1994.

Groves, G. V. and J. M. Forbes, Equinox tidal heating of the upper atmosphere. *Planet. Space Sci.*, 32, 447–456, 1984.

Hardy, D. A., M. S. Gussenhoven, and E. Holeman, A statistical model of auroral electron precipitation. *J. Geophys. Res.*, 90, 4229–4248, 1985.

Hays, P. B., T. L. Killeen, N. W. Spencer, L. E. Wharton, R. G. Roble, T. J. Fuller-Rowell, D. Rees, L. A. Frank, and J. D. Craven, Jr., Observations of the dynamics of the polar thermosphere. *J. Geophys. Res.*, 89, 5547–5612, 1984.

Hays, P. B., UARS HRDI Report, paper presented at CEDAR Meeting, Boulder, Colorado, June 21–25, 1993.

Hedin, A. E., E. L. Fleming, A. H. Manson, F. J. Schmidlin, S. K. Avery, and S. J. Franke, *Empirical Wind Model for Middle and Lower Atmosphere—Part I: Local Time Average*, NASA Technical Memorandum 104581, 1993.

Heppner, J. P. and M. L. Miller, Thermospheric winds at high latitudes from chemical release observations. *J. Geophys. Res.*, 87, 1633–1647, 1982.

Heppner, J. P. and N. C. Maynard, Empirical high-latitude electric field models. J. Geophys. Res., 92, 4467–4490, 1987.

Hines, C. O., Internal atmospheric gravity waves at ionospheric heights. *Can, J. Phys.*, 38, 1441–1481, 1960.

Johnson, R. M., V. B. Wickwar, R. G. Roble, and J. G. Luhmann, Lower thermospheric winds at high latitude: Chatanika radar observations. *Ann. Geophys.*, 5A, 6, 383–404, 1987.

Killeen, T. L. and R. G. Roble, An analysis of the high-latitude thermospheric wind and temperature structure using a thermospheric general circulation model: I, Momentum forcing. *J. Geophys. Res.*, 89, 7509–7522, 1984.

Killeen, T. L., B. Nardi, P. N. Purcell, R. G. Roble, T. J. Fuller-Rowell and D. Rees, Neutral winds in the lower thermosphere from Dynamics Explorer 2. *Geophys. Res. Lett.*, 19, 1093–1096, 1992.

Lindzen, R. S., Tides and gravity waves in the upper atmosphere, in *Mesospheric Models and Related Experiments*, edited by G. Fiocco, D. Reidel, Hingham, Mass., 1971.

Lindzen, R. S., Turbulence and stress owing to gravity wave and tidal breakdown. *J. Geophys. Res.*, 86, 9707–9714, 1981.

Lindzen, R. S. and D. Blake, Internal gravity waves in atmospheres with realistic dissipation and temperature, II, Thermal tides excited below the mesopause. *Geophys. Fluid Dyn.*, 2, 31–61, 1971.

Manson, A. H., C. E. Meek, and J. B. Gregory, Winds and waves (10 min–30 days) in the mesosphere and lower thermosphere at Saskatoon (52°N, 107°W, L=4.3) during the year October 1979 to July 1980. *J. Geophys. Res.*, 86, 9615–9625, 1981.

Manson, A. H., C. E. Meek, J. B. Gregory, and D. K. Chakrabarty, Fluctuations in tidal (24, 12h) characteristics and oscillations (8h–5d) in the mesosphere and lower thermosphere (70–110km): Saskatoon (52°N, 107°W), 1979–1981. *Planet. Space Sci.*, 30, 1283–1294, 1982.

Manson, A. H., C. E. Meek, S. K. Avery, G. J. Fraser, R. A. Vincent, A. Phillips, R. R. Clark, R. Schminder, D. Kurschner, and E. S. Kasimirovsky, Tidal winds from the mesosphere, lower thermosphere global radar network during the second LTCS campaign: December 1988. *J. Geophys. Res.*, 96, 1117–1127, 1991.

Parish, H. F., T. J. Fuller-Rowell, D. Rees, T. S. Virdi, and P. J. S. Williams, Numerical simulations of the seasonal response on the thermosphere to propagating tides. *Adv. Space Sci.*, 10, 287–291, 1990.

Pereira, E., M.C. Kelley, D. Rees, I. S. Mikkelsen, T. S. Jorgensen, and T. J. Fuller-Rowell, Observations of neutral wind profiles between 115 and 175 km altitude in the dayside auroral oval. *J. Geophys. Res.*, 85, 2935–2940, 1980.

Rees, D., Ionospheric winds in the auroral zone. *J. Br. Interplanet. Soc.*, 24, 233–246, 1971.

Richmond, A. D., E. C. Ridley, and R. G. Roble, A thermosphere/ionosphere general circulation model with coupled electrodynamics. *Geophys. Res. Lett.*, 19, 601–604, 1992.

Rishbeth, H., Polarization fields produced by winds in the equatorial F-region, *Planet. Space Sci.,* 19, 357–369, 1971.

Roble, R. G., E. C. Ridley, and R. E. Dickinson, On the global mean structure of the thermosphere, *J. Geophys. Res.*, 92, 8745–8758, 1987.

Roble, R. G., E. C. Ridley, A. D. Richmond, and R. E. Dickinson, A coupled thermosphere/ionosphere general circulation model. *Geophys. Res. Lett.*, 15, 1325–1328, 1988.

Salah, J. E., R. M. Johnson, and C. A. Tepley, Coordinated incoherent scatter radar observations of the semidiurnal tide in the lower thermosphere. *J. Geophys. Res.,* 96, 1071–1080, 1991.

Salby, M. L., Survey of planetary-scale traveling waves: The state of theory and observations. *Rev. Geophys. Space Phys.*, 22, 209–236, 1984.

Schoeberl, M. R., D. F. Strobel, and J. P. Apruzese, A numerical model of gravity wave breaking and stress in the mesosphere, *J. Geophys. Res.*, 88, 5249–5259, 1983.

Spiro, R. W., P. H. Reiff, and L. J. Maher, Precipitating electron energy flux and auroral zone conductances: An empirical model. *J. Geophys. Res.*, 87, 8215–8227, 1982.

Vincent, R. A., Planetary and gravity waves in the mesosphere and lower thermosphere. *Adv. Space Res.*, 10, 93–101, 1990.

T. J. Fuller-Rowell, NOAA Space Environment Laboratory R/E/SE, 325 Broadway, Boulder, CO 80303

The Lower Ionosphere at High Latitudes

R. W. Schunk and J. J. Sojka

Center for Atmospheric and Space Sciences
Utah State University
Logan, UT

The lower ionosphere is a particularly difficult region to both observe and model. Although radars and rockets have probed this region for more than two decades, our overall understanding of the interplay between radiative, chemical, dynamical, and electrodynamical processes in the lower ionosphere is relatively poor in comparison to the other regions of the solar-terrestrial system. Part of the problem is that the various radar and rocket campaigns have focused on different scientific issues, have been of limited duration, or have been restricted to specific geographical locations. However, the lower ionosphere is a complex region, being acted upon by magnetospheric processes from above and stratospheric processes from below. Within the lower ionosphere are chemical reactions involving negative, positive, and cluster ions; transport processes that sometimes involve ordinary diffusion, turbulence, and wave-particle interactions due to plasma instabilities; radiative processes that could involve multiple scattering effects; and energetics that could result in non-Maxwellian ion velocity distribution functions. A further complication arises in that the processes acting on and within the lower ionosphere do so on widely different spatial and temporal scales, and these scales are directly reproduced in the medium. An overview of our current knowledge of the lower ionosphere is presented in this brief review, with the emphasis on the high latitude region.

1. INTRODUCTION

The Earth's mesosphere-lower thermosphere-ionosphere (MLTI) region is poorly understood relative to some of the other regions in the solar-terrestrial environment, i.e., the upper thermosphere, and the upper ionosphere. One of the reasons for this lack of knowledge of the MLTI system is that it is acted upon by magnetospheric processes from above and stratospheric processes from below, and their effects have not been fully elucidated. Another reason is that there are complex chemical reactions, transport processes, and electrodynamical effects that occur in the MLTI region. A further complication arises because the time constant for the chemical reactions is short (of the order of minutes) and significant spatial structure, both in the vertical and horizontal directions, occurs in this region. Hence, it is difficult to interpret measurements made at a specific location and time in terms of the basic physics and chemistry that are thought to operate in the MLTI region. It is also difficult to model this region because of the lack of sufficient spatial and temporal resolutions concerning the measured inputs needed by the models.

Figure 1 is a schematic diagram that shows the different processes acting on the MLTI region, which encompasses the $F1$, E, and D regions of the ionosphere and the mesosphere-lower thermosphere ($\approx$ 50-200 km). At high latitudes, the dominant momentum and energy sources for the MLTI region derive from magnetospheric particle precipitation and electric fields. The plasma production due to auroral precipitation and the Joule heating associated with convection electric fields maximize at MLTI altitudes. These energy and momentum sources directly affect the densities, temperatures and composition of both the ionized and neutral components in this region which, in turn, affects the coupling between the ionosphere and lower thermosphere. The effects of these magnetospheric processes can be felt at mid- and low latitudes during

The Upper Mesosphere and Lower Thermosphere:
A Review of Experiment and Theory
Geophysical Monograph 87
Copyright 1995 by the American Geophysical Union

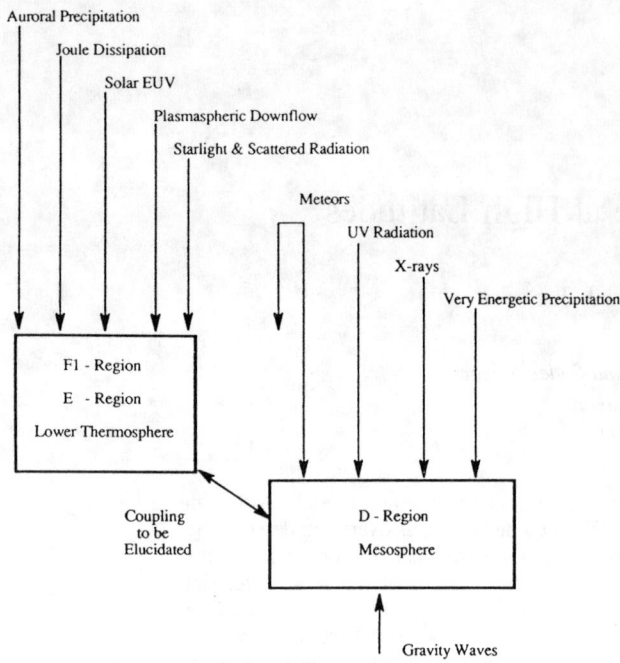

Figure 1. External drivers for the MLTI system.

magnetic storms and substorms. The magnetosphere also affects the MLTI region via very energetic particle precipitation from the radiation belts, which can produce ionization at all latitudes in the D region/mesosphere, and via a downward plasmaspheric flow, which helps maintain the nocturnal $F1$ region at mid-latitudes. With regard to solar radiation, ion production and photoelectron heating occurs in the lower thermosphere due to EUV radiation, while similar processes occur in the D region/mesosphere due to UV radiation. These processes occur globally on the sunlit side of the Earth. On the nightside, ion production and electron heating result from starlight and resonantly scattered solar radiation. In a sporadic fashion, ablation of impacting meteors acts to produce long-lived metallic ions. As far as the stratosphere is concerned, it has a significant effect on the MLTI region in that the upward propagating tides and gravity waves that originate in this region deposit most of their energy at MLTI altitudes owing to wave breaking and dissipation.

Unfortunately, the physical processes shown in Figure 1 that operate on and within the MLTI region do so on widely different spatial and temporal scales. To study the long-term effect of these processes on the MLTI region (i.e., climate) requires many months of data, while diurnal/UT variations occur over a one-day period. Propagating tides and magnetic storm effects manifest themselves over a 0.5-1 day period, while substorm, SAR-arc, and ring current precipitation act on the MLTI region over a period of hours. Penetration electric fields during substorm expansions and meteor ablation act on the MLTI system with a characteristic time measured in minutes, while very energetic particle precipitation bursts can be as short as seconds.

Not only do the processes operating on the MLTI region display widely different time scales, but the horizontal spatial scale associated with them varies from tens of meters to global dimensions. Starlight and resonantly scattered solar radiation operate over the entire globe, although their effects are mainly important at night. The effects associate with solar EUV and UV radiation, diffuse auroral precipitation, plasmaspheric downflows, and upward propagating gravity waves occur over a horizontal scale of thousands of kilometers. Processes like Joule heating, auroral arc precipitation, ring current precipitation, and sub-auroral red arcs have a horizontal scale length of from 100 to 500 km. On a still smaller spatial scale, polar cap arcs and sub-auroral electric field spikes typically have horizontal dimensions in the 10-200 km range. Finally, field-aligned current structures in the auroral oval have dimensions as small as tens of meters.

The above discussion only provides a brief description of the various processes acting on and within the MLTI region as well as their spatial and temporal scales. Further details can be found in recent texts and review articles [*Sojka*, 1989; *Tsunoda*, 1988; *Rees*, 1991; *Kelley*, 1989; *Hargreaves*, 1992].

2. LOWER IONOSPHERIC PROCESSES

The main emphasis of this review paper is on the lower ionosphere, including the processes that operate within and on this region. Some of the topics to be discussed are the chemical reaction scheme, the effect of minor neutral species, frictional heating and ion composition changes, non-Maxwellian ion velocity distributions, anomalous electron temperatures, and the effect of propagating tides and gravity waves. A discussion of the spatial and temporal scales that exist in the lower ionosphere is also given.

2.1. *Chemical Reaction Scheme*

The chemical reactions that need to be included in the description of the ionospheric D and E regions are considerably more involved than those needed to describe the F1 and F2 regions. This is because of the need to include positive ions, negative ions, cluster ions, and three-body chemical reactions, many of which have uncertain rate coefficients. The cluster ions dominate the D region at altitudes below about 85 km and their formation occurs via hydration starting from the primary ions NO^+ and O_2^+.

Also, in addition to the usual neutrals (N_2, O_2, O, N, He, H) that are considered in F region studies, it is necessary to take account of a number of important minor neutral species [NO, CO_2, H_2O, O_3, OH, NO_2, HO_2, $O_2(^1\Delta g)$], with nitric oxide playing a crucial role in the D region chemistry.

Numerous chemical models of the D region have been developed since the first rocket-borne mass spectrometer measurements by *Narcisi and Bailey* [1965] revealed the presence of cluster ions. Several relatively simple 6-ion models have been developed [*Mitra and Rowe*, 1972; *Friedrich et al.*, 1979; *Tomko et al.*, 1980], and more recently, some sophisticated many-ion models have been developed [*Devlin et al.*, 1986; *Burns et al.*, 1991]. In all cases, when the model densities were compared with data, there were several 'free' parameters that could be adjusted to get the models to agree with the data. Also, the chemical reaction schemes were found to be sensitive to geophysical conditions (especially the temperature) and, hence, the D region is expected to be extremely variable.

In the study by *Burns et al.* [1991], both the 6-ion chemical scheme developed by *Mitra and Rowe* [1972] and the 35-ion chemical model developed at the Sodankylä Geophysical Observatory were compared with electron density profiles measured by the EISCAT incoherent scatter radar over the altitude range from 80-120 km. The study was restricted to the daytime, summer ionosphere and 'quiet' geomagnetic conditions (no production due to energetic electron precipitation). The overall strategy underlying the modeling/data comparisons was to use the measured electron density profiles to constrain the models. To get agreement with the measurements, the solar flux was adjusted to obtain a fit at E region altitudes and then the nitric oxide profile was adjusted to get a better fit at all altitudes. Once a fit was obtained at a given solar zenith angle, the adjustments were kept fixed and the models were then compared to data taken at other solar zenith angles.

Figure 2 shows the Mitra-Rowe (M-R) 6-ion chemical scheme for the D and E regions. The main reactions and the rates for the M-R model are given in *Burns et al.* [1991] and are not repeated here. Some of the important features of this simplified model are (1) NO^+ and O_2^+ are the precursor ions; (2) Clustering occurs through NO^+ above about 70 km and through O_2^+ below this altitude; (3) O_4^+ is included explicitly because the back reaction to O_2^+ inhibits clustering from the O_2^+ channel above about 85 km; (4) All cluster ions are lumped under a common ion called Y^+ (main simplifying assumption); and (5) All negative ions, except O_2^-, are lumped under X^-. The main advantage of the 6-ion model is its computational efficiency, but another advantage is that numerous reactions with uncertain rate coefficients are lumped together.

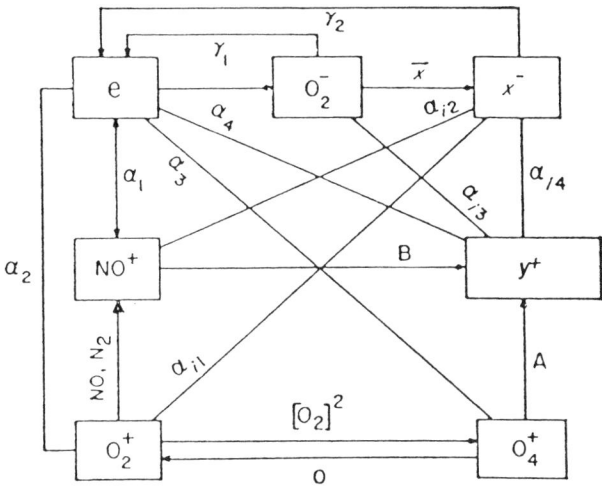

Figure 2. The 6-ion chemical scheme for the D and E regions suggested by *Mitra and Rowe* [1972].

The more sophisticated Sodankylä ion chemistry (SIC) model for the D and E regions includes 24 positive ions and 11 negative ions. The chemical reaction schemes associated with both the positive and negative ions are shown in Figure 3. Note that the water cluster ions are primarily of the form $H^+(H_2O)n$, $NO^+(H_2O)n$ and $O_2^+(H_2O)n$, where n can be as large as 8. In the chemical scheme, account must be taken of both 2-body and 3-body positive ion-neutral reactions, recombination of positive ions with electrons, photodissociation of positive ions, both 2-body and 3-body negative ion-neutral reactions, electron photodetachment of negative ions, photodissociation of negative ions, electron attachment to neutrals, and ion-ion recombination. In all there are 174 reactions in the SIC chemical scheme. The specific reactions and their rates are given in the appendix of *Burns et al.* [1991].

To start the model/data comparisons, *Burns et al.* [1991] adopted a reference solar spectrum, a representative Lyman α flux, the MSIS model densities for the main neutral species (N_2, O_2, O, N, He, H), and altitude profiles for the minor neutral species [NO, CO_2, H_2O, O_3, OH, NO_2, HO_2, $O(^1\Delta g)$], which were obtained from a variety of 'separate' measurements. As expected, the initial runs of both the 6-ion and 35-ion models yielded fairly large discrepancies between the models and data. Subsequently, the solar EUV flux was adjusted to get better agreement between the measured and modeled electron densities at E region altitudes and then the NO density profile was adjusted to get a better fit at all altitudes. With the adjustments, both the 6-ion and 35-ion models were in good agreement with the measurements at several solar zenith angles (Figure 4). However, the NO and solar flux adjustments needed by the

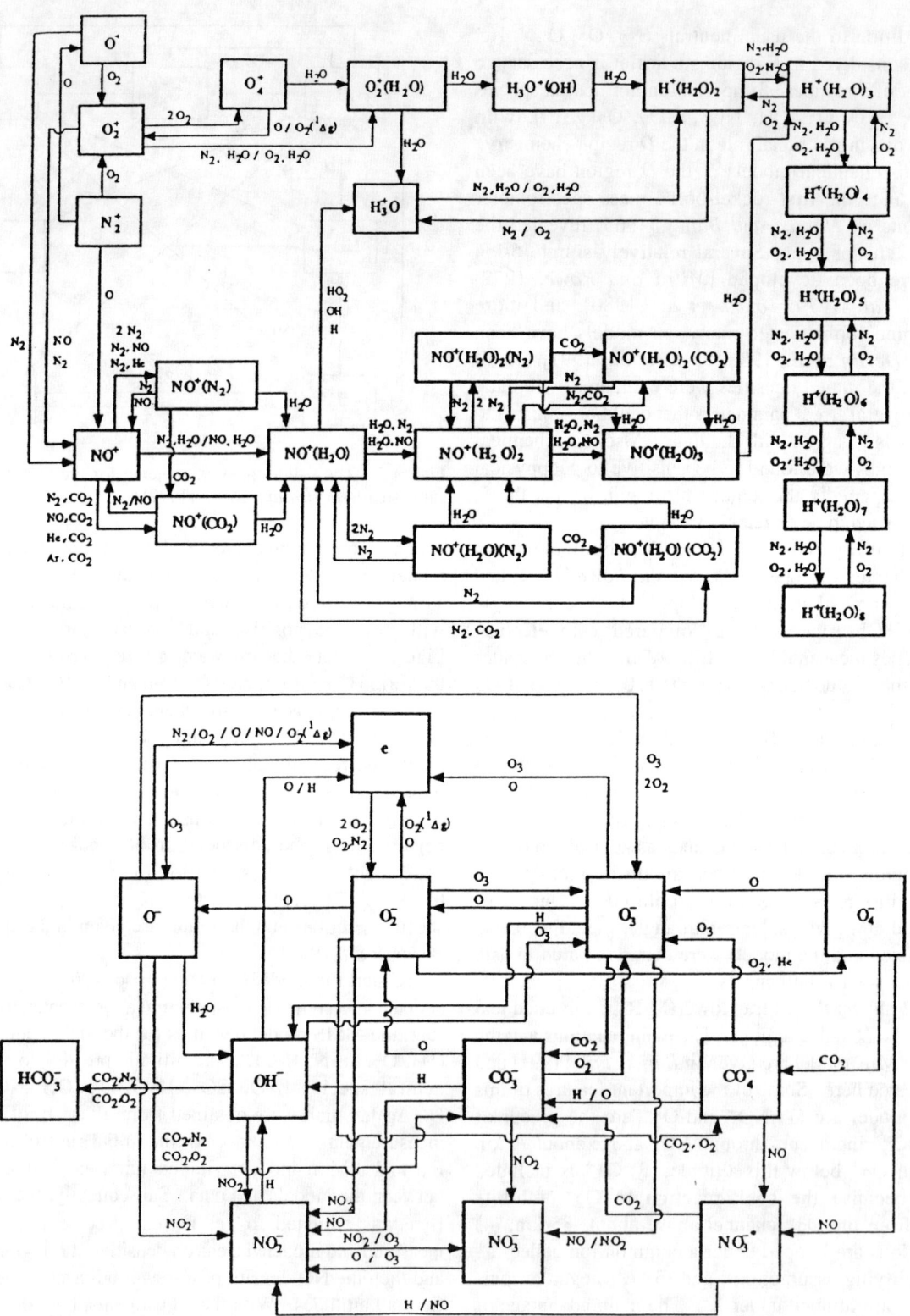

Figure 3. The 35-ion chemical scheme developed by the Sodankylä Geophysical Observatory for the D and E regions. From *Burns et al.* [1991].

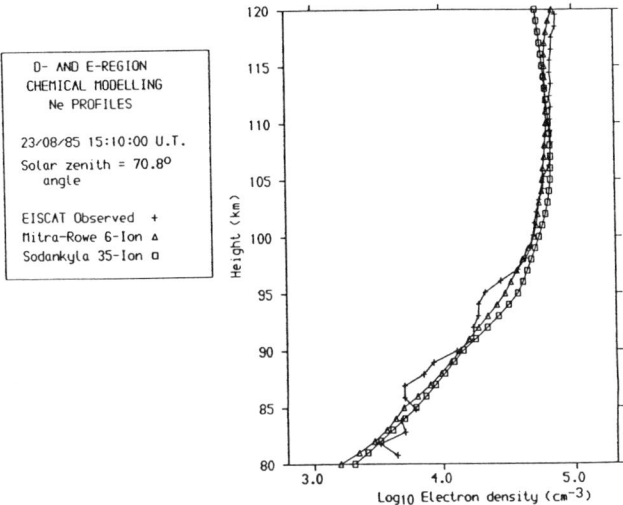

Figure 4. The M-R and SIC models compared with the EISCAT electron density profile measured at 1510 UT. From *Burns et al.* [1991].

two models 'were significantly different.' The EUV fluxes had to be multiplied by a factor of 2.5 for the M-R model and 1.3 for the SIC model. Large, but different, adjustments also had to be made to the NO profile to get the agreement shown in Figure 4. Based on this study, *Burns et al.* [1991] point out the need to further constrain the inputs required by the chemical models. What this means is that the fundamental chemical reactions that control the electron density behavior in the MLTI region have not, as yet, been clearly established.

2.2. Minor Neutral Species

As noted above, minor neutral species play an important role in the MLTI region. Of the odd nitrogen components, $N(^2D)$, $N(^4S)$, and NO, nitric oxide is especially important for several reasons: it has a major effect on the *D* region chemistry; infrared radiation of NO is an important cooling mechanism for the thermosphere; and it is speculated that the downward transport of NO might affect the chemistry and energetics in the mesosphere and stratosphere. In recent years, these minor neutral species have received considerable attention, primarily because of the data available via the Solar Mesospheric Explorer (SME) satellite [*Rusch et al.*, 1991; *Roble et al.*, 1987; *Fesen et al.*, 1990; *Rees and Fuller-Rowell*, 1990; *Siskind and Rusch*, 1992].

The most comprehensive modeling of the odd nitrogen constituents is described in the papers by *Roble et al.* [1987], *Rees and Fuller-Rowell* [1990], and *Fesen et al.* [1990]. However, all three papers are based on the same formulation, i.e., the one developed by *Roble et al.* [1987].

To model the dynamics and energetics of the minor neutral species, it is necessary to solve coupled continuity and diffusion equations, including various production, loss, and transport processes. These equations must be solved in parallel with the major species continuity, momentum, and energy equations. As far as the production and loss processes are concerned, the main sources of $N(^2D)$ and $N(^4S)$ are dissociation of N_2 by solar radiation and auroral electron impact. However, the reactions of $NO^+ + e$ and $N_2^+ + O$ also yield $N(^2D)$ and these could be important in the auroral oval. The main source of NO comes from the reaction of $N(^2D)$ with O_2, and a secondary source comes from the reaction of $N(^4S)$ with O_2. The primary loss mechanism for NO occurs via a reaction of NO with $N(^4S)$. All three odd nitrogen components undergo vertical transport via both molecular and eddy diffusion, and they tend to be transported in both the vertical and horizontal directions owing to the thermospheric winds that are associated with the major neutral species [cf. *Roble et al.*, 1987]. Also, with regard to NO, a downward flow condition must be imposed at the lower boundary of the simulation domain to allow escape to the stratosphere.

The minor neutral species play an important role in the energy balance that determines the neutral temperature of the mesosphere/lower thermosphere region, which in turn affects the ionospheric chemistry. Overall, the energy balance equation must include photoelectron heating, heating from precipitating auroral electrons, Joule heating, heating via collisional coupling to the thermal electrons and ions, O_2 absorption of solar UV radiation in the Schumann-Runge continuum and bands, excess energy from exothermic ion-neutral and neutral-neutral chemical reactions, O_3 absorption of solar UV radiation in the Hartley bands, and heating via atomic oxygen recombination. The main cooling processes are associated with infrared radiation of NO at 5.3-μm, CO_2 at 15-μm, and O fine structure cooling at 63-μm. Also, account must be taken of thermal conduction via both molecular and eddy processes [cf. *Roble et al.*, 1987]. In general, tidal and gravity wave dissipation must be included when the forcing functions are known [cf. *Fesen et al.*, 1986].

As the various authors clearly point out, the existing models of the lower thermosphere/mesosphere region are very uncertain for several reasons. First, the NO and CO_2 infrared cooling rates are not well known, yet they are important components of the neutral gas energy balance. Also, the eddy diffusion and thermal conductivity coefficients are basically 'free parameters' that can be adjusted to 'correct' for uncertain rate coefficients. Furthermore, there is little to no information on the global distribution of tides and gravity waves, yet these waves play an important role in the MLTI region. The downward transport of NO to the stratosphere is another free parameter. In addition, the important auroral source of NO due to electron precipitation is highly variable, both spatially and temporally, and this aspect has not been rigorously included in the models. The net effect of all of the competing processes is that the important minor neutral species NO is extremely variable. Even for very quiet geomagnetic conditions, *Fesen et al.* [1990] note that "the nitric oxide densities vary considerably from day to day, suggesting that the concept of an average distribution is of limited usefulness in understanding nitric oxide." Only after the behavior of the important minor neutral species is clearly established can progress be made in elucidating the essential chemical reactions in the lower ionosphere.

2.3. *Tidal Forcing*

Tides and gravity waves play an important role in the dynamics and energetics of the MLTI region. Unfortunately, it is difficult to include gravity wave and tidal effects in a realistic manner owing to the lack of global measurements of the forcing function. For example, tides are excited in the middle and lower atmosphere due to the insolation absorption by O_3 and H_2O and then the tides propagate upward into the thermosphere. Both diurnal and semidiurnal tidal components exist, but the semidiurnal component is the most important in the lower thermosphere.

The effect of semidiurnal tides has been calculated recently by *Crowley et al* [1989] using the NCAR thermospheric general circulation model (TGCM). The simulations were for the September 18-19, 1984 period of the Equinox Transition Study. For the magnetically quiet time period, calculations were performed both with and without the semidiurnal tides. Figure 5 shows the variation of the meridional neutral wind versus altitude and latitude for 70° W at 18 UT. The top panel shows the wind without semidiurnal tides, while the bottom panel shows the wind with tidal effects. It is apparent that semidiurnal tides are very important in the lower thermosphere. There is a complex wind structure below about 300 km, with reversals of the wind direction clearly evident. The semidiurnal tides also have a similar effect on the neutral temperature and densities. The tidal-induced structure in the neutral parameters then affects the ionospheric densities at D and E region altitudes because of the short time constant for chemical reactions.

2.4. *Convection Electric Fields*

Convection electric fields have an important effect on the

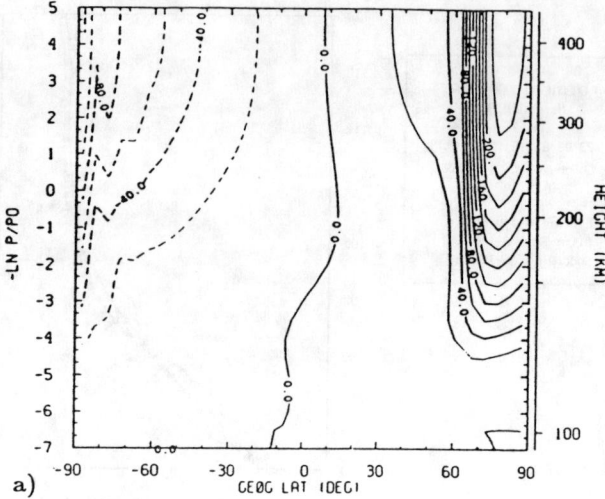

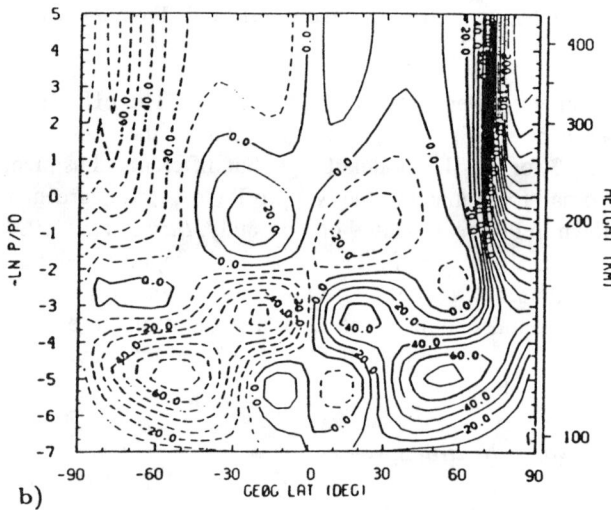

Figure 5. Variation of the meridional neutral wind versus altitude and latitude for 70° W at 18 UT on a quiet day. The variation is shown both without (a) and with (b) tidal effects. Solid contours are for winds blowing toward the south and dashed contours correspond to northward winds. The contour interval is 10 m s^{-1}. From *Crowley et al.* [1989].

lower ionosphere at high latitudes. The electric fields are generated by a dynamo action as the solar wind plasma interacts with the Earth's strong magnetic field. The dynamo electric fields are then mapped down along the geomagnetic field lines to the high latitude ionosphere because the ionospheric plasma is a good conductor. These electric fields, which are perpendicular to the geomagnetic field, cause the high latitude ionosphere to drift horizontally across the polar region. The electrons execute an **E** x **B** (Hall) drift at altitudes above about 90 km, whereas the ions

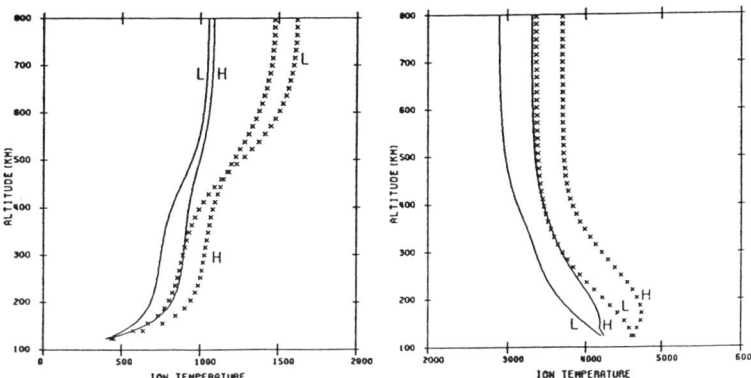

Figure 6. Daytime O$^+$ temperature profiles for 0 (left panel) and 100 mV/m (right panel) meridional electric fields. The profiles were calculated for solar minimum conditions, for summer (x curves) and winter (solid curves), and for high (H) and low (L) geomagnetic activity. From *Schunk and Sojka* [1982].

execute an **E** x **B** drift only above about 150 km. Below this altitude, the ion drift velocity rotates from the **E** x **B** to the **E** direction as altitude decreases.

As the ions drift through the neutrals, they are frictionally heated, and depending on the electric field strength, the ion temperature can be elevated to values well in excess of the electron temperature. The ion heating peaks in the lower ionosphere where the neutrals are abundant, as shown in Figure 6. The ion temperature profiles shown in this figure were calculated for daytime, steady state conditions at solar minimum for both summer and winter solstices and for both high and low geomagnetic activity [*Schunk and Sojka*, 1982]. The left panel shows the results when the electric field is absent, while the right panel is for a moderate-high meridional electric field strength of 100 mVm^{-1}. Without the electric field, T_i is equal to T_n at low altitudes. However, with a 100 mVm^{-1} electric field, T_i is greater than 4000 K in the lower ionosphere ($\approx$120 km).

The elevated ion temperatures shown in Figure 6 act to alter the ion composition in the lower ionosphere through temperature-dependent chemical reaction rates. For example, the O$^+$ + N$_2$ $\rightarrow$ NO$^+$ + N reaction rate increases rapidly with T_i, and therefore, in regions where the convection electric field is large, the associated frictional heating should lead to a rapid conversion of O$^+$ into NO$^+$. This effect is shown in Figure 7, where ion and electron density profiles are shown for convection electric fields of 0 and 100 mVm^{-1}. The profiles were calculated for daytime steady state conditions by *Schunk et al.* [1975]. For no electric field, the molecular ions dominate in the E-region, while O$^+$ is the major ion in the F-region. The transition from molecular to atomic ion dominance occurs at about 225 km. For a 100 mVm^{-1} electric field, on the other hand, the elevated ion temperatures lead to an enhanced conversion of

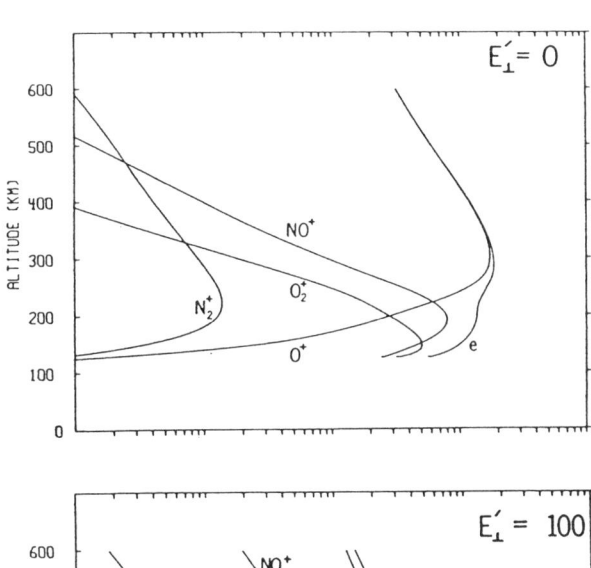

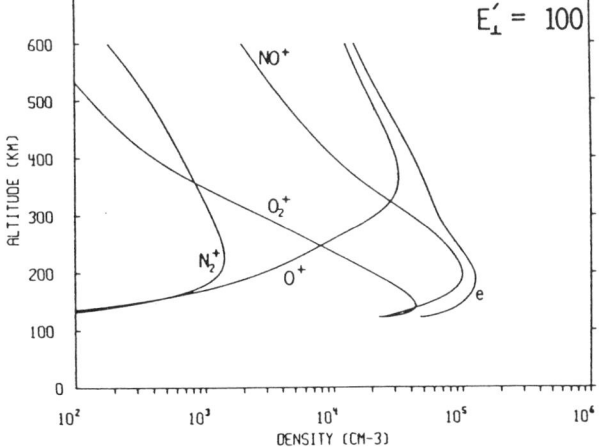

Figure 7. Ion and electron density profiles calculated for the daytime high-latitude ionosphere and for meridional electric fields of 0 (top panel) and 100 mVm^{-1} (bottom panel). From *Schunk et al.* [1975].

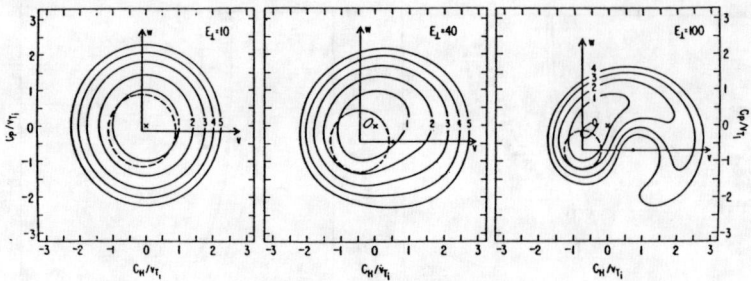

Figure 8. Ion velocity distributions in the E-region for meridional electric fields of 10 (left panel), 40 (middle panel), and 100 mVm^{-1} (right panel). The contours are shown in the plane perpendicular to **B**. They indicate the points where the ion distribution has decreased by a factor of e^{α}, where α is the number attached to each curve. The x marks the location of the ion drift velocity, and the dashed circle shows the region of velocity space occupied by most of the ions after collisions. From *St.-Maurice and Schunk* [1974].

O^+ into NO^+, which significantly alters the ion composition at altitudes below 330 km.

The ion frictional heating discussed above is a manifestation of changes in the ion velocity distribution due to ion-neutral collisions. For small electric fields, the ion-neutral relative drift is small and ion-neutral collisions do not appreciably alter the ion velocity distribution. In this case, the ion distribution is basically a drifting Maxwellian with an enhanced temperature, as shown in Figure 8 for an altitude of about 120 km [*St.-Maurice and Schunk*, 1974]. However, when the electric field is greater than about 40 mVm^{-1}, the ion drift exceeds the neutral thermal speed and the ion velocity distribution becomes non-Maxwellian. For large electric fields (≥ 100 mVm^{-1}), the ion distribution tends to become bean-shaped in the lower ionosphere. Such highly non-Maxwellian distributions are unstable, and the resulting wave-particle interactions have a significant effect on the ion energetics. Note that the non-Maxwellian features shown in Figure 8 relate to an altitude of about 120 km. At higher altitudes, the non-Maxwellian features change markedly, while at lower altitudes they rapidly disappear owing to the decrease in the ion drift velocity as the ions try to penetrate a more dense atmosphere [cf. *St-Maurice and Schunk*, 1979].

Large electric fields also lead to anomalous electron temperatures in the E-region owing to the excitation of plasma instabilities [*Schlegel and St.-Maurice*, 1981; *St.-Maurice et al.*, 1981; *Williams et al.*, 1992]. Specifically, in the auroral E-region the electrons drift in the **E** x **B** direction, while the ions drift in the **E** direction. This ion-electron relative drift excites a modified two-stream instability when the electric field exceeds a threshold. The subsequent interaction of the plasma waves and the electrons acts to heat the electron gas. For large electric fields, T_e can be much greater than T_n in the lower ionosphere. This is illustrated in Figure 9, where EISCAT radar measurements

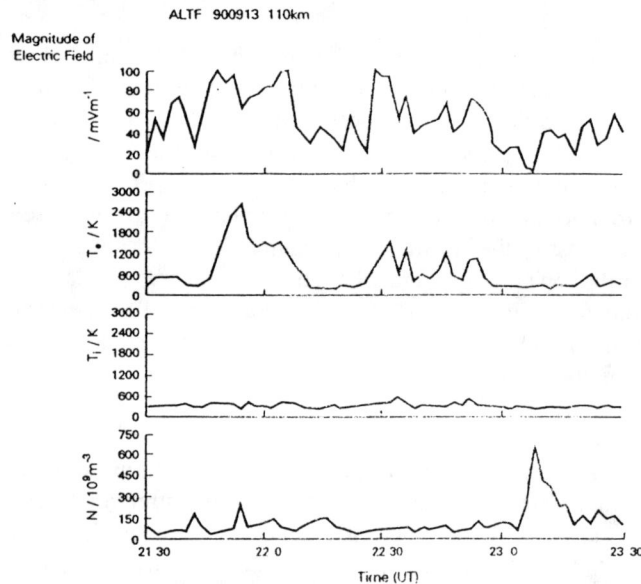

Figure 9. EISCAT measurements of electric field strength, electron and ion temperatures, and electron density as a function of time. The measurements were made at an altitude of 110 km on 13 September 1990 between 2130 and 2330 UT. From *Williams et al.* [1992].

of the electric field, electron and ion temperatures, and electron density are shown versus time at an altitude of 110 km. The radar measurements were made on 13 September between 2130 and 2330 UT [*Williams et al.*, 1992]. Note that T_i, and probably T_n, remain below 600 K throughout the observing period, but that T_e is significantly enhanced at certain times. The peaks in the electron temperature coincide with electric field enhancements. However, not all of the electric field enhancements produce T_e increases, but this is probably due to the need to satisfy certain threshold conditions for the plasma instability. Note that this area of research is still controversial and more work is needed before definitive conclusions can be drawn about the cause of the T_e increases [cf. *Williams et al.*, 1992; and references therein].

2.5. *Particle Precipitation*

Auroral particle precipitation, both electrons and protons, has a significant impact on the lower ionosphere. The main effect of the precipitation is to cause ionization when the energetic particles strike the ambient neutrals. However, the precipitation also leads to electron, ion, and neutral gas heating as well as airglow. The ionization that is produced depends on both the energy flux and the characteristic energy of the precipitation. Only the so-called "hard" precipitation penetrates to the D and E regions. This is shown in Figure 10, where theoretical O^+ production rates are plotted as a function of altitude. The solid line is for a hard auroral precipitation spectrum, with an energy flux of 0.92 ergs cm^{-2} s^{-1} sr^{-1} and a characteristic energy of 2 keV. The other three production rate profiles are for soft auroral spectra. These production rate profiles were used by *Sojka and Schunk* [1986] to study the effect on convecting plasma flux tubes of different types of electron precipitation that persist for only a brief period of time (~ 10 minutes). With regard to the lower ionosphere, the study obtained the well-known result that the E-region density enhancement associated with the precipitation occurs only in the precipitation region and it only lasts as long as the precipitation lasts because of the rapid dissociative recombination rate.

3. SPATIAL STRUCTURE AND TIME VARIATIONS

The interplay between the various chemical, radiative, dynamical, and electrodynamical processes that operate in the lower ionosphere, coupled with the widely different spatial and temporal scales of the processes, act to create a considerable amount of structure in this region. The structure can result from spatial structure in the magnetospheric parameters, including convection electric fields, particle precipitation, and field-aligned currents. It can also result from the structure associated with upward propagating tides and gravity waves, as noted earlier. In addition, small-scale structure can result from a variety of plasma instabilities, including the modified two-stream, gradient-drift, and velocity space instabilities. Finally, structure can develop when the magnetospheric and thermospheric parameters undergo significant time variations, as occurs during magnetic storms and substorms [c.f., *Weber et al.*, 1984; *Robinson et al.*, 1985; *Tsunoda*, 1988; and references therein].

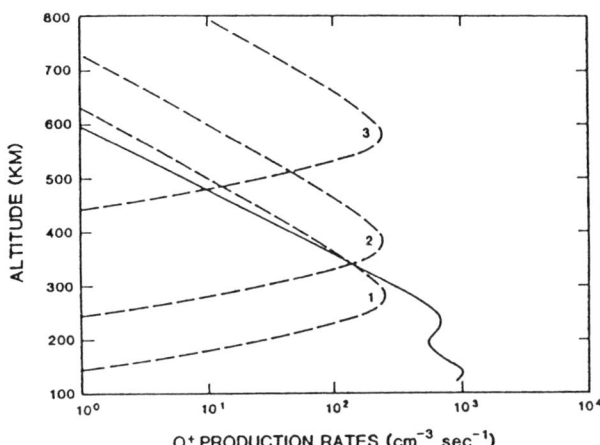

Figure 10. Theoretical O^+ production rates as a function of altitude. The solid curve corresponds to an O^+ production rate for a typical hard auroral spectrum, while the dashed curve labeled 1 corresponds to a soft cusp spectrum. Curves 2 and 3 correspond to soft spectra with peak O^+ production rates at higher altitudes. From *Sojka and Schunk* [1986].

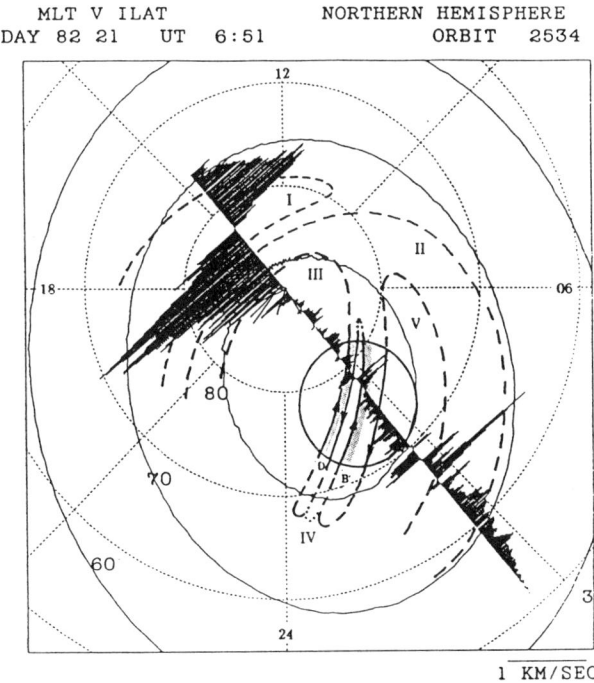

Figure 11. An implied convection pattern consistent with observed optical emissions and plasma convection velocities measured by DE 2. The flow lines describing cells III, IV, and V are at the same potential and may be connected "fingers" or separate convection cells. From *Carlson et al.* [1988].

Figure 11 shows an example of spatial structure in the magnetospheric electric field and particle precipitation for northward interplanetary magnetic field. The structure was observed by *Carlson et al.* [1988] using a ground-based all-sky imaging photometer in combination with simultaneous ion convection velocities and precipitation fluxes measured by DE 2. The authors found that in most cases the velocity structure was associated with subvisual sun-aligned polar cap arcs. The sun-aligned arcs were of the order of 100 km in width and 1000-2000 km in length. The locations of two sun-aligned arcs are shown in Figure 11 via the areas shaded in gray. Using the optical and satellite data, *Carlson et al.* [1988] were able to deduce a possible convection signature in the vicinity of the arcs. The lines perpendicular to the satellite track correspond to the measured ion drift velocities, while the shaded regions marked D and B correspond to sun-aligned arcs. The dashed lines correspond to possible convection paths. The cells labeled I and II can be identified as those usually expected. However, within the large dawn convection cell there appears to be at least three discrete convection cells or large-scale 'fingers' (labeled III, IV, V).

The spatial structure in the convection and precipitation shown in Figure 11 will produce spatial structure in the lower ionosphere with the same scale-lengths. The structure should also develop fairly rapidly and track the changes in the magnetospheric parameters. The rapid build-up of ionization in response to on-going auroral precipitation is shown in Figure 12. The measurements were made using the Chatanika incoherent scatter radar when it was in the auroral oval. Two altitude-latitude scans are shown, separated by about 10 minutes. Note the rapid build-up of ionization and the structure in the lower ionosphere.

The electron density structure that occurs at F-region altitudes will not be reflected in the thermospheric parameters with the same scale length. At these altitudes, viscous dissipation in the neutral gas acts to smooth velocity gradients and, hence, density gradients. However, viscous dissipation is not important at MLTI altitudes and there the structure that occurs in the ionospheric and magnetospheric parameters will be reflected in the neutral parameters. Evidence for this fact comes from a number of 'high-resolution' modeling studies [*St.-Maurice and Schunk*, 1981; *Fuller-Rowell*, 1984; *Walterscheid and Lyons*, 1989]. In particular, *St.-Maurice and Schunk* [1981] used a two-dimensional MHD model and studied the 'steady state' momentum coupling between the ionosphere and neutral atmosphere in the vicinity of discrete high-latitude structures, such as convection channels and plasma density troughs. After a systematic and extensive parameter study, these authors concluded that ionospheric structure and structured electric fields may have an important effect on the global thermospheric circulation. However, this conclusion is speculative, and to date, all of the model studies have been

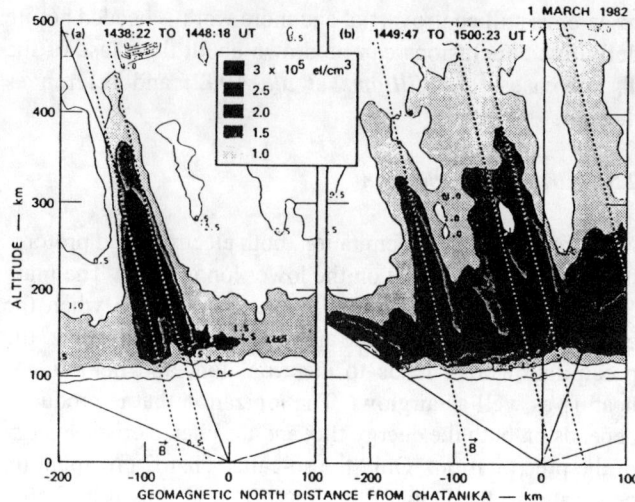

Figure 12. Electron densities measured with the Chatanika radar when it was in the auroral oval. The two altitude-latitude scans are separated by about 10 minutes. From *Tsunoda* [1988].

concerned with 'single' ionospheric structures and how they may affect the 'local' thermospheric circulation.

4. SUMMARY

The lower ionosphere is a particularly difficult region to both probe and model. Numerous chemical, radiative, dynamical, and electrodynamical processes operate in this region. The region is also influenced by solar, magnetospheric and stratospheric processes. To properly describe the lower ionosphere, account must be taken of ionization due to solar EUV, UV and x-ray radiation; ionization due to meteor ablation; magnetospheric electric fields, auroral precipitation, and Birkeland currents; plasmaspheric downflows and radiation belt precipitation; and upward propagating tides and gravity waves from the stratosphere. In the lower ionosphere are chemical reactions involving positive, negative, and cluster ions; transport processes involving ordinary diffusion and wave-particle interactions; and non-Maxwellian ion velocity distributions. In addition, the various processes act on widely different temporal and spatial scales, and these scales are directly reproduced in the medium. The net effect is that "structure" is probably the normal state for the lower ionosphere. Hence, it is highly doubtful that climatological studies will be useful for elucidating the various processes that operate on and within this region.

Acknowledgment. This research was supported by NASA grant NAG5-1484 and NSF grant ATM-89-13230 to Utah State University.

REFERENCES

Burns, C. J., E. Turunen, H. Matveinen, H. Ranta, and J. K. Hargreaves, Chemical modeling of the quiet summer D and E regions using EISCAT electron density profiles, *J. Atmos. Terr. Phys., 53*, 115-134, 1991.

Carlson, H. C., R. A. Heelis, E. J. Weber, and J. R. Sharber, Coherent mesoscale convection patterns during northward interplanetary magnetic field, *J. Geophys. Res., 93*, 14,501-14,514, 1988.

Crowley, G., B. A. Emery, R. G. Roble, H. C. Carlson and D. J. Knipp, Thermospheric dynamics during September 18-19, 1984. 1. Model simulations, *J. Geophys. Res., 94*, 16925-16944, 1989.

Devlin, T., J. K. Hargreaves, and P.N. Collis, EISCAT observations of the ionospheric D region during auroral radio absorption events, *J. Atmos. Terr. Phys., 48*, 795-805, 1986.

Fesen, C. G., R. E. Dickinson, and R. G. Roble, Simulation of thermospheric tides at equinox with the NCAR thermospheric general circulation model, *J. Geophys. Res., 91*, 4471-4489, 1986.

Fesen, C. G., D. W. Rusch, and J.-C. Gerard, The latitudinal gradient of the NO peak density, *J. Geophys. Res., 95*, 19,053-19,059, 1990.

Friedrich, M., K. M. Torkar, K. Spenner, G. Rose, and H. V. Widdel, Electron densities during winter anomalous absorption of different intensity, *J. Atmos. Terr. Phys., 41*, 1121-1125, 1979.

Fuller-Rowell, T. J., A two-dimensional, high-resolution, nested-grid model of the thermosphere. 1. Neutral response to an electric field "spike", *J. Geophys. Res., 89*, 2971-2990, 1984.

Hargreaves, J. K., *The Solar-Terrestrial Environment*, Cambridge University Press, 1992.

Kelley, M. C., *The Earth's Ionosphere*, Academic Press, 1989.

Mitra, A. P. and J. N. Rowe, Ionospheric effects of solar flares – VI. Changes in D region ion chemistry during solar flares, *J. Atmos. Terr. Phys., 34*, 795-806, 1972.

Narcisi, R. S. and A. D. Bailey, Mass spectrometric measurements of positive ions at altitudes from 64 to 112 kilometers, *J. Geophys. Res., 70*, 3687, 1965.

Rees, D., and T.-J. Fuller-Rowell, Numerical simulation of the seasonal/latitudinal variations of atomic oxygen and nitric oxide in the lower thermosphere and mesosphere, *Adv. Space Res., 10*, 83-102, 1990.

Rees, M. H., *Physics and Chemistry of the Upper Atmosphere*, Cambridge University Press, 1991.

Robinson, R. M., R. T. Tsunoda, J. F. Vickrey, and L. Guerin, Sources of F region ionization enhancements in the nighttime auroral zone, *J. Geophys. Res., 90*, 7533, 1985.

Roble, R. G., E. C. Ridley, and R. E. Dickinson, On the global mean structure of the thermosphere, *J. Geophys. Res., 92*, 8745-8758, 1987.

Rusch, D. W., J.-C. Gerard and C. G. Fesen, The diurnal variation of NO, N(^{2}D) and ions in the thermosphere: A comparison of satellite measurements to a model, *J. Geophys. Res., 96*, 11,331-11,339, 1991.

Schlegel, K. and J.-P. St.-Maurice, Anomalous heating of the polar E region by unstable plasma waves, I., Observations, *J. Geophys. Res., 86*, 1447-1452, 1981.

Schunk, R. W. and J. J. Sojka, Ion temperature variations in the daytime high-latitude F region, *J. Geophys. Res., 87*, 5169-5183, 1982.

Schunk, R. W., W. J. Raitt and P. M. Banks, Effect of electric fields on the daytime high-latitude E and F regions, *J. Geophys. Res., 80*, 3121-3130, 1975.

Siskind, D. E., and D. W. Rusch, Nitric oxide in the middle to upper thermosphere, *J. Geophys. Res., 97*, 3209-3217, 1992.

Sojka, J. J., Global scale, physical models of the F region ionosphere, *Rev. Geophys., 27*, 371-403, 1989.

Sojka, J. J. and R. W. Schunk, A theoretical study of the production and decay of localized electron density enhancements in the polar ionosphere, *J. Geophys. Res., 91*, 3245-3253, 1986.

St-Maurice, J.-P. and R. W. Schunk, Behavior of ion velocity distributions for a simple collision model, *Planet. Space Sci., 22*, 1-18, 1974.

St-Maurice, J.-P. and R. W. Schunk, Ion velocity distribution in the high latitude ionosphere, *Rev. Geophys. Space Phys., 17*, 99-134, 1979.

St-Maurice, J.-P. and R. W. Schunk, Ion-neutral momentum coupling near discrete high-latitude ionospheric features, *J. Geophys. Res., 86*, 11,299-11,321, 1981.

St-Maurice, J.-P., K. Schlegel and P.M. Banks, Anomalous heating of the polar E region by unstable plasma waves, 2, Theory, *J. Geophys. Res., 86*, 1453-1462, 1981.

Tomko, A. A., A. J. Ferraro, H. S. Lee, and A. P. Mitra, A theoretical model of D region ion chemistry modifications during high power radio wave heating, *J. Atmos. Terr. Phys., 42*, 275-285, 1980.

Tsunoda, R. T., High-latitude F region irregularities: A review and synthesis, *Rev. Geophys., 26*, 719-760, 1988.

Walterscheid, R. L. and L. R. Lyons, The neutral E region zonal winds during intense postmidnight diffuse aurora: Response to observed particle fluxes, *J. Geophys. Res., 94*, 3703-3712, 1989.

Williams, P. J. S., B. Jones, and G. O. L. Jones, The measured relationship between electric field strength and electron temperature in the auroral E region, *J. Atmos. Terr. Phys., 54*, 741-748, 1992.

Weber, E. J., J. Buchau, J. G. Moore, J. R. Sharber, R. C. Livingston, J. D. Winningham, and B. W. Reinisch, F layer ionization patches in the polar cap, *J. Geophys. Res., 89*, 1683, 1984.

R. W. Schunk and J. J. Sojka, Center for Atmospheric and Space Sciences, Utah State University, Logan, UT 84322-4405

The Ionospheric Wind Dynamo: Effects of its Coupling With Different Atmospheric Regions

A. D. Richmond

High Altitude Observatory, National Center for Atmospheric Research, Boulder, Colorado

Effects of the ionospheric wind dynamo related to its coupling with different atmospheric regions are reviewed. Early studies showed that height-averaged tidal-like winds of order 30 m/s must be present in the dynamo region (90–200 km height) to produce the observed geomagnetic daily variations. They also indicated that diurnal solar tides drive more current than semidiurnal tides, although the latter are important in explaining the observed electric fields and asymmetries of the currents about the equator. Lunar tidal winds of order 10 m/s in the dynamo region, roughly opposite in phase to the tides at ground level, are required to explain the observed lunar geomagnetic variations. There is strong seasonal and longitudinal variability in lunar geomagnetic variations that indicates similar variability in the lunar tidal winds. Day-to-day variability in ionospheric dynamo effects points to short-term variability in the global winds, probably due to variability in tidal propagation conditions through the middle atmosphere and/or to penetration of planetary waves into the dynamo region. However, attempts to correlate observed upper-mesospheric winds with geomagnetic variations have had only limited success. Searches for two-day variations in geomagnetic data associated with the two-day wave in mesospheric winds have not consistently found a significant signal. Suggestions of a 16-day variation in geomagnetic and ionospheric data associated with 16-day planetary waves remain to be verified, as do suggestions of possible associations between the stratospheric quasi-biennial oscillation and geomagnetic variations. Coupling among dynamo electric fields and currents, ionospheric plasma variations, and thermospheric dynamics have been shown to be important in a number of situations. Recently, global simulation models that take these mutual-coupling effects into account have been developed. Coupling between the ionospheric and magnetospheric dynamos is also significant, though the quantitative importance and full implications of the "flywheel" effect, "fossil-wind" effect, and "disturbance-dynamo" effect remain to be determined. Fruitful areas of future research will be the further exploitation of observed geomagnetic and ionospheric phenomena to study tidal and planetary-wave propagation conditions in the middle atmosphere, including possible long-term changes associated with a changing atmospheric state; exploitation of simulation models of coupled thermosphere/ionosphere dynamics and electrodynamics; and further investigation of effects associated with coupling between the ionospheric and magnetospheric dynamos.

1. INTRODUCTION

Earth's upper atmosphere is electrically conducting, owing to the presence of free electrons and ions that constitute the ionosphere. Electric fields and currents are produced in this conducting medium by various processes (see *Volland* [1984], *Kelley* [1989], and *Roble* [1991]). At middle and low latitudes the ionospheric wind dynamo (or "ionospheric dynamo" for short) is especially important. Winds in the lower thermosphere move the conducting medium through Earth's geomagnetic field, generating an electromotive force that leads to current flow, the build-up of polarization charges, and resultant electric fields and magnetic perturbations. The state of the middle atmosphere affects the ionospheric dynamo by modulating tides and planetary waves that propagate up into the dynamo region. Consequently, studies of dynamo ef-

The Upper Mesosphere and Lower Thermosphere:
A Review of Experiment and Theory
Geophysical Monograph 87
Copyright 1995 by the American Geophysical Union

fects can shed light on certain properties of the middle atmosphere. The dynamo electric fields and currents interact with the dynamics of the ionosphere and thermosphere: the electric field contributes to plasma drifts that alter the ionospheric electron density distribution, and the ionospheric current exerts a force on the thermospheric medium that affects its motion. Modeling this coupled system has presented a major challenge to the upper-atmospheric research community. At high latitudes, the ionosphere is electrically coupled with the magnetosphere, where an important additional source of electric fields and currents is present due to the interaction of the solar wind with the magnetosphere. This source is particularly important during magnetic disturbances, of which the most dramatic form is the magnetic storm. The ionospheric and magnetospheric dynamos are electrically coupled. For completeness, it should also be noted that electric currents can penetrate into the ionosphere from below, most strongly so over thunderclouds. Although the total current flowing between the lower atmosphere and the upper atmosphere is relatively small [e.g., *Hays and Roble*, 1979; *Makino and Takeda*, 1984; *Kuznetsov et al.*, 1992] it is possible that significant local effects on the ionospheric electric fields and currents can be produced. Electric fields of ionospheric origin are also large enough to be detectable in the lower atmosphere [e.g., *Roble and Hays*, 1979]. Some recent reviews of the ionospheric dynamo and its effects are given by *Volland* [1984], *Kelley* [1989], and *Richmond* [1989, 1991].

The present paper reviews the ionospheric dynamo and effects related to tides and planetary waves propagating up from the middle atmosphere, to the role of the dynamo in coupling the dynamics of the thermosphere and the ionosphere, and to the coupling of the ionospheric dynamo with the magnetosphere.

The ionospheric conductivity is highly anisotropic, being many orders of magnitude greater in the direction of the geomagnetic field than perpendicular to it. Consequently, ionospheric electrodynamics is strongly organized by the geomagnetic field. The large-scale electric field is essentially electrostatic, and the high field-aligned conductivity usually ensures that geomagnetic field lines are nearly equipotential. Figure 1 shows the average potential distribution at 300 km altitude, at middle and low magnetic latitudes [*Richmond et al.*, 1980]; this figure does not include the much stronger potential variations that are present above 60° associated with field-aligned currents from the magnetosphere. At middle latitudes the potential is nearly constant in height. Near the magnetic equator, where ge-

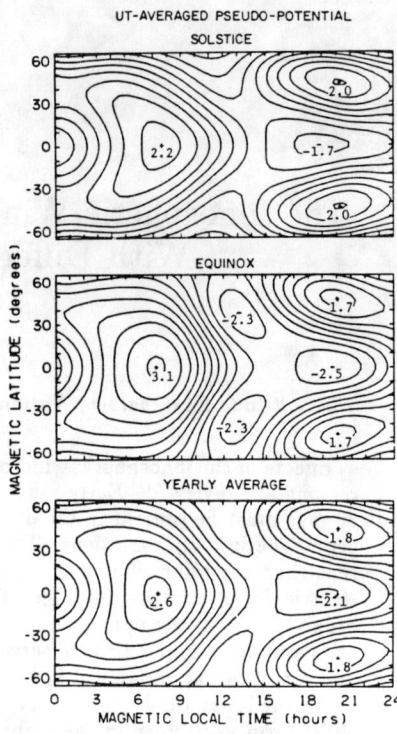

Fig. 1. Magnetic-local-time/magnetic-latitude plots of the UT-averaged electric pseudo-potential (that is, the potential normalized to the longitudinally averaged geomagnetic-field strength) at 300 km altitude. Contour spacing is 500 V. Maximum and minimum values are in kilovolts. [*Richmond et al.*, 1980]

omagnetic field lines become horizontal, significant altitude variations of the potential and electric field can occur. The most dramatic manifestation of this is the so-called "equatorial electrojet" (see reviews by *Forbes* [1981], *Reddy* [1989], and *Rastogi* [1989]), consisting of a relatively intense current along the magnetic equator between 100 km and 110 km, caused by a strong vertical polarization electric field that can be established there. Figure 2 shows a simulation of the height-integrated horizontal current for the December solstice [*Takeda*, 1990], again neglecting high-latitude coupling with the magnetosphere. Two large current vortices exist on the dayside of Earth, clockwise in the southern hemisphere and counterclockwise in the northern hemisphere. The equatorial electrojet is also apparent. These currents produce perturbations in the geomagnetic field that are easily measured at ground level. In fact, it was on the basis of observed geomagnetic variations that the existence of electric currents in the upper atmosphere was first postulated (see *Chapman and Bartels* [1940]). The

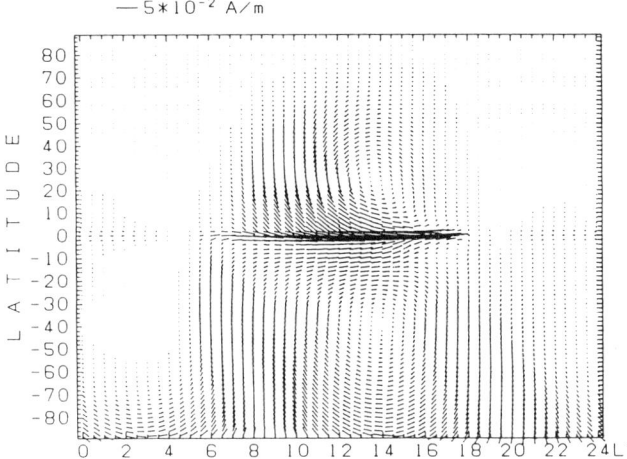

Fig. 2. Global distribution of the height-integrated ionospheric currents at the December solstice, in geomagnetic coordinates. (Reprinted from *Takeda* [1990], with kind permission from Pergamon Press Ltd, Headington Hill Hall, Oxford OX3 0BW, UK.)

average quiet-day geomagnetic perturbations are commonly labeled S_q, for "solar quiet." The perturbations vary on a day-to-day basis as well as with season and solar cycle, and the quiet-day variations on a single day are sometimes labeled S_R [*Mayaud*, 1965], to distinguish them from the average S_q.

2. INFERENCES CONCERNING LOWER-THERMOSPHERIC WINDS

Models of the ionospheric-wind-dynamo process, when combined with observed electrodynamic effects produced by the dynamo, have provided a considerable amount of information concerning the characteristics of winds in the lower thermosphere. *J. Fejer* [1953], using a dynamo model with conductivities of the correct order of magnitude, showed that the height-averaged wind over the thickness of the dynamo layer is of the order of 30 m/s. The first dynamo calculations that used realistic day-night variations of the electrical conductivity, by *Maeda* [1955, 1957], *Hirono and Kitamura* [1956], and *Kato* [1956, 1957], showed that the winds are predominantly diurnal, that is, they have a dominant 24-hour period. Later studies [*Stening*, 1969, 1981; *Tarpley*, 1970b; *Volland*, 1971; *Schieldge et al.*, 1973; *Murata*, 1974; *Richmond et al.*, 1976; *Möhlmann*, 1976, 1977; *Takeda and Maeda*, 1980, 1981; *Takeda et al.*, 1986] have tended to confirm this fact, but some have also shown that 12-hour winds play an important role in explaining observed ionospheric electric fields [*Richmond et al.*, 1976; *Hanuise et al.*, 1983; *Richmond and Roble*, 1987; *Takeda and Yamada*, 1987; *Takeda*, 1990]. However, an unresolved problem is that several studies have found the modeled magnitude of semidiurnal dynamo effects, using wind amplitudes based on observations at a few locations, tends to be too strong when compared with observations [e.g., *Forbes and Lindzen*, 1976; *Forbes and Garrett*, 1979; *Takeda and Maeda*, 1981; *Richmond and Roble*, 1987; *Takeda and Yamada*, 1987]. *Stening* [1969] and *Takeda and Maeda* [1981] also found that the phase of semidiurnal dynamo effects in their simulations did not agree with observations.

Figure 3 [*Richmond and Roble*, 1987] compares observed equinoctial S_q variations (points) at various stations, from a northern magnetic upper-midlatitude site (Fredericksburg) to a southern magnetic lower-midlatitude site (Trelew), with results from two dynamo simulations. One simulation (dashed line) uses winds from the National Center for Atmospheric Research thermospheric general circulation model (NCAR TGCM) [*Dickinson et al.*, 1981] as driven solely by thermospheric absorption of solar radiation, without allowance for tides propagating up from the middle atmosphere. The other simulation (solid line) uses the same model, but includes upward-propagating semidiurnal tides as estimated by *Fesen et al.* [1986]. Whereas the simulation without tides generally underestimates the observed magnitude of the S_q variations, the simulation that includes propagating tides tends to overestimate the variations somewhat, suggesting that the model of semidiurnal tides used in the simulation needs improvement.

It is to be expected that tidal components antisymmetric about the equator will be present at the solstices, but there have been only a few attempts to include the dynamo effects of the asymmetric propagating tides [*Schieldge et al.*, 1973; *Marriott et al.*, 1979; *Stening*, 1989a]. These studies have established the likely importance of such tides but have not yet uniquely defined their probable characteristics. An unexpected result of early studies of S_q variations was the fact that significant asymmetry exists even at equinox, strongly suggesting the importance of lower-thermospheric winds that are asymmetric about the equator at equinox. Figure 4 [*van Sabben*, 1964] shows that the northern-hemisphere vortex tends to be situated to the west of the southern vortex at the September equinox; this behavior was observed at all universal times. Such observations have been verified by other studies [e.g., *Suzuki*, 1978]. *Wulf* [1963, 1965a, b] has pointed out that S_q variations can differ significantly between dif-

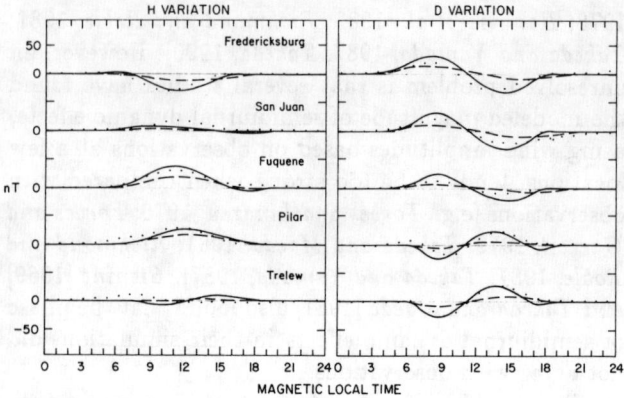

Fig. 3. Comparisons of average observed equinoctial solar-minimum northward (H) and eastward (D) magnetic perturbations (points) at Fredericksburg (magnetic latitude 51°), San Juan (30°), Fuquene (18°), Pilar (-17°), and Trelew (-28°) with perturbations computed from two simulations: without tidal forcing at the lower boundary (dashed lines) and with tidal forcing (solid lines). The computations show perturbations at 1700 local time as a function of magnetic local time (representing magnetic longitude), rather than actual local-time variations for these stations. [*Richmond and Roble*, 1987]

ferent times of the year when the solar position is similar (e.g., April and August), and that this should have implications for seasonal variations in the structure and circulation of the middle atmosphere.

Lunar tidal effects were detected early in geomagnetic variations (see reviews by *Stewart* [1882], *Chapman and Bartels* [1940], and *Matsushita* [1967a]) and in ionospheric plasma densities (see review by *Matsushita* [1967b]). Model calculations [*Maeda and Fujiwara*, 1967; *Tarpley*, 1970a] verified that lunar atmospheric tides are capable of generating the observed effects, and showed that the amplitude of the lunar semidiurnal tidal winds must be of the order of 10 m/s in the dynamo region, while the phase of the tide is roughly opposite to that at the ground. There are strong indications that the lunar tide is quite variable with respect to latitude, longitude, and season. Global analyses of geomagnetic lunar effects [*Malin*, 1973; *Stening and Winch*, 1979; *Winch*, 1981] have found significant longitudinal variations. Figure 5 [*Stening and Winch*, 1979] shows the longitudinal variation of phase of the geomagnetic lunar variation across Europe. *Stening and Winch* [1979] and *Gupta* [1982] found strong seasonal variations of the lunar tide, having different forms at different locations: sometimes an annual variation dominates, while at other sites the main seasonal variation is semiannual. In the equatorial electrojet, the lunar geomagnetic effect is considerably stronger around the December solstice than around the June solstice, on a global-average basis [*Rastogi and Trivedi*, 1970; *Gupta*, 1973]. *Schlapp and Malin* [1979] showed that a tendency for a January maximum is present not only for equatorial and southern-hemisphere stations, but extends well into the northern hemisphere as well. They also found that the phase of the tide around January is frequently considerably different from the phase in other seasons. Similar results were found by *Gupta* [1982] and *Matsushita and Xu* [1984]. *Stening* [1977b] modeled the seasonal variations in currents driven by a (2,2) tidal mode of fixed amplitude and phase within the dynamo region, and found that the calculated changes are much too small to explain the observed seasonal variations in lunar geomagnetic variations. He concluded that there must be a significant seasonal variation of lunar tidal winds in the dynamo region. *Stening* [1989b] also found that the amplitude of lunar geomagnetic variations is often larger in the morning than in the afternoon; the explanation for this effect is not clear. *Chapman et al.* [1971] found that the amplitude of the lunar geomagnetic perturbation, when considered on a global basis, usually changes more strongly with the solar cycle around the December solstice than around the June solstice, an effect that cannot be simply explained solely by solar-cycle changes in ionospheric conductivity.

Irregular day-to-day variations in ionospheric electrodynamics are often observed. One source of the irregular variations is known to be irregular magnetospheric activity that couples into the ionosphere via electric fields and magnetic-field-aligned currents at high latitudes, spreading to lower latitudes. However, a large part of the day-to-day variability in ionospheric electrodynamics is not directly associated with magnetic activity [e.g., *Chapman and Stagg*, 1929, 1931; *Schlapp*, 1968; *Gupta*, 1970; *Kane*, 1976; *Mayaud*, 1977; *Greener and Schlapp*, 1979; *Clauer et al.*, 1980; *Sastri*, 1982; *G. Walker and Kannangara*, 1982; *Schlapp and Mann*, 1983; *Briggs*, 1984; *Mann and Schlapp*, 1985; *Hibberd and Davidson*, 1988; *Butcher*, 1989], and is believed to be linked to irregular variability of winds in the dynamo region. The causes of the wind variability are only poorly understood. Since simulation models of global thermospheric dynamics do not reveal any intrinsic variability of thermospheric dynamics associated with solar forcing when a fixed lower boundary is employed [e.g., *Fuller-Rowell and Rees*, 1980; *Dickinson et al.*, 1981], it is believed that variability in the winds must be as-

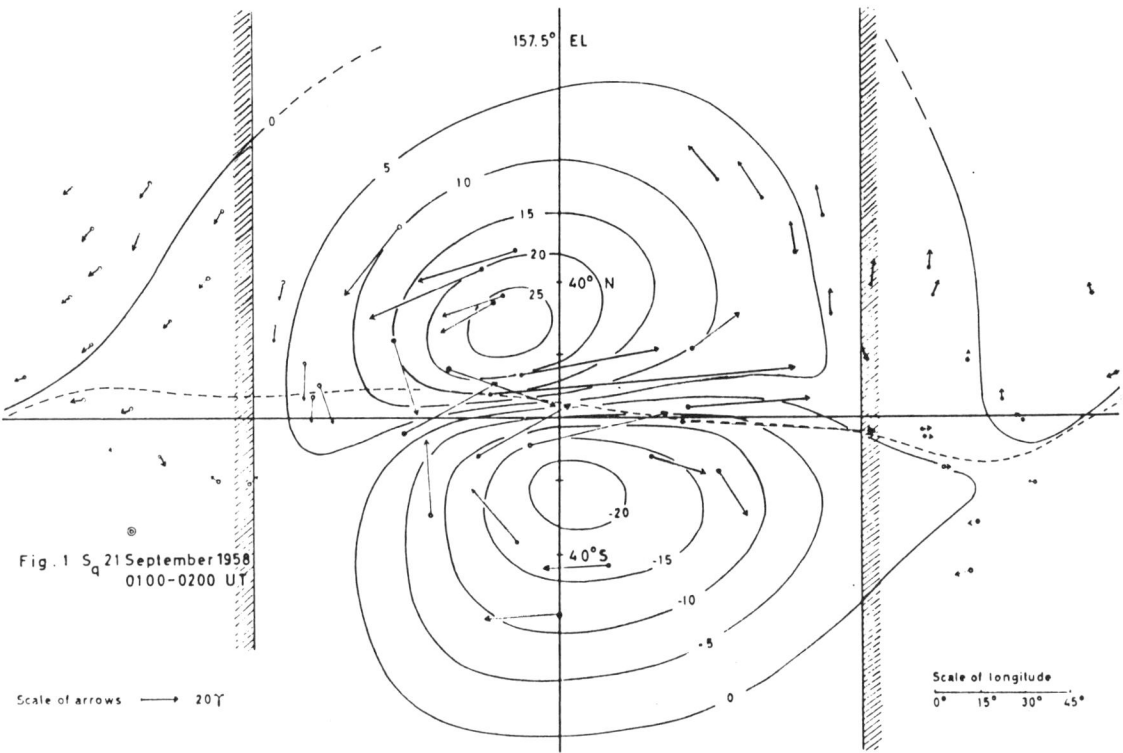

Fig. 4. Representative current system of the S_R variation on 1958 September 21, 0130 UT. The arrows give the horizontal S_R vectors, turned clockwise by 90°. The streamlines are drawn at intervals of 50 kA. Geographical latitudes, magnetic dip equator and ionospheric twilight zones are indicated. (Reprinted from *van Sabben* [1964], with kind permission from Pergamon Press Ltd, Headington Hill Hall, Oxford OX3 0BW, UK.)

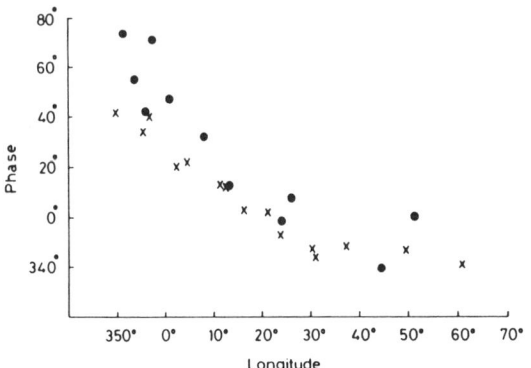

Fig. 5. Variation with longitude of the phase of the semidiurnal lunar variation in the the magnetic horizontal component H in the European sector in August. Symbols: • stations in the geographic latitude range 30°-45°; × latitude range 45°-55°. (Reprinted from *Stening and Winch* [1979], with kind permission from Pergamon Press Ltd, Headington Hill Hall, Oxford OX3 0BW, UK.)

sociated with variability in the dynamic conditions at the base of the thermosphere, that is, in the characteristics of the upward-propagating tides and other wave features.

Figure 6 [*Mayaud*, 1977] shows an extreme example of day-to-day variability in quiet-day variations of the horizontal magnetic component at Huancayo, Peru, which is under the equatorial electrojet. The ten magnetically disturbed days of the month have been removed, although some of the remaining variability apparent in Figure 6 could be due to residual activity, especially the more-rapid fluctuations. The day-to-day variability is found to be only weakly correlated between stations widely separated in longitude (e.g., *Kane* [1972]). *Bartels and Johnston* [1940] noted that the daily geomagnetic variation at Huancayo often showed an abnormally large lunar-like variation that is in phase with the average lunar variation, especially around January. *Onwumechilli* [1963, 1964] found that the anomalous daily-variation curves could have various shapes

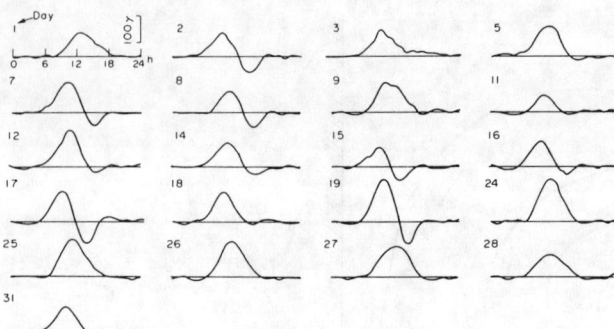

Fig. 6. S_R variations in the magnetic horizontal component H at Huancayo during the month of January 1923. All the quiet days of the month, as classified by the average aa indices for the local day, are displayed. (Reprinted from *Mayaud* [1977], with kind permission from Pergamon Press Ltd, Headington Hill Hall, Oxford OX3 0BW, UK.)

for a given lunar phase, casting doubt on whether the cause is really lunar. *Stening* [1975] suggested that this variability may be due more to variability in the solar semidiurnal tide. Simulations by *Marriott et al.* [1979] and *Hanuise et al.* [1983] supported the idea that anomalous semidiurnal tides, whether of solar or lunar origin, could cause the global changes in currents that are associated with equatorial counterelectrojet events, when the daytime electrojet current reverses from its usual eastward direction to a westward direction. However, *Stening* [1977a] examined simultaneous changes in the S_R current system at different latitudes, and came to the conclusion that, although the world-wide changes are sometimes consistent with a change in the solar semidiurnal tide penetrating into the dynamo region, often the changes do not have the characteristics expected of tidal generation of currents. He proposed that planetary waves might often be responsible.

Ito et al. [1986], *Takeda and Yamada* [1989], and *Chen* [1992] have modeled the dynamo effects of quasi-two-day planetary waves that have been observed at mesopause heights, and that they proposed could extend into the dynamo region. *Ito et al.* [1986] examined geomagnetic data to test the model predictions. They found a minor peak in the power spectrum of magnetic perturbations at 0.5 cycles per day, simultaneous with meteor-wind observations of a large two-day wave below 100 km in late July and early August, 1979. They concluded that the available wind and geomagnetic observations point to the dominance of a two-day wave mode that is antisymmetric about the equator and has a longitudinal wave number of 3, and that can produce a geomagnetic variation at midlatitudes of the order of 10% the magnitude of S_q. *Takeda and Yamada* [1989], performing a spectral analysis on geomagnetic data at a number of low-latitude stations for May–August 1958, sometimes found peaks around 45 hours and 58 hours, which they attributed to the two-day wave. In contrast, *Parkinson* [1982] computed spectra of geomagnetic variations by processing 15 days at a time, in order to allow for the limited duration of the two-day wave seen in mesospheric-wind measurements, and found that the occurrences of two-day variations have low statistical significance when compared with a random time series. It is possible that the alternating large and small daily ranges of equatorial electrojet amplitude reported by *Fambitakoye* [1971] could be associated with such planetary waves, and *Chen* [1992] and *Yi and Chen* [1993] have attributed an apparent 2-day variation of the equatorial ionospheric anomaly to the dynamo effects of this wave. There is a hint of a possible two-day variation in Figure 6, on the days January 16–19.

By correlating anomalous S_R perturbations with stratospheric pressure variations, *Brown* [1975] found evidence that the S_R variations at midlatitudes are influenced by planetary waves of 15- to 20-day periods in the winter. *Forbes and Leveroni* [1992] have noted a 16-day variation in the equatorial electrojet amplitude in January 1979 that they presented as evidence of the 16-day planetary wave. (However, *MacDougall* [1979] and *Burrows* [1970] found 15-day variations not much smaller in amplitude during the winters of 1964 and 1965 that they attributed to the semimonthly lunar modulation of the electrojet.) *Yi and Chen* [1993] proposed that variations in the strength of the equatorial ionization anomaly with periods ranging from 2 to 18 days could be associated with planetary waves. If the effects are indeed due to planetary waves, it is not known whether the planetary waves themselves penetrate well into the dynamo region, or whether they might act indirectly through modulation of the amplitudes and phases of upward-propagating tides. Some studies have found a weak geomagnetic variation at a period around 26 months, approximately that of the quasi-biennial oscillation in the stratospheric wind [*Stacey and Wescott*, 1962; *Yacob and Bhargava*, 1968; *Currie*, 1973; *Nastrom and Belmont*, 1976], while other studies have found no such geomagnetic variation [*London and Matsushita*, 1963; *Shapiro and Ward*, 1964; *Currie*, 1966]. *Yacob and Bhargava* [1968] found a minor spectral peak of the same 26-month period in sunspot-number data processed for the same sixty-year interval. The 26-month spectral peak in the geomagnetic data could therefore have been due to variations in ionospheric conductivity rather than winds.

An attempt by *Phillips and Briggs* [1991] to relate day-to-day variations in geomagnetic data with observed mesospheric winds over Australia found no significant correlation. The authors discussed several possible explanations for this, one of which is that local wind measurements are not adequate representations of the global conditions responsible for ionospheric currents. It will be important to pursue this line of research.

3. INTERACTIVE DYNAMO MODELS

The dynamics of thermospheric winds, ionospheric conductivities, and electric fields and currents are mutually coupled (see *Kelley* [1989]). The winds generate electric fields and currents through dynamo action. Both the winds and the electric fields produce plasma drifts that can alter the ionospheric conductivity, especially at night [e.g., *Footitt et al.*, 1983; *Singhal*, 1991]. The currents, flowing through the geomagnetic field, exert an Ampère force on the medium of $\mathbf{J} \times \mathbf{B}$ (where $\mathbf{J}$ is current density and $\mathbf{B}$ is magnetic-field strength) that can significantly affect the dynamics of the winds [e.g., *Baker and Martyn*, 1953; *Piddington*, 1954; *Axford and Hines*, 1961]. Because the Ampère force results primarily from the collisional drag force acting between differentially drifting ions and neutrals, it is commonly called the "ion-drag" force.

There have been several studies to investigate the interactive effects of these coupling processes. *Baker and Martyn* [1953] and *Akasofu and DeWitt* [1965] showed how the neutral air tends to be accelerated by the electric currents towards the ion drift velocity $\mathbf{E} \times \mathbf{B}/B^2$ (where $\mathbf{E}$ is electric field) such that the magnitude of the electric field in the frame of the moving neutrals, $\mathbf{E} + \mathbf{V}_n \times \mathbf{B}$ (where $\mathbf{V}_n$ is neutral velocity), is reduced, and hence the currents are reduced. The effect is the same as if the conductivity itself were reduced, so the feedback of the winds onto the currents is such as to reduce the effective conductivity. *Axford and Hines* [1961] and *Forbes and Harel* [1989] also discussed this effect, and *Saveliev and Zheleznjakov* [1991] have taken this and other effects into account in deriving a generalized version of Ohm's law. *Dougherty* [1961] noted that the neutral winds set into motion by ion drag will in turn influence the distribution of plasma by inducing magnetic-field-aligned motions. *Fuller-Rowell et al.* [1984, 1987, 1988, 1991], *Rees et al.* [1988], *Rees and Fuller-Rowell* [1989, 1991], and *Millward et al.* [1993] simulated the effects of coupling between neutral winds and plasma density at high magnetic latitudes in response to magnetospheric electric fields imposed on the ionosphere, and found a significant degree of feedback affecting both the winds and the plasma density. *Rishbeth* [1971b, 1981] pointed out that zonal winds in the low-latitude F-region evening ionosphere generate nearly vertical electric fields that cause the ionospheric plasma to drift in the zonal direction with a velocity approaching that of the neutral wind. Consequently, the ion drag on the wind becomes relatively small, so that high wind speeds (and zonal ion velocities) can develop. Indeed, a strong correlation of zonal neutral and ion motions at equatorial latitudes is observed [e.g., *Herrero and Mayr*, 1986]. *Heelis et al.* [1974] modeled the coupled neutral and ion dynamics for assumed inputs of neutral pressure gradients and of ionospheric densities. *Matuura* [1974] presented another model with coupled neutral and ion motions. *Volland* [1976a, b, 1984] and *Volland and Grellmann* [1978] discussed and modeled dynamo feedback on neutral dynamics for mid-latitude conditions, including the self-consistent thermodynamic feedback of temperature and pressure changes associated with the wind system. Even in their simplified models of the "hydromagnetic dynamo" in which height variations of the various parameters were not explicitly taken into account and in which the conductivity was a given input, the coupled neutral-ion electrodynamic response showed complex behavior. *Glushakov et al.* [1980, 1981] and *Mayr et al.* [1990] have modeled the interacting neutral and ion dynamics on a global scale, and have shown that the dynamo feedback on the neutral thermospheric motions is a major effect. In addition, *Mayr et al.* [1990] showed that most of the dynamo generation of diurnal electric fields occurs above 150 km, in the F region. *Anderson and Roble* [1974] showed that the changes in ion drag produced by electrodynamic uplifting of the ionosphere at low latitudes can also have an important influence on the neutral dynamics. *Rishbeth and Walker* [1982] and *Rishbeth* [1983] discussed how winds and electric fields in the nighttime ionosphere, when they are in certain directions, can produce layers of enhanced conductivity where the current tends to flow in a preferred direction. *Haerendel and Eccles* [1992] and *Haerendel et al.* [1992] presented models of the evening equatorial ionosphere that demonstrate how conductivity changes associated with ion drifts can significantly feed back onto the distribution and strength of the electric fields, possibly in an unstable fashion. *Crain et al.* [1993a, b] modeled the coupling between the ionospheric dynamo and electron density variations for a given neutral wind distribution, showing that F-region dynamo effects are important at all local times at low latitudes, and that variations of plasma density due to the electric fields have an appreciable feedback on the dynamo generation of these fields. *Rishbeth* [1979] and *J. Walker* [1988] have reviewed and discussed many of the interactive phenomena occurring in the thermosphere-ionosphere system.

Recently, two simulation models have been developed that take into account the mutual coupling and feedback effects among neutral and plasma dynamics, conductivities, and electric fields and currents [*Namgaladze et al.*, 1990, 1991; *Richmond et al.*, 1992]. The model of *Namgaladze et al.* [1990, 1991] is based on the earlier work of *Namgaladze et al.* [1988]. It has relatively low spatial resolution, but it additionally models the plasmasphere, so that the exchange of plasma and energy between the ionosphere and plasmasphere can be self-consistently simulated. *Namgaladze et al.* [1990] stated that their fully-coupled model tended to underestimate thermospheric neutral temperatures and electron densities; results presented by *Namgaladze et al.* [1990, 1991] therefore relied on an empirical model of thermospheric temperature and composition in order to examine other coupling effects. The NCAR thermosphere-ionosphere-electrodynamics general circulation model (TIE-GCM) [*Richmond et al.*, 1992] is an extension of the NCAR thermosphere-ionosphere general circulation model (TIGCM) of *Roble et al.* [1988]. Like its predecessor, the TIE-GCM incorporates detailed chemistry and radiation processes, so that few empirical inputs are required. It also has relatively high spatial resolution, such that it can readily handle modes of thermospheric motion like atmospheric tides, and can resolve the equatorial electrojet.

Figure 7 [*Richmond et al.*, 1992] illustrates some of the effects of electrodynamical coupling between the neutral and plasma motions as modeled in the TIE-GCM. This example is for solar-cycle maximum conditions, when electron densities are large so that coupling effects are strongest. The simulation is for equinox, with upward-propagating semidiurnal tides imposed at the lower boundary (97 km) as in *Fesen et al.* [1991], and with an imposed magnetospheric electric field pattern at high magnetic latitudes, with a 40 kV potential drop that corresponds to conditions of less-than-average magnetic activity. Figure 7a shows the neutral temperature and wind at 350 km altitude for 0 UT, while Figure 7b shows the electric potential and horizontal component of $\mathbf{E} \times \mathbf{B}$ ion drift at the same altitude and time. The identical scale for vector velocities is used to facilitate comparison. (The very strong high-latitude ion velocities are not shown in order to avoid cluttering the di-

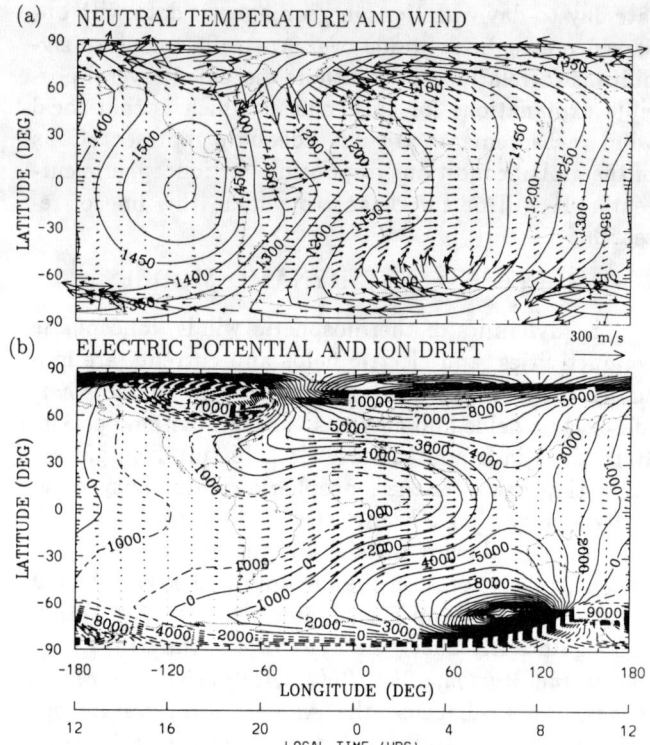

Fig. 7. (a) Contours of neutral temperature (in kelvins) and vectors of horizontal wind at a pressure level of 6.8 μPa (approximately 350 km) at 0000 UT for solar-maximum equinox conditions. The contour interval is 50 K. (b) Contours of electric potential (in volts) and vectors of the horizontal component of the electrodynamic drift for the same conditions as (a). The contour interval is 1000 V. Vector scales are the same for (a) and (b). Very large velocity vectors at high latitudes are not displayed. [*Richmond et al.*, 1992].

agram.) Midnight is at the center of the diagrams, and noon is at the edges. The electric potential is the same at the northern and southern ends of every geomagnetic field line, but the tilt and distortion of Earth's quasi-dipolar magnetic field results in asymmetry of the potential about the geographic equator. The ionospheric plasma density (not shown) tends to be organized with respect to the magnetic field, so that ion-drag effects on the neutral dynamics are also asymmetric about the geographic equator, leading to some asymmetry in the wind and temperature distributions, even for the symmetric solar-UV forcing present at equinox. The strong high-latitude winds are produced by the rapid ion convection, which is also organized with respect to the geomagnetic field.

At low latitudes there is a striking degree of similarity between the neutral and ion motions, particularly at night. This can be attributed to the F-region dynamo effects proposed by *Rishbeth* [1971a]. A further noteworthy effect is the existence of relatively high-speed winds near the equator in the evening, over Brazil and the Atlantic Ocean. The maximum winds trace fairly closely the position of the magnetic equator, which is distorted considerably with respect to a great circle in this region of Earth. Examination of the electron density distribution (not shown) reveals that these high-speed winds are flowing just beneath a vault of relatively large electron densities. The ionospheric plasma has been raised along the magnetic equator by $\mathbf{E} \times \mathbf{B}$ drifts, but it diffuses downward along magnetic field lines away from the equator on either side, so that plasma densities are considerably greater at 350 km a thousand kilometres either side of the magnetic equator than they are along the magnetic equator. The reduced ion drag near the equator allows the winds to be accelerated eastward by the strong horizontal pressure gradient existing at this local time, an effect first discussed by *Anderson and Roble* [1974]. Observations [*Raghavarao et al.*, 1991] have also detected this enhanced wind velocity near the magnetic equator.

4. HIGH-LATITUDE DYNAMO EFFECTS

The dynamics of the thermosphere is strongly influenced by inputs of energy and momentum from the magnetosphere at high magnetic latitudes. Through ion drag, the rapidly drifting ions impart momentum to the neutral molecules and set up a neutral circulation above about 125 km that tends to have characteristics similar to those of the ion circulation. The strong currents, especially those during magnetic storms, dissipate electrical energy in the form of Joule heat. This can be an important source of thermospheric heating at high latitudes, and can affect thermospheric dynamics even at midlatitudes. The consequences of these effects for the ionospheric dynamo have been investigated from a number of perspectives.

Axford and Hines [1961] pointed out that winds set into motion by the convecting plasma will in turn produce dynamo effects. It was noted by *Banks* [1972] that these winds, once set into motion, will generate an electric field that operates in the sense so as to continue the plasma motion, even if the source of magnetospheric field-aligned currents into the ionosphere were cut off. This effect would act as a "flywheel," tending to maintain magnetospheric plasma convection in the aftermath of stronger magnetospheric activity that occurred earlier. *Lyons et al.* [1985], *Deng et al.* [1991, 1993], and *Thayer and Vickrey* [1992] examined this phenomenon from a different perspective. They supposed that, instead of a stoppage of field-aligned current (an open circuit), there is a cancellation of the electric field (a shorted circuit) following storm activity. In this case, significant field-aligned current would flow, roughly in the opposite direction to the current that flowed earlier in association with strong magnetospheric convection.

Figure 8 illustrates the flywheel effect proposed by *Banks* [1972] with a numerical simulation. For this example, the TIE-GCM was run under conditions of steady high-latitude convection, of the typical two-cell configuration (electric-potential high on the morning side of the polar cap and low on the evening side), with a potential drop of 30 kV. The distribution of electric potential and associated plasma $\mathbf{E} \times \mathbf{B}/B^2$ drifts are shown in Figure 8a. Field-aligned currents were allowed to flow, of whatever configuration and strength would be necessary to maintain this electric-potential distribution. The distribution of winds that were established at 145 km is shown in Figure 8b. The two-cell circulation pattern of the imposed plasma drifts makes its mark on the wind distribution, although the wind velocities are weaker than the plasma velocities, and the entire high-latitude portion of the wind pattern appears to be rotated counterclockwise by about two hours in local time with respect to the plasma drifts. This rotation is explained by the rotation of Earth, taking into account the fact that the inertia of the neutral gas causes it to lag by about two hours behind the ion-drag forcing. Within one model time-step (5 minutes), the field-aligned currents were arbitrarily stopped, and the polarization electric field produced by the dynamo action of the full three-dimensional wind distribution was calculated. The resultant electric potential and $\mathbf{E} \times \mathbf{B}/B^2$ plasma drifts are shown in Figure 8c. The drifts are similar to the high-latitude winds at 145 km. The total potential drop generated by these winds is 7.5 kV, or 25% of what was originally imposed. This simulation was representative of equinox, solar-cycle-minimum conditions. At solar-cycle maximum the ionospheric plasma density is greater, so the the ion-drag acceleration of the wind is stronger, and one might expect even stronger flywheel effects.

Forbes and Harel [1989] considered the influence of the "flywheel" winds upon magnetosphere-ionosphere coupling, using a simplified model wherein the wind speeds were assumed to be a fixed fraction of the ion ve-

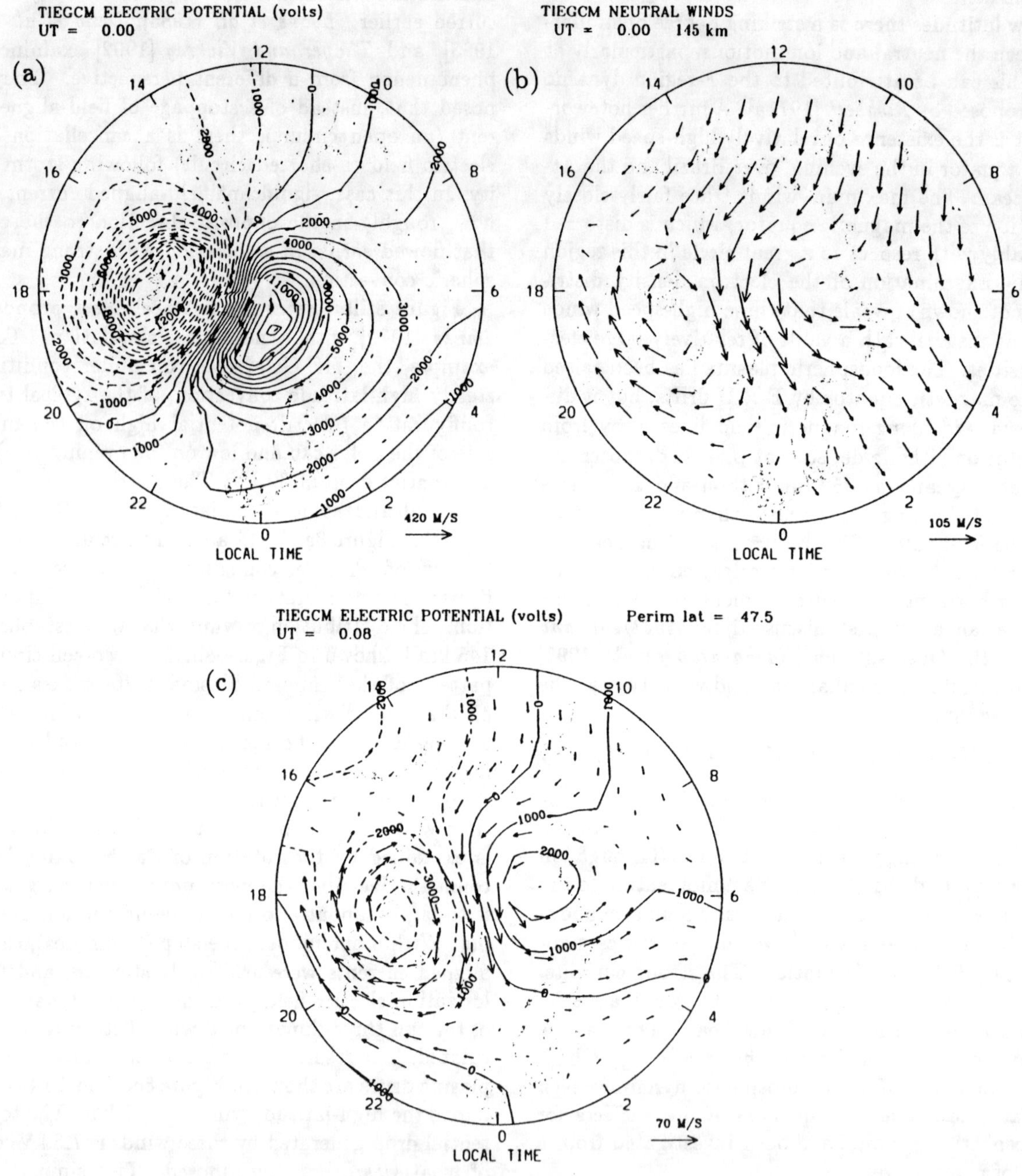

Fig. 8. Results of a TIE-GCM simulation illustrating the "flywheel" effect for solar-minimum equinox conditions. Contour intervals are 1000 V; vector velocity scales vary, as shown at the lower right of each plot. (a) Electric-potential contours and ion-drift vectors in geographic coordinates between 47.5° and the North Pole, at 0000 UT, for a diurnally reproducible simulation with an imposed cross-polar-cap potential of 30 kV. (b) Corresponding neutral wind vectors at 145 km altitude. (c) Electric potential contours and ion-drift vectors one time-step later, after field-aligned current between the ionosphere and the outer magnetosphere has been cut off.

locities. Since the dynamo effect of the winds tends to produce a net reduction in the current that would flow in their absence, the effect is somewhat similar to a reduction in the Pedersen conductivity. One consequence of this is a tendency for the so-called "shielding" effect associated with energetic magnetospheric plasma to be enhanced. Under steady-state conditions this plasma tends to produce a quasi-equipotential layer at the inner edge of the ring current, such that ionospheric regions equatorward of this inner edge, as mapped along geomagnetic field lines to the ionosphere, tend to be electrically shielded from high-latitude dynamo sources [e.g., *Wolf et al.*, 1986]. The time constant for shielding to become established varies inversely as the ionospheric Pedersen conductance, so that a reduction in the conductance tends to increase the shielding effect.

A variant of the "flywheel" concept is what has been called the "fossil-wind" effect [*Spiro et al.*, 1988]. During a magnetic storm, the auroral oval and the pattern of high-latitude convection expand, so that the wind system accelerated by ion drag extends down in latitude to upper mid-latitudes. As the storm calms, this wind system will continue to exert a dynamo influence. The additional feature of the "fossil-wind" concept is the relation of this dynamo influence to the shielding layer. In the calming phase of a magnetic storm, the ionospheric projection of the ring-current inner edge retreats to higher magnetic latitudes, thereby exposing the lower-latitude ionosphere to the full dynamo effects of the winds that had been previously accelerated by ion convection. Some of the polarization electric field associated with this dynamo effect can spread all the way to the magnetic equator. *Spiro et al.* [1988] and *B. Fejer et al.* [1990] modeled the fossil-wind effect under simplified assumptions about the nature of the winds that are accelerated and about the rate of retreat of the inner edge of the ring current. They found that the effect could potentially help explain certain long-lived electric-field perturbations seen at the magnetic equator. This effect has been further discussed by *B. Fejer* [1991].

Winds at midlatitudes can also be altered by the Joule heating that occurs during a storm. The high-latitude heating causes an upwelling of the air and an outflow towards lower latitudes at altitudes above 120 km. There is a subsidence of the air at midlatitudes and a return flow to high latitudes below 120 km, though to conserve mass flux the return flow proceeds at much smaller velocities than the outflow, because of the much larger air density at the lower altitudes. The outflow generally decreases in magnitude away from the region of heating. The Coriolis force acting on this equatorward wind produces a westward wind. Whereas the outflow will cease fairly quickly after the heating subsides, the westward wind can continue to flow in the absence of vertical motions long after the end of the storm. The dynamo effects of the globally altered wind system can thus persist for a day or more beyond the storm, as noted by *Blanc and Richmond* [1980], who called the dynamo effect during and following the storm the "disturbance dynamo." One prediction from their model is that the disturbance-dynamo contribution to low-latitude east-west electric fields is opposite to the regular daily variation. *Sastri* [1989] invoked the disturbance dynamo mechanism to explain the variation in the equatorial-electrojet strength with magnetic activity associated with changes in the interplanetary-magnetic-field sector structure; there was some suggestion of a delay in the electrojet response to the activity. *Mazaudier et al.* [1987] attempted to simulate observed storm events with a two-dimensional model of thermospheric dynamics and electrodynamics, and found at least some qualitative agreement. Low-latitude observations of the electrojet and of the equatorial ionization anomaly sometimes support the disturbance-dynamo concept [*B. Fejer et al.*, 1983; *Sastri*, 1988], but there are also many examples that show no clear effect following magnetic storms [*Kane*, 1978; *B. Fejer et al.*, 1983; *Sastri*, 1988]. Analyses of electric-field perturbations during disturbed periods generally show a daily-mean westward plasma drift with respect to quiet times at midlatitudes [*Blanc*, 1978; *Wand*, 1981; *Ganguly et al.*, 1987; *Heelis and Coley*, 1992; *Buonsanto et al.*, 1993; *B. Fejer*, 1993], compatible with the *Blanc and Richmond* [1980] model, but the westward shift of the drift tends to be strongest at night and relatively small at day, which is not a feature predicted by Blanc and Richmond. Figure 9 [*B. Fejer*, 1993], shows the difference between average disturbed and quiet eastward plasma drifts, indicating the equatorward electric field, for two midlatitude data sets that illustrate this feature. Possibly, the nighttime westward drifts may be due more to direct penetration of disturbed electric fields from high latitudes than to the disturbance dynamo [e.g., *Heelis and Coley*, 1992]. The quantitative importance of both the disturbance dynamo and the fossil wind effect remain to be determined.

5. FUTURE DIRECTIONS IN IONOSPHERIC DYNAMO RESEARCH

The ability of observed dynamo effects, when combined with modeling studies, to provide useful informa-

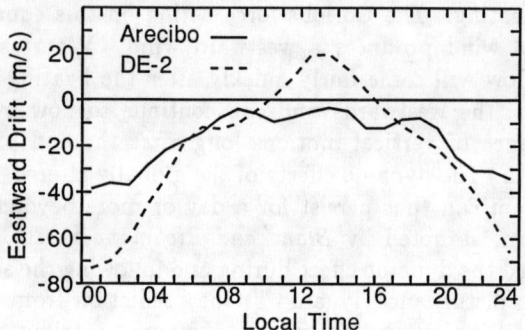

Fig. 9. Average disturbance zonal ion drifts obtained by subtracting the quiet-time drift patterns from the Arecibo and Dynamics Explorer-2 disturbed drifts. [*B. Fejer*, 1993]

tion about the global wind systems in the lower thermosphere has not been fully exploited. These observations can potentially tell us a great deal about the nature of tides and planetary waves propagating up from the middle atmosphere. Since magnetometers are spread over the globe, and some of them have very long series of nearly continuous observations, their measurements can complement the much more limited direct observations of winds in this region, when interpreted with the aid of models. For example, the apparent sensitivity of geomagnetic lunar tidal effects to atmospheric conditions may allow us to investigate possible long-term changes in the middle and upper atmosphere [e.g., *Roble and Dickinson*, 1989] by looking for changes in the amplitudes and phases of the lunar geomagnetic effects. Much remains to be done before a full quantitative understanding of winds in the dynamo region is achieved, especially of the solar and lunar semidiurnal tides and of planetary waves. It is important that further observational studies be carried out to determine the empirical relations between global mesospheric winds and geomagnetic variations, building on preliminary work like that of *Ito et al.* [1986] and *Phillips and Briggs* [1991].

Now that models have been developed that can simulate the mutually coupled dynamics and electrodynamics of the thermosphere and ionosphere, it is possible to investigate more realistically many aspects of the neutral winds, ionospheric variations, and dynamo effects under a wide variety of conditions. The mutual interactions are especially important for the nighttime dynamo, where conductivities tend to be dominated by the highly variable electron densities above 180 km. The initial work that has been done on examining the mutual coupling between the ionospheric and magnetospheric dynamos, such as the flywheel effect, the fossil-wind effect, and the disturbance-dynamo effect, has also indicated that considerably more research will be required in this field before a satisfactory understanding of all the implications of this coupling is achieved.

Acknowledgments. This research was supported by the NASA Space Physics Theory Program (S-97239-E). I am grateful to J. M. Forbes for helpful comments.

REFERENCES

Akasofu, S.-I., and R. N. DeWitt, Dynamo action in the ionosphere and motions of the magnetospheric plasma, 3, The Pedersen conductivity, generalized to take account of acceleration of the neutral gas, *Planet. Space Sci., 13*, 737-744, 1965.

Anderson, D. N., and R. G. Roble, The effect of vertical $\mathbf{E} \times \mathbf{B}$ ionospheric drifts on F region neutral winds in the low-latitude thermosphere, *J. Geophys. Res., 79*, 5231-5236, 1974.

Axford, W. I., and C. O. Hines, A unifying theory of high-latitude geophysical phenomena and geomagnetic storms, *Can. J. Phys., 39*, 1433-1464, 1961.

Baker, W. G., and D. F. Martyn, Electric currents in the ionosphere, I, The conductivity, *Phil. Trans. R. Soc., A246*, 281-294, 1953.

Banks, P. M., Magnetospheric processes and the behaviour of the neutral atmosphere, *Space Res., 12*, 1051-1067, 1972.

Bartels, J., and H. F. Johnston, Geomagnetic tides in horizontal intensity at Huancayo, *Terr. Magn. Atmos. Elect., 45*, 269-308, 1940.

Blanc, M., Midlatitude convection electric fields and their relation to ring current development, *Geophys. Res. Lett., 5*, 203-206, 1978.

Blanc, M., and A. D. Richmond, The ionospheric disturbance dynamo, *J. Geophys. Res., 85*, 1669-1686, 1980.

Briggs, B. H., The variability of ionospheric dynamo currents, *J. Atmos. Terr. Phys., 46*, 419-429, 1984.

Brown, G. M., Sq variability and aeronomic structure, *J. Atmos. Terr. Phys., 37*, 107-117, 1975.

Buonsanto, M. J., M. E. Hagan, J. E. Salah, and B. G. Fejer, Solar cycle and seasonal variations in F region electrodynamics at Millstone Hill, *J. Geophys. Res.*, in press, 1993.

Burrows, K., The day-to-day variability of the equatorial electrojet in Peru, *J. Geophys. Res., 75*, 1319-1323, 1970.

Butcher, E. C., Abnormal Sq behaviour, *Pure Appl. Geophys., 131*, 463-483, 1989.

Chapman, S., and J. Bartels, *Geomagnetism*, Clarendon, Oxford, 1940.

Chapman, S., and J. M. Stagg, On the variability of the quiet-day diurnal magnetic variation at Eskdalemuir and Greenwich, *Proc. R. Soc., 123*, 27-53, 1929.

Chapman, S., and J. M. Stagg, On the variability of the quiet-day diurnal magnetic variation, II, *Proc. R. Soc., 130*, 668-697, 1931.

Chapman, S., J. C. Gupta, and S. R. C. Malin, The sunspot cycle influence on the solar and lunar daily geomagnetic variations, *Proc. Roy. Soc., A324*, 1-15, 1971.

Chen, P.-R., Two-day oscillation of the equatorial ionization anomaly, *J. Geophys. Res., 97*, 6343-6357, 1992.

Clauer, C. R., R. L. McPherron, and M. G. Kivelson, Uncertainty in ring current parameters due to the quiet magnetic field variability at mid-latitudes, *J. Geophys. Res., 85*, 633-643, 1980.

Crain, D. J., R. A. Heelis, and G. J. Bailey, Effects of electrical coupling on equatorial ionospheric plasma motions: When is the F region a dominant driver in the low-latitude dynamo?, *J. Geophys. Res., 98*, 6033-6037, 1993a.

Crain, D. J., R. A. Heelis, G. J. Bailey, and A. D. Richmond, Low-latitude plasma drifts from a simulation of the global atmospheric dynamo, *J. Geophys. Res., 98*, 6039-6046, 1993b.

Currie, R. G., The geomagnetic spectrum - 40 days to 5.5 years, *J. Geophys. Res., 71*, 45-4598, 1966.

Currie, R. G., Geomagnetic line spectra - 2 to 70 years, *Astrophys. Space Sci., 21*, 425-438, 1973.

Deng, W., T. L. Killeen, A. G. Burns, and R. G. Roble, The flywheel effect: Ionospheric currents after a geomagnetic storm, *Geophys. Res. Lett., 18*, 1845-1848, 1991.

Deng, W., T. L. Killeen, A. G. Burns, R. G. Roble, J. A. Slavin, and L. E. Wharton, The effects of neutral inertia on ionospheric currents in the high-latitude thermosphere following a geomagnetic storm, *J. Geophys. Res., 98*, 7775-7790, 1993.

Dickinson, R. E., E. C. Ridley, and R. G. Roble, A three-dimensional general circulation model of the thermosphere, *J. Geophys. Res., 86*, 1499-1512, 1981.

Dougherty, J. P., On the influence of horizontal motion of the neutral air on the diffusion equation of the F-region, *J. Atmos. Terr. Phys., 20*, 167-176, 1961.

Fambitakoye, O., Variabilité jour-à-jour de la variation journalière du champ magnétique terrestre dans la région de l'électrojet équatorial, *C. R. Acad. Sci. Paris, 272*, 637-640, 1971.

Fejer, B. G., Low latitude electrodynamic plasma drifts: A review, *J. Atmos. Terr. Phys., 53*, 677-693, 1991.

Fejer, B. G., F-region plasma drifts over Arecibo: Solar cycle, seasonal, and magnetic activity effects, *J. Geophys. Res., 98*, 13,645-13,652, 1993.

Fejer, B. G., M. F. Larsen, and D. T. Farley, Equatorial disturbance dynamo electric fields, *Geophys. Res. Lett., 10*, 537-540, 1983.

Fejer, B. G., R. W. Spiro, R. A. Wolf, and J. C. Foster, Latitudinal variation of perturbation electric fields during magnetically disturbed periods: 1986 SUNDIAL observations and model results, *Ann. Geophysicae, 8*, 441-454, 1990.

Fejer, J. A., Semidiurnal currents and electron drifts in the ionosphere, *J. Atmos. Terr. Phys., 4*, 184-203, 1953.

Fesen, C. G., R. E. Dickinson, and R. G. Roble, Simulation of thermospheric tides at equinox with the National Center for Atmospheric Research thermospheric general circulation model, *J. Geophys. Res., 91*, 4471-4489, 1986.

Fesen, C. G., R. G. Roble, and E. C. Ridley, Thermospheric tides at equinox: Simulations with coupled composition and auroral forcings, 2, Semidiurnal component, *J. Geophys. Res., 96*, 3663-3677, 1991.

Footitt, R. J., G. J. Bailey, and R. J. Moffett, Ion transport in the mid-latitude F1-region, *Planet. Space Sci., 31*, 671-687, 1983.

Forbes, J. M., The equatorial electrojet, *Rev. Geophys. Space Phys., 19*, 469-504, 1981.

Forbes, J. M., and H. B. Garrett, Solar tidal wind structures and the E-region dynamo, *J. Geomagn. Geoelectr., 31*, 173-182. 1979.

Forbes, J. M., and M. Harel, Magnetosphere-thermosphere coupling: An experiment in interactive modeling, *J. Geophys. Res., 94*, 2631-2644, 1989.

Forbes, J. M., and S. Leveroni, Quasi 16-day oscillation in the ionosphere, *Geophys. Res. Lett., 19*, 981-984, 1992.

Forbes, J. M., and R. S. Lindzen, Atmospheric solar tides and their electrodynamic effects, I, The global Sq current system, *J. Atmos. Terr. Phys., 38*, 897-910, 1976.

Fuller-Rowell, T., and D. Rees, A three-dimensional time-dependent global model of the thermosphere, *J. Atmos. Sci., 37*, 2545-2567, 1980.

Fuller-Rowell, T. J., D. Rees, S. Quegan, G. J. Bailey, and R. J. Moffett, The effect of realistic conductivities on the high-latitude neutral thermospheric circulation, *Planet. Space Sci., 32*, 469-480, 1984.

Fuller-Rowell, T. J., S. Quegan, D. Rees, R. J. Moffett, and G. J. Bailey, Interactions between neutral thermospheric composition and the polar ionosphere using a coupled ionosphere-thermosphere model, *J. Geophys. Res., 92*, 7744-7748 and 7775-7777, 1987.

Fuller-Rowell, T. J., D. Rees, S. Quegan, R. J. Moffett, and G. J. Bailey, Simulations of the seasonal and universal time variations of the high-latitude thermosphere and ionosphere using a coupled, three-dimensional model, *Pure Appl. Geophys., 127*, 189-217, 1988.

Fuller-Rowell, T. J., D. Rees, S. Quegan, and R. J. Moffett, Numerical simulations of the sub-auroral F-region trough, *J. Atmos. Terr. Phys., 53*, 529-540, 1991.

Ganguly, S., R. A. Behnke, and B. A. Emery, Average electric field behavior in the ionosphere above Arecibo, *J. Geophys. Res., 92*, 1199-1210, 1987.

Glushakov, M. L., V. N. Dul'kin, and A. I. Ivanovskiy, Model of the daily variations of thermospheric characteristics, III, Self-consistent method of accounting for the electrostatic field in the daily variations of thermospheric composition, *Geomagn. Aeron., 20*, 39-41, (Engl. trans.), 1980.

Glushakov, M. L., V. N. Dul'kin, and A. I. Ivanovskiy, Model of diurnal variations in the variables of the thermosphere, IV, Results of calculations upon self-consistent correction for the effect of the electrostatic polarization field, *Geomagn. Aeron., 21*, 629-631, (Engl. trans.), 1981.

Greener, J. G., and D. M. Schlapp, A study of day-to-day variability of S_q over Europe, *J. Atmos. Terr. Phys., 41*, 217-223, 1979.

Gupta, J. C., Daily variability of the equatorial electrojet current system, *J. Atmos. Terr. Phys., 32*, 1159-1164, 1970.

Gupta, J. C., On solar and lunar equatorial electrojets, *Ann. Géophys., 29*, 49-60, 1973.

Gupta, J. C., Solar and lunar seasonal variations in the American sector, *Ann. Géophys., 38*, 255-265, 1982.

Haerendel, G., and J. V. Eccles, The role of the equatorial electrojet in the evening ionosphere, *J. Geophys. Res., 97*, 1181-1192, 1992.

Haerendel, G., J. V. Eccles, and S. Çakir, Theory for modeling the equatorial evening ionosphere and the origin of the shear in the horizontal plasma flow, *J. Geophys. Res., 97*, 1209-1223, 1992.

Hanuise, C., C. Mazaudier, P. Vila, M. Blanc, and M. Crochet, Global dynamo simulation of ionospheric currents and their connection with the equatorial electrojet and counter electrojet: A case study, *J. Geophys. Res., 88*, 253-270, 1983.

Hays, P. B., and R. G. Roble, A quasi-static model of global atmospheric electricity, 1, The lower atmosphere, *J. Geophys. Res., 84*, 3291-3305, 1979.

Heelis, R. A., and W. R. Coley, East-west ion drifts at midlatitudes observed by Dynamics Explorer 2, *J. Geophys. Res., 97*, 19461-19469, 1992.

Heelis, R. A., P. C. Kendall, R. J. Moffett, D. W. Windle, and H. Rishbeth, Electrical coupling of the E- and F-regions and its effect on F-region drifts and winds, *Planet. Space Sci., 22*, 743-756, 1974.

Herrero, F. A., and H. G. Mayr, Tidal decomposition of zonal neutral and ion flows in the earth's upper equatorial thermosphere, *Geophys. Res. Lett., 13*, 359-362, 1986.

Hibberd, F. H., and R. E. Davidson, Global scale of the day-to-day variability of S_q, *Geophys. J., 92*, 315-321, 1988.

Hirono, M., and T. Kitamura, A dynamo theory in the ionosphere, *J. Geomagn. Geoelectr., 8*, 9-23, 1956.

Ito, R., S. Kato, and T. Tsuda, Consideration of an ionospheric wind dynamo driven by a planetary wave with a two-day period, *J. Atmos. Terr. Phys., 48*, 1-13, 1986.

Kane, R. P., Longitudinal spread of equatorial Sq variability, *J. Atmos. Terr. Phys., 34*, 1425-1430, 1972.

Kane, R. P., Geomagnetic field variations, *Space Sci. Rev., 18*, 413-540, 1976.

Kane, R. P., S_q variations at low latitudes during geomagnetic storms, *J. Geophys. Res., 83*, 5312-5314, 1978.

Kato, S., Horizontal wind systems in the ionospheric E-region deduced from the dynamo theory of the geomagnetic S_q variations, II, Rotating earth, *J. Geomagn. Geoelectr., 8*, 24-37, 1956.

Kato, S., Horizontal wind systems in the ionospheric E-region deduced from the dynamo theory of the geomagnetic S_q variations, IV, *J. Geomagn. Geoelectr., 9*, 107-115, 1957.

Kelley, M. C., *The Earth's Ionosphere: Plasma Physics and Electrodynamics*, Academic Press, San Diego, 1989.

Kuznetsov, V. V., V. V. Plotkin, I. I. Nesterova, and N. I. Izraileva, Universal geomagnetic variation, *J. Geomagn. Geoelectr., 44*, 481-494, 1992.

London, J., and S. Matsushita, Periodicities of the geomagnetic variation field at Huancayo, Peru, *Nature, 198*, 374, 1963.

Lyons, L. R., T. L. Killeen, and R. L. Walterscheid, The neutral wind flywheel as a source of quiet-time polar-cap currents, *Geophys. Res. Lett., 12*, 101-104, 1985.

MacDougall, J., Equatorial electrojet and Sq current system: I, *J. Geomagn. Geoelectr., 31*, 341-357, 1979.

Maeda, H., Horizontal wind systems in the ionospheric E-region deduced from the dynamo theory of the geomagnetic S_q variations, I, Non-rotating earth, *J. Geomagn. Geoelectr., 7*, 121-132, 1955.

Maeda, H., Horizontal wind systems in the ionospheric E region deduced from the dynamo theory of the geomagnetic S_q variations, III, *J. Geomagn. Geoelectr., 9*, 86-93, 1957.

Maeda, H., and M. Fujiwara, Lunar ionospheric winds deduced from the dynamo theory of geomagnetic variations, *J. Atmos. Terr. Phys., 29*, 917-936, 1967.

Makino, M., and M. Takeda, Three-dimensional ionospheric currents and fields generated by the atmospheric global circuit current, *J. Atmos. Terr. Phys., 46*, 199-206, 1984.

Malin, S. R. C., Worldwide distribution of geomagnetic tides, *Phil. Trans. R. Soc. London, A274*, 551-594, 1973.

Mann, R. J., and D. M. Schlapp, The effect of disturbance on the day-to-day variability of S_q, *Geophys. J. R. astr. Soc., 80*, 535-540, 1985.

Marriott, R. T., A. D. Richmond, and S. V. Venkateswaran, The quiet-time equatorial electrojet and counter-electrojet, *J. Geomagn. Geoelectr.*, *31*, 311-340, 1979.

Matsushita, S., Solar quiet and lunar daily variation fields, in *Physics of Geomagnetic Phenomena*, edited by S. Matsushita and W. H. Campbell, pp. 301-427, Academic Press, NY, 1967a.

Matsushita, S., Lunar tides in the ionosphere, *Handbuch der Physik*, *49/2*, pp. 547-602, Springer, Berlin, 1967b.

Matsushita, S., and W.-Y. Xu, Seasonal variations of L equivalent current systems, *J. Geophys. Res.*, *89*, 285-294, 1984.

Matuura, N., Electric fields deduced from the thermospheric model, *J. Geophys. Res.*, *79*, 4679-4689, 1974.

Mayaud, P.-N., Analyse morphologique de la variabilité jour-a-jour de la variation journalière ‹régulière› S_R du champ magnétique terrestre, I. - Le système de courants C_P (régions polaires et sub-polaires), *Ann. Géophys.*, *21*, 369-401, 1965.

Mayaud, P. N., The equatorial counter-electrojet - a review of its geomagnetic aspects, *J. Atmos. Terr. Phys.*, *39*, 1055-1070, 1977.

Mayr, H. G., I. Harris, and F. A. Herrero, The dynamo of the diurnal tide and its effect on the thermospheric circulation, *Planet. Space Sci.*, *38*, 301-309, 1990.

Mazaudier, C., A. D. Richmond, and D. Brinkman, On thermospheric winds produced by auroral heating during magnetic storms and associated dynamo electric fields, *Ann. Geophysicae*, *5A*, 443-448, 1987.

Millward, G. H., S. Quegan, R. J. Moffett, R. J. Fuller-Rowell, and D. Rees, A modelling study of the coupled ionospheric and thermospheric response to an enhanced high-latitude electric field event, *Planet. Space Sci.*, *41*, 45-56, 1993.

Möhlmann, D., Two wind systems causing the global ionospheric dynamo-electric field, *Gerlands Beitr. Geophysik*, *85*, 343-344, 1976.

Möhlmann, D., Ionospheric electrostatic fields, *J. Atmos. Terr. Phys.*, *39*, 1325-1332, 1977.

Murata, H., An estimation of the electric potential field generated by the diurnal atmospheric tide with the first negative mode excited in the lower ionosphere, *Planet. Space Sci.*, *22*, 569-582, 1974.

Namgaladze, A. A., Yu. N. Koren'kov, V. V. Klimenko, I. V. Karpov, F. S. Bessarab, V. A. Surotkin, T. A. Glushchenko, and N. M. Naumova, A global model of the thermosphere-ionosphere-protonosphere system, *Pure Appl. Geophys.*, *127*, 219-254, 1988.

Namgaladze, A. A., Yu. N. Koren'kov, V. V. Klimenko, I. V. Karpov, F. S. Bessarab, V. A. Surotkin, T. A. Glushchenko, and N. M. Naumova, Global numerical model of the thermosphere, ionosphere, and protonosphere of the earth, *Geomagn. Aeron.*, *30*, 515-521, (Engl. trans.), 1990.

Namgaladze, A. A., Yu. N. Koren'kov, V. V. Klimenko, I. V. Karpov, V. A. Surotkin, and N. M. Naumova, Numerical modelling of the thermosphere-ionosphere-protonosphere system, *J. Atmos. Terr. Phys.*, *53*, 1113-1124, 1991.

Nastrom, G. D., and A. D. Belmont, The influence of long-period dynamo region winds on the surface geomagnetic field elements, *J. Geophys. Res.*, *81*, 4800-4804, 1976.

Onwumechilli, A., Lunar effects on the diurnal variation of the geomagnetic horizontal field near the magnetic equator, *J. Atmos. Terr. Phys.*, *25*, 55-70, 1963.

Onwumechilli, A., On the existence of days with extraordinary geomagnetic lunar tide, *J. Atmos. Terr. Phys.*, *26*, 729-748, 1964.

Parkinson, W. D., Bi-diurnal geomagnetic variations, *Ann. Géophys.*, *38*, 327-329, 1982.

Phillips, A., and B. H. Briggs, The day-to-day variability of upper atmosphere tidal winds and dynamo currents, *J. Atmos. Terr. Phys.*, *53*, 39-47, 1991.

Piddington, J. H., The motion of ionized gas in combined magnetic electric and mechanical fields of force, *Mon. Not. R. astr. Soc.*, *114*, 651-663, 1954.

Raghavarao, R., L. E. Wharton, N. W. Spencer, H. G. Mayr, and L. H. Brace, An equatorial temperature and wind anomaly (ETWA), *Geophys. Res. Lett.*, *18*, 1193-1196, 1991.

Rastogi, R. G., The equatorial electrojet: Magnetic and ionospheric effects, in *Geomagnetism*, vol. 3, edited by J. A. Jacobs, pp. 461-525, Academic Press, San Diego, 1989.

Rastogi, R. G., and N. B. Trivedi, Luni-solar tides H at stations within the equatorial electrojet, *Planet. Space Sci.*, *18*, 367-377, 1970.

Reddy, C. A., The equatorial electrojet, *Pure Appl. Geophys.*, *131*, 485-508, 1989.

Rees, D., and T. J. Fuller-Rowell, The response of the thermosphere and ionosphere to magnetospheric forcing, *Phil. Trans. R. Soc. London*, *A328*, 139-171, 1989.

Rees, D., and T. J. Fuller-Rowell, Thermospheric response and feedback to auroral inputs, in *Auroral Physics*, edited by C.-I. Meng, M. J. Rycroft, and L. A. Frank, pp. 51-65, Cambridge UP, 1991.

Rees, D., T. J. Fuller-Rowell, S. Quegan, R. J. Moffett, and G. J. Bailey, Simulations of the seasonal variations of the thermosphere and ionosphere using a coupled, three-dimensional, global model, including variations of the interplanetary magnetic field, *J. Atmos. Terr. Phys.*, *50*, 903-930, 1988.

Richmond, A. D., Modeling the ionospheric wind dynamo:

A review, *Pure Appl. Geophys.*, *131*, 413-435, 1989.

Richmond, A. D., The ionospheric wind dynamo: Recent progress and remaining research problems, *J. Geomagn., Geoelectr.*, *43*, Suppl., 433-440, 1991.

Richmond, A. D., and R. G. Roble, Electrodynamic effects of thermospheric winds from the NCAR thermospheric general circulation model, *J. Geophys. Res.*, *92*, 12365-12376, 1987.

Richmond, A. D., S. Matsushita, and J. D. Tarpley, On the production mechanism of electric currents and fields in the ionosphere, *J. Geophys. Res.*, *81*, 547-555, 1976.

Richmond, A. D., M. Blanc, B. A. Emery, R. H. Wand, B. G. Fejer, R. F. Woodman, S. Ganguly, P. Amayenc, R. A. Behnke, C. Calderon, and J. V. Evans, An empirical model of quiet-day ionospheric electric fields at middle and low latitudes, *J. Geophys. Res.*, *85*, 4658-4664, 1980.

Richmond, A. D., E. C. Ridley, and R. G. Roble, A thermosphere/ionosphere general circulation model with coupled electrodynamics, *Geophys. Res. Lett.*, *19*, 601-604, 1992.

Rishbeth, H., The F-layer dynamo, *Planet. Space Sci.*, *19*, 263-267, 1971a.

Rishbeth, H., Polarization fields produced by winds in the equatorial F-region, *Planet. Space Sci.*, *19*, 357-369, 1971b.

Rishbeth, H., Ion-drag effects in the thermosphere, *J. Atmos. Terr. Phys.*, *41*, 885-894, 1979.

Rishbeth, H., The F-region dynamo, *J. Atmos. Terr. Phys.*, *43*, 387-392, 1981.

Rishbeth, H., Further studies of directional E-layer currents, *Planet. Space Sci.*, *31*, 1177-1180, 1983.

Rishbeth, H., and J. C. G. Walker, Directional currents in nocturnal E-region layers, *Planet. Space Sci.*, *30*, 209-214, 1982.

Roble, R. G., On modeling component processes in the earth's global electric circuit, *J. Atmos. Terr. Phys.*, *53*, 831-847, 1991.

Roble, R. G., and R. E. Dickinson, How will changes in carbon dioxide and methane modify the mean structure of the mesosphere and thermosphere?, *Geophys. Res. Lett.*, *16*, 1441-1444, 1989.

Roble, R. G., and P. B. Hays, A quasi-static model of global atmospheric electricity, 2, Electrical coupling between the upper and lower atmosphere, *J. Geophys. Res.*, *84*, 7247-7256, 1979.

Roble, R. G., E. C. Ridley, A. D. Richmond, and R. E. Dickinson, A coupled thermosphere/ionosphere general circulation model, *Geophys. Res. Lett.*, *15*, 1325-1328, 1988.

Sastri, J. H., Phase variability of $Sq(H)$ on normal quiet days in the equatorial electrojet region, *Geophys. J. R. astr. Soc.*, *71*, 187-197, 1982.

Sastri, J. H., Equatorial electric fields of ionospheric disturbance dynamo origin, *Ann. Geophysicae*, *6*, 635-642, 1988.

Sastri, J. H., Response of equatorial electric field to polarity of interplanetary magnetic field, *Planet. Space Sci.*, *37*, 1403-1408, 1989.

Saveliev, V. L., and E. V. Zheleznjakov, Ohm's law for multicomponent moving ionospheric plasma, conductivity tensors and their eigenvalues, *Planet. Space Sci.*, *39*, 1133-1137, 1991.

Schieldge, J. P., S. V. Venkateswaran, and A. D. Richmond, The ionospheric dynamo and equatorial magnetic variations, *J. Atmos. Terr. Phys.*, *35*, 1045-1061, 1973.

Schlapp, D. M., World-wide morphology of day-to-day variability of S_q, *J. Atmos. Terr. Phys.*, *30*, 1761-1776, 1968.

Schlapp, D. M., and S. R. C. Malin, Some features of the seasonal variation of geomagnetic lunar tides, *Geophys. J. R. astr. Soc.*, *59*, 161-170, 1979.

Schlapp, D. M., and R. J. Mann, The spatial scale of correlation of the day to day variability of Sq, *Geophys. J. R. astr. Soc.*, *73*, 671-673, 1983.

Shapiro, R., and F. Ward, Possibility of a 26- or 27-month periodicity in the equatorial geomagnetic field, *Nature*, *201*, 909, 1964.

Singhal, R. P., The effect of the electric field and neutral winds on E-region ion densities and conductivities at low latitudes, *J. Atmos. Terr. Phys.*, *53*, 949-957, 1991.

Spiro, R. W., R. A. Wolf, and B. G. Fejer, Penetration of high-latitude-electric-field effects to low latitudes during SUNDIAL 1984, *Ann. Geophysicae*, *6*, 39-50, 1988.

Stacey, F. D., and P. Wescott, Possibility of a 26- or 27-month periodicity in the equatorial geomagnetic field and its correlation with stratospheric winds, *Nature*, *196*, 730-732, 1962.

Stening, R. J., An assessment of the contributions of various tidal winds to the Sq current system, *Planet. Space Sci.*, *17*, 889-908, 1969.

Stening, R. J., Problems of identifying lunar geomagnetic effects at Huancayo, *J. Geomagn. Geoelectr.*, *27*, 409-424, 1975.

Stening, R. J., Magnetic variations at other latitudes during reverse equatorial electrojet, *J. Atmos. Terr. Phys.*, *39*, 1071-1077, 1977a.

Stening, R. J., Ionospheric dynamo calculations with semidiurnal winds, *Planet. Space Sci.*, *25*, 1075-1080, 1977b.

Stening, R. J., A two-layer ionospheric dynamo calculation, *J. Geophys. Res.*, *86*, 3543-3550, 1981.

Stening, R. J., A calculation of ionospheric currents due to semidiurnal antisymmetric tides, *J. Geophys. Res.*, *94*, 1525-1531, 1989a.

Stening, R. J., A diurnal modulation of the lunar tide in the upper atmosphere, *Geophys. Res. Lett.*, *16*, 307-310,

1989b.

Stening, R. J., and D. E. Winch, Seasonal changes in the global lunar geomagnetic variation, *J. Atmos. Terr. Phys.*, *41*, 311-323, 1979.

Stewart, B., Terrestrial magnetism, *Encyclopaedia Britannica*, 9th ed., *16*, 159-184, 1882.

Suzuki, A., Geomagnetic S_q field at successive universal times, *J. Atmos. Terr. Phys.*, *40*, 449-463, 1978.

Takeda, M., Geomagnetic field variation and the equivalent current system generated by an ionospheric dynamo at the solstice, *J. Atmos. Terr. Phys.*, *52*, 59-67, 1990.

Takeda, M., and H. Maeda, Three-dimensional structure of ionospheric currents, 1, Currents caused by diurnal tidal winds, *J. Geophys. Res.*, *85*, 6895-6899, 1980.

Takeda, M., and H. Maeda, Three-dimensional structure of ionospheric currents, 2, Currents caused by semidiurnal tidal winds, *J. Geophys. Res.*, *86*, 5861-5867, 1981.

Takeda, M., and Y. Yamada, Simulation of ionospheric electric fields and geomagnetic field variation by the ionospheric dynamo for different solar activity, *Ann. Geophysicae*, *5A*, 429-434, 1987.

Takeda, M., Y. Yamada, and T. Araki, Simulation of ionospheric currents and geomagnetic field variations of S_q for different solar activity, *J. Atmos. Terr. Phys.*, *48*, 277-287, 1986.

Takeda, M., and Y. Yamada, Quasi two-day period of the geomagnetic field, *J. Geomagn. Geoelectr.*, *41*, 469-478, 1989.

Tarpley, J. D., The ionospheric wind dynamo, I, Lunar tide, *Planet. Space Sci.*, *18*, 1075-1090, 1970a.

Tarpley, J. D., The ionospheric wind dynamo, I, Solar tides, *Planet. Space Sci.*, *18*, 1091-1103, 1970b.

Thayer, J. P., and J. F. Vickrey, On the contribution of the thermospheric neutral wind to high-latitude energetics, *Geophys. Res. Lett.*, *19*, 265-268, 1992.

van Sabben, D., North-south asymmetry of Sq, *J. Atmos. Terr. Phys.*, *26*, 1187-1195, 1964.

Volland, H., A note on the height of the geomagnetic S_q current, *J. Geomagn. Geoelectr.*, *23*, 117-121, 1971.

Volland, H., Coupling between the neutral wind and the ionospheric dynamo current, *J. Geophys. Res.*, *81*, 1621-1628, 1976a.

Volland, H., The atmospheric dynamo, *J. Atmos. Terr. Phys.*, *38*, 869-877, 1976b.

Volland, H., *Atmospheric Electrodynamics*, Springer, Berlin, 1984.

Volland, H., and L. Grellmann, A hydromagnetic dynamo of the atmosphere, *J. Geophys. Res.*, *83*, 3699-3708, 1978.

Walker, G. O., and S. I. Kannangara, A study of quiet day magnetic field variations in East Asia at sunspot minimum, *Ann. Géophys.*, *38*, 271-282, 1982.

Walker, J. C. G., The mid-latitude thermosphere, *Planet. Space Sci.*, *36*, 1-10, 1988.

Wand, R. H., A model representation of the ionospheric electric field over Millstone Hill ($\Lambda = 56°$), *J. Geophys. Res.*, *86*, 5801-5808, 1981.

Winch, D. E., Spherical harmonic analyses of geomagnetic tides, 1964-1965, *Phil. Trans. R. Soc.*, *A303*, 1-104, 1981.

Wolf, R. A., G. A. Mantjoukis, and R. W. Spiro, Theoretical comments on the nature of the plasmapause, *Adv. Space Res.*, *6*, 177-186, 1986.

Wulf, O. R., A possible effect of atmospheric circulation in the daily variation of the earth's magnetic field, *Mon. Weather Rev.*, *91*, 520-526, 1963.

Wulf, O. R., A possible effect of atmospheric circulation in the daily variation of the earth's magnetic field, II, *Mon. Weather Rev.*, *93*, 127-132, 1965a.

Wulf, O. R., On winds in the lower ionosphere and variations of the earth's magnetic field, *Mon. Weather Rev.*, *93*, 655, 1965b.

Yacob, A., and B. N. Bhargava, On 26-month periodicity in quiet-day range of geomagnetic horizontal force and in sunspot number, *J. Atmos. Terr. Phys.*, *30*, 1907-1911, 1968.

Yi, L., and P.-R. Chen, Long period oscillations in the equatorial ionization anomaly correlated with the neutral wind in the mesosphere, *J. Atmos. Terr. Phys.*, *55*, 1317-1323, 1993.

A. D. Richmond, High Altitude Observatory, National Center for Atmospheric Research, P.O. Box 3000, Boulder, CO 80307-3000.

Tidal and Planetary Waves

Jeffrey M. Forbes[1]

High Altitude Observatory, National Center for Atmospheric Research, Boulder, Colorado

This is a graduate student level tutorial on atmospheric tidal and planetary waves for non-dynamicists. The intent is to provide a basic understanding of the physics, mathematics, and observational evidence pertaining to solar tidal period (12 and 24 hours) and planetary wave period (near 2, 5, 10, and 16 days) oscillations in the mesopause region (ca. 95±15 km) of the upper atmosphere. Analytic solutions for "free" and "forced" oscillations in an isothermal background atmosphere are derived. The basic characteristics of these solutions, and anticipated modifications due to non-isothermality, dissipation, and mean winds, are described. Observations and numerical simulations are interpreted using these results, and some perspective is provided regarding future studies of tidal and planetary waves in the mesosphere/lower thermosphere.

1. INTRODUCTION

Atmospheric tides are global-scale oscillations in temperature, wind, density, and pressure at periods which are subharmonics of a solar or lunar day. Strictly speaking, atmospheric tides may be either eastward or westward propagating, but by far the largest components are those which are westward propagating or *migrating* with the apparent motion of the sun or moon. Planetary waves are longer-period global oscillations which are either stationary (i.e., fixed to the Earth) or zonally propagating in either direction. Without intending to diminish the importance of the other wave components, in the interest of brevity the present tutorial will mainly concentrate on migrating solar tides and the westward propagating family of traveling planetary waves. Except possibly at equatorial latitudes, existing evidence suggests these to be the dominant planetary-scale wave components in the mesosphere and lower thermosphere.

---

[1] Permanent affiliation: Department of Aerospace Engineering Sciences, University of Colorado, Boulder, Colorado

A brief view of typical observations provides adequate motivation for the present tutorial. Figure 1 illustrates height/local time contours representing average meridional wind patterns between 80 and 100 km over Townsville, Australia (19°S, 147°E) and Saskatoon, Canada (54°N, 107°W) during the period March 18–27, 1979. Note first of all that the character is mainly diurnal over Townsville (24-hour harmonic dominates), and mainly semidiurnal over Saskatoon (12-hour harmonic dominates). Why do you suppose this is? Why is it that phase progression is downward (i.e., the wind contours tilt to the left in Figure 1)? And, given that there are no significant heat sources at these heights, why is it that these "tidal" oscillations assume such a prominent role in the meteorology of the mesosphere and lower thermosphere?

Instead of the average local time (day/night) wind structure examined in Figure 1, now suppose that we compute the daily mean wind (24-hour average) each day at a single height and form a time series of the daily values. The spectral density curve corresponding to daily mean winds measured over Obninsk, Russia (54°N, 38°E) during January through February, 1979, are illustrated in Figure 2. Note that prominent peaks occur near 5, 9, and 16 days period; a simple bandpass/IFT analysis demonstrates that these peaks each correspond to some 5–10 ms^{-1} oscillation in the wind, a substantial fraction of the total wind at any given time. Assuming this example is representative, why is it that

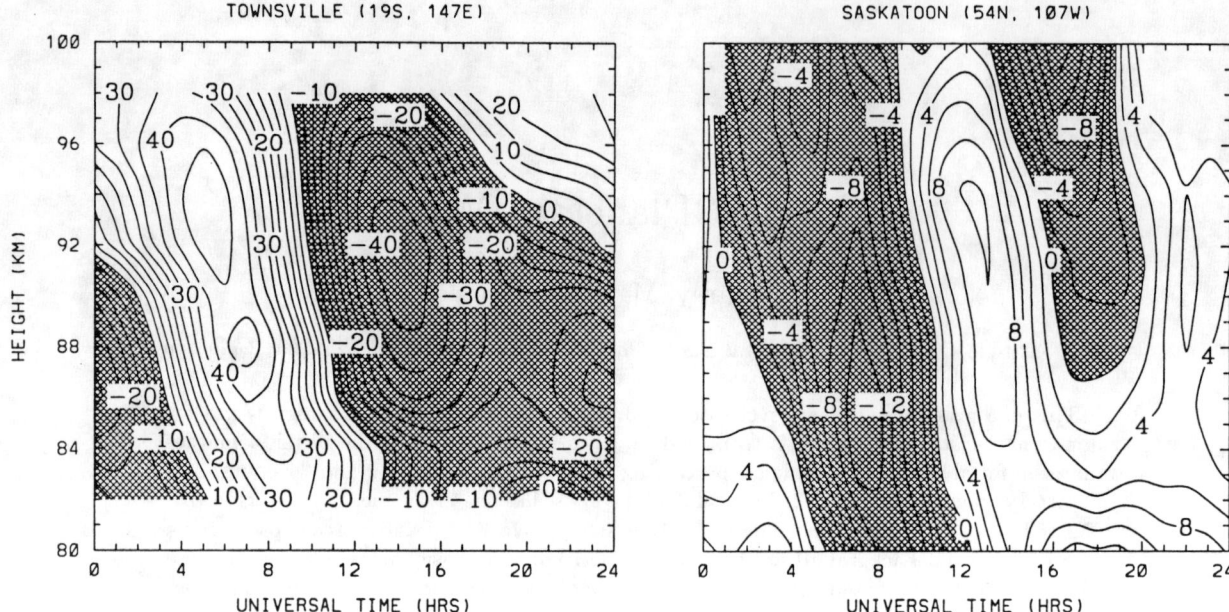

Fig. 1. Height-local time contours depicting average northward winds during the period March 18–27, 1979, over Townsville, Australia (19°S, 147°E) and Saskatoon, Canada (54°N, 107°W), as measured by the spaced antenna drift method. Data courtesy of Prof. R. A. Vincent and Prof. A. H. Manson.

the spectral peaks should fall at these specific periods?

Perhaps the most well-known long-period oscillation is the so-called 2-day wave. A history of 2-day wave amplitudes determined from mesopause winds measured over Adelaide, Australia (35°S, 138°E) is illustrated in Figure 3 [*Harris*, 1993]. Note that amplitudes of order 20–40 ms^{-1} episodically occur. Why should a prominent oscillation at this period exist near the mesopause?

This tutorial is motivated by the simple fact that tidal and planetary waves often dominate the meteorology of the atmospheric region between 80 and 150 km. Students and scientists engaged in studies of this regime should have some rudimentary understanding of the origins, characteristics, and governing mechanisms pertinent to these oscillations. The present tutorial seeks to impart this basic understanding.

Given that the present work is a tutorial rather than a comprehensive review of research in the field, I have not provided extensive referencing to the huge body of published literature on tidal and planetary waves. Some exceptions are works of historical importance, or recent papers which are particularly illustrative or instructive. For more extensive expositions than provided here, including references to the literature, the reader is referred to *Chapman and Lindzen* [1970] and *Forbes* [1982a, b] for solar and lunar atmospheric tides, and to *Walterscheid* [1980] and *Salby* [1984] for traveling planetary waves. There also exists an extensive literature on stationary planetary waves. A recent paper which emphasizes the vertical extension of stationary planetary waves into the mesosphere/lower thermosphere is that of *Pogorel'tsev and Sukhanova* [1993].

In the following section, the mathematics governing free and forced oscillations in a horizontally stratified isothermal atmosphere is developed. The resulting analytic solutions provide a reference framework for interpreting observations and numerical simulations. Anticipated modifications to this simple theory due to nonisothermality, mean winds, and dissipation are also discussed. Thermal forcing of atmospheric tides is covered in section 3. In section 4 several examples of numerical simulations of tidal and planetary waves are presented and interpreted. A brief outlook of potentially fruitful areas of research is provided in section 5.

2. MATHEMATICAL BASIS

2.1 Governing Equations

The general dynamical equations governing atmospheric motions are nonlinear and contain a variety of dissipative terms. These equations are often linearized, leading to coupled equations describing the zonal mean state and perturbations upon this 'background' atmosphere. If the background atmosphere is assumed to be horizontally stratified, the zonal mean winds are zero and the equations are decoupled. Assumption of an isothermal background atmosphere leads to further simplification without loss of instructional value. The resulting perturbation equations represent the starting point for the present mathematical analysis. The *linearized* equations, then, for perturbations on a spherical isothermal atmosphere are [*Holton*, 1975]:

$$\frac{\partial u}{\partial t} - 2\Omega \sin\theta v + \frac{1}{a\cos\theta}\frac{\partial \Phi}{\partial \lambda} = 0 \quad (1)$$

$$\frac{\partial v}{\partial t} + 2\Omega \sin\theta u + \frac{1}{a}\frac{\partial \Phi}{\partial \theta} = 0 \quad (2)$$

$$\frac{\partial}{\partial t}\Phi_z + N^2 w = \frac{\kappa J}{H} \quad (3)$$

$$\frac{1}{a\cos\theta}\left[\frac{\partial u}{\partial \lambda} + \frac{\partial}{\partial \theta}(v\cos\theta)\right] + \frac{1}{\rho_o}\frac{\partial}{\partial z}(\rho_o w) = 0 \quad (4)$$

where

u	eastward velocity
v	northward velocity
w	upward velocity
Φ	perturbation geopotential
N^2	buoyancy frequency squared $= \kappa g/H$
Ω	angular velocity of Earth
ρ_o	basic state density $\propto e^{-z/H}$
z	altitude
λ	longitude
θ	latitude
κ	$R/c_p \approx 2/7$
J	heating per unit mass
a	radius of Earth
g	acceleration due to gravity
H	constant scale height
t	time

We will now follow the spirit of the development in *Holton* [1975], although some normalization factors will differ. Assume the perturbations to consist of longitudinally propagating waves of zonal wavenumber s and frequency σ:

$$\{u, v, w, \Phi\} = \{\hat{u}, \hat{v}, \hat{w}, \hat{\Phi}\}\exp[i(s\lambda - \sigma t)] \quad (5)$$

The zonal wavenumber is a positive integer ($s = 0$ permitted) that gives the number of maxima of the sinusoidal oscillation in longitude. The $(s\lambda - \sigma t)$ form for the phase is chosen so that positive values for σ correspond to eastward propagating waves and negative values to westward propagating waves (i.e., the real part of equation (5) is $\cos(s\lambda - \sigma t)$ and the crest of the wave occurs where $\lambda = \sigma t/s$). Substituting equation (5) into equations (1)–(4) eliminates derivatives with respect to t and λ, permitting consolidation into a single second-order partial differential equation for Φ in z and θ. Separable solutions of the following form exist:

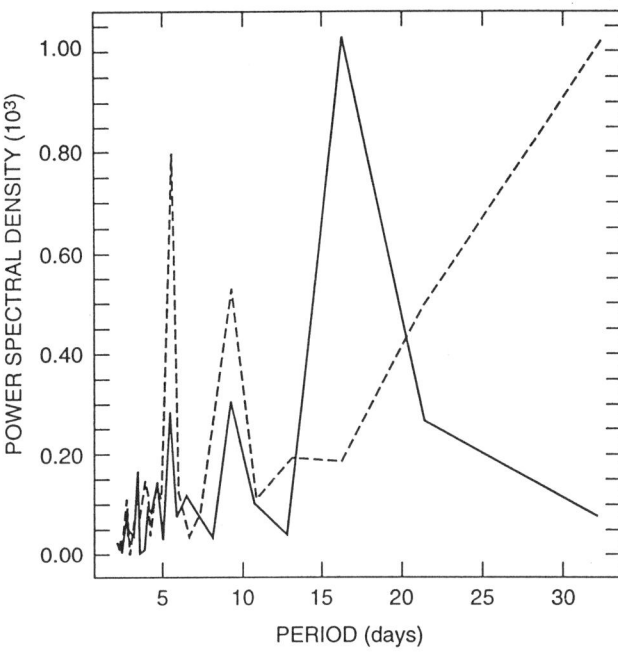

Fig. 2. Power spectrum of daily mean meridional (dashed line) and zonal (solid line) winds observed near the mesopause over Obninsk, Russia (54°N, 38°E) for January through February, 1979. Data courtesy of Dr. Yu.I. Portnyagin.

$$\hat{\Phi} = \sum_n \Theta_n(\theta) G_n(z) \quad (6)$$

where $\{\Theta_n\}$ is a complete orthogonal set and $G_n(z)$ is defined below. Since Θ_n is a complete orthogonal set, the thermal excitation can be expanded in the following form:

$$\hat{J} = \sum_n \dot{\Theta}_n(\theta) J_n(z) \quad (7)$$

The $J_n(z)$ functions are discussed in section 3. From equations (1), (2), and (6), expressions for the horizon-

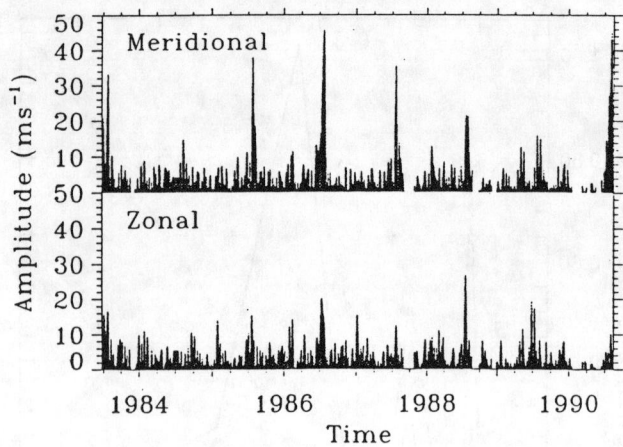

Fig. 3. The amplitude of the quasi-two-day wave for January, 1984, through January, 1991, near the mesopause over Adelaide, South Australia (35°S, 138°E). These amplitudes were determined using a complex demodulation procedure, with an effective bandpass of 44 to 53 hours. Meridional winds are shown in the top panel with the zonal winds below. From *Harris* [1993].

tal velocity components in terms of Θ_n and G_n may be derived:

$$\hat{u} = \frac{\sigma}{4\Omega^2 a} \sum_n U_n(\theta) G_n(z) \qquad (8)$$

$$\hat{v} = \frac{-i\sigma}{4\Omega^2 a} \sum_n V_n(\theta) G_n(z) \qquad (9)$$

where

$$U_n = \frac{1}{(f^2 - \sin^2\theta)} \left[\frac{s}{\cos\theta} + \frac{\sin\theta}{f} \frac{d}{d\theta} \right] \Theta_n \qquad (10)$$

$$V_n = \frac{1}{(f^2 - \sin^2\theta)} \left[\frac{s\tan\theta}{f} + \frac{d}{d\theta} \right] \Theta_n \qquad (11)$$

The following expression is a consequence of separation:

$$i\sigma H \left[\frac{1}{\rho_o} \frac{\partial}{\partial z} \rho_o \frac{\partial}{\partial z} G_n \right] + \frac{1}{\rho_o} \frac{\partial}{\partial z} (\rho_o \kappa J_n) = -\frac{i\sigma\kappa}{h_n} G_n \qquad (12)$$

where h_n arises as the separation constant. Defining $G'_n = G_n \rho_o^{1/2} N^{-1}$, taking $N^2 = \kappa g/H$ for an isothermal atmosphere where H = constant = 7.5 km (corresponding to $T_o = 256K$), and letting $x = z/H$, results in the canonical form for the *vertical structure equation* (for an isothermal atmosphere):

$$\frac{d^2 G'_n}{dx^2} + \left[\frac{\kappa H}{h_n} - \frac{1}{4} \right] G'_n = -\frac{\rho_o^{-1/2}}{i\sigma N} \frac{d}{dx}(\rho_o J_n) \qquad (13)$$

The θ-dependent part of the solution is embodied in *Laplace's tidal equation* [*Laplace*, 1799, 1825]:

$$\frac{d}{d\mu} \left[\frac{(1-\mu^2)}{(f^2 - \mu^2)} \frac{d\Theta_n}{d\mu} \right] - \frac{1}{f^2 - \mu^2}$$

$$\left[-\frac{s}{f} \frac{(f^2 + \mu^2)}{(f^2 - \mu^2)} + \frac{s^2}{1-\mu^2} \right] \Theta_n + \epsilon \Theta_n = 0 \qquad (14)$$

where $\mu = \sin\theta$ and $\epsilon_n = (2\Omega a)^2/gh_n$. Thus, the equations governing atmospheric perturbations are now formulated in terms of an eigenfunction-eigenvalue problem. Solutions to equations (13) and (14) must be sought subject to certain boundary conditions; these solutions are described in sections 2.2 and 2.3, respectively. Note that equations (13) and (14) are linked through h_n, the set of eigenvalues, which is referred to as the "equivalent depth." This nomenclature originates from the first appearance of equation (14) in connection with the ocean tide problem where h is the ocean depth [*Laplace*, 1799, 1825; *Taylor*, 1936].

2.2 Vertical Structure Equation: Forced and Free Solutions

Rewriting equation (13) as follows

$$\frac{d^2 G'_n}{dx^2} + \alpha^2 G'_n = F(x) \qquad (15)$$

where $\alpha^2 = \kappa H/h_n - 1/4$, the form of the solution is

$$G'_n \sim A e^{i\alpha x} + B e^{-i\alpha x} \qquad (16)$$

Now we will examine the cases where $F(x) \neq 0$ ("forced" solution) and where $F(x) = 0$ ("free" solution). When $F(x) \neq 0$ there are two possibilities. If $h_n < 0$ or $h_n > 4\kappa H$, then $\alpha^2 < 0$ and

$$G'_n \sim e^{-|\alpha|x} \qquad (17)$$

above the source region for a bounded solution. In this case the solutions are referred to as "evanescent" or "trapped" since the wave oscillations are more or less confined to the region of excitation. If $0 < h_n < 4\kappa H$, then $\alpha^2 > 0$ and a "radiation condition" ($C_{gx} > 0$) at $x = \infty$ implies

$$G'_n \sim e^{i\alpha x} \quad (18)$$

where the plus (minus) sign in the expression for $\alpha = \pm(\kappa H/h_n - 1/4)^{1/2}$ is chosen for westward (eastward) propagating waves (see section 2.4). This is the so-called propagating solution, where the wave propagates away from the source region.

When $F(x) = 0$ the only nontrivial solution satisfying boundedness and $w = 0$ at $z = 0$ is:

$$G'_n \sim e^{(\kappa - \frac{1}{2})x} \quad (19)$$

and

$$h_n = \frac{H}{1 - \kappa} \quad (20)$$

where $h_n = 10.5$ km for $H = 7.5$ km. This free (unforced) solution corresponds to a resonant response of the atmosphere. Note that the above solution implies

$$u \sim e^{\kappa x} \quad (21)$$

corresponding to energy ($\rho_o u^2$) decay away from the surface while horizontal velocity and other wave fields increase exponentially (by a factor of 40 from the surface to 100 km). These waves are sometimes called "Lamb" or "edge" waves. Furthermore, for $h_n = 10.5$ km α^2 is negative, implying no vertical flux of energy out of the atmosphere ($w = 0$) and no phase change with height. Without dissipation, such free oscillations would continue indefinitely without forcing [*Lindzen and Blake*, 1972].

2.3 Laplace's Tidal Equation: Nomenclature and Classification of Wave Modes

Laplace's tidal equation is often written as follows to emphasize the explicit dependences on s, σ, and ϵ_n:

$$F_{s,\sigma}(\Theta_n^{s,\sigma}) = \epsilon_n^{s,\sigma} \Theta_n^{s,\sigma} \quad (22)$$

For each choice of s and σ, there exist sets of ϵ_n and Θ_n which satisfy (22) and the condition of boundedness at the poles. The ϵ_n^s and σ are generally related parametrically for a given s in diagrams like the ones comprising Figure 4 for $s = 1$. (Diagrams for $s = 2$ and $s = 3$ are very similar to Figure 4, and are not shown here to conserve space.) Two families of curves are evident for both eastward propagating ($\sigma > 0$) and westward propagating ($\sigma < 0$) solutions. These families are sometimes referred to as "Class I" or "Solutions of the First Kind" and "Class II" or "Solutions of the Second Kind." A more common usage is to refer to the first class as "gravity modes" and the second class as "Rossby," "rotational," or "planetary wave" modes. Note that for $\sigma < 0$ there is one so-called "mixed" or "Rossby-Gravity" mode that belongs to Class I for ϵ_n large and positive and Class II for ϵ_n large and negative. There are energy partitionings and other properties which differ between the two classes of solutions [*Longuet-Higgins*, 1968], but we will not concern ourselves with these issues. The remainder of this tutorial will mainly use the terms "gravity" and "Rossby" to distinguish the wave types.

We note from Figure 4 some general features and properties. For instance, gravity (Class I) modes always have $\epsilon_n > 0$, whether they are westward propagating or eastward propagating. On the other hand, Rossby (Class II) modes possess $\epsilon_n > 0$ only for westward propagating waves; all eastward propagating Rossby modes have $\epsilon_n < 0$. From the expression for α^2 in equation (15), for negative or sufficiently small ϵ_n (or large h_n) the vertical structures are "trapped" or "evanescent," whereas for ϵ_n greater than about 100 solutions are propagating with vertical wavelengths less than 100 km. Note that the (1,-1) mode in Figure 4 belongs to the Rossby mode class for $\epsilon_n^1 < 10$ but joins the Gravity mode family of curves for $\epsilon_n^1 > 10$. This is the so-called mixed Rossby-gravity wave. This mode exists for higher wavenumbers as well. The gravest ($n = 1$) of the eastward propagating ($\sigma > 0$) gravity modes in Figure 4 is referred to as the Kelvin wave.

The collection of all Θ_n are the eigenfunctions of Laplace's tidal equation, and are called Hough functions in honor of the individual who pioneered in their numerical computation [*Hough*, 1897, 1898]. Either the ϵ_n or the h_n (where $\epsilon_n = 4\Omega^2 a^2/gh_n \approx 88$ km/h_n) are referred to as eigenvalues of the system. Each eigenfunction/eigenvalue pair constitutes a "mode." A common nomenclature in identifying modes is to explicitly express s, the zonal wavenumber, and n, the meridional index (so-named since it provides information on the number of latitudinal nodes and symmetry characterizing Θ_n). It is common therefore to refer to a particular mode as the Θ_n^s mode or just the (s,n) mode, and to add some information on wave period, as in the "(1,-2) diurnal tide." The (1,-2) diurnal mode might also be referred to as the "first symmetric trapped diurnal tide" and the "(1,1) mode" as the "first symmetric propagating diurnal tide." Note also from Figure 4 that the (1,-2) mode can assume other periods; at the free mode

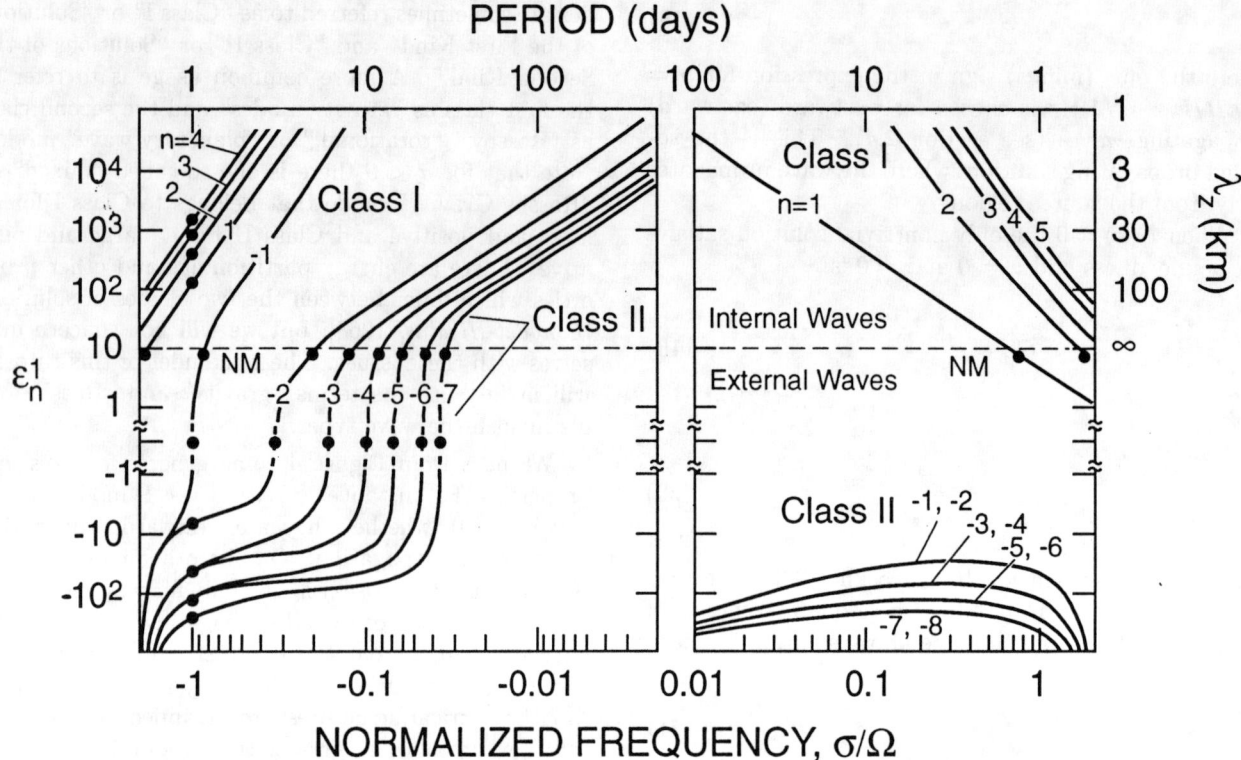

Fig. 4. Eigenvalues ϵ_n^s of wave modes of zonal wavenumber $s = 1$ vs. normalized frequency σ/Ω. Waves with positive (negative) frequencies propagate to the east (west). The dots corresponding to $\epsilon_n^1 = 0$ denote the so-called Rossby-Haurwitz waves. The dots corresponding to "NM" refer to the normal modes ($\epsilon_n^1 \approx 8.4$). The vertical series of dots at $\sigma/\Omega = -1.0$ define the ϵ_n^1 for the diurnal tide. The gravest ($n = 1$) of the eastward propagating gravity-type (Class I) modes is the Kelvin Wave. Figure and caption adapted from *Volland* [1988].

value of $\epsilon_n = 8.4$ ($h_n = 10.5$ km) for a 256 K isothermal atmosphere, the (1,-2) mode would represent the "5-day wave" ($\sigma/\Omega \approx$ -0.20).

The above experience in locating the "5-day wave" alludes to two possible ways in which diagrams like Figure 4 can be utilized. For forced modes we generally know the frequency of forcing, σ; by drawing a vertical line at σ/Ω on Figure 4, the points of intersection define the ϵ_n^1 values corresponding to the modes which comprise the response at that frequency. This provides information on the vertical structure of the forced response. The points of intersection corresponding to the diurnal tide ($\sigma/\Omega = -1.0$) are indicated in Figure 4. We see that the response consists of a mixture of trapped ((1,-1), (1,-2) ......) and propagating ((1,1), (1,2), ...) modes, the latter with vertical wavelengths between 15 and 50 km. This means that some localized heating in the lower atmosphere will result in (a) several modes which propagate to higher levels; and (b) a response partially contained at the levels of excitation. (The e-folding distance of the latter will depend on the value of ϵ_n). The degree to which the response falls into either of these categories is determined by how well the horizontal and vertical structures of these modes matches that of the forcing. Examination of the analog of Figure 4 for $s = 2$ (not shown here) would show the semidiurnal response (σ/Ω = -2.0) to consist only of propagating modes ($\epsilon_n > 0$); Rossby modes at frequencies higher than 2Ω do not exist.

For free (unforced) modes, we know that $\epsilon_n = 8.4$ for an isothermal atmosphere at 256 K. In Figure 4 the horizontal line defines the free or normal modes that exist for $s = 1$. Looking down from the points of intersection (labeled "NM"), we can then infer the frequencies or periods of the normal modes. For $s = 1$, these occur approximately at periods of 28 hours, 5 days, 8 days, and 12 days, and so on. According to our present nomenclature, we may refer to the last three of these, respectively, as the (1,-2), (1,-3), and (1,-4) westward propagating Rossby modes of zonal wavenumber one.

Table 1. Nomenclatures and other data for various common westward propagating waves in the middle and upper atmosphere. The column (s,n) indicates the nomenclature used in the present work, and in *Volland* [1988] and *Chapman and Lindzen* [1970]. The column $(s, |n| - s)$ indicates the nomenclature, generally restricted to planetary-wave usage, used by *Salby* [1981a, b; 1984] and *Longuet-Higgins* [1968]. Also provided are the equivalent depth for each mode [*Chapman and Lindzen*, 1970], h_n, propagating-mode vertical wavelengths λ_z for an isothermal atmosphere at 256 K, and further descriptors pertaining to the wave. Note that many of the values of λ_z in the real atmosphere vary significantly from the isothermal values given below, especially above and below the mesopuase where the dT/dz term in equation (26) (cf. equation (23)) plays an important role.

Wave	(s,n)	$(s, |n| - s)$	h_n (km)	λ_z (km)	Additional Descriptors
Diurnal tide	(1,1)		0.6909	27.9	Gravity; first symmetric propagating
Diurnal tide	(1,2)		0.2384	15.9	Gravity; first asymmetric propagating
Diurnal tide	(1,3)		0.1203	11.2	Gravity; second symmetric propagating
Diurnal tide	(1,-1)		803.356		Rotational; first asymmetric trapped
Diurnal tide	(1,-2)		-12.2703		Rotational; first symmetric trapped
Diurnal tide	(1,-4)		-1.7581		Rotational; second symmetric trapped
Semidiurnal tide	(2,2)		7.8519	311.	Gravity; first symmetric (propagating)
Semidiurnal tide	(2,3)		3.6665	81.4	Gravity; first asymmetric (propagating)
Semidiurnal tide	(2,4)		2.1098	53.8	Gravity; second symmetric (propagating)
Semidiurnal tide	(2,5)		1.3671	41.0	Gravity; second asymmetric (propagating)
Semidiurnal tide	(2,6)		0.9565	33.4	Gravity; third symmetric (propagating)
5-day wave	(1,-2)	(1,1)	10.5		Rotational; Rossby; first symmetric
10-day wave	(1,-3)	(1,2)	10.5		Rotational; Rossby; first asymmetric
16-day wave	(1,-4)	(1,3)	10.5		Rotational; Rossby; second symmetric
4-day wave	(2,-3)	(2,1)	10.5		Rotational; Rossby; first symmetric
2-day wave	(3,-3)	(3,0)	10.5		Mixed Rossby-Gravity; asymmetric

The 28-hour mode is a mixed Rossby-gravity mode, and is designated (1,-1). If one examines the $s = 3$ family of curves (not shown here), we would find that the mixed Rossby-gravity normal mode for $s = 3$ occurs close to $\sigma/\Omega = -.5$, corresponding to the "2-day wave."

Table 1 lists some of the more common westward propagating modes and their nomenclatures, with approximate values of h_n and the corresponding vertical scale in an isothermal atmosphere calculated from

$$\lambda_z = \frac{2\pi}{\alpha} = \frac{2\pi H}{\sqrt{\frac{\kappa H}{h_n} - \frac{1}{4}}} \quad (23)$$

The present nomenclature is consistent with that of *Volland* [1988] and *Chapman and Lindzen* [1970]. In this nomenclature a mode is symmetric about the equator if $(n + s)$ is even (odd) and antisymmetric if $(n + s)$ is odd (even) for gravity (Rossby) solutions. The mixed Rossby-gravity modes obey the Rossby mode symmetry conditions. For symmetric modes Θ_n (and hence all variables δp, $\delta \rho$, δT, w, and u) are mirror images with respect to the equator, whereas v is antisymmetric; for antisymmetric modes v is symmetric and the other variables change sign at the equatorial node. Another commonly used nomenclature used for planetary waves is due to *Longuet-Higgins* [1968], and is based on the value of $(|n| - s)$. This notation is also provided in Table 1.

Figure 5 illustrates the Θ_n for the first three westward propagating free Rossby modes for $s = 1$. Note

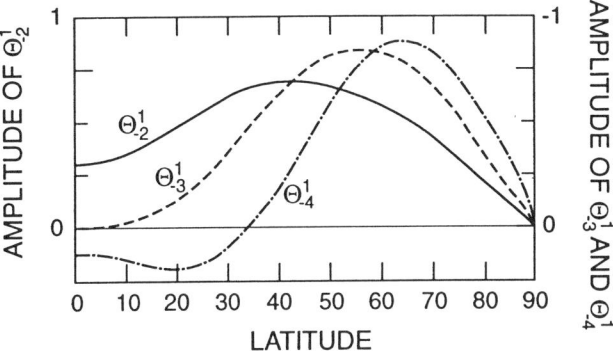

Fig. 5. Hough modes corresponding to the first three free Rossby modes of zonal wavenumber one. Adapted from *Walterscheid* [1980].

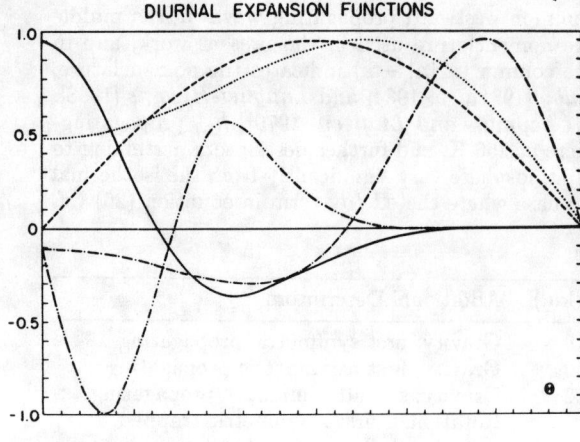
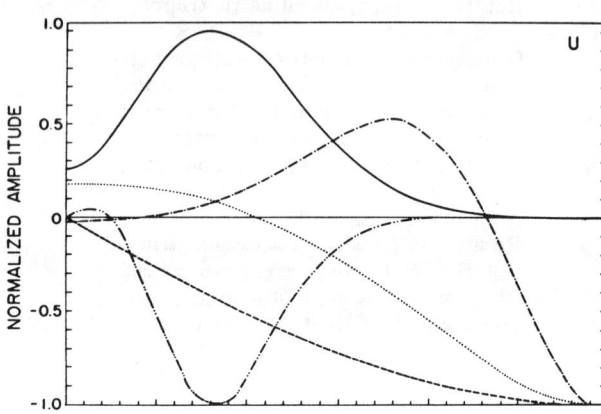
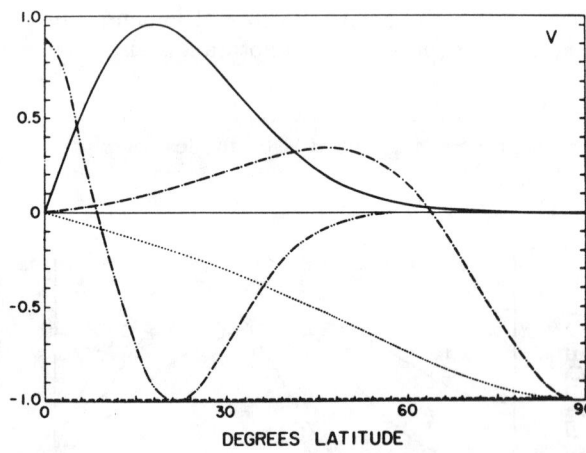

Fig. 6. Normalized expansion functions for the solar diurnal tide. Top: Hough Function. Middle: Eastward wind expansion function. Bottom: Northward wind expansion function. Solid line, (1,1); dashed, (1,-1); dashed-double dot, (1,2); dashed, (1,-2); dashed-dot, (1,-4). From *Forbes* [1982a].

that these are global scale modes with maximum amplitudes at middle and high latitudes. Figures 6 and 7 illustrate the corresponding Θ_n for the diurnal and semidiurnal tides, respectively. Also shown are the velocity expansion functions U_n and V_n defined by equations (10) and (11). For the diurnal tide, note the relative concentration of Θ_n, U_n, and V_n at low latitudes for the propagating ($h_n > 0$) modes and high latitudes for the trapped ($h_n < 0$) modes. The propagating modes are also more oscillatory in character. The semidiurnal wind expansion functions, on the other hand, tend to maximize at middle to high latitudes, increasingly so as the meridional index of the mode increases. This provides the first hint of why wind observations around the mesopause should appear predominantly semidiurnal in character at middle to high latitudes, and more diurnal in character at low latitudes (cf. Figure 1).

2.4 Group and Phase Velocity

Let us now return to our "propagating" solution to the vertical structure equation at the beginning of section 2.2. If $0 < h_n < 4\kappa H$, then $\alpha^2 > 0$ and the form of the solution of equation (16) consists of an upgoing and downgoing wave. Imposition of a "radiation condition" at the top of our domain determines which term in equation (16) to retain. The radiation condition demands that at sufficiently high altitudes the energy is upgoing, i.e., the vertical group velocity is positive, or $C_{gx} > 0$. To derive this condition, note that since

$$\alpha_n^2 = \frac{\kappa H}{h_n} - \frac{1}{4} = \frac{\kappa H g \epsilon_n}{(2\Omega a)^2} - \frac{1}{4} \qquad (24)$$

then

$$C_{gx} \equiv \frac{\partial \sigma}{\partial \alpha} = 2\alpha \frac{\partial \sigma}{\partial \epsilon} \bigg/ \frac{\partial \alpha^2}{\partial \epsilon} = \alpha \frac{8\Omega^2 a^2}{\kappa g H} \frac{\partial \sigma}{\partial \epsilon} \qquad (25)$$

[cf. *Andrews et al.*, 1987, p.164]. The choice of sign of $\alpha = \pm(\kappa H/h_n - 1/4)^{1/2}$ must be consistent with that in equation (18). From Figure 4, we see that $\partial \sigma/\partial \epsilon > 0$ for westward propagating waves and $\partial \sigma/\partial \epsilon < 0$ for eastward propagating waves. Therefore, to maintain $C_{gx} > 0$, in (25) we must choose $\alpha > 0$ for westward propagating waves and $\alpha < 0$ for eastward propagating waves.

Now, let us see what this implies in terms of phase progression in height and longitude. Our solution for propagating modes is of the form

$$e^{i(s\lambda + \alpha x - \sigma t)}$$

The equation $s\lambda + \alpha x - \sigma t = K$ defines the line of constant phase, e.g., the crest of the oscillation if $K = 0$. At a fixed λ, $\alpha x - \sigma t = K'$. Therefore, for either westward propagating ($\sigma < 0, \alpha > 0$) or eastward propagating ($\sigma > 0, \alpha < 0$) waves we have $x = \sigma t/\alpha + K''$ with $\sigma/\alpha < 0$, i.e., downward phase progression as time increases. (We see now why the downward phase progressions characterizing Figure 1 are consistent with a wave source at lower heights, i.e., O_3 and H_2O insolation absorption, and upward propagation through the mesopause.) Continuing, for a fixed t we have $s\lambda + \alpha x = K'$ or $x = -s\lambda/\alpha + K''$ implying westward phase tilt ($-s/\alpha < 0$) for westward propagating ($\alpha > 0$) waves and eastward phase tilt ($-s/\alpha > 0$) for eastward propagating ($\alpha < 0$) waves. Therefore, westward (eastward) phase tilt for westward (eastward) propagating waves is consistent with downward phase progression and upward energy propagation. These are important features to look for in observational data to verify theoretical interpretations.

2.5 Effects of Temperature Structure, Dissipation, and Mean Winds

In section 2.3 we showed that free oscillations exist in an isothermal, dissipationless atmosphere. For such oscillations and $T_o = 256$ K, h_n and ϵ_n had to assume values of 10.5 km and 8.4, respectively. In section 2.2 we were able to find the points of intersection corresponding to $\epsilon_n = 8.4$, and to infer the periods and horizontal structures of the various free modes. In this section we will discuss in simple terms how the additional complexities of vertical temperature structure, mean zonal winds, and dissipation modify our concepts about free atmospheric oscillations. In section 4 we will examine the effects of more complicated distributions of winds and temperatures that necessitate comprehensive numerical treatment of the problem.

For a non-isothermal atmosphere, the derivation leading to equations (13), (14), and (15) from equations (1)–(4) proceeds exactly as before, except the resulting expression for α^2 in equation (15) appears as follows:

$$\alpha^2 = \frac{\kappa H + dH/dx}{h_n} - \frac{1}{4} \qquad (26)$$

Above ≈ 90 km $\alpha^2 > 0$, implying propagating solutions, energy leakage into the thermosphere, and a finite time for the oscillations in the absence of continual forcing. A true resonance (infinite response) no longer exists. *Lindzen and Blake* [1972] assumed a mean dis-

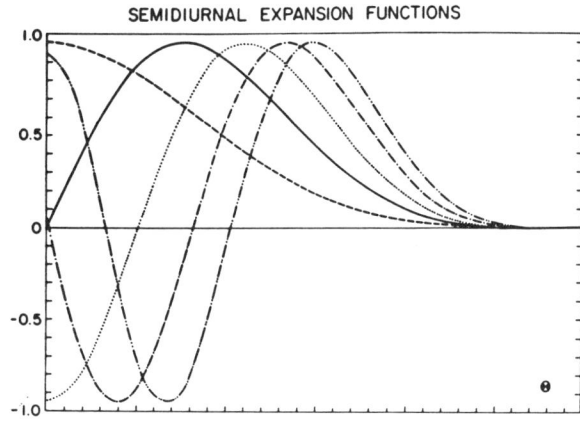
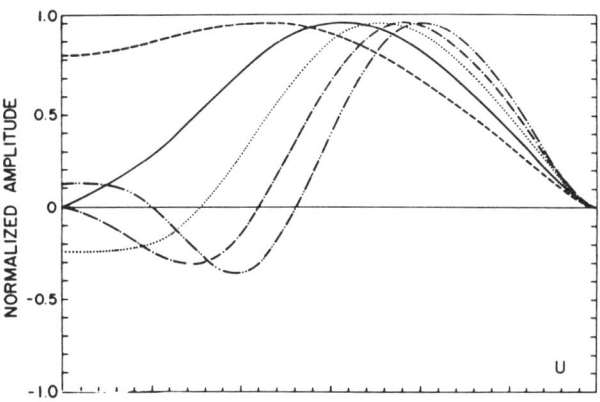
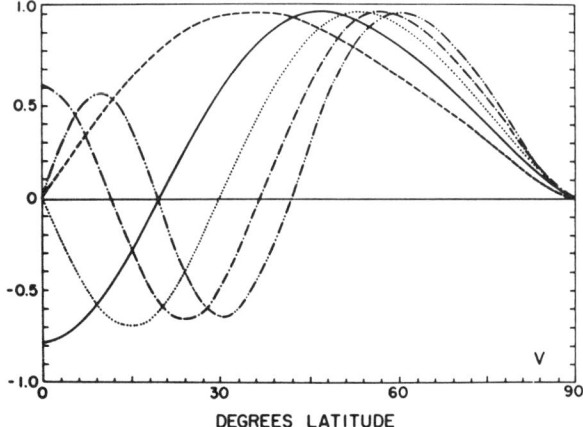

Fig. 7. Same as Figure 6, except for the semidiurnal tide. Dashed line, (2,2); solid, (2,3); dotted, (2,4); dashed-dot, (2,5); dashed-double dot, (2,6). From *Forbes* [1982b].

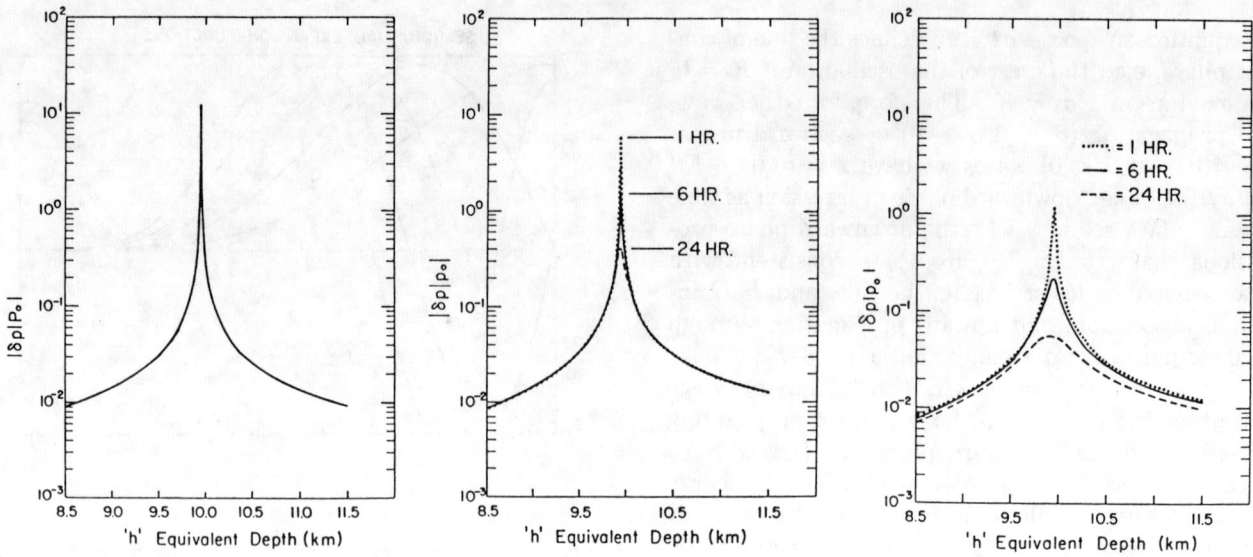

Fig. 8. Fractional response in surface pressure $|\delta p/p_o|$ as a function of equivalent depth. (a), left, without dissipation and without surface friction; (b), middle, with dissipation and without surface friction; and (c), right, with dissipation and with surface friction. From *Lindzen and Blake* [1972].

tribution of temperature with height, solved equations (15) and (26) subject to a specification of tropospheric heating, and examined the response (surface perturbation pressure) as a function of the equivalent depth (h_n in equation (26)). Their result is shown in Figure 8a, and illustrates a sharp but finite maximum at $h = 9.95$ km. Their solutions also exhibited amplitude growth and phase tilt with height above 90 km, and nonzero vertical velocities when T_o varies with height.

Lindzen and Blake [1972] also examined the influences of eddy and molecular dissipation and surface friction on the Lamb modes. In this case the response is dependent on wave period. Figure 8b illustrates the analog to Figure 8a, except that dissipation is taken into account. Figure 8c illustrates the additional effects of including surface friction. We see that the effects of dissipation and surface friction are to reduce the magnitude and significantly broaden the response, with increasing effects as the period becomes longer. It is also evident that surface friction dominates over internal dissipation, and is therefore the determining factor in limiting the "lifetime" of free modes. Lindzen and Blake estimate lifetimes on the order of 10 to 100 wave cycles for periods between 24 hours and 2 hours, respectively.

Salby [1979, 1980] examined the resonance characteristics of Lamb modes in the presence of vertical temperature structure and dissipation, with emphasis on the longer-period 'Lamb' waves (2 to 20 days). His results for the $s = 1$ Rossby-gravity mode are illustrated in Figure 9. He notes the secondary peak occurring near $h = 6.4$ km, which was discovered by *Pekeris* [1937]. This secondary peak is due to the stratospheric temperature duct, and was apparently overlooked by Lindzen and Blake who only took their calculations down to $h = 8.5$ km. This secondary peak disappears in the presence of realistic dissipation [*Salby*, 1979; see Figure 9]. In honor of the original discoverer, *Platzman* [1988] has suggested referring to this as the 'Pekeris' mode.

Lindzen and Blake [1972] did not find any noticeable effects on amplitude and phase structures due to dissipation below 100 km for Lamb periods < 24 hours. However, for the longer-period modes examined by Salby, increased amplitude reduction and phase tilt with height accompany an increase in wave period (see Figure 10). Figure 10 also implies that phase tilt with height in the real atmosphere is not inconsistent with the concept of a 'quasi'-free atmospheric mode. Enhanced vertical leakage of energy should diminish wave lifetimes, but the dominant dissipative effect remains surface friction. Salby's work also suggests free mode lifetimes to be on the order of tens of wave cycles, and also discusses the role of variations in dissipation in reflection of wave modes, particularly when the doppler-shifted frequency becomes small.

The zero-order effects of non-zero winds can be ascertained quite easily. If we suppose that the troposphere is characterized by a mean eastward wind $\overline{U}\sin\theta$

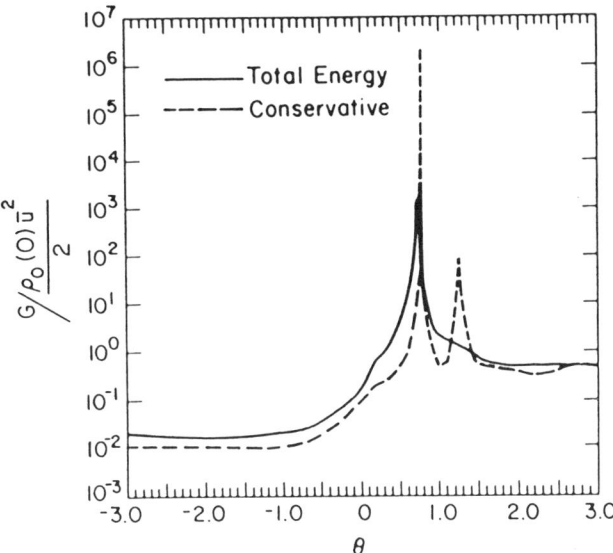

Fig. 9. Total energy as a function of $\theta = H/h_n$ for the $s = 1$ mixed Rossby-gravity mode, with dissipation (solid line) and without dissipation (dashed line). The secondary peak for the conservative case (dashed line) is due to buoyancy trapping at upper levels [*Salby*, 1979]. The absence of a secondary peak in the presence of dissipation is due to the reduced energy flux reaching these levels from the surface [*Salby*, 1980]. Figure and caption adapted from *Salby* [1980].

(effectively a uniform superrotation of the atmosphere), then the σ appearing in our equations

$$\frac{\partial}{\partial t} \to -i\sigma$$

should be replaced by the Doppler-shifted or intrinsic frequency, σ_D:

$$\frac{\partial}{\partial t} + \frac{\overline{U}}{a}\frac{\partial}{\partial \lambda} \to -i(\sigma - k\overline{U}) = -i\sigma_D$$

where $k = s/a$. In this case, the horizontal scale in Figure 4 is σ_D/Ω, not σ/Ω. To an observer on the ground, however, the wave frequency would be

$$\sigma_{obs} = \sigma_D + k\overline{U},$$

or equivalently, the observed period is

$$T_{obs} = \frac{2\pi}{|\sigma_D + k\overline{U}|}$$

For the (1,-2), (1,-3), and (1,-4) normal modes for which $\epsilon = 8.4$, we infer from Figure 4 the corresponding normalized frequencies of about -0.2, -0.12, -0.08, or periods of 5, 8.3, and 12.5 days, respectively. If we interpret these to be Doppler-shifted frequencies, then for a nominal value of $\overline{U} = 10$ ms^{-1}, the *observed* periods ought to be about 5.6, 10.2, and 17.1 days. Therefore, we expect the actual atmospheric manifestations of free Rossby modes to be Doppler-shifted to longer periods. This is why we associate, for instance, the observed 2–3 week oscillation referred to as the "quasi 16-day wave" [*Madden*, 1979] with the (1,-4) Rossby mode possessing an eigenperiod of only 12.5 days.

3. FORCING OF ATMOSPHERIC TIDES

Atmospheric tides represent an obvious example of "forced" atmospheric waves for which we know the wave periods quite well. Lunar tides are of course determined by the period of the moon's *apparent* rotation around the Earth. Here we will be mainly concerned with solar or thermally-forced tides, which are excited by the periodic absorption of solar radiation connected with the *apparent* motion of the Sun around the Earth. Figure 11 is a schematic of the main points: various parts of the solar spectrum are absorbed by tropospheric water vapor (near-IR), stratospheric ozone (UV), and major atmospheric constituents (O_2 and N_2) in the lower thermosphere (Figure 11a). (Note that the region around the mesopause, where most meteor and MF radar measurements provide wind data, cf., Figures 1–3, are in a region of "no excitation." At any given height, the day-night variation of absorbed radiation (and hence heating) gives rise to Fourier components which are integral subharmonics of a solar day: 24 hours, 12 hours, 8 hours, etc.; Figure 11c). Each of these harmonic components (referred to as the diurnal tide, semidiurnal tide, terdiurnal tide, respectively) possess a height-latitude distribution (we are ignoring longitude dependences for the moment). Near the height of maximum heating, the latitudinal distribution for a given harmonic might look something like Figure 11b (i.e., maximum at low latitudes and minimum at the poles, in concert with the solar zenith angle influence).

Now, given that the Θ_n form a complete orthogonal set, we can expand the height-latitude distribution of heating for a given frequency component, $J^\sigma(z,\theta)$ (cf. equation (7)):

$$J^\sigma(z,\theta) = \sum \Theta_n(\theta) J_n(z)$$

Each "mode" defined by its eigenfunction-eigenvalue pair (Θ_n, h_n) now possesses its own vertical profile of heating $J_n(z)$. The vertical structure of each mode is determined by $J_n(z)$, h_n, and the mean thermal structure of the atmosphere vis-a-vis equations (15) and (6). Typical examples of $J_n(z)$ for diurnal and semidiurnal tides are provided in Figure 12 [*Forbes and Garrett*, 1978]. Note that most of the heating goes into the (1,-2) mode for the diurnal tide, and into the (2,2) mode for the semidiurnal tide, as the Θ_n for these modes (cf. Figures 6 and 7) most closely correspond with the latitudinal distribution of heating (cf. Figure 11b).

4. NUMERICAL MODEL RESULTS

4.1 *Atmospheric Tides*

In section 2 we discussed the eigenfunction-eigenvalue problem corresponding to forced and free atmospheric oscillations in an isothermal, dissipationless atmosphere. By virtue of separability a set of independent horizontal modal structures existed, each with its own vertical structure. Separability of height and latitude dependences also exists in a non-isothermal atmosphere, and additionally for special treatments of height-dependent dissipation [*Lindzen and McKenzie*, 1967] or for constant Coriolis parameter [*Lindzen*, 1970]. However, for vertical diffusion of heat and momentum in a spherical rotating atmosphere, or for latitude-dependent mean winds in a diffusive or non-diffusive atmosphere, the equations for an oscillation with specified frequency and zonal wavenumber are nonseparable. This necessitates a numerical approach to the problem wherein a universal set of modal structures does not naturally emerge. However, it is commonplace to use modal terminology from the separable problem nonetheless, as many observed features of prominent oscillations exhibit characteristics very similar to what would be expected on the basis of "classical" theory. In fact, it is commonplace to decompose the thermal forcing in numerical models into Hough modes (as in Figure 12), even though the solution is nonseparable; and in fact, the solutions are sometimes decomposed into Hough modes to facilitate interpretation of the results. In this context mode-mode coupling is a way of expressing the consequences of nonseparability. We will now briefly review some of these numerical models.

Forbes and Garrett [1979] review basically two types of numerical models which take into account dissipation, mean winds, and other processes in simulations of middle and upper atmosphere tides. The first genre neglect eddy and molecular dissipation, but include mean winds and meridional temperature gradients, Newtonian cooling, and perhaps a Rayleigh friction (linear damping) term to filter out small-scale noise or to facilitate application of upper boundary conditions. Dispensing with diffusion allows one to derive a sin-

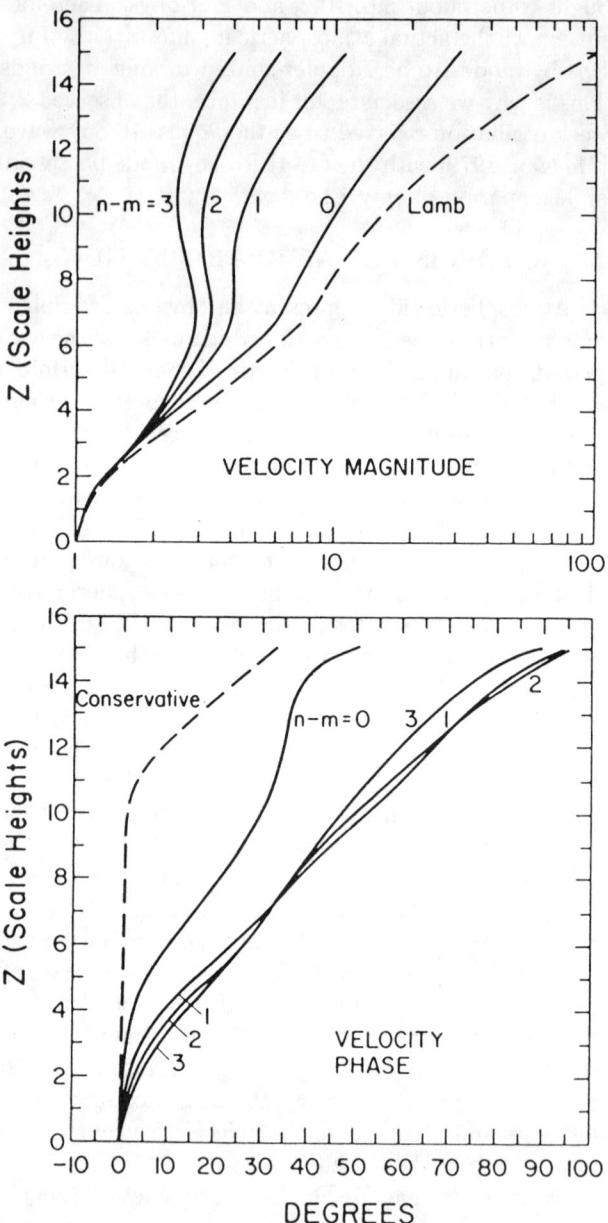

Fig. 10. Normalized velocity magnitude (top) and phase (bottom) for the lowest order $s = 1$ westward propagating waves in the presence of dissipation. The notation $n - m = 0,1,2,3$ refers, respectively, to the mixed Rossby-gravity mode, and the "5-day," "10-day," and "16-day" waves. Vertical structures for the Lamb mode in a non-isothermal atmosphere without dissipation are shown as dashed lines. From *Salby* [1980].

gle second-order partial differential equation in height and latitude for the perturbation geopotential [*Lindzen and Hong*, 1974; *Aso et al.*, 1981; *Walterscheid and Venkateswaran*, 1979a, b; *Walterscheid et al.*, 1980; *Vial*, 1987; *Forbes and Vial*, 1989]. In the context of the solution of these nonseparable equations, the terminology of "mode coupling" has arisen. This refers to the generation of tidal modes (determined through an orthogonal expansion of the calculated response) which are not forced directly by thermal excitation, but which arise because of the nonseparability of the governing equation. For instance, if only the (2,2) mode is excited in these models, the response at say 90 km consists of many modes ((2,2), (2,3) , ...) due to the "distorting" effects of the mean wind distribution. In the above models the (2,4) mode appears to receive about equal contributions from direct thermal forcing and mode coupling via the (2,2) mean wind interactions which tend to add in phase. On the other hand, for the (2,3) mode the effect of mode coupling is to interfere with the directly forced component and thereby reduce the (2,3) response above the level of ozone heating. In the case of (2,5), excitation appears to arise almost exclusively due to direct thermal forcing (mode coupling is weak). More recent studies by *Forbes and Hagan* [1987] and *Vial* [1986] address the diurnal tide, and utilize a Rayleigh friction (linear damping) term to parameterize turbulent diffusion of momentum. For the dominant diurnal propagating (1,1) mode, latitudinal broadening (or leakage to high latitudes) due to dissipation near 90 km can be viewed as a coupling into the trapped or evanescent (1,-2) mode, whereas the asymmetries in the modified modal shape induced by the global mean wind distribution (particularly around solstice) can be interpreted as a coupling into the (1,2) and (1,-1) asymmetric modes.

At the next hierarchal level of modeling pertaining to atmospheric tides, *Forbes* [1982a, b] includes eddy and molecular diffusion of momentum and heat so as to properly address the structural modification of tides in the 80–150 km region and their penetration to higher altitudes. This requires numerical solution of the four coupled partial differential equations in the three velocity components and temperature, as opposed to a single equation for the geopotential as in the above studies. *Forbes* [1982a, b] provides explicit simulations from the surface to 400 km for the solar diurnal, solar semidiurnal, and lunar semidiurnal tides due to realistic thermal and gravitational forcing, as well as normalized thermospheric extensions of solar semidiurnal modes above 80 km for use in the fitting, extrapolation and interpolation of observational data [*Forbes and Hagan*, 1982].

Illustrations of amplitude and phase vertical structures for the solar semidiurnal and diurnal tides from the *Forbes* [1982a, b] model are shown in Figures 13 and 14, respectively. In Figure 13, note the relatively long vertical wavelength characterizing the response below 50 km; this is consistent with most of the heating going into the long-wavelength (2,2) mode (Figure 12, Table 1); above about 50 km, the wavelength decreases, due to the increased presence of short-wavelength modes induced by "mode coupling" due to the strong mesospheric jets. The region between 70 and 90 km is a region of evanescence for the (2,2) mode, due to the combined effects of its large h_n and the negative temperature gradient (cf. equations (15) and (6)). However, in this region the higher-order modes are growing exponentially with height, and soon begin to dominate the solution in the lower thermosphere. However, as molecular viscosity begins to dominate in the 120–150 km region, these shorter vertical wavelength modes (cf. Table 1) are more susceptible to dissipation, and the longer wavelength modes begin to dominate at higher altitudes. In the upper thermosphere, molecular diffusion of heat and momentum are so efficient that it is difficult to maintain vertical shears in the wind and temperature fields, and the tidal fields asymptote to constant values above about 200 km.

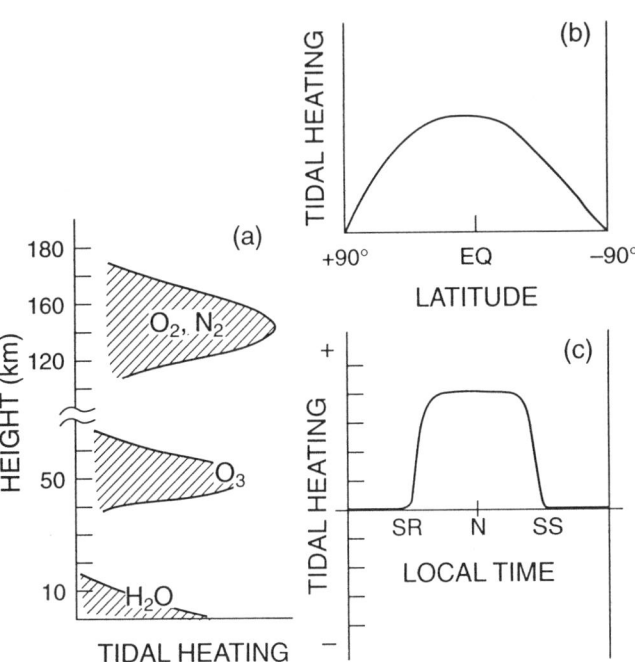

Fig. 11. Schematic of (a) vertical (left), (b) latitudinal (top), and (c) diurnal (bottom) variations in tidal heating.

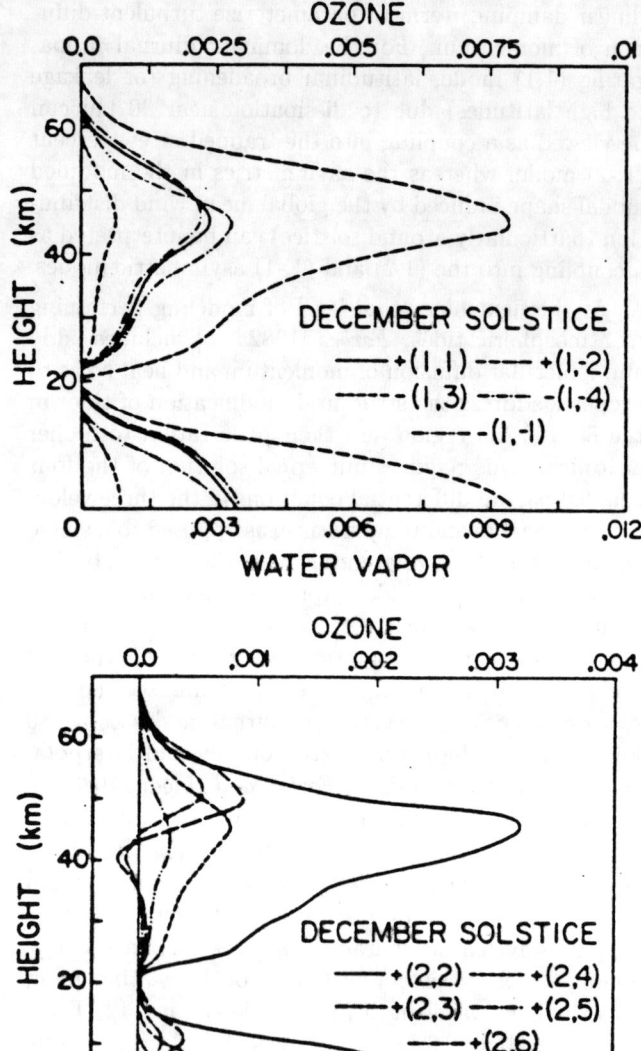

Fig. 12. Vertical profiles of diurnal (top) and semidiurnal (bottom) heating, $e^{-x/2}J_n$ where $x = -ln(p/p_o)$, due to insolation absorption by ozone and water vapor, corresponding to various solar tidal modes. The units are Joules kg^{-1} sec^{-1}. Adapted from *Forbes and Garrett* [1978].

The situation is similar for the diurnal tide, illustrated in Figure 14. Note that at high latitudes (60° in Figure 14) the phase is more or less constant with height, consistent with the dominance of trapped modes whose maxima are at high latitudes (cf. Figure 6). At low latitudes the solution is dominated by the (1,1) mode, with a vertical wavelength of order 30 km (cf. Table 1).

At this point we should comment on the characteristics noted in reference to Figure 1 in section 1. In section 2.4 the downward phase progression with height was shown to be consistent with a positive (upward) group velocity, consistent with the excitation sources being located somewhere below 80 km. At high latitudes, the tidal fields near the mesopause are dominated by the semidiurnal propagating modes, particularly higher order modes than (2,2); since there is relatively little in-situ heating, the diurnal tide is weak at latitudes greater than about 40°; furthermore, the higher-order semidiurnal tides are growing exponentially with height in this regime where the (2,2) mode is quasi-evanescent. This accounts for the predominance of the semidiurnal tides at Saskatoon in Figure 1. At Townsville (19°S), much closer to the equator, the propagating semidiurnal tides are relatively small, and the diurnal tide enjoys its maximum amplitudes (cf. Figure 6).

Atmospheric tides are beginning to be addressed in general circulation models [*Hunt*, 1990; *Miyahara et al.*, 1993; *Roble and Ridley*, 1994]. These models are capable of investigating a variety of nonlinear interactions, and in some cases related chemical transport effects. For instance, disspation of the diurnal propagating tide is found to produce a lower thermosphere mean zonal jet ($\sim$ 20–40 ms^{-1}) over the equator [*Miyahara et al.*, 1991] with accompanying consequences for the mean zonal distributions of O, NO, and a number of other minor constituents [*Forbes et al.*, 1993]. Present and future contributions of GCMs to the study of planetary waves and tide/gravity-wave interactions are discussed further in sections 4.2 and 5, respectively.

4.2 Planetary Waves

Although the above mode coupling effects are important in the context of atmospheric tides, the tidal wave phase speeds are generally large compared to the mean flow speed U; the resulting effects do not represent *drastic* consequences. In effect, these "fast" waves do not strongly feel the effects of the relatively slow background flow. However, as the wave periods increase from 2 to 20 days for planetary (Rossby) waves, the phase speeds get smaller and the effects of mean winds assume much greater importance. The above arguments can be made more quantitative by noting the Doppler-shifting effects of mean winds that appear when assuming solutions of the form $e^{i(s\lambda - \sigma t)}$:

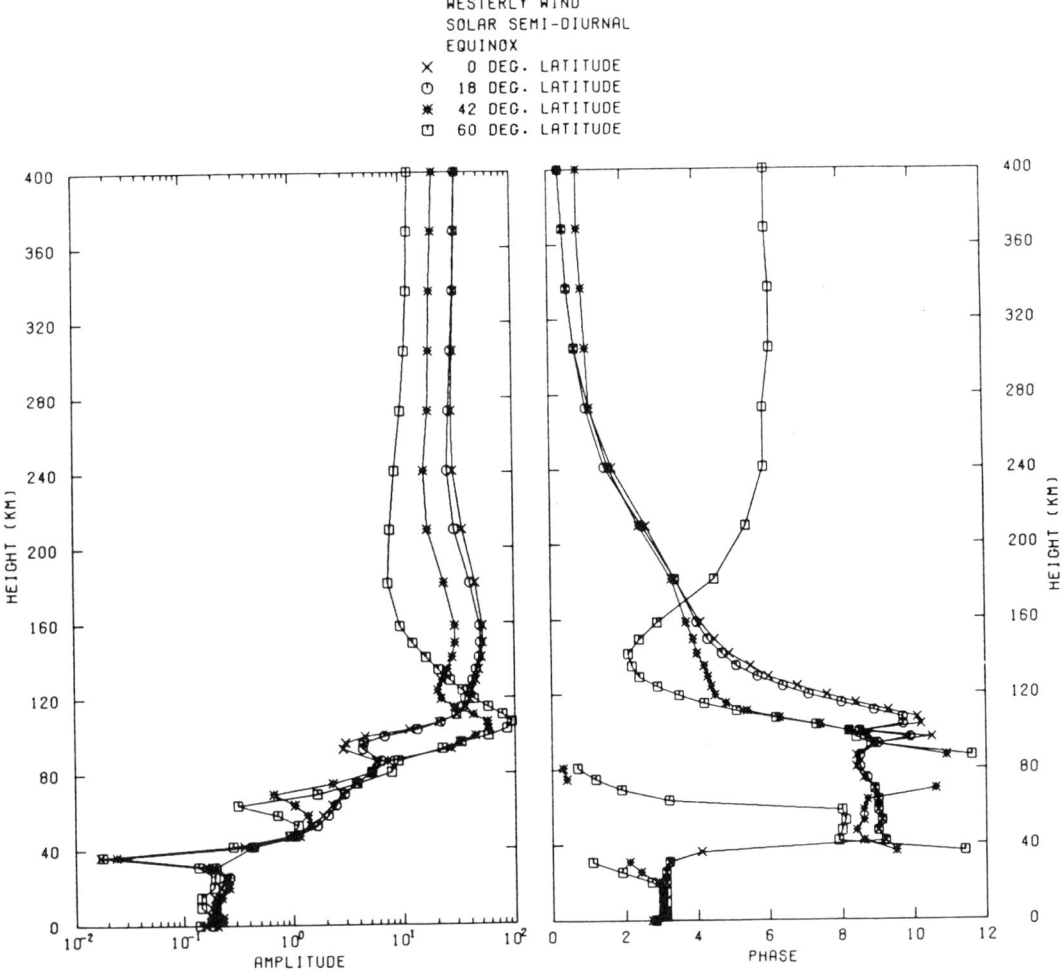

Fig. 13. Amplitude (left) and phase (right) for solar semidiurnal eastward winds at 0°, ±30°, and ±60° latitude for December solstice conditions. From *Forbes* [1982b].

$$\frac{\partial}{\partial t} + \frac{\overline{U}}{a\sin\theta}\frac{\partial}{\partial \lambda} \to ik(-C_{ph} + \overline{U})$$

where $k = s/(a\sin\theta)$ and the zonal phase speed is $C_{ph} = \sigma/k$. For the migrating tides $\sigma = s\Omega$, so that $C_{ph} = \Omega a \sin\theta$ or about 464 ms^{-1} at the equator and 232 ms^{-1} at 60° latitude. If T is the period in days, then for the westward propagating planetary waves,

$$C_{ph} = -\frac{\Omega a \sin\theta}{sT}$$

and the above tidal phase speeds are reduced by the factor sT. For the $s = 1$ 10-day wave $C_{ph} \approx 23$ ms^{-1} at 60° latitude, and for the $s = 3$ 2-day wave $C_{ph} \approx 39$ ms^{-1}. In either case, and for other planetary waves as well, summer easterlies of order -20 to -60 ms^{-1} can obviously have drastic effects on the propagation of planetary waves. When the condition $\sigma_D = -C_{ph} + \overline{U} = 0$ is satisfied, we refer to this as a critical line, and anticipate that this must imply drastic effects (N.B. for stationary planetary waves $C_{ph} = 0$, and this condition reduces to the zero wind line). Moreover, when σ_D becomes small we intuitively expect the wave to be more sensitive to dissipative processes. Below, we will examine the role of zonal mean winds in greater detail.

Salby [1981a, b] has utilized the first genre of model described above to investigate the behavior of planetary waves in the presence of realistic background winds. In this work he forced the lower boundary with a constant vertical velocity with respect to latitude (with a change of sign at the equator for asymmetric forcing), and ex-

82   TIDAL AND PLANETARY WAVES

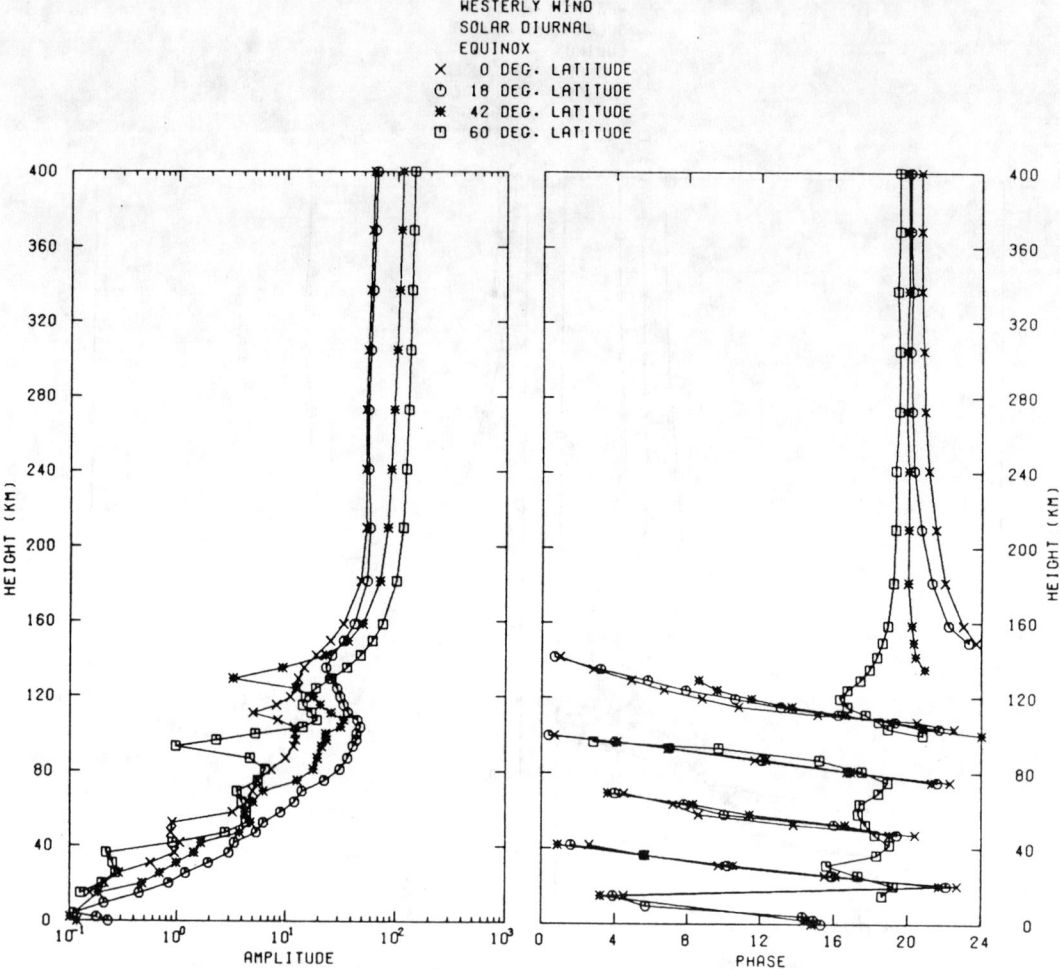

Fig. 14. Same as Figure 13, except for the solar diurnal tide. From *Forbes* [1982a].

amined the response as a function of frequency, with the zonal wavenumber and background wind configuration fixed. An example of his results for $s = 1$ westward propagating modes is illustrated in Figure 15. Note that the response is very structured, and differs considerably between typical equinoctial and solsticial conditions. These results reflect the extreme sensitivity of the planetary wave response to the background wind field. Note, however, that the responses tend to maximize near periods of 5, 9, and 16 days. As Salby shows, the tropospheric and lower stratospheric responses near these periods are structurally similar (i.e., latitude dependence of amplitude and phase) to what we would expect on the basis of "classical theory" presented in section 2. Therefore, it appears that even in realistic wind configurations that it is valid to speak in terms of a resonant response of the atmosphere, and to associate these responses with the free Rossby modes of Laplace's tidal equation. However, above the lower stratosphere, the atmospheric response is further complicated by high wind speeds and shears. We will now briefly discuss some of these effects.

Dickinson [1968] has performed an analytic investigation of the vertical propagation of stationary planetary waves through a background wind field consisting of significant vertical and horizontal shears. This theory is applicable to the long-period oscillations investigated here, and provides a framework for interpreting the results. Dickinson's work represents an extension of *Charney and Drazin* [1961], who limit their analysis to a mean zonal wind independent of latitude with constant Coriolis parameter. Charney and Drazin conclude that vertical propagation of stationary planetary waves is only possible in westerly wind regimes,

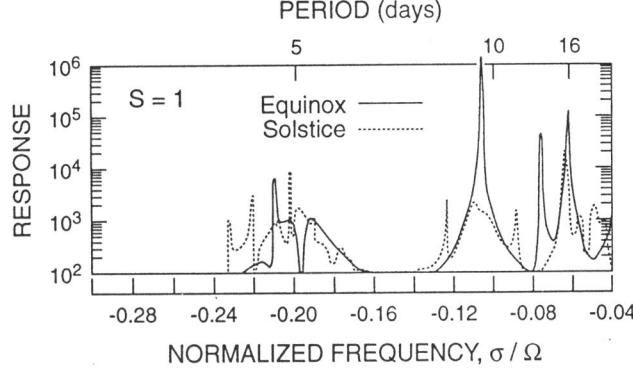

Fig. 15. Simulated atmospheric response as a function of normalized frequency for $s = 1$ westward propagating waves, for typical solstice and equinox background wind conditions. From *Salby* [1981b].

when the westerly wind speed is below some upper limit (sometimes referred to as the "Charney-Drazin critical speed"). The basic idea of Dickinson's work can be extended to traveling planetary waves if we simply replace "westerly wind" ($\overline{U} > 0$) with "westerly wind with respect to the wave" (($\overline{U} - C_{ph}) > 0$). The main conceptual results of *Dickinson* [1968] are summarized in his Figure 1, which is reproduced here as Figure 16. Assuming winter solstice conditions and a mid-latitude source of wave energy, vertical propagation of planetary waves is affected as follows. At middle latitudes, the westerly jet is sufficiently strong to preclude efficient propagation above the stratopause. Poleward of the westerly jet, a wave guide is formed which traps waves between the strong westerlies and the geometric pole; Dickinson refers to this as the *polar cap wave guide*. This wave guide provides a ducting channel through which planetary waves can penetrate to the mesosphere and lower thermosphere. (A similar ducting channel can in principle be realized between two regions of high westerly wind separated by weak westerlies.) Planetary wave disturbances can also be diffracted into an *equatorial wave guide* formed between the westerly jet of the winter hemisphere and the zero-wind line (or critical line in the case of traveling waves) transition to stratospheric summer easterlies. Dickinson's analysis indicates the planetary disturbances would be absorbed rather than reflected along such zero-wind lines, providing an impediment to significant vertical penetration (i.e., to the mesopause). Dickinson therefore suggests that whatever stationary (and therefore long-period) planetary wave disturbances might be realized at the equatorial mesopause would probably originate from leakage connected with the polar wave guide. It seems reasonable to assume, however, that the degree of attenuation is dependent on the strength of the westerly jet, and the separation distance between the jet and the zero-wind (or critical) line.

A steady-state linear numerical simulation [*Salby*, 1981c] of the 2-day wave under northern hemisphere winter solstice conditions is presented in Figure 17. This is a case of 'moderate' mean wind effects, i.e., not so extreme as the $s = 1$ 10-day and 16-day waves. Nevertheless, the tendency for exclusion of the solution from the strong winter westerly and summer easterly jets is evident. In this case the equatorial waveguide is very broad (wide separation between the critical line in the summer hemisphere and the winter jet maximum). There is also a tendency for the wave maxima to shift to the summer hemisphere, i.e., to the region of weak westerlies with respect to the wave. Note that the equatorial amplitudes of meridional wind at 35°S are of the same order ($\approx 20 - -30 ms^{-1}$) as the episodically large 2-day wave amplitudes observed during local summer over Adelaide, South Australia, (Figure 3). Therefore, Figure 17 provides some measure of the true height/latitude temperature and meridional wind distributions for the episodically large 2-day wave.

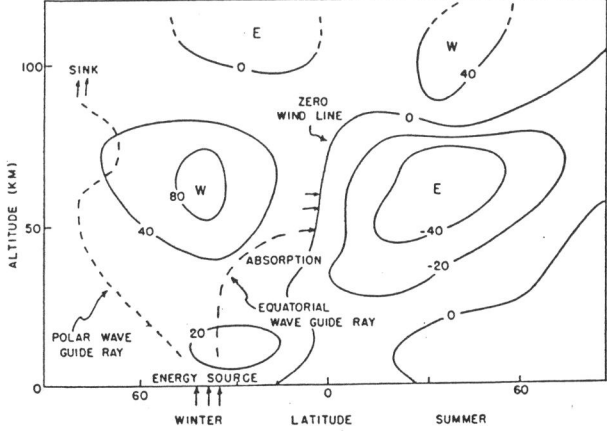

Fig. 16. Schematic of propagation paths for stationary planetary waves excited in the mid-latitude northern hemisphere during winter conditions. From *Dickinson* [1968]. For the traveling planetary waves, the barrier represented here by the zero wind line would be replaced by the frequency-dependent critical line; for periods greater than about 10 days and small zonal wavenumbers, the critical line is close to the zero wind line, and for progressively smaller periods the critical line recedes up into the mesosphere and towards high latitudes.

Despite considerable simplifications underlying the theory of so-called "free" or "normal" modes (cf. section 2), atmospheric manifestations of these modes apparently exist, and many salient features are captured with steady-state linear models. However, the mechanisms responsible for the appearance of normal mode signatures in the atmosphere remain nebulous. As noted by *Salby* [1984], spatially and temporally broadband forcing in the troposphere is expected to preferentially result in atmospheric responses corresponding to the normal mode spectrum. General circulation models in fact reveal structures in close agreement with atmospheric normal modes, which appear to be connected with inclusion of regional gravity wave activity and the inherent nonlinearity of the system [*Jakobs and Hass*, 1987; *Manzini and Hamilton*, 1993]. Baroclinic instability in the summer mesosphere is another mechanism proposed for triggering of traveling planetary waves, at least for the quasi-two-day wave [*Plumb*, 1983].

5. CONCLUDING REMARKS AND OUTLOOK FOR THE FUTURE

This tutorial has sought to expose the nondynamicist to the fundamental theory, observational evidence, and numerical modeling results pertaining to tides and planetary waves in the mesosphere and lower thermosphere. At this point, there remains much to be done. While radars are capable of providing long time series and therefore identifying the presence of planetary wave periodicities, they are distributed too sparsely to provide adequate information on zonal wavenumbers. On the other hand, satellites are now capable of providing good spatial coverage with marginally useful temporal information. Moreover, the region between 100 and 150 km is practically devoid of any measurements capable of delineating planetary waves. This combination of capabilities and circumstances represents an ideal situation for joint ground-based/satellite observations of the mesosphere/lower thermosphere (MLT), a task promised to be accomplished by the NASA TIMED mission.

There is also considerable room for theoretical and numerical modeling advances. The question of nonmigrating (i.e., longitude-dependent) diurnal tides needs to be addressed, and the work of *Forbes and Groves* [1987] improved upon with greater attention to various tropospheric excitation sources. The pioneering work of *Salby* [1981a, b, c] needs to be extended, particularly with regard to inclusion of more realistic mean wind distributions and dissipative processes, both of which are essential to understanding the propagation charac-

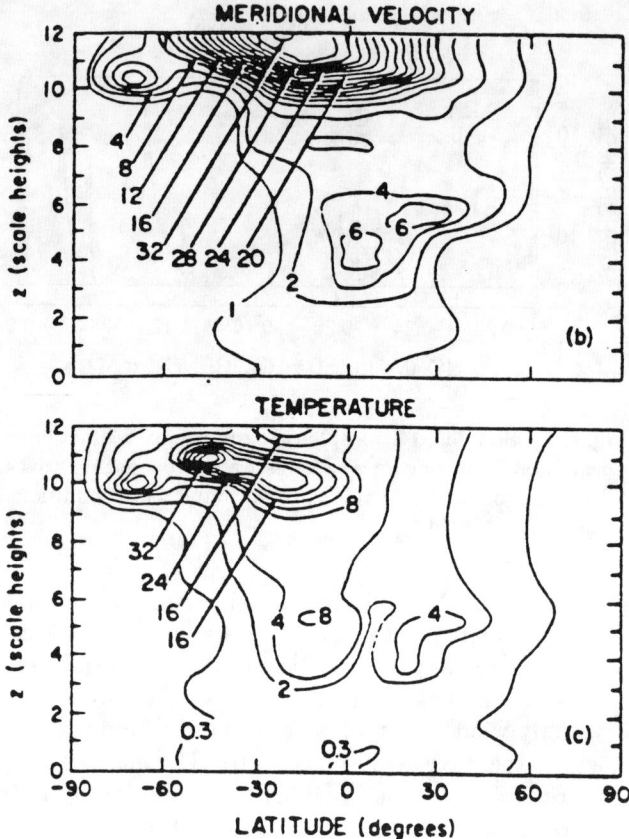

Fig. 17. Simulated meridional wind (ms^{-1}) and temperature (K) as a function of height and latitude for the quasi-two-day wave. Typical December solstice conditions are assumed. From *Salby* [1981 c].

teristics of planetary waves in the mesosphere and lower thermosphere. Work along these lines is now being realized [*Hagan et al.*, 1993; *Forbes et al.*, 1994]. The possible in-situ generation of Rossby-like modes in the MLT regime, possibly due to solar radiation or Joule heating variations, or the periodic filtering of gravity waves originating in the lower atmosphere and depositing heat and momentum in the upper atmosphere, warrant investigation. Some initial work on gravity-type normal modes of the thermosphere has been accomplished by *Larsen and Mikkelsen* [1987], but no work has been done on possible Rossby-like normal modes of the thermospheric regime.

Nonlinear interactions between gravity waves and planetary-scale waves will occupy a significant component of forthcoming research activities devoted to MLT dynamics. General circulation models are now beginning to include parameterized interactions between tides and gravity waves [*Hunt*, 1986, 1990; *Roble and*

Ridley, 1994], with significant consequences: While the tides serve to modulate the propagation of gravity waves through the MLT region and heterogenize the global spatial production of turbulence, the variable gravity wave momentum deposition feeds back to significantly modify the tidal structures. A secondary effect is to change the wave drag on the zonal mean circulation. A similar mechanism has been discovered in connection with the wintertime stationary planetary wave [McLandress and McFarlane, 1993]. These authors demonstrate that deceleration of the mesospheric mean zonal wind due to Eliassen-Palm Flux Divergence of the modified planetary wave is comparable in magnitude to the deceleration induced by gravity wave drag. It remains to be seen whether similar consequences hold for the the migrating solar tides and traveling planetary waves discussed in this tutorial.

Figure 18 illustrates one potential influence of planetary waves on the MLT region, even when they may not penetrate beyond the mesosphere. This is a plot of power spectral densities constructed from daily values of the *semidiurnal tidal amplitude* derived from wind observations near 95 km over Obninsk, Russia, during January through February, 1979 (cf. Figure 2). We see that the semidiurnal tide is modulated at periods near 10 days and 20 days, possible due to interactions with mean winds of these periodicities in the mesosphere. (Unusually large 10-day and 18-day oscillations were known to characterize the stratosphere and mesosphere during the same time period [Gille and Lyjak, 1984; Smith, 1985].) The modulations are significant; the 10-day modulation amounts to about ± 7 ms^{-1} about a mean value of ≈ 20 ms^{-1} in the semidiurnal wind amplitude. The modulation of gravity wave fluxes may also may be involved in explaining this dynamical feature of the mesopause region.

Heretofore, studies of tides and planetary waves have considered these wave components to be linearly independent. To advance our understanding we must now pursue the consequences of nonlinear interactions between these wave components [cf. Teitelbaum and Vial, 1991], and interactions with with the mean dynamics, thermodynamics, and compositional state of the MLT region. General circulation models will undoubtedly play an increasingly important role in elucidating planetary wave triggering mechanisms, and the nonlinear interactions between planetary waves, tides, gravity waves, and the mean circulation. The data in Figure 18 demonstrate the potential influence of planetary waves on the day-to-day variability of atmospheric tides, and underscores the importance of continuous wind and temperature observations. New methods of data analysis must also be explored, such as bispectral estimation which may provide greater insight into the interactions between waves. Finally, considerable progress will not be made until a combined ground-based and satellite-based effort is launched, hopefully in connection with the TIMED mission, to provide the necessary space-time coverage to disentangle the wavenumber/frequency spectra of large-scale waves in the MLT regime.

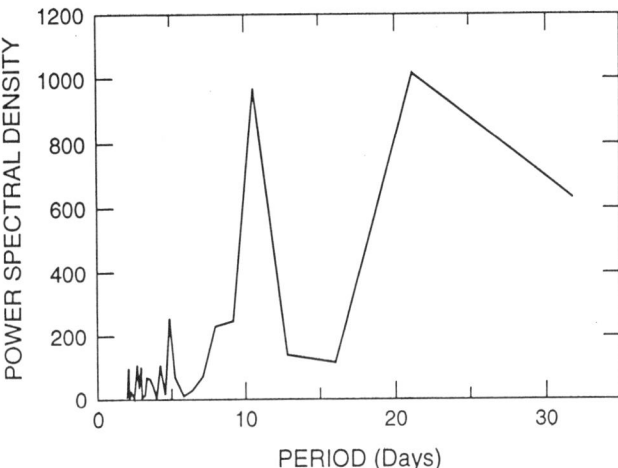

Fig. 18. Power spectrum of daily amplitudes of semidiurnal meridional wind observed near the mesopause over Obninsk, Russia (54°N, 38°E) for January through February, 1979. Data courtesy of Dr. Yu.I. Portnyagin.

Acknowledgments. I am indebted to the reviewers for careful reading of the manuscript; their comments improved its quality considerably. I thank Dr. Roberta Johnson and Dr. Timothy Killeen for inviting me to submit this work, and for their patience in accepting a tardy delivery of final copy. The efforts of Louise Beierle in preparing the camera-ready manuscript under tight schedule is greatly appreciated. The National Center for Atmospheric Research is sponsored by the National Science Foundation. This work was sponsored by Grant ATM-9102200 from the National Science Foundation to Boston University.

REFERENCES

Andrews, D. G., J. R. Holton, and C. B. Leovy, *Middle Atmosphere Dynamics*, p. 164, Academic Press, New York, N.Y., 1987.

Aso, T., T. Nonoyama, and S. Kato, Numerical simulation of semidiurnal atmospheric tides, *J. Geophys. Res., 86,*

11,388-11,400, 1981.

Chapman, S., and R. S. Lindzen, *Atmospheric Tides*, D. Reidel, Hingham, MA, 1970.

Charney, J. G., and P. G. Drazin, Propagation of planetary-scale disturbances from the lower into the upper atmosphere, *J. Geophys. Res.*, *66*, 83-109, 1961.

Dickinson, R. E., Planetary Rossby waves propagating vertically through weak westerly wind wave guides, *J. Atmos. Sci.*, *25*, 984-1002, 1968.

Forbes, J. M., Atmospheric tides, I, Model description and results for the solar diurnal component, *J. Geophys. Res.*, *87*, 5222-5240, 1982a.

Forbes, J. M., Atmospheric tides, II, The solar and lunar semidiurnal components, *J. Geophys. Res.*, *87*, 5241-5252, 1982b.

Forbes, J. M., and H. B. Garrett, Thermal excitation of atmospheric tides due to insolation absorption by H_2O and O_3, *Geophys. Res. Lett.*, *5*, 1013-1016, 1978.

Forbes, J. M., and H. B. Garrett, Theoretical studies of atmospheric tides, *Rev. Geophys. Space Phys.*, *17*, 1951-1981, 1979.

Forbes, J. M., and G. V. Groves, Diurnal propagating tides in the low-latitude middle atmosphere, *J. Atmos. Terr. Phys.*, *49*, 153-164, 1987.

Forbes, J. M., and M. E. Hagan, Thermospheric extensions of the classical expansion functions for semidiurnal tides, *J. Geophys. Res.*, *87*, 5253-5259, 1982.

Forbes, J. M., and M. E. Hagan, Diurnal propagating tide in the presence of mean winds and dissipation: A numerical investigation, *Planet. Space Sci.*, *36*, 579-590, 1987.

Forbes, J. M., and F. Vial, Monthly simulations of the solar semidiurnal tide in the mesosphere and lower thermosphere, *J. Atmos. Terr. Phys.*, *51*, 649-661, 1989.

Forbes, J. M., R. G. Roble, and C. G. Fesen, Acceleration, heating and compositional mixing of the thermosphere due to upward-propagating tides, *J. Geophys. Res.*, *98*, 311-321, 1993.

Forbes, J. M., M. E. Hagan, S. Miyahara, F. Vial, A. Manson, and Yu. I. Portnyagin, Quasi 16-day oscillation in the mesosphere and lower thermosphere, *J. Geophys. Res.*, (in press), 1994.

Gille, J. C., and L. V. Lyjak, An overview of wave-mean flow interactions during the winter of 1978-1979 derived from LIMS observations, in *Dynamics of the Middle Atmosphere*, edited by J. R. Holton and T. Matsuno, Terr Sci. Publ. Co., 1984.

Hagan, M. E., J. M. Forbes, and F. Vial, Numerical investigation of the propagation of the quasi-two-day wave into the lower thermosphere, *J. Geophys. Res.*, *98*, 23,193-23,205, 1993.

Harris, T. J., A long-term study of the quasi-two-day wave in the middle atmosphere, *J. Atmos. Terr. Phys.*, in press, 1994.

Holton, J. R., *The Dynamic Meteorology of the Stratosphere and Mesosphere*, Meteor. Monog. 15(37), Amer. Met. Soc., MA, 1975.

Hough, S. S., On the application of harmonic analysis to the dynamical theory of tides, Part I, On Laplace's 'Oscillations of the First Species,' and on the dynamics of ocean currents, *Phil. Trans. Roy. Soc. London*, *A189*, 201-257, 1897.

Hough, S. S., On the application of harmonic analysis to the dynamical theory of tides, Part II, On the general integration of Laplace's dynamical equations, *Phil. Trans. Roy. Soc. London*, *A191*, 139-185, 1898.

Hunt, B. G., The impact of gravity wave drag and diurnal variability on the general circulation of the middle atmosphere, *J. Meteor. Soc. Japan*, *64*, 1-16, 1986.

Hunt, B. G., A simulation of the gravity wave characteristics and interactions in a diurnally varying model atmosphere, *J. Meteor. Soc. Japan*, *68*, 145-161, 1990.

Jakobs, H. J., and H. Hass, Normal modes as simulated in a three-dimensional circulation model of the middle atmosphere including regional gravity wave activity, *Ann. Geophys.*, *5A(3)*, 103-114, 1987.

Laplace, P. S., *Mechanique Celeste*, 2, pp. 294-298, Paris, France, 1799.

Laplace, P. S., *Mechanique Celeste*, 5, pp. 145-169, Paris, France, 1825.

Larsen, M. F., and I. S. Mikkelsen, The normal modes of the thermosphere, *J. Geophys. Res.*, *92*, 6023-6043, 1987.

Lindzen, R. S., Internal gravity waves in atmospheres with realistic dissipation and temperature, I, Mathematical development and propagation of waves into the thermosphere, *Geophys. Fl. Dyn.*, *1*, 303-355, 1970.

Lindzen, R. S., and D. Blake, Lamb waves in the presence of realistic distributions of temperature and dissipation, *J. Geophys. Res.*, *77*, 2166-2176, 1972.

Lindzen, R. S., and S.-S. Hong, Effects of mean winds and horizontal temperature gradients on solar and lunar tides in the atmosphere, *J. Atmos. Sci.*, *31*, 1421-1466, 1974.

Lindzen, R. S., and D. J. McKenzie, Tidal theory with Newtonian cooling, *Pageoph.*, *66*, 90-96, 1967.

Longuet-Higgins, M. S., The eigenfunctions of Laplace's tidal equation over a sphere, *Phil. Trans. Roy. Soc. London*, *A262*, 511-607, 1968.

Madden, R. A., Observations of large-scale traveling Rossby waves, *Rev. Geophys. Space Phys.*, *17*, 1935-1950, 1979.

Manzini, E., and K. Hamilton, Middle atmosphere traveling waves forced by latent and convective heating, *J. Atmos. Sci.*, *50*, 2180-2200, 1993.

McLandress, C., and N. A. McFarlane, Interactions between orographic gravity wave drag and forced stationary planetary waves in the winter northern hemisphere middle atmosphere, *J. Atmos. Sci.*, *50*, 1966-1990, 1993.

Miyahara, S., Y. Yoshida, and Y. Miyoshi, Dynamic coupling between the lower and upper atmosphere by tides and gravity waves, *J. Atmos. Terr. Phys.*, *55*, 1039-1053, 1993.

Pekeris, C. L., Atmospheric oscillations, *Proc. Roy. Soc. London*, *A158*, 650-671, 1937.

Platzman, G. W., The atmospheric tide as a continuous

spectrum: Lunar semidiurnal tide in surface pressure, *Meteorol. Atmos. Phys.*, *38*, 70-88, 1988.

Plumb, R. A., Baroclinic instability of the summer mesosphere: A mechanism for the quasi-two-day wave?, *J. Atmos. Sci.*, *40*, 262-270, 1983.

Pogorel'tsev, A. I., and S. A. Sukhanova, Simulation of the global structure of stationary planetary waves in the mesosphere and lower thermosphere, *J. Atmos. Terr. Phys.*, *55*, 33-40, 1993.

Roble, R. G., and E. C. Ridley, A thermosphere-ionosphere-mesosphere-electrodynamics general circulation model (TIME-GCM): Equinox solar cycle minimum simulations, *Geophys. Res. Lett.*, *21*, 417-420, 1994.

Salby, M. L., On the solution of the homogeneous vertical structure problem for long-period oscillations, *J. Atmos. Sci.*, *36*, 2350-2359, 1979.

Salby, M. L., The influence of realistic dissipation on planetary normal structures, *J. Atmos. Sci.*, *37*, 2186-2199, 1980.

Salby, M. L., Rossby normal modes in nonuniform background configurations, Part I, Simple fields, *J. Atmos. Sci.*, *38*, 1803-1826, 1981a.

Salby, M. L., Rossby normal modes in nonuniform background configurations, Part II, Equinox and solstice conditions, *J. Atmos. Sci.*, *38*, 1827-1840, 1981b.

Salby, M. L., The 2-day wave in the middle atmosphere: Observations and theory, *J. Geophys. Res.*, *86*, 9654-9660, 1981c.

Salby, M. L., Survey of planetary-scale traveling waves: The state of theory and observation, *Rev. Geophys. Space. Phys.*, *22*, 209-236, 1984.

Smith, A. K., Wave transience and wave-mean flow interactions caused by the interference of stationary and traveling waves, *J. Atmos. Sci.*, *42*, 529-535, 1985.

Taylor, G. I., The oscillations of the atmosphere, *Proc. Roy. Soc. London*, *A156*, 318-326, 1936.

Teitelbaum, H., and F. Vial, On tidal variability induced by non-linear interaction with planetary waves, *J. Geophys. Res.*, *96*, 14,169-14,178, 1991.

Vial, F., Numerical simulations of atmospheric tides for solstice conditions, *J. Geophys. Res.*, *91*, 8955-8969, 1986.

Volland, H., *Atmospheric Tidal and Planetary Waves*, Kluwer Academic Publ., Boston, MA, 1988.

Walterscheid, R. L., Traveling planetary waves in the stratosphere, *Pageoph.*, *118*, 239-265, 1980.

Walterscheid, R. L., and S. V. Venkateswaran, Influence of mean zonal motion and meridional temperature gradients on the solar semidiurnal atmospheric tide: A spectral study, Part 1, Theory, *J. Atmos. Sci.*, *36*, 1623-1635, 1979a.

Walterscheid, R. L., and S. V. Venkateswaran, Influence of mean zonal motion and meridional temperature gradients on the solar semidiurnal atmospheric tide: A spectral study, Part 2, Numerical results, *J. Atmos. Sci.*, *36*, 1636-1662, 1979b.

Walterscheid, R. L., J. G. DeVore, and S. V. Venkateswaran, Influence of mean zonal motion and meridional temperature gradients on the solar semidiurnal atmospheric tide: A revised spectral study with improved heating rates, *J. Atmos. Sci.*, *37*, 455-470, 1980.

J. M. Forbes, Department of Aerospace Engineering Sciences, Campus Box 429, University of Colorado, Boulder, CO 80309-0429.

Gravity Wave Forcing and Effects in the Mesosphere and Lower Thermosphere

DAVID C. FRITTS

Laboratory for Atmospheric and Space Physics and Department of Electrical and Computer Engineering,
University of Colorado, Boulder, CO 80309

Gravity waves are now widely recognized to play a major role in the forcing, structure, and variability of the mesosphere and lower thermosphere. The intent of this paper is to provide a simple understanding of the reasons for their importance, the processes accounting for their observed characteristics and effects, and their forcing of and variability at these altitudes. Gravity waves are important dynamically, not because they are dominant energetically, but because they contribute the majority of the energy and momentum fluxes above the stratosphere. Wave dissipation causes flux divergences which lead to local heating, turbulent diffusion, and accelerations of the local mean flow. Observations suggest an approximately universal gravity wave spectral shape and systematic variations in spectral energy with height. Together, these permit an assessment of mean and variable wave forcing of the middle atmosphere. Also discussed are the factors believed to contribute most to gravity wave variability at these heights.

1. INTRODUCTION

Gravity waves in the mesosphere and lower thermosphere (MLT) have been a subject of study since they were first suggested to account for observed structure in this region by *Hines* [1960]. More recently, they have been recognized to contribute significantly both to the large-scale circulation and the thermal and constituent structures of the MLT and to the variability of this region on many temporal and spatial scales. The importance of gravity waves in the MLT is due to a variety of factors, including their excitation by a wide range of sources throughout the lower and middle atmosphere, their rapid vertical propagation and corresponding large vertical fluxes of energy and momentum, their potential for strong wave-wave and wave-mean flow interactions, and their dissipation via nonlinear interactions and wave breaking processes which cause flux divergences, turbulent heating and diffusion, and accelerations of the local mean flow.

The above processes have important consequences for the mean MLT circulation and structure, including large departures from geostrophic balance, closure of the mesospheric jets in the summer and winter hemispheres, a strong induced residual (meridional and vertical) circulation, and reversal of the mean meridional temperature gradient near the mesopause. These effects are due primarily to gravity wave momentum transports, dissipation, and their induced zonal mean accelerations. On smaller spatial and temporal scales, the same processes account for large variability of the tidal and planetary wave fields, highly variable turbulent diffusion and mixing, and potentially large forcing of the local mean flow in response to strong episodic and spatially localized gravity wave sources at lower altitudes.

The purposes of this paper are to outline as simply as possible 1) the reasons for the importance of gravity waves in the MLT, 2) the spectral character of the motion field and its implications for wave transports of energy and momentum, and 3) the consequences of wave dissipation for the wave field evolution and wave forcing of the circulation and structure of the MLT. We will begin by considering in section 2 the implications of the zonal-mean momentum and continuity equations. These reveal a direct link between gravity wave momentum fluxes and the induced residual circulation. We then examine simple gravity wave structure and the spectral character of gravity wave motions in section 3. These show the wave fluxes of energy and momentum to be dominated by motions with high intrinsic frequencies and imply a potential for interactions and variability on many time scales. The magnitude of mean wave forcing is reviewed in section 4

The Upper Mesosphere and Lower Thermosphere:
A Review of Experiment and Theory
Geophysical Monograph 87
Copyright 1995 by the American Geophysical Union

and seen largely to be consistent with the observed mean circulation and thermal structure. Finally, we examine in section 5 the most likely sources of variability of wave forcing and effects in the MLT. These suggest strong vertical coupling and sensitivity of the MLT both to gravity wave sources at lower levels and to the intervening wind and thermal fields imposed by the mean structure and low-frequency wave motions. Our conclusions are presented in section 6.

2. Zonal-Mean Equations

We begin by considering the steady, mean zonal momentum and continuity equations in spherical coordinates. These may be written in the approximate form

$$\overline{v}^*[\frac{1}{cos\phi}\frac{\partial}{\partial y}(\overline{u}cos\phi)-f]+\overline{w}^*\overline{u}_z = -\frac{1}{\rho_0(z)}\frac{\partial}{\partial z}(\rho_0 S) \quad (1)$$

with

$$S = \overline{u'w'} + (\overline{u}_y - f)\overline{v'\theta'}/\overline{\theta}_z \quad (2)$$

and

$$\frac{\rho_0(z)}{cos\phi}\frac{\partial}{\partial y}(\overline{v}^* cos\phi) + \frac{\partial}{\partial z}(\rho_0 \overline{w}^*) = 0, \quad (3)$$

where $\mathbf{u} = (u,v,w)$, θ is potential temperature, the right hand side of eq. (1) is the zonal body force per unit mass induced by the vertical flux of zonal momentum and the meridional flux of heat by the wave field, $\rho_0(z)$ is the mean state density at height z, $f = 2\Omega sin\phi$ is the Coriolis parameter, ϕ is latitude, primes and overbars denote perturbation and mean quantities, subscripts denote derivatives, and the * superscripts denote the residual diabatic circulation induced by wave forcing [*Andrews and McIntyre*, 1976].

We now assume for simplicity that $\overline{u}_y$ and $\overline{u}_z \simeq 0$ and that gravity waves satisfy a dispersion relation of the form

$$m^2 = k^2(N^2 - \omega^2)/(\omega^2 - f^2), \quad (4)$$

where k and m are the horizontal and vertical wavenumbers and N and $\omega = k(c - \overline{u})$ are the buoyancy and intrinsic frequencies of the wave motion. This form assumes that motions may be described using the Boussinesq approximation and that the wave environment is uniform. Also using the linearized adiabatic energy equation,

$$\theta'_t + \overline{u}\theta'_x + w'\overline{\theta}_z = 0, \quad (5)$$

eqs. (1) and (2) may be written

$$-f\overline{v}^* \simeq -\frac{1}{\rho_0(z)}\frac{\partial}{\partial z}(\rho_0 \overline{u'w'})(1 - f^2/\omega^2). \quad (6)$$

As will be seen below, the vertical fluxes of horizontal momentum are dominated by wave motions at high intrinsic frequencies so that we may neglect the term in eq. (5) varying as f^2/ω^2 to a good approximation [*Fritts and Vincent*, 1987].

Then substituting for $\overline{v}^*$ from eq. (6) into eq. (3) and integrating from height z to ∞ yields approximately

$$\overline{w}^* \simeq \frac{-1}{cos\phi}\frac{\partial}{\partial y}(\frac{cos\phi}{f}\overline{u'w'}). \quad (7)$$

This relation, known as the "downward control principle" following *McIntyre* [1989] and *Haynes et al.* [1991], represents a powerful constraint on extratropical motions and provides a direct link between the strength of wave forcing and the magnitude of the wave-induced diabatic circulation. Specifically, the induced vertical motion is related to the meridional gradient of the wave momentum flux (per unit mass) at that level and is independent of the height or manner of wave dissipation above.

We now consider, as an example, the implications of eq. (7) for the high-latitude summer mesopause region, where observations reveal momentum fluxes of ~ 10 to $20\ m^2 s^{-2}$ over a depth of $\sim 10\ km$ about the mesopause [*Reid et al.*, 1988; *Fritts and Yuan*, 1989; *Wang and Fritts*, 1990]. At these latitudes $sin\phi \simeq 1$, $cos\phi \simeq \phi$, and $\overline{u'w'}$ must vanish at the pole, so that

$$\overline{w}^* \simeq \frac{(\overline{u'w'})_0}{a\phi_0\Omega}, \quad (8)$$

where a is the radius of the earth and subscripts denote values at the latitude at which the momentum flux is specified. Using values appropriate for Poker Flat, Alaska ($\phi_0 = 65°$), we infer a vertical velocity that is approximately uniform at polar latitudes of $\overline{w}^* \simeq 2$ to $4\ cm\ s^{-1}$, in reasonable agreement with the values needed to account for the observed thermal structure of this region [*Garcia*, 1989].

3. Gravity Wave Spectra and Implications

Gravity wave structure is well known and provides useful insights into those motions likely to contribute preferentially to wave energy and momentum fluxes in the atmosphere. Assuming, for example, that motions are hydrostatic and two dimensional ($f^2 << \omega^2 << N^2$), the horizontal and vertical perturbation velocities are related through the polarization relations [*Gossard and Hooke*, 1975] and given by

$$u' = -mw'/k = Nw'/\omega, \quad (9)$$

where we have taken k and m positive eastward and upward for $\omega > 0$ corresponding to downward energy propagation. Likewise, the upward energy flux and zonal momentum flux (per unit mass) for a single wave motion may be written

$$F_E = c_{gz}E = -\frac{\omega}{m}E \quad (10)$$

and

$$F_{Px} = \overline{u'w'} = \frac{\omega}{N}\overline{u'^2}, \qquad (11)$$

where c_{gz} and E are the vertical group velocity and total (kinetic plus potential) energy density of the wave. For a given kinetic energy density, $E \simeq \rho\overline{u'^2}/2$, then, these fluxes clearly are dominated by those motions with large intrinsic frequencies and small vertical wavenumbers (large vertical scales, $\lambda_z = 2\pi/m$).

In order to quantify these arguments, we must consider the spectral character of gravity wave motions in the atmosphere. Fortunately, considerable efforts have been devoted to observational and theoretical studies of these spectra, and their mean forms are now reasonably well defined. There is even some evidence suggesting that the intrinsic frequency and vertical wavenumber spectra are approximately separable [*Fritts and Chou*, 1987], though there is no reason to expect this to be the case on theoretical grounds. Nevertheless, this proves to be a useful assumption and we adopt it for our purposes, writing the spectral energy density following *Garrett and Munk* [1975] and *VanZandt* [1982] as

$$E(\mu, \omega) = E_0 A(\mu) B(\omega). \qquad (12)$$

Here E_0 is the total energy for the complete gravity wave spectrum, $\mu = m/m_*$ is the vertical wavenumber scaled by a characteristic value (see below), and $A(\mu)$ and $B(\omega)$ express the dependencies on m and ω and are each normalized to unity after integration over the spectrum.

3.1. Vertical Wavenumber Spectrum

Of the two spectra, the vertical wavenumber spectrum is best understood. Numerous observational studies have shown the m spectrum of horizontal velocity (or equivalently temperature, potential temperature, or density) to exhibit a saturated form at large m, with a slope and amplitude given approximately by

$$E_u(m) \sim N^2/6m^3 \qquad (13)$$

[*Dewan*, 1979; *Fritts and Chou*, 1987; *Fritts et al.*, 1988; *Tsuda et al.*, 1989; *Tsuda et al.*, 1991; *Wu and Widdel*, 1989a; *Kwon et al.*, 1990; *Wu and Widdel*, 1992; *Kuo et al.*, 1992]. These observations have received some theoretical confirmation, with various wave saturation theories predicting generally comparable slopes and amplitudes, despite widely varying assumptions and formulations [*Dewan and Good*, 1986; *Smith et al.*, 1987; *Sidi et al.*, 1988; *VanZandt and Fritts*, 1989; *Hines*, 1991]. There remains at this time considerable uncertainty over the processes contributing most to spectral evolution and maintenance of the mean spectral shape. Recent numerical studies, however, have demonstrated the validity of saturation (instability) processes and the associated wave amplitude constraints and may also provide a means of examining more directly the competing saturation theories [*Andreassen et al.*, 1994; *Fritts et al.*, 1994].

In contrast to the approximately constant amplitude of the vertical wavenumber spectrum at large m, that at small m displays a considerable increase with increasing altitude, growing only slightly less than implied for conservative wave propagation. The net effect of these disparate behaviors at large and small m is a vertical wavenumber spectrum that remains nearly invariant in shape, but with a characteristic, or transition, wavenumber separating these two ranges. This transition wavenumber decreases with increasing altitude with a scale height substantially larger than that describing the growth of energy density (per unit mass) with altitude, $E_0(z) \sim e^{z/H_E}$, with $H_E > H$ and H the density scale height. An example of this behavior observed with the MU radar in Shigaraki, Japan ($35^\circ N$) by *Tsuda et al.* [1989] is shown for reference in Figure 1. Note here, despite the decrease in density from the troposphere to the mesosphere by $\sim 10^4$, that the spectral energy density at large m remains essentially unchanged, apart from the variations with N anticipated by eq. (13). At small m, in contrast, there is evidence of a considerable increase in spectral energy density (per unit mass) with altitude and a corresponding decrease in the transition wavenumber, m_*.

Theory does not as yet provide a description of the shape of the vertical wavenumber spectrum, apart from the saturated range at large m. However, observations suggest consistency with a wavenumber spectrum of the form

$$A(\mu) = A_0 \frac{\mu^s}{1 + \mu^{s+t}}, \qquad (14)$$

where again $\mu = m/m_*$, A_0 is a coefficient normalizing the integrated spectrum to unity, m_* describes the transition between the ranges with $A(\mu) \sim \mu^s$ and μ^{-t}, $t \simeq 3$, and a smaller, positive slope of $s \sim 1$ is inferred at $\mu \ll 1$ ($m \ll m_*$) from limited observations [*Fritts and Chou*, 1987; *Fritts et al.*, 1988; *Tsuda et al.*, 1991] and the requirement that the vertical wave energy (or wave action) flux remain finite [*VanZandt and Fritts*, 1989]. It must be noted, however, that there are at present much firmer observational bounds on the value of t than of s. With these limiting slopes, m_* represents the most energetic wavenumber and varies with altitude as $m_* \sim e^{-z/H_*}$, with $H_* = 2H_E$ [*Fritts and VanZandt*, 1993]. In general, both E_0 and m_* also vary somewhat with variations in N due to the tendency for increased (decreased) dissipation when N is increasing (decreasing) with altitude [*VanZandt and Fritts*, 1989; *Fritts and VanZandt*, 1993]. These variations represent subtleties in the theory, however, and will not be discussed here. The interested reader is referred instead to the papers cited above.

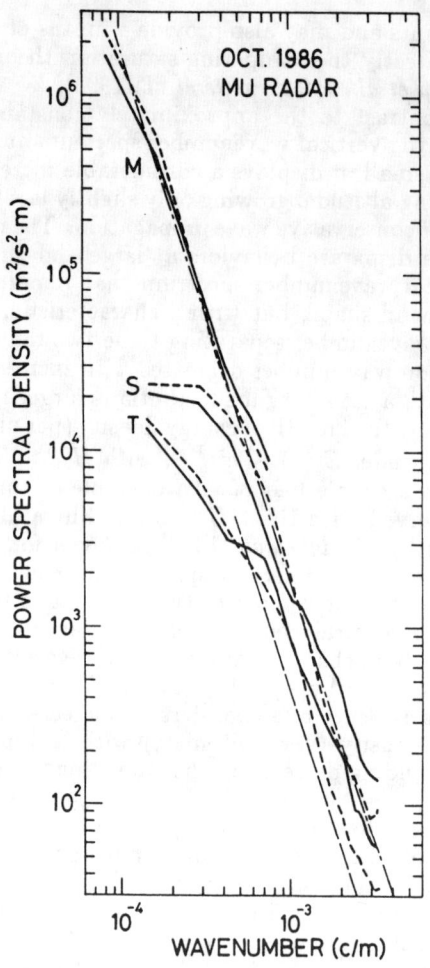

Figure 1. Vertical wavenumber spectra in the troposphere (T), stratosphere (S), and mesosphere (M) obtained with the MU radar. Straight dashed lines show the saturation limit given by eq. (13) for each range.

3.2. Frequency Spectra

Frequency spectra of atmospheric gravity waves are usually obtained at fixed locations and are therefore observed rather than intrinsic spectra. This causes ambiguities in the inference of the intrinsic properties of the motion field. There is, nevertheless, some evidence that the intrinsic frequency spectra of gravity waves are approximately uniform with altitude in the mean. Relatively few studies of the effects of Doppler shifting of frequency spectra have been performed to date. Those that have suggest that frequency spectra of horizontal velocities are much less sensitive to Doppler shifting than those of vertical velocities [*Scheffler and Liu*, 1986; *Fritts and VanZandt*, 1987]. The reasons become obvious when we consider the forms of these spectra. For convenience, we will discuss observed frequency spectra first.

Observed frequency spectra of horizontal, vertical, and more commonly oblique (or radial) velocities have been compiled for a number of years using a variety of measurement systems. In most cases, the oblique spectra have been more representative of horizontal than of vertical velocities because of the dominance of the wave field by horizontal motions. For the most part, horizontal or oblique spectra have displayed clear tidal peaks and an apparent gravity wave continuum at higher frequencies with typical slopes of ~ -3/2 to -2 [*Balsley and Carter*, 1982; *Vincent*, 1984; *Balsley and Garello*, 1985; *Fritts and Chou*, 1987; *Vincent and Lesicar*, 1991; *Fritts and Isler*, 1992]. Examples of the zonal and meridional velocity spectra obtained with the MF radar at Kauai, Hawaii ($22°N$) are shown for reference in Figure 2. These display significant tidal and planetary wave structure at lower frequencies and a gravity wave continuum at higher frequencies.

Observed frequency spectra of vertical velocities, in contrast, have been found to be highly variable, with slopes and amplitudes dependent on several factors. Under strong wind conditions (large Doppler shifting effects), these spectra display a negative slope comparable to that of horizontal velocity spectra [*Ecklund et al.*, 1986; *Fritts et al.*, 1990]. A similar behavior is generally noted in close proximity to mountains, which likely act as a strong source of wave activity at phase speeds near zero [*Ecklund et al.*, 1981]. Under light wind conditions, however, the spectral shape is very different and generally exhibits a weak, positive slope with a maximum near or slightly below the local buoyancy frequency [*Ecklund et al.*, 1986; *Fritts et al.*, 1990].

It was noted by *VanZandt et al.* [1991] that observed tropospheric vertical velocity spectra were in good agreement with predictions of the Doppler-shifting model by *Fritts and VanZandt* [1987], suggesting that the assumption of a constant intrinsic frequency spectrum by those authors was at least partially justified. Similar results were obtained in a study of horizontal and vertical velocity spectra and Doppler shifting near the mesopause by *Fritts and Wang* [1991]. It was found here that observed vertical velocity spectra were highly variable, but consistent with the predictions of Doppler shifting by *Fritts and VanZandt* [1987] assuming a constant intrinsic frequency spectrum and wave motions propagating primarily against the local mean flow. Fortunately, this latter assumption is itself consistent with numerous observations of gravity wave momentum fluxes (see section 4 below). Frequency spectra of horizontal velocity, by comparison, were found to be nearly invariant with Doppler shifting. Thus, there is clear, but limited, evidence at present suggesting that a model that assumes uniform intrinsic frequency spectra for gravity wave motions has some observational validity. It should be noted again, however, that there is no theoretical ba-

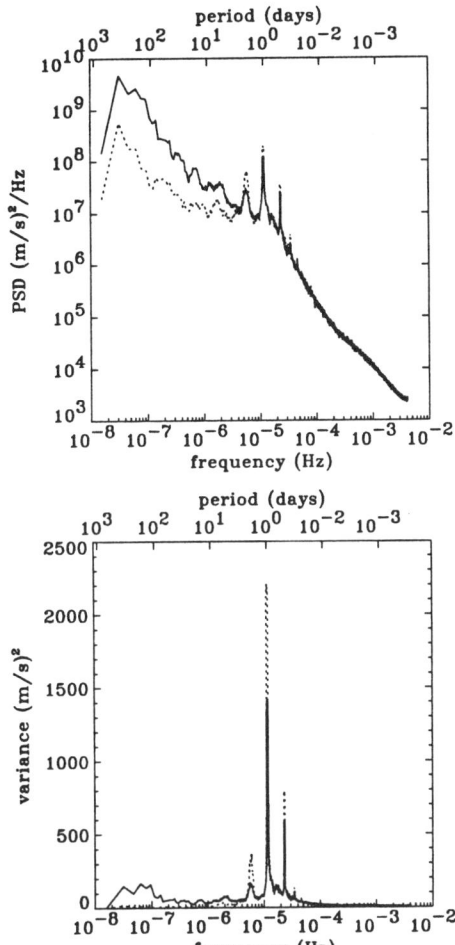

Figure 2. Frequency spectra of zonal (solid) and meridional (dashed) velocities in standard (a) and energy-content (b) forms for a 2-year data set obtained with the Hawaii MF radar.

sis at present for the universality of these spectra or for their apparent separability from the vertical wavenumber dependence.

To understand the different responses of the horizontal and vertical frequency spectra to Doppler shifting, we refer to eq. (4) and note that this implies a dominance of the motion field by the horizontal velocity for $\omega \ll N$ and by the vertical velocity for $\omega \sim N$. Despite their different intrinsic frequencies, all motions with a common vertical scale have comparable characteristic intrinsic phase speeds, $c_* \simeq N/m_*$, because of separability and are thus equally subject to Doppler shifting. But a mean wind $\overline{u} = c_*$, implying a change in frequency comparable to ω, can Doppler shift intrinsic frequencies near N, where most of the energy in the intrinsic frequency spectrum of vertical velocity resides, to observed frequencies near ~ 0 or $2N$, resulting in dramatic influences on the shape of the observed frequency spectrum. In contrast, a frequency shift of $\sim f$, where the maximum energy in the intrinsic frequency spectrum of horizontal velocity resides, can Doppler shift wave energy to observed frequencies of only ~ 0 or $2f \ll N$, resulting in relative insensitivity of this spectrum to Doppler shifting.

A model spectrum which appears to describe well the form of the inferred intrinsic frequency spectrum of total wave energy is

$$B(\omega) = B_0 \omega^{-p}, \qquad (15)$$

where $p = 5/3$ is assumed to provide the best fit to observational data following *VanZandt and Fritts* [1989] and *Fritts and VanZandt* [1993]. As in eq. (14) above, B_0 is a coefficient normalizing the integrated spectrum to unity. Together, the vertical wavenumber and intrinsic frequency spectra of total energy and the polarization relations for gravity waves fully specify the dependence of all components of the wave field on m and ω [*VanZandt and Fritts*, 1989; *Fritts and VanZandt*, 1993].

3.3. Spectral Implications for Wave Fluxes

Following the development by *Fritts and VanZandt* [1993], but assuming only a single direction of wave propagation for simplicity, the spectral energy and momentum fluxes may be expressed in terms of the above spectra as

$$F_E(\mu, \omega) = c_{gz} E(\mu, \omega) = -\frac{\omega}{m} E(\mu, \omega) \delta_- \gamma \qquad (16)$$

and

$$F_{Px} = \frac{\omega}{N} E(\mu, \omega)(\delta_+ \delta_- \gamma)^{1/2}, \qquad (17)$$

where $\delta_\pm(\omega) = 1 \pm (f/\omega)^2$, $\gamma(\omega) = 1 - (\omega/N)^2$, and we have neglected terms of order $(f/N)^2$.

Assuming canonical values of f and N with $f/N = 1/200$ and integrating over all m and ω, we obtain the approximate relations

$$F_E = E_0 \overline{(1/m)\omega} \simeq \frac{N E_0}{18 m_*} \qquad (18)$$

and

$$F_{Px} = E_0 \frac{\overline{\omega}}{N} \simeq \frac{E_0}{22}. \qquad (19)$$

The factors $N/18m_*$ and $N/22$ in each case may be interpreted as an effective vertical group velocity and an effective intrinsic frequency for wave fluxes of energy and momentum, respectively. Thus, the dominant fluxes are accomplished by gravity waves with characteristic periods of ~ 1 to 2 hours and vertical scales representative of the most energetic spectral components. It should also be emphasized here that the wave frequencies characteristic of the dominant fluxes are substantially larger than those most representative of the dominant wave energy densities. As an example, $\sim 70\%$ of wave momentum fluxes occur at wave periods less than 1 hour, while $\sim 90\%$ of wave energy occurs at periods longer than 1 hour. It is also interesting to note that the energy flux scales as E_0/m_* and thus increases more rapidly than the momentum flux with increasing altitude, due to the simultaneous increase of E_0 and decrease of m_*. This is because a fixed frequency spectrum and a decreasing m_* imply an increasing effective vertical group velocity with altitude.

Observations suggest values for E_0 and m_* near the tropopause and mesopause of $E_0 \sim 10$ and $10^3 \; m^2 s^{-2}$ and $m_* \sim 1/3$ and $1/30 \; km^{-1}$, respectively. These imply corresponding energy and maximum momentum fluxes of $F_E \sim 5$ and $5 \times 10^3 \; m^3 s^{-3}$ and $F_{Px} \sim 0.5$ and $50 \; m^2 s^{-2}$ which are in reasonable agreement with observed values (see below).

4. Gravity Wave Forcing of the MLT

Returning to eq. (6) and assuming a constant $\overline{u'w'}$ with height for simplicity, we see that a maximum momentum flux of $\sim 0.5 \; m^2 s^{-2}$ implies a compensating Coriolis torque (or an equivalent acceleration) of $\sim 6 \; m \; s^{-1} day^{-1}$. In reality, gravity waves propagate in all directions, so the mean flux and implied acceleration are likely to be somewhat smaller. These magnitudes are easily balanced by a weak diabatic circulation in association with the tropospheric jet structure. At mesospheric heights, however, the maximum momentum flux inferred from spectral estimates implies a Coriolis torque (or acceleration) of $\sim 600 \; m \; s^{-1} day^{-1}$ and dramatic influences on the circulation and structure of this region. Gravity wave fluxes of energy are likewise anticipated to be large in the MLT and are expected to contribute significantly to turbulent mixing and diffusion. The purpose of this section is to review the existing evidence of gravity wave forcing of the MLT and compare observed values with those inferred from the spectral form of the motion field above.

Measurements of momentum fluxes in the MLT are relatively limited at this time, due to the existence of only a few systems able to perform such studies. Nevertheless, these measurements display a high degree of consistency in the mean and maximum values observed and suggest that gravity wave forcing is indeed strong and highly variable. In all observations, momentum fluxes increase with altitude and exhibit maximum values in the MLT, due to the increasing gravity wave scales and energies (per unit mass). Typical mean momentum fluxes vary from ~ 5 to $15 \; m^2 s^{-2}$ and appear to be somewhat larger at low and high than at middle latitudes [*Vincent and Reid*, 1983; *Fritts and Vincent*, 1987; *Reid and Vincent*, 1987; *Fritts and Yuan*, 1989; *Meyer et al.*, 1989; *Rüster and Reid*, 1990; *Hitchman et al.*, 1992]. An example of the momentum flux pro-

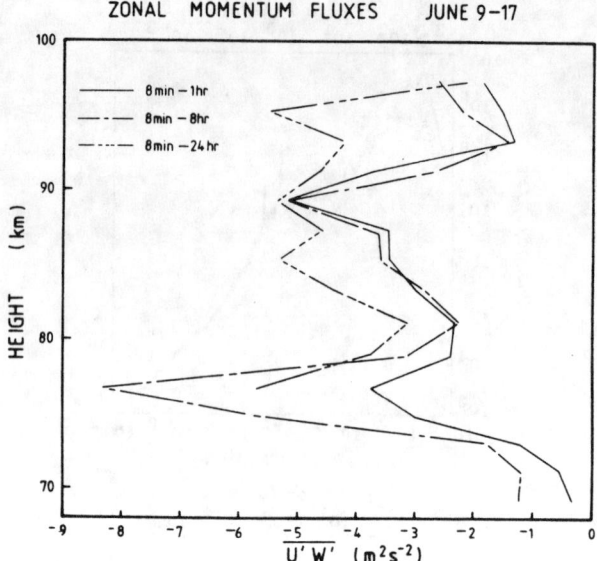

Figure 3. Zonal momentum flux profiles obtained for 8 days during June 1984 (winter) with the MF radar at Adelaide, Australia. Note that periods less than 1 hour account for $\sim 70\%$ of the total flux.

file obtained with the MF radar at Adelaide, Australia ($35°S$) during June 1984 (winter) is displayed in Figure 3 and shows clearly the dominance of the total flux by those motions with periods less than 1 hour. A corresponding mean profile obtained during summer with the MST radar at Poker Flat, Alaska ($65°N$) is shown in Figure 4 and exhibits the mesopause maximum which drives the strong residual circulation discussed in section 2 above.

At middle and high latitudes, momentum fluxes tend to peak during summer near the mesopause in the region of strong eastward zonal mean shear above the westward jet maximum. During winter, the maximum fluxes occur at comparable heights, but appear to be somewhat smaller in magnitude [*Tsuda et al.*, 1990; *Wang and Fritts*, 1990]. In both seasons, there is clear evidence that the mean momentum fluxes are acting in the manner required to close the mesospheric jets and account for the observed residual cirulation and thermal structure (positive fluxes in summer and negative in winter). Finally, there is some evidence that gravity wave energies and momentum fluxes and the wave-driven residual circulation are smaller on average in the Southern Hemisphere than in the Northern Hemisphere [*Vincent et al.*, 1988]. At this stage, however, it is not possible to say with any certainty that these trends will persist when more data are available.

Daily and hourly fluxes may achieve much larger values, exceeding ~ 30 and $60 \; m^2 s^{-2}$, respectively, and exhibiting consistency between adjacent profiles and an anti-phase relationship with the local mean winds

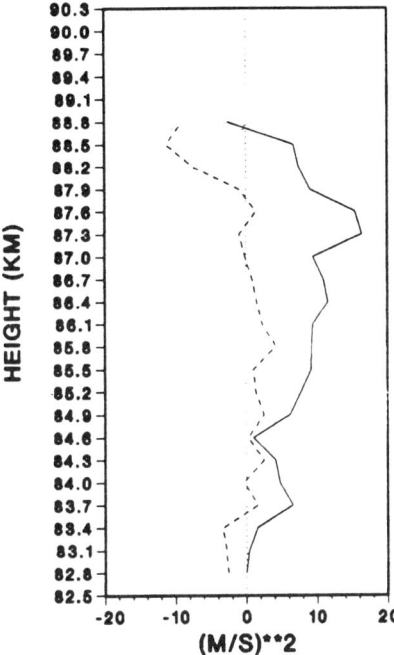

Figure 4. Zonal (solid) and meridional (dashed) momentum flux profiles obtained for 8 days during July 1986 with the MST radar at Poker Flat, Alaska. Note the limited depth over which fluxes are large.

[*Fritts and Vincent*, 1987; *Reid et al.*, 1988; *Fritts and Yuan*, 1989; *Fritts and Wang*, 1991; *Fritts et al.*, 1992]. An example of this variability obtained during summer using the Poker Flat MST radar in Alaska is displayed in Figure 5 and shows large positive (meridional) fluxes at times of strong southward winds and large vertical wind shears. Similar, though smaller, positive zonal fluxes are also seen to accompany westward mean winds during the same interval.

Energy fluxes and divergences are generally more difficult to quantify and have, at present, poorly understood influences on the mean profiles of temperature and constituents due to the uncertainties concerning the processes contributing most to turbulence production and mixing. Nevertheless, it is possible to compare the spectral predictions discussed above with those observations of energy dissipation rates that are available. The spectral model by *Fritts and VanZandt* [1993] predicts maximum values of the energy dissipation rate of $\epsilon \sim 0.1$ to $0.3 \ m^2 s^{-3}$, depending on the energy assigned to the initial wave spectrum and its variations with altitude. Estimates based on $in-situ$ rocket measurements of fine structure and radar spectral widths are typically smaller, but with maxima in this range [*Hocking*, 1985; *Thrane et al.*, 1985; *Lübken et al.*, 1987; *Watkins et al.*, 1988; *Wu and Widdel*, 1989b; *Blix et al.*, 1990]. It must be noted, however, that the spectral estimates are almost certainly high, as there are various processes that will compete with turbulence generation in removing available energy from the wave field, but which were not accounted for in the spectral model. These include the vertical (downward) reflection of wave energy, transfers of energy to other components of the motion spectrum that do not result in small-scale turbulence, i.e., 2-D or geostrophic turbulence, wave-mean flow interactions which may reduce wave energy prior to instability, and radiative damping of wave motions occurring at lower frequencies but having substantial energy densities. With these effects taken into account, there is general consistency between the spectral model of gravity wave effects and observations in the MLT.

5. VARIABILITY OF WAVE FORCING AND EFFECTS

It should now be clear that gravity wave influences are strong and highly variable in the MLT region. Perhaps less obvious are the sources of this variability. Nevertheless, observations and theory have suggested a number of candidates. One alluded to above is the filtering of gravity waves, and thus a modulation of gravity wave momentum fluxes, by background winds or low-frequency motions. A second is the variability imposed at MLT altitudes by temporally and spatially variable sources of gravity wave activity at lower levels of the atmosphere.

Gravity wave propagation is strongly influenced by the local shear environment, resulting in significant anisotropy (preferred directions of propagation and momentum fluxes) of the motion field. Because tidal structures achieve large amplitudes in the MLT (see Figure 2) and occur at frequencies substantially below those characteristic of gravity wave energy and momentum transports, they have the potential to modulate strongly these gravity wave fluxes. Observations have provided evidence of this modulation, with short-period momentum fluxes large and anticorrelated with tidal wind fields when tidal amplitudes are large [*Fritts and Vincent*, 1987; *Wang and Fritts*, 1991]. Parallel theoretical and modeling efforts have confirmed the importance of these interactions, but have yet to account fully for the observed variations in forcing, turbulent diffusion, and influences on the mean and tidal structures themselves [*Walterscheid*, 1981; *Fritts and Vincent*, 1987; *Forbes et al.*, 1991; *Miyahara and Forbes*, 1991; *Fritts and Lu*, 1993; *Lu and Fritts*, 1993]. A more complete discussion of gravity wave-tidal interactions is provided by *Fritts* [1994] separately in this volume.

Gravity wave sources in the lower atmosphere have received considerable observational and theoretical attention in recent years due to the increasing awareness

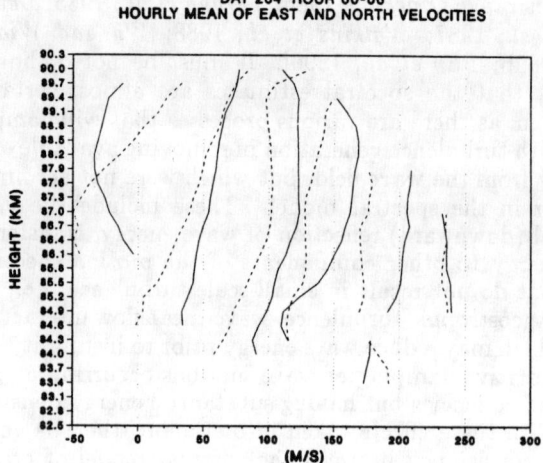

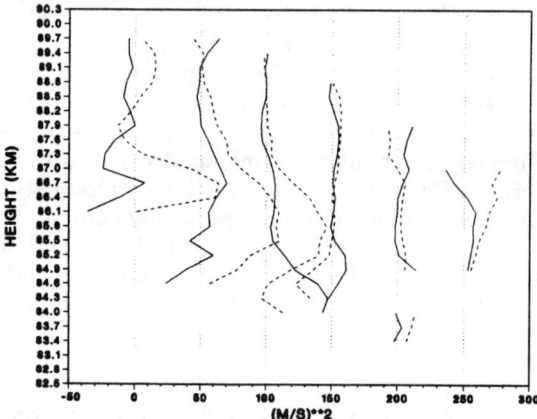

Figure 5. Hourly zonal (solid) and meridional (dashed) winds (upper) and momentum fluxes (lower) obtained during July 1986 at Poker Flat, Alaska.

of their importance in a variety of atmospheric processes. As a result, the literature is too extensive to review here. We will focus instead on those observational studies that delineate specific source strengths in a statistical manner and refer the interested reader to the more complete discussions of gravity wave sources provided by *Fritts and Nastrom* [1992] and *Fritts* [1993].

From an MLT perspective, we are interested both in the gravity wave strengths and characteristics imposed by various wave sources at lower levels and in the ability of those gravity waves to propagate vertically and influence the atmosphere above. The major gravity wave sources at lower levels are believed to include topography, frontal and convective activity, and wind shear. Of these, topography has received the greatest attention, with extensive theoretical and observational efforts over several decades (see the discussion

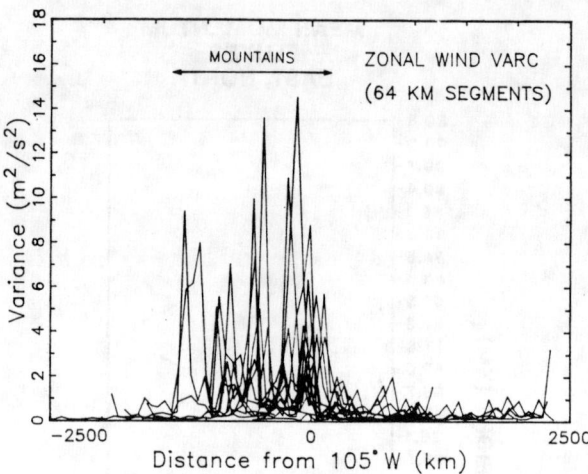

Figure 6. Zonal wind variances for 64-km data segments obtained from GASP data and showing large enhancements over topography in the Western U.S.

by *Fritts and Lu* [1993]). These have revealed considerable enhancements in gravity wave energy densities and momentum fluxes under conditions of strong forcing, but have not addressed the statistical effects of this source. Likewise, individual studies of frontal and convective excitation have shown strong gravity wave responses [*Nastrom et al.*, 1990; *Fovell et al.*, 1992; *Eckermann and Vincent*, 1993] under specific conditions. It has proven difficult, however, to assess relative source strengths with measurements at a single location. To do this, it has been necessary to consider other, more global, data sets or intercompare observations at various locations.

Evidence of possible hemispheric differences in the strengths of gravity wave sources is provided indirectly by the tidal observations at conjugate sites by *Vincent et al.* [1988], which show larger tidal amplitudes in the Southern than in the Northern Hemisphere and suggest less gravity wave-induced diffusion and damping in the Southern Hemisphere as a result. Other observations comparing the energy in the motion spectrum over oceans and continental land masses also suggest statistical enhancements over land at MLT altitudes [*Fritts et al.*, 1989; *Vincent and Lesicar*, 1991]. Whether these differences reflect variable mean source strengths at lower levels or differing influences by the intervening atmospheric structure or wave interactions in the two hemispheres is not yet known.

Finally, a global aircraft data set (GASP) has been used to assess individual source responses, focussing initially on topography and then expanding the analysis to include frontal/convective excitation and wind shear [*Nastrom et al.*, 1987; *Jasperson et al.*, 1990; *Nastrom and Fritts*, 1992; *Fritts and Nastrom*, 1992]. These efforts have helped to clarify the relative strengths of these three sources of wave activity at small horizontal scales. An example of the statistical response of the motion field to topography over the Western U.S. assembled by *Nastrom and Fritts* [1992] is displayed in Figure 6 and reveals a dramatic enhancement in the wave variances over significant topography relative to those measured elsewhere. Similar enhancements were also noted in association with frontal and convective systems and regions of strong wind shear [*Fritts and Nastrom*, 1992].

The collective influences of these sources averaged over many flights are summarized and compared with no-source regions using mean wind and temperature variances for both 64- and 256-km segments in Figure 7. This figure shows topography to be the dominant source in terms of the mean strength of wave forcing, with the other sources smaller but comparable, and with all source responses showing enhancements relative to the no-source segments of $\sim$ 3 to 10. Individual segments were found to achieve variances as high as $\sim$ 30 to 100 times background levels in response to strong wave forcing events and suggest considerable variability in the atmospheric response. However, no study to date has addressed the distribution and relative occurrence of strong forcing events such as those evaluated by these authors. Nor do we know at this stage to what extent these variable wave energies imposed at lower levels persist to greater altitudes. Yet these issues are important if we are to understand gravity wave effects in the MLT and describe them quantitatively in large-scale models of this region.

6. SUMMARY AND CONCLUSIONS

We have reviewed the evidence for and effects of gravity wave forcing in the MLT. Gravity waves achieve major importance in this region because their energy (per unit mass) increases by a factor of $\sim$ 100 from the tropopause to the mesopause. The corresponding energy and momentum fluxes and the heating, turbulent diffusion, and accelerations resulting from flux divergences due to wave dissipation have large effects on the circulation and structure of the MLT.

Theory provides a simple link between the magnitude of wave forcing and the strength and form of the residual circulation. This "downward control principle" and observations of wave momentum fluxes together imply maximum mean vertical motions at high latitudes and large departures from radiative equilibrium conditions during solstice periods.

The spectral character of the mean gravity wave field is reasonably well known based on numerous remote-sensing and *in–situ* observations. These data suggest that the intrinsic frequency and vertical wavenumber dependencies are approximately separable and that the forms of the individual spectra are approximately invariant with altitude. Estimates of wave energy and momentum fluxes based on these spectral forms are in good agreement with observed values, imply that the majority of wave energy and momentum fluxes are accomplished by motions at high intrinsic frequencies and large vertical scales, and provide a convenient means of describing systematic variations of wave forcing and effects.

Gravity wave variability appears to arise primarily in response to two factors. One is the variability of sources, including source type, strength, wave character, and temporal and spatial distribution. The second factor is modulation of the wave field by its local environment, including mean and low-frequency wave structures, due to the predominant wave fluxes occurring at higher frequencies.

Acknowledgments. This work was supported by the

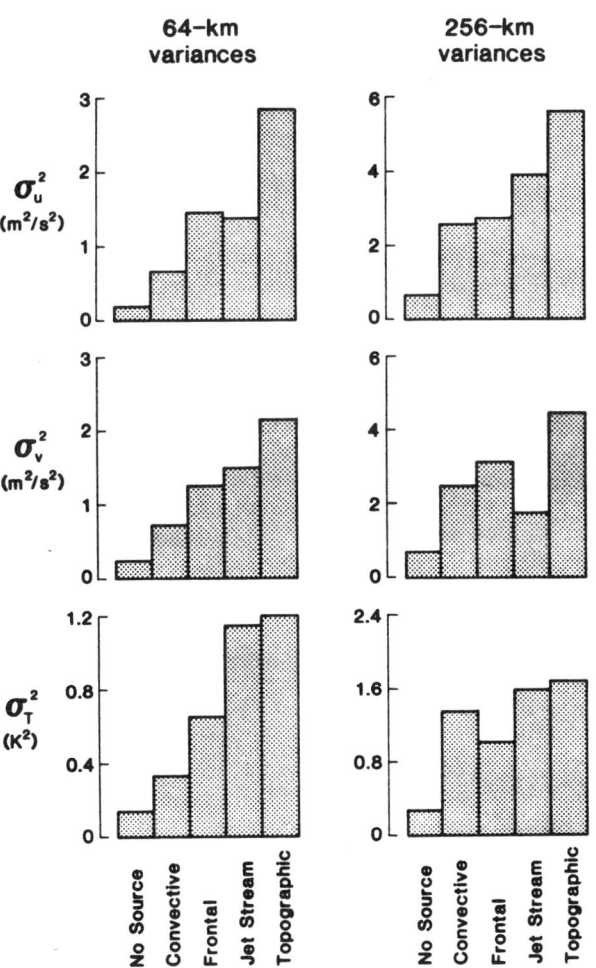

Figure 7. Variances of horizontal wind and temperature computed from GASP data in 64- and 256-km segments and categorized by source.

Air Force Office of Scientific Research (AFSC) under grant F49620-92-J-0138 and by the SDIO/IST and managed by the Naval Research Laboratory under grant N00014-92-J-2005.

REFERENCES

Andreassen, Ø., Wasberg, C. E., Fritts, D. C., Isler, J. R., Gravity wave breaking in two and three dimensions, part i: Model description and comparison of two-dimensional evolutions, *J. Geophys. Res.*, in press, 1994.

Andrews, D. G., McIntyre, M. E., Planetary waves in horizontal and vertical shear: the generalized eliassen-palm relation and the mean zonal acceleration, *J. Atmos. Sci.*, 33, 2031–2048, 1976.

Balsley, B. B., Carter, D. A., The spectrum of atmospheric velocity fluctuations at 8 and 86 km, *Geophys. Res. Lett.*, 9, 465–468, 1982.

Balsley, B. B., Garello, R., The kinetic energy density in the troposphere, stratosphere, and mesosphere: A preliminary study using the poker flat radar in alaska, *Radio Sci.*, 20, 1355-1362, 1985.

Blix, T. A., Thrane, E. V., Andreassen, Ø., *in – situ* measurements of the fine-scale structure and turbulence in the mesosphere and lower thermosphere by means of electrostatic positive ion probes, *J. Geophys. Res.*, 95, 5533–5548, 1990.

Dewan, E. M., Stratospheric wave spectra resembling turbulence, *Science*, 204, 832–835, 1979.

Dewan, E. M., Good, R. E., Saturation and the universal spectrum for vertical profiles of horizontal scalar winds in the atmosphere, *J. Geophys. Res.*, 91, 2742–2748, 1986.

Eckermann, S. D., Vincent, R. A., Vhf radar observations of gravity-wave production by cold fronts over southern australia, *J. Atmos. Sci.*, 50, 785–806, 1993.

Ecklund, W. L., Gage, K. S., Nastrom, G. D., Balsley, B. B., A preliminary climatology of the spectrum of vertical velocity observed by clear-air doppler radar, *J. Climate Appl. Meteor.*, 25, 885–892, 1986.

Ecklund, W. L., Gage, K. S., Riddle, A. C., Gravity wave activity in vertical winds observed by the poker flat mst radar, *Geophys. Res. Lett.*, 8, 285–288, 1981.

Forbes, J. M., Gu, J., Miyahara, S., On the interaction between gravity waves and the propagating diurnal tide, *Planet. Space Sci.*, 39, 1249–1257, 1991.

Fovell, R., Durran, D. R., Holton, J. R., Numerical simulation of convectively generated gravity waves, *J. Atmos. Sci.*, 49, 1427–1442, 1992.

Fritts, D. C., Gravity wave sources, variability, and lower and middle atmosphere effects, *Coupling Processes in the Lower and Middle Atmosphere*, 387, 191–208, 1993.

Fritts, D. C., Gravity wave-tidal interactions in the middle atmosphere: Observations and theory, this volume, 1994.

Fritts, D. C., Blanchard, R. C., Coy, L., Gravity wave structure between 60 and 90 km inferred from space shuttle reentry data, *J. Atmos. Sci.*, 46, 423–434, 1989.

Fritts, D. C., Chou, H.-G., An investigation of the vertical wavenumber and frequency spectra of gravity wave motions in the lower stratosphere, *J. Atmos. Sci.*, 44, 3610–3624, 1987.

Fritts, D. C., Hoppe, U.-P., Inhester, B., A study of the vertical motion field near the high-latitude summer mesopause during mac/sine, *J. Atmos. Terres. Phys.*, 52, 927–938, 1990.

Fritts, D. C., Isler, J. R., First observations of mesospheric dynamics with a partial reflection radar in hawaii ($22°n, 160°w$), *Geophys. Res. Lett.*, 19, 409–412, 1992.

Fritts, D. C., Isler, J. R., Andreassen, Ø., Gravity wave breaking in two and three dimensions, part ii: Three dimensional evolution and instability structure, *J. Geophys. Res.*, in press, 1994.

Fritts, D. C., Lu, W., Spectral estimates of gravity wave energy and momentum fluxes, ii: Parameterization of wave forcing and variability, *J. Atmos. Sci.*, 50, 3695–3713, 1993.

Fritts, D. C., Nastrom, G. D., Sources of mesoscale variability of gravity waves. part ii: Frontal, convective, and jet-stream excitation, *J. Atmos. Sci.*, 49, 111–127, 1992.

Fritts, D. C., Tsuda, T., Sato, T., Fukao, S., Kato, S., Observational evidence of a saturated gravity wave spectrum in the troposphere and lower stratosphere, *J. Atmos. Sci.*, 45, 1741–1759, 1988.

Fritts, D. C., VanZandt, T. E., Effects of doppler shifting on the frequency spectra of atmospheric gravity waves, *J. Geophys. Res.*, 92, 9723–9732, 1987.

Fritts, D. C., VanZandt, T. E., Spectral estimates of gravity wave energy and momentum fluxes, i: Energy dissipation, acceleration, and constraints, *J. Atmos. Sci.*, 50, 3685–3694, 1993.

Fritts, D. C., Vincent, R. A., Mesospheric momentum flux studies at adelaide, australia: Observations and a gravity wave-tidal interaction model, *J. Atmos. Sci.*, 44, 605–619, 1987.

Fritts, D. C., Wang, D.-Y., Doppler-shifting effects on frequency spectra of gravity waves observed near the summer mesopause at high latitude, *J. Atmos. Sci.*, 48, 1535–1544, 1991.

Fritts, D. C., Yuan, L., Measurement of momentum fluxes near the summer mesopause at poker flat, alaska, *J. Atmos. Sci.*, 46, 2569–2579, 1989.

Fritts, D. C., Yuan, L., Hitchman, M. H., Coy, L., Kudeki, E., Woodman, R. F., Dynamics of the equatorial mesosphere observed using the jicamarca mst radar during june and august 1987, *J. Atmos. Sci.*, 49, 2353–2371, 1992.

Garcia, R. R., Dynamics, radiation, and photochemistry in the mesosphere: Implications for the formation of noctilucent clouds, *J. Geophys. Res.*, **94**, 14605–14615, 1989.

Garrett, C. J. R., Munk, W. H., Space-time scales of internal waves: A progress report, *J. Geophys. Res.*, **80**, 291–297, 1975.

Gossard, E. E., Hooke, W. H., Waves in the Atmosphere, Elsevier, New York, 1975.

Haynes, P. H., Marks, C. J., McIntyre, M. E., Shepard, T. G., Shine, K. P., On the downward control of extratropical diabatic circulations by eddy-induced mean zonal forces, *J. Atmos. Sci.*, **48**, 651–678, 1991.

Hines, C. O., Internal gravity waves at ionospheric heights, *Can. J. Phys.*, **38**, 1441–1481, 1960.

Hines, C. O., The saturation of gravity waves in the middle atmosphere. part ii: Development of doppler-spread theory, *J. Atmos. Sci.*, **48**, 1360–1379, 1991.

Hitchman, M. H., Bywaters, K. W., Fritts, D. C., Coy, L., Kudeki, E., Surucu, F., Mean winds and momentum fluxes over jicamarca, peru during june and august 1987, *J. Atmos. Sci.*, **49**, 2372–2383, 1992.

Hocking, W. K., Measurement of turbulent energy dissipation rates in the middle atmosphere by radar techniques, *Radio Sci.*, **20**, 1403–1422, 1985.

Jasperson, W. H., Nastrom, G. D., Fritts, D. C., Further study of terrain effects on the mesoscale spectrum of atmospheric motions, *J. Atmos. Sci.*, **47**, 979–987, 1990.

Kuo, F. S., Lue, H. Y., Huang, C. M., Lo, C. L., Liu, C. H., Kukao, S., A study of velocity fluctuation spectra in the troposphere and lower stratosphere using mu radar, *J. Atmos. Terres. Phys.*, **54**, 31–48, 1992.

Kwon, K. H., Senft, D. C., Gardner, C. S., Airborne sodium lidar observations of horizontal and vertical wavenumber spectra of mesopause density and wind perturbations, *J. Geophys. Res.*, **95**, 13723–13736, 1990.

Lu, W., Fritts, D. C., Spectral estimates of gravity wave energy and momentum fluxes, iii: Gravity wave-tidal interactions, *J. Atmos. Sci.*, **50**, 3714–3727, 1993.

Lübken, F.-J., von Zahn, U., Thrane, E. V., Blix, T. A., Kokin, G. A., Pakhomov, S. V., in – situ measurements of turbulent energy dissipation rates and eddy diffusion coefficients during map/wine, *J. Atmos. Terres. Phys.*, **49**, 763–775, 1987.

McIntyre, M. E., On dynamics and transport near the polar mesopause in summer, *J. Geophys. Res.*, **94**, 14617–14628, 1989.

Meyer, W., Siebenmorgen, R., Widdel, H.-U., Estimates of gravity wave momentum fluxes in the winter and summer high mesosphere over northern scandinavia, *J. Atmos. Terres. Phys.*, **51**, 311–319, 1989.

Miyahara, S., Forbes, J. M., Interactions between gravity waves and the diurnal tide in the mesosphere and lower thermosphere, *J. Meteor. Soc. Japan*, **69**, 523–531, 1991.

Nastrom, G. D., Fritts, D. C., Sources of mesoscale variability of gravity waves. part i: Topographic excitation, *J. Atmos. Sci.*, **49**, 101–110, 1992.

Nastrom, G. D., Fritts, D. C., Gage, K. S., An investigation of terrain effects on the mesoscale spectrum of atmospheric motions, *J. Atmos. Sci.*, **44**, 3087–3096, 1987.

Nastrom, G. D., Peterson, M. R., Green, J. L., Gage, K. S., VanZandt, T. E., Sources of gravity wave activity seen in the vertical velocities observed by the flatland vhf radar, *J. Appl. Meteor.*, **29**, 783–792, 1990.

Reid, I. M., Rüster, R., Czechowsky, P., Schmidt, G., Vhf radar measurements of momentum flux in the summer polar mesosphere over andenes ($69°n, 16°e$), norway, *Geophys. Res. Lett.*, **15**, 1263–1266, 1988.

Reid, I. M., Vincent, R. A., Measurements of mesospheric gravity wave momentum fluxes and mean flow accelerations at adelaide, australia, *J. Atmos. Terres. Phys.*, **49**, 443–460, 1987.

Rüster, R., Reid, I. M., Vhf radar observations of the dynamics of the summer polar mesopause region, *J. Geophys. Res.*, **95**, 10005–10016, 1990.

Scheffler, A. O., Liu, C. H., The effects of doppler shift on the gravity wave spectra observed by mst radar, *J. Atmos. Terres. Phys.*, **48**, 1225–1231, 1986.

Sidi, C., Lefreve, J., Dalaudier, F., Barat, J., An improved atmospheric buoyancy wave model, *J. Geophys. Res.*, **93**, 774–790, 1988.

Smith, S. A., Fritts, D. C., VanZandt, T. E., Evidence for a saturated spectrum of atmospheric gravity waves, *J. Atmos. Sci.*, **44**, 1404–1410, 1987.

Thrane, E. V., Andreassen, Ø., Blix, T. A., Grandal, B., Brekke, A., Philbrick, C. R., Schmidlin, F. J., Widdel, H.-U., von Zahn, U., Lübken, F.-J., Neutral air turbulence in the upper atmosphere, *J. Atmos. Terres. Phys.*, **47**, 243–265, 1985.

Tsuda, T., Inoue, T., Fritts, D. C., VanZandt, T. E., Kato, S., Sato, T., Fukao, S., Mst radar observations of a saturated gravity wave spectrum, *J. Atmos. Sci.*, **46**, 2440–2447, 1989.

Tsuda, T., Murayama, Y., Yamamoto, M., Kato, S., Fukao, S., Seasonal variation of momentum flux in the mesosphere observed with the mu radar, *Geophys. Res. Lett.*, **17**, 725–728, 1990.

Tsuda, T., VanZandt, T. E., Mizumoto, M., Kato, S., Fukao, S., Spectral analysis of temperature and brunt-vaisala frequency fluctuations observed by radiosondes, *J. Geophys. Res.*, **96**, 17265–17278, 1991.

VanZandt, T. E., A universal spectrum of buoyancy waves in the atmosphere, *Geophys. Res. Lett.*, **9**, 575–578, 1982.

VanZandt, T. E., Fritts, D. C., A theory of enhanced

saturation of the gravity wave spectrum due to increases in atmospheric stability, *Pure Appl. Geophys.*, **130**, 399-420, 1989.

VanZandt, T. E., Nastrom, G. D., Green, J. L., Frequency spectra of vertical velocity from flatland vhf radar data, *J. Geophys. Res.*, **96**, 2845–2855, 1991.

Vincent, R. A., Gravity wave motions in the mesosphere, *J. Atmos. Terres. Phys.*, **46**, 119–128, 1984.

Vincent, R. A., Lesicar, D., Dynamics of the equatorial mesosphere: First results with a new generation partial reflection radar, *Geophys. Res. Lett.*, **18**, 825–828, 1991.

Vincent, R. A., Reid, I. M., Hf doppler measurements of mesospheric momentum fluxes, *J. Atmos. Sci.*, **40**, 1321–1333, 1983.

Vincent, R. A., Tsuda, T., Kato, S., A comparative study of mesospheric solar tides observed at adelaide and kyoto, *J. Geophys. Res.*, **93**, 699–708, 1988.

Walterscheid, R. L., Inertio-gravity wave induced accelerations of mean flow having an imposed periodic component: Implications for tidal observations in the meteor region, *J. Geophys. Res.*, **86**, 9698–9706, 1981.

Wang, D.-Y., Fritts, D. C., Mesospheric momentum fluxes observed by the mst radar at poker flat, alaska, *J. Atmos. Sci.*, **47**, 1512–1521, 1990.

Wang, D.-Y., Fritts, D. C., Evidence of gravity wave-tidal interaction observed near the summer mesopause at poker flat, alaska, *J. Atmos. Sci.*, **48**, 572–583, 1991.

Watkins, B. J., Philbrick, C. R., Balsley, B. B., Turbulent energy dissipation rates and inner scale sizes from rocket and radar data, *J. Geophys. Res.*, **93**, 7009-7014, 1988.

Wu, Y.-F., Widdel, H.-U., Observational evidence of a saturated gravity wave spectrum in the mesosphere, *J. Atmos. Terres Phys.*, **51**, 991–996, 1989a.

Wu, Y.-F., Widdel, H.-U., Turbulent energy dissipation rates and eddy diffusion coefficients derived from foil cloud experiments, *J. Atmos. Terres Phys.*, **51**, 497–506, 1989b.

Wu, Y.-F., Widdel, H.-U., Saturated gravity wave spectrum in the polar lower thermosphere observed by foil chaff during campaign sodium 88, *J. Atmos. Terres Phys.*, **49**, 1781–1789, 1992.

Observations of the Meridional Quasi Two-Day Wave in the Mesosphere and Lower Thermosphere at Christmas Island

S.E. PALO AND S.K. AVERY

Department of Electrical and Computer Engineering, University of Colorado, Boulder, Colorado

Cooperative Institute for Research in Environmental Sciences, University of Colorado, Boulder, Colorado

Observations of the quasi two-day wave at Christmas Island (1.95°N, 157.30°W) made with a MEDAC meteor scatter system are presented. The data were collected from September 1988 through June 1992. These data shows the presence of a quasi two-day wave in the meridional wind field maximizing in late January with an amplitude of ∼50 m s^{-1}. The period of the quasi two-day wave from December through February varies between 46 hours and 52 hours. The temporal and vertical structure of the observed quasi two-day wave indicate similarities with observations of the quasi two-day wave in the northern and southern hemispheres.

INTRODUCTION

The first observations of the quasi two-day wave in the northern hemisphere were reported at Sheffield (53°N, 2°W) by *Muller* [1972]. Muller noted that a significant peak very close to 51 hours was present in the Fourier decomposition of meteor wind data collected during August 1966. Recent ground based observations of the quasi two-day wave (hereafter referred to as the two-day wave) in the northern hemisphere [*Clark et al.*, 1994; *Fritts and Isler*, 1992; *Williams and Avery*, 1992] exhibit the salient features of the two-day wave. The observed period in the northern hemisphere is near 52 hours during maximum wave activity in late July. The frequency and amplitude of the observed two-day wave exhibit moderate temporal variability. Observations in the northern hemisphere indicate the period of the two-day wave is between 48 and 55 hours. Medium frequency (MF) radar data collected by *Craig et al.* [1980] indicated the two-day wave is also present in the southern hemisphere. *Harris* [1994] has recently published a study of the two-day wave over Adelaide during twelve summers. These results are representative of previous observations in the southern hemisphere and show the two-day wave in the southern hemisphere to have a period near 48 hours. Peak two-day wave activity at Adelaide (35°S, 138°E) occurs in late January with an amplitude between 50 and 60 m s^{-1}.

Ground based wind [*Glass et al.*, 1975; *Muller and Nelson*, 1978] and satellite temperature measurements [*Rodgers and Prata*, 1981] provided data indicating the two-day wave is a westward traveling global scale oscillation with zonal wave number 3. The results of *Rodgers and Prata* [1981] also show the maximum amplitude associated with the two-day wave occurs in the mesosphere at low latitudes of the summer hemisphere. *Salby* [1981] suggests the two-day wave to be the third Rossby-gravity normal mode of a windless isothermal atmosphere. Using a model based upon the linearized primitive equations and realistic mean winds, the frequency and spatial amplitude response of the (3,0) normal model were computed. These results indicate the (3,0) mode has a resonant period near 2.27 days (equinox) and 2.22 days (solstice). The spatial amplitude response exhibits a maximum in the mesospheric meridional wind field at low to mid latitudes in the southern hemisphere. The regions in which the quasi two-day response was large correspond to regions of time mean eastward flow. *Hagan et al.* [1993] have extended the results of *Salby* [1981] by changing the mean wind field to include a summer mesospheric jet during December solstice conditions. By changing the stratospheric and mesospheric mean winds, *Hagan et al.* [1993] found the spatial amplitude, vertical phase structure and resonant frequency of the (3,0) mode changed. This indicates the stratospheric and mesospheric zonal mean flow impact the salient features of the two-day wave observed in the mesosphere.

Observations of the two-day wave at mid latitude northern and southern hemisphere sites have indicated the two-day wave behaves differently in the northern and southern

The Upper Mesosphere and Lower Thermosphere:
A Review of Experiment and Theory
Geophysical Monograph 87
Copyright 1995 by the American Geophysical Union

hemispheres. The results of *Hagan et al.* [1993] would indicate the variability in frequency and amplitude of the two-day wave is a result of differences in the zonal mean circulation between hemispheres. Early observational results of *Kal'chanko and Bulgakov* [1973] at Mogadishu (2°N, 45°E) indicated that the two-day wave is also present at the equator. Recently *Harris and Vincent* [1993] have presented observations made at Christmas Island (1.95°N, 157.30°W) using a MF radar from January 1990 to April 1992. The results of *Harris and Vincent* [1993] are the first comprehensive observational results of the two-day wave in the equatorial mesosphere. These results will be compared with the MEDAC data collected at Christmas Island from August 1988 to June 1992. A comprehensive overview of the two-day wave from a historical perspective can be found in *Hagan et al.* [1993] and the references therein.

DATA ANALYSIS

The data used in this analysis were collected using the meteor echo detection and collection system (MEDAC) [*Avery et al.*, 1990] located on Christmas Island (1.95°N, 157.30°W), between August 1988 and June 1992. The MEDAC system operates in conjunction with the 50 MHz stratosphere-troposphere (ST) radar located on Christmas Island. The Christmas Island ST radar operates one antenna with a 2.7° two way beamwidth. This results in a typical echo rate of 150-200 echoes/day [*Palo and Avery*, 1993]. Estimates of the horizontal wind are only computed when sufficient ionization is detected, which results in a data set that contains wind measurements at random time intervals. To reduce the complexity required in directly analyzing a randomly spaced time series the data is temporally and spatially averaged, to yield a time series containing values at uniform time intervals.

Signal Processing

In the analysis of the Christmas Island mesospheric data four separate signal processing techniques were utilized to gain insight into the characteristics of the two-day wave. These consisted of the discrete spectrogram [*Palo and Avery*, 1993], time domain filtering with finite impulse response (FIR) filters [*Roberts and Mullis*, 1987], complex demodulation [*Bloomfield*, 1976; *Campbell and Walker*, 1977; *Bingham et al.*, 1967] and least squares fitting [*Palo and Avery*, 1993]. Before proceeding it should be stated that we have made a number of implicit assumptions about the temporal structure of the quasi two-day wave. From the results of *Hagan et al.* [1993] it is obvious the amplitude and frequency characteristics of the two-day wave are temporally variable. This is a direct result of any temporal variations in intervening stratospheric and mesospheric wind fields. Therefore, to account for this amplitude and frequency variability, the two-day wave is modelled as

$$x(t) = A(t)\cos(\omega_o t + \phi(t)) \quad t \in [t_1, t_2] \quad (1)$$

where ω_o is the dominant resonant frequency of the (3,0) normal Rossby-gravity mode, $A(t)$ is the linear, stationary, time varying amplitude, and $\phi(t)$ is the time varying phase.

Assuming the two-day wave is modelled by Equation 1, then the two-day wave cannot be considered stationary unless $\phi(t)$ is a linear function of time. Therefore standard Fourier methods cannot be used to compute the power spectral density of the two-day wave. However, if we assume that $x(t)$ is locally stationary on time scales less than a month (eg. $\phi(t)$ is approximately linear on these time scales) then the discrete spectrogram can be used to compute the time varying power spectral density. The discrete spectrogram is implemented by computing fast Fourier transforms (FFT) for 30 days of data. Each consecutive FFT is shifted in time by 2 days relative to its predecessor. Prior to computation of the FFT these data were temporally averaged into 12 hour bins yielding a Nyquist period of 1 day.

While the spectrogram yields coarse information pertaining to the temporal distribution of power in the frequency domain, time domain filtering can be utilized to better illustrate the temporal evolution of the two-day wave. In our analysis we have chosen to use FIR filters because they are simple to design and have a linear phase response. In determining our filter bandwidths a number of considerations had to be taken into account. To accurately represent $x(t)$ (Equation 1) the filter bandwidth (ω_B) must meet the condition

$$\omega_B > \omega_A + \max_t \left[\frac{\partial}{\partial t}\phi(t)\right] - \min_t \left[\frac{\partial}{\partial t}\phi(t)\right] \quad (2)$$

where ω_A is the bandwidth of $A(t)$. In effect the filter must designed such that the information contained in the amplitude and phase of the wave is not destroyed. *Clark et al.* [1994] have indicated the two-day wave may undergo 180° phase shifts. This shift has the undesirable property of increasing the effective bandwidth of $A(t)$. In order to represent these phase changes in the filtered time series, the filter must have a bandwidth large enough to encompass the maximum fundamental frequency of the phase shifts. From empirical results we have determined a filter centered at 0.5 cycles per day (cpd) with a bandwidth between 0.2 cpd and 0.3 cpd will adequately represent the salient features of the two-day wave. Figure 1 shows the magnitude and phase response of the FIR filter used in our analysis.

Complex demodulation is a technique used to determine the local amplitude and phase characteristics of a nonstationary function such as that described in Equation 1. Complex demodulation is implemented simply by multiplying a given time series by a complex sinusoid ($e^{-j\omega_d t}$ where $j = \sqrt{-1}$ and ω_d is the demodulation frequency) and then passing the result through a low pass filter. If $\omega_d = \omega_o$ then applying the technique of complex demodulation to $x(t)$ (Equation 1) will result in a complex time series of the form

$$x_d(t) = x_{lp}(t) + x_{hp}(t) \quad (3)$$

$$x_{lp}(t) = \frac{A(t)}{2}e^{j\phi(t)} \qquad (4)$$

$$x_{hp}(t) = \frac{A(t)}{2}e^{-j(2\omega_o t + \phi(t))} \qquad (5)$$

where the time varying amplitude $A(t) = 2|x_{lp}(t)|$ and the time varying phase $\phi(t) = \text{Arg}(x_{lp}(t))$. In demodulating $x(t)$ it is assumed that

$$\omega_o > \frac{\omega_B}{2} \qquad (6)$$

otherwise aliasing will occur. Passing the demodulated signal, $x_d(t)$, through a low pass filter which meets the requirements described in Equation 2, will result in

$$x_{lp}(t) = h_{lp}(t) * x_d(t) \qquad (7)$$

where $h_{lp}(t) * x_d(t)$ denotes the operation of passing $x_d(t)$ through a lowpass filter with frequency response $H_{lp}(j\omega)$.

Additionally, complex demodulation can be used to determine the instantaneous frequency of a signal. If $\phi(t)$ is modelled as approximately linear over some time interval, such that

$$\phi(t) = \omega_\Delta t + \phi_o \qquad t \in [t_1, t_2] \qquad (8)$$

then

$$\frac{\partial}{\partial t}\text{Arg}(x_{lp}(t)) = \omega_\Delta \qquad t \in [t_1, t_2]. \qquad (9)$$

Therefore, $\frac{\partial}{\partial t}\text{Arg}(x_{lp}(t))$ represents the frequency difference between $x(t)$ and $x_d(t)$. If $\phi(t)$ is nonlinear in time then equation 9 will be time varying, and have the interpretation of an instantaneous frequency difference between $x(t)$ and $x_d(t)$ where the frequency of $x(t)$ is defined as

$$\frac{\partial}{\partial t}(\omega_o t + \phi(t)). \qquad (10)$$

By using a narrow band low pass filter, with bandwidth ω_{nb}, in the complex demodulation technique, the center of mass (ω_{cm}) for the power spectral density in this narrow spectral band can be estimated. The results are as follows

$$\frac{\partial}{\partial t}\text{Arg}(x_{lp}(t)) \begin{cases} > 0 & \Rightarrow \quad \omega_d < \omega_{cm} \leq \omega_d + \frac{\omega_{nb}}{2} \\ = 0 & \Rightarrow \quad \omega_{cm} = \omega_d \\ < 0 & \Rightarrow \quad \omega_d + \frac{\omega_{nb}}{2} \leq \omega_{cm} < \omega_d \end{cases} \qquad (11)$$

Therefore, complex demodulation can be used to estimate the instantaneous frequency of the two-day wave during periods of high activity.

The three signal processing techniques described above are utilized in an effort to more accurately determine the temporal variability associated with the amplitude and frequency of the observed two-day wave. To determine the vertical structure of the two-day wave during its maximum in late January a linear least squares fit was employed. The least squares fit was computed for a model consisting of a mean, 12 hour, 24 hour, and two-day period (P_{2d}). The two-day wave period for January was estimated at five heights using complex demodulation. The median period over the

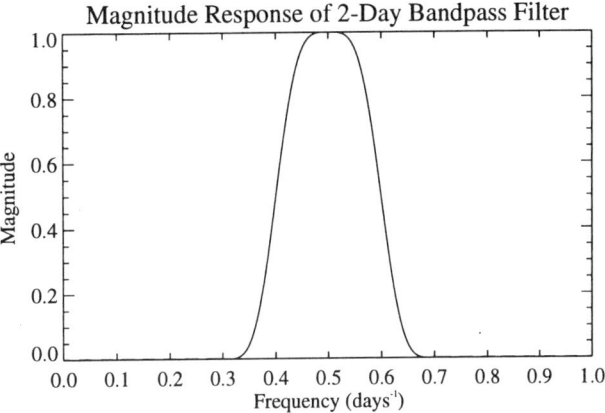

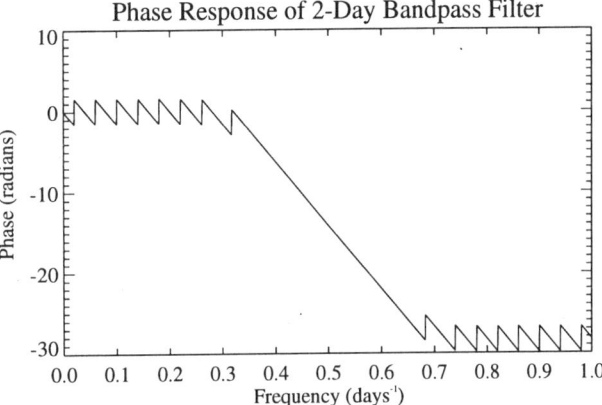

Fig. 1. The (top) magnitude and (bottom) phase response of the finite impulse response (FIR), bandpass, two-day wave filter.

five heights was assumed to be P_{2d} for the fitted time interval.

RESULTS

The data collected at Christmas Island from August 1988 to June 1992 show evidence of a strong mesospheric oscillation in the meridional wind field near a period of 48 hours. Using only measurements from a single site the zonal wave number associated with this oscillation cannot be determined although the amplitude and frequency structure are consistent with models and observations of the two-day wave. The basic characteristics of the two-day wave at Christmas Island indicate it is closely associated with the manifestation of the two-day wave in the southern hemisphere. These characteristics include maximum amplitudes of 35-50 m s^{-1}, a period of 48 $\pm$ 1.5 hours in late January and sporadic enhancements throughout the remainder of the year. From December 1989 to January 1991 the radar and MEDAC systems experienced intermittent failures that produced large gaps in the data set. Therefore, only data collected from August 1988 to December 1989 and January 1991 to June 1992 will be presented.

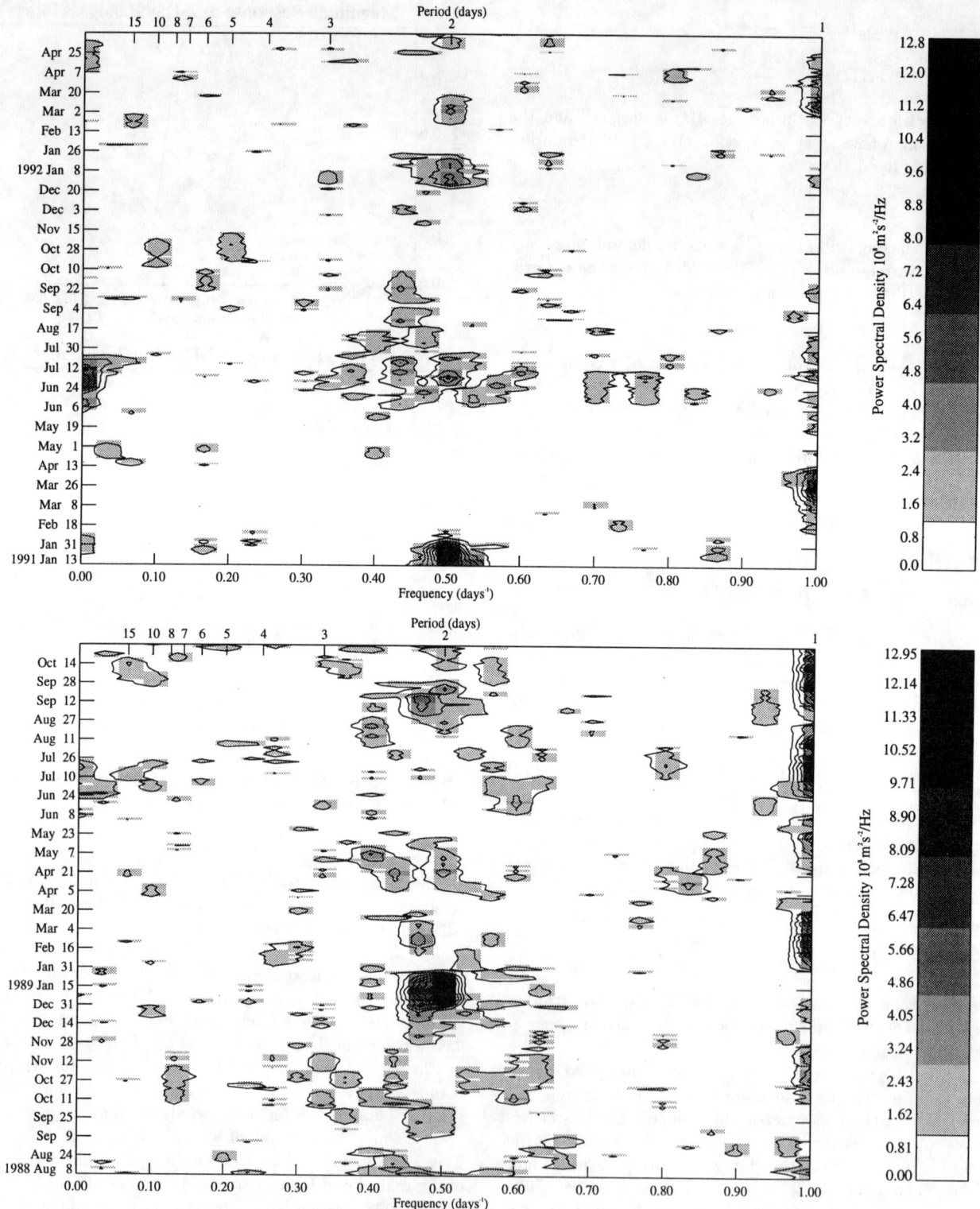

Fig. 2. The spectrogram of the meridional wind field from 88 to 91.5 km for (top) 1991-92 and (bottom) 1988-89. The spectrum is computed for a 30-day interval and shifted by 2 days.

Spectral Analysis

The spectrogram of the meridional wind field from August 1988 through October 1989 at 90 km (88-91.5 km) is shown in the lower panel of Figure 2. The spectral energy in the meridional wind field during 1988-89 is concentrated around the diurnal tide (1.0 cpd) and the quasi two day wave (0.5 cpd), with only sporadic enhancements for wave periods greater than 3 days. This is in contrast to the zonal spectrogram for the same time period [*Palo and Avery*, 1993] where the spectral power is concentrated in the waves with periods greater than 3 days. The two-day wave exhibits a dominant peak in January concentrated near 48 hours, although significant signal power between 44 and 48 hours is present before and after the January maximum. In addition to the primary peak in January there are sporadic enhancements near 48 hours in September and May. The amplitude and frequency variability of the two-day can be clearly seen by the variability of the spectrogram in the 40 hour (0.6 cpd) to 60 hour (0.4 cpd) spectral band.

The meridional power spectral distribution for January 1991 through April 1992, shown in the upper panel of Figure 2, exhibits characteristics similar to those seen in 1988-89. Maximum two-day wave activity during January 1991 is concentrated around a period of 48 hours. The peak in the spectrogram associated with the January 1992 maximum is concentrated at 48 hours although it is weaker than peaks observed in January 1989 and 1991. A possible explanation for the weaker peak in January 1992 is the propagation effect due to the quasi biennial oscillation in the intervening stratospheric wind field. A comparison of the spectrograms for 1988-89 and 1991-92 indicate less variability in the 0.4 to 0.6 cpd spectral bands for 1991-92. However, a prominent wideband event occurs between June and September 1992 with frequency components ranging from 0.31 cpd to 0.62 cpd. The mean period during this event is near 54 hours indicating it is similar to the manifestation of the two-day wave observed in the northern hemisphere. The spectrograms also indicate the amplitude of the diurnal tide during maximum two-day wave activity is low.

Time Series Analysis

The amplitude variability of the two-day wave in the meridional wind field is evident in the filtered time series at 90 km shown in Figure 3. The filtered time series was computed in the time domain by convolving the impulse response of the FIR filter, shown in Figure 1, with the data collected at 90 km. The filtered time series show maxima in late January 1989 and 1991 of 50 m s^{-1} and late January 1992 of 40 m s^{-1}. On average the two-day wave in the meridional wind field at Christmas Island exhibits persistent amplitudes of 5 to 10 m s^{-1}. The wideband event seen in the spectrogram during July and August 1991 is also present in the filtered time series (bottom panel Figure 3). Additionally, there is increased two-day wave activity during September and October 1989 which is similar in magnitude to the January 1992 event. The wave event maintains an amplitude greater that 20 m s^{-1} for five cycles. The two peaks occurring in early and late July 1991 associated with this event have amplitudes of 35 m s^{-1}. The time interval between these peaks is ~15 days, indicating the amplitude

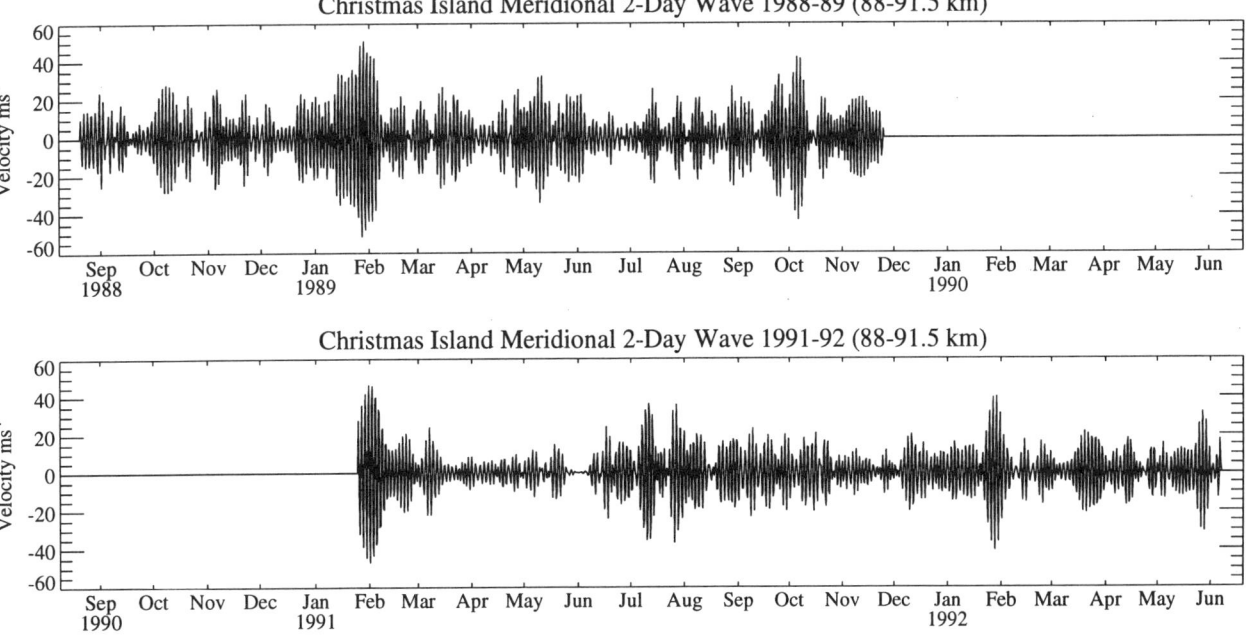

Fig. 3. The meridional filtered time series of the two-day wave from 88 to 91.5 km for (top) 1988-89 and (bottom) 1991-92.

of the two-day wave during July 1991 is dominated by a 15 day oscillation. Peaks in the amplitude of the two-day wave separated by 12-18 days are also evident during November 1988, March 1989 and July 1989. This could indicate a possible interaction between the two-day wave and the quasi sixteen-day wave. It cannot be determined if the variation in the amplitude of the two-day wave is due to a mechanism similar to modulation (wave-wave interaction) or superposition of closely spaced independent sinusoidal oscillations (beating frequencies) without the use of higher order spectral estimation techniques.

The filtered zonal time series at 90 km shown in Figure 4 exhibits sporadic enhancements in amplitude similar to those seen in the meridional time series. On average the amplitude of the zonal two-day oscillation is 50% to 75% weaker than the two-day oscillation seen in the meridional time series at 90 km. Throughout the year the two-day oscillation in the zonal direction is between 5 m s^{-1} and 10 m s^{-1} with enhanced amplitudes up to 25 m s^{-1} occurring in November 1988.

Case Study

Examination of the power spectral density and filtered time series clearly exhibit maximum two-day wave activity in late January. In an effort to better understand the equatorial two-day wave we have chosen to focus on the data collected from December 15 - February 15 during 1988-89, 1991-92, and January 15 - February 15 during 1991.

The dominant periods (P_{2d}) present during maximum two-day wave activity were computed by demodulating the meridional time series and analyzing the resulting phase structure. P_{2d} was computed for five independent heights between December and February 1988-89, 1991-92 and January to February 1991. P_{2d} was not computed for December 1990 since the system was inoperable during this period. Figure 5 shows the resulting dominant periods during these time intervals. In the left panel, the two-day wave period is shown for December 15, 1988 to January 15, 1989 and January 15, 1989 to February 15, 1989. During the January 15 to February 15 time interval the two-day wave exhibited a monochromatic behavior. P_{2d} varied between 48.5 hours and 50 hours, depending on height, with a median period of 49.5 hours. During the interval between December 15, 1988 and January 15, 1989, prior to maximum activity, the two-day wave exhibited substantial spectral power at multiple periods between 43 hours and 53 hours. Both 1988-89 and 1991-92 exhibited similar characteristics where spectral power was present at multiple periods prior to January 15 evolving into a monochromatic spectrum between January 15 and February 15. The center panel of Figure 5 shows P_{2d} for January 15 to February 15 1991. P_{2d} varied between 47 hours and 51.5 hours with a median period of 47 hours. The right panel shows P_{2d} for December 15, 1991 to February 15 1992. Again, prior to January 15, there is power located at multiple periods between 44 hours and 52 hours evolving into a monochromatic spectrum from January 15 to February 15 with P_{2d} between 46 hours and 51 hours.

The amplitude and phase profiles shown in Figure 6 were computed using a least squares fitting algorithm. The meridional time series were fit to a model consisting of a mean, semidiurnal, diurnal and two-day component. The

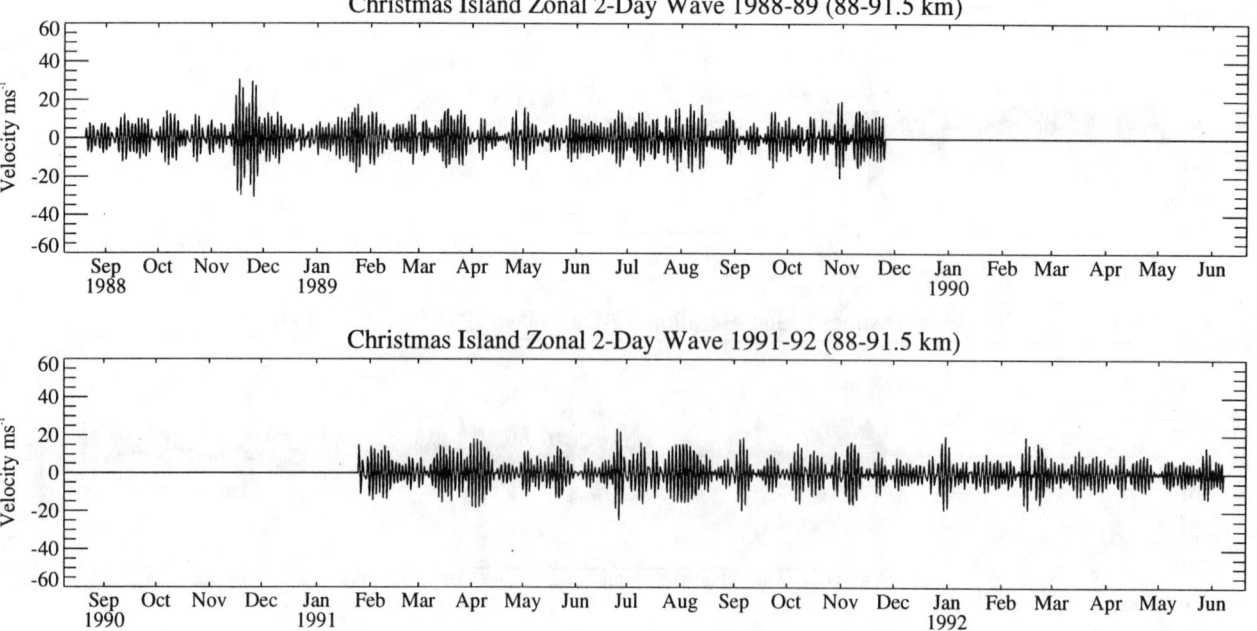

Fig. 4. The zonal filtered time series of the two-day wave from 88 to 91.5 km for (top) 1988-89 and (bottom) 1991-92.

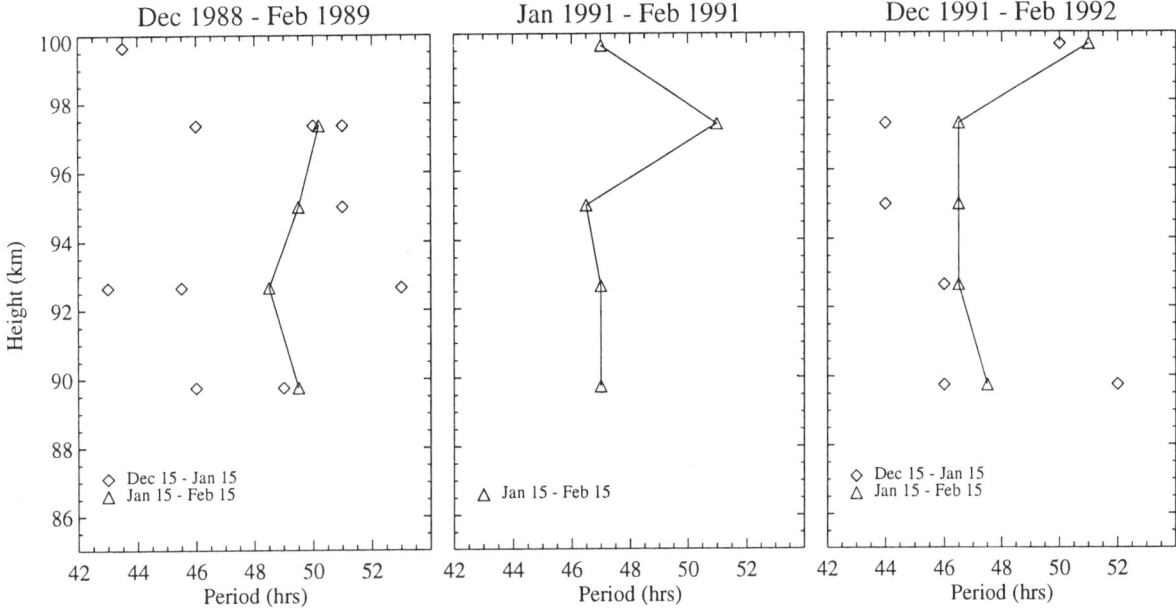

Fig. 5. The dominant wave periods present between 42 and 54 hours in the meridional wind field.

horizontal error bars shown on the amplitude and phase plots are the 66% confidence intervals associated with each fitted parameter [*Palo and Avery*, 1993]. The left panels of Figure 6 are the amplitude (upper panel) and phase (lower panel) of the 49.5 hour component during January and February 1989 as a function of height. The amplitude of the two-day wave during January and February 1989 decreases as a function of increasing height. The phase of the 49.5 hour component during this time period indicates an upward transport of energy with an estimated vertical wavelength of 40 km. The fit for January and February 1991 was computed for a harmonic component of 47 hours. The amplitude and phase characteristics are shown in the middle panel of Figure 6. The amplitude (upper panel) of the two-day wave during this time interval is nearly constant (20-25 m s^{-1}) between 85 km and 100 km. Between 80 km and 84 km the amplitude is increasing with a jump of 40 m s^{-1} at 84 km to 10 m s^{-1} at 85 km. The phase (lower panel) of the two-day wave is similar to the phase seen in 1989 with an estimated vertical wavelength of 44 km. The right panel shows the amplitude (upper panel) and phase (lower panel) of a harmonic fit to a component of 47 hours during January and February 1992. Again, the amplitude is relatively constant (15-20 m s^{-1}) between 85 km and 100 km with a small bite-out at 87 km. The phase during January 1992 indicates an upward transport of energy with an estimated vertical wavelength of 36 km. The hour of maximum at 90 km during all three years is between 3 am and 6 am solar local time, indicating similarities to the phase locking phenomenon associated with observation of the two-day wave [*Craig et al.*, 1980; *Craig and Elford*, 1981; *Harris*, 1994; *Clark et al.*, 1994].

DISCUSSION

The two-day wave is a feature that dominates the summer hemisphere mesosphere and lower thermosphere for nearly a month shortly after the summer solstice. Observations made in the southern hemisphere [*Harris and Vincent*, 1993; *Poole*, 1990; *Phillips*, 1989; *Craig and Elford*, 1981; *Craig et al.*, 1983] indicate that the two-day wave maximizes about one month after the summer solstice in late January with amplitudes of 10-50 m s^{-1} depending upon latitude. During this period of high two-day wave activity in the southern hemisphere the period of the two-day wave is close to 48 hours. Observations made in the northern hemisphere [*Clark*, 1989; *Tsuda et al.*, 1988; *Clark et al.*, 1994; *Williams and Avery*, 1992; *Fritts and Isler*, 1992; *Kingsley et al.*, 1978; *Muller and Nelson*, 1978; *Glass et al.*, 1975; *Reddi et al.*, 1988; *Salby and Roper*, 1980] indicate that the quasi two-day wave maximizes in late summer and rarely exceeds 30 m s^{-1}. In contrast to the two-day wave observed in the southern hemisphere the period of the two-day wave observed in the northern hemisphere tends towards 52 hours.

The two-day wave at Christmas Island (1.95°N, 157.30°W) exhibits characteristics of both the northern and southern hemisphere manifestations of the two-day wave. The primary maxima during 1989, 1991, and 1992, of the two-day wave in the meridional wind field occurs within five days of February 1 each year. The maximum amplitudes at 90 km during 1989, 1991 and 1992 all exceed 40 m s^{-1} (see Figure 3). During these periods of maximum two-day wave activity a method of complex demodulation was used to compute the dominant wave periods that were present between 44 hours and 54 hours. The results, shown in Figure 5,

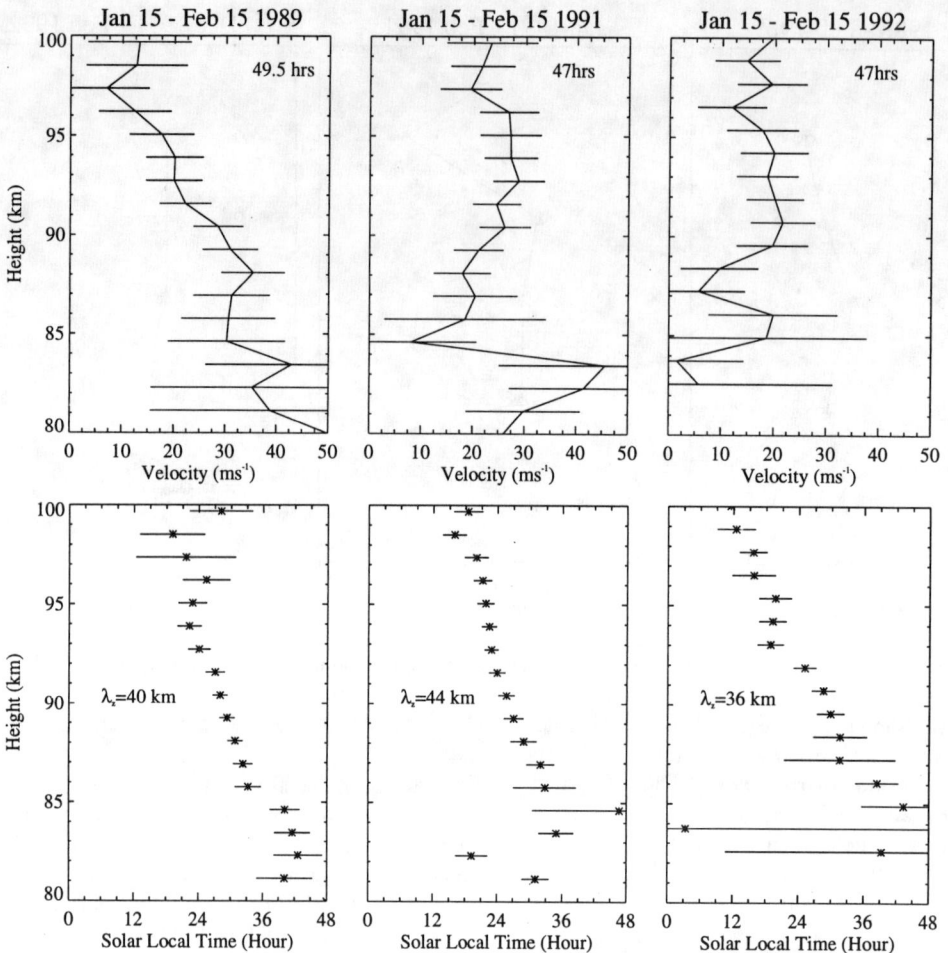

Fig. 6. Least squares (top) amplitude and (bottom) phase fits for periods of strong two-day wave activity. Error bars indicate the 66% confidence intervals.

indicate that the dominant periods during late January were typically within 1.5 hours of 48 hours. These characteristics compare favorably to observations made at Grahamstown (33°S) [*Poole*, 1990], Adelaide (35°S) [*Phillips*, 1989] and Christmas Island [*Harris and Vincent*, 1993]. The results of *Harris and Vincent* [1993] for December-February show the two-day wave has period between 47 and 50 hours depending upon height.

The two-day wave in the meridional wind field maintains an amplitude of 10 m s^{-1} throughout the year, although there are periods of enhancement where the two-day wave has amplitudes exceeding 20 m s^{-1}. During October 1988 and 1989, May 1989 and 1992, and July 1991 the two-day wave has amplitudes exceeding 20 m s^{-1} for more than 2 cycles. Examination of the spectrograms (Figure 2) during these events indicates the period of the quasi two-day wave tends towards periods greater than 48 hours. *Harris and Vincent* [1993] have indicated the presence of two spectral peaks during the July-August 1991 event. The largest peak during July is near 51 hours while a secondary peak is present close to 44 hours. *Harris and Vincent* [1993] suggest the peak near 44 hours to be a manifestation of the (2,0) Rossby normal mode. The longer periods present during these events are indicative of the two-day wave observed in the northern hemisphere summer at Durham (43°N) [*Clark*, 1989], Saskatoon (52°N) [*Clark et al.*, 1994] and Sheffield (53°N) [*Kingsley et al.*, 1978]. The filtered meridional time series during these events also exhibits amplitude variations with a period between 12 and 18 days. Examination of the filtered meridional time series presented in *Harris and Vincent* [1993] (Figure 3) exhibits a similar phenomenon during July and August 1991 from 84 to 92 km. There are five peaks in the meridional amplitude (Figure 3 of *Harris and Vincent* [1993]) over an interval of 60 days indicating an oscillation near 15 days is present in the amplitude of the two-day wave.

The vertical structure of the two-day wave at Christmas Island during January 1989, 1991 and 1992 is shown in Figure 6. During January 1989 the amplitude exhibits a vertical structure that is decaying with increasing height while both

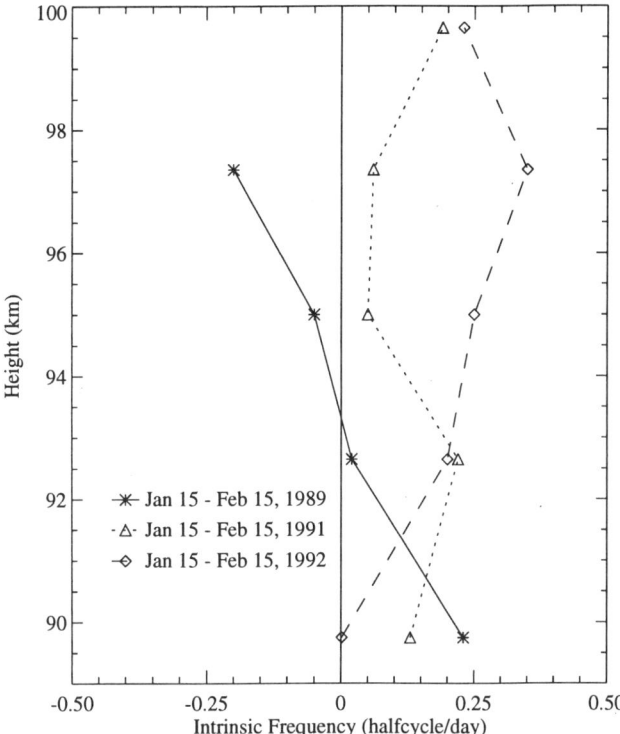

Fig. 7. The quasi two-day wave intrinsic frequency for (solid) Jan-Feb 1989, (short dash) Jan-Feb 1991 and (long dash) Jan-Feb 1992. Mean wind profiles taken from *Palo and Avery* [1993].

January 1991 and 1992 exhibit an amplitude structure that is relatively constant with height. A possible explanation for the variability in amplitude structure between years can be inferred from Figure 7. In Figure 7, the normalized intrinsic frequency has been computed for each interval of data that the fits were performed. The normalized intrinsic frequency is defined as

$$\tilde{\omega} = \frac{\omega}{2\Omega} \qquad (12)$$

where

$$\omega = \sigma + m \frac{\bar{u}}{a \cos \phi}, \qquad (13)$$

Ω is the angular velocity of the earth, σ is frequency, m is the zonal wave number, $\bar{u}$ is the time mean zonal velocity, a is the radius of the earth and ϕ is latitude [*Salby*, 1981]. Examination of Figure 7 indicates that during January 1989 vertical propagation of the two-day wave was prohibited above 92 km whereas there were no critical regions present between 90 km and 100 km during 1991 or 1992. Similar amplitude characteristics were observed by *Craig et al.* [1980]. At Townsville and Adelaide the amplitude of the two-day wave grew with height below 90 km where a maximum was reached and the amplitude decayed with increasing height above 90 km. In contrast, *Craig et al.* [1980] also observed amplitudes that grew with increasing height at Christchurch until the radar ceiling was reached at 98 km.

The vertical phase structure of the two-day wave during January 1989, 1991 and 1992 shown in Figure 6 indicates the upward transport of energy. The vertical wavelength of the two-day wave during these time intervals are between 36 km and 44 km. This is shorter than the 100 km wavelengths observed at Kyoto (35°N, 136°E) [*Tsuda et al.*, 1988] and Adelaide (35°S, 138°E) [*Craig et al.*, 1980], however a recurring vertical wavelength of 50 km was observed at Townsville (19°S, 147°E) [*Craig et al.*, 1983]. *Harris and Vincent* [1993] have estimated the vertical wavelength to be 70 km at Christmas Island during the period from December 1990 - February 1991. The difference between the vertical wavelength estimated by *Harris and Vincent* [1993] and the vertical wavelength presented herein may be a result of the time intervals analyzed. Our estimate of the vertical wavelength was only computed for the intervals of maximum two-day activity from January 15 to February 15, whereas the estimate of vertical wavelength cited in *Harris and Vincent* [1993] was an average computed for three months. Additionally, *Harris and Vincent* [1993] discuss three case studies of the two-day wave conducted for two week time intervals indicate the vertical wavelength to be shorter during these periods. Examination of the peaks present in the filtered time series from February 1 to February 5, 1991 (Figure 6 of [*Harris and Vincent*, 1993]) indicate a vertical wavelength close to 50 km.

A feature of the two-day wave, similar to phase locking [*Poole*, 1990; *Craig et al.*, 1980], is observed at Christmas Island. The maximum wave amplitudes during January and February occur between 3 am and 6 am solar local time at 90 km (Figure 6). *Clark et al.* [1994] have suggested the phase locking of the two-day wave to a preferred local time indicates an interaction with the solar tides. Moreover, during periods of strong two-day wave activity the diurnal tide is weak. *Harris and Vincent* [1993] have also observed suppressed diurnal tidal amplitudes during periods of high two-day wave activity and suggest a nonlinear interaction between the diurnal tide and the quasi two-day wave.

CONCLUSIONS

Observations of a persistent oscillation in the meridional wind field at Christmas Island exhibit temporal and spatial characteristics indicating it is a manifestation of the two-day wave. The amplitude of the wave in the meridional wind field is between 5 and 50 m s^{-1} with a period of 46 to 54 hours. During maximum wave activity in January the spectrogram indicates the period to be near 48 hours. Although the zonal wave number cannot be determined from a single site the lack of power in zonal wind field near 0.5 cpd is consistent with the (3,0) Rossby-gravity normal mode associated with the two-day wave.

Maximum two-day wave amplitudes occur in late January with sporadic enhancements throughout the remainder of the year. During July and August 1991 a two-day wave event was observed with a mean period near 54 hours. These observations indicate the two-day wave at the equator exhibits characteristics associated with the two-day wave in both the northern and southern hemispheres.

Acknowledgements. This work was supported by the National Science Foundation grant ATM-8918669. The Christmas Island wind profiler is supported by the U.S. Tropical Oceans and Global Atmosphere project office. The Christmas Island profiler is part of the trans-Pacific network which is supported by the National Science Foundation under grants ATM-8720797 and ATM-8720812. The authors would also like to thank John Bryden for supporting the MEDAC system at Christmas Island and the reviewers for their comments and suggestions.

REFERENCES

Avery, S. K., J. P. Avery, T. A. Valentic, S. E. Palo, M. J. Leary, and R. L. Obert, A new meteor echo detection and collection system: Christmas Island mesospheric measurements, *Radio Sci.*, 25, 657–669, 1990.

Bingham, C., M. D. Godfrey, and J. W. Tukey, Modern techniques of power spectrum estimation, *IEEE Trans. Audio Electroacoustics*, AU-15, 56–66, 1967.

Bloomfield, P., *Fourier Analysis of Time Series: An Introduction*, Wiley, New York, 1976.

Campbell, M. J., and A. M. Walker, A survey of statistical work on the MacKenzie River series of annual Canadian lynx trappings for the years 1821-1934, and a new analysis, *J. Roy. Statist. Soc. Ser. A*, 140, 411–431, 1977.

Clark, R. R., The quasi 2-day wave at Durham (43°N): solar magnetic effects, *J. Atmos. Terr. Phys.*, 51, 617–622, 1989.

Clark, R. R., A. C. Current, A. H. Manson, C. E. Meek, S. K. Avery, S. E. Palo, and T. Aso, Global properties of the 2-day wave from mesosphere-lower thermosphere radar observations, *J. Atmos. Terr. Phys.*, in press, 1994.

Craig, R. L., and W. G. Elford, Observations of the quasi 2-day wave near 90 km altitude at Adelaide (35°S), *J. Atmos. Terr. Phys.*, 43, 1051–1056, 1981.

Craig, R. L., R. A. Vincent, G. J. Fraser, and M. J. Smith, The quasi 2-day wave in the southern hemisphere mesosphere, *Nature*, 287, 319–320, 1980.

Craig, R. L., R. A. Vincent, S. P. Kingsley, and H. G. Muller, Simultaneous observations of the quasi 2-day wave in the northern and southern hemispheres, *J. Atmos. Terr. Phys.*, 45, 539–541, 1983.

Fritts, D. C., and J. R. Isler, First observations of mesospheric dynamics with a partial reflection radar in Hawaii (22°N, 160°W), *Geophys. Res. Lett.*, 19, 409–412, 1992.

Glass, M., J. L. Fellous, M. Massebeuf, A. Spizzichino, I. A. Lysenko, and Y. I. Portniaghin, Comparison and interpretation of the results of simultaneous wind measurements in the lower thermosphere at Garchy (France) and Obinsk (U.S.S.R.) by meteor radar technique, *J. Atmos. Terr. Phys.*, 37, 1077–1087, 1975.

Hagan, M. E., J. M. Forbes, and F. Vial, Numerical investigation of the propagation of the quasi-two-day wave into the lower thermosphere, *J. Geophys. Res.*, 98, 23193–23205, 1993.

Harris, T. J., A long-term study of the quasi 2-day wave in the middle atmosphere, *J. Atmos. Terr. Phys.*, in press, 1994.

Harris, T. J., and R. A. Vincent, The quasi 2-day wave observed in the equatorial middle atmosphere, *J. Geophys. Res.*, 98, 10481–10490, 1993.

Kal'chanko, B. V., and S. V. Bulgakov, Study of periodic components of wind velocity in the lower thermosphere above the equator, *Geomag. Aeronomy*, 13, 955–956, 1973.

Kingsley, S. P., H. G. Muller, L. Nelson, and A. Scholefield, Meteor winds over Sheffield (53°N, 2°W), *J. Atmos. Terr. Phys.*, 40, 917–922, 1978.

Muller, H. G., Long-period meteor wind oscillations, *Phil. Trans. R. Soc. Lond. Ser. A.*, 271, 585–598, 1972.

Muller, H. G., and L. Nelson, A travelling quasi 2-day wave in the meteor region, *J. Atmos. Terr. Phys.*, 40, 761–766, 1978.

Palo, S. E., and S. K. Avery, Mean Winds and the Semiannual Oscillation in the Mesosphere and Lower Thermosphere at Christmas Island, *J. Geophys. Res.*, 98, 20385–20400, 1993.

Phillips, A., Simultaneous observations of the quasi 2-day wave at Mawson, Antartica, and Adelaide, South Australia, *J. Atmos. Terr. Phys.*, 51, 119–124, 1989.

Poole, L. M. G., The characteristics of the mesospheric two-day wave as observed at Grahamstown (33.3°S, 26.5°E), *J. Atmos. Terr. Phys.*, 52, 259–268, 1990.

Reddi, C. R., A. Geetha, and K. R. Lekshmi, Quasi 2-day wave in the middle atmosphere over Trivandrum, *Annales Geophysicae*, 6, 231–238, 1988.

Roberts, R. A., and C. T. Mullis, *Digital Signal Processing*, Addison-Wesley, Reading, Mass, 1987.

Rodgers, C. D., and A. J. Prata, Evidence for a traveling two-day wave in the middle atmosphere, *J. Geophys. Res.*, 86, 9661–9664, 1981.

Salby, M. L., The 2-day wave in the middle atmosphere: observations and theory, *J. Geophys. Res.*, 86, 9654–9660, 1981.

Salby, M. L., and R. G. Roper, Long-period oscillations in the meteor region, *J. Atmos. Sci.*, 37, 237–244, 1980.

Tsuda, T., S. Kato, and R. A. Vincent, Long period wind oscillations observed by the Kyoto meteor radar and comparisons of the quasi-2-day wave with Adelaide HF radar observations, *J. Atmos. Terr. Phys.*, 50, 225–230, 1988.

Williams, C. R., and S. K. Avery, Analysis of long-period waves using the MST radar at Poker Flat, Alaska, *J. Geophys. Res.*, 97, 20,855–20,861, 1992.

S. E. Palo and S. K. Avery, Department of Electrical and Computer Engineering, University of Colorado, Campus Box 425, Boulder, CO 80309-0425.

Analysis of Wave Signatures in the Equatorial Ionosphere

H. F. Parish[1]

Center for Space Physics, Boston University, Boston, Massachusetts

J. M. Forbes[2] and F. Kamalabadi

Center for Space Physics and Department of Electrical, Computer and Systems Engineering, Boston University, Boston, Massachusetts

Analyses of magnetic intensity measurements taken over several years indicate fluctuations in the E and F regions of the equatorial ionosphere with periods in the range of 2 to 45 days. The periods of some fluctuations are characteristic of resonant, or Rossby normal mode oscillations, probably excited in the lower atmosphere. Observations from the Huancayo Observatory in Peru (12.00°S, 75.30°W geographic; 0.72°S, 4.78°W geomagnetic) are examined from early 1979 to the end of 1986. Measurements give hourly values of the perturbation of the ground horizontal magnetic field intensity associated with the equatorial electrojet, ΔH. Fluctuations in the magnetic field intensity at ground level can be interpreted in terms of oscillations of wind fields in the E-region (ca. 100-160 km) which cause perturbations in the electric field produced by the wind-driven atmospheric dynamo. Oscillations which are significant in magnitude, are found with periods close to 2, 4, 6, 9 and 16 days throughout 1979 to 1986 and are found to persist with nearly constant periods for intervals of days to weeks. The interpretation of the oscillations in terms of planetary wave variations is supported by observational evidence of a 2 day wave in the mesosphere at the same time as a significant 2 day oscillation in the ΔH measurements. Modulations of diurnal and semidiurnal variations are investigated through bispectrum analysis of ΔH during July to September 1985. The results of this analysis suggest non-linear interactions between oscillations with periods in the range of 12 to 16 days and the diurnal and semidiurnal variations at this time.

1. INTRODUCTION

Oscillations with periods in the range of 2 days to a few weeks are frequently observed in the troposphere and stratosphere and have been associated with resonant, Rossby normal modes (see reviews by *Salby*, 1984; *Madden*, 1979). There is evidence that these oscillations are present in the 80 to 100 km altitude range [*Manson and Meek*, 1986; *Pancheva et al.*, 1989; *Cevolani and Kingsley*, 1992] and that oscillations in this altitude range are coherent with planetary waves observed in the stratosphere [*Fraser*, 1977]. The propagation of planetary waves above 80 km is not well understood, but recent studies show evidence of 5, 10, and 16 day oscillations in the E-region equatorial ionosphere [*Forbes and Leveroni*, 1992] and there is also evidence of a 2-day modulation of the equatorial F-region anomaly [*Chen*, 1992]. Investigations by *Parish et al.*, [1994] show that

---

[1] Now at Radio and Space Plasma Physics Group, Department of Physics and Astronomy, University of Leicester, Leicester, LE1 7RH, United Kingdom.
[2] Now at Department of Aerospace Engineering Sciences, University of Colorado, Boulder, Colorado.

The Upper Mesosphere and Lower Thermosphere:
A Review of Experiment and Theory
Geophysical Monograph 87
Copyright 1995 by the American Geophysical Union

waves with periods characteristic of planetary waves were present in equatorial ionospheric measurements throughout the year of 1979.

This prompts further investigation into the presence of oscillations in the equatorial ionosphere and thermosphere over a number of years. In this analysis, measurements from the magnetic observatory at Huancayo, Peru, which is close to the geomagnetic equator (12.00° S, 75.30° W geographic; 0.72° S, 4.78° W geomagnetic), have been studied for the interval from early 1979 to the end of 1986. The data which are studied are values of the ground horizontal perturbation magnetic field intensity associated with the equatorial electrojet current, ΔH, which is available at hourly intervals. Variations in ΔH are related to changes in currents flowing in the dynamo region. Oscillations in the neutral winds in the dynamo region will produce oscillations in the induced electric fields, $\mathbf{E}$, and hence in ΔH.

Spectral analysis is performed on the ΔH measurements to look for fluctuations with periods of a few days to around two weeks, and the relationship of these fluctuations to planetary wave oscillations is assessed.

In Sections 2.1 and 2.2, the occurrence of different components in the ΔH spectra and the importance of solar flux variations are discussed. In Sections 2.3 and 2.4 the persistence over time of spectral components and their magnitude is examined. In Sections 2.5 and 2.6 evidence of a non-linear interaction between waves with periods in the range of 12 to 16 days and the diurnal and semidiurnal oscillations is investigated through bispectrum analysis.

2. RESULTS

2.1. *Power Spectra*

Spectral analysis techniques were applied to the ΔH measurements to determine the periodicities present in the data and their variations with time. A time-frequency analysis technique known as the short-time Fourier transform (STFT) [*Nawab and Quatieri*, 1988] was used to provide a time-variant measure of frequency content. The STFT can be viewed as a set of Fourier transforms each corresponding to a short-time section of the analysed signal. In this way, the STFT provides a local measure of frequency content, one which changes with time. The ΔH measurements were taken once per hour, giving a Nyquist period of two hours, so that periods greater than two hours are valid in these analyses.

The three-dimensional spectrographs in Figures 1 and 2 show examples of the spectral variation of ΔH with time, for periods in the range of 1.25 to 40 days. Figure

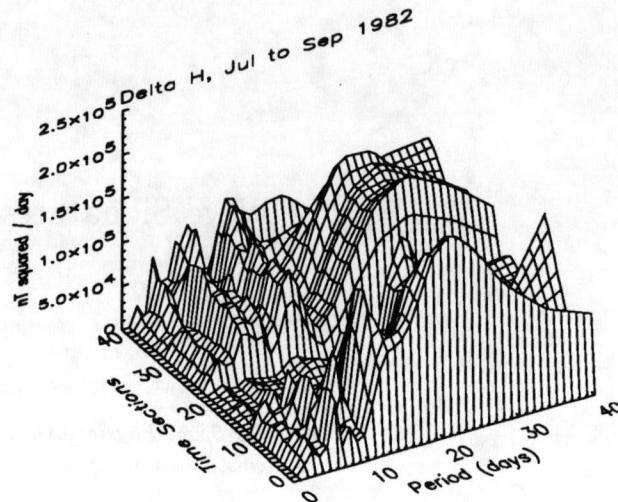

Fig. 1. Three-dimensional spectrograph showing the magnitude squared of the STFT for ΔH during July to September 1982 with a window length of 42.7 days and a decimation interval of 28 hours.

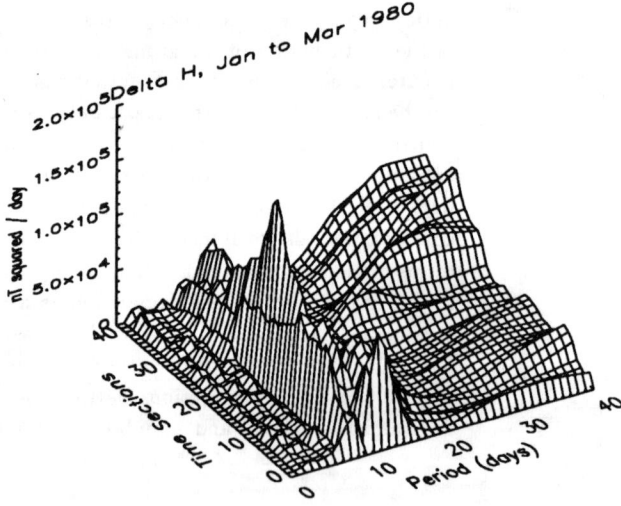

Fig. 2. Three-dimensional spectrograph showing the magnitude squared of the STFT for ΔH during January to March 1980 with a window length of 42.7 days and a decimation interval of 28 hours.

1 corresponds to the interval July to September 1982 (JS82) and Figure 2 represents January to March 1980 (JM80). The three-dimensional spectrographs show the magnitude squared of the STFT for ΔH, computed using 1024-point time windows, each window corresponding to 42.7 days of data, with a decimation interval between consecutive spectra of 28 hours. Each spectrum is produced using an FFT size of four times the length of the time window, i.e. 4096 points, which is

four times the minimum required length and therefore yields better frequency sampling. The period resolution is 3.7 days near the 25 day period, 35.1 hours near the 16 day period, and 32.7 minutes near the 2 day period.

There are several prominent peaks in the spectrograph for JS82 (Figure 1). There is a peak close to 3 days early in the interval, and a peak close to 4 days at the end of the interval. There is also a peak close to 6 days throughout the entire interval. During JM80 (Figure 2), prominent peaks are seen close to 2 days and 3 days throughout the interval and there is also a large peak close to 9 days which is especially prominent in the middle of the interval.

The periodicity at 6 days in Figure 1 may be related to a quasi 5 day normal mode, which has been observed in the troposphere and stratosphere with periods between 4.5 and 6.2 days [*Rodgers*, 1976; *Madden*, 1978; *Madden and Stokes*, 1975]. Peaks close to 3 days and 4 days may be associated with a quasi 4 day wave which has been measured in the stratosphere with periods in the range of 3.8 to 4.5 days [*Hirota and Hirooka*, 1984]. The periodicity close to 9 days in JM80 (Figure 2) may be connected with the quasi 10 day wave, which has been observed in the stratosphere with periods in the range of 7.5 to 12 days and has been associated with the (1,2) normal mode [*Hirooka and Hirota*, 1985; *Salby*, 1984]. Periods close to 2 days in JM80 may be related to a "2-day" wave, which has been measured in the upper stratosphere and mesosphere with periods in the range of 46 to 55 hours and has been associated with the (3,0) normal mode [*Craig and Elford*, 1981; *Kingsley et al*, 1978; *Salby and Roper*, 1980]. The (2,0) normal mode has a 1.8 day period predicted by modelling studies [*Salby*, 1981]. The observed periodicity close to 2 days may be associated with one of these 2-day normal modes.

There are peaks between 12 and 17 days throughout JS82 and JM80. Peaks with periods in the range of 12 to 20 days have frequently been observed in the stratosphere, and have been termed the "16-day" wave [*Madden*, 1978; 1979] and associated with the (1,3) Rossby normal. Several other normal modes may also have periods in this range, for example, the (1,4) mode has a predicted period of 17.5 days, and the (2,4) mode a period of 11.5 days [*Hirooka and Hirota*, 1989]. Periods close to 16 days may also be affected by periodicities due to 13-14 day solar flux variations. This will be discussed further in the next section.

In a previous study *Forbes and Leveroni*, [1992] demonstrate the presence of oscillations in ΔH and foF2 characteristic of the 16-day wave. The noon-time values of foF2 and ΔH analysed by *Forbes and Leveroni* suffer from contamination by lunar aliasing, introducing a periodicity close to 14.75 days, which they correct for empirically. Since our analyses concern hourly data only, any lunar contamination will be greatly reduced.

Large magnitudes are seen for periods between 20 and 30 days, in both intervals. There is an inherent ambiguity in the exact period of any peak, due to the limits of the period resolution (around 4 days in the region of the 25 day period). These peaks may be related to a 27-day solar flux variation (see next section).

The overall magnitude of the spectra for JS82 decreases significantly for a wide range of periods, in the middle of the interval, between the beginning and the middle of August 1982. Some of the periodicites persist over this part of the interval, but with much reduced magnitudes, for example the 6-day period and a period close to 13 days. Since the decrease in magnitude occurs nearly simultaneously at all periods in the middle part of the interval, this suggests that oscillations with a wide range of periods may be excited or damped by a common source at this time.

2.2. *Effects of Solar Flux Variations*

A complication in interpreting spectra such as those illustrated in Figure 1 arises from periodicities in solar emissions which may be reflected in variations in E-region conductivity, and hence in ground magnetic variations. Much of the work on periodicities in UV, EUV, and X-ray emissions has been reviewed by *Donnelly and Puga*, [1990]. Periodicities close to 27 days and sometimes 13.5 days are prominent in the solar flux variations. The influence of solar flux variations on the spectrum of ΔH measurements was investigated in *Parish et al.*, [1994] for the year of 1979. *Parish et al.* conclude that solar flux variations dominate periodicities in the range of 20 to 35 days and that periods between 10 and 20 days may be due to a combination of solar flux variations and other variations such as planetary waves. For periods below 10 days, the effects of solar flux variations are found to be negligible relative to the effects of the other variations, which are probably due to planetary waves. 1979 was a year of high solar activity, and during years of lower solar activity, solar flux variations might be less important. This is investigated further elsewhere (see H.F. Parish, J.M. Forbes, and F. Kamalabadi, A Spectral Analysis of ΔH measurements at Huancayo, Peru, submitted to *Journal of*

Atmospheric and Terrestrial Physics, 1994).

2.3. *Period Peaks*

For the two intervals discussed above, the persistence of the main peaks was investigated. Peaks are plotted as a function of period and time in Figures 3 and 4 for JS82 and JM80 respectively, for successive 42.7 day time windows, with a shift of 28 hours between each time window. Within a given time window the five largest peaks are shown. The day number, in days from 1st July or 1st January respectively, defines the centre of a time window of width 42.7 days.

For both intervals a number of peaks are found to have nearly constant periods over intervals of days to weeks. Peaks close to 6 days, and 12 to 15 days, are significant over the entire JS82 interval and may be related to quasi 5-day and 16-day normal modes respectively. Peaks close to 4 days, 5 days, and 7 to 8 days are present over intervals of several weeks in JS82, and may be related to the "4-day" and "10-day" normal modes.

For JM80 peaks are seen between 8 and 9 days for intervals of several weeks. There is also a periodicity between 2 and 3 days throughout JM80. The interpretation of the 2 to 3 day variations in JM80 in terms of wind oscillations is supported by observations of a significant quasi 2-day wave at altitudes between 65 and 100 km during January to February 1980, as shown by partial reflection drift measurements at Adelaide, Townsville, and Christchurch [*Harris*, 1993; *Craig et al.*, 1980].

There is some evidence during JS82 that waves drift in period with time. For example, the wave close to 5 days appears to drift downwards in period between day 20 and day 45 (Figure 3). The waves close to 6 and 7 days appear to drift to higher periods around day 60.

If the periodicities in ΔH are associated with normal modes in the E-region, then drifts in the periods of the waves may be related to changes in the excitation mechanisms, or changes in the background atmosphere, since the periods and horizontal and vertical structures of normal modes are strongly affected by the wind and temperature structure through which they propagate [*Salby*, 1984].

2.4. *Significance of Peaks*

In order to determine whether the oscillations in the 1.25 to 40 day period range are significant in magnitude, the measurements have been band pass filtered around the main peaks and the resulting amplitudes are given as a percentage of the average noon to midnight range of ΔH.

During JS82 one of the most prominent peaks is close

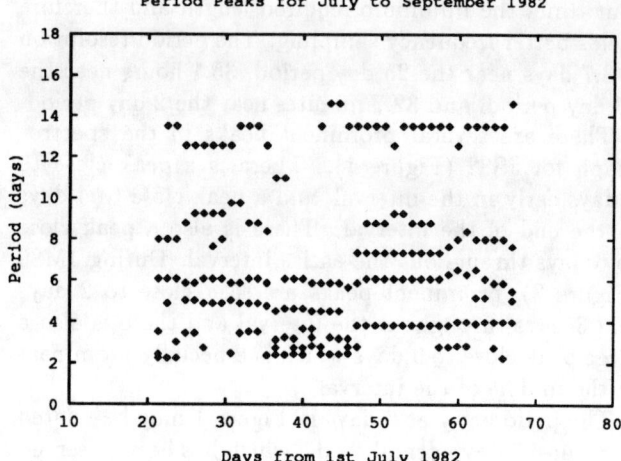

Fig. 3. Peaks in the power spectra of ΔH, for July to September 1982, plotted as a function of period and time for successive 42.7 day time windows starting on July 1st, with a shift of 28 hours between each time window. Within a given time window the 5 largest peaks are shown.

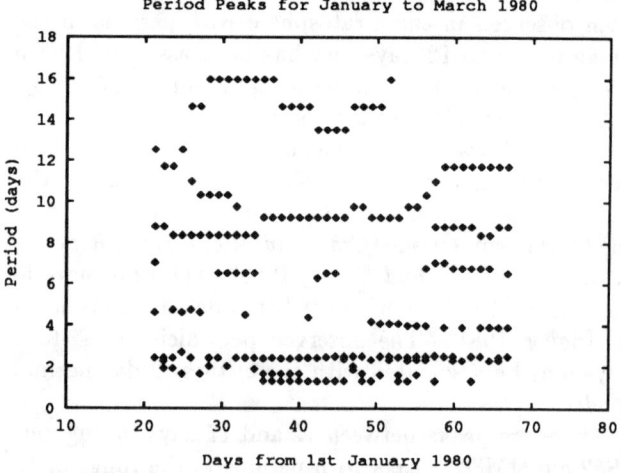

Fig. 4. Peaks in the power spectra of ΔH, for January to March 1980, plotted as a function of period and time for successive 42.7 day time windows starting on January 1st, with a shift of 28 hours between each time window. Within a given time window the 5 largest peaks are shown.

to 6 days. A band pass filter with a bandwidth of 2 days, centred at 5.5 days, was used to investigate the magnitude of the "5-day" wave, based on the range of periods close to the 6 day peak in the ΔH measurements and consistent with the periods suggested by observations [*Madden*, 1978, 1979] and theory [*Salby*, 1984]. A wave with a period close to 2 days is prominent throughout JM80. A band pass filter of width 0.4 days, centred at 2 days, was therefore used to determine the significance of the 2-day wave in JM80, based on values found in the ΔH measurements, and on theory [*Salby*, 1984] and observations [*Kingsley et al.*, 1978].

The data filtered for the quasi 5-day wave in JS82 are shown in Figure 5, as a function of time from July 1st 1982, where values are expressed as a percentage of the average daily noon to midnight range of ΔH, which is assumed to have a nominal value of 140 nT for 1982.

Variations up to 25% of the noon to midnight range of ΔH are seen, showing that the 5 day oscillation in the ΔH measurements is significant in magnitude over this interval. The ΔH measurements vary in magnitude with time, with largest amplitudes at the beginning and end of this interval, consistent with the overall decrease in magnitude over a wide range of periods in the middle of the interval (see Figure 1).

The JM80 data filtered for the 2 day wave, are shown in Figure 6. Variations up to 9 % of the noon to midnight range of ΔH are seen, showing that the 2 day variation is significant in magnitude over the interval. The magnitude is large in the middle of January and the middle of February. This is consistent with the observations of *Craig et al.*, [1980], and *Harris*, [1993] who noted significant magnitudes for a quasi 2-day wave at Adelaide from mid January to the beginning of February 1980. The results in Figures 5 and 6 therefore suggest that the 5 day and 2 day variations, which may be associated with normal modes, are significant in magnitude in the lower thermosphere during the JS82 and JM80 intervals respectively.

2.5. Non-linear Wave Interactions

Close to the magnetic equator the noon to midnight range of ΔH, $r(\Delta H)$, is a measure of the height-integrated current flowing in the electrojet, and $r(\Delta H) \sim \Sigma \mathbf{E}$, where Σ is the height-integrated Cowling conductivity and $\mathbf{E}$ is the zonal electric field (other numerical factors would take into account the external and internal induction ratio and the limited north-south extent of the electrojet). The conductivity, Σ, is strongly influenced by solar variations, and has significant diurnal and semidiurnal harmonics [*Forbes and Lindzen*, 1976], which will appear as corresponding diurnal and semidiurnal variations in ΔH. Oscillations in the neutral wind in the dynamo region caused, for example, by planetary wave or tidal variations, will generate oscillations in the induced electric field, $\mathbf{E}$, and hence in ΔH. Due to the strong influence of solar variations on the

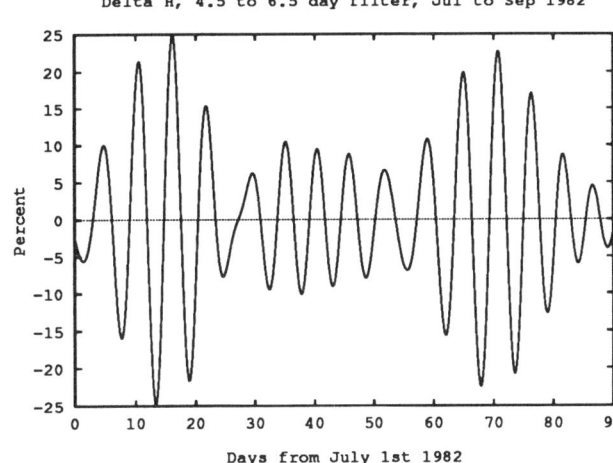

Fig. 5. ΔH measurements filtered with a band-pass filter between 4.5 and 6.5 days, for the interval July to September 1982.

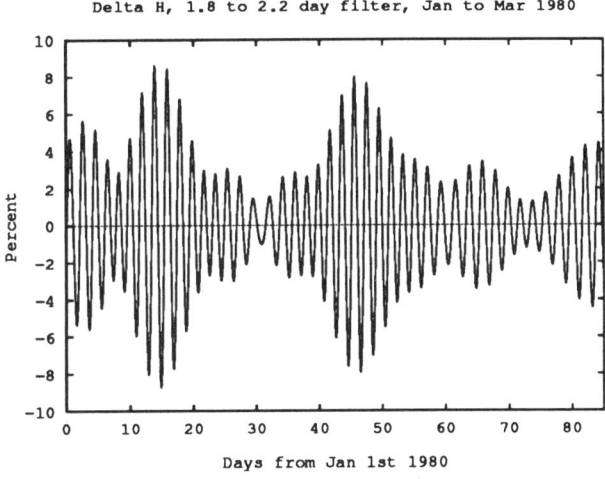

Fig. 6. ΔH measurements filtered with a band-pass filter between 1.8 and 2.2 days, for the interval January to March 1980.

conductivity, the diurnal and semidiurnal variations in ΔH may be mainly related to the Fourier components of the conductivity, rather than the tidal variations of the neutral wind. Oscillations in $\mathbf{E}$ may generate sidebands due to non-linear interactions with the strong diurnal and semidiurnal harmonic components of the conductivity. The frequencies of the sidebands will be given by the sum and difference of the frequencies of the primary waves [*Teitelbaum and Vial*, 1991]. We next examine possible wave-wave interaction through a quantitative analysis.

2.6. Bispectrum Analysis

Observed subsidiary peaks around the diurnal and semidiurnal periods, prompt further investigation of the origin of mechanisms generating such periodicities. Conventional spectrum analysis techniques which are based on the power spectrum, fall short of providing adequate information due to the fact that phase information is suppressed in the power spectrum (autocorrelation) domain. Consequently one must look beyond the power spectrum for identifying phase relations. A powerful technique based on third-order moments, known as *bispectrum analysis*, retains phase information and is therefore capable of detecting and identifying possible nonlinear processes that might generate phase coupling. The bispectrum of a discrete-time signal $x[k]$ is defined [*Nikias and Petropulu*, 1993] as

$$B(w_1, w_2) = \sum_{m=-\infty}^{\infty} \sum_{n=-\infty}^{\infty} R(m,n) e^{-j(w_1 m + w_2 n)} \quad (1)$$

where $R(m,n)$ is the third-order moment sequence given by

$$R(m,n) = E\{x[k]x[k+m]x[k+n]\} \quad (2)$$

and $E\{.\}$ denotes the expectation operator. Note that the bispectrum is equivalent to the two-dimensional Fourier transform of the third-order moment sequence $R(m,n)$. Our implementation of the bispectrum, which belongs to the indirect class of bispectrum estimators, utilizes this fact by computing the two-dimensional fast Fourier transform (FFT) of the third-order moment sequence.

It can be shown [*Nikias and Raghuveer*, 1987] that the bispectrum represents the contribution to the mean product of three Fourier components where one frequency equals the sum of the other two. Consequently the bispectrum will only be non-zero at locations (w_1, w_2) where $w_3 = w_1 \pm w_2$ and $\phi_3 = \phi_1 \pm \phi_2$ where ϕ denotes the phase of the components. Such a phase relation is refered to as *quadratic phase coupling*. In cases where ϕ_3 is random and independent of ϕ_1 and ϕ_2, the bispectrum will be zero. In this way the bispectrum determines if components at harmonically related positions in the power spectrum are, in fact, phase coupled. The non-zero points (t_1, t_2) in the two-dimensional bispectrum period plane indicate periods that are non-linearly interacting and result in quadratic phase coupling.

In order to determine whether such non-linear wave interactions may be occurring, we examine the interval July to September 1985 as an example. First, consider the power spectrum computed for this interval. Figure 7 shows the corresponding power spectral density of ΔH which was computed using a 8192-point FFT. The PSD in Figure 7a is plotted against frequency while the PSD in Figure 7b is depicted against period in the range 1.25 to 45 days. Prominent peaks are seen close to 6 days, 9 days, and 16 days, which may correspond to quasi 5 day, 10 day and 16 day normal modes respectively, and there are also periodicities greater than 20 days, which may be related to solar flux variations, and a large number of peaks with periods less than 5 days, where there is greater period sampling resolution than for longer periods. The most prominent feature in the 1.25 to 45 day spectrum during this interval is the peak close to 16 days. Periodicities between 10 and 20 days may be due to a combination of solar flux and other variations, such as planetary waves, as discussed above. Since 1985 is a year of low solar activity, the periodicities in ΔH due to solar flux variations around 13.5 days may be smaller than in a year of high solar activity. The importance of solar flux variations in a year of low solar activity is investigated further in Parish et al., submitted manuscript, 1994.

Next we compute the bispectrum estimate of ΔH for the same interval. Figure 8 shows the contour plot for a section of the bispectrum plane of ΔH values. The non-zero points in this plane indicate quadratic phase coupling of various degrees between waves with periods corresponding to the x and y values of these points. Note that only the relative magnitudes are significant and that the true values of the bispectrum do not provide any additional information in this context. The two narrow strips in the Figure depict phase coupling between oscillations with periods in the range of 2 to 19 days and the diurnal and semidiurnal components. These quadratic phase couplings generate subsidiary peaks of proportional magnitude around the diurnal and the semidiurnal variations. The two highest regions in the bispectrum, which indicate the strongest interaction, are detected between (a) a wave with period close to 16 days and the diurnal variation, indicating that peaks at 22.59 and 25.60 hours would result from such phase coupling, and (b) the 12-day wave and the semidiurnal component, indicating coupling at periods 11.52 and 12.52 hours.

The spectrum in the region of the 24 hour and 12 hour peaks is shown in Figures 9 and 10 respectively. It is indeed interesting to see sidebands around the 24 hour peak at 0.94 days (22.59 hours) and 1.07 days (25.6 hours), corresponding to the predicted interaction with a 16-day wave. In Figure 11, sidebands can be seen at 0.48 days (11.52 hours) and 0.52 days (12.52 hours), suggesting interaction of the 12-hour oscillation with a

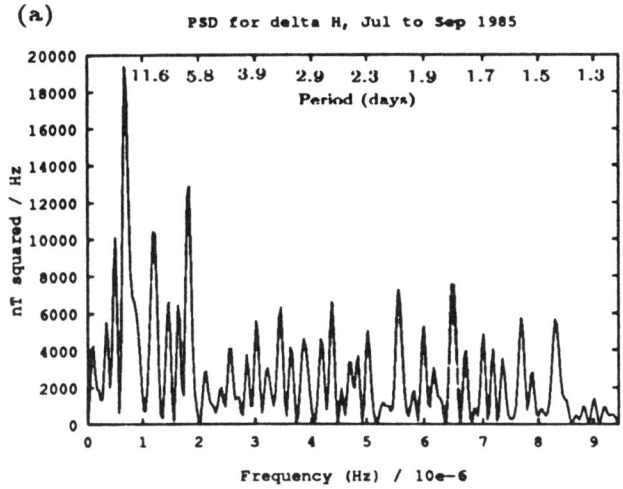

Fig. 7a. Power spectral density as a function of frequency for ΔH, for the interval July to September 1985.

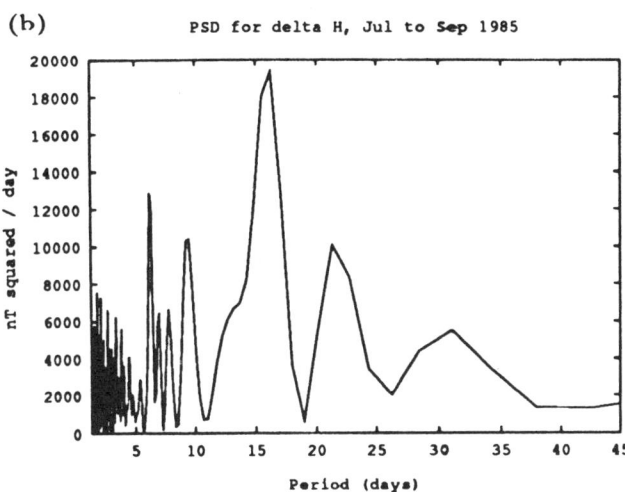

Fig. 7b. Power spectral density as a function of period for ΔH, for the interval July to September 1985, for periods between 1.25 and 45 days.

12-day wave, as predicted. The sideband at 0.52 days on the 12-hour peak is considerably larger than the sideband at 0.48 days. This may be due to the presence of a lunar semidiurnal tidal variation, which would cause oscillations in the neutral winds in the dynamo region and hence in the electric field, and ΔH. A peak is seen close to the expected period of the lunar tide, at 0.518 days, and this peak may contaminate the 12-day sideband of the semidiurnal variation at 0.516 days. Note that the other peaks around the diurnal and the semidiurnal comonents may be partially due to contributions from other quadratic phase interactions indicated by the bispectrum.

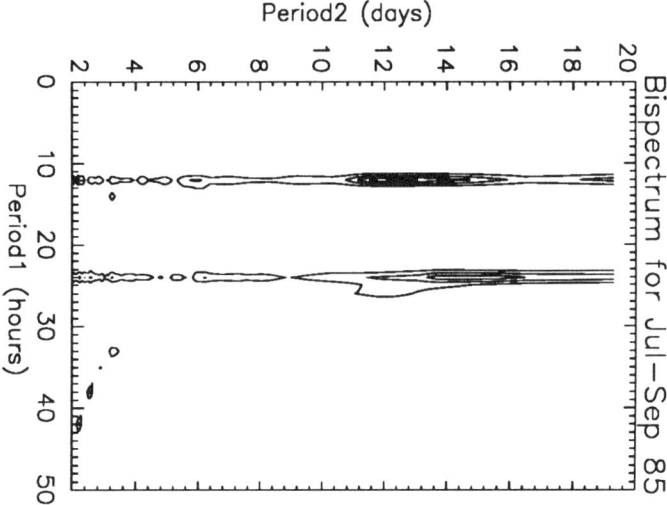

Fig. 8. Contour plot of a section of the bispectrum estimate for ΔH during the interval July to September 1985.

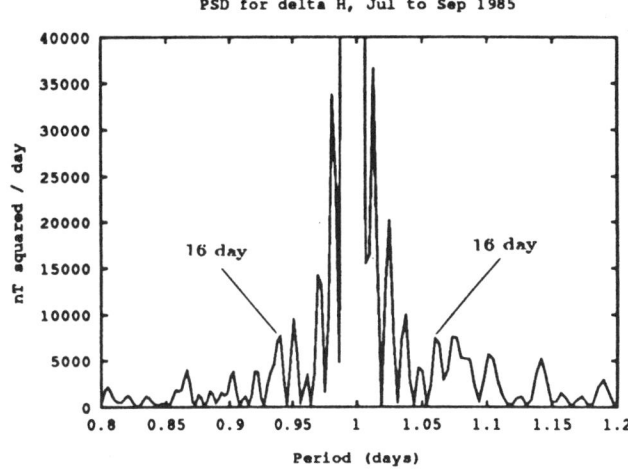

Fig. 9. Power Spectral Density as a function of period for ΔH, for the interval July to September 1985, centred around the 24-hour peak.

3. CONCLUSIONS

Oscillations with periods characteristic of planetary waves are observed in the ground magnetic intensity measured at the Huancayo Observatory in Peru between early 1979 and the end of 1986. Oscillations of significant magnitude are frequently observed with periods close to 16 days, 9 days, 6 days, 4 days and 2 days. Oscillations are found to persist with nearly constant periods over intervals of weeks to months, although there is evidence that some of the variations with periods less than 8 days may drift in period with time. The simulta-

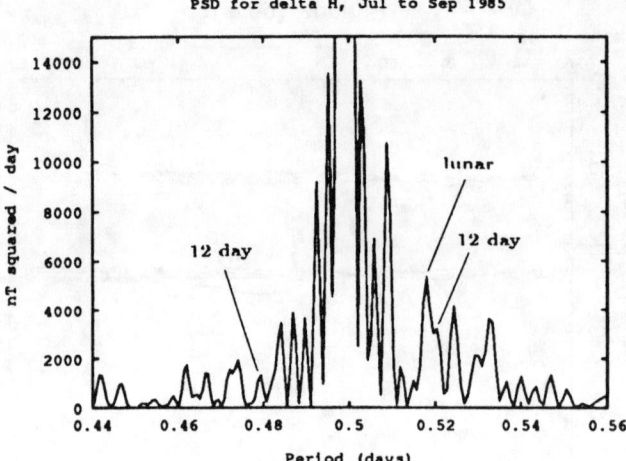

Fig. 10. Power Spectral Density as a function of period for ΔH, for the interval July to September 1985, centred around the 12-hour peak.

neous variation of the magnitude of the ΔH spectrum for a wide range of periods during July to September 1982 suggests that planetary waves may be damped or excited by a common source at this time.

The interpretation of the fluctuations in magnetic intensity in terms of planetary wave variations is supported by observations of a quasi 2 day wave in January to February 1980 at the same time as there was a large 2 day variation in the ΔH measurements for January to March 1980. There is evidence from bispectrum analysis of the July to September interval of 1985, which suggests a non-linear interaction between a quasi 16-day wave and the diurnal oscillation as well as another interaction between a period close to 12 days and the semidiurnal oscillation at this time. Other subsidiary peaks around the diurnal and semidiurnal variations may receive contributions from coupling of these oscillations with waves having periods in the range of 2 to 19 days.

Investigations by *Parish et al.*, [1994] for a year of high solar activity (1979) suggest that EUV flux variations may dominate periodicities above 20 days, and may also be important relative to possible planetary wave effects for periods in the range of 10 to 20 days. This prompts further analysis to determine the importance of solar flux variations during years of lower solar activity, and this is investigated in Parish et al., submitted manuscript, 1994.

Acknowledgments. The authors gratefully acknowledge helpful assistance and discussions with the following individuals, with respect to the importance of EUV flux influences on E-region conductivity: R.Roble (NCAR), R.White (NCAR), R. Donnelly and L. Puga (NOAA/SEL) and K. Tobiska (JPL). This work was supported by grants ATM-9302258 and ATM-9102200 from the National Science Foundation to Boston University.

REFERENCES

Cevolani G., and S.P. Kingsley, Non-linear Effects on Tidal and Planetary Waves in the Lower Thermosphere: Preliminary Results, *Adv. Space Res., 12,* 77-80, 1992.

Chen P.-R., Two day Oscillation of the Equatorial Ionization Anomaly, *J. Geophys. Res., 97,* 6343-6357, 1992.

Craig R.L., and W.G. Elford, Observations of the quasi two-day wave near 90 km altitude at Adelaide (35 S), *J. Atmos. Terr. Phys., 43,* 1051-1056, 1981.

Craig R.L., R.A. Vincent, G.J. Fraser and M.J. Smith, The Quasi 2-day Wave in the Southern Hemisphere Mesosphere, *Nature, 287,* 319-320, 1980.

Donnelly, R.F., and L.C. Puga, Thirteen-day periodicity and the center-to-limb dependence of UV, EUV, and X-ray emission of solar activity, *Solar Physics, 130,* 369-390, 1990.

Forbes J.M., and S. Leveroni, Quasi 16-day Oscillation in the Ionopshere, *Geophys. Res. Lett., 19,* 981-984, 1992.

Forbes J.M., and R.S. Lindzen, Atmospheric Solar Tides and their Electrodynamic Effects - I. The global Sq current system, *J. Atmos. Terr. Phys., 38,* 897-910, 1976.

Fraser G.J., The 5-day Wave and Ionospheric Absorption, *J. Atmos. Terr. Phys., 39,* 121-124, 1977.

Harris T.J., A long term study of the Quasi-Two-Day Wave in the Middle Atmosphere, submitted to *J. Atmos. Terr. Phys.,* 1993.

Hirooka T., and I. Hirota, Normal Mode Rossby Waves Observed in the Upper Stratosphere. Part II: Second Anitsymmetric and Symmetric Modes of Zonal Wave Numbers 1 and 2, *J. Atmos. Sci., 42,* 536-548, 1985.

Hirooka T., and I. Hirota, Further Evidence of Normal Mode Rossby Waves, *Pure and Appl. Geophys., 130,* 277-289, 1989.

Hirota I., and T. Hirooka, Normal Mode Rossby Waves Observed in the Upper Stratosphere. Part I: First Symmetric Modes of Zonal Wavenumbers 1 and 2, *J. Atmos. Sci., 41,* 1253-1267, 1984.

Kingsley S.P., H.G. Muller, L. Nelson, and A. Scholfield, Meteor winds over Sheffield, *J. Atmos. Terr. Phys., 40,* 917-922, 1978.

Madden R.A., Further Evidence of Traveling Planetary Waves, *J. Atmos. Sci., 35,* 1605-1618, 1978.

Madden R.A., Observations of Large-Scale Traveling Rossby waves, *Rev. Geophys. Space Phys., 17,* 1935-1949, 1979.

Madden R.A., and J. Stokes, Evidence of Global-Scale 5-

day Waves in the 73-year Pressure Record, *J. Atmos. Sci., 32*, 831-836, 1975.

Manson A., and C. Meek, Dynamics of the Middle Atmosphere at Saskatoon (52°N, 107°W): A Spectral Study During 1981,1982, *J. Atmos. Terr. Phys., 48*, 1039-1055, 1986.

Nawab S.H., and T.F. Quatieri, Short-Time Fourier Transform, in *Advanced Topics in Signal Processing*, chapter 6, ed. Lim J.S., and Oppenheim A.V., Prentice Hall, Englewood Cliffs, New Jersey, 1988.

Nikias C.L., and A.P. Petropulu, *Higher-order Spectra Analysis: A Nonlinear Signal Processing Framework*, Prentice Hall, Englewood Cliffs, New Jersey, 1993.

Nikias C.L., and M.R. Raghuveer, Bispectrum Estimation: A Digital Signal Processing Framework, *Proc. of IEEE., 75, No. 7*, 869-891, 1987.

Pancheva D., E. Apostolov, J. Laštovička, and J. Boška, Long-period fluctuations of meteorological origin observed in the lower ionosphere, *J. Atmos. Terr. Phys., 51*, 381-388, 1989.

Parish H.F., J.M. Forbes, and F. Kamalabadi, Planetary Wave and Solar Emission Signatures in the Equatorial Electrojet, *J. Geophys. Res., 99*, 355-368, 1994.

Rodgers C.D., Evidence for the Five-day Wave in the Upper Stratosphere, *J. Atmos. Sci., 33*, 710-711, 1976.

Salby M.L., Rossby Normal Modes in Nonuniform Background Configurations. Part 1: Simple Fields, *J. Atmos. Sci., 38*, 1803-1826, 1981.

Salby M.L., Survey of Planetary-Scale Traveling Waves: The State of Theory and Observations, *Rev. Geophys. and Space Phys., 22*, 209-236, 1984.

Salby M.L., and R. Roper, Long Period Oscillations in the Meteor Region, *J. Atmos. Sci., 37*, 237-244, 1980.

Teitelbaum H. and F. Vial, On Tidal Variability Induced by Nonlinear Interaction with Planetary Waves, *J. Geophys. Res., 96*, 14169-14178, 1991.

J. M. Forbes, Department of Aerospace Engineering Sciences, University of Colorado, Boulder, Colorado.

F. Kamalabadi, Center for Space Physics, 725, Commonwealth Avenue, Boston University, Boston, MA 02215.

H. F. Parish, Radio and Space Plasma Physics Group, Department of Physics and Astronomy, University of Leicester, Leicester, LE1 7RH, United Kingdom.

Gravity Wave-Tidal Interactions in the Middle Atmosphere: Observations and Theory

DAVID C. FRITTS

Laboratory for Atmospheric and Space Physics and Department of Electrical and Computer Engineering, University of Colorado, Boulder, CO 80309

Evidence has emerged in recent years of strong interactions between gravity waves and tides in the middle atmosphere. These interactions are likely to be important both in determining the mean and variable wave structures in the mesosphere and lower thermosphere and in modulating wave forcing of the large-scale circulation and thermal structure of this region. Gravity waves are influenced by large tidal wind shears, which filter and impose directional anisotropy on the gravity wave spectrum. Gravity waves may also be sensitive to stability variations induced by tidal thermal fluctuations. Together, these processes modulate the gravity wave fluxes of energy and momentum and the turbulent diffusion, heating, and accelerations accompanying wave dissipation. Diurnal variability imposed by tidal filtering also influences the tidal structures themselves, contributing expected variations in tidal amplitudes and phases. This paper reviews the observational evidence of gravity wave-tidal interactions available at this time and the modeling studies that have been performed of these processes to date.

1. INTRODUCTION

Gravity waves and tides are ubiquitous features of the motion spectrum in the mesosphere and lower thermosphere (MLT) due to their amplitude growth with increasing altitude. The primary sources of gravity wave activity in the lower atmosphere are believed to be topography, frontal and convective activity, and wind shear. Tidal structures are excited primarily through absorption of solar radiation by water vapor in the troposphere and by ozone in the stratosphere. Both gravity waves and tides experience considerable growth with altitude and achieve their maximum amplitudes in the MLT. Because these motions become large at comparable altitudes, they are also subject to strong mutual interactions which have important implications for each other and for the MLT.

Our purpose in this paper is to review the evidence of gravity wave-tidal interactions available in observational data as well as the modeling studies that have been performed and their implications for modulation and variability of both motion fields. We begin by reviewing in section 2 the characteristics of the gravity wave and tidal fields which enable them to interact strongly. The factors most important to the interaction appear to be the disparate time scales of the tidal motions and of the gravity wave motions accounting for the majority of the momentum flux. Evidence of gravity wave-tidal interactions in observational data is presented in section 3. This reveals strong modulation of gravity wave momentum fluxes at tidal periods, a characteristic phase relationship between momentum fluxes and tidal winds, and a tendency for substantial tidal variability in response to gravity wave modulation. Modeling studies of gravity wave-tidal interactions are considered in section 4 and are seen to exhibit several features in common with observations, despite very different model formulations. The various models also make predictions for gravity wave and tidal behaviour which will have important implications for the variability and effects of these motions in the MLT and numerical models, but which are poorly defined at present. Our conclusions are presented in section 5.

2. CHARACTERISTICS OF GRAVITY WAVE AND TIDAL MOTIONS IN THE MLT

Numerous observations are now available which enable us to describe the gravity wave and tidal fields in the MLT in a statistical manner. Despite considerable variability, both gravity wave and tidal motions exhibit certain characteristics and/or mean structures that vary little with altitude or season.

2.1. Gravity Wave Structure and Fluxes

Gravity waves display a growth with altitude of mean wave energy density (per unit mass) given approximately by

$$E(z) \sim e^{z/H_E}, \qquad (1)$$

where $H_E \simeq 2.3H$, H is the density scale height, and H_E varies only slightly with altitude and season below $\sim 100\ km$ [Fritts and VanZandt, 1993]. At greater altitudes, wave energy densities eventually decay, due to increasing molecular diffusion and heat conduction, but there is only limited data on the mean variations at present [Fritts et al., 1993].

Due to saturation processes that act to constrain wave amplitudes at large vertical wavenumbers, $m \gg m_*$, where m_* is the characteristic (or most energetic) vertical wavenumber, this characteristic scale also exhibits a systematic variation with altitude given by

$$m_* \sim e^{-z/H_*}, \qquad (2)$$

where $H_* \simeq 2H_E$, due to the relationship between E_0 and m_* implied by the mean spectral form [Fritts and VanZandt, 1993].

Because the forms of the gravity wave vertical wavenumber and inferred intrinsic frequency spectra are nearly invariant with altitude, gravity wave fluxes of energy and momentum may likewise be expressed approximately as

$$F_E \simeq \frac{NE_0}{18m_*} \qquad (3)$$

and

$$F_{Px,y} \simeq b_{x,y}\frac{E_0}{22}, \qquad (4)$$

where the coefficients $N/18m_*$ and $1/22$ may be regarded as an effective vertical group velocity and an effective frequency (normalized by N) for gravity wave transports of energy and momentum and $-1 \leq b_x + b_y \leq 1$ express the degree of anisotropy (or preferential direction of propagation) in the zonal and meridional directions within the gravity wave field due to source or filtering conditions [Fritts and VanZandt, 1993].

With $E_0 \sim 10\ m^2 s^{-2}$ near the tropopause, eqs. (3) and (4) imply energy and momentum fluxes of $\sim 5 \times 10^3\ m^3 s^{-3}$ and $50 b_{x,y}\ m^2 s^{-2}$ near the mesopause. In regions where the environment and wave field anisotropy are approximately constant with altitude, wave dissipation leads to an energy dissipation rate and accelerations given approximately by

$$\epsilon \simeq F_E(\frac{1}{H} - \frac{3}{2H_E}) \simeq 0.28 m^2 s^{-3}, \qquad (5)$$

and

$$D_{Fx,y} \simeq F_{Px,y}(\frac{1}{H} - \frac{1}{H_E}) \simeq 350 ms^{-1} day^{-1}, \qquad (6)$$

where we have assumed that all the energy removed from the wave field has been dissipated, representing an upper limit on the energy dissipation rate.

The above flux and flux divergence estimates were found to be in good agreement with the atmospheric observations reviewed by *Fritts and VanZandt* [1993] and also by *Fritts* [1994] in this volume, and the interested reader is referred to those papers for the relevant references. What is important for our purposes here is to recognize that these gravity wave flux divergences represent a substantial forcing of the mean MLT circulation and structure and a potential for substantial interaction with and modulation of the MLT tidal structures because of the magnitude of the forcing and the time scales characteristic of gravity wave energy and momentum fluxes.

As reviewed by *Fritts* [1994], mean gravity wave momentum fluxes achieve maximum values of ~ 5 to $15\ m^2 s^{-2}$ in the MLT, with $\sim 70\%$ of this flux associated with gravity wave motions with periods less than 1 hour. This suggests small anisotropies of $b_{x,y} \sim 0.1$ to 0.3 in the mean. Maximum flux estimates of $\sim 50\ m^2 s^{-2}$ for individual hourly estimates, on the other hand, imply highly anisotropic motions with $|b_{x,y}| \sim 1$. Corresponding flux divergences likewise imply potentially strong and variable turbulent diffusion and local accelerations which may influence strongly the local mean circulation and thermal and constituent distributions as well as the MLT tidal fields.

2.2. Tidal Structure

Mean tidal structures in the MLT are becoming better known due to advances in tidal modeling and extensive observations [*Forbes*, 1982a; *Forbes*, 1982b; *Vial*, 1986; *Avery et al.*, 1989; *Forbes and Vial*, 1989; *Fraser et al.*, 1989; *Manson et al.*, 1989; *Vial and Forbes*, 1989; *Vincent et al.*, 1989]. These suggest mean tidal structures that vary in systematic ways with season, altitude, and latitude, with maximum amplitudes of the propagating modes in the MLT. Also improving is the agreement with observed amplitudes and phases when realistic wind and thermal fields and damping schemes are included in tidal models.

Mean tidal amplitudes vary from ~ 10 to $40\ m\ s^{-1}$ and vertical wavelengths range from $\sim 25\ km$ upward, depending on latitude and season. Like gravity waves, however, the tidal fields are found to be highly variable on short and long time scales. Variability on short time scales may arise in response to transience in tidal forcing and propagation [*Vial et al.*, 1991] or to the gravity wave-tidal interactions that are the subject of this paper. On longer time scales, tidal variability appears to be due to systematic seasonal variations in tidal forcing and the circulation and thermal structure of the lower and middle atmosphere and to interactions with planetary wave structures occurring at somewhat lower frequencies.

An example of the short- and long-term variability in the diurnal amplitude and phase observed in 4-*day* and 30-*day* fits to the MLT wind field measured with the MF radar at Kauai, Hawaii ($22°N$) by *Fritts and Isler* [1993] is shown in Figure 1. The 30-*day* fits were performed at 10-*day* intervals and exhibit longer-term and seasonal variations in amplitude and phase which are approximately reproducible from 1991 to 1992. The 4-*day* fits at 2-*day* intervals, however, exhibit substantially greater variability in amplitude and phase, with amplitudes varying by as much as ~ 5 times at periods of ~ 10 to 20 days. These latter variations are suggestive of strong interactions of the tides with or modulation by the planetary wave motions at these periods. Indeed, power spectra of the 4-*day* tide fits compiled by *Fritts and Isler* [1993] reveal a maximum response at a 16-*day* period, corresponding to a dominant period of the planetary wave field.

The above discussion has made it clear that both gravity waves and tides achieve large amplitudes and are expected to have large effects as a consequence of wave-wave and wave-mean flow interactions in the MLT region. The remainder of this paper will focus more specifically on the interactions and mutual influences among these motions.

3. OBSERVATIONAL EVIDENCE OF GRAVITY WAVE-TIDAL INTERACTIONS

There have been many observational studies of gravity wave and tidal propagation, forcing, and effects in the MLT to date. Only two of these studies have specifically addressed gravity wave-tidal interactions, however. This is due, in part, to the need both for detailed information on the direction of gravity wave propagation and for full diurnal definition of the large-scale motion field in order to assess the strength of the interaction and the filtering effects of the tidal wind fields. Information on propagation direction (or wave field anisotropy) is most easily obtained with measurement systems sensitive to gravity wave momentum fluxes (see section 2), while full diurnal measurements require either sensitivity to the motion field during nighttime, when electron densities are far below daytime values, or continuous daytime conditions. The two systems with which such measurements have been possible are the MF Doppler radar at Adelaide, Australia and the MST radar at Poker Flat, Alaska during summer.

The Adelaide MF radar was used by *Fritts and Vincent* [1987] to examine the mean and variable momentum fluxes due to gravity waves in the winter MLT. It was found, consistent with theoretical expectations, that there was a negative mean momentum flux and an associated westward acceleration that opposed the mean zonal eastward wind, resulting in closure of the winter mesospheric jet structure and a residual circulation with subsidence near the winter polar mesosphere and mean poleward motions at mesopause altitudes and lower latitudes (see the discussion of the zonal-mean equations and the "downward control principle" by *McIntyre* [1989] or by *Fritts* [1994] in this volume).

Not expected, but also observed in this data set was a strong modulation of the gravity wave momentum flux computed for 1-*h* and 8-*h* intervals having a diurnal period. These data, averaged in 8-*h* intervals, are displayed at nine altitudes for three days in Figure 2. What is most noteworthy here is that the maximum momentum fluxes for these 8-*h* intervals exceed $\sim 30 \ m^2 s^{-2}$ in magnitude, with individual hourly values (not shown) as high as $\sim 60 \ m^2 s^{-2}$. The mean momentum flux for the entire observation period, in contrast, was $\sim -5 \ m^2 s^{-2}$. Thus, it was obvious that some process was acting to modulate strongly the measured momentum fluxes.

The source of the variable momentum fluxes proved to be a highly variable mean and tidal wind field, with the maximum negative fluxes occurring during the interval in which the tidal wind shears reversed the mean wind shear and resulted in positive zonal shears and a maximum positive zonal wind above 90 *km*. Local mean (mean plus tidal) zonal wind profiles centered at 0000, 0800, and 1600 local time (LT) and averaged over the three-day interval shown in Figure 2 are displayed in Figure 3. The corresponding gravity wave behavior during this interval is shown with vertical profiles of west-beam variance and momentum flux averaged for the 8-*h* blocks centered at the above times and over three days in Figure 4. Examining these figures, we note that the momentum fluxes appear to have been constrained near zero for the intervals with local mean winds having a negative zonal shear with height. At lower altitudes, the momentum fluxes were comparable to the mean negative value inferred for the full data set (see *Fritts* [1994]). At greater altitudes, however, these fluxes increased to smaller negative or positive values, suggesting removal or filtering of wave motions with zonal phase speeds less than the local mean winds. Only when the local mean shear was reversed did west-beam variance and net westward momentum flux increase approximately exponentially with altitude, suggesting in this case approximately conservative wave propagation at these altitudes and times. The interpretation and implications of these results will be discussed at greater length in the following section.

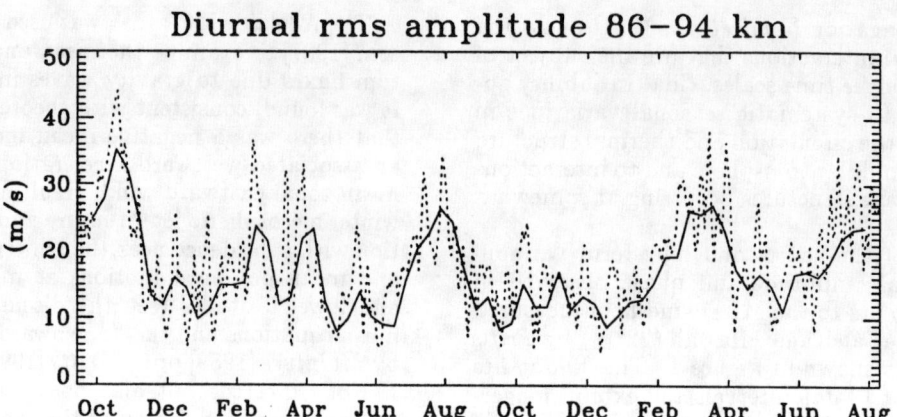
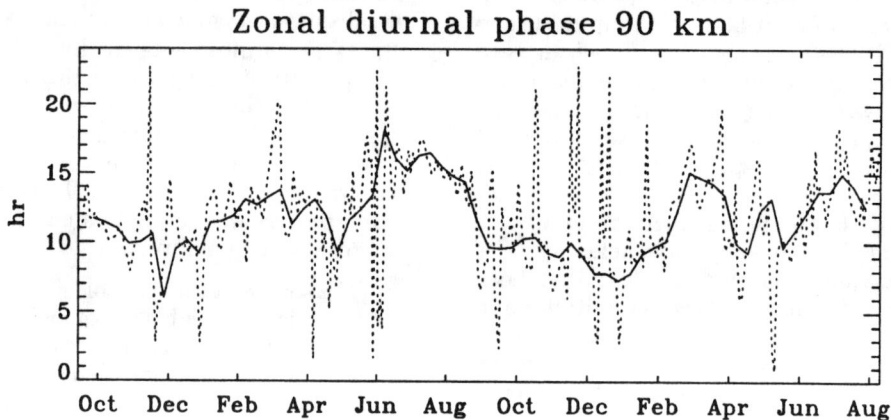
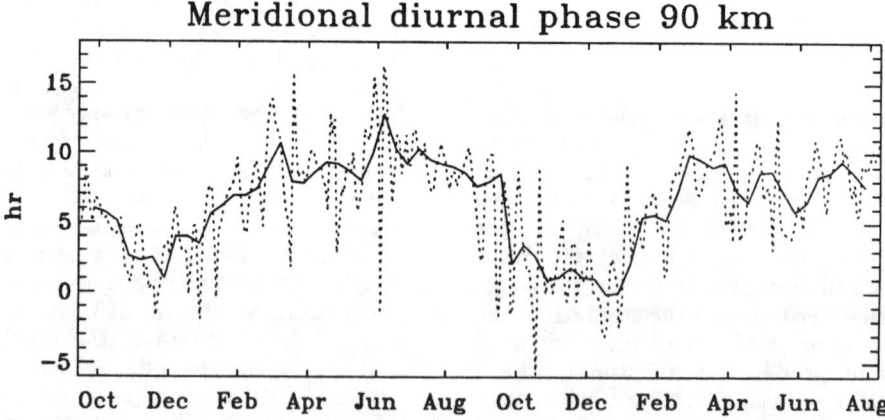

Figure 1. RMS diurnal amplitude computed for 30-*day* (solid) and 4-*day* (dashed) data segments at 10-*day* and 2-*day* intervals and averaged from 86 to 94 *km* (top) and zonal and meridional phases at 90 *km* for the same intervals and averages (middle and bottom).

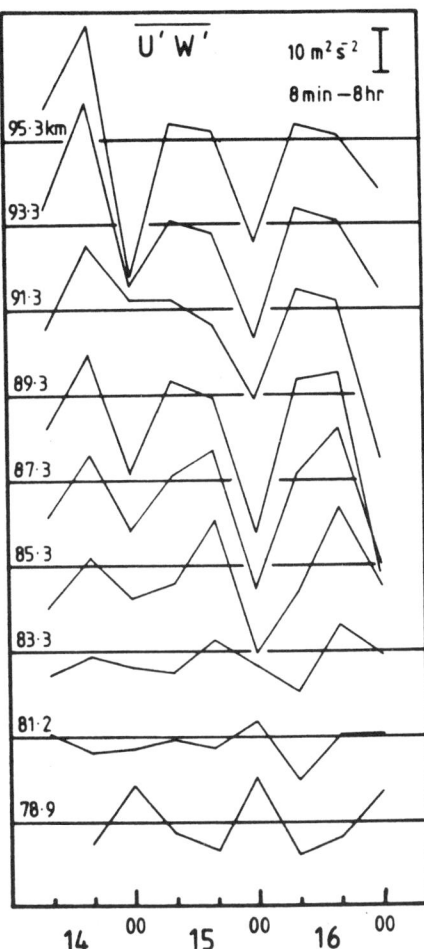

Figure 2. Momentum fluxes measured with the Adelaide MF radar and averaged in 8-h intervals for three days in June 1984 with large diurnal tidal motions.

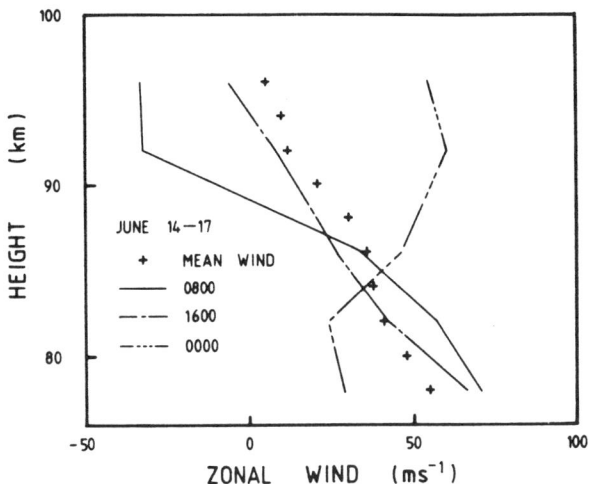

Figure 3. Local mean zonal wind profiles for the times and averaging intervals for which momentum fluxes were computed. The local mean shear is reversed by the diurnal tide near 0000 LT.

A similar, more systematic, study of possible gravity wave-tidal interactions using the Poker Flat MST radar was performed by *Wang and Fritts* [1991] using data collected during 16 days in July 1986 when the mesosphere was continuously sunlit. These data were analyzed by computing hourly mean winds and momentum fluxes and averaging the results by hour-of-day for 16-*day* and 4-*day* segments. The resulting data at each altitude were then fitted to obtain the diurnal and semi-diurnal amplitudes and phases for each series.

The diurnal and semi-diurnal fits for the 4-*day* interval of 15 to 18 July, which experienced larger than average tidal amplitudes, are shown in Figures 5 and 6. Those profiles in Figure 5 reveal diurnal tidal amplitudes of $\sim$ 10 to 30 $m\ s^{-1}$, with larger amplitudes below than above the mesopause ($\sim$ 86 km). Diurnal tidal phases exhibit a general descent with time, but with nearly evanescent behavior in the zonal component above the mesopause. Diurnally-varying momentum fluxes achieved maximum values of $\sim$ 10 and 5 $m^2 s^{-2}$ for the zonal and meridional directions, respectively, with the maximum values occurring coincident with the larger tidal amplitudes at lower levels. What is most relevant to this discussion, however, is the relationship between the diurnal wind and momentum flux phases observed at altitudes where both amplitudes are large. Referring to the zonal profiles below $\sim$ 84 km and above $\sim$ 87 km and to the meridional profiles from $\sim$ 82 to 85 km in Figure 5, we note a tendency for the tidal winds and the gravity wave momentum fluxes to be in approximate anti-phase, with the momentum fluxes typically leading the tidal winds by $\sim$ 9 to 12 h.

The semi-diurnal tidal wind and gravity wave momentum flux amplitudes and phases displayed in Figure 6 exhibit behavior very similar to that seen in the diurnal fits. Semi-diurnal winds are smaller, in general, than the diurnal components, with the maximum response ($\sim$ 20 $m\ s^{-1}$) in the zonal component at lower altitudes. But like the diurnal fits, the semi-diurnal winds and momentum fluxes achieve large amplitudes at common altitudes and exhibit a strong anti-correlation in phase, with a phase advance of the momentum flux of very nearly 6 h at altitudes with significant wind and momentum flux amplitudes (see especially below $\sim$ 85 km in the zonal component and below $\sim$ 83 km in the meridional component).

The Poker Flat observations are in good agreement with those by *Fritts and Vincent* [1987] discussed above and with other observations of local mean wind and momentum flux anti-correlations which did not address gravity wave-tidal interactions specifical-

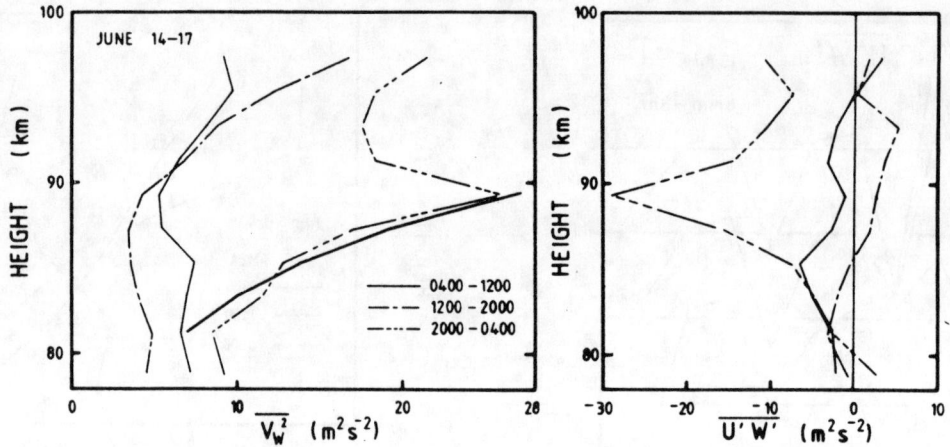

Figure 4. West-beam variance and zonal momentum flux averaged for the intervals displayed in Figures 2 and 3. The heavy solid line with the variance profiles shows an exponential growth with altitude for conservative wave motions.

ly [*Reid and Vincent*, 1987; *Reid et al.*, 1988; *Fritts and Yuan*, 1989; *Rüster and Reid*, 1990; *Fritts et al.*, 1992]. Similar anti-correlations of long-term mean winds and momentum fluxes have also been observed [*Tsuda et al.*, 1990; *Wang and Fritts*, 1990] and are consistent with those needed to account for the large-scale circulation and thermal structure of the MLT.

4. Modeling Studies of Gravity Wave-Tidal Interactions

Several modeling studies have addressed the mutual interactions among gravity waves and tidal motions in the MLT. The first, by *Walterscheid* [1981], considered the accelerations implied by the dissipation of gravity waves with equal and opposite zonal phase speeds near critical levels induced by the semi-diurnal tide. This study found such forcing to lead, in the case of larger gravity wave phase speeds (and momentum fluxes), to induced mean motions that were competitive with the imposed tidal amplitudes. It was concluded that gravity wave forcing could account for much of the variability noted in observations of the semi-diurnal tidal motion.

A subsequent conceptual model proposed by *Fritts and Vincent* [1987] was motivated by the observations described in section 3 above and relied on gravity wave saturation to limit wave amplitudes consistent with linear theory ($u' \leq c - \overline{u}$). This represents a generalization of the dissipation preferentially near critical levels assumed by *Walterscheid* [1981] and enables significant accelerations to occur where wave phase speeds, amplitudes, and momentum fluxes are large, but constrained by saturation conditions. The effects can be quantified by assuming a single gravity wave component with zonal phase speed c, a saturated amplitude $u' = c - \overline{u}$, a vertical velocity $w' = -ku'/m$, with k and m the horizontal and vertical wavenumbers of the motion, and a momentum flux (per unit mass)

$$\overline{u'w'} = -k(\overline{u} - c)^3/2N, \qquad (7)$$

where N is the buoyancy frequency.

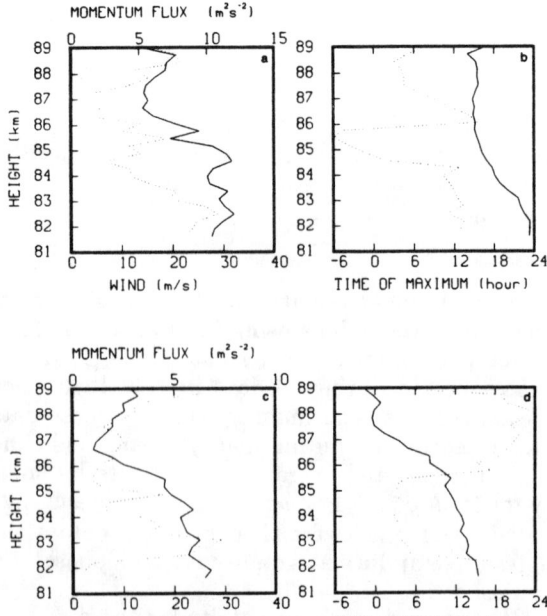

Figure 5. Diurnal wind (solid) and momentum flux (dotted) amplitudes and phases for the zonal (upper) and meridional (lower) components during 15 to 18 July 1986 over Poker Flat, Alaska. Note the anti-phase relationship where amplitudes are large.

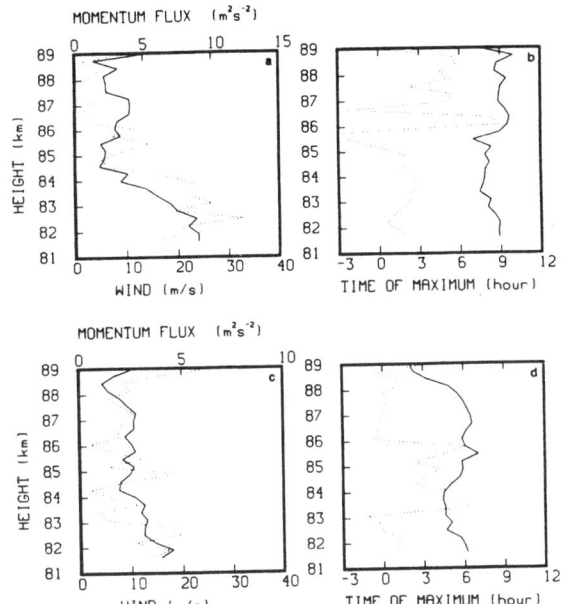

Figure 6. As in Figure 5, but for the semi-diurnal wind and momentum flux fits.

Then, to a good approximation (see *Fritts* [1994]), the implied zonal acceleration and/or balancing Coriolis torque in regions of wave dissipation are given by

$$\overline{u}_t - f\overline{v}^* = -\frac{k(\overline{u}-c)^2}{2N}[\frac{(\overline{u}-c)}{H} - 3\overline{u}_z], \quad (8)$$

where subscripts denote differentiation, $\overline{v}^*$ is the meridional component of the wave-driven residual (or diabatic) circulation, and we have assumed that

$$3\overline{u}_z \leq (\overline{u}-c)/H. \quad (9)$$

When this condition is violated, the intrinsic phase speed increases more rapidly with height than allowed for u' assuming conservative wave propagation and the momentum flux is constrained to grow as

$$\overline{u'w'} \sim e^{z/H}. \quad (10)$$

Implied variations in momentum flux for $c < \overline{u}$ with two mean profiles representative of the variations observed by *Fritts and Vincent* [1987] due to tidal motions are shown in Figure 7. These suggest substantially greater accelerations where gravity wave amplitudes are large, but constrained than where wave amplitudes have already become small due to saturation during critical level approach. Specifically, this conceptual model appears to capture the structure and correlations between local mean winds and gravity wave momentum fluxes observed by *Fritts and Vincent* [1987] and described in section 3. The model also describes the manner in which a gravity wave field comprised of many components with different phase speeds will

be modulated or filtered by the mean and tidal wind fields. Conversely, the momentum flux profiles imply forcing of the mean and tidal structures given by eq. (8) where the divergence of the momentum flux is large. In these respects, this formulation is equivalent to that by *Walterscheid* [1981].

The implications of gravity wave forcing expressed by eq. (8) are illustrated qualitatively in Figure 8. Assuming that the largest momentum fluxes accompany the maximum tidal winds, the implied accelerations will also be aligned closely with the wind maxima, suggesting a gravity wave enhancement of the tidal accelerations. This would result in an advance of the tidal phase (towards earlier times) during periods of strong forcing, consistent with the observations by *Fritts and Vincent* [1987]. Also proposed by these authors was a reduction in apparent tidal amplitudes. This prediction is more sensitive, however, to the phase relation between the tidal wind field and the resulting gravity wave momentum fluxes and, as will be seen below, has not yet been resolved in more quantitative models of gravity wave-tidal interactions.

The study by *Wang and Fritts* [1991] posed two simple, analytic models of the gravity wave-tidal interaction in order to assess the degree of wave field anisotropy required to account for observed momentum flux variations. These authors concluded that relatively small anisotropies could account for the weak momentum flux modulations noted in long-term fits to observed data. Large and variable momentum flux-

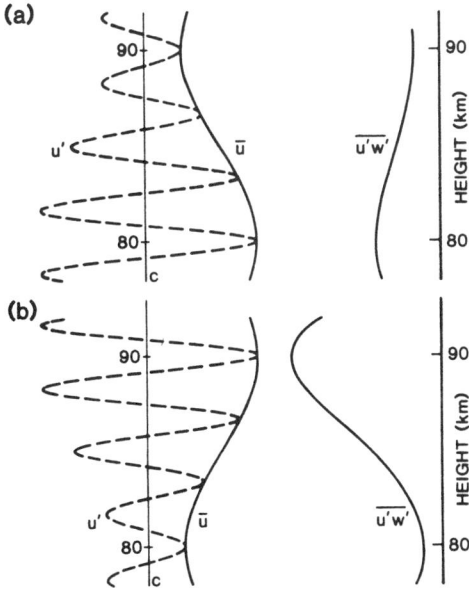

Figure 7. Schematic of a saturated gravity wave amplitude and momentum flux with $\overline{u}-c > 0$ and decreasing (top) or increasing (bottom) with altitude.

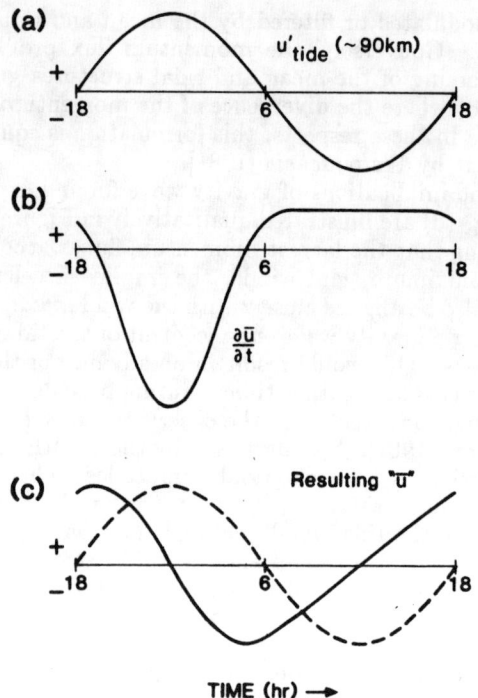

Figure 8. Schematic of the temporal variations of a tidal wind component (a), the acceleration induced by gravity wave dissipation due to tidal filtering (b), and the resulting apparent tidal structure (c).

es, on the other hand, with $\overline{u'w'} \sim \pm 50 \ m^2 s^{-2}$ [*Fritts and Vincent*, 1987; *Reid et al.*, 1988; *Fritts and Yuan*, 1989; *Rüster and Reid*, 1990; *Fritts et al.*, 1992], were found to imply considerable anisotropy and significant momentum fluxes at small horizontal scales ($\lambda_z \sim 100$ km or less).

A different approach was taken by *Forbes et al.* [1991] and *Miyahara and Forbes* [1991] to assess the role of gravity wave-tidal interactions in the MLT. These authors used global tidal models to specify the tidal structures and the Lindzen parameterization to incorporate the effects of a spectrum of gravity waves with discrete phase speeds and tidal interactions. Their results provide support for the predictions by *Fritts and Vincent* [1987] of an advance of the apparent tidal phase and a suppression of tidal amplitudes during strong forcing. Also found by these authors was a gravity wave-induced diffusion which was a function of tidal phase and highly variable in space and time due to tidal filtering of the wave spectrum. It was suggested that this diurnally-variable diffusion might account for the diurnal variability observed in photochemically active atmospheric constituents.

Most recently, *Lu and Fritts* [1993] employed a spectral parameterization of gavity wave propagation, filtering, and energy and momentum fluxes in a study of gravity wave-tidal interactions with canonical tidal wind fields. The parameterization was specified to reflect, as closely as possible, observed mean gravity wave spectral characteristics and energy variations with altitude [*Fritts and VanZandt*, 1993; *Fritts and Lu*, 1993]. As such, gravity wave energy and momentum fluxes were constrained by those anticipated based on the mean spectral forms (see *Fritts* [1994] in this volume), with predicted momentum flux profiles in general agreement with observed values at various locations. This scheme was applied to superposed mean and tidal wind fields for various tidal phases. Examples of the profiles obtained with CIRA 1986 winter and summer mean winds at 40°N, tidal structures with $\lambda_z = 25 \ km$ and an amplitude increasing to 50 $m \ s^{-1}$ at 120 km, and a characteristic intrinsic phase speed of the gravity wave spectrum of 5 $m \ s^{-1}$ at the surface are shown in Figure 9.

The inferred momentum flux profiles exhibit a number of interesting characteristics. First, there is a tendency for anisotropy to develop in response to the mean wind fields, resulting in gravity wave propagation preferentially opposed to the mean motion and corresponding momentum fluxes which act to decelerate the zonal jets in regions of wave dissipation. As noted above, these flux profiles are in general agreement with those observed to close the mesospheric jets and drive the strong residual circulation in the MLT. More important, perhaps, for our purposes here is an approximate anti-correlation of the momentum fluxes with the tidal wind fields, in close agreement with the observations discussed above. It is also of interest that the mean of the tidally-modulated momentum flux profiles differs from that obtained in the absence of tidal variations. If this is representative of gravity wave-tidal interactions in the MLT, it implies that the mean gravity wave forcing of the MLT is itself dependent on the strength (and variability) of these interactions. Other experiments with larger and smaller gravity wave energies (or characteristic intrinsic phase speeds) and maximum anisotropies showed these results to be fairly robust, with momentum fluxes generally consistent with observed values.

Because of the observed anti-correlation of tidal winds and gravity wave momentum fluxes, *Lu and Fritts* [1993] presented a simple analytic model of the effects of gravity wave forcing of the tidal structures. This model suggested a tendency, as noted above, for a phase advance of the tidal motion. It also predicted an increase in apparent tidal amplitude, however, in contrast to earlier studies. These predictions were tested with a simple, coupled model of gravity wave-tidal interactions which assumed relaxation of the tidal structure to its unforced state in the absence of gravity wave forcing. Simulations for two tidal wavelengths confirmed the tendency for phase advances and amplitude increases during intervals of strong forcing, but

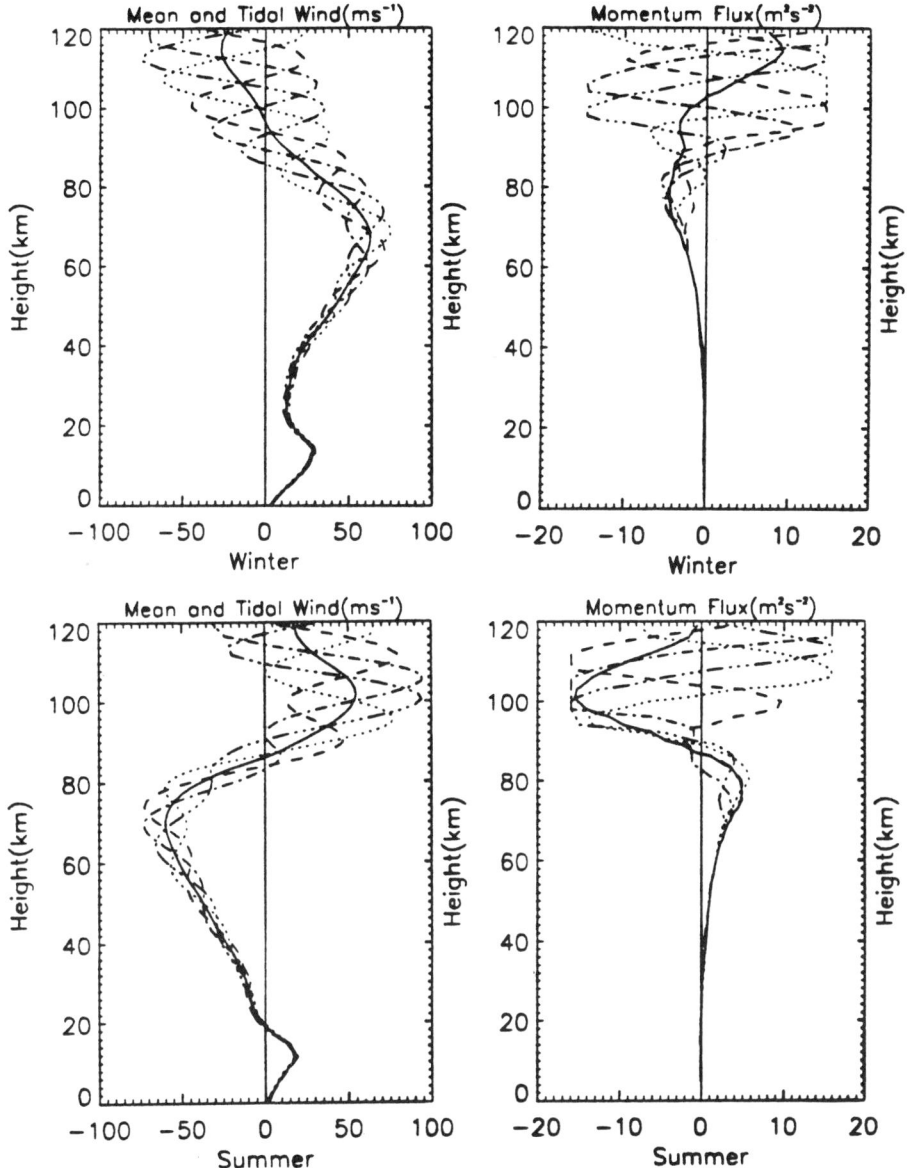

Figure 9. Winter (upper) and summer (lower) mean wind profiles from the CIRA 1986 model at 40°N (left, solid) and corresponding mean gravity wave momentum fluxes (right, solid) obtained with the parameterization by *Lu and Fritts* [1993] with an intrinsic phase speed of $5\ m\ s^{-1}$. Dashed and dotted lines show four phases of a superposed canonical tidal wind field (left) and the corresponding momentum fluxes (right).

also revealed that the prediction of phase advances is more robust and less dependent on wave parameters than that of amplitude increases. Thus, the issue of gravity wave influences on tidal amplitudes remains uncertain at this time. It should be noted that this gravity wave parameterization ultimately results in the suppression of tidal amplitudes in the NCAR TGCM, which describes the evolution of the tidal fields more accurately than the simple coupled model (R. Roble, private communication, 1993).

5. SUMMARY AND CONCLUSIONS

Gravity wave-tidal interactions have been found during the last decade to contribute importantly to the observed variability of gravity wave and tidal motions in the MLT. These interactions arise due to a number of factors. Primary among these factors are the sensitivity of gravity wave propagation to the local environment (shear and stability) and the disparate periods of tidal motions and the gravity wave motions accounting for the majority of energy and momentum fluxes in the MLT. The variable wind and stability environments created locally by the large-scale tidal structures result in filtering and differential propagation of gravity waves, depending on their orientation. This filtering induces strong anisotropy and potentially large momentum fluxes within the gravity wave field which in turn influence the tidal and mean fields through which they propagate.

Mean momentum fluxes at MLT altitudes have typical magnitudes of $\sim$ 5 to 15 $m^2 s^{-2}$ and vary relatively little with latitude and season. Maximum momentum fluxes accompanying strong tidal modulation, on the other hand, are often $\sim$ 50 $m^2 s^{-2}$ or larger for short intervals, with mean tidally-modulated fluxes smaller, but still comparable to mean values. Thus, variable fluxes due to tidal modulation are at least comparable to, and may far exceed, the fluxes required to account for the mean circulation and thermal structure. As such, gravity wave-tidal interactions may have important, but as yet unknown, influences on the mean and the variable structure of the MLT.

Observations suggest an anti-correlation between tidal winds and gravity wave momentum fluxes due to tidal filtering of the wave spectrum. A similar correlation is also suggested by the simple gravity wave-tidal interaction models that have been posed to date. Both theory and observations suggest a tendency for tidal phase advances in response to strong gravity wave modulation and forcing. Influences on tidal amplitudes are less well defined, with observations generally suggesting amplitude reductions, models based on the observed phase relation between winds and fluxes suggesting amplitude increases, and other studies suggesting decreases. Additional influences of variable gravity wave fluxes are likely to include diurnal variations in turbulent heating, diffusion, and mean transports and in the distributions of photochemically active species affected by these processes. Yet, there is too little data at present to assess the importance of these responses to gravity wave-tidal interactions.

Acknowledgments. This work was supported by the Air Force Office of Scientific Research (AFSC) under grant F49620-92-J-0138 and by the SDIO/IST and managed by the Naval Research Laboratory under grant N00014-92-J-2005.

References

Avery, S. K., Vincent, R. A., Phillips, A., Manson, A. H., Fraser, G. J., High-latitude tidal behavior in mesosphere and lower thermosphere, *J. Atmos. Terres. Phys.*, 51, 595–608, 1989.

Forbes, J. M., Atmospheric tides, 1, model description and results for solar diurnal component, *J. Geophys. Res.*, 87, 5222–5240, 1982a.

Forbes, J. M., Atmospheric tides, 2, the solar and lunar semidiurnal components, *J. Geophys. Res.*, 87, 5241–5252, 1982b.

Forbes, J. M., Gu, J., Miyahara, S., On the interaction between gravity waves and the propagating diurnal tide, *Planet. Space Sci.*, 39, 1249–1257, 1991.

Forbes, J. M., Vial, F., Monthly simulations of the solar semidiurnal tide in the mesosphere and lower thermosphere, *J. Atmos. Terres. Phys.*, 51, 649–661, 1989.

Fraser, G. J., Vincent, R. A., Manson, A. H., Meek, C. E., Clark, R. R., Interannual variability of tides in the mesosphere and lower thermosphere, *J. Atmos. Terres. Phys.*, 51, 555–567, 1989.

Fritts, D. C., Gravity wave forcing and effects in the mesosphere and lower thermosphere, this volume, 1994.

Fritts, D. C., Isler, J. R., Mean motions and tidal and two-day structure and variability in the mesosphere and lower thermosphere over hawaii, *J. Atmos. Sci.*, 51, in press, 1993.

Fritts, D. C., Lu, W., Spectral estimates of gravity wave energy and momentum fluxes, ii: Parameterization of wave forcing and variability, *J. Atmos. Sci.*, 50, 3695–3713, 1993.

Fritts, D. C., VanZandt, T. E., Spectral estimates of gravity wave energy and momentum fluxes, i: Energy dissipation, acceleration, and constraints, *J. Atmos. Sci.*, 50, 3685–3694, 1993.

Fritts, D. C., Vincent, R. A., Mesospheric momentum flux studies at adelaide, australia: Observations and a gravity wave-tidal interaction model, *J. Atmos. Sci.*, 44, 605–619, 1987.

Fritts, D. C., Wang, D.-Y., Blanchard, R. C., Gravity wave and tidal structures between 60 and 140 km inferred from space shuttle reentry data, *J. Atmos. Sci.*, 50, in press, 1993.

Fritts, D. C., Yuan, L., Measurement of momentum fluxes near the summer mesopause at poker flat, alaska, *J. Atmos. Sci.*, 46, 2569–2579, 1989.

Fritts, D. C., Yuan, L., Hitchman, M. H., Coy, L., Kudeki, E., Woodman, R. F., Dynamics of the equatorial mesosphere observed using the jicamarca mst radar during june and august 1987, *J. Atmos. Sci.*, 49, 2353–2371, 1992.

Lu, W., Fritts, D. C., Spectral estimates of gravity wave energy and momentum fluxes, iii: Gravity wave-tidal interactions, *J. Atmos. Sci.*, 50, 3714–3727, 1993.

Manson, A. H., Meek, C. E., Teitelbaum, H., Vial, F., Schminder, R., Kurschner, D., Smith, M. J., Fraser, G. J., Clark, R. R., Climatologies of semidiurnal and diurnal tides in the middle atmosphere (70 - 110 km) at middle latitudes (40 0 55°), *J. Atmos. Terres. Phys.*, 51, 579–593, 1989.

McIntyre, M. E., On dynamics and transport near the polar mesopause in summer, *J. Geophys. Res.*, 94, 14617–14628, 1989.

Miyahara, S., Forbes, J. M., Interactions between

gravity waves and the diurnal tide in the mesosphere and lower thermosphere, *J. Meteor. Soc. Japan*, **69**, 523–531, 1991.

Reid, I. M., Rüster, R., Czechowsky, P., Schmidt, G., Vhf radar measurements of momentum flux in the summer polar mesosphere over andenes ($69°n, 16°e$), norway, *Geophys. Res. Lett.*, **15**, 1263-1266, 1988.

Reid, I. M., Vincent, R. A., Measurements of mesospheric gravity wave momentum fluxes and mean flow accelerations at adelaide, australia, *J. Atmos. Terres. Phys.*, **49**, 443–460, 1987.

Rüster, R., Reid, I. M., Vhf radar observations of the dynamics of the summer polar mesopause region, *J. Geophys. Res.*, **95**, 10005–10016, 1990.

Tsuda, T., Murayama, Y., Yamamoto, M., Kato, S., Fukao, S., Seasonal variation of momentum flux in the mesosphere observed with the mu radar, *Geophys. Res. Lett.*, **17**, 725–728, 1990.

Vial, F., Numerical simulations of atmospheric tides for solstice conditions, *J. Geophys. Res.*, **91**, 8955–8969, 1986.

Vial, F., Forbes, J. M., Recent progress in tidal modeling, *J. Atmos. Terres. Phys.*, **51**, 663–671, 1989.

Vial, F., Forbes, J. M., Miyahara, S., Some transient aspects of tidal propagation, *J. Geophys. Res.*, **96**, 1215–1224, 1991.

Vincent, R. A., Tsuda, T., Kato, S., Asymmetries in mesospheric tidal structure, *J. Atmos. Terres. Phys.*, **51**, 609–616, 1989.

Walterscheid, R. L., Inertio-gravity wave induced accelerations of mean flow having an imposed periodic component: Implications for tidal observations in the meteor region, *J. Geophys. Res.*, **86**, 9698–9706, 1981.

Wang, D.-Y., Fritts, D. C., Mesospheric momentum fluxes observed by the mst radar at poker flat, alaska, *J. Atmos. Sci.*, **47**, 1512–1521, 1990.

Wang, D.-Y., Fritts, D. C., Evidence of gravity wave-tidal interaction observed near the summer mesopause at poker flat, alaska, *J. Atmos. Sci.*, **48**, 572–583, 1991.

Gravity Wave Mean State Interactions in the Upper Mesosphere and Lower Thermosphere

Richard L. Walterscheid

Space and Environment Technology Center, The Aerospace Corporation, Los Angeles, California

Nonlinear interactions between gravity waves and the mean state of the upper atmosphere can have important consequences for the mean state. The mean state refers to variations having temporal or spatial scales that are much greater than the gravity wave scales. Gravity waves can alter the mean-state winds, temperature and constituent concentrations when the nonacceleration conditions are violated. The nonacceleration conditions are satisfied when the waves are linear, conservative, steady-state and do not encounter a critical level. In addition, waved-perturbed chemistry can also induce changes in the mean-state constituent distributions even when the nonacceleration conditions are satisfied. The best known gravity wave-mean flow effect is the role that wave drag is believed to have in maintaining the cold polar summer mesopause. Gravity waves can also be effective in forcing temporal variations in the mean winds at tidal frequency, altering the mean state during large magnetic storms, cooling the lower thermosphere, and altering the mean profiles of minor constituent concentrations. Conversions of wave energy from the wave to the mean state can deplete wave energy and significantly diminish the growth of wave amplitude with altitude while causing a build-up of the mean wind. Wave stresses induced by wave transience might contribute significantly to the wave drag required to produce the cold summer polar mesopause.

1. INTRODUCTION

Atmospheric gravity waves are quasi-periodic fluctuations that may be most notable for the nonfluctuating changes that they induce. Gravity waves can induce changes in the mean-state wind, temperature and minor constituent distribution when the so called nonacceleration conditions are not satisfied [Eliassen and Palm, 1961; Andrews and McIntyre, 1976]. These conditions are that the waves be linear, conservative, steady and not encounter a critical level. Wave breakdown leads to nonconservative processes, and, as we shall see, has a strong association with critical level encounters. When the nonacceleration conditions are violated, waves induce mean-state changes through wave fluxes of momentum, sensible heat and minor constituent concentrations. In addition, waves can induce a flux of minor constituents concentrations when the nonacceleration conditions are satisfied if chemistry can act to change minor constituent concentrations on the time scale of the wave [Strobel, 1981; Walterscheid and Schubert, 1989]

Gravity wave mean-state interactions can have a profound influence on the mean state. The best example of this influence in the upper mesosphere and lower thermosphere is the cold summer polar mesopause. This is the coldest region in the atmosphere. It is believed that the summer mesopause is maintained out of radiative equilibrium by gravity wave drag [Holton, 1982, 1983; Garcia and Solomon, 1985, Miyahara et al., 1986]. In the summer hemisphere, wave drag opposes the acceleration of the mid-latitude upper mesospheric easterlies by the Coriolis force associated with poleward meridional winds. This causes the easterlies to decrease with altitude in the upper mesosphere. On seasonal time scales, an approximate balance is maintained in the meridional momentum equation with the result that the zonal wind is close to geostrophic balance with the meridional pressure gradient (Leovy, 1964). According to the thermal wind relation, decreasing easterlies

The Upper Mesosphere and Lower Thermosphere:
A Review of Experiment and Theory
Geophysical Monograph 87
Copyright 1995 by the American Geophysical Union

(westerly shear) implies that temperature must decrease poleward.

The upper mesosphere and lower thermosphere is the region where wave-mean flow interactions involving upward propagating gravity wave are apt to be most significant. This region is where waves achieve their largest amplitudes owing, on the one hand, to the exponential growth of waves as they propagate up from below, and, on the other, to the dissipation of waves by various mechanisms as they propagate through the region. The dissipation mechanisms include wave breakdown and viscous dissipation [Hodges, 1967; Lindzen, 1981; Pitteway and Hines, 1963]. Wave breakdown may proceed by convective, dynamic (Kelvin-Helmholtz), or parametric instability [Fritts, 1984, Yeh and Liu, 1981, Dunkerton, 1987].

In the remainder of this paper I shall discuss some phenomena of the upper mesosphere and lower thermosphere related to gravity wave-mean state interactions. These are temporal variations in the mean winds having tidal frequency, changes in the mean state temperature during large magnetic storms, cooling of the lower thermosphere, and changes in the mean profiles of minor constituent concentrations. Additional topics are depletion of wave energy by conversions from the wave to the mean state, and the build-up of the mean wind by wave transience.

2. THEORY

For simplicity we consider wave-mean state interactions for vertically propagating two-dimensional gravity waves and ignore the Earth's rotation. For these waves, the wave induced change in the mean state of an arbitrary quantity ψ is given by

$$\frac{\partial \overline{\psi}}{\partial t} = -\frac{1}{\overline{\rho}} \frac{\partial \overline{\rho w' \psi'}}{\partial z} \qquad (1)$$

where ψ may be a quantity such as the zonal wind u, temperature T or the mixing ratio r of a minor constituent, and w is the vertical velocity, z is the vertical coordinate, and ρ is mass density. The overbar denotes a mean-state value and primes denote deviations therefrom (waves). The quantity $\overline{\rho w' \psi'}$ is the wave flux of $\overline{\psi}$. For the special case of $\psi = u$ the wave flux is the wave stress.

According to (1) any process that induces a height variation in the wave flux of $\overline{\psi}$ induces a change in $\overline{\psi}$. When the nonacceleration conditions apply $\overline{\rho w' \psi'}$ is constant in altitude, or nil. The processes that induce a height variation in $\overline{\rho w' \psi'}$ are those which cause the nonacceleration conditions to be violated. These are wave dissipation, transience, nonlinearity and critical level interactions. The formal development of the nonacceleration conditions is given by Eliassen and Palm [1961], Andrews and McIntyre [1976], Boyd [1976] and others. Qualitative discussions of wave-induced changes of the mean state forced by wave dissipation, wave transience and critical levels is presented below.

2.1. Wave Dissipation

The integral form of (1) is

$$\left\langle \overline{\rho} \frac{\partial \overline{\psi}}{\partial t} \right\rangle = \left(\overline{\rho w' \psi'} \right)_{z_1} - \left(\overline{\rho w' \psi'} \right)_{z_2} \qquad (2)$$

Where the angle brackets denote a height integration between z_1 and z_2. The terms on the right hand side represent fluxes through the boundaries of the layer. If the fluxes are positive (upward), the first term on the right represents the flux into the region from below (import) and the second term represents the flux out of the region from within (export). For the special case of zonal wind acceleration, $\overline{\psi} = \overline{u}$ and $\overline{\rho w' \psi'} = \overline{\rho w' u'}$, and $\overline{\rho w' u'}$ is the momentum flux (wave stress). When the nonacceleration conditions apply $\left(\overline{\rho w' u'} \right)_{z_1} = \left(\overline{\rho w' u'} \right)_{z_2}$.

The magnitude of the stress can be related to kinetic energy density. The kinetic energy density per unit mass is

$$\frac{1}{2} \left(\overline{u'^2} + \overline{w'^2} \right) \propto a^2 \qquad (3a)$$

where a is a measure of wave amplitude. Likewise,

$$\overline{w' u'} \propto a^2 \qquad (3b)$$

Consider a steady-state wave with upward energy transfer subject to dissipation in the region between z_1 and z_2, and consider the forcing of the mean wind by wave stresses. Assume also $\overline{\rho w' u'} > 0$. Because the wave is dissipating, the kinetic energy is diminished between z_1 and z_2. According to (3) this implies that $\left(\overline{\rho w' u'} \right)_{z_1} > \left(\overline{\rho w' u'} \right)_{z_2}$ and $\langle \overline{\rho} \partial \overline{u}/\partial t \rangle = \left(\overline{\rho w' u'} \right)_{z_1} - \left(\overline{\rho w' u'} \right)_{z_2} > 0$; the gravity wave is importing more westerly momentum into the layer through the bottom than it is exporting through the top. Since total momentum is conserved, the momentum accumulated in the layer is manifested as an acceleration of the mean wind. This is illustrated schematically in Figure 1. The mean wind is accelerated toward the horizontal phase velocity of the wave because for upward energy propagation the intrinsic

Forcing of Mean Wind by Momentum Flux

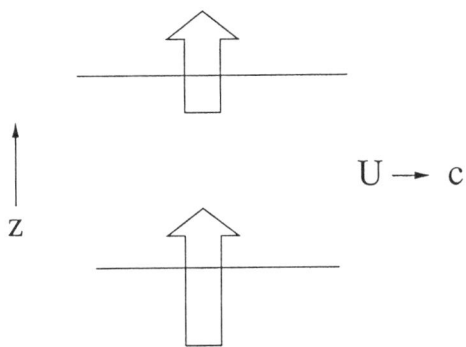

Figure 1. The thick open arrows denote the flux of momentum into and out of a layer. The horizontal lines denote the boundaries of the layer. The symbol z denotes the vertical coordinate, U denotes the mean wind, and c denotes the wave phase velocity. The situation depicted is a net import of momentum into the layer because of dissipation within the layer. The mean wind is accelerated toward the phase velocity of the wave.

phase velocity and momentum flux have the same sense (i. e., westward moving waves transport westward momentum upward).

Modification of the mean flow through viscous dissipation of the wave can feed back on the wave and lead to its rapid demise. Dissipating waves cause the mean flow to be accelerated toward the wave's phase speed, reducing the intrinsic frequency of the wave, with the result that $\bar{u} - c \to 0$. This causes the group velocity to slow and the vertical scale to shrink. Both of these processes favor the rapid absorption of the wave by scale-dependent viscosity. The slower the group velocity, the more time a wave spends in a dissipative layer. The shorter the vertical scale, the faster viscosity diffuses momentum. This means that the wave-caused acceleration has a positive feedback on the rate of wave dissipation. Jones and Houghton [1972] simulated the self-destruction of a gravity wave subject to viscous dissipation. These authors found that with wave-mean flow interactions included, dissipating waves rapidly give up their energy to the mean flow.

The feedback is strongly altered when wave breaking is the dissipation mechanism. In this case, the wave stress is reduced by a factor of $(\bar{u} - c)^3$ [Lindzen, 1981; Walterscheid, 1984] as the mean wind is accelerated. Thus, the tendency to accelerate the mean flow to the phase speed is strongly mitigated. Weinstock [1982] has obtained a similar expression for wave stresses in which the dissipative mechanism is the diffusive-like action of wave-wave interactions involving a spectrum of waves.

2.2. Transience

Consider the same wave as in the previous section except that here the wave is a transient, conservative wave propagating up from below. Since z_1 is closer to the wave source than z_2, and the wave reaches z_1 before it reaches z_2, the build-up of wave energy density is greater at z_1 than at z_2 during the approach to steady-state conditions. By virtue of (2) and (3), this implies that $\left(\overline{\rho w'u'}\right)_{z_1} > \left(\overline{\rho w'u'}\right)_{z_2}$ and $\langle \bar{\rho} \partial \bar{u}/\partial t \rangle > 0$.

Dunkerton [1981, 1982] has examined transience-induced wave-mean flow interactions in a compressible atmosphere. In a process that is similar to the one described by Jones and Houghton [1972], Dunkerton found that vertically propagating waves leads to mean-flow accelerations through group velocity feedback and that this process is enhanced by the exponential growth of the waves in a compressible atmosphere. A simple analytical model of transience and wave saturation indicates that transience in a compressible atmosphere can lower the level of wave breaking on the order of a scale height. This should reduce the mean wind acceleration because momentum deposition occurs where the atmosphere is more dense.

A number of investigators have numerically examined wave mean flow interactions forced by breaking waves. Schoeberl et al. [1983] employed a wave saturation scheme following Lindzen [1981] in a time dependent mean-state model with wave fluxes derived at each time step from a steady-state wave model. Walterscheid [1984] incorporated the Lindzen parameterization into a quasi-linear model with a time-dependent wave model. Fritts and Dunkerton [1984] and Dunkerton and Fritts [1984] incorporated a convective adjustment scheme to limit wave growth in a quasi-compressible nonlinear model. Walterscheid and Schubert [1990] simulated wave breakdown, convective adjustment and mean-state acceleration with a fully compressible nonlinear model. All of these studies have indicated a strong forcing of the mean wind by breaking waves.

2.3. Critical levels

Here we consider the case where there is a critical level between z_1 and z_2. According to an Eliassen-Palm relation [Eliassen and Palm, 1961]

$$\overline{p'w'} = \bar{\rho}(c - \bar{u})\overline{u'w'} \qquad (4)$$

If we assume that the energy flux is upward at z_1 and z_2, then $\left(\overline{\rho w'u'}\right)_{z_1} > \left(\overline{\rho w'u'}\right)_{z_2}$ because $(c - \bar{u}) > 0$ at z_1 and $(c - \bar{u}) < 0$ at z_2. This means that $\langle \bar{\rho} \partial \bar{u}/\partial t \rangle > 0$ [Holton, 1975].

The earliest quantitative study to examine the behavior in the neighborhood of the critical level was Booker and Bretherton [1967]. They showed that in the inviscid case the momentum fluxes have a discontinuity at the critical level implying an infinite acceleration of the mean state. The discontinuity is removed by dissipation, but not the implication of strong wave forcing of the mean flow.

Critical level interactions have been studied extensively by a number of authors since Booker and Bretherton's analysis. An issue of particular interest is whether critical levels are reflecting or absorbing and to what extent transience modifies the critical level interaction. Reflecting critical layers are associated with Kelvin-Helmholtz modes of instability and nonlinear effects associated with wave overturning [Davies and Peltier, 1979; Peltier and Clark, 1979; Clark and Peltier, 1984, Fritts, 1978, 1979]. The reflecting properties of the interaction are sensitive to the amount of viscous wave dissipation; with modest dissipation the interaction becomes absorbing. Rather large values of eddy dissipation are inferred for the upper mesosphere, at least subsequent to wave overturning [Allen et al., 1981; Lindzen, 1981]. Dunkerton and Fritts [1984] simulated the interaction between transient gravity waves and a critical level and found that transience causes the region of momentum deposition to thicken considerably.

The occurrence of critical levels and wave breaking are closely connected. The region just below a critical level should be a favored region for wave breaking [Geller et al., 1975; Walterscheid, 1984]. This can be understood in terms of the Orlanski and Bryan [1969] criterion for the onset of overturning. The onset of overturning leading to breakdown occurs when $u' \sim |\bar{u} - c|$. Near critical levels this is easily satisfied because of large u' and small $|\bar{u} - c|$.

2.4. Wave Forcing of Temperature and Mean-State Minor Constituent Distributions

When the nonacceleration conditions apply, the wave transport of heat by gravity waves is essentially nil [Andrews and McIntyre, 1976; Walterscheid, 1981b]. Likewise the flux of minor constituents is nil when chemical production and loss are zero. Breaking waves, in contrast, induce a flux of heat and constituents by means of convection and turbulence generated during breakdown [Lindzen, 1981; Fritts and Dunkerton, 1985; Coy and Fritts, 1988; Walterscheid and Schubert, 1990]. The concern here is with the fluxes induced by chemical production and loss.

It is well-known that viscously dissipating gravity waves heat the mean state by means of frictional heating [Hines, 1965]. Additionally, gravity waves can affect the mean-state temperature by the heat fluxes they induce. In a stably stratified atmosphere, dissipating gravity waves induce a downward flux of sensible heat that cools the upper regions affected by the wave, and warms the lower region [Walterscheid, 1981b]. Since the upper region is less dense, the heat removed from this cools this region more strongly than the lower region is warmed by the addition of heat from above. The cooling in the upper region can be quite significant. Walterscheid [1981b] used a WKB model of wave dissipation in the lower thermosphere and found that cooling rates of ~ 20K per day are possible. Weinstock [1983] applied a similar model to waves dissipating by wave-wave interactions in the mesosphere and estimated significant cooling rates in this region as well. Schoeberl et al. [1983] evaluated wave fluxes induced by viscously dissipating wave in their numerical model and found these fluxes contributed to a net cooling above ~ 20 km.

As we have mentioned, chemistry can induce a constituent flux even when the nonacceleration conditions are satisfied. If one multiplies the linearized species continuity equations by w' and averages one obtains the following result when the nonacceleration conditions are satisfied

$$\overline{w'r'} = (\partial \bar{r}/\partial z)^{-1} \left\{ \frac{\overline{r'(P' - L')}}{\bar{N}} - (\bar{P} - \bar{L}) \frac{\overline{N'r'}}{\bar{N}} \right\} \qquad (5)$$

where r is the minor constituent mixing ratio, P is chemical production and L is loss. In most cases the second term in curly brackets on the right side is negligible. The chemically induced fluxes of minor constituents, like turbulent fluxes and transport by the mean circulation, can alter the mean-state distribution of minor constituents.

3. PSEUDOTIDES

In this section, the theory and observations of gravity wave-induced mean wind fluctuations having tidal period are discussed.

3.1. Theory

Atmospheric tides modulate the rate at which gravity waves dissipate near critical levels formed by tidal winds. This modulation occurs, because, as we have mentioned, altitudes near critical levels are favored regions for viscous dissipation and wave breakdown.

Because tides modulate wave dissipation they also modulate wave-mean flow interactions. Consider the modulation due to zonally propagating gravity waves. At altitudes where the tide is westerly, westerly propagating gravity waves are preferentially absorbed because these waves are Doppler shifted to smaller intrinsic frequencies, and can encounter critical levels. Westerly propagating gravity waves that transfer energy upwards also transfer westerly momentum upwards. The absorption of these waves imparts a westerly acceleration. Similar arguments indicate an easterly acceleration when the tidal winds are easterly. This means that the mean flow is subject to an acceleration having tidal period because of wave-mean-flow interactions that are tidally modulated. A similar argument applies when the dissipation mechanism is wave breakdown.

The wave-induced acceleration will produce a component of the mean flow that fluctuates with tidal period. We refer to this component as a pseudotide. It can be argued that the tides are the total periodic response of the atmospheric system to periodic forcing. However, it is conceptually useful to distinguish between the large-scale response to external forcing and the secondary response owing to local nonlinear interactions modulated by tides.

3.2. Observations

The first experimental evidence for a wave-driven modulation of the wind at tidal frequencies was provided by Fritts and Vincent [1987]. These authors observed a strong diurnal variation of the gravity wave momentum fluxes observed by a dual-beam Doppler radar at Adelaide, Australia. The diurnal variation was correlated with the phase of the diurnal tidal motions, with an out of phase relation between tidal amplitudes and momentum fluxes. They proposed a qualitative gravity-wave interaction model in which wave dissipation occurs as a consequence of wave breakdown. Using the wave stress parameterization of Lindzen [1981] they reasoned that the maximum easterly (westerly) acceleration is expected where $\bar{u} - c$ is most negative (or most positive) and the largest accelerations should occur where wave amplitudes are greatest. They were able to qualitatively account for the observed phase relation between the diurnal tide and the diurnal modulation of the momentum fluxes, and explain a large reduction in the amplitude of the diurnal tide below 90 km.

Figure 2 shows the vertical flux of westerly momentum $\overline{u'w'}$ derived from dual beam Doppler radar measurements at Adelaide, Australia [Fritts and Vincent, 1987]. The momentum fluxes are shown as a function of altitude and time. The momentum fluxes are derived from 8-h blocks of data. Of interest is the large diurnal variation in the upper levels and the phase lag at lower levels. The phase lag was correlated with the phase of the diurnal tide.

Fritts and Yuan [1989] examined the momentum fluxes derived from the Poker Flat, Alaska MST radar and found a tendency for large fluxes to occur at times of, and in

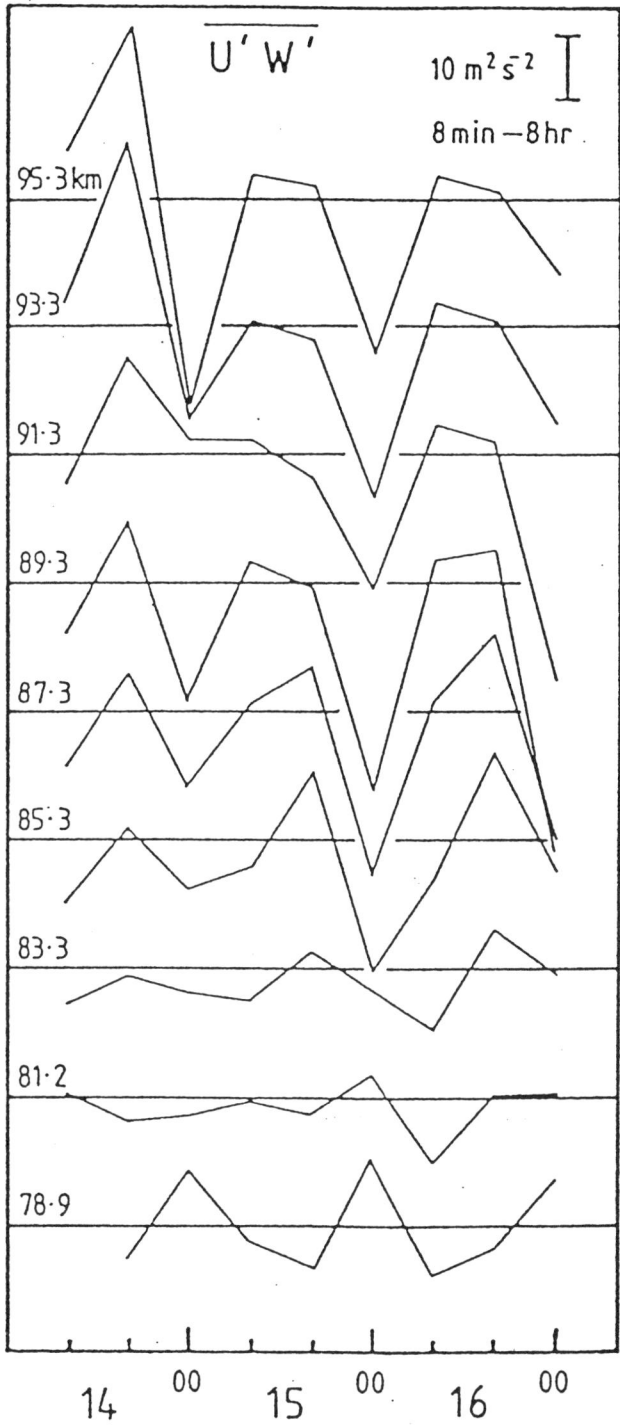

Figure 2. The vertical flux of westerly momentum $\overline{u'w'}$ derived from dual beam Doppler radar measurements at Adelaide, Australia (Fritts and Vincent, 1987). The momentum fluxes derived from 8-h blocks of data are shown as a function of altitude and time.

directions opposed to, the local mean flow maximum, in accord with the tendency noted at Adelaide. This was examined in more detail by Wang and Fritts [1991]. They noted variations in the amplitude and phase of momentum fluctuations that supported the model proposed by Fritts and Vincent [1987] and agreed with the earlier observations. The interpretation of the observation was given further quantitative supported by an elaboration of the earlier proposed model.

The steady-state pseudotidal response is proportional to the factor $\omega^2(f^2-\omega^2)^{-1}$, where $f=2\Omega \sin(\varphi)$ is the inertial frequency and ω is the tidal frequency, and where Ω is the Earth's angular frequency and φ is latitude [Walterscheid et al., 1986]. The response is also proportional to the vertical gradient (divergence) of the tidally modulated wave stresses. The amplitude of the wave driving depends upon the amplitude of the stresses, and the amplitude and vertical wavelength of tide. As the poles are approached $f^2-\omega^2 \to 0$ for the semidiurnal tide. Thus the very high latitude semidiurnal pseudotidal response should be quite significant. At the poles the singularity in the amplitude factor is mitigated by the nonlinear amplitude modulation seen in the simulations shown in Figure 3 and the vanishing of the main (migrating) semidiurnal tide. Walterscheid et al. [1986] interpreted an observed large-amplitude semidiurnal temperature variation in the mesopause rotational OH airglow temperature at Spitzbergen (78°N) as a pseudotide. The amplitude of the semidiurnal temperature variation was ~ 10 K, whereas model predictions for the migrating semidiurnal tide at the latitude of Spitzbergen are only ~ 2K or less [Walterscheid et al., 1980; Forbes, 1982]. At the latitude of Spitzbergen the amplification factor $\omega^2(f^2-\omega^2)^{-1}$ is 23, which represents a very significant degree of amplification for tidal forcing. Collins et al. [1992] observed a strong semidiurnal oscillation in sodium density with a sodium lidar located at the south pole. These authors noted a correlation between this oscillation and gravity wave amplitudes that is consistent with a pseudotide interpretation. At the pole, the underling semidiurnal tide would have to be a nonmigrating tide (i. e., zonal wavenumber different from 2) since the migrating tide vanishes at the pole.

It is possible that the observed very-high latitude semidiurnal fluctuations are due to zonally symmetric tides. These tides may have their largest amplitudes at the pole, and are the likely cause of observed wave-like temperature variations in the very high latitude mesopause region with periods of 6 and 8 hours [Longuet-Higgins, 1968; Sivjee and Walterscheid, 1994; Sivjee et al., 1994]. These variations are not apt to be pseudotidal since the amplification factors for these frequencies are order unity; nor are they apt to be zonally propagating tides, since these tides have small temperature amplitudes at very high latitudes [Longuet-Higgins, 1968].

3.3. Numerical Simulations

Numerical studies of pseudotides are distinguished by the dissipation mechanism invoked (viscous dissipation or wave breakdown) and the scheme for simulating wave dissipation and wave-mean flow effects (explicitly modeled or parameterized).

3.3.1 Quasi-linear models. The earliest simulations were by Walterscheid [1981a]. In Walterscheid [1981a] the dissipation mechanism was viscous absorption near critical levels. A time-dependent wave-mean-flow (quasilinear) model was employed to simulate gravity waves, dissipation and wave-induced acceleration of the mean flow. The mean state consisted of an imposed background component and a wave-induced component. The properties of the imposed background component (period, vertical wavelength, amplitude) were more or less characteristic of the semidiurnal tide. At every time step, the quasilinear model first solves the time dependent gravity wave equations with the mean flow as a basic state, and then solves time-dependent mean-state equations of the form (1), where the gravity wave solutions are used to evaluate the flux terms. The wave field consists of two gravity waves forced at the lower boundary by two disturbances propagating in opposite

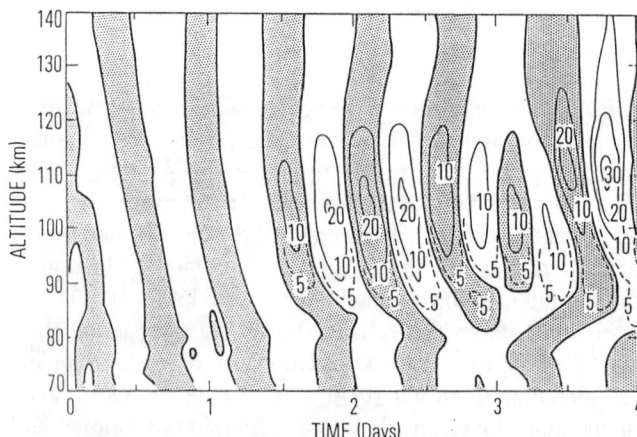

Figure 3. The wave-induced mean wind in an altitude versus time plot for gravity waves excited at the lower boundary with horizontal phase speeds of ± 37 m s^{-1} (Walterscheid, 1981a). The amplitude of the forcing was adjusted to give a wave amplitude of ~ 10 m s^{-1} at 80 km altitude. The shaded and unshaded areas correspond to regions of opposite sense, the unshaded regions being for $\bar{u}>0$.

directions with the same amplitudes, period and horizontal wavelength. These waves propagate into the upper mesosphere where the tidal winds are comparable to the phase speeds of the waves.

Figure 3 from Walterscheid [1981a] shows the wave-induced mean wind in an altitude versus time plot for gravity waves forced at the lower boundary by traveling disturbances with horizontal phase speeds of $\pm$ 37 m s^{-1}, corresponding to a horizontal wavelength of 400 km and a period of 3 h. The amplitude of the forcing was adjusted to give a wave amplitude of ~ 10 m s^{-1} at 80 km altitude. The shaded and unshaded areas correspond to regions of opposite sense, the unshaded regions being for $\bar{u} > 0$. We note the establishment of a well-defined oscillation with a predominant semidiurnal period. The amplitude of the fluctuating mean wind vacillates between about 10 and 30 m s^{-1} between about 100 and 110 km altitude. This magnitude is comparable to the amplitude of the background tide in this region. Waves with slower phase speeds are absorbed at lower altitudes where the atmosphere is denser. As a result, the wave forcing causes a weaker pseudotide. On the other hand, faster waves are absorbed where the atmosphere is less dense and the pseudotide is stronger. For example waves with phase speeds of $\pm$ 74 m s^{-1} generate a pseudotide with an amplitude of ~ 100 m s^{-1}.

3.3.2 Parameterized models with wave breakdown. More recent simulations of pseudotides all invoke wave breakdown as the dissipation mechanism and use parameterized wave stresses following Lindzen [1981], or elaborations thereon. Lu and Fritts [1993] have calculated the periodic wave-mean flow forcing at tidal frequency using a generalization of the Lindzen parameterizations developed by Fritts and VanZandt [1993]. This parametization is based on linear saturation theory [Smith et al., 1987] and incorporates a spectrum of gravity waves. The results are parameterization dependent, but the inclusion of a spectrum of waves represents a significant advance. The calculations support a significant pseudotidal response with amplitudes on the order of tens of meters per second.

In a somewhat different vein, Forbes et al. [1991] have simulated the effects of gravity wave stresses on the diurnal tide. They employed a time-dependent equivalent gravity wave approximation and used Lindzen's parameterization to specify the wave stresses for a spectrum of waves. The stresses are included in the momentum equations for the tides, and the effects of the wave stresses on the tide itself are simulated. This is a different manifestation of wave-mean flow interaction from a pseudotide. They found that the effect of the waves was to damp the amplitude of the tide.

4. WAVE MEAN FLOW INTERACTIONS IN THE STORM-TIME THERMOSPHERE:

During geomagnetically active periods intense time-variable heating in the auroral zones can launch large-amplitude equatorward-propagating gravity waves [Richmond and Matsushita, 1975; Fuller-Rowell and Rees, 1981; Larsen and Mikkelsen, 1983; Roble et al., 1987]. The response to the zonally-averaged storm-time heating can be separated into two components: the storm-averaged response and the deviations therefrom (eddies). The storm-averaged response is a large-scale convection cell in the meridional plane, with strong flow from high to low latitudes throughout the main part of lower thermosphere, and a weak return flow in the lowest part of the affected region. The eddies comprise contributions from the time dependence of convection and from waves. When the heating is characterized by strong transients the waves dominate the eddy contribution.

Richmond [1979] simulated the response of the thermosphere to a magnetic storm with a two-dimensional dynamical model and analyzed the low-latitude heating. He analyzed the combined effects of the adiabatic heating in the descending branch of the storm-averaged convection and the viscous heating due to dissipating waves. He found that the low-latitude heating caused by the storm was dominated by the adiabatic heating, with viscous heating accounting for about one-third of the total.

However, the viscous heating does not account fully for effects of gravity waves. Waves can cause heating and cooling through the transports they cause [Walterscheid, 1981b] and through a wave-induced contribution to the storm-averaged circulation. Brinkman et al. [1992] employed the model of Richmond and Matsushita [1975] generalized to extend the domain from pole to pole and include the seasonal background winds. These authors simulated the response to a model geomagnetic storm with four distinct pulses over a 12-h period. They found that when all wave effects were accounted for, the total low-latitude heating due to the waves (including the viscous heating previously evaluated by Richmond [1979]) was roughly comparable to the heating caused by the thermally direct circulation.

5. MEAN FLOW ACCELERATION AND WAVE SATURATION BY TRANSIENCE

We examine two aspects of wave transience: the forcing of the mean state by wave stresses, and, the reduction in wave growth in height relative to linear steady conservative waves.

5.1. Mean-Wind Acceleration.

While a wave is undergoing a transient build-up, kinetic energy is transferred from the wave to the background state via wave-mean flow interactions, forcing the mean flow and depleting the energy of the wave [Grimshaw, 1975; Dunkerton, 1982]. For conservative waves, the mean wind forced by transience is reversible [Boyd, 1976; Andrews and McIntyre, 1976] and is nil when integrated over the passage of a wave packet. When breakdown occurs, the wave-driven mean flow is not completely reversible.

Walterscheid and Schubert [1990] simulated the nonlinear evolution and breakdown of an upward propagating gravity wave with a time-dependent nonlinear model of gravity wave propagation. The wave was forced by imposing a traveling disturbance at the lower boundary. The wave propagated up into decreasing ambient density until it achieved large amplitude, overturned and broke down via a convective instability.

Figure 4 shows a time versus altitude cross-section of the build-up of the mean wind. All times are the elapsed time after the onset of wave forcing. Wave overturning begins at ~ 390 minutes and breakdown begins at ~ 435 minutes. The acceleration of the mean wind between 80 and 100 km is fairly constant until ~ 430 minutes when the mean wind has reached a maximum value of ~ 50 m s^{-1}. After ~ 430 minutes, a jet forms centered near 85 km. Near the end of the simulation the maximum wind is ~ 70 m s^{-1}. The main part of the mean wind build-up is due to wave transience. The acceleration of the mean wind following breakdown is not much greater than the acceleration before breakdown, probably because the effect of wave dissipation is diminished as a result of smaller wave amplitude.

Since the wave is dissipating in the region of breakdown because of the combined effects of convection and turbulence, it seems unlikely that the acceleration caused by transience would be essentially reversible integrated over the life of the wave. We speculate that a significant fraction of the Eliassen-Palm flux divergence [Eliassen and Palm, 1961] required to maintain the mean mesospheric circulation, including the cold summer mesopause discussed above, is due to wave transience.

5.2. Saturation.

Transience is also important with regard to the growth of large-amplitude waves with altitude. It is important for two reasons: first, during the transient build-up of the wave nonlinear processes inhibit the build-up of wave kinetic energy, and, second, breakdown occurs above the altitude $z_g = c_g \Delta t$ where steady state has been achieved (c_g is the

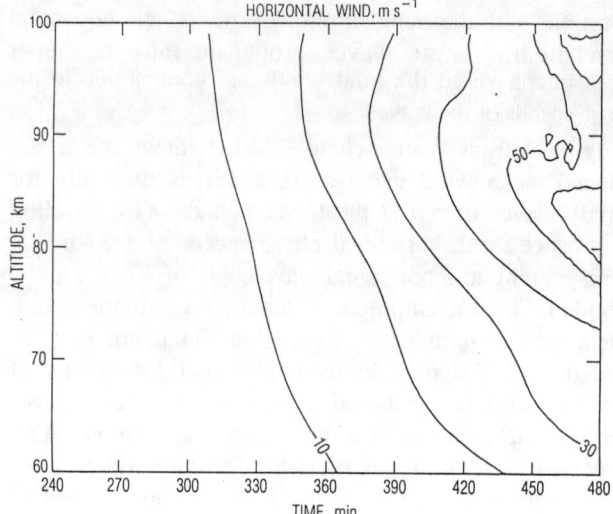

Figure 4. A time versus altitude cross-section of the build-up of the mean wind forced by transience. Contour interval in 10 m s^{-1}. Overturning commences at around 390 minutes and breakdown (i. e., onset of strong convection) commences at around 435 minutes (Walterscheid and Schubert, 1990).

vertical group velocity of the wave and Δt is the time elapsed since the onset of wave excitation).

Large-amplitude transient waves do not necessarily obey equipartition of wave energy. This is explained in terms of the wave kinetic energy budget as follows. The wave energy flux, which is the direct source of wave kinetic energy, is approximately balanced by three processes: the conversion of wave kinetic energy to wave available potential energy, the conversion to mean-state kinetic energy, and the eddy flux of wave kinetic energy. This balance allows wave kinetic energy to remain relatively constant while total wave energy continues to increase [Schubert and Walterscheid, 1990].

The rate of growth of steady-state linear waves with altitude is inferred theoretically from the result that wave energy flux is be constant in altitude without sources and sinks of wave energy. This condition is not met when the wave energy flux in a region is driving the conversions mentioned above. As long as this continues there is a sink of wave energy flux and the growth inferred from linear steady-state considerations is an overestimate of wave growth.

Even if nonlinear processes did not reduce growth relative to a linear steady state, growth can be reduced because breakdown can occur in the region above z_g where steady-state amplitudes have not been achieved. This is because wave amplitudes still increase above z_g despite decreasing energy density. The increase occurs because, although energy density decays away above z_g, it does so more slowly than $\bar{\rho}$. We believe that the increase in wave amplitude above z_g is an inevitable feature of waves excited by realistic forcing [Walterscheid, 1984].

The observed rate of growth of wave amplitude with altitude is substantially less than expected for linear unsaturated waves [Manson et al., 1974; Vincent, 1984]. The processes discussed above are potentially important contributors to the observed decrease of wave kinetic energy density with altitude. The observed decrease is usually attributed to saturation processes such as wave breakdown and wave-wave interactions [Weinstock, 1976; Fritts, 1984].

Figure 5 from Walterscheid and Schubert [1990], shows the dependence of the amplitude of the primary wave horizontal velocity on altitude at 386, 448 and 464 minutes after onset for the same simulation that pertains to Figure 4. (The primary wave is the component of the total disturbance having the horizontal wavelength of the traveling disturbance imposed at the lower boundary.) The sloping line represents the growth of linear, conservative, steady waves. The linear growth rate is seen only at altitudes below about 50 km where wave amplitudes are not too great. At greater altitudes the increase in wave velocity with altitude is much slower and the decrease in kinetic energy density is correspondingly large. The slower growth above 50 km is largely a consequence of the waves not having time to establish a steady state. The vertical group speed of waves at the dominant frequency is ~3 km h^{-1}, and the forcing not able to establish steady state above ~ 50 km before breakdown occurs.

6. CHEMICALLY INDUCED CONSTITUENT FLUXES

Strobel [1981] developed a parameterized approach to computing chemically induced constituent fluxes by linear waves. He developed an eddy diffusion parametization for transport by two-dimensional waves within a residual Eulerian framework, and demonstrated that strong coupling of chemistry and temperature through the rate coefficients requires additional terms proportional to the mean mixing ratio rather than its gradient. He also demonstrated that the usual one-dimensional vertical eddy diffusion coefficient is rigorously valid only when the temperature dependence of the rate constants is negligible, and the constituent in question has strong vertical stratification relative to its horizontal variation.

Schoeberl et al. [1983] modified Strobel's parameterization to include the Lindzen [1981] parameterization of turbulent mixing by breaking gravity waves. The modified parameterization was used in chemical-dynamical models to quantify the combined effect

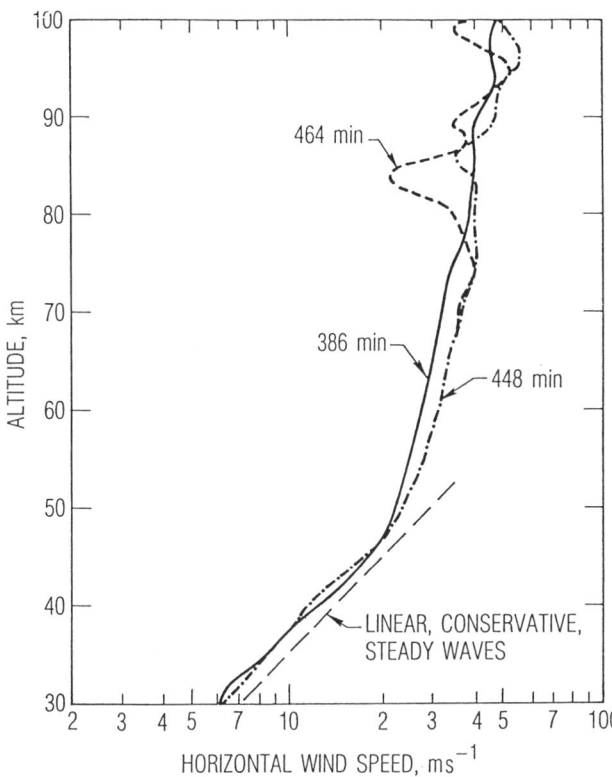

Figure 5. The dependence of the amplitude of the primary wave horizontal velocity on altitude at 386, 448 and 464 minutes after onset (Walterscheid and Schubert, 1990). The sloping line represents the growth of linear, conservative, steady waves.

of chemistry and wave-generated turbulence on constituent transport in the middle atmosphere [Garcia and Solomon, 1985; Bjarnason et al., 1987; LeTexier et al., 1987].

Walterscheid and Schubert [1989] used a chemical-dynamical model to directly calculate the chemically-induced gravity wave fluxes of OH, O_3, HO_2, H and O at mesopause heights. The model incudes a five-reaction chemical scheme for the production and loss of the minor species and the complete dynamics of linear gravity waves in a motionless background atmosphere. Fluxes of OH and O_3 number density were evaluated by means of relations similar to (5). Dwelling on the chemically induced flux of O_3, the strongest flux occurs in the region where the odd hydrogen chemistry is most effective in controlling O_3 abundance and where the constituent scale heights are smallest (vertical gradients are greatest). For a given vertical velocity perturbation the wave-induced constituent perturbations are greatest where the scale heights are smallest.

Figure 6 shows the chemically-induced number density fluxes of O_3 and OH as a function of altitude for a wave with a period of 3 h, a horizontal wave length of 1000 km and an amplitude of $T'/\overline{T} = 0.15$ at 80 km altitude [Walterscheid and Schubert, 1989]. The waves satisfy the nonacceleration conditions. Two sets of results are shown corresponding to fluxes induced by chemistry and dynamics (solid curves) and by dynamics alone (dashed curves). The curves are annotated with the scaling applied to each. (Note that while the mixing ratio flux vanishes for dynamics alone, the number density flux does not. However, it tends to be small.)

The flux of O_3 is essentially zero except for a region about 5 km thick centered at about 80 km wherein the fluxes are large and downward. The flux tends to deplete the upper part of the region and enhance the lower part. The flux divergence computed from the flux profile (not shown) indicates that the time scale for significantly changing the O_3 profile is ~ 1000 s. This is comparable to or shorter than the chemical time constant for ozone in this region [Walterscheid et al., 1987]. This means that chemically induced fluxes may be an important contributor to the chemical-dynamical balance that determines the background O_3 concentration in the mesopause region. The results of the calculations also indicate that the fluxes are not simply related to the background minor constituent gradients and can be up as well as down gradient.

7. SUMMARY

The role of gravity-wave-mean-state interactions is discussed for tidally modulated mean-flow accelerations, wave drag due to wave transience, chemically induced wave transport of minor constituents, and wave saturation due to wave-mean-flow energy conversions. The major findings discussed are

• Gravity waves can be effective in forcing large variations in the mean winds at tidal frequency.
• Wave drag due to wave transience coupled with wave dissipation or breakdown may be important and play a significant role in maintaining the cold polar summer mesopause.
• Conversions of wave energy from the wave to the mean state can deplete wave energy, significantly diminish the growth of wave amplitude with altitude, and thereby cause the wave to exhibit saturation like growth with altitude. Wave breakdown occurring before the establishment of steady-state amplitude growth with altitude can have a similar effect.
• During magnetic storms, gravity waves can induce a warming of the low-latitude thermosphere which competes with the warming due the large-scale thermally direct pole to equator circulation.

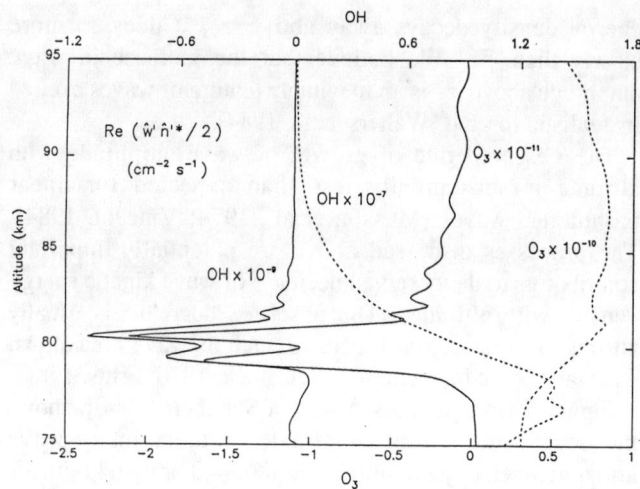

Figure 6. The wave induced number density fluxes of O_3 and OH as a function of altitude for a wave with a period of 3 h, a horizontal wave length of 1000 km and an amplitude of $T'/\overline{T} = 0.15$ at 80 km altitude (Walterscheid and Schubert, 1989). Two sets of results are shown corresponding to fluxes induced by chemistry and dynamics (solid curves) and by dynamics alone (dashed curves). The curves are annotated with scaling factors.

• Chemically-induced wave transports can alter the mean profile of ozone in the mesopause region.
• Removal of heat by wave transports can cool the lower thermosphere.

Acknowledgments. This work was supported by The Aerospace Sponsored Research Program of the Aerospace Corporation, by National Science Foundation Grant ATM-900216 and by NASA Space Physics Grant NAGW 2887.

REFERENCES

Allen, M., Y. L. Yung, and J. W. Waters, Vertical transport and photochemistry in the terrestrial mesosphere and lower thermosphere (50-120 km), J. Geophys. Res., 86, 3617-3627, 1981

Andrews, D. G., and M. E. McIntyre, Planetary waves in horizontal and vertical shear: the generalized Eliassen and Palm relation and the mean zonal acceleration, J. Atmos. Sci., 33, 2031-2048, 1976

Boyd, J. P., The noninteraction of waves with the zonally averaged flow on a spherical earth and interrelationships of eddy fluxes of energy, heat and momentum, J. Atmos. Sci, 33, 2285-2291, 1976

Booker J. R., F. P. Bretherton, The critical layer for gravity waves in shear flow, J. Fluid Mech., 27, 513-530, 1967

Brinkman,. D. G., R. L. Walterscheid, A. D. Richmond, and S. V.

Venkateswaran, Wave-mean flow interaction in the storm-time thermosphere: A two-dimensional model simulation, *J. Atmos. Sci., 49,* 660-680, 1992

Collins, R. L., D. C. Senft and C. S. Gardner, Observations of a 12 h wave in the mesopause region at the south pole, *Geophys. Res. Lett, 19,* 67-60, 1992

Clark, T. L., and W. R. Peltier, Critical level reflection and the resonant growth of nonlinear mountain waves, *J. Atmos. Sci, 41,* 3122-3134, 1984

Coy, L., and D. C. Fritts, Gravity wave heat fluxes: A Lagrangian approach, *J. Atmos. Sci., 45,* 1770-1780, 1988

Davies, P. A., and W. R. Peltier, Some characteristics of the Kelvin-Helmholtz and resonant over-reflection modes of shear flow instability and of their interaction through vortex pairing, *J. Atmos., Sci, 36,* 2394-2412, 1979

Dunkerton, T. J., Wave transience in a compressible atmosphere, part I, Transient internal wave, mean-flow interaction, *J. Atmos, Sci., 38,* 281-297, 1981

Dunkerton, T. J., Wave transience in a compressible atmosphere. Part III: The saturation of internal gravity waves in the mesosphere, *J. Atmos. Sci., 39,* 1042-1051, 1982

Dunkerton, T. J., and D. C. Fritts, The transient gravity wave critical layer, Part I: Convective adjustment and the mean zonal acceleration, *J. Atmos. Sci., 41,* 992-1007, 1984

Dunkerton, T. J., Effect of nonlinear instability on gravity-wave momentum transport, *J. Atmos. Sci., 44,* 3188-3209, 1987

Eliassen, A., and E. Palm, On the transfer of energy in stationary mountain waves, *Geophys. Publ., 22(3),* 1-13, 1961

Forbes, J. M., Atmospheric tides, 2, the solar and lunar semidiurnal components, *J. Geophys. Res., 87,* 5241-5252, 1982

Forbes, J. M., J. Gu, and S. Miyahara, On the interactions between gravity waves and the propagating diurnal tide, *Planet. Space Sci., 39,* 1249-1257, 1991

Fritts, D. C, The nonlinear gravity wave-critical level interaction, *J. Atmos. Sci, 35,* 12-23, 1978

Fritts, D. C, The excitation of radiating waves and Kelvin-Helmholtz instabilities by the gravity wave-critical level interaction, *J. Atmos. Sci, 36,* 12-23, 1979

Fritts, D. C., Gravity wave saturation in the middle atmosphere: A review of theory and observations, *Rev. Geophys. Space Phys., 22,* 275-308, 1984

Fritts, D. C., and T. J. Dunkerton, A quasi-linear study of gravity wave saturation and self-acceleration, *J. Atmos. Sci., 41,* 3272-3289, 1984

Fritts, D. C., and T. J. Dunkerton, Fluxes of heat and constituents due to convectively unstable gravity waves, *J. Atmos. Sci., 42,* 549-556, 1985

Fritts D. C., and R. A. Vincent, Mesospheric momentum flux studies at Adelaide, Australia: observations and a gravity wave/tidal interaction model, *J. Atmos. Sci., 44,* 605-619, 1987

Fritts, D. C., and L. Yuan, Measurements of momentum fluxes near the summer mesopause at Poker Flat, Alaska, *J. Atmos. Sci., 46,* 2569-2569, 1989

Fritts, D. C., and T. E. VanZandt, Spectral estimates of gravity wave energy and momentum fluxes, I: Energy dissipation, acceleration, and constraints, *J. Atmos. Sci., 50,* submitted, 1993

Fuller-Rowell, T. J., and D. Rees, A three-dimensional time-dependent simulation of the global dynamical response of the thermosphere to a geomagnetic substorm, *J. Atmos. Terr. Phys., 43,* 701-721, 1981

Garcia, R. R., and S. Solomon, The effect of breaking gravity waves on the dynamics and chemical composition of the mesosphere and lower thermosphere, *J. Geophys. Res., 90,* 3850-3868, 1985

Geller M. A., A. H. Tanaka, and D. C. Fritts, Production of turbulence in the vicinity of critical levels for internal gravity waves, *J. Atmos. Sci., 32,* 2125-2135, 1975

Grimshaw, R., Nonlinear internal gravity waves and their interaction with the mean wind. *J. Atmos. Sci., 32,* 1779-1793, 1975

Hines, C. O., Dynamical heating of the upper atmosphere, *J. Geophys. Res., 70,* 177-183, 1965

Holton, J. R., *The Dynamic Meteorology of the Stratosphere and Mesosphere,* 218pp., American Meteorological Society, Boston, 1975

Holton, J. R., The role of gravity wave-induced drag and diffusion in the momentum budget of the mesosphere, *J. Atmos. Sci., 39,* 781-799, 1982

Holton, J. R., The influence of gravity wave-breaking on the general circulation of the middle atmosphere, *J. Atmos. Sci., 40,* 2497-2507, 1983

Hodges, R. R., Jr., Generation of turbulence in the upper atmosphere by internal gravity waves, *J. Geophys. Res., 72,* 3455-3458, 1967

Jones, W. L., and D. D. Houghton, The self-destructing internal gravity wave, *J. Atmos. Sci, 29,* 844-849, 1972

Larsen, M. F., and I. S. Mikkelsen, The dynamic response of the high-latitude thermosphere and geostrophic adjustment, *J. Geophys. Res., 88,* 3158-3168, 1983

Leovy, C., Simple models of thermally driven mesospheric circulation, *J. Atmos. Sci., 21,* 327-341, 1964

Lindzen, R. S., Turbulence and stress due to gravity wave and tidal breakdown, *J. Geophys. Res., 86,* 9707-9714, 1981

Longuet-Higgins, M. S., The eigenfunctions of Laplace's tidal equations over a sphere, *Phil. Trans. Roy. Soc. London A262,* 511-607, 1968.

Lu, W., and D. C. Fritts, Spectral Estimates of Gravity Wave Energy and Momentum Fluxes, III: Gravity Wave-Tidal Interactions, *J. Atmos. Sci,* in press, 1993

Manson, A. H., J. B. Gregory, and D. G. Stephenson, Winds and wave motions to 110 km at mid-latitudes I. Partial reflection radio wave soundings 1972-73. *J. Atmos. Sci., 31,* 2207-2215, 1974

Miyahara, S., Y. Hayashi, and J. D. Mahlman, Interactions between gravity waves and planetary scale flow simulated by the GFDL "SKYHI" general circulation model, *J. Atmos. Sci., 43,* 1844-1861, 1986

Orlanski, I., and K. Bryan, Formation of the thermocline step structure by large-amplitude internal gravity waves, *J. Geophys. Res., 74,* 6975-6983, 1969

Peltier, W. R., and T. L. Clark, The evolution and stability of finite-amplitude mountain waves, part II: surface wave drag and severe downslope windstorms, *J. Atmos. Sci., 36*, 1498-1529, 1979

Pitteway, J. L. V., and C. O. Hines, The viscous damping of atmospheric gravity waves, *Can. J. Phys., 41*, 1935-1948, 1963

Richmond A. D., Thermospheric heating in a magnetic storm: Dynamic transport of energy from high to low latitudes, *J. Geophys. Res., 84*, 5259-5266, 1979

Richmond, A. D., and S. Matsushita, Thermospheric response to a magnetic substorm, *J. Geophys. Res., 80*, 2839-2850, 1975

Roble, R. G., J. M. Forbes, and F. A. Marcos, Thermospheric dynamics during the March 22, 1979, magnetic storm, 1: model simulations, *J. Geophys. Res., 92*, 6045-6068, 1987

G. G. Sivjee, and R. L. Walterscheid, R. L., Six-hour zonally symmetric tidal oscillations of the winter mesopause over the south pole station, *Plan. Space Sci.*, in press, 1994

Sivjee, G. G., R. L. Walterscheid, and D. J. McEwan, Planetary wave disturbances in the arctic winter mesopause over Eureka (80°N), *Plan. Space Sci.*, in press, 1994

Schoeberl, M. R., D. F. Strobel and J. P. Apruzese, A numerical model of gravity wave breaking and stress in the mesosphere, *J. Geophys. Res., 88*, 5249-5259, 1983

Smith, S. A., D. C. Fritts, and T. E. VanZandt, Evidence for a saturated spectrum of atmospheric gravity waves, *J. Atmos. Sci., 44*, 1404-1410, 1987

Strobel, D. F., Parameterizations of linear wave chemical transport in planetary atmospheres by eddy diffusion, *J. Geophys. Res., 86*, 9806-9810, 1981

Vincent, R. A., Gravity wave motions in the mesosphere, *J. Atmos. Terr. Phys., 46*, 119-128, 1984

Walterscheid, R. L., J. G. DeVore, and S. V. Venkateswaran, Influence of mean zonal motion and meridional temperature gradients on the solar semidiurnal atmospheric tide: A revised spectral study with improved heating rates, *J. Atmos. Sci, 37*, 455-470, 1980

Walterscheid, R. L., Inertio-gravity wave induced accelerations of mean flow having an imposed periodic component: Implications for tidal observations in the meteor region, *J. Geophys. Res., 86*, 9698-9706, 1981a

Walterscheid, R. L., Dynamical cooling induced by dissipating internal gravity waves, *Geophys. Res., Lett., 8*, 1235-1238, 1981b

Walterscheid, R., Gravity wave attenuation and the evolution of the mean state following wave breakdown. In *Dynamics of the Middle Atmosphere*, ed. by J. R. Holton and T. Matsuno, pp. 19-43, Terra, Tokyo, 1984:

Walterscheid, R. L., G. G. Sivjee, G. Schubert and R. M. Hamwey, Large-amplitude semidiurnal temperature variations in the polar mesopause: evidence of a pseudotide, *Nature, 324*, 347-349, 1986

Walterscheid, R. L., G. Schubert, and J. M. Straus, A dynamical-chemical model of wave driven fluctuations in the OH nightglow, *J. Geophys. Res., 92*, 1241-1254, 1987

Walterscheid, R. L. and G. Schubert, Gravity wave fluxes of O_3 and OH at the nightside mesopause, *Geophys. Res. Lett., 16*, 719-722, 1989

Walterscheid, R. L., and G. Schubert, Nonlinear evolution of an upward propagating wave: overturning, convection, transience and turbulence. *J. Atmos. Sci., 47*, 101-125, 1990

Wang, D.-Y., and D. C. Fritts, Evidence of gravity wave-tidal interaction observed near the summer mesopause at Poker Flat, Alaska, *J. Atmos. Sci., 48*, 572-583, 1991

Weinstock, J., Nonlinear theory of acoustic-gravity waves, 1, Saturation and enhanced diffusion. *J. Geophys. Res., 81*, 633-652, 1976

Weinstock, J., Nonlinear theory of gravity waves: momentum deposition, generalized Rayleigh friction, and diffusion, *J. Geophys. Res., 81*, 633-652, 1976

Weinstock, J., Heat flux induced by gravity waves, *Geophys. Res. Lett., 10*, 165-167, 1983

Yeh, K. C., and C. H. Liu, The instability of gravity waves through wave-wave interactions, *J. Geophys. Res., 86*, 9722-9728, 1981

R. L. Walterscheid, The Aerospace Corporation, P. O. Box 92957, Los Angeles, CA 9009

Modulation of Gravity Wave Drag in the Mesosphere and Lower Thermosphere by Tides and the Effects on the Time Mean Flow - Model Results

Charles McLandress and W. E. Ward

Institute for Space and Terrestrial Science, York University,
4700 Keele St., North York, Canada, M3J 1P3

It is now generally accepted that gravity waves play a significant role in the large scale dynamics of the mesosphere and lower thermosphere. In this paper, the question addressed concerns how the inclusion of the semidiurnal tidal winds and temperature modifies the temporal and spatial variation of the forcing associated with breaking orographically generated gravity waves. For this investigation a modified version of the model developed by *McLandress and McFarlane* [1993] is used. In the particular case studied, the results indicate that the presence of the tide enhances the time-averaged gravity wave drag at northerly latitudes by about 20 to 30% which in turn generates weaker zonal mean westerlies and larger stationary planetary waves.

1. INTRODUCTION

Since the development of the first simple gravity wave drag parameterization scheme by *Lindzen* [1981], a number of investigations have been undertaken to understand the effects of these waves on the large scale dynamics of the middle atmosphere. Early successes include the simulation of the zonally averaged circulation [*Holton*, 1983] and an explanation for the observed variation of atomic oxygen and ozone with season [*Garcia and Solomon*, 1985]. More recently, parameterizations have been developed for gravity waves generated by flow over orography [*Palmer et al.*, 1986; *McFarlane*, 1987].

Recently, two papers have been published in which the interaction between the spatial distribution of orographic gravity wave sources, the wind field and the spatial distribution of momentum deposition in the middle atmosphere is explored. *Bacmeister* [1993] pays special attention to the filtering properties of the specific wind fields on the resulting deposition pattern and points out that use of daily wind fields as opposed to temporally averaged fields leads to significantly different results. In a complementary study, *McLandress and McFarlane* [1993] (henceforth MM) investigate the climatological effects of this localized forcing. They demonstrate that mesospheric planetary scale disturbances are generated in response to the spatial inhomogeneity of the gravity wave breaking (a feature first noted by *Holton* [1984]). MM also show that these disturbances can exert a drag on the zonal mean wind via the Eliassen-Palm flux divergence with a magnitude similar to that of the zonal mean gravity wave drag (henceforth GWD).

None of the above papers, however, include tides as part of the background field through which the gravity waves propagate. Nevertheless, it is evident that the tidal wind field will significantly modify the location at which momentum deposition due to gravity waves occurs [*Lu and Fritts*, 1993; and references therein]. *Hunt* [1990] used a general circulation model which generated tidal motions and showed that significant modulation of the parameterized gravity wave momentum flux in the mesosphere arose as a result of the presence of tides.

In this paper we report on the result of including tidal

winds and temperatures in the calculation of the GWD. As in MM, the approach is to determine the climatological effects of such a forcing with emphasis put on the influence of orographically forced gravity waves. This differs from *Lu and Fritts* [1993] where individual wind profiles are considered and the spatial variation of the forcing is not, and from *Hunt* [1990] by our emphasis of the specific effects of the inclusion of the tidal winds. Moreover, with the top of Hunt's model at 100 km, the tides in the upper mesosphere cannot be properly simulated since reflections at the boundary will be significant.

2. MODEL DESCRIPTION

The model is a global quasi-geostrophic (QG) model, modified slightly from the hemispheric version described in MM. The reader is referred to that paper for more details. It extends from 16 to 140 km and simulates both the zonal mean and planetary wave components (waves 1-3) of the circulation. The observed monthly mean climatological 100 mb geopotential fields are used as the lower boundary condition. Sponge layers are inserted in the equatorial regions to maintain the low latitude winds at realistic values and at the upper boundary to prevent spurious reflections.

The heating in the 16-120 km regime is modeled using the radiation code of *Fomichev and Shved* [1988], modified using the approach outlined by *Fomichev et al.* [1993]. Above 120 km, where the radiation code is not used, a momentum sponge layer is inserted and the heating is reduced to zero at 140 km where an upper boundary condition of vanishing residual vertical velocity is applied. In the results which will be discussed shortly, the 120 to 140 km region is not plotted.

The effects of breaking gravity waves are parameterized in a manner similar to *Lindzen* [1981] using the scheme discussed in detail in MM. A simple spectrum of gravity waves, consisting of both orographic (i.e., the same as as in MM) and a single eastward propagating non-orographic wave, is assumed. A similar approach has been used by *Jackson* [1993] in a study using a middle atmosphere general circulation model. For mechanistic studies, like ours, in which the interaction between tides and gravity waves is explored, such a choice of gravity waves is justifiable but is not intended to discount the possible importance of waves with other phase speeds. [*Rind et al.*, 1988; *Miyahara et al.*, 1986].

The use of only a single non-orographic wave whose phase speed c is 25 m/s is based simply on the need for a momentum sink in the summer hemisphere. This choice of c can be loosely justifed as the typical phase speed of a wave generated by systems moving with the speed of the tropospheric flow [*Holton*, 1983]. It was found that a value of 2.5×10^{-6} N/m^2 for the lower boundary vertical momentum flux τ_B is adequate to generate wave breaking at mesospheric heights. In addition, to prevent wave breaking in equatorial regions where the model dynamics are not valid, a $\sin^2\theta$ latitudinal dependence is assumed for τ_B. The remaining adjustable parameter is the product of the gravity wave efficiency factor and horizontal wavenumber. This is given values of 6×10^{-6} and 3×10^{-6} m^{-1} for the orographic and non-orographic waves, respectively.

The tidal wind and temperature fields must be specified a priori since they cannot be generated internally using a QG model. For this particular study the effects of the semidiurnal tide is examined using January data from the Vial model [*Vial*, 1986]. It should be stressed that in this study the tidal fields are affected by neither the model mean zonal winds nor by the GWD - their only effect is to temporally modulate the gravity wave fluxes that emanate from the stratosphere. The amplitude and phase of the zonal wind component is shown in Figure 1. Although the large amplitude diurnal tide may be important, it is generally restricted to subtropical latitudes in contrast to the semidiurnal component which is dominant at higher latitudes [*Forbes*, 1982a,b]. Consequently the results presented here should not be significantly modified outside the 30°N to 30°S latitude band by the presence of the diurnal tide.

A Lindzen-like GWD parameterization has been used in several studies concerning tidal/gravity wave interactions. [e.g., *Miyahara and Forbes*, 1991]. A few remarks should, however, be made about the applicability of the scale separation assumption inherent in the GWD parameterization. Inspection of Figure 1 indicates that in the region of large tidal amplitude at 60°N wind speeds attain a maximum of about 50 m/s and the vertical wavelength is about 40 km. The vertical wavelength of the gravity wave can be estimated from linear theory and is proportional to the mean wind speed. Thus wind speeds in the 30 to 50 m/s range yield wavelengths from 10 to 15 km, indicating that a spatial scale separation between the tide and the gravity wave exists.

Concerning the temporal scale, the vertical component of the gravity wave group velocity can easily be shown to be proportional to the square of the Doppler-shifted wind. A typical horizontal wavelength of 100 km, a Brunt-Vassala frequency of 2×10^{-2} s^{-1} and $(c - U)$ of 40 m/s yield a group velocity of approximately 3 m/s. This implies that it takes less than 2 hours for a gravity

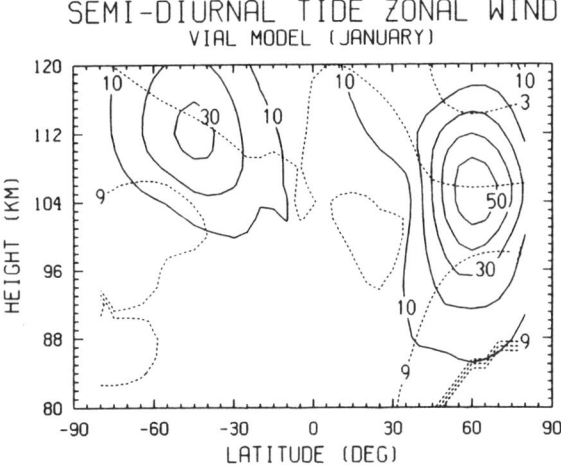

Fig. 1. Amplitude in m/s (solid) and phase in hours (dashed) of semidiurnal tide zonal wind component from the *Vial* [1986] model.

wave packet to propagate through the region of largest semidiurnal tidal amplitudes from 90 to 120 km (see Figure 1). In this time the tidal phase has changed by 1/6 of a period which is a relatively small but not insignificant amount.

One final point concerns the regions near critical lines at $c = U$. Below these levels gravity wave breaking must take place due to wind shear effects and so the eddy diffusion which has been invoked to maintain the wave at its saturated amplitude strongly damps the wave so that its momentum flux is completely absorbed as $(U - c) \to 0$.

The GWD is calculated on a longitude by latitude by height grid each hour using the combined zonal mean, planetary and tidal wind and temperature fields as the background flow field. The resulting drag field is then averaged over a tidal period (12 hours) to obtain the forcing of the geostrophic flow and finally projected onto the zonal mean and planetary wave components. The model is integrated forward in time from December 1 until mid January using the monthly varying boundary and radiative forcing fields. A time step of 1 hour and a spatial grid of 4 km by 10 degrees are used.

3. MODEL RESULTS

To evaluate the impact of the presence of the semidiurnal tide on the time mean GWD and wind fields, two separate experiments were performed in which the tidal winds and temperatures were first included and then omitted from the drag calculations. In panels (a) to (d) of Figure 2 are shown height vs latitude plots of the zonally averaged zonal wind and GWD components for the case with the tide present and the corresponding differences between this case and the no tide case. In extratropical regions where the tidal winds are strongest (see Figure 1), the zonal mean drag is enhanced by the presence of the tide, which in turn corresponds to reduced wind speeds in both hemispheres. In the northern hemisphere lower thermosphere, above the tidal wind amplitude maximum, the drag is reduced. The magnitude of the differences in panel (b), being less than 5 m/s, is not large, however, and indicates that the presence of the tide in the GWD calculation does not substantially alter the zonal and time mean wind field.

In panels (e) and (f) of the same figure are shown the stationary zonal wavenumber 1 geopotential height and the associated amplitude differences with the no tide case. In the lower stratosphere the planetary wave is forced from below by the climatological boundary condition while in the mesosphere and lower thermosphere it is generated in situ by the localized orographic GWD. Panel (f) indicates that in northerly latitudes this in situ forcing of wave 1 is further enhanced by the presence of the tide as is evident from the 30% increase in amplitude at 120 km. Since the zonal wind component is proportional to the latitudinal gradient of the geopotential, the dipole structure in panel (f) indicates that the wave 1 zonal wind component (not shown) is increased by about 5 m/s at 100 km at 60°N.

As was discussed by *Hunt* [1990] the presence of tides significantly modifies the vertical momentum flux of gravity waves. The next series of diagrams illustrates this point. In Figure 3 northern hemisphere polar stereographic maps of the zonal component of the GWD at a height of 100 km is shown at 3 hour intervals. The large temporal variation in the strength and location of the regions of wave breaking are strikingly clear and represent the modulation of the drag by the semidiurnal tide. For example, the strong centre over North America in panel (b) has all but vanished 6 hours later in panel (c).

The time averaged GWD and the corresponding differences with the no tide case are shown in Figure 4. The inclusion of the tide is seen to increase the time mean drag at all locations by as much as 10 m/s/day in northerly latitudes. At mid-latitudes, however, the differences are quite small despite the substantial semidiurnal signal in the GWD (see over eastern Asia, for instance).

The modifications of the time mean GWD that arise due to the presence of the tides results from the nonlinear dependence of the gravity wave momentum flux on the local wind speed. This dependence is, in fact,

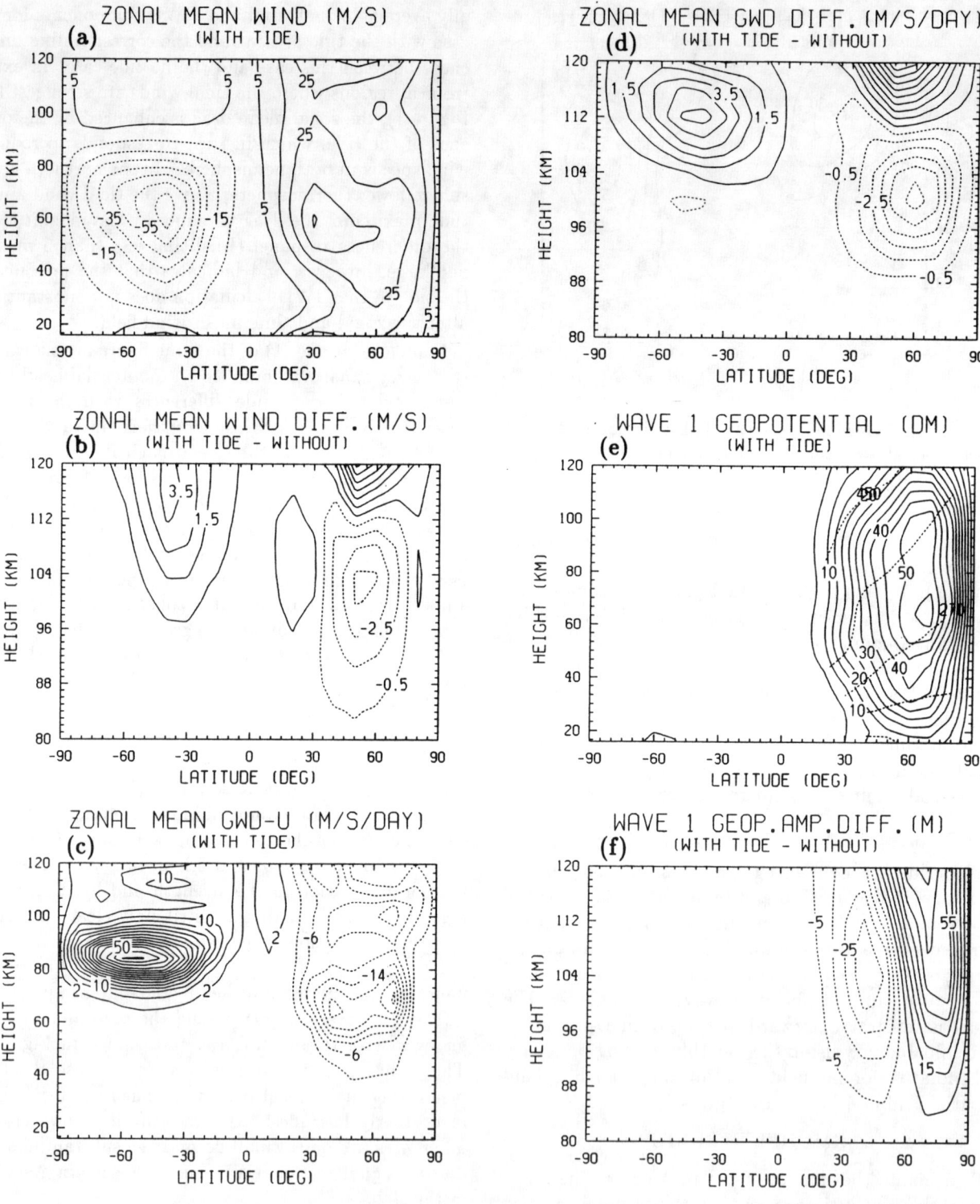

Fig. 2. Model results for the case with the tide included in the GWD calculation of (a) zonal mean wind, (c) zonal and time mean GWD, and (e) stationary planetary wavenumber one geopotential height amplitude in decameters (solid) and phase in degrees (dashed using a 90° contour interval). The corresponding differences with the no tide case are shown in (b), (d) and (f).

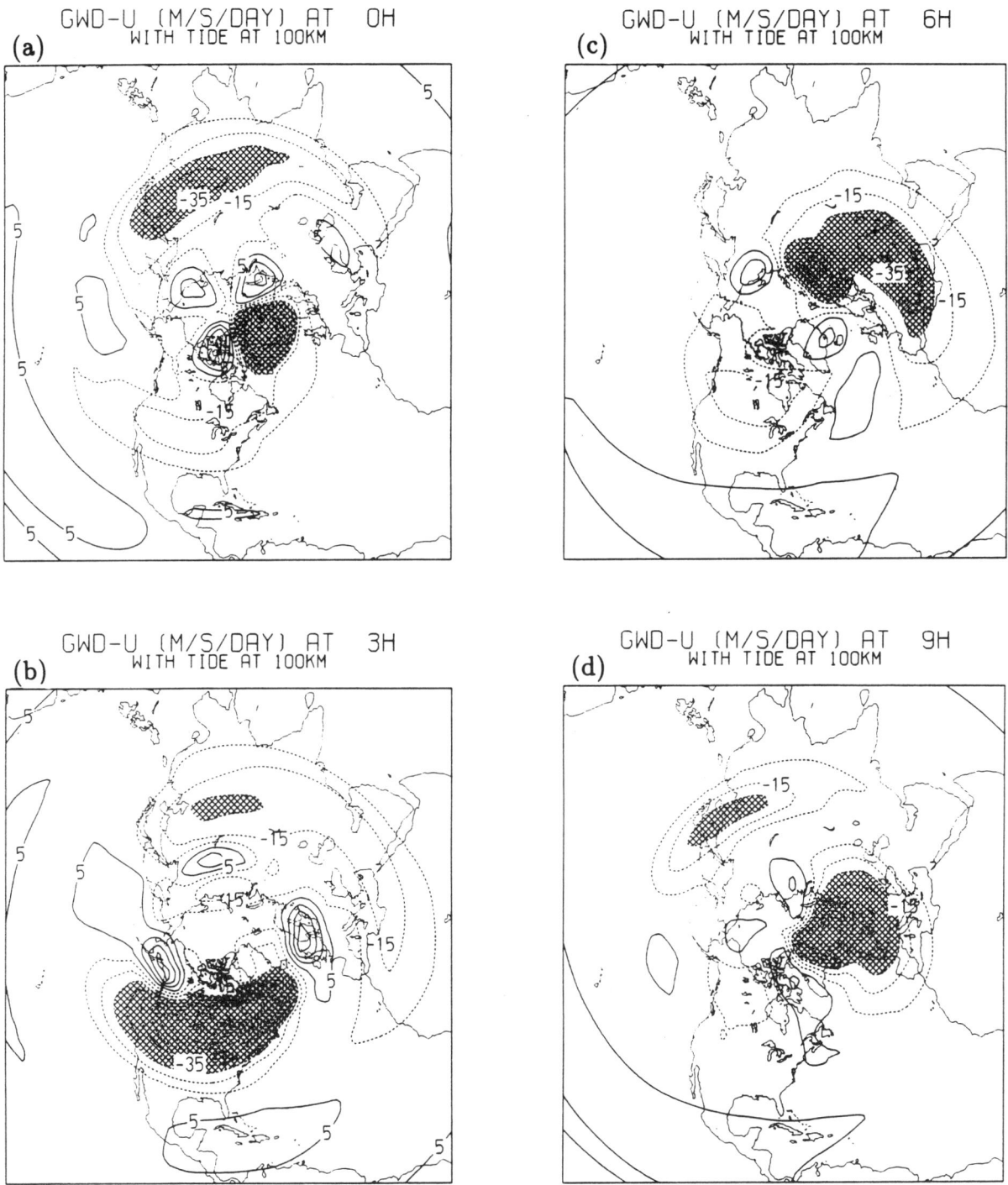

Fig. 3. Northern hemisphere polar stereographic maps at 100 km of zonal GWD component at (a) 0 hours, (b) 3 hours, (c) 6 hours and (d) 9 hours. A contour interval of 10 m/s/day is used and values < -25 m/s/day are shaded.

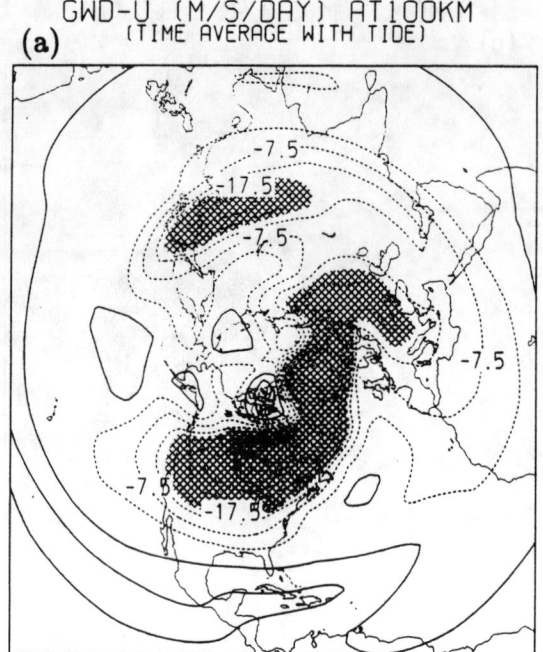

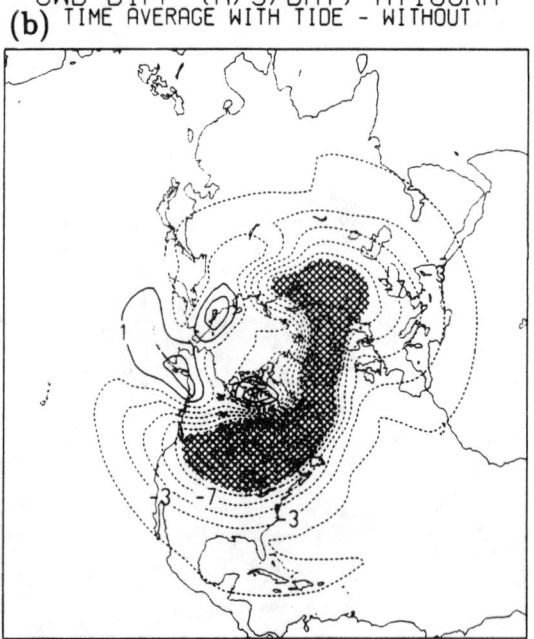

Fig. 4. Same as Figure 3 but for (a) time averaged GWD and (b) difference with no tide case. In panel (a) a 5 m/s/day contour interval is used and values < -15 m/s/day are shaded. In panel (b) a 2 m/s/day contour interval is used and values < -10 m/s/day are shaded.

cubic and occurs as a direct consequence of the gravity wave saturation hypothesis inherent in the parameterization scheme. In our model this results in a drag term proportional to $(U_o + U_T)^3$, where U_o and U_T denote the stationary (zonal mean plus planetary waves) and semidiurnal tidal wind fields, respectively. It can easily be seen that this yields the term $3U_oU_T^2$ which projects onto the time mean component. Thus in regions where the tidal amplitudes are large one would expect larger differences of the time mean GWD than in the case without tides.

In Figure 5 is shown time vs height plots of the zonal components of GWD and velocity at 50°N, 90°W. Both the semidiurnal variation in the GWD and the importance of the zero wind line in blocking gravity wave activity are evident in these figures. In addition, on the right hand side of these panels are shown the corresponding semidiurnal tidal, stationary planetary wave one and zonal mean components of the GWD and wind at 50°N, which have been calculated from consideration of the drag and wind fields around the entire latitude circle. The tide is clearly seen to dominate both the wind and drag at 100 km, while the stationary planetary effects are largest around 60 km.

Although our model does not permit a modification of the tidal structures by the GWD, the possible effects of GWD on the tides can be inferred from Figure 5. Clearly, a large semidiurnal component in the zonal GWD is seen in Figure 5(a) and the strong correlation between regions of strong drag and strong westerlies indicates that at these locations the breaking gravity waves would in fact damp the tidal winds.

4. SUMMARY

A quasi-geostrophic model has been used to examine the impact of tidal motions on the time mean flow of the extratropical middle atmosphere that occurs as a result of the modulation of gravity wave drag by the semidiurnal tide. Since tides cannot be generated in the model, previously calculated semidiurnal tidal wind and temperature fields were included along with the model-simulated planetary wave and zonal mean fields in the determination of the GWD.

It was found that the presence of the tide significantly alters the instantaneous GWD fields in mid and northerly latitudes where the tidal winds are strongest. While causing large temporal modulations of the drag field, their presence does not, however, cause large departures of the time-mean drag from the values calculated without the tide present. This result indicates

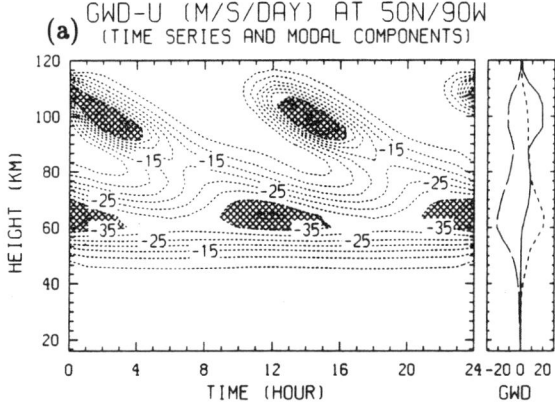

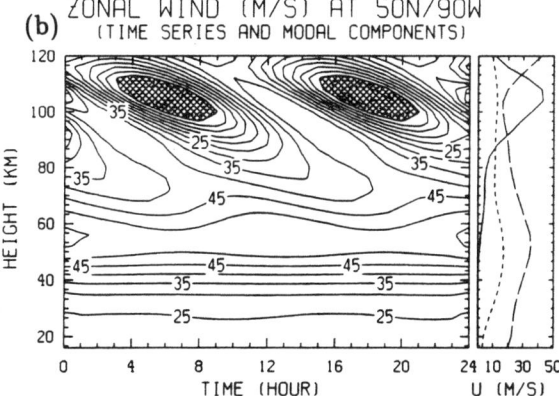

Fig. 5. Temporal variation of zonal components of (a) GWD and (b) velocity at 50°N, 90°W with tide present. Values of GWD < -35 m/s/day and of wind < 0 are shaded. On right side of figures are vertical profiles of semidiurnal tide (solid), stationary planetary wave one (short dashes) and zonal mean (long dashes) components of the GWD in m/s/day and wind in m/s.

that in this case the tidally modulated GWD projects predominently onto the tidal components of the drag. Although the overall pattern of the time averaged drag fields remained qualitatively similar, the inclusion of tides did result in some enhancement of the orographic drag in northerly latitudes in the upper mesosphere. The tidal modulation of the GWD fields resulted in a 10% decrease in zonal mean wind speeds and a 20-30% increase in planetary wave amplitudes in the mesosphere, both of which are consistent with the enhanced drag in this region.

Some caution must be used in interpreting these results since they depend on the particular choice of the gravity wave spectrum, on the particular structure of the tides, and on the use of only the semidiurnal tide. Presently we are studying the effects of both the diurnal and semidiurnal tides and our results indicate that in mid and polar latitudes the dominant GWD modulation is by the semidiurnal tide. A paper on this work is in preparation. Thus, while the results from these model runs are dependent on the specific details of the model used, it is clear that tidal motions can modify the structure of the time-averaged GWD and so must be included to accurately model the large scale dynamics of the mesosphere and lower thermosphere.

Acknowledgments. The authors would like to express their thanks to Dr. Francois Vial for his support in the development of this paper and to two anonymous reviewers for their constructive suggestions. This work was funded by the Institute for Space and Terrestrial Science, which is a designated Centre of Excellence supported by the Province of Ontario's Technology Fund.

REFERENCES

Bacmeister, J. T., Mountain-wave drag in the stratosphere and mesosphere inferred from observed winds and a simple mountain-wave parameterization scheme, *J. Atmos. Sci.*, 50, 377-399, 1993.

Fomichev, V. I., and G. M. Shved, Net radiative heating in the middle atmosphere, *J. Atmos. Terr. Phys.*, 50, 671-688, 1988.

Fomichev, V. I., A. A. Kutepov, R. A. Akmaev, and G. M. Shved, Parameterization of the 15 μm CO_2 band cooling in the middle atmosphere (15-115 km), *J. Atmos. Terr. Phys.*, 55, 7-18, 1993.

Forbes, J. M., Atmospheric tides, 1: Model description and results for the solar diurnal component, *J Geophys. Res.*, 87, 5222-5240, 1982a.

Forbes, J. M., Atmospheric tides, 2: The solar and lunar semidiurnal components, *J Geophys. Res.*, 87, 5241-5252, 1982b.

Garcia, R. R., and S. Solomon, The effect of breaking gravity waves on the dynamics and chemical composition of the mesosphere and lower thermosphere, *J. Geophys. Res.*, 90, 3850-3868, 1985.

Holton, J. R., The influence of gravity wave breaking on the general circulation of the middle atmosphere, *J. Atmos. Sci.*, 40, 2497-2507, 1983.

Holton, J. R., The generation of mesospheric planetary waves by zonally asymetric gravity wave breaking, *J. Atmos. Sci.*, 41, 3427-3430, 1984.

Hunt, B. G., A simulation of the gravity wave characteristics and interactions in a diurnally varying model atmosphere, *J. Meteor. Soc. Japan*, 68, 145-161, 1990.

Jackson, D. R., Sensitivity of the Extended UGAMP General Circulation Model to specification of gravity-wave phase speeds, *Quart. J. R.Met. Soc.*, 119, 457-468, 1993.

Lindzen, R. S., Turbulence and stress owing to gravity wave and tidal breakdown, *J. Geophys. Res.*, 86, 9707-9714, 1981.

Lu, W., and D. C. Fritts, Spectral estimates of gravity wave energy and momentum fluxes, III: gravity wave-tidal interactions, *J. Atmos. Sci.*, in press, 1993.

McFarlane, N. A., The effect of orographically excited gravity wave drag on the general circulation of the lower stratosphere and troposphere, *J. Atmos. Sci.*, 44, 1775-1800, 1987.

McLandress, C., and N. A. McFarlane, Interactions between orographic gravity wave drag and forced stationary planetary waves in the winter northern hemisphere middle atmosphere, *J. Atmos. Sci.*, 50, 1966-1990, 1993.

Miyahara, S., Y. Hayashi, and J. D. Mahlman, Interactions between gravity waves and planetary-scale flow simulated by the GFDL "SKYHI" General Circulation Model, *J. Atmos. Sci.*, 43, 1844-1861, 1986.

Miyahara, S., and J. M. Forbes, Interactions between gravity waves and the diurnal tide in the mesosphere and lower thermosphere, *J. Met. Soc. Japan*, 69, 523-531, 1991.

Palmer, T. N., G. J. Shutts, and R. Swinbank, Alleviation of a systematic westerly bias in general circulation and numerical weather prediction models through an orographic gravity wave parameterization, *Quart. J. R.Met. Soc.*, 112, 1001-1039, 1986.

Rind, D., R. Suozzo, N.K. Balachandran, A. Lacis, and G. Russell, The GISS global climate-middle atmosphere model. Part I: Model structure and climatology, *J. Atmos. Sci.*, 45, 329-370, 1988.

Vial, F., Numerical simulations of atmospheric tides for solstice conditions, *J. Geophys. Res.*, 91, 8955-8969, 1986.

C. McLandress and W. E. Ward, Institute for Space and Terrestrial Science, Centre for Research in Earth and Space Science, York University, North York, Ontario, Canada M3J 1P3.

Scale-Independent Diffusive Filtering Theory of Gravity Wave Spectra in the Atmosphere

Chester S. Gardner

Department of Electrical and Computer Engineering, University of Illinois at Urbana-Champaign, Urbana, Illinois

Shear and convective instabilities, Doppler spreading and non-linear wave-wave interactions are mechanisms which have been proposed to explain the form of the gravity wave vertical wave number spectrum of horizontal winds. In this paper we present an alternative explanation by assuming that the damping effects of molecular viscosity, turbulence and off-resonance wave-wave interactions can all be characterized in terms of a scale-independent diffusivity (D) which increases with altitude. The components of the gravity wave source spectrum are assumed to grow exponentially with altitude in response to decreasing atmospheric density until they are removed by diffusive damping. A wave of intrinsic frequency ω and vertical wave number m is assumed to be completely damped when the effective vertical velocity of momentum diffusion (mD) exceeds the vertical phase velocity of the wave (ω/m). Only waves satisfying $mD < \omega/m$, or equivalently $m^2D < \omega$ and $m < (\omega/D)^{1/2}$ are permitted to grow in amplitude as they propagate upward in the atmosphere. If the gravity wave temporal spectrum varies as ω^{-p}, we show that the vertical wave number spectrum must vary as m^{-2p+1} and the zonal (or meridional) wave number spectrum must vary as $k^{-(2p+1)/3}$. For p near 2, the diffusion theory predicts that the spectra are proportional to ω^{-2}, m^{-3} and $k^{-5/3}$. Because the joint (m, ω) intrinsic spectrum for scale-independent diffusive filtering is not separable, the theory predicts that the m-spectrum of vertical winds is proportional to m^{5-2p}. The model spectra compare favorably with recent lidar and radar observations of middle atmosphere density, temperature and horizontal wind fluctuations.

1. INTRODUCTION

The spectra of atmospheric velocity, temperature and density fluctuations extend from the sub-meter scale perturbations associated with turbulence to global scale disturbances associated with planetary waves and tides. Gravity waves are the primary source of mesoscale fluctuations throughout the atmosphere. Several theories have been proposed to explain the gravity wave vertical wave number spectrum of horizontal wind fluctuations. One group of researchers has argued that shear and convective instabilities control the shape and magnitude of the spectrum. Commonly referred to as the linear instability theory, this model was initially proposed by Dewan and Good [1986] and is based on earlier work of Hodges [1967] who showed that shear and convective instabilities limit the horizontal wind velocities of quasi-monochromatic waves to values approximately equal to the horizontal phase speed of the waves. Another group has argued that the spectrum is controlled by nonlinear effects arising from strong wave-wave interactions. Hines [1991] proposed that the form of the vertical wave number spectrum is a consequence of Doppler spreading of the vertical wavelengths by the irregular winds of the wave system itself. Weinstock [1990] has suggested that the amplitudes of waves of all scales are limited by diffusion-like processes related to the chaotic motions imposed by all smaller scale waves.

In this paper we also assume that wave amplitudes are controlled by nonlinear interactions of the wave field in a way which can be modeled as a diffusion process. However, our approach differs from that of Weinstock in that the diffusion effects are modeled as a filtering process which selectively removes waves from the source spectrum as the wave field propagates upward. Diffusion degrades the organized bulk motions imparted to the atmosphere by the propagating wave and eventually leads to attenuation of the wave. Dissipation is most severe when the effective vertical velocity of momentum diffusion is comparable to or larger than the vertical phase speed of the wave. Thus the gravity wave spectrum is selectively filtered with the slowest phase speed waves experiencing the most severe damping. To model these effects we assume all components of the wave spectrum grow exponentially with altitude in response to decreasing density until they are damped by diffusion. In this way the spectrum is partitioned into regions corresponding to damped or undamped waves. This partitioning is related to the effective wave induced eddy diffusivity. By assuming an appropriate source spectrum in the lower atmosphere, the spectrum in the upper atmosphere is computed by including only the wave components lying in the undamped region. Because the diffusivity is related to the wave amplitudes and increases with altitude, the wave spectrum also changes with altitude.

We ignore the effects of the mean background wind field and wave-induced Doppler effects and employ simplified forms of the gravity wave polarization and dispersion relations to derive the wave spectra. We restrict our attention to this simplified model so that the physical consequences of diffusive filtering are not obscured by excessively complicated mathematics. Doppler shifting effects, shears and critical layers associated with both the mean and wave-induced wind fields are important in the real atmosphere. Fortunately our spectral models can be extended in a straightforward manner to include the effects of a mean horizontal wind but their treatment is beyond the scope of this paper. The use of the simplified polarization and dispersion relations yield results that differ only slightly from those obtained using the Boussinesq approximations [e.g., *Gardner et al.*, 1993a, b]. The rather simple mathematical formulas derived in this paper are useful because they provide considerable physical insight into the processes controlling the forms of the gravity wave spectra. This paper is based largely on a more extensive treatment of diffusive filtering which was recently published by Gardner [1994].

2. DIFFUSIVE FILTERING THEORY

It has long been recognized that the energy dissipating processes that occur in the real atmosphere due to its viscosity play a crucial role in gravity wave propagation. Viscous effects arise from the diffusion of energetic molecules and from turbulence. The early work of Hines [1960, 1964, 1968, 1970] and Pitteway and Hines [1963] showed that molecular viscosity and thermal conduction play important roles in gravity wave dissipation, particularly in the upper atmosphere. Superimposed on the coherent motions of the atmospheric wave are the random thermal motions of individual molecules. The diffusion of these molecules as a result of their thermal motion, transports energy and momentum from one region of the wave system to another in a partially chaotic way. This process degrades the organized bulk motions imparted to the atmosphere by the propagating wave and eventually leads to attenuation of the wave. The degradation proceeds more vigorously in the upper atmosphere where the mean-free paths of the molecules are large. This chaotic diffusive motion can be described either by a coefficient of molecular diffusion D_{mole} (subscript "mole" denotes molecular diffusion) or by a coefficient of viscosity μ. If we include the effects of both thermal conduction and molecular viscosity which are physically different processes but mathematically similar, then in the middle and upper atmosphere D_{mole} can be expressed in terms of μ as follows [*Hines*, 1974]

$$D_{mole} = 2.4\mu/\rho \qquad (1)$$

where ρ is the atmospheric density. For an ideal gas, μ is independent of both pressure and density and is only a function of temperature.

Eddy diffusion caused by turbulence also contributes to attenuation by inhibiting organized wave motions via processes analogous to molecular diffusion. This mechanism is characterized by a coefficient of eddy diffusion D_{eddy}. Breaking or saturating waves contribute to eddy diffusion by creating turbulence [e.g., *Hodges*, 1967, 1969; *Hines*, 1970]. The most extensive theories, which have been developed by Weinstock [1976; 1982; 1984a, b; 1990], include non-linear wave effects and off-resonance wave-wave interactions. Weinstock extended the early work of Hodges and Hines by showing quantitatively that because of non-linear interactions the small scale, high frequency wind perturbations associated with the gravity wave field can also contribute to eddy diffusion. He has shown that D_{eddy} is related to the gravity wave spectrum and, in general, its value increases with altitude. Below the turbopause, the eddy diffusion coefficient is much larger than the molecular diffusion coefficient, while above the turbopause, molecular diffusion dominates. The total effective diffusivity of the atmosphere D, is just the sum of the eddy and molecular diffusivities.

$$D = D_{\text{eddy}} + D_{\text{mole}} \qquad (2)$$

It is convenient to distinguish between the turbulence and wave induced eddy diffusion by expressing D_{eddy} as the sum of the wave and turbulence contributions.

$$D_{\text{eddy}} = D_{\text{wave}} + D_{\text{turb}} \qquad (3)$$

Wave induced diffusivity is in general anisotropic and representable only by a tensor [see e.g., *Weinstock*, 1976 and 1982]. For gravity waves having vertical wave numbers large compared to their horizontal wave numbers, the important component of the tensor may be expected to be the zz component which we associate with D_{wave}. We consider the case of scale-independent diffusion where the effective wave induced eddy diffusivity at a given altitude is assumed to be the same for each wave regardless of its vertical wavelength and period. Scale-dependent diffusivity, which depends upon wavelength and period, is discussed by Gardner [1994].

Wave damping due to scale-independent diffusion is frequency and wave number dependent. Dissipation is most severe when the vertical diffusive transport, during a time comparable to the wave period, is a significant fraction of the vertical wavelength. Consequently, the gravity wave spectrum is selectively filtered as the wave field propagates upward in the atmosphere with the long period, short vertical wavelength waves (i.e., slow vertical phase speed waves) experiencing the most severe damping. To model these effects, we assume all the components of the gravity wave spectrum grow exponentially with altitude in response to decreasing atmospheric density until they are each in turn damped by diffusion. A wave of intrinsic frequency ω and vertical wave number m is assumed to be severely damped when the effective vertical diffusion velocity (mD) of particles experiencing the wave motion exceeds the vertical phase velocity of the wave (ω/m). Thus only waves satisfying

$$mD \leq \omega/m \qquad (4)$$

or equivalently

$$Dm^2 \leq \omega \qquad (5)$$

$$m \leq (\omega/D)^{1/2} \qquad (6)$$

are permitted to grow in amplitude with increasing altitude and in the present model, are assumed to do so as if quite free from any dissipative effects. Waves not satisfying (4), (5), or (6) experience significant damping and are assumed to be eliminated from the spectrum. Since D increases with altitude, progressively more waves are removed as the wave field propagates upward. This simple damping criterion can also be derived directly from the equations of motion by comparing the relative magnitudes of the main viscous and inertial terms [e.g., *Gossard and Hooke*, 1975, pp. 218-219].

The assumed abrupt discontinuous cutoff in the wave spectrum is a crude approximation to the physical conditions existing in the real atmosphere. In the real atmosphere, waves in the undamped region near the cutoff limit $\omega = Dm^2$, will grow more slowly with increasing altitude than waves deep in the undamped regions, while waves in the damped region near the cutoff will decay with increasing altitude but not be completely eliminated from the spectrum. The transition from undamped to completely damped waves will, in reality, be smooth and continuous. However, as will be shown later, simple spectral models which are in excellent agreement with observations, can be derived by employing the approximate cutoff criterion.

The selective filtering effects described by (5) and (6) are more easily visualized by examining the (m, ω) diagram plotted in Fig. 1(a). The gravity wave frequency spectrum extends from the inertial frequency f to the buoyancy frequency N, while the vertical wave number spectrum extends from 0 to the buoyancy wave number m_b, which marks the transition between waves and turbulence. Because of diffusive damping, the low frequency, high wave number waves to the right of the diagonal line in Fig. 1(a) are removed from the spectrum. On the log-log plot, this line represents the $\omega = Dm^2$ damping limit given by (5). The intersections of this line with the $\omega = f$ and $\omega = N$ lines occur at the respective wave numbers $m_* = (f/D)^{1/2}$ and $m_d = (N/D)^{1/2}$. As D increases with increasing altitude, more and more waves are eliminated from the spectrum as the $\omega = Dm^2$ line moves upward and to the left in Fig. 1(a). The cutoff wave numbers m_* and m_d also decrease as D increases. Thus, even though the undamped wave energies per unit mass increase in proportion to $e^{z/H}$ where z is the altitude and H is the atmospheric scale height, the total wave energy increases more slowly than $e^{z/H}$ because successively more waves are removed from the spectrum as D also increases with altitude. Notice that the resulting spectrum cannot be separable for $m > m_*$ since the cutoff wave number $m_c = (\omega/D)^{1/2}$ depends on ω. The simplified gravity wave dispersion relation $m = Nh/\omega$ can be used to express the damping limits given by (5) and (6) in terms of the horizontal wave number h. The damped and undamped regions in the (h, ω) and (h, m) joint spectra are illustrated in Figs. 1(b) and 1(c). Because the simplified dispersion relation fails at ω near f and N, the actual

Scale-Independent Diffusion Theory

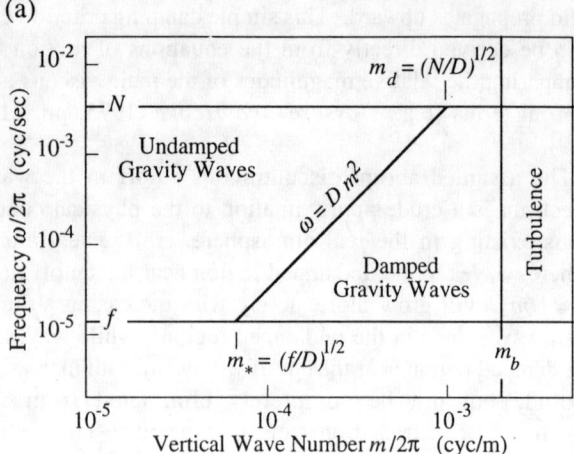

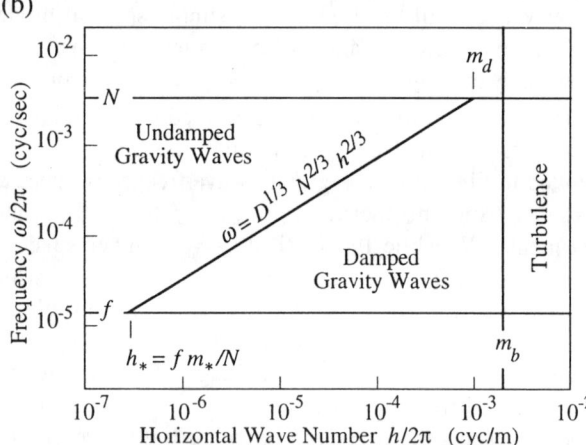

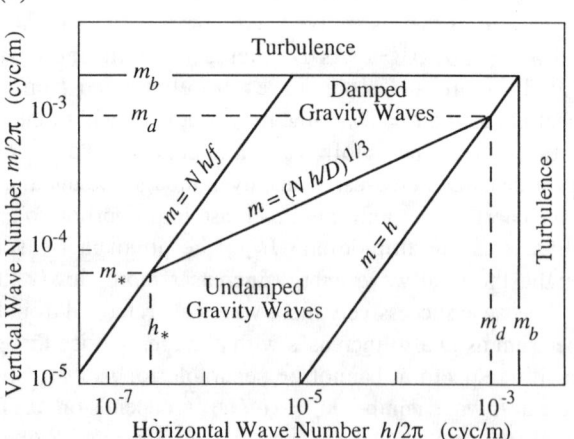

Fig. 1. Scale-independent diffusive damping limits for atmospheric gravity waves. The plotted limits are typical of the mid-latitude mesopause region where $\omega_{sc} = f \simeq 2\pi/(20\ h)$, $N \simeq 2\pi/(5\ min)$, $D \simeq 500\ m^2/s$, $m_* = m_{sc} \simeq 2\pi/(15\ km)$, $h_* \simeq 2\pi/(3600\ km)$ and $m_d \simeq 2\pi/(1\ km)$.

boundaries between the damped and undamped regions depart from those plotted in Figs. 1(b) and 1(c) near h_* and m_d. These figures are discussed in Section 4 where we derive expressions for the horizontal wave number spectra.

Let $F_u^{sc}(m, \omega)$ denote the source spectrum of gravity wave horizontal winds (u) and $F_u(m, \omega)$ denote the spectrum at an altitude z above the source where the atmosphere is characterized by an effective diffusivity D. We assume all spectral components satisfying (4)–(6) grow unimpeded in response to decreasing density while the remaining components are severely damped so that

$$F_u(m, \omega) = \begin{cases} e^{z/H} F_u^{sc}(m, \omega) & m \leq (\omega/D)^{1/2} \\ & \text{and}\ f \leq \omega \leq N \quad (7) \\ 0 & \text{otherwise.} \end{cases}$$

The temporal frequency spectrum is obtained by integrating the joint (m, ω) spectrum over m subject to the damping limit given by (6), that is, from $m = 0$ to the $m = (\omega/D)^{1/2} = m_*(\omega/f)^{1/2}$ limit delineated by the diagonal line in Fig. 1(a).

$$F_u(\omega) = \frac{e^{z/H}}{2\pi} \int_0^{(\omega/D)^{1/2}} F_u^{sc}(m, \omega)\ dm. \quad (8)$$

The vertical wave number spectrum is obtained by integrating the joint spectrum over ω subject to (5). When $m \leq m_*$, the lower limit is the $\omega = f$ horizontal line in Fig. 1a. When $m > m_*$, the lower limit is the $\omega = Dm^2$ diagonal line. In both cases the upper limit is N.

$$F_u(m) = \frac{e^{z/H}}{2\pi} \int_{Dm^2}^{N} F_u^{sc}(m, \omega)\ d\omega. \quad (9)$$

Notice that we have used the standard engineering convention of writing the joint spectrum in units of $(m^2/s^2)/(cyc/m)/(cyc/s)$. The 1-D spectra and wave variance are obtained by integrating $F_u(m, \omega)$ with respect to $dm/(2\pi)$, $d\omega/(2\pi)$ or $dmd\omega/(2\pi)^2$.

To proceed we need to make some assumptions about the source spectrum in the lower atmosphere. It seems reasonable to characterize each source by a vertical wave number m'_{sc} related to its vertical size and a temporal frequency ω_{sc} related to its lifetime. We assume the 1-D vertical wave number and temporal frequency spectra are maximum near m'_{sc} and ω_{sc} as illustrated by the hypothetical spectra plotted in Fig. 2. $F_u^{sc}(m)$ is not expected to remain finite

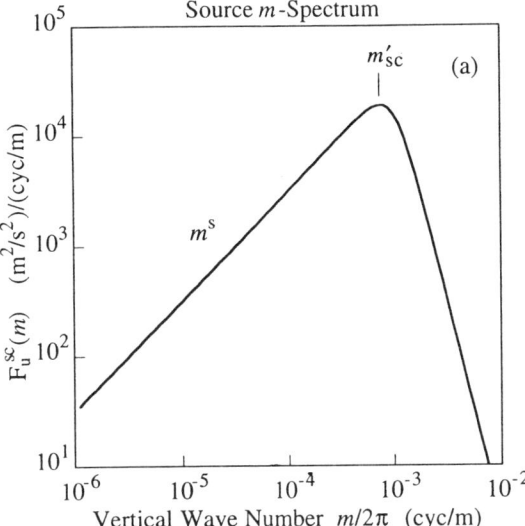

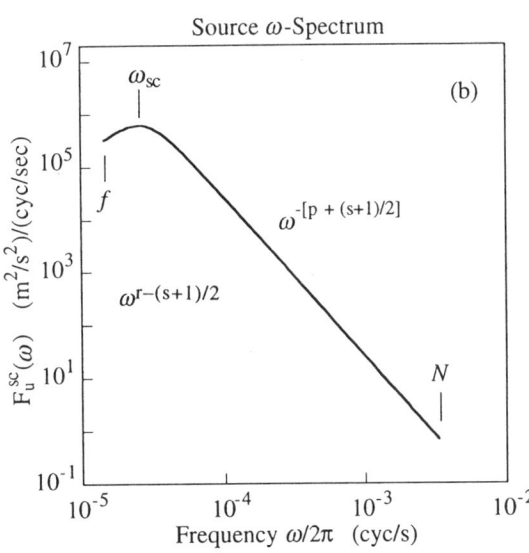

Fig. 2. Hypothetical models for the 1-D gravity wave source spectrum. m'_{sc} is related to the physical size of the source while ω_{sc} is related to the source lifetime.

as m approaches zero, because otherwise this would imply the existence of sources capable of generating infinitely long vertical wavelengths. $F_u^{sc}(\omega)$ goes to zero at the gravity wave cutoff frequencies f and N. As the wave field propagates upward the high m, low ω components of the source spectrum are eliminated by diffusive damping. Therefore, we only need to model the low m, high ω behavior of the sources in order to evaluate (8) and (9). While it is unlikely that $F_u^{sc}(m, \omega)$ is separable for m near m'_{sc} and ω near ω_{sc}, to simplify the mathematics we will

assume that the source spectrum is separable in the low m, high ω region of interest and that both the m and ω dependencies are power-laws of the form

$$F_u^{sc}(m, \omega) \sim \begin{cases} m^s \omega^{r-(s+1)/2} & f \leq \omega \leq \omega_{sc} \\ \dfrac{m^s}{\omega^{p+(s+1)/2}} & \omega_{sc} \leq \omega \leq N \end{cases} \quad (10)$$

where $s > 0$, $r \geq (s+1)/2$, and $p > -1/2$. The separability assumption seems reasonable since at high altitudes many statistically independent low-altitude sources contribute to F_u^{sc}. The low m behavior of one source is not expected to be related to the high ω behaviors of all the other sources and vice versa. Even though we have written the spectral indices of the ω dependences in terms of s, the model given by (10) is completely general since p and r are arbitrary. The mathematical convenience of this representation will become apparent shortly.

If we denote $\langle (u')^2 \rangle$ as the total wave variance at an altitude z where the atmosphere is characterized by an effective diffusivity D, then the joint (m, ω) spectrum consistent with (7) and (10) is

$$F_u(m, \omega) =$$

$$\begin{cases} (2\pi)^2 \langle (u')^2 \rangle \dfrac{(s+1)}{m_{sc}} \left(\dfrac{m}{m_{sc}}\right)^s \\ \quad \dfrac{(r+1)(p-1)}{(p+r)\omega_{sc}} \left(\dfrac{\omega}{\omega_{sc}}\right)^{r-(s+1)/2} \\ \qquad f \leq \omega \leq \omega_{sc} \quad m \leq (\omega/D)^{1/2} \\ \\ (2\pi)^2 \langle (u')^2 \rangle \dfrac{(s+1)}{m_{sc}} \left(\dfrac{m}{m_{sc}}\right)^s \\ \quad \dfrac{(r+1)(p-1)}{(p+r)\omega_{sc}} \left(\dfrac{\omega_{sc}}{\omega}\right)^{p+(s+1)/2} \\ \qquad \omega_{sc} \leq \omega \leq N \quad m \leq (\omega/D)^{1/2} \end{cases} \quad (11)$$

where $m_{sc} = (\omega_{sc}/D)^{1/2}$. The joint spectrum given by (11) is valid for $p > 1$ and $(f/N)^{p-1} \ll 1$. It has the same shape as the source spectrum in the low m, high ω region, is limited to the undamped region in the (m, ω) plane defined by (5) and (6) and has a total integrated variance of $\langle (u')^2 \rangle$. We emphasize that (11) is an approximation that ignores the departure of the source ω-spectrum from a pure power law near ω_{sc}. The 1-D spectra are easily computed by substituting (11) in (8) and (9) using (7) so that

$$F_u(\omega) = \frac{1}{2\pi} \int_0^{(\omega/D)^{1/2}} F_u(m, \omega) \, dm \simeq$$

$$\begin{cases} 2\pi <(u')^2> \dfrac{(r+1)(p-1)}{(p+r)\omega_{sc}} \left(\dfrac{\omega}{\omega_{sc}}\right)^r & f \leq \omega \leq \omega_{sc} \\ 2\pi <(u')^2> \dfrac{(r+1)(p-1)}{(p+r)\omega_{sc}} \left(\dfrac{\omega_{sc}}{\omega}\right)^p & \omega_{sc} \leq \omega \leq N \end{cases} \quad (12)$$

and

$$F_u(m) = \frac{1}{2\pi} \int_{m^2 D}^{N} F_u(m, \omega) \, d\omega \simeq$$

$$\begin{cases} 2\pi <(u')^2> \dfrac{2(s+1)(r+1)(p-1)}{(p+r)(2p+s-1)} \\ \left(\dfrac{(2p+s-1)}{(2r-s+1)}\left[1-\left(\dfrac{f}{\omega_{sc}}\right)^{r-(s-1)/2}\right]+1\right)\dfrac{1}{m_{sc}}\left(\dfrac{m}{m_{sc}}\right)^s \\ \qquad m \leq m_* \\[1em] 2\pi <(u')^2> \dfrac{2(s+1)(r+1)(p-1)}{(p+r)(2p+s-1)} \\ \left(\dfrac{(2p+s-1)}{(2r-s+1)}\left[1-\left(\dfrac{m}{m_{sc}}\right)^{2r-s+1}\right]+1\right)\dfrac{1}{m_{sc}}\left(\dfrac{m}{m_{sc}}\right)^s \\ \qquad m_* \leq m \leq m_{sc} \\[1em] 2\pi <(u')^2> \dfrac{2(s+1)(r+1)(p-1)}{(p+r)(2p+s-1)} \\ \left[1-\left(\dfrac{\omega_{sc} m^2}{N m_{sc}^2}\right)^{p+(s-1)/2}\right]\dfrac{1}{m_{sc}}\left(\dfrac{m_{sc}}{m}\right)^{2p-1} \\ \qquad m_{sc} \leq m \leq m_d \end{cases} \quad (13)$$

The scale-independent diffusive filtering theory spectra are summarized in Table 1. The vertical wave number and temporal frequency spectra are plotted in Fig. 3. The temporal spectrum has a shallower slope than the source, because low frequency energy, due to high m waves, is removed by diffusive damping as the wave field propagates upward. In the region $m > m_*$, even though the source spectrum increases with increasing m, the vertical wave number spectrum decreases with increasing m because energy at high wave numbers due to energetic low frequency waves is removed by diffusive damping. However, for $m < m_*$ none of the frequencies are damped (see Fig. 1) so the m-spectrum retains the shape of the source. Notice that for $m_{sc} < m << m_d$, the m spectrum index $-(2p-1)$ is related to the ω-spectrum index $-p$. When $p = 2$ the temporal spectrum is proportional to ω^{-2} and the vertical wave number spectrum is proportional to m^{-3}. Diffusive damping eliminates all waves with $m > m_d = (N/D)^{1/2}$.

Although the vertical wave number spectrum plotted in Fig. 3 exhibits many of the features predicted by various saturation theories [e.g., *Dewan and Good*, 1986; *Hines*, 1991], the physics controlling the shape of $F_u(m)$ is considerably different from these other theories. The spectrum for $m \geq m_{sc}$ is not saturated. We have assumed that all components of the spectrum, including those waves contributing to the region $m_* < m_{sc} < m$, grow with increasing altitude until they are removed by diffusive damping. However, for $m_{sc} < m$ successively more energy is removed as D increases with altitude because successively more low frequency waves with $\omega < Dm^2$ are eliminated. Depending on the growth rate of D, the spectral magnitude for $m_{sc} < m$ can increase, remain constant or even decrease with increasing altitude. Since diffusive damping is unimportant for $m < m_*$ (i.e., all frequencies between f and N contribute, see Fig. 1), $F_u(m)$ is proportional to $e^{z/H}$ in this region.

Shear and convective instabilities, related critical layer effects, and wave-induced Doppler spreading effects can also contribute to wave damping. Although our model does not specifically incorporate these effects in the spectrum, their contributions are not precluded. For example, if a source in the lower atmosphere generates an especially energetic wave, it is possible for the amplitude of the wave to reach the shear or convective instability limit before the wave is eliminated from the spectrum by diffusive damping. The relative importance of linear instabilities and nonlinear diffusion depends in large measure on the strengths of the sources. If the sources are so strong that waves regularly reach their shear or convective limits before diffusion becomes important, then instability effects will control the dissipation of those waves. Otherwise, we believe wave-induced diffusion will be the controlling process.

3. WAVE INDUCED DIFFUSIVITY

To completely characterize the gravity wave spectrum, it is necessary to determine the relationship between D_{wave} and the spectrum. One approach is to relate D_{wave} to other characteristics of the wave field such as the vertical shear variance of horizontal winds $<(\partial u'/\partial z)^2>$ which can also be expressed as a form of the Richardson number

$$\bar{Ri} = \frac{N^2}{<\left(\dfrac{\partial u'}{\partial z}\right)^2>} \quad (14)$$

TABLE 1. Spectra Models: Scale-Independent Diffusive Filtering Theory ($f < \omega_{sc} < N$)

$$F_u(m, \omega) = \begin{cases} (2\pi)^2 <(u')^2> \dfrac{(s+1)}{m_{sc}} \left(\dfrac{m}{m_{sc}}\right)^s \dfrac{(r+1)(p-1)}{(p+r)\omega_{sc}} \left(\dfrac{\omega}{\omega_{sc}}\right)^{r-(s+1)/2} & \begin{array}{l} f \leq \omega \leq \omega_{sc} \\ m \leq (\omega/D)^{1/2} \end{array} \\ (2\pi)^2 <(u')^2> \dfrac{(s+1)}{m_{sc}} \left(\dfrac{m}{m_{sc}}\right)^s \dfrac{(r+1)(p-1)}{(p+r)\omega_{sc}} \left(\dfrac{\omega_{sc}}{\omega}\right)^{p+(s+1)/2} & \begin{array}{l} \omega_{sc} \leq \omega \leq N \\ m \leq (\omega/D)^{1/2} \end{array} \end{cases}$$

$$F_u(m) = \begin{cases} 2\pi <(u')^2> \dfrac{2(s+1)(r+1)(p-1)}{(p+r)(2p+s-1)} \left(\dfrac{(2p+s-1)}{(2r-s+1)}\left[1-\left(\dfrac{f}{\omega_{sc}}\right)^{r-(s-1)/2}\right]+1\right) \dfrac{1}{m_{sc}} \left(\dfrac{m}{m_{sc}}\right)^s & m \leq m_* \\ 2\pi <(u')^2> \dfrac{2(s+1)(r+1)(p-1)}{(p+r)(2p+s-1)} \left(\dfrac{(2p+s-1)}{(2r-s+1)}\left[1-\left(\dfrac{m}{m_{sc}}\right)^{2r-s+1}\right]+1\right) \dfrac{1}{m_{sc}} \left(\dfrac{m}{m_{sc}}\right)^s & m_* \leq m \leq m_{sc} \\ 2\pi <(u')^2> \dfrac{2(s+1)(r+1)(p-1)}{(p+r)(2p+s-1)} \left[1-\left(\dfrac{\omega_{sc} m^2}{N m_{sc}^2}\right)^{p+(s-1)/2}\right] \dfrac{1}{m_{sc}} \left(\dfrac{m_{sc}}{m}\right)^{2p-1} & m_{sc} \leq m \leq m_d \end{cases}$$

$$F_u(\omega) = \begin{cases} 2\pi <(u')^2> \dfrac{(r+1)(p-1)}{(p+r)\omega_{sc}} \left(\dfrac{\omega}{\omega_{sc}}\right)^r & f \leq \omega \leq \omega_{sc} \\ 2\pi <(u')^2> \dfrac{(r+1)(p-1)}{(p+r)\omega_{sc}} \left(\dfrac{\omega_{sc}}{\omega}\right)^p & \omega_{sc} \leq \omega \leq N \end{cases}$$

$$D_{wave} = \dfrac{(s+1)\tilde{R}i <(u')^2> \omega_{sc}}{(s+3)N^2} \xi(p,r)$$

$$<(u')^2> = \dfrac{(s+3)}{(s+1)\tilde{R}i\,\xi(p,r)} \dfrac{N^2}{m_{sc}^2}$$

$$m_* = (f/D_{wave})^{1/2}$$

$$m_{sc} = (\omega_{sc}/D_{wave})^{1/2}$$

$$m_d = (N/D_{wave})^{1/2}$$

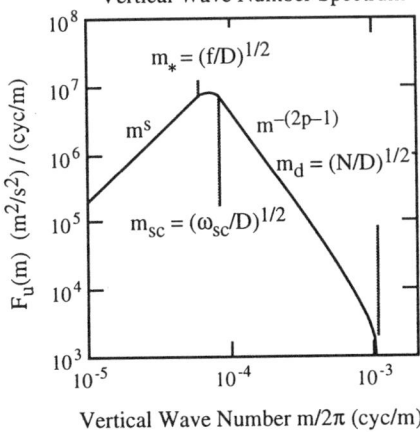 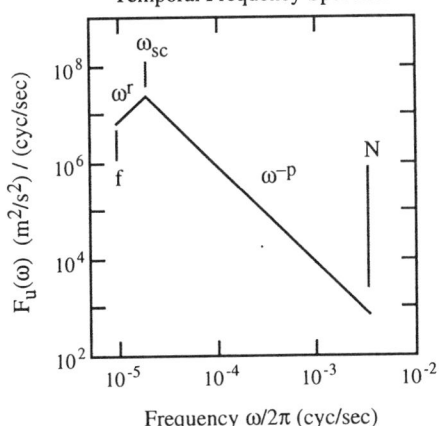

Fig. 3. Gravity wave vertical wave number, temporal frequency and horizontal wave number spectra predicted by scale-independent diffusive filtering theory plotted for the special case $p = r = s = 2$. These spectra are typical of the mid-latitude mesopause region where $f \simeq 2\pi/(30\ h)$, $\omega_{sc} \simeq 2\pi/(15\ h)$, $N \simeq 2\pi/(5\ min)$, $<(u')^2> \simeq (25\ m/s)^2$, $D \simeq 400\ m^2/s$, $m_* \simeq 2\pi/(16.5\ km)$, $m_{sc} \simeq 2\pi/(11.6\ km)$ and $m_d \simeq 2\pi/(870\ m)$.

$\tilde{R}i$ is easily calculated using (11) and noting

$$\left\langle\left(\frac{\partial u'}{\partial z}\right)^2\right\rangle = \frac{1}{(2\pi)^2}\int_f^N d\omega \int_0^{m_{sc}(\omega/\omega_{sc})^{1/2}} m^2 F_u(m,\omega)dm$$

$$= \frac{(s+1)}{(s+3)}\xi(p,r)\langle(u')^2\rangle m_{sc}^2 \qquad (15)$$

so that

$$\tilde{R}i = \frac{(s+3)}{(s+1)\xi(p,r)\langle(u')^2\rangle}\frac{N^2}{m_{sc}^2}$$

$$= \frac{(s+3)}{(s+1)\xi(p,r)\langle(u')^2\rangle}\frac{N^2}{\omega_{sc}}D_{\text{wave}} \qquad (16)$$

where

$$\xi(p,r) = \begin{cases} \frac{(r+1)(p-1)}{(p+r)(r+2)}\left[\left(1-\left(\frac{f}{\omega_{sc}}\right)^{r+2}\right)+\frac{(r+2)}{(2-p)}\left(\left(\frac{N}{\omega_{sc}}\right)^{2-p}-1\right)\right] \\ \qquad\qquad\qquad\qquad\qquad\qquad p\neq 2 \quad (17) \\ \frac{(r+1)}{(r+2)}\left[\frac{1}{(r+2)}\left(1-\left(\frac{f}{\omega_{sc}}\right)^{r+2}\right)+\ln(N/\omega_{sc})\right] \\ \qquad\qquad\qquad\qquad\qquad\qquad p=2 \end{cases}$$

and we have used the fact that $m_{sc}^2 = \omega_{sc}/D_{\text{wave}}$ in (16). By rearranging terms in (16) we obtain the final results

$$D_{\text{wave}} = \frac{(s+1)\tilde{R}i\langle(u')^2\rangle\omega_{sc}}{(s+3)N^2}\xi(p,r) \qquad (18)$$

$$\langle(u')^2\rangle = \frac{(s+3)}{(s+1)\tilde{R}i\,\xi(p,r)}\frac{N^2}{m_{sc}^2} \qquad (19)$$

and

$$\lambda_z^* = \sqrt{\frac{\omega_{sc}}{f}}\lambda_z^{sc} = \sqrt{\frac{\omega_{sc}}{f}\frac{(s+1)}{(s+3)}\tilde{R}i\,\xi(p,r)}\,u_{\text{rms}}\,T_B \qquad (20)$$

where T_B is the buoyancy period.

An alternate approach suggested by R. Walterscheid (private communication), employs the first law of thermodynamics, to derive an expression for D_{wave}.

$$\frac{\partial T'}{\partial t} + \Gamma_a w' = D_{\text{wave}}\frac{\partial^2 T'}{\partial z^2} \qquad (21)$$

$$\Gamma_a = -N^2\,\bar{T}/g \qquad (22)$$

where Γ_a is the adiabatic lapse rate for dry air (~9.8 K/km), $\bar{T}$ is the mean temperature, and g is the gravitational acceleration. After multiplying both sides of (21) by T', computing the expectation and recognizing that

$$\left\langle T'\frac{\partial T'}{\partial t}\right\rangle = 0 \qquad (23)$$

$$\left\langle T'\frac{\partial^2 T'}{\partial z^2}\right\rangle = -\left\langle\left(\frac{\partial T'}{\partial z}\right)^2\right\rangle \qquad (24)$$

we obtain

$$\Gamma_a\langle w'T'\rangle = D_{\text{wave}}\left\langle\left(\frac{\partial T'}{\partial z}\right)^2\right\rangle. \qquad (25)$$

Upon rearranging terms and recognizing that $|T'| \simeq |\Gamma_a u'/N|$, we obtain the final result

$$D_{\text{wave}} = \frac{\Gamma_a\langle w'T'\rangle}{\left\langle\left(\frac{\partial T'}{\partial z}\right)^2\right\rangle} \simeq \frac{\tilde{R}i\langle w'T'\rangle}{\Gamma_a}, \qquad (26)$$

where as before $\tilde{R}i$ is a form of the Richardson number defined by (14). According to (26) the effective wave-induced diffusivity is proportional to the gravity wave vertical thermal flux $\langle w'T'\rangle$.

Unfortunately, Eq. (26) is not a convenient expression for relating D_{wave} to more commonly observed parameters such as the total wave variance or to the spectrum of horizontal winds. From the gravity wave polarization and dispersion relations we have

$$w'T' \propto \frac{\Gamma_a}{N^2}\omega(u')^2. \qquad (27)$$

By using (27) in (26), D_{wave} can be expressed in terms of the joint (m,ω) spectrum of horizontal winds given by (11) or in terms of the total horizontal wind variance

$$D_{\text{wave}} = \frac{a\tilde{R}i}{(2\pi N)^2}\int_f^N d\omega\int_0^{(\omega/D)^{1/2}}\omega F_u(m,\omega)dm$$

$$= \frac{a\tilde{R}i\langle(u')^2\rangle\omega_{sc}}{N^2}\xi(p,r) \qquad (28)$$

where α is a dimensionless constant of proportionality which relates the thermal flux, horizontal wind variance, and D_{wave}. By comparing (28) to (18), we conclude that

$$\alpha = \frac{(s+1)}{(s+3)}. \quad (29)$$

Extensive lidar observations of the mesopause region reported by Collins [1994] suggest that $s \simeq 1$, $\tilde{R}i \simeq 1$ and $T_{sc} = 2\pi/\omega_{sc} \simeq 15$ h. At mid-latitude mesopause heights where $T_i \simeq 0$ h, $T_B \simeq 5$ min and $<(u')^2> \simeq (25 \text{ m/s})^2$, these equations predict $D_{wave} \simeq 330$ m^2/s, $\lambda_z^* \simeq 12$ km, $\lambda_{sc} = 10.4$ km and $\lambda_d = 2\pi/m_d \simeq 0.8$ km. To compute these values we assumed $r \simeq 2$. For comparison, at these same heights the molecular diffusivity D_{mole} is approximately 5 m^2/s and the turbulence contribution to eddy diffusivity D_{turb} is approximately 100 m^2/s. In the middle stratosphere $<(u')^2>$ decreases to about $(6 \text{ m/s})^2$ so that if $\tilde{R}i = 1$ then $D_{wave} \simeq 20$ m^2/s, $\lambda_z^* \simeq 3$ km and $\lambda_d \simeq 200$ m.

The altitude variations of D_{wave}, m_*, total wave variance and spectral magnitudes can now be determined by noting that the magnitudes of all components of the joint (m, ω) spectrum are assumed to vary in proportion to $e^{z/H}$. By assuming $D_{wave} \gg D_{turb} \gg D_{mole}$ and substituting (19) in (11) we obtain

$$F_u(m, \omega) = \begin{cases} (2\pi)^2 \dfrac{(s+3)}{\tilde{R}i\,\xi(p,r)} \dfrac{N^2}{m_{sc}^3} \left(\dfrac{m}{m_{sc}}\right)^s \\ \quad \dfrac{(r+1)(p-1)}{(p+r)\omega_{sc}} \left(\dfrac{\omega}{\omega_{sc}}\right)^{r-(s+1)/2} \\ \quad f \leq \omega \leq \omega_{sc} \qquad m \leq (\omega/D)^{1/2} \\[6pt] (2\pi)^2 \dfrac{(s+3)}{\tilde{R}i\,\xi(p,r)} \dfrac{N^2}{m_{sc}^3} \left(\dfrac{m}{m_{sc}}\right)^s \\ \quad \dfrac{(r+1)(p-1)}{(p+r)\omega_{sc}} \left(\dfrac{\omega_{sc}}{\omega}\right)^{p+(s+1)/2} \\ \quad \omega_{sc} \leq \omega \leq N \qquad m \leq (\omega/D)^{1/2} \end{cases} \quad (30)$$

If we assume $\tilde{R}i$ does not vary with altitude, then from (30) we have

$$m_* \sim m_{sc} \sim \exp\left[-\frac{z}{(s+3)H}\right] \quad (31)$$

$$D_{wave} \sim <(u')^2> \sim \exp\left[\frac{2z}{(s+3)H}\right]. \quad (32)$$

It is especially interesting to examine the altitude behavior of the vertical wave number spectrum in the region $m_{sc} < m \ll m_d$. By substituting (19) in (13) we find

$$F_u(m) = 2\pi \frac{2(s+3)(r+1)(p-1)}{(p+r)(2p+s-1)} \frac{N^2}{\tilde{R}i\,\xi(p,r)} \frac{m_{sc}^{2(p-2)}}{m^{2p-1}}$$
$$m_{sc} \leq m \ll m_d \quad (33)$$

and when p = 2, (33) reduces to

$$F_u(m) \simeq \frac{4\pi}{\tilde{R}i\,\ln(N/\omega_{sc})} \frac{N^2}{m^3} = \frac{4\pi}{\ln(N/\omega_{sc})} \frac{<\left(\frac{\partial u'}{\partial z}\right)^2>}{m^3}$$
$$m_{sc} \leq m \ll m_d \quad (34)$$

and (12) becomes

$$F_u(\omega) \simeq \frac{2\pi(s+3)D_{wave}}{(s+1)\,\tilde{R}i\,\ln(N/\omega_{sc})} \frac{N^2}{\omega^2}$$
$$= \frac{2\pi(s+3)D_{wave}}{(s+1)\,\ln(N/\omega_{sc})} \frac{<\left(\frac{\partial u'}{\partial z}\right)^2>}{\omega^2} \quad \omega_{sc} \leq \omega \leq N. \quad (35)$$

When p = 2, the temporal spectrum is proportional to ω^{-2} for $\omega > \omega_{sc}$, the vertical wave number spectrum is proportional to m^{-3} for $m > m_{sc}$ and the m-spectrum magnitudes in this region depend on altitude only weakly through N (assuming $\tilde{R}i$ does not vary with altitude). This behavior has also been predicted by several of the saturation theories [*Dewan and Good*, 1986; *Smith et al.*, 1987; *Weinstock*, 1990; *Hines*, 1991]. However, again we emphasize that the spectra in our model are not saturated. When p = 2, the elimination of waves through increased diffusive damping is just sufficient to counteract the exponential growth of the remaining wave components in the region $m > m_{sc}$ so that the spectral magnitude remains constant with increasing altitude. Notice also from (33), because m_{sc} decreases with increasing altitude, the spectral magnitudes increase with increasing altitude when p < 2, remain constant when p = 2, and decrease with increasing altitude when p > 2. Thus the behavior of the spectra appear to be far more complex and influenced to a greater extent by the gravity wave sources than that predicted by the saturation theories. The predicted altitude variations of the spectra and various wave parameters are summarized in Table 2 for the special case $\omega_{sc} = f$. The altitude variations of the spectral magnitudes are illustrated in Figure 4 where the vertical wave number and temporal frequency spectra are

TABLE 2. Predicted Altitude Variations of Gravity Wave Parameters and Spectral Magnitudes for the Scale-Independent Diffusion Theory ($\omega_{sc} = f, m_{sc} = m_*$)

Parameter	Region	Altitude Dependence	Scale Height General	Scale Height p = 2, s = 2, H = 7 km
λ_z^*		$\exp\left[\dfrac{z}{(s+3)H}\right]$	$(s+3)H$	$5H = 35$ km
$<(u')^2>$		$(\lambda_z^*)^2$	$\dfrac{(s+3)}{2}H$	$2.5H = 17.5$ km
D		$(\lambda_z^*)^2$	$\dfrac{(s+3)}{2}H$	$2.5H = 17.5$ km
$<w'u'>$		$(\lambda_z^*)^2$	$\dfrac{(s+3)}{2}H$	$2.5H = 17.5$ km
$<w'T'>$		$(\lambda_z^*)^2$	$\dfrac{(s+3)}{2}H$	$2.5H = 17.5$ km
$F_u(m)$	$m < m_*$	$(\lambda_z^*)^{s+3}$	H	$H = 7$ km
	$m_* < m$	$(\lambda_z^*)^{2(2-p)}$	$\dfrac{(s+3)}{2(2-p)}H$	∞
$F_u(\omega)$	$f \leq \omega \leq N$	$(\lambda_z^*)^2$	$\dfrac{(s+3)}{2}H$	$2.5H = 17.5$ km
$F_u(h, \phi)$	$h < h_*$	$(\lambda_z^*)^{s+3}$	H	$H = 7$ km
	$h_* < h \ll m_d$	$(\lambda_z^*)^{2(4-p)/3}$	$\dfrac{3(s+3)}{2(4-p)}H$	$\dfrac{15}{4}H = 26.2$ km
$F_u(k)$	$k < h_*$	$(\lambda_z^*)^3$	$\dfrac{(s+3)}{3}H$	$5H/3 = 11.7$ km
	$h_* < k \ll m_d$	$(\lambda_z^*)^{2(4-p)/3}$	$\dfrac{3(s+3)}{2(4-p)}H$	$\dfrac{15}{4}H = 26.2$ km

plotted for several values of λ_z^* for the special case of $p = r = s = 2$ and $\omega_{sc} = f$.

4. HORIZONTAL WAVE NUMBER SPECTRA

Because the waves in the diffusive filtering model are not saturated, the gravity wave dispersion relation can be used to transform the joint (m, ω) spectrum into various joint spectra involving the horizontal wave numbers. The approach has been described in detail by Gardner et al. [1993a] who derived models for the horizontal wave number spectra by assuming the joint (m, ω) spectrum is separable and the wave field is horizontally isotropic. Their final results are easily modified to accommodate the nonseparable model given by (11). From Table 1 of Gardner et al. [1993a] and (11), we obtain for the 3-D joint (k, l, ω) and (k, l, m) spectra for the special case $\omega_{sc} = f$

$$F_u(k, l, \omega) = F_u(h, \phi, \omega) =$$

$$\begin{cases} \dfrac{4N}{h\omega} F_u(Nh/\omega, \omega) \\ \quad h \leq \omega^{3/2}/(ND^{1/2}) = h_*(\omega/f)^{3/2} \\ \quad \text{and } f \leq \omega \leq N \\ 0 \quad \text{otherwise} \end{cases}$$

$$= \begin{cases} (2\pi)^3 <(u')^2> \dfrac{2}{\pi}\dfrac{(s+1)}{h_*^2}\left(\dfrac{h}{h_*}\right)^{s-1} \\ \quad \dfrac{(p-1)}{f}\left(\dfrac{f}{\omega}\right)^{p+3(s+1)/2} \\ \quad h \leq h_*(\omega/f)^{3/2} \\ \quad \text{and } f \leq \omega \leq N \\ 0 \quad \text{otherwise} \end{cases} \quad (36)$$

$$F_u(k, l, m) = F_u(h, \phi, m) =$$

$$\begin{cases} \dfrac{4N}{hm} F_u(m, Nh/m) \\ \quad h \leq m \leq (Nh/D)^{1/3} \\ \quad = m_*(h/h_*)^{1/3} \leq Nh/f \\ \quad \text{and } f \leq \omega \leq N \\ 0 \quad \text{otherwise} \end{cases}$$

$$= \begin{cases} (2\pi)^3 <(u')^2> \dfrac{2}{\pi}\dfrac{(s+1)}{m_*}\left(\dfrac{m}{m_*}\right)^{p+(3s-1)/2} \\ \quad \dfrac{(p-1)}{h_*^2}\left(\dfrac{h_*}{h}\right)^{p+(s+3)/2} \\ \quad h \leq m \leq m_*(h/h_*)^{1/3} \leq Nh/f \\ \quad \text{and } f \leq \omega \leq N \\ 0 \quad \text{otherwise} \end{cases} \quad (37)$$

where

$$h_* = f m_*/N \quad (38)$$

and $h = (k^2 + l^2)^{1/2}$, $k = h \cos\phi$, $l = h \sin\phi$ and $0 \leq \phi \leq \pi/2$. The parameters k and l are respectively the zonal and meridional wave numbers and h is the magnitude of the horizontal wave number vector. The joint 3-D spectra are constant with respect to ϕ because the wave field is assumed to be horizontally isotropic.

The frequency and wave number limits in (36) and (37) are a direct consequence of the damping limit given by (5) subject to the simplified dispersion relation $m = Nh/\omega$ with $f \leq \omega \leq N$. These spectral limits are illustrated in Figs. 1(b) and 1(c). They partition the (h, ω) and (h, m) planes into regions associated with damped and undamped gravity waves. We see that diffusive damping eliminates the small horizontal scale, long period waves to the right and below the $\omega = D^{1/3}N^{2/3}h^{2/3}$ diagonal line in Fig. 1(b). Equivalently, damping eliminates the small vertical scale waves lying above the $m = (Nh/D)^{1/3}$ line in Fig. 1(c).

The 2-D $(k, l) = (h, \phi)$ spectrum is obtained by integrating (36) over ω

$$F_u(k, l) = F_u(h, \phi) = \dfrac{1}{2\pi}\int_{D^{1/3}N^{2/3}h^{2/3}}^{N} F_u(h, \phi, \omega)d\omega =$$

$$\begin{cases} (2\pi)^2 <(u')^2> \dfrac{4(p-1)(s+1)}{\pi(2p+3s+1)} \\ \quad \left[1-(f/N)^{p+(3s+1)/2}\right]\dfrac{1}{h_*^2}\left(\dfrac{h}{h_*}\right)^{s-1} \\ \qquad\qquad h \leq h_* \\ (2\pi)^2 <(u')^2> \dfrac{4(p-1)(s+1)}{\pi(2p+3s+1)} \quad (39) \\ \quad \left[1-\left(\dfrac{f}{N}\dfrac{h^{2/3}}{h_*^{2/3}}\right)^{p+(3s+1)/2}\right]\dfrac{1}{h_*^2}\left(\dfrac{h_*}{h}\right)^{(2p+4)/3} \\ \qquad\qquad h_* \leq h \leq m_d \end{cases}$$

The 1-D zonal or meridional wave number spectrum is obtained by integrating (39) over l or k. By evaluating the integral over l approximately, we obtain the following approximate expression for the zonal wave number spectrum.

$$F_u(k) = \dfrac{1}{2\pi}\int_0^{m_d} F_u(k, l)\, dl \simeq$$

$$\begin{cases} 2\pi<(u')^2> \dfrac{4(p-1)(s+1)}{\pi(2p+1)s}\dfrac{1}{h_*} \\ \qquad\qquad k << h_* \\ 2\pi<(u')^2> \dfrac{2(p-1)(s+1)}{\pi(2p+3s+1)} \quad (40) \\ \quad \dfrac{\Gamma(1/2)\Gamma\left(\dfrac{p+2}{3}-\dfrac{1}{2}\right)}{\Gamma\left(\dfrac{p+2}{3}\right)}\dfrac{1}{h_*}\left(\dfrac{h_*}{k}\right)^{(2p+1)/3} \\ \qquad\qquad h_* << k << m_d \end{cases}$$

The 2-D horizontal wave number spectrum given by (39) and the 1-D zonal wave number spectrum obtained by integrating (39) numerically with respect to the meridional wave number are plotted respectively in Figures 3(c) and 3(d). For $h < h_*$ all frequencies between f and N contribute to the 2-D horizontal wave number spectrum (see Fig. 1(b)). In this source region the spectrum is proportional to h^{s-1}. For $h > h_*$, diffusive damping eliminates the energetic low frequency waves between f and $D^{1/3}N^{2/3}h^{2/3}$ so that the spectrum decreases with increasing h. In this region the spectrum is proportional to $h^{-(2p+4)/3}$ and falls to zero at $h = m_d$, i.e., at the intersection of the $\omega = N$ and $\omega = D^{1/3}N^{2/3}h^{2/3}$ limits plotted in Fig. 1(b). The zonal

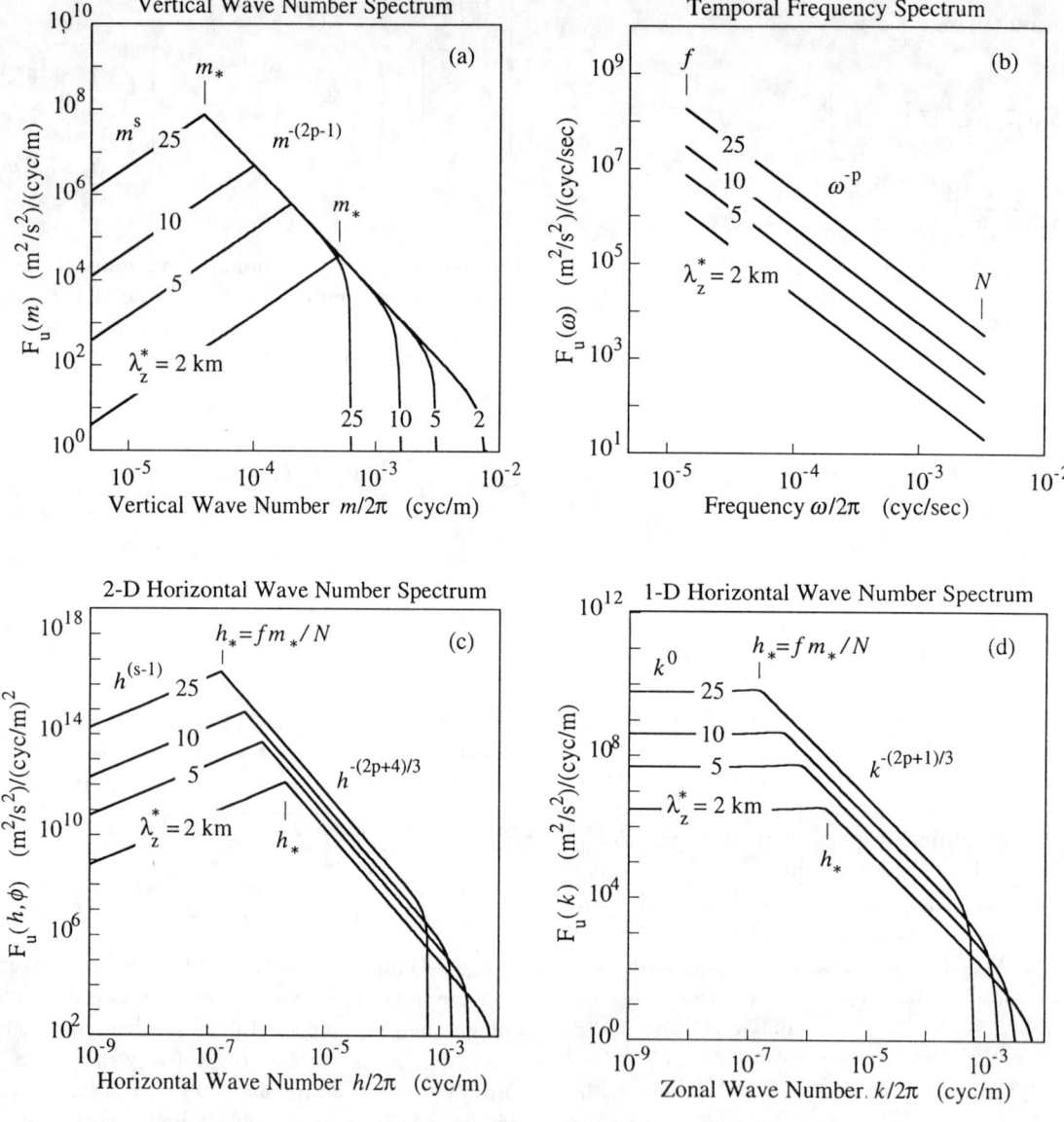

Fig. 4. Altitude dependencies of the gravity wave vertical wave number, temporal frequency and horizontal wave number spectra predicted by scale-independent diffusive filtering theory for the special case $p = s = 2$ and $\omega_{sc} = f$. λ_z^* is proportional to the rms horizontal wind velocity (see Eq. (20)) and increases with altitude.

wave number spectrum is flat (i.e., constant) for $k \ll h_*$. The energy at the lowest zonal wave numbers is associated with waves propagating primarily in the meridional direction. For $h_* \ll k \ll m_d$, the zonal spectrum is proportional to $k^{-(2p+1)/3}$ and falls to zero at $k = m_d$. For p = 2 and $h, k, l > h_*$, the horizontal wave number spectra are proportional to $h^{-8/3}$, $k^{-5/3}$ and $l^{-5/3}$.

By expressing the total variance in terms of m_* using (19) and noting the altitude variation of m_* given by (31), the altitude dependencies of the spectral magnitudes can be determined. The results are summarized in Table 2. The general formulas for all the spectra are summarized in Table 3. The altitude variations of the horizontal wave number spectra are illustrated in Figure 4 for the special case p = s = 2.

5. QUASI-MONOCHROMATIC GRAVITY WAVES

Although the random perturbations in wind, density and temperature characterized by wave spectra have received considerable attention in recent years, quasi-monochromatic wave perturbations are frequently seen in radar and lidar profiles and in airglow images. Several studies of the middle atmosphere using radars [e.g., *Reid and Vincent*, 1987; *Manson and Meek*, 1988], lidars [e.g., *Gardner and Voelz*, 1987; *Gardner et al.*, 1989; *Beatty et al.*, 1992; *Collins*, 1994] and airglow imagers [e.g., *Taylor and Hill*, 1991; *Taylor and Edwards*, 1991] have provided detailed information on the horizontal and vertical scales, periods and phase speeds of individual wave motions. Reid [1986] and Manson [1990] provide excellent summaries and reviews of many of these previous observations.

A curious feature of the Na and Rayleigh lidar observations reported by Gardner and Voelz [1987], Gardner et al. [1989], Beatty et al. [1992], and Collins [1994] is the remarkably systematic variations of the intrinsic vertical wavelengths and observed horizontal wavelengths with the observed periods (T_{ob}). Although Doppler shifting of the intrinsic periods by the mean winds can be substantial, Beatty et al. [1992] showed that their measured relationships were statistically significant and reflected an underlying relationship between the intrinsic vertical and horizontal wavelengths and the intrinsic periods (T) of the waves. Gardner and Voelz [1987] showed that the measured λ_z vs T_{ob} relationship was consistent with the measured wave amplitude dependencies on λ_z and T_{ob}, provided the kinetic energy distribution of monochromatic waves was not separable. They also suggested that the observed relationship was a consequence of viscous damping.

The relationships between the wave amplitudes, wavelengths and periods reported by Gardner and Voelz [1987] and Beatty et al. [1992] are readily explained by diffusive filtering theory. The horizontal velocity variance of an average wave comprising the spectrum can be estimated by multiplying the joint (m, ω) spectrum by the wave bandwidth.

$$<u'^2(m, \omega)> \simeq F_u(m, \omega) \, \Delta m \Delta \omega / (2\pi)^2 \quad (41)$$

From the joint (m, ω) spectrum given by (11) and Fig. 1(a), for the case $\omega_{sc} = f$ where $m_{sc} = m_*$, we see that the most energetic waves in the spectrum, i.e., the waves with the largest amplitudes and therefore the ones most easily observed, lie along the $\omega = f$ line for $m < m_*$ and along the $\omega = Dm^2$ line for $m > m_*$ because these are the regions where the magnitude of $F_u(m, \omega)$ is largest. Thus in the region $m > m_*$ for scale-independent diffusion, the relationship between the vertical wavelength and intrinsic period of the largest amplitude waves is

$$\lambda_z = (2\pi D T)^{1/2} = \lambda_z^* (T/T_i)^{1/2}, \quad (42)$$

where T_i is the inertial period.

From the joint (h, ϕ, ω) spectrum given by (36) and the joint (h, ϕ, m) spectrum given by (37), we see that the largest amplitude waves also lie along the $\omega = D^{1/3} N^{2/3} h^{2/3}$ line in Fig. 1b and the $m = (Nh/D)^{1/3}$ line in Fig. 1c so that for these waves

$$\lambda_h = N(D/2\pi)^{1/2} T^{3/2} = \lambda_h^* (T/T_i)^{3/2} \quad (43)$$

$$\lambda_z = (2\pi)^{2/3} (D/N)^{1/3} \lambda_h^{1/3} = \lambda_z^* (\lambda_h/\lambda_h^*)^{1/3}. \quad (44)$$

These are precisely the relationships measured by Gardner and Voelz [1987] and Beatty et al. [1992] for quasi-monochromatic waves observed in the upper stratosphere and upper mesosphere with Rayleigh and Na lidars. Similar relationships can be derived using the scale-dependent diffusion model. The results for scale-independent diffusion are summarized in Table 4.

The mean-square amplitude of a wave packet can be determined by multiplying the spectrum by the bandwidth of the wave packet (Eq. (41)). The bandwidth is determined from Fourier transform theory. In our model, the wind variances of all of the waves are growing exponentially with altitude in proportion to $e^{z/H}$ until they are damped by diffusion. Consider the following model for the horizontal velocity profile $u_0(z)$ of a wave packet with a vertical extent

TABLE 3. Spectra Models: Scale-Independent Diffusive Filtering Theory ($\omega_{sc} = f$, $m_{sc} = m_*$)

Spectrum	Model	
$F_u(m, \omega) =$	$(2\pi)^2 <(u')^2> \dfrac{(s+1)}{m_*} \left(\dfrac{m}{m_*}\right)^s \dfrac{(p-1)}{f} \left(\dfrac{f}{\omega}\right)^{p+(s+1)/2}$	$m \leq m_*(\omega/f)^{1/2}$ $f \leq \omega \leq N$
$F_u(m) =$	$\begin{cases} 2\pi <(u')^2> \dfrac{2(p-1)(s+1)}{(2p+s-1)} \dfrac{1}{m_*} \left(\dfrac{m}{m_*}\right)^s \\ 2\pi <(u')^2> \dfrac{2(p-1)(s+1)}{(2p+s-1)} \dfrac{1}{m_*} \left(\dfrac{m_*}{m}\right)^{2p-1} \end{cases}$	$m \leq m_*$ $m_* \leq m \ll m_d$
$F_u(\omega) =$	$2\pi <(u')^2> \dfrac{(p-1)}{f} \left(\dfrac{f}{\omega}\right)^p$	$f \leq \omega \leq N$
$F_u(k, l, \omega) = F_u(h, \phi, \omega) =$	$(2\pi)^3 <(u')^2> \dfrac{2}{\pi} \dfrac{(s+1)}{h_*^2} \left(\dfrac{h}{h_*}\right)^{s-1} \dfrac{(p-1)}{f} \left(\dfrac{f}{\omega}\right)^{p+3(s+1)/2}$	$h \leq h_*(\omega/f)^{3/2}$ $f \leq \omega \leq N$
$F_u(k, l, m) = F_u(h, \phi, m) =$	$(2\pi)^3 <(u')^2> \dfrac{2}{\pi} \dfrac{(s+1)}{m_*} \left(\dfrac{m}{m_*}\right)^{p+(3s-1)/2} \dfrac{(p-1)}{h_*^2} \left(\dfrac{h_*}{h}\right)^{p+(s+3)/2}$	$h \leq m \leq m_*(h/h_*)^{1/3} \leq Nh/f$
$F_u(k, l) = F_u(h, \phi) =$	$\begin{cases} (2\pi)^2 <(u')^2> \dfrac{4(p-1)(s+1)}{\pi(2p+3s+1)} \dfrac{1}{h_*^2} \left(\dfrac{h}{h_*}\right)^{s-1} \\ (2\pi)^2 <(u')^2> \dfrac{4(p-1)(s+1)}{\pi(2p+3s+1)} \dfrac{1}{h_*^2} \left(\dfrac{h_*}{h}\right)^{(2p+4)/3} \end{cases}$	$h \leq h_*$ $h_* \leq h \ll m_d$
$F_u(k) =$	$\begin{cases} 2\pi <(u')^2> \dfrac{4(p-1)(s+1)}{\pi(2p+1)s} \dfrac{1}{h_*} \\ 2\pi <(u')^2> \dfrac{2(p-1)(s+1)}{\pi(2p+3s+1)} \dfrac{\Gamma(1/2)\,\Gamma\left(\dfrac{p+2}{3}-\dfrac{1}{2}\right)}{\Gamma\left(\dfrac{p+2}{3}\right)} \dfrac{1}{h_*} \left(\dfrac{h_*}{k}\right)^{(2p+1)/3} \end{cases}$	$k \ll h_*$ $h_* \ll k \ll m_d$
$<(u')^2> =$	$\begin{cases} \dfrac{(s+3)}{(s+1)\,\tilde{R}i\,\ln(N/f)} \dfrac{N^2}{m_*^2} & p = 2 \\ \dfrac{(2-p)(s+3)}{(s+1)\,\tilde{R}i\,(p-1)[(N/f)^{2-p}-1]} \dfrac{N^2}{m_*^2} & p \neq 2 \end{cases}$	
$m_* = (f/D_{\text{wave}})^{1/2}$		
$h_* = fm_*/N$		
$m_d = (N/D_{\text{wave}})^{1/2}$		

L centered about altitude z_0, amplitude u'_0 at z_0, and the vertical wave number m_0.

$$u_0(z) = \begin{cases} u'_0 \, e^{(z-z_0)/2H} \cos[m_0(z-z_0)] & -L/2 \leq (z-z_0) \leq L/2 \\ 0 & \text{otherwise} \end{cases} \quad (45)$$

This wave packet is assumed to have been launched abruptly at $z_0 - L/2$, then grows exponentially with increasing altitude in response to decreasing atmospheric density, and finally is abruptly damped at $z_0 + L/2$. This approximate model ignores the details of the source profile near $z_0 - L/2$ and the damping profile near $z_0 + L/2$. The power spectral density of this wave packet, $F_{u_0}(m)$, is defined as the magnitude squared Fourier transform of the velocity profile divided by L and is easily calculated from (45).

$$F_{u_0}(m) = \frac{u'^2_0}{L} \frac{4H^2 \left\{ \sinh^2 \frac{L}{4H} \cos^2\left[\frac{(m-m_0)L}{2}\right] + \cosh^2 \frac{L}{4H} \sin^2\left[\frac{(m-m_0)L}{2}\right] \right\}}{[4H^2(m-m_0)^2 + 1]} \quad (46)$$

TABLE 4. Relationships Between the Wave Variances, Wavelengths and Intrinsic Periods ($\omega_{sc} = f$, $m_{sc} = m_*$)

Parameter	Scale-Independent Diffusion
λ_z	$\lambda_z^* (T/T_i)^{1/2}$
λ_h	$\lambda_h^* (T/T_i)^{3/2}$
λ_z	$\lambda_z^* (\lambda_h/\lambda_h^*)^{1/3}$
$\langle u^2(\lambda_z) \rangle$	$c_0 \dfrac{2(p-1)(s+3)}{(2p+s-1)\, \tilde{R}i\, \xi(p)} \dfrac{\lambda_z^{*3}}{HT_B^2} \left(\dfrac{\lambda_z}{\lambda_z^*}\right)^{2p-1}$
$\langle u^2(\lambda_h) \rangle$	$c_0 \dfrac{2(p-1)(s+3)}{(2p+3s-1)\, \tilde{R}i\, \xi(p)} \dfrac{\lambda_z^{*3}}{HT_B^2} \left(\dfrac{\lambda_h}{\lambda_h^*}\right)^{(2p-1)/3}$
$\langle u^2(T) \rangle$	$c_0 \dfrac{(p-1)(s+3)}{s\, \tilde{R}i\, \xi(p)} \dfrac{\lambda_z^{*3}}{HT_B^2} \left(\dfrac{T}{T_i}\right)^{(2p-1)/2}$
λ_z^*	$[\langle (u')^2 \rangle (s+1)\, \tilde{R}i\, \xi(p)/(s+3)]^{1/2}\, T_B$
λ_h^*	$T_i \lambda_z^*/T_B$

If the vertical extent of the wave packet is much less than the atmospheric scale height, i.e., $L \ll H$, then the hyperbolic functions in (46) can be approximated as $\cosh(L/4H) \simeq 1$ and $\sinh(L/4H) \simeq L/4H$ so that

$$F_{u_0}(m) \simeq \frac{u'^2_0}{4} L \frac{\sin^2\left[\frac{(m-m_0)L}{2}\right] + \left(\frac{L}{4H}\right)^2 \cos^2\left[\frac{(m-m_0)L}{2}\right]}{[(m-m_0)^2 L^2/4 + (L/4H)^2]}$$

$$\simeq \frac{u'^2_0}{4} L \left\{ \frac{\sin[(m-m_0)L/2]}{(m-m_0)L/2} \right\}^2 \quad (47)$$

In this case the bandwidth, Δm, of the spectrum is determined by the $\sin^2(\bullet)$ term in (47) and is proportional to $\Delta m \sim 1/L$. If the vertical extent of the wave packet is much larger than the scale height, i.e., $L \gg H$, then $\cosh(L/4H) \simeq \sinh(L/4H) \simeq 1/2\, e^{L/4H}$ so that

$$F_{u_0}(m) \simeq \frac{u'^2_0 H^2}{L} \frac{e^{L/2H}}{[4H^2(m-m_0)^2 + 1]}. \quad (48)$$

In this case the bandwidth is determined by the denominator in (48) and is proportional to $\Delta m \sim 1/H$. Thus, if a wave packet is damped after propagating a vertical distance L above the source that is less than the atmospheric scale height, then the vertical wave number bandwidth is proportional to $1/L$. However, if the wave is damped after propagating a vertical distance larger than the scale height, then the bandwidth is proportional to $1/H$. Including more realistic launching and damping profiles in the wave packet model will not alter these bandwidth dependencies on L and H, although the shape of the actual wave packet spectrum will be different from that given by (46).

We assume that all waves propagate a vertical distance at least equal to or larger than the atmospheric scale height (~6-7 km) so that the vertical wave number bandwidth is

$$\Delta m = \frac{2\pi}{H} c_0 \quad (49)$$

where the precise value of c_0 is determined experimentally. The mean-square amplitude $\langle u^2(m) \rangle$ of the average wave with wave number m is obtained by integrating (41) over ω using (49) and (11)

$$\langle u^2(m)\rangle = \int_{D^2m}^{N} F_u(m,\omega)\frac{\Delta m}{2\pi}\frac{d\omega}{2\pi}$$

$$\simeq \langle u'^2\rangle \frac{2(p-1)(s+1)}{(2p+s-1)}\frac{2\pi c_o}{Hm_*}\left(\frac{m_*}{m}\right)^{2p-1}$$

$$= c_o \frac{2(p-1)(s+3)}{(2p+s-1)\,\tilde{R}i\,\xi(p)}\frac{2\pi N^2}{Hm_*^3}\left(\frac{m_*}{m}\right)^{2p-1} \quad (50)$$

where we have used (28) to express the total variance in terms of m_*. The temporal frequency and horizontal wave number bandwidths are related to Δm through the dispersion relation

$$\Delta\omega = \frac{\omega}{m}\Delta m \simeq \frac{\omega}{m}\frac{2\pi}{H}c_o \quad (51)$$

$$\Delta h = \frac{\omega}{N}\Delta m \simeq \frac{\omega}{N}\frac{2\pi}{H}c_o \quad (52)$$

The mean-square amplitudes of the average wave of frequency ω or horizontal wave number h are given by

$$\langle u^2(\omega)\rangle = \int_0^{(\omega/D)^{1/2}} F_u(m,\omega)\frac{\Delta\omega}{2\pi}\frac{dm}{2\pi}$$

$$\simeq \langle u'^2\rangle \frac{(p-1)(s+1)}{s}\frac{2\pi c_o}{Hm_*}\left(\frac{f}{\omega}\right)^{(2p-1)/2}$$

$$= c_o \frac{(p-1)(s+3)}{s\,\tilde{R}i\,\xi(p)}\frac{2\pi N^2}{Hm_*^3}\left(\frac{f}{\omega}\right)^{(2p-1)/2} \quad (53)$$

$$\langle u^2(h)\rangle = \frac{1}{2\pi}\int_0^{\pi/2}d\phi\int_{D^{1/3}N^{2/3}h^{2/3}}^{N} hF_u(h,\phi,\omega)\frac{\Delta h}{2\pi}\frac{d\omega}{2\pi}$$

$$\simeq \langle u'^2\rangle \frac{2(p-1)(s+1)}{(2p+3s-1)}\frac{2\pi c_o}{Hm_*}\left(\frac{h_*}{h}\right)^{(2p-1)/3}$$

$$= c_o \frac{2(p-1)(s+3)}{(2p+3s-1)\,\tilde{R}i\,\xi(p)}\frac{2\pi N^2}{Hm_*^3}\left(\frac{h_*}{h}\right)^{(2p-1)/3}. \quad (54)$$

The results for scale-independent diffusion are summarized in Table 4. These predicted relationships are compared with observations in Section 6.

6. COMPARISONS WITH OBSERVATIONS AND SATURATION THEORIES

Senft and Gardner [1991] analyzed more than 450 h of high resolution Na lidar observations obtained at Urbana, IL (40°N, 88°W) on 60 different nights distributed throughout the year. The data were used to compute the vertical wave number and temporal frequency spectra as well as a variety of related parameters including the values of the spectral indices, wave variance and λ_z^*. These authors reported considerable variability in the spectral indices and magnitudes of both the measured m- and ω-spectra. Gardner et al. [1993a] reported similar airborne lidar data obtained during the ALOHA-90 Campaign from which the horizontal wave number spectra and related parameters were computed. The measured and predicted values of the spectral-indices, spectral magnitudes and the values of λ_z^* and λ_h^* are summarized in Table 5. The spectral magnitudes for the scale-independent diffusion theory are less than these observed values, but the differences are not substantial especially considering the large variability in the measured values. The predicted magnitudes of the temporal spectra are a factor of three less than the mean values reported by Senft and Gardner [1991]. This difference may be partly explained by the fact that Doppler shifting caused by the background wind field is known to enhance the magnitude of the temporal spectrum at high frequencies [Gardner et al., 1993b].

Beatty et al. [1992] and Senft et al. [1993] reported simultaneous Na and Rayleigh lidar measurements of the m-spectrum in the upper stratosphere (25-45 km) and upper mesosphere (80-105 km) at Arecibo. Their data show that the m-spectral indices in the upper stratosphere are approximately –2.5 and the spectral magnitudes are between 5 and 10 times smaller than the mesopause region magnitudes. The scale-independent diffusion theory predicts that the m-spectrum magnitudes for $m > m_*$ will increase with increasing altitude when $p < 2$. The observations of Beatty et al. and Senft et al. suggest that $p = 1.75$ (i.e., $2p-1 = 2.5$), so the altitude variations in the m-spectrum appear to be qualitatively consistent with the scale-independent diffusion theory.

Senft and Gardner [1991] reported substantial nightly and seasonal variations in the spectral indices and magnitudes of the m- and ω-spectra. For scale-independent diffusion, the spectral indices of both spectra are related to the sources through p and are expected to vary as the source characteristics change. However, Senft and Gardner found no systematic relationship between m- and ω-spectral indices of the type predicted. It is possible that the combined effects of Doppler shifting by the background wind field and

TABLE 5. Measured and Predicted Values of λ_z^*, λ_h^* and Spectrum Magnitudes at Mesopause Heights

Parameter	Measured	Predicted (p = 2, s = 1, $\tilde{R}i$ = 1)
	Senft and Gardner [1991] (annual mean)	Scale-Independent Diffusion
$<(u')^2>$	$(25.3 \text{ m/s})^2$	
λ_z^*	14.1 km	12 km
m-spectrum index	-2.9	-3
$F_u(m)$ @ $2\pi/(4 \text{ km})$	4.4×10^5 (m²/s²)/(cyc/m)	2.9×10^5 (m²/s²)/(cyc/m)
ω-spectrum index	-1.8	-2
$F_u(\omega)$ @ $2\pi/(1 \text{ h})$	3.6×10^5 (m²/s²)/(cyc/s)	1.2×10^5 (m²/s²)/(cyc/s)
	Gardner et al. [1993a] (ALOHA-90 27 March)	
$<(u')^2>$	$(18.6 \text{ m/s})^2$	
λ_z^*	10.3 km	9.2 km
λ_h^*	4140 km	3680 km
m-spectrum index	-3.1	-3
$F_u(m)$ @ $2\pi/(4 \text{ km})$	2.9×10^5 (m²/s²)/(cyc/m)	2.7×10^5 (m²/s²)/(cyc/m)
k-spectrum index	-2	$-5/3 = -1.67$
$F_u(k)$ @ $2\pi/(500 \text{ km})$	2.2×10^7 (m²/s²)/(cyc/m)	1.6×10^7 (m²/s²)/(cyc/m)

data filtering may have obscured the relationship between the spectral indices [*Gardner et al.*, 1993b]. This issue certainly deserves further study since this relationship is a key prediction of the scale-independent diffusion theory.

Figures 5 and 6 are comparisons of the predicted diffusion theory spectra with airborne and groundbased Na lidar observations made during the ALOHA-90 campaign. These data are taken from Hostetler and Gardner [1994]. The predicted vertical wave number, zonal wave number and temporal frequency spectra agree remarkably well with the observations. In all cases, the models were computed by assuming s = 2, $\omega_{sc} = f$, $m_{sc} = m_*$, and by determining p from the m-spectra.

From the discussion in Section 3 and Table 2, we find that for scale-independent diffusion the spectral magnitudes are related to the source variance $<(u'_0)^2>$ at altitude z_0 as follows.

$$F_u(\omega) \sim (p-1) <(u'_0)^2> e^{2(z-z_0)/(s+3)H} \quad (55)$$

$$F_u(m) \sim \frac{(p-1)}{(2p+s-1)} <(u'_0)^2>^{(2-p)} e^{(4-2p)(z-z_0)/(s+3)H}$$
$$m > m_* \quad (56)$$

These equations show that the magnitudes of both the m- and ω-spectra can vary substantially as the strength and spectral characteristics (i.e., s and p) of the sources change from day to day and seasonally. We note that this behavior for the m-spectrum is not predicted by scale-dependent diffusion theory [*Weinstock*, 1990; *Gardner*, 1994] or linear instability theory [*Dewan and Good*, 1986]. For these theories, the m-spectrum is proportional to N^2/m^3 in the region $m > m_*$ and does not vary with altitude or the strengths of the lower atmospheric sources.

Gardner and Voelz [1987], Gardner et al. [1989] and Beatty et al. [1992] reported observations of the wavelengths, observed periods and amplitudes of quasi-monochromatic gravity waves using Na and Rayleigh lidars. Their data exhibited remarkably systematic relationships between these parameters. Theoretical relationships between the monochromatic wave parameters were derived using the scale-independent diffusion theory spectra. The results are listed in Table 4. The relationships measured by Gardner and Voelz [1987] at Urbana, IL, are summarized in Table 6 along with the relationships predicted by the scale-independent diffusive filtering theory. The theoretical formulas were evaluated assuming p = 2, s = 1, H = 7 km,

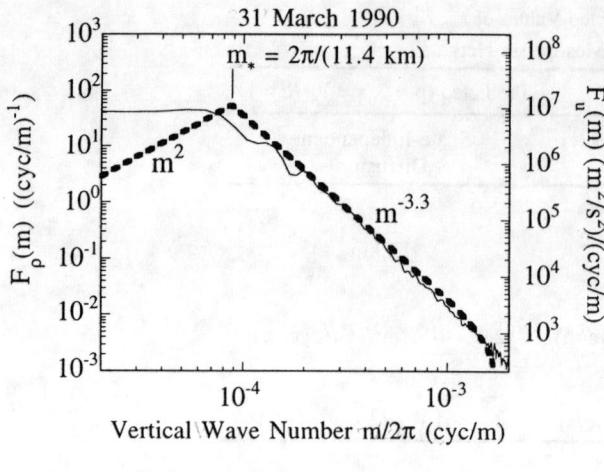
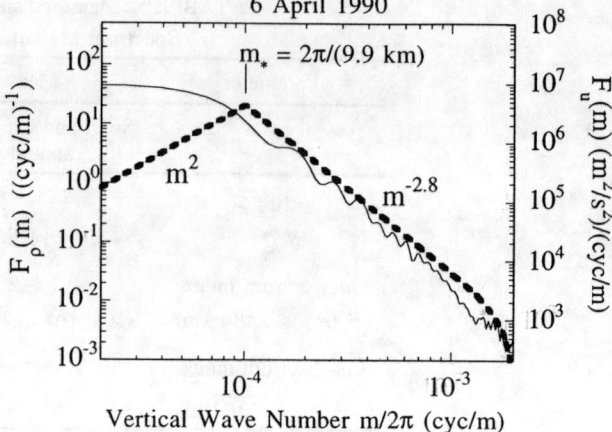
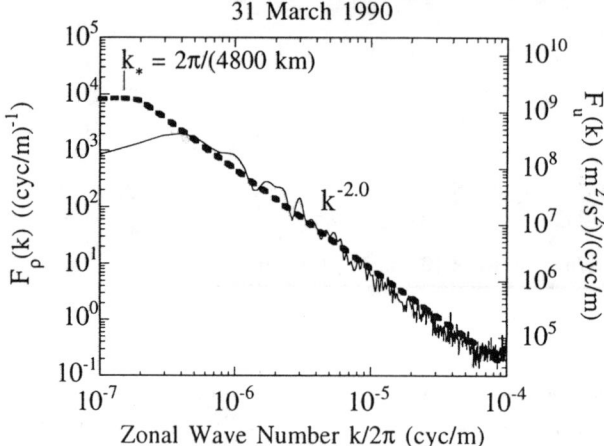
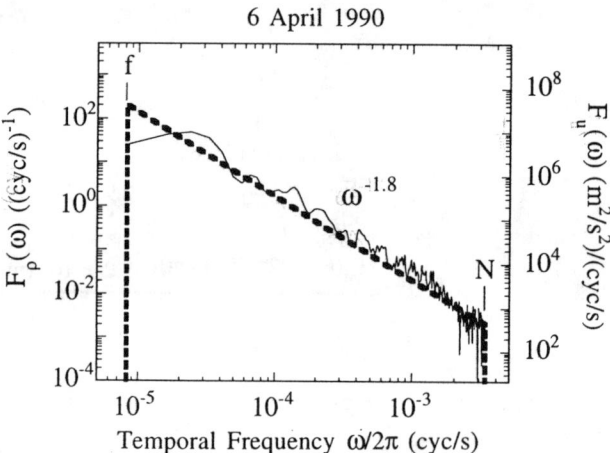

Fig. 5. (a) Vertical and (b) zonal wave number spectra measured with an airborne Na lidar during the ALOHA-90 campaign. The dashed curves are the spectra predicted by the scale-independent diffusion theory. The measured spectral indices are noted in the figures. F_ρ is the relative density spectrum and F_u is the horizontal wind spectrum.

Fig. 6. (a) Vertical and (b) temporal frequency spectra measured with a groundbased Na lidar during the ALOHA-90 campaign. The dashed curves are the spectra predicted by the scale-independent diffusion theory. The measured spectral indices are noted in the figures.

$\omega_{sc} = f$, $m_{sc} = m_*$, and $T_i = 18.67$ h. The values for λ_z^* were calculated using the value of $(25.3 \text{ m/s})^2$ for the mean wave variance at Urbana reported by Senft and Gardner [1991]. The measured and predicted relationships are in good agreement.

Similar relationships were reported by Beatty et al. [1992] for mesopause region waves observed at Arecibo (18°N) during the AIDA-89 campaign with a Na lidar. Figure 7 includes plots of the kinetic energies versus vertical wavelength and observed period of the quasi-monochromatic waves measured by Beatty et al. [1992]. For p = 2, diffusion theory predicts that the kinetic energies are proportional to λ_z^3 and $T_{intrinsic}^{1.5}$, as observed in Figure 7.

The relationships between the vertical wave numbers, horizontal wave numbers and temporal frequencies of the waves observed by Beatty et al. [1992] are illustrated in Figure 8 (pluses). The solid diagonal lines are the damping limits $\omega = Dm^2$ and $\omega = D^{1/3} N^{2/3} h^{2/3}$ predicted by scale-independent diffusion theory (assuming $\omega_{sc} \leq f$ which is probably not valid at the low latitudes of Arecibo and Haleakala where $T_i \sim 40$ h). The dashed diagonal lines are the maximum likelihood regression fits to the data. All of the waves lie just to the left of the damping limits. As discussed in Section 5, these are the largest amplitude waves, which are the most easily observed. Also plotted are waves observed by an O_2 airglow imager near Arecibo

TABLE 6. Measured and Predicted Relationships Between the Wave
Variances, Wavelengths and Periods at Mesopause Heights

Parameter	Measured	Predicted (p = 2, s = 1, $\tilde{R}i = 1$)
	Gardner and Voelz [1987]	Scale-Independent Diffusion
λ_z (km)	0.40 $T_{ob}^{0.55}$(min)	0.35 $T^{0.5}$(min)
λ_h (km)	0.093 $T_{ob}^{1.52}$(min)	0.071 $T^{1.5}$(min)
$<u^2(\lambda_z)>$ (m²/s²)	0.78 $\lambda_z^{2.95}$(km)	0.66 $c_o \lambda_z^3$(km)
$<u^2(\lambda_h)>$ (m²/s²)	0.75 $\lambda_h^{1.05}$(km)	0.41 $c_o \lambda_h$(km)
$<u^2(T)>$ (m²/s²)	0.060 $T_{ob}^{1.59}$(min)	0.058 $c_o T^{1.5}$(min)

during AIDA-89 (closed circles) [*Zhang et al.*, 1993] and waves observed by an OH imager at Haleakala (20°N) during ALOHA-90 (open circles) [*Taylor and Edwards*, 1991; *Taylor and Hill*, 1991]. The airglow waves are all deep within the undamped region. Significant intensity modulation of the 10 km thick airglow layers is possible only by gravity waves with vertical wave lengths comparable to or larger than the layer thickness. At mesopause region altitudes, λ_z^* is comparable to the airglow layer thickness. Thus the waves observed by the imagers all fall in the undamped region $m \lesssim m_*$. Notice that in the (ω, h) plane, the airglow waves are distributed well within the undamped region—roughly parallel to the lidar wave distribution.

So far we have restricted our attention to the spectra of horizontal winds. Vertical wind spectra can also be derived by applying the polarization relation $(w')^2 = (\omega/N)^2 (u')^2$ to the joint (m, ω) spectra. For scale-independent diffusion, we obtain from (11) for the case $\omega_{sc} = f$ and $m_{sc} = m_*$

$$F_w(m, \omega) = \left(\frac{\omega}{N}\right)^2 F_u(m, \omega) =$$

$$\begin{cases} (2\pi)^2 <(w')^2> \frac{(s+1)}{m_*}\left(\frac{m}{m_*}\right)^s \\ \left(\frac{f}{N}\right)^{3-p} \frac{(3-p)}{f}\left(\frac{f}{\omega}\right)^{p-2+(s+1)/2} \\ \qquad m \leq m_*(\omega/f)^{1/2} \\ \text{and } f \leq \omega \leq N \\ 0 \qquad \text{otherwise} \end{cases} \quad (57)$$

so that

$$F_w(\omega) = 2\pi <(w')^2>\left(\frac{f}{N}\right)^{3-p}\frac{(3-p)}{f}\left(\frac{f}{\omega}\right)^{p-2}$$

$$= 2\pi <(\omega')^2> \frac{(3-p)}{N}\left(\frac{N}{\omega}\right)^{p-2} \quad f \leq \omega \leq N \quad (58)$$

and

$$F_w(m) = \begin{cases} 2\pi <(w')^2> \frac{2(p-3)(s+1)}{(2p+s-5)} \\ \left[1-(f/N)^{p-2+(s-1)/2}\right]\frac{1}{m_*}\left(\frac{m}{m_*}\right)^s \\ \qquad m \leq m_* \\ 2\pi <(w')^2> \frac{2(p-3)(s+1)}{(2p+s-5)} \\ \left[1-\left(\frac{fm^2}{Nm_*^2}\right)^{p-2+(s-1)/2}\right]\frac{1}{m_*}\left(\frac{m}{m_*}\right)^{5-2p} \\ \qquad m_* \leq m \leq m_d \end{cases} \quad (59)$$

When p = 2, (58) predicts $F_w(m) \sim m$ in the region $m_* \leq m \ll m_d$. Diffusive filtering theory predicts that the vertical wave number spectrum of vertical winds is considerably shallower than horizontal wind spectrum and in fact, the w' spectrum will increase in magnitude with increasing m for p < 5/2. This feature is a direct consequence of the inseparability of the (m, ω) spectrum enforced by the diffusive damping limits.

Because vertical winds are difficult to measure, very little data have been reported on vertical wind spectra. Kuo et al. [1985] and Larsen et al. [1986 and 1987] reported m-spectra

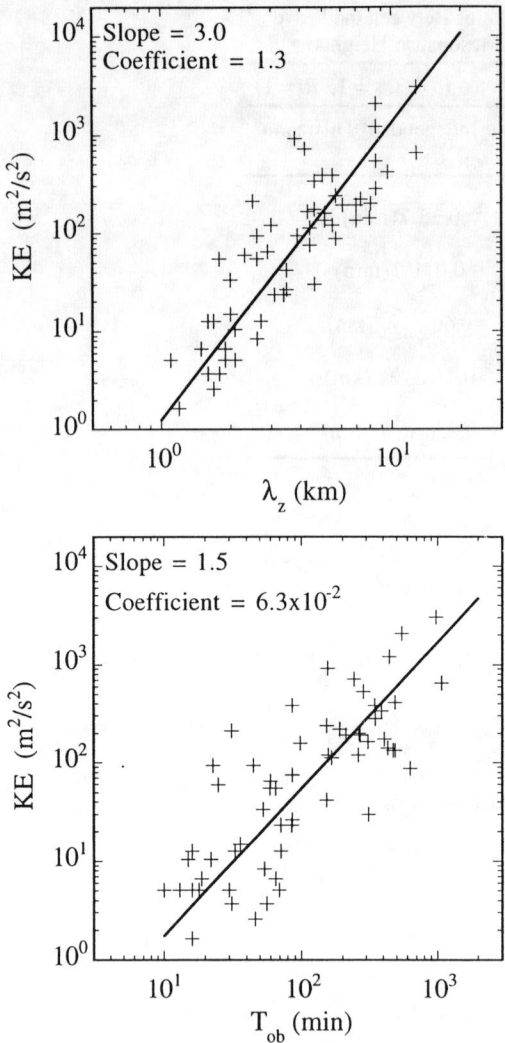

Fig. 7. Kinetic energies of quasi-monochromatic gravity waves observed by Beatty et al. [1992] at Arecibo (18°N) plotted versus (a) vertical wavelength and (b) observed period. The diagonal lines are the maximum likelihood power-law fits to the data. Scale independent diffusive filtering theory predicts $KE \propto \lambda_z^3$ and $KE \propto T^{1.5}$ as observed.

of vertical velocities measured with ST radars in the troposphere and stratosphere at Arecibo and at Bad Lauterberg, Germany. The spectral indices varied between −0.3 and −1.7. Although these slopes are steeper than those predicted by diffusion theory (~ +1), they are considerably shallower than the −3 value predicted by linear instability theory which invokes separability of the joint (m, ω) spectrum. Because the gravity wave horizontal winds are much larger than the vertical winds, even a small contamination of the vertical velocity measurement by the horizontal velocities would cause the corresponding m-spectra to exhibit characteristics similar to the horizontal wind spectra. Also, Lintleman and Gardner [1993] have shown that the combined effects of Doppler shifting by the mean background winds, and low-pass temporal filtering of the vertical wind data, can result in substantially steeper slopes for the measured m-spectra. The existing data on vertical wind spectra appear qualitatively consistent with diffusive filtering theory, although the observations are not conclusive. This area deserves further study. The vertical wind spectrum predictions of diffusion theory are considerably different from those of the separable linear instability model. Accurate simultaneous measurement of the vertical wave number and temporal frequency spectra of vertical and horizontal winds may provide a crucial test of the linear saturation and nonlinear diffusion theory.

CONCLUSIONS

The gravity wave horizontal wind spectra predicted by the scale-independent diffusive filtering theory compare favorably with existing observations. The m, k and ω spectral magnitudes and indices are comparable to those observed using a variety of lidar, radar and in situ measurement techniques. Although the vertical wave number spectrum predicted by the diffusive filtering theory exhibits many of the features predicted by various saturation theories, the physics controlling the shape of the m-spectrum is considerably different from these other theories. The waves in the region $m > m_*$ are not saturated. All components of spectrum, including those contributing to the region $m > m_*$, grow with increasing altitude in response to decreasing density until they are eliminated by diffusive damping. For scale-independent diffusion, the spectral magnitude for $m > m_*$ is very sensitive to the strengths and spectral characteristics of the gravity wave sources in the lower atmosphere. The magnitudes can increase, remain constant or even decrease with increasing altitude. Recent lidar observations of wave spectra reported by Senft and Gardner [1991], Beatty et al. [1992] and Senft et al. [1993] suggest that the magnitudes and indices of the m-spectra are highly variable and the magnitudes can increase substantially with altitude. These observations are consistent with the scale-independent diffusion theory.

Because the joint (m, ω) spectrum predicted by diffusive filtering theory is not separable, the shape of the m-spectrum of vertical winds is considerably different from the shape of the corresponding horizontal wind spectrum. Diffusion theory predicts that the vertical wind spectrum will typically increase with increasing m. The spectral index is predicted to be near + 1. This prediction is perhaps

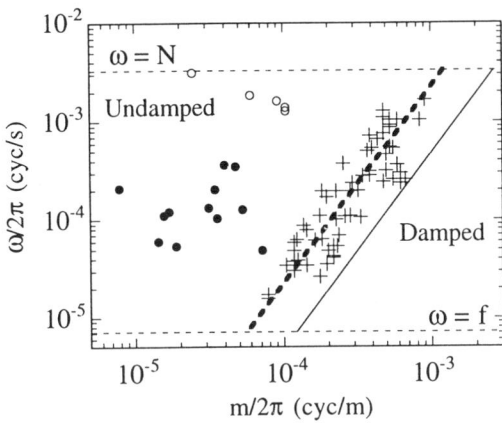

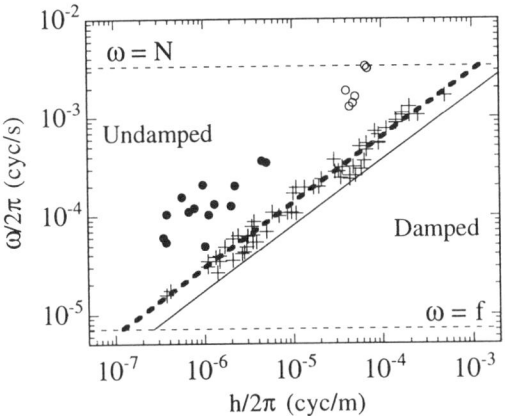

Fig. 8. Relationship between observed gravity waves and the damping limits (solid diagonal lines) predicted by scale-independent diffusive filtering theory. The + denotes waves observed at Arecibo (18°N) during AIDA 89 with a Na lidar, [*Beatty et al.*, 1992], • denotes waves observed at Arecibo (18°N) during AIDA 89 with an oxygen airglow imager [*Zhang et al.*, 1993] and o denotes waves observed at Haleakala (20°N) during ALOHA-90 with an OH imager [*Taylor and Hill*, 1991; *Taylor and Edwards*, 1991].

the most significant because it provides a robust test of diffusive filtering theory. The linear instability model predicts that the *m*-spectrum of vertical winds will have the same shape as the horizontal wind spectrum (i.e., spectral index equals –3). Existing radar observations of vertical winds show spectral indices between –0.3 and –1.7. However, vertical winds are difficult to measure with radars and even a small contamination by the strong horizontal winds as well as data filtering effects can seriously distort measurements of the vertical wind spectrum. Although these existing observations suggest that the vertical wind spectrum is considerably different from the horizontal wind spectrum and similar to the predictions of diffusive filtering

theory, we do not consider the observations to be conclusive.

In the real atmosphere the cutoff conditions for diffusive filtering will not be as abrupt or as absolute as that implied by the derivations leading to the spectra listed in Tables 1 and 3. The altitude growth characteristics of the spectrum components will, in reality, transition smoothly from $e^{z/H} F_u^{sc}(m, \omega)$ deep in the undamped region to zero deep in the damped region. Even so, it does not seem productive to modify the theory to include these subtle details since the accuracies of the final spectrum models will be limited ultimately, by the accuracy of the assumed source spectrum. The power-law models we employed provided good agreement with existing experimental data but are at best an idealized description. Deviations of the source spectrum from a pure power-law behavior will be reflected in the spectra observed at higher altitudes.

The analysis presented in this paper is incomplete because of several other deficiencies. All of the calculations used the simplified dispersion and polarization relations. While the changes will be relatively minor, more accurate spectral models can be derived using the more accurate Boussinesq approximations for the gravity wave polarization and dispersion relation. We also ignored the effects of the mean wind field which can be very important in the real atmosphere. Doppler shifting of the intrinsic frequencies by the mean horizontal wind can substantially alter the ω-spectrum and the joint (m, ω) spectrum. Fortunately it is a rather straightforward task to model the influence of the mean wind field and this has been done elsewhere [*Lintleman and Gardner*, 1993].

Finally, we note the significance of ω_{sc}. When $f < \omega_{sc} < N$, the *m*-spectrum magnitude in the region $m_{sc} < m$ for the case where p = 2 is given by (34)

$$F_u(m) \simeq \frac{4\pi}{\tilde{R}i \ln(N/\omega_{sc})} \frac{N^2}{m^3} = \frac{4\pi}{\ln(N/\omega_{sc})} \frac{\left<\left(\frac{\partial u'}{\partial z}\right)^2\right>}{m^3} \quad (60)$$

If $\tilde{R}i$ or equivalently the vertical shear variance of horizontal winds is independent of altitude and geographic location, then the spectrum magnitudes are constant. However, if $\omega_{sc} \leq f$, the *m*-spectrum magnitude for p = 2 is given by [*Gardner*, 1994]

$$F_u(m) = \frac{4\pi}{\tilde{R}i \ln(N/f)} \frac{N^2}{m^3} = \frac{4\pi}{\ln(N/f)} \frac{\left<\left(\frac{\partial u'}{\partial z}\right)^2\right>}{m^3} \quad (61)$$

In this case, the *m*-spectrum magnitude depends on latitude through *f*. The magnitudes should decrease at lower latitudes where *f* is smaller. The existing mesopause region data suggest $F_u(m)$ is independent of latitude [*Collins*, 1994] with $\omega_{sc} \sim 2\pi/(10\ h)$, highlighting the importance of modeling accurately the low ω dependence of the source spectrum.

Acknowledgments. The author thanks Chris Hostetler for preparing the figures included in this paper. The author also acknowledges with pleasure the stimulating discussions with Colin Hines, Ed Dewan, Jerry Weinstock and Chris Hostetler during the preparation of the manuscript. This work was supported in part by NSF Grant ATM90-24367.

REFERENCES

Beatty, T. J., C. A. Hostetler and C. S. Gardner, Lidar observations of gravity waves and their spectra near the mesopause and stratopause at Arecibo, *J. Atmos. Sci., 49*, 477-496, 1992.

Collins, R. L., Middle atmosphere structure and dynamics: Lidar observations at the South Pole, Syowa, and Urbana, Ph.D. Dissertation, University of Illinois, January 1994.

Dewan, E. M., and R. E. Good, Saturation and the "universal" spectrum for vertical profiles of horizontal scalar winds in the atmosphere, *J. Geophys. Res., 91*, 2742-2748, 1986.

Gardner, C. S., Diffusive filtering theory of gravity wave spectra in the atmosphere, *J. Geophys. Res.*, in press February 1994.

Gardner, C. S., and D. G. Voelz, Lidar studies of the nighttime sodium layer over Urbana, Illinois, 2, Gravity waves, *J. Geophys. Res., 92*, 4673-4694, 1987.

Gardner, C. S., C. A. Hostetler and S. J. Franke, Gravity wave models for the horizontal wave number spectra of atmospheric velocity and density fluctuations, *J. Geophys. Res., 98*, 1035-1049, 1993a.

Gardner, C. S., C. A. Hostetler and S. Lintleman, Influence of the mean wind field on the separability of atmospheric perturbation spectra, *J. Geophys. Res., 98*, 8859-8872, 1993b.

Gardner, C. S., D. C. Senft, T. J. Beatty, R. E. Bills and C. A. Hostetler, Rayleigh and sodium lidar techniques for measuring middle atmospheric density, temperature and wind perturbations and their spectra, in *World Ionosphere/Thermosphere Study Handbook, 2*, edited by C. H. Liu and B. Edwards, International Congress of Scientific Unions, Urbana, pp. 148-187, 1989.

Gossard, E. E., and W. H. Hooke, *Waves in the Atmosphere*, Developments in Atmospheric Science 2, Elsevier Scientific Publishing, New York, 1975.

Hines, C. O., Internal atmospheric gravity waves at ionospheric heights, *Can. J. Phys., 38*, 1441-1481, 1960.

Hines, C. O., Minimum vertical scale sizes in the wind structure above 100 kilometers, *J. Geophys. Res., 69*, 2847-2848, 1964.

Hines, C. O., An effect of molecular dissipation in upper atmospheric gravity waves, *J. Atmos. Terr. Phys., 30*, 845-849, 1968.

Hines, C. O., Eddy diffusion coefficients due to instabilities in internal gravity waves, *J. Geophys. Res., 75*, 3937-3939, 1970.

Hines, C. O., *The Upper Atmosphere in Motion*, Geophys. Monogr. Ser., vol. 18, AGU, Washington, DC, pg. 426, 1974.

Hines, C. O., The saturation of gravity waves in the middle atmosphere, Part II, Development of Doppler-spread theory, *J. Atmos. Sci., 48*, 1360-1379, 1991.

Hodges, R. R., Generation of turbulence in the upper atmosphere by internal gravity waves, *J. Geophys. Res., 72*, 3455, 1967.

Hodges, R. R., Jr., Eddy diffusion coefficients due to instabilities in internal waves, *J. Geophys. Res., 74*, 4087-4090, 1969.

Hostetler, C. A., and C. S. Gardner, Observations of horizontal and vertical wave number spectra of gravity wave motions in the stratosphere and mesosphere over the mid-Pacific, *J. Geophys. Res., 99*, 1283-1302, 1994.

Kuo, F. S., et al., Altitude dependence of vertical velocity spectra observed by VHF radar, *Radio Sci., 20*, 1349-1354, 1985.

Larsen, M. F., et al., Power spectra of vertical velocities in the troposphere and lower stratosphere observed at Arecibo, Puerto Rico, *J. Atmos. Sci., 43*, 2230-2240, 1986.

Larsen, M. F., J. Rottger and D. N. Holden, Direct measurements of vertical-velocity power spectra with the Sousy-VHF-Radar wind profiler system, *J. Atmos. Sci., 44*, 3442-3448, 1987.

Lindzen, R. S., Turbulence and stress owing to gravity wave and tidal breakdown, *J. Geophys. Res., 86*, 9707-9714, 1981.

Lintleman, S. A., and C. S. Gardner, Influence of the mean wind field on observed gravity wave spectra: Nonseparable scale independent diffusive filtering models, *J. Geophys. Res.*, Submitted, July 1993.

Manson, A. H., and C. E. Meek, Gravity wave propagation characteristics (60-120 km) as determined by the Saskatoon MF radar (Gravnet) system: 1983-85 at 52°N, 10°W, *J. Atmos. Sci., 45*, 932-946, 1988.

Manson, A. H., Gravity wave horizontal and vertical wave lengths; An update of measurements in the mesopause region (~ 80-100 km), *J. Atmos. Sci., 47*, 2765-2773, 1990.

Pitteway, M. L. U., and C. O. Hines, The viscous damping of atmospheric gravity waves, *Can. J. Phys., 41*, 1935-1948, 1963.

Reid, I. M., and R. A. Vincent, Measurements of the horizontal scales and phase velocities of short period mesospheric gravity waves at Adelaide, Australia, *J. Atmos. Terr. Phys., 49*, 1033-1048, 1987.

Reid, I. M., Gravity wave motions in the upper middle atmosphere (60-110 km), *J. Atmos. Terr. Phys., 48*, 1057-1072, 1986.

Senft, D. C., and C. S. Gardner, Seasonal variability of gravity wave activity and spectra in the mesopause region at Urbana, *J. Geophys. Res., 96*, 17,229-17,264, 1991.

Senft, D. C., C. A. Hostetler and C. S. Gardner, Characteristics of gravity wave activity and spectra in the upper stratosphere and upper mesosphere at Arecibo during early April 1989, *J. Atmos. Terr. Phys., 55*, 499-511, 1993.

Smith, S. A., D. C. Fritts and T. E. Van Zandt, Evidence for a saturated spectrum of atmospheric gravity waves, *J. Atmos. Sci., 44*, 1404-1410, 1987.

Taylor, M. J., and M. J. Hill, Near infrared imaging of hydroxyl wave structure over an ocean site at low latitudes, *Geophys.*

Res. Lett., 18, 1333-1336, 1991.

Taylor, M. J., and R. Edwards, Observations of short period mesospheric wave patterns: In situ or tropospheric wave generation? *Geophys. Res. Lett., 18*, 1337-1340, 1991.

Weinstock, J., Nonlinear theory of acoustic-gravity waves, 1, Saturation and enhanced diffusion, *J. Geophys. Res., 81*, 633-652, 1976.

Weinstock, J., Nonlinear theory of gravity waves: Momentum deposition, generalized Rayleigh friction, and diffusion, *J. Atmos. Sci., 39*, 1698-1710, 1982.

Weinstock, J., Simplified derivation of an algorithm for nonlinear gravity waves, *J. Geophys. Res., 89*, 345-350, 1984a.

Weinstock, J., Gravity wave saturation and eddy diffusion in the middle atmosphere, *J. Atmos. Terr. Phys. 46*, 1069-1082, 1984b.

Weinstock, J., Saturated and unsaturated spectra of gravity waves and scale-dependent diffusion, *J. Atmos. Sci., 47*, 2211-2225, 1990.

Zhang, S. P., R. N. Peterson, R. H. Wiens and G. G. Shepherd, Gravity waves from O_2 nightglow during the AIDA '89 Campaign I: emission rate/temperature observations, *J. Atmos. Terr. Phys., 55*, 355-375, 1993.

C. S. Gardner, Department of Electrical and Computer Engineering, University of Illinois, Everitt Laboratory, 1406 W. Green Street, Urbana, IL 61801.

Note: The work described in the paper was first presented at the November 1992 AGU Chapman Conference at Asilomar. A more extensive version of this paper was submitted to *JGR–Atmospheres* in October 1992 and was accepted for publication in February 1994.

An Investigation of Thunderstorms as a Source of Short Period Mesospheric Gravity Waves

M. J. Taylor

Space Dynamics Laboratory, Utah State University, Logan, UT

V. Taylor

Logan, UT

R. Edwards

Physics Department, The University of Southampton, U.K.

For three months during the spring and early summer of 1988, low-light TV images showing wave structure in the near infrared hydroxyl (OH) nightglow emission (peak altitude ~87 km) were recorded from the Mountain Research Station near Nederland, Colorado (40.0°N, 105.6°W) as part of the AFOSR MAPSTAR'88 campaign. Well-defined, coherent wave patterns associated with the passage of short period (<1 hour) gravity waves were observed on a total of 22 occasions. One potential source of these waves has been studied using radar summary charts to identify regions of strong convection associated with the existence or development of thunderstorms. Comparison of the "storm" positions with the location and direction of motion of the OH patterns shows that there was always at least one disturbance suitably located in both space and time to have been the source. The analysis presented here is qualitative, but the large number of wave events associated with favorably located convective activity provides strong evidence for a relationship between the observed waves and storms. This result, although preliminary, suggests that thunderstorms are an important source of mesospheric gravity waves at this site and time of year.

1. INTRODUCTION

Internal atmospheric gravity waves, generated by sources in the troposphere, play an important role in governing the dynamical coupling between the lower and the upper atmospheric regions. In particular, small scale gravity waves with periods of <1 hour are now recognized to be a ubiquitous feature of the upper middle atmosphere [*Reid*, 1986]. These waves are responsible for as much as 70% of the wave induced transport that occurs in the mesosphere and lower thermosphere [*Fritts and Vincent*, 1987]. At such heights wave energy may be absorbed gradually into the mean flow or it may be deposited at a single "critical layer" if the background winds Doppler shift the wave frequency to zero [e.g., *Booker and Bretherton*, 1967]. Waves propagating through this region may also become unstable and break, depositing their energy in the form of small-scale turbulence. Variations in the seasonal and latitudinal abundance of short period gravity waves therefore can have a marked effect on the dynamics of the upper atmospheric circulation.

Observations of the hydroxyl (OH) nightglow emission, which originates at a mean altitude of ~87 km and has a night-time halfwidth of 5-8 km [*Baker and Stair*, 1988], provide an excellent method for remote sensing mesospheric gravity waves [e.g., *Armstrong*, 1982; *Takahashi et al.*, 1985; *Taylor et al.*, 1987]. Imaging studies are important as they provide unique information on the two-dimensional

The Upper Mesosphere and Lower Thermosphere:
A Review of Experiment and Theory
Geophysical Monograph 87
Copyright 1995 by the American Geophysical Union

horizontal gravity wave parameters (λ_x, v_x, τ_{obs}), over a large geographic area (up to ~10^6 km^2) and with a high spatial and temporal resolution [*Taylor et al.*, 1993]. In particular, image data are valuable for investigating gravity wave sources as they contain unambiguous information on the frequency of occurrence, geographic location, orientation, and the horizontal direction of propagation of the waves [*Taylor and Hapgood*, 1988; *Taylor and Edwards*, 1991].

Any disturbance that introduces a change in the atmosphere on a time scale of a few minutes to several hours may be capable of generating gravity waves. Tropospheric sources are thought to be important at all latitudes as waves created near the earth's surface may grow considerably in amplitude as they propagate energy upwards. Weather related disturbances such as jet streams, fronts, depressions, severe storms and winds blowing over prominent topographic features therefore may be responsible for many of the wave-like motions that frequently permeate the upper neutral atmosphere and the ionosphere. Several attempts have been made in the past to associate tropospheric sources with various upper atmospheric wave phenomena [e.g., *Hines*, 1968; *Röttger*, 1977; *Freund and Jacka*, 1979; *Hung et al.*, 1979]. To date, most of these studies have proven inconclusive and only on rare occasions has an individual source been positively identified. This is because there are often several potential sources present and because the effects of wind filtering on the gravity waves as they propagate upwards are usually unknown. As a consequence, the spatial distribution and the temporal variability of the wave sources primarily responsible for the transport of momentum into the upper atmosphere remain obscure.

One potential source mechanism that is thought to be quite efficient at generating short period gravity waves is strong tropospheric convection that often culminates as thunderstorms [*Pierce and Coroniti*, 1966]. In a previous study we successfully identified an isolated thunderstorm as a source of short period (~15 min) gravity waves imaged in the OH, Na (589.2 nm) and OI (557.7 nm) nightglow emissions [*Taylor and Hapgood*, 1988]. In this report we build on this result by using a series of all-sky OH data recorded over a three month period to determine the relationship between the occurrence of regions of strong convection and the appearance of short period mesospheric gravity waves.

2. INSTRUMENTATION

Measurements of the near infrared (NIR) OH nightglow emission were made using a low light, Image Isocon TV system capable of obtaining good quality images of wave structure with an integration time of typically 2–5 s [*Taylor et al.*, 1993]. The TV camera was fitted with a Nikon 8 mm f/2.8 "fish eye" lens, giving an all-sky (180°) field of view, and a Schott RG715 glass filter limiting the bandwidth of the observations to 715-810 nm. This region of the nightglow spectrum is dominated by several OH Meinel band emissions which have an integrated intensity of ~2 kR. Wave structure was detectable at all azimuths up to a maximum range of ~600 km (limited by the local horizon) which corresponds to a nominal search area ~10^6 km^2 (for an emission altitude of 87 km).

3. MEASUREMENTS

To investigate any association between the occurrence of "storms" and the subsequent appearance of short period, mesospheric gravity waves, data from a three month campaign conducted during spring and early summer of 1988 (new moon periods May 11-22, June 7-22, and July 9-18) were analyzed. The measurements were made from the Mountain Research Station (MRS) near Nederland, Colorado (40.0°N, 105.6°W, 3050m) which is located near the eastern edge of the Rocky Mountains. This region is well known for its strong convective activity and thunderstorms were observed to occur with increasing frequency during this time. Well defined OH wave patterns were detected on a total of 22 occasions, 18 of which have yielded accurate measurements (see Table 1).

Figure 1 is an example of a coherent OH wave display extending over the entire camera field [*Taylor et al.*, 1993]. The image was obtained on July 10 at 05:10 UT during an unusually high contrast event and shows numerous near east-west aligned wave fronts. Data were obtained for ~4 hours during which time the wave forms were observed to move uniformly towards the north. Several convective regions were detected on this occasions but only one occurred at an appropriate azimuth to have been an acceptable candidate source (see Results). Most of the wave patterns were less conspicuous than this display but all of them appeared as a set of well-formed elongated structures, usually extending over a limited area of the camera's field.

4. ANALYSIS PROCEDURE

The data have been analyzed to determine the geographic location and orientation of the wave patterns, their time of occurrence and where possible their horizontal wavelengths (λ_x), phase speeds (v_x), and observed periods (τ_{obs}). The images were calibrated first in terms of elevation and azimuth using the right ascension and declination of several known stars within the all-sky field [*Hapgood and Taylor*, 1982]. Ground maps showing the position and motion of the wave forms as a function of time were then calculated

TABLE 1. Summary of Image Measurements

UT Date 1988	Data Interval (UT)	λ_x (km)	v_x (m/s)	τ_{obs} (min)	Wave Azi (±5°)
May 13	7:50-8:30	38±3	30±2	21.0	90
May 14	6:00	56±5	-	-	120
May 16	8:30-10:00	70±2	21±2	55.5	20
May 17	7:00-9:00	26±2	28±2	15.5	40
May 21	9:35-10:05	30±6	22±4	23.0	125
Mean	-	44	25.3	28.8	-
Jun 12	6:54	41±2	-	-	10
Jun 17	7:40-8:50	41±14	22±1	31	15
Jun 19	8:30	19±5	-	-	55
Jun 20a	8:00-9:00	35±5	5±2	117.0	15
Jun 20b	8:00-9:00	21±2	42±2	8.0	340
Jun 21a	6:10-8:45	50±5	18±3	46	80
Jun 21b	6:10-8:45	35±9	14±3	42	20
Mean	-	34.5	29.6	48.8 (32)*	-
Jul 10	5:10-6:40	21±5	20±1	23.0	15
Jul 12	4:50-5:50	30±5	23±3	22.0	30
Jul 15	6:42	30±3	-	-	345
Jul 16	5:20-8:40	45±2	28±2	27	140
Jul 17	5:30-7:30	21±4	32±4	11	345
Jul 18	5:30-6:40	25±5	37±3	11	0
Mean	-	28.7	28	18.8	-

* Without June 20a data

for each display (assuming an emission altitude of 87 km), from which its average horizontal wavelength and velocity was determined. Only those displays that exhibited well-defined structures subtending elevations >5° have been used in this study.

Radar summary charts covering the continental USA (one per hour) were used to identify regions of strong convection associated with thunderstorm activity. In May the number of convection zones on any one day were few, but in June and July the summary charts were considerably more complex. Measurements of the geographic location, altitude and time of occurrence of only those disturbances that achieved ~10 km in height (i.e. storms approaching the tropopause region) were made. However, on some occasions, mainly during May, lower altitude isolated disturbances were also considered (see Discussion).

Linear gravity wave theory indicates that the range from the tropospheric source to the region of observation of the wave in the upper atmosphere is strongly dependent on its periodicity [Hines, 1967]. For an isothermal, stationary atmosphere, the minimum horizontal range (R_{min}) from the source to the point of intersection with an airglow layer is given by:

$$R_{min} = (h_a - h_s)\sqrt{\frac{\tau^2}{\tau_b^2} - 1} \qquad (1)$$

where h_a and h_s are the heights of the airglow layer and the gravity wave source, τ_b is the Brunt-Väisälä period, and τ is the intrinsic period of the wave (which equals its observed period in a stationary atmosphere) [Taylor and Hapgood, 1988]. This relationship holds provided the vertical extent of the wave source is not large (i.e. typically <8 km). For a convective source such as a thunderstorm h_s would be about 5-15 km. Using a height-averaged value of $\tau_b = 5.5 \pm 0.5$ min and a wave period of 1 hour (which includes all the observations listed in Table 1, with the exception of June 20a display), an upper limit for R_{min} would be ~1,000 km. Many of the observed wave motions exhibited τ_{obs} << 1 hour indicating ground ranges of only a few hundred kilometers. However, as wave propagation can be affected significantly by background winds in the stratosphere and mesosphere (which are unknown), the actual horizontal range may vary considerably. Thus, to encompass as many candidate sources as practical, a search region 1,000 km in radius (extending from the Mexican to the Canadian border), was used for all wave events.

The time taken for wave energy emanating from a tropospheric source to reach the upper mesosphere is dependent upon the wave's group velocity. Assuming an isothermal, stationary atmosphere Taylor and Hapgood [1988] calculated a time of flight of 6 ± 0.5 hrs for a wave of period 17 min, originating from a storm at a range of ~500 km, and propagating with a horizontal group velocity of 23 ± 2 ms^{-1} to reach the OH layer. Waves generated near the limit of our search region, or waves exhibiting significantly lower group velocities would be expected to take several hours longer. As background winds also affect the group velocity of the waves as well as their path length through the intervening atmosphere, radar summary charts were investigated for

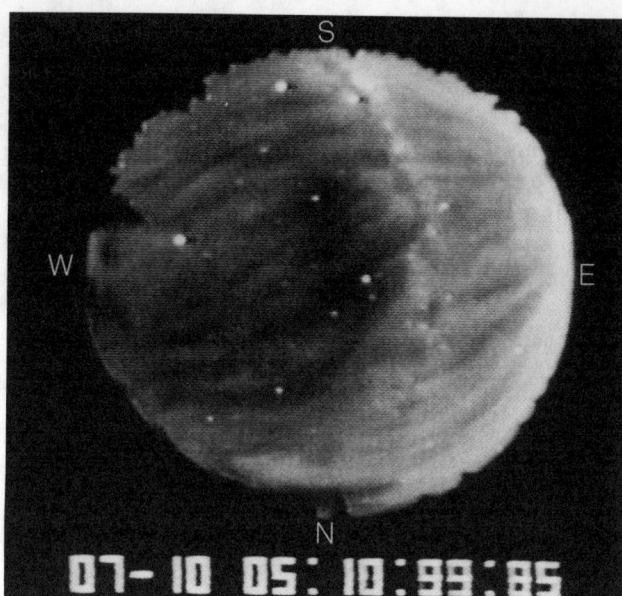

Figure 1. All-sky image showing coherent NIR OH wave structure extending over the entire field of view. The data were obtained on 10 July at 05:10 UT using an integration time of 1.5 s [*Taylor et al., 1993*]. The bright band intersecting the wave display at an acute angle is the Milky Way. The apparent increase in sky brightness to the east is due to scattered light from cities on the Colorado plains.

intervals up to 14 hours prior to each display in an attempt to ensure sufficient time for wave propagation (equivalent to a horizontal group speed of 20 ms^{-1} for a source located at the limit of the search area).

5. RESULTS

5.1. Wave Measurements

Table 1 summarizes the results of the image analysis. A total of 18 events were recorded where λ_x and the wave azimuth (i.e. the horizontal direction of motion) were measured. However, only 14 of these displays gave accurate measurements of v_x and hence τ_{obs}. The horizontal wavelengths ranged from 19-70 km yielding a mean value of 36 km for the campaign. The spread in the individual measurements suggests no major difference from month to month, but the average value was found to decrease systematically from 44 km in May to 29 km in July. The horizontal phase speeds ranged from 14 to 42 ms^{-1} and within the limits of the measurements showed no month to month variation (average value = 28 ms^{-1}). The observed periods of the wave patterns (with the exception of June 20a event), were all less than 1 hour. The average period for May (29 min) and June (32 min; excluding June 20a) were similar, but somewhat higher than the mean for July (19 min), reflecting the lower average λ_x for this month. These parameters are typical of many of the short period wave events reported in the literature [e.g., *Reid*, 1986].

Table 1 shows that a distinct preference for wave propagation towards the north existed during this campaign (with 68% of the wave azimuths within ±40°N). In particular, it was found that none of the OH displays exhibited a significant westward component of motion suggesting that the gravity waves were subject to considerable directional filtering. *Taylor et al.* [1993] have shown that this anisotropy in wave propagation can be attributed solely to the effects of "critical layer" filtering of the gravity waves by background winds in the stratosphere and lower mesosphere. This result is used in the next section to help discriminate between potential convective sources.

5.2. Thunderstorm Comparison

The OH patterns were compared with the ensemble of "storm" maps to determine any obvious candidate sources. (Note, the term storm refers to any region of strong convection as determined from the radar summary charts.) Figure 2 shows four example maps comparing OH wave data with radar summary data; for two displays in May, and one each in June and July. Figure 2a shows the wave display of 14 May recorded at 06:00 UT superimposed on the storm map for 13 May at 22:35 UT (time interval 7.5 hours). On this occasion, meteorological cloud limited the measurements of the wave parameters to λ_x only. However, a sufficient number of images were recorded to determine the wave's direction of motion. Three storm centers existed within the search area at this time, two to the west/northwest of MRS and one to the east. The storm to the west at a range of about 650 km (shaded area) appears to be located well in both position and time to be a candidate source. The disturbance to the east was potentially more powerful (cloud tops reaching 11.3 km), but its position was not consistent with the observed direction of wave motion (indicated by the arrow). This fact, together with the result that the prevailing background winds tend to limit westward wave propagation, indicates that this storm was not the source of the waves. In total, five regions of strong convection occurred within the search area/interval on this occasion, but only one (the storm to the west) was suitably located to have been the source of the gravity waves.

Figure 2b shows a more extensive wave pattern recorded on 17 May at 07:30 UT (τ_{obs} = 15.5 min). The direction of motion of the waves (towards the northeast) is quite

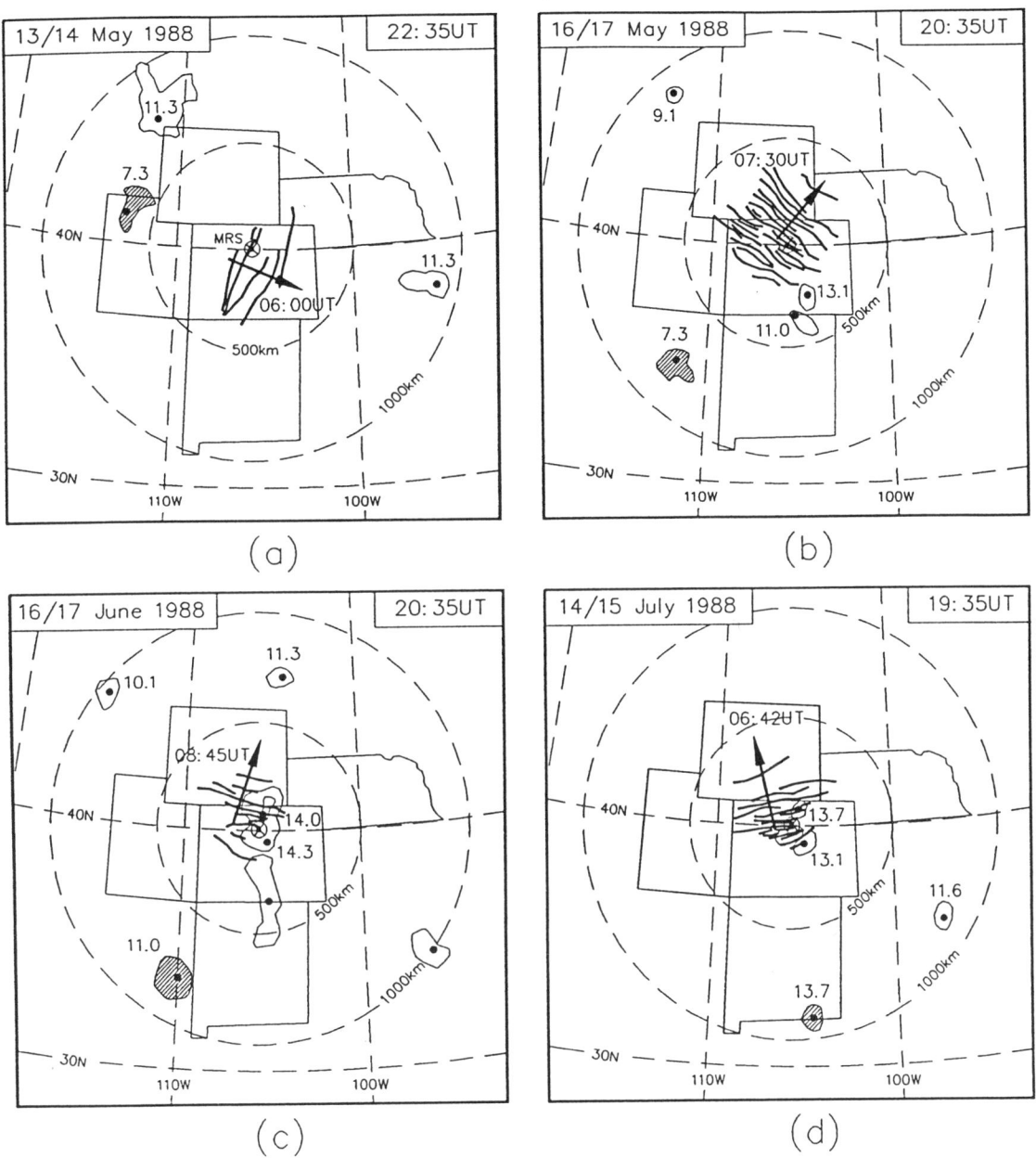

Fig. 2. Four example maps comparing the OH wave structure (bold lines) with radar summary data. The "storm" positions are indicated by the irregular areas marked with a black spot, the numbers give maximum cloud height in km. The dashed outer circle indicates the 1,000 km radius search area. In each example the "candidate storm" is marked by the hatched area. The arrows show the direction of motion of the waves.

different from the previous display. In this example, the radar summary map is plotted for 20:35 UT (time interval 11 hours) and shows two storm regions to the south, one to the northwest, and one to the southwest. This latter disturbance (shaded area) occurred at a range of ~800 km and was favorably oriented to be a candidate source. A second candidate source was also found on this occasion at a larger range (~1,000 km), but at a shorter average time interval (~7 hours). Figures 2c and 2d show similar situations for two more displays observed on 17 June and 15 July. On each of these nights several storms occurred within the source area/interval. Both wave displays exhibited north-

TABLE 2. Results of Thunderstorm Comparison Study for May

UT Date 1988	Maximum Height (km)	Range to MRS (km)	Time Interval (hrs)	Deviation From Wave Azi (±5°)
May 13	6.1	1400	8 ± 3	15
May 14	7.3	650	9 ± 1	0
May 16	12.2	450	10 ± 2	10
May 17	7.3	800	12 ± 2	5
	-	1000	7 ± 2	0
May 21	8.5	550	8 ± 2	20

ward progression and in each case more than one candidate source was identified.

A summary of the results of this study, for each month, is given in Tables 2, 3 and 4, which include information on the candidate storms: their altitudes, range from MRS and the average time intervals. This latter parameter is a measure of the mean time between the storm occurrence and the time of the wave observations listed in Table 1. As meteorological cloud often limited the available observing time this number is only an approximation for the actual time of separation. The deviation in azimuth of each of these candidate sources from the direction of wave propagation is also listed. Because of uncertainties in positioning individual storms (due to their evolution and motion), all storms within ± 20° of the nominal wave azimuth were considered as candidate sources. However, the majority of these disturbances (77%), occurred within ± 10°. As expected the number of candidate sources and their heights were observed to increase considerably from May to July.

6. DISCUSSION

On every occasion when OH wave measurements were possible (a total of 18 displays) there was always at least one storm favorably located in both space and time to have been the source. This is a remarkable result considering the observed distribution of the potential sources. During the course of the campaign it would not have been surprising to find one or two favorably located candidate sources by chance, given the selection criteria used. However, the large number of wave events that can be associated with regions of strong convection suggests that this correlation was most probably not due to chance alone. On a case study basis the data recorded in May provided the most critical test of this result as far fewer thunderstorms occurred during this month yet acceptable candidate storms were found on each of the five nights when wave structure was observed (see Table 2).

Candidate storms were found within the nominal 1,000 km radius search area for all but one of the 18 events. For the display of 13 May (τ_{obs} = 21 min) a disturbance located at a range of 1,400 km was identified. On this occasion the background winds may have substantially increased the path length of the gravity waves through the intervening atmosphere. For this reason several other potential candidate sources outside the nominal search area have been included in the summary Tables where appropriate. The mean time interval from storm occurrence to wave observation varied from about 3-12 hours and the ground distance separating the storm centers from MRS (over which the wave structures were measured) ranged from 200-1,400 km. In June and July more than one candidate storm was often identified. Although estimates of the ground range, time of flight, and deviation from the nominal wave azimuth may be used to help choose between these sources, considerable uncertainty would remain due to the unknown influence of the background winds. A better method of distinguishing between several potential sources would be to trace the path of the waves through the atmosphere for each event. Dr T.F. Tuan (University of Cincinnati) is currently developing a ray tracing model for a realistic atmosphere, based on climatological background winds and numerical tidal modes, to investigate further the candidate storms identified here. The

TABLE 3. Results of Thunderstorm Comparison Study for June

UT Date 1988	Maximum Altitude (km)	Range to MRS (km)	Time Interval (hrs)	Deviation from Wave Azi (±5°)
Jun 12	-	600	6 ± 3	10
	14.0	900	3 ± 2	15
Jun 17	11.0	750	10 ± 3	10
	12.1	200	7 ± 2	20
	-	600	5 ± 1	5
Jun 19	-	750	12 ± 1	5
	9.8	1150	11 ± 1	10
	-	600	5 ± 1	5
Jun 20a	12.2	300	12 ± 7	0
	11.6	900	5 ± 1	20
Jun 20b	12.2	300	9 ± 4	5
Jun 21a	13.1	800	11 ± 3	20
	11.9	750	12 ± 3	5
	12.8	1300	8 ± 3	10
Jun 21b	11.0	300	11 ± 3	25

TABLE 4. Results of Thunderstorm Comparison Study for July

UT Date 1988	Maximum Height (km)	Range to MRS (km)	Time Interval (hrs)	Deviation From Wave Azi (±5°)
Jul 10	-	800	10 ± 2	0
Jul 12	11.3	800	8 ± 3	5
	-	400	7 ± 3	0
Jul 15	15.2	1150	9 ± 2	5
	13.7	950	9 ± 2	5
	14.6	950	4 ± 2	0
Jul 16	17.1	700	7 ± 3	10
	11.3	800	7 ± 3	20
Jul 17	12.8	1000	11 ± 2	10
	17.7	750	8 ± 5	0
	12.2	300	9 ± 3	5
Jul 18	15.5	1200	7 ± 4	10
	16.2	800	7 ± 3	15
	13.1	300	7 ± 3	0

results of this modelling study will help significantly to quantify this apparent correlation.

Several researchers have published convincing evidence in support of the generation of short period gravity waves by thunderstorms [e.g., *Curry and Murty* 1974; *Balachandran* 1980; *Larsen et al.*, 1982]. However, evidence establishing the propagation of these storm-induced waves into the upper neutral atmosphere and the ionosphere is far less common [*Röttger*, 1977; *Taylor and Hapgood*, 1988]. This study indicates that thunderstorms, or convective regions associated with their development, are a potentially important source of short period wave energy reaching the upper atmosphere at this time of the year. Because cumulus convection associated with the development of thunder cells (typical radius ~10 km at mid-latitudes) produces strong vertical updrafts, thunderstorms are particularly well suited to the generation of short horizontal wavelength (a few tens of km) gravity waves. *Pierce and Coroniti* [1966] proposed penetrative convection of the thunder cell at the tropopause as one mechanism for generating waves with periods a few tens of minutes (dependent upon the temperature gradient at the tropopause), similar to those observed in the OH emission. The candidate storms identified in this study were not always the highest, most prominent disturbances present on any one occasion (see Tables) and their altitudes varied considerably over the three month period (ranging from 6 to 18 km). This suggests that many vertically developing, convective systems may be capable of generating gravity

azimuths agrees well with the hypothesis that waves were launched by storm systems that were subject to considerable "critical layer" directional filtering [*Taylor et al.*, 1993]. This result may be used to account for the apparent absence of wave structure associated with potentially powerful convective sources that often developed to the east of the optical site. However, on several occasions storms were also observed at azimuths which generally should not have been restricted by the background winds, and yet little evidence of wave structure associated with these potential sources was found. Indeed, on only two occasions (June 20 and 21) were more than one OH wave pattern discerned. This is a surprising observation and may be connected with the sensitivity of the TV system. Recent all-sky OH observations using a new generation of solid state (CCD) imagers indicate that it is not uncommon to detect several different wave patterns during the course of a night. However, the brightness and contrast of these displays can vary considerably. Thus it seems plausible that the Isocon camera used for this study recorded only the most prominent wave motions on any one occasion, and that gravity waves generated by other storms may have gone undetected.

The result of this qualitative study suggest a direct link between storm related convection and the appearance of short period OH wave structure, but should be viewed as preliminary. Other tropospheric sources, thought to be capable of generating gravity waves (e.g. fronts, depressions, jet streams and wind flow over topography), also existed during this campaign. The all-sky observations were made from a mountainous site, but orographically generated waves (stationary waves) can be discounted on most occasions as significant phase progression, of typically a few tens of ms^{-1} was observed in each case (with the exception of the June 20a display which moved at a speed of 5 ± 2 ms^{-1}). In-situ sources such as the breakdown of long period tidal-type waves may also have been present [*Taylor and Edwards*, 1991]. The near linear appearance of several wave displays suggests gravity wave generation by an extended source region (e.g a front) rather than a discrete "point-like" convective region, but this argument is conditional on the range of the source from MRS, which we have shown may be quite large, depending upon the wave period and the prevailing atmospheric conditions. A detailed investigation of the occurrence of these potential sources and their association (if any) with the OH wave events is in progress.

7. SUMMARY

Short period (<1 hour) gravity waves of varying scale sizes and phase speeds were imaged on nearly every night that the sky was clear during the three month campaign. The wave waves of period <1 hour and not just those storm cells reaching tropopause heights.

The anisotropy in the observed distribution of wave forms often appeared as coherent, quasi-linear patterns with lifetimes of a few hours. A striking result of this investigation is that at least one candidate storm was found for every occasion that OH wave structure was detected. The observed distribution and abundance of the storms does not indicate that this result was due to chance alone. These observations, although preliminary, support the hypothesis that thunderstorms may be an important and abundant source of short period gravity waves capable of reaching mesospheric heights. The characterization of such wave events (λ_x, v_x, τ_{obs}) and the investigation of seasonal variations in source anisotropy is of considerable importance for modelling studies of the global-scale circulation.

Acknowledgements. We are most grateful to the director and staff of the Mountain Research Station, University of Colorado for allowing us the use of their facilities for the all-sky measurements. We thank M.J. Hill and P. Mace for their considerable help during the campaign and E. A. Dewan for procuring large numbers of radar summary charts from the USAF Environmental Technical Applications Center (MAC). Support for this research was provided by the USAF Office of Scientific Research MAPSTAR program and by the USAF Phillips Laboratory SOAR program.

REFERENCES

Armstrong, E. B., The association of visible airglow features with a gravity wave, *J. Atmos. Terr. Phys., 44,* 325, 1982.

Baker, D. J., and A. T. Stair, Jr., Rocket measurements of the altitude distributions of the hydroxyl airglow, *Phys. Scr., 37,* 611, 1988.

Balachandran, N. K., Gravity waves from thunderstorms. *Mon. Weath. Rev., 108,* 804, 1980.

Booker, J. R., and F. P. Bretherton, The critical layer for internal gravity waves in a shear flow, *J. Fluid Mech., 27,* 3, 513, 1967.

Curry, M. J. and R. C. Murty, Thunderstorm-generated gravity waves. *J. Atmos. Sci., 31,* 1402, 1974.

Freund J. T., and F. Jacka, Structure in the 557.7 nm (OI) airglow, *J. Atmos. Terr. Phys., 41,* 25, 1979.

Fritts, D. C., and R. A. Vincent, Mesospheric momentum flux studies at Adelaide, Australia: Observations and a gravity wave-tidal interaction model, *J. Atmos. Sci., 44,* 605, 1987.

Hapgood M. A., and M. J. Taylor, Analysis of airglow image data, *Ann. Geophys., 38,* 805, 1982.

Hines, C. O., Internal atmospheric gravity waves, *Can J. Phys., 38,* 1441, 1960.

Hines, C.O., On the nature of travelling ionospheric disturbances launched by low-altitude nuclear explosions, *J. Geophys. Res., 72,* 1877, 1967.

Hines, C. O., A possible source of waves in noctilucent clouds, *J. Atmos. Sci., 25,* 937, 1968.

Hung, R. J., T. Phan, and R. E. Smith, Coupling of ionosphere and troposphere during the occurrence of isolated tornadoes on November 20, 1973, *J. Geophys. Res., 84,* 1261, 1979.

Larsen, M. F., W. E. Swartz and R. F. Woodman, Gravity-wave generation by thunderstorms observed with a vertically-pointing 430 MHz radar, *Geophys. Res. Lett., 9,* 571, 1982.

Pierce, A. D., and S. C. Coroniti, A mechanism for the generation of acoustic gravity waves during a thunderstorm formation, *Nature, 210,* 1209, 1966.

Reid, I. M., Gravity wave motions in the upper middle atmosphere (60-110km), *J. Atmos. Terr. Phys., 48,* 1057, 1986.

Röttger, J., Traveling disturbances in the equatorial ionosphere and their association with penetrative cumulus convection, *J. Atmos. Terr. Phys., 39,* 987, 1977.

Takahashi, H., P. P. Batista,, Y. Sahai, and B. R. Clemesha, Atmospheric wave propagation in the mesopause region observed by the OH (8,3) band, NaD, O_2 At (8645 Å) band and OI 5577 Å nightglow emissions, *Planet. Space Sci. 33,* 381, 1985.

Taylor, M. J., M. A. Hapgood, and P. Rothwell, Observations of gravity wave propagation in the OI (557.7 nm), Na (589.2 nm) and the near infrared OH nightglow emissions, *Planet. Space Sci., 35,* 413, 1987.

Taylor, M. J., and M. A. Hapgood, Identification of a thunderstorm as a source of short period gravity waves in the Upper atmospheric nightglow emissions, *Planet. Space Sci., 36,* 975, 1988.

Taylor, M. J., and R. Edwards, Observations of short period mesospheric wave patterns: In situ or tropospheric wave generation?, *Geophys. Res. Lett., 18,* 1337, 1991.

Taylor M. J., E. H. Ryan, T. F. Tuan and R. Edwards, Evidence of preferential directions for gravity wave propagation due to wind filtering in the middle atmosphere, *J. Geophys. Res., 98,* 6047, 1993.

M. J. Taylor, Science Division, Space Dynamics Laboratory, Utah State University, Logan, UT 84322-4437.

V. Taylor, 91 Quail Way, Logan, UT 84321.

R. Edwards, Physics Department, University of Southampton, Southampton, SO9 5NH, U.K.

Climatology of Polar Mesospheric Clouds: Interannual Variability and Implications for Long-Term Trends

Gary E. Thomas

Laboratory for Atmospheric and Space Physics and Department of Astrophysical, Planetary and Atmospheric Sciences University of Colorado, Boulder, Colorado

The climatology of Polar Mesospheric Clouds (PMC) is studied using data from two ultraviolet spectrometers over the period 1978-1986. The basic quantities measured are data from the UVS instrument on board the Solar Mesosphere Explorer (SME) satellite, the scattered limb albedo at 265 nm; and from the SBUV instrument on the Nimbus-7 satellite, the nadir albedo at 273.5 nm. The statistical distributions of albedo over each PMC summer season (both north and south), and over the latitude range 50° to 75°, are derived. Empirical climatological measures of the albedo distribution are: (1) the 'half-power' value of the limb albedo, A_o, the median value dividing the clouds into equal numbers of dimmer and brighter albedo values; (2) the slope of the nadir albedo distribution β_n; and (3) the cumulative number of clouds $g(A)$ exceeding a certain albedo A, either at the limb or at the nadir. Solar cycle variations from season to season are examined by studying the correlations of the climatological measures with seasonally-averaged solar Lyman-α flux. To explain the nearly two orders of magnitude range of PMC albedo our microphysical cloud model requires seasonally-averaged mesopause temperature variations δT (in the presence of clouds) to be $\approx$ 6K in the north and 3.5K in the south. The estimated seasonal mean mesopause temperature in the presence of clouds averages 126K in the north, and 130K in the south.

Factor-of-ten changes in the cumulative cloud number $g(A_n > 1 \times 10^{-5})$ occur over the solar cycle. The results suggest the following: first, that temperature variability is the dominant factor in setting the large brightness of PMC observed within a given season. Second, the longer-period (11-year) changes in the number of bright clouds over the solar cycle are controlled by the mean water vapor amounts resulting from the changing UV flux (mainly Ly-α). The inferred higher temperatures and lower temperature variances in the south are consistent with recent reports of less gravity wave activity and smaller meriodional wind speed in the Antarctic than in the Arctic; and also the apparent absence of Polar Mesospheric Scattering Echoes (PMSE) in the south. These considerations suggest that *both* solar-cycle and secular changes of noctilucent cloud occurrence over the past thirty years are manifested in the number of the brightest clouds. These longer-term changes in turn appear to be driven primarily by long-term variability of middle atmosphere water content.

INTRODUCTION

The presence of clouds high in the mesosphere has been known for over a century. Unlike more subtle sky phenomena like airglow, and the faint apparitions of zodiacal light and the gegenschein, they can occasionally provide a dramatic spectacle. During these displays, an observer north of about 50° latitude and south of the arctic circle can witness silvery-blue cloud features of astonishing lustre and brilliance. Typically located low on the horizon above the sunset arch, they extend upward to the arc of the earth's shadow where they vanish into the blackness of the night sky. Their most common form is that of wispy cirrus-like clouds, rippled by complex waves and billows. The ordinary process of scattering of sunlight from small particles suspended against a deep blue backdrop can be transformed in the mind's eye to what an early observer described "as that which shines from white phosphor paint". Hence the term 'night-luminous' clouds, or 'noctilucent clouds' (NLC).

The subject of my paper is a phenomenon which, though probably closely related to NLC, has perhaps never been seen from the ground. Polar Mesospheric Clouds (PMC) have only been observed from earth orbit, and for reasons related to the specific objectives of the various satellite missions, these have always been made during daylight hours. To a ground observer the brightness of the sky background obscures these optically-thin clouds, and no coincident measurements from the ground have been possible to date. Almost certainly the same general phenomenon is involved in both PMC and NLC, that of small ice particles formed at the cold mesopause at 86-88 km heights. I will be considering mainly the satellite data in the present study, but the conclusions drawn should apply to mesospheric clouds, no matter how they are viewed or where they are located. (*Thomas*, [1991] discusses this point in more detail.)

Among the advantages of observing clouds from space is that they are detectable even in full sunlight provided the background radiance can be suppressed. There are two ways to achieve this when viewing from space. The first way is to view the earth's limb where the background scattering is very low as a result of the low atmospheric density at the height of PMC (83 km). (An adequate optical baffling is essential to eliminate stray-light contributions.) The second way is to observe the clouds at a wavelength of sunlight which scatters efficiently from small particles and which is absorbed strongly by the atmosphere beneath the clouds. The ultraviolet wavelength range between about 250 and 280 nm is ideal for this purpose, situated in the very strong Hartley absorption band of ozone where the earth's albedo is extraordinarily low. The Ultraviolet Spectrometer (UVS) on board the Solar Mesosphere Explorer satellite (SME) uses both the above advantages [*Thomas*, 1984]. The nadir-viewing Solar Backscatter Spectrometer (SBUV) on board the Nimbus 7 satellite uses the same ultraviolet wavelengths as the UVS but, as far as mesospheric cloud detection is concerned, does not have the advantage of limb pointing. The pioneering measurements by *Donahue et a.* [1972] were made by a visible-light airglow photometer on board the OGO-6 satellite, which possessed only the single advantage of limb viewing.

The suppression of sky background allowed all three of the above-mentioned satellite experiments to view PMC over their entire domain of existence in space and time. The space era opened up the known cloud zone from a narrow band of latitudes (50°-65°) and local times (10 PM to 2AM) to a domain extending (in principle) over

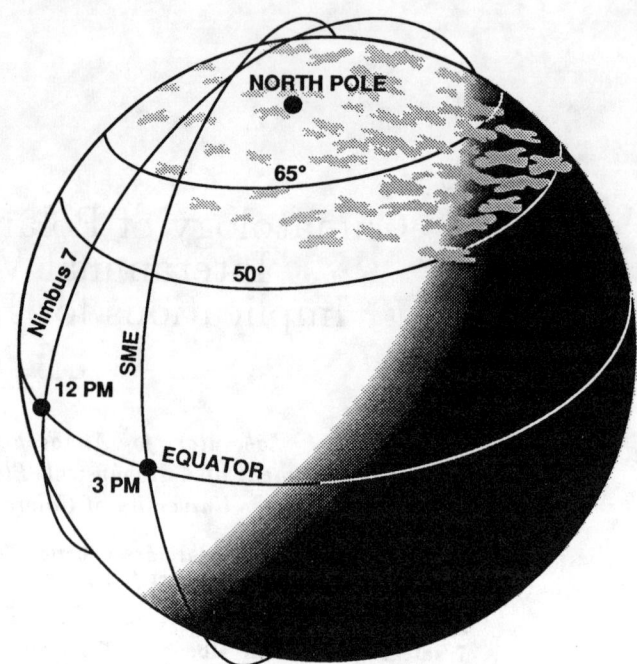

Fig. 1. Schematic appearance of mesospheric clouds as seen from space at middle UV wavelengths. The nightside is shown as a dark crescent. The variation of PMC occurrence with local time within the NLC observing zone (between 50° and 65° latitude) is hypothetical. Shown also are the locations of the orbits of the SME and Nimbus 7 spacecraft, and the local times of their node crossings.

the entire polar cap and all local times during summer. The timing of the 'cloud season' is tied to the summer solstice, in both polar caps, presumably due to the occurrence of supersaturation conditions during that time of year [*Thomas and Olivero*, 1989]. Our current understanding of the physics of ice-crystal formation at the cold summertime mesopause leads us to believe, as *Donahue et al.* [1972] first put it 20 years ago, that NLC are the "ragged edge" of a much more extensive polar cap phenomenon. This concept is based on the fact that high-latitude summertime mesopause (HLSM) temperatures are the lowest on the planet (in the range 110-140K). Figure 1 illustrates how mesospheric clouds might be distributed over the north polar regions, if the entire earth could be imaged at middle UV wavelengths. Figure 2 shows a four-year (plus) times series of PMC data acquired by the SME mission. This figure shows that the equatorward cloud boundary in the SME data is at about 60° latitude, which may seem hard to understand, given that the NLC observing zone is 50-65°. This may be a local time effect due to the tidal forcing of temperature [*Jensen et al*; 1989; *Thomas*, 1991] which causes more and brighter clouds near the evening terminator as illustrated in Figure 1.

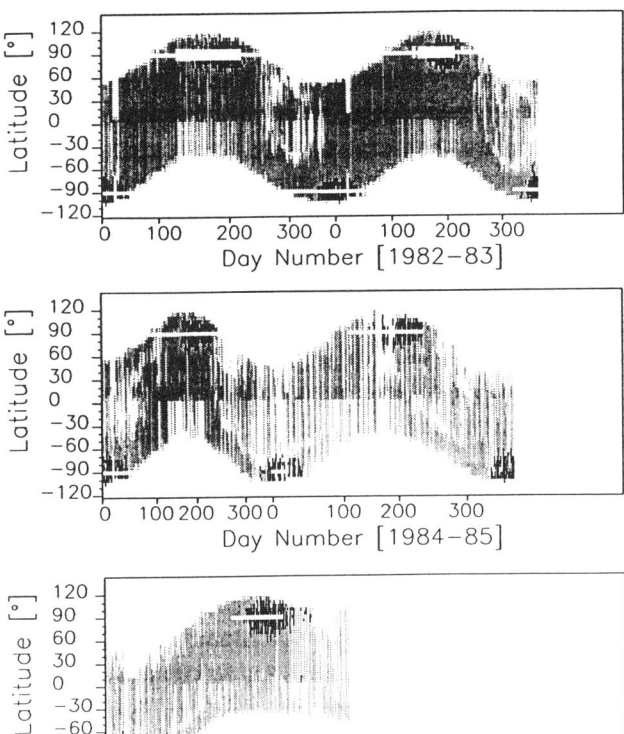

Fig. 2. Summary of SME data coverage (grey vertical stripes) with latitude and day number. Only the daytime portions are shown. Latitudes greater than 90° indicate coverage on the AM side of the sun-synchronous orbit. The local time of the node crossing gradually changed from 3PM early in the mission to 4PM in 1986. The darker stripes show presence of PMC with limb albedos greater than 1×10^{-3}. Note missing coverage of the 80-85° N and 95-100° N regions in 1982 and 1983. The horizontal scales are of variable length because of the varying number of orbits for which data were available. Note in particular the high amount of coverage in early 1986.

Several recent reviews provide more background information on mesospheric clouds [*Gadsden and Schröder*, 1989; *Thomas*, 1991; 1993; and *Avaste*, 1993]. In this paper I will focus on the subject of long-term variability of mesospheric cloud activity. In particular, I begin with two questions: are mesospheric clouds undergoing long-term changes? If so, is it possible to understand this behavior in terms of climate changes brought on by human activities? These ideas have been raised recently as a result of two developments. First, the apparently abrupt beginning of NLC sightings in 1885 suggested to us [*Thomas et al.*, 1989] that the existence of NLC came about as a result of the century-long rise of atmospheric methane CH_4. This molecule is oxidized to water vapor in the middle atmosphere, and is known to have doubled its concentration since pre-industrial times. Second, increasing observations of NLC since the 1960's are evident in reports from two independent observing networks, one in northeastern Europe and the other in the USSR [*Gadsden*, 1990]. It is important to point out that NLC observations provide us with the longest record available (since 1885) of possible changes in the middle atmosphere. (A possible exception are wintertime stratospheric mother-of-pearl cloud sightings *Stanford*, 1974; *Stanford and Davis*, 1974]. Unfortunately, despite their earlier discovery in 1870, there are far fewer numbers of recorded sightings of mother-of-pearl clouds than of NLC.)

My approach in this paper will be first to analyze interannual changes of PMC which have occurred during the last decade of space observations. For this purpose the SME and Nimbus 7 PMC data are superior to NLC data in that the satellite observations are well-calibrated, essentially free of observing gaps, immune to changes in lower atmospheric transparency, and apply to a greater fraction of the polar cap. Second, I will compare these data with known changes in solar UV irradiance over that same period. Finally, I will describe a microphysical theory of PMC albedo based upon current knowledge of the physics of ice formation, assuming that variability of PMC within a season is due to variability of one of two atmospheric 'forcings', either mesopause temperature or water vapor amount. The parameters of the theory are related to the climatological factors derived from the satellite data. Interannual changes in these parameters appear to depend upon solar activity. From the expected atmospheric changes over a solar cycle I conclude with some speculation as to how these results relate to possible longer-term variability of mesospheric clouds.

PMC CLIMATOLOGY FROM THE SME AND NIMBUS SATELLITE DATA

Details concerning the UVS and SBUV instruments, the orbits of the SME and Nimbus 7 spacecraft, and the type of data collected by these two instruments are described for SME by *Thomas* [1984], *Thomas and McKay* [1985], *Olivero and Thomas* [1986], and *Thomas and Olivero* [1989]; and for Nimbus 7 by *Heath et al.* [1975] and *Thomas et al.* [1991]. Both these instruments were devoted to measuring ozone—however I will be concerned here with their serendipitous measurements of PMC presence. The basic quantity of interest is the brightness (or more precisely, albedo which is used interchangeably with brightness) of UV sunlight scattered by cloud layers as seen at the limb (for SME) or in the nadir (for Nimbus 7). My goal is to describe the statistical ensemble of measurements over an entire season. The two PMC data bases provide a total of 12 complete PMC seasons extending from the northern summer of 1978 to the southern season 1986-87. Unfortunately Nimbus 7 provided only partial seasonal coverage in the North (1982, 1983 and 1985). I will first describe the SME data base, which includes 9 seasons in all.

I will use a newly-analyzed SME data base (1982-1986) which is superior to earlier PMC data sets in that it references the scattered radiance to that of the background atmosphere. Thus temporal changes in absolute sensitivity are automatically removed. Details of the inversion algorithm are found in the references by *Clancy and Rusch* [1989] and *Thomas and Clancy* [1994]. The derived quantity is the *scattering ratio*, commonly used in lidar studies and defined as the scattered light from the cloud divided by the scattered light from the atmospheric background, both applying to the same atmospheric volume. This ratio is available at two wavelengths (265 and 296 nm) at intervals of 5° in latitude and 3 km in height (up to 95 km). In this paper I will report only the 265 nm data, since the second channel contains largely redundant information. The scattering ratios have been averaged over 6 consecutive spacecraft spins (limb scans), corresponding to a motion of the spacecraft through 5° of latitude. This acknowledges the basic limitation of limb viewing which, because of the long intersection of the line of sight within the mesopause region, effectively averages over several hundred kilometers. Nominally, the pointing direction is along the orbit track. An individual data point will apply to the scattering ratio averaged over a rectangular area which is 550 km along the orbit track, and about 35 km normal to the track. The pointing direction slowly changes with time from along the orbit plane to an orientation at small angles with this plane. This circumstance helped extend the SME latitudinal coverage closer to the geographic poles. In this paper I will use the *directional limb albedo*, $A_l(\lambda) = 4\pi I_\lambda^l/F_\lambda$, where I_λ^l is the limb intensity and F_λ is the extraterrestrial solar flux at wavelength λ. I will use the symbol A_l for this quantity whose name I will shorten to simply *limb albedo*. $A_l(\lambda = 265$ nm) is obtained from the scattering ratio by multiplication by the background sky limb albedo at 83 km tangent height (1×10^{-3} at 265 nm). This allows the reader to compare with our previous publications.

Figure 2 provides an overview of the new three-year SME data base showing the occurrence of clouds for every orbit, and for every 5° of latitude. Latitudes greater than 90° indicate the AM (descending) portion of the sun-synchronous orbit (inclination=97°). Clouds brighter than 25% of the sky background may be distinguished from background by the cloud detection algorithm. This figure illustrates the seasonal variation of the latitude of the terminator, marking the boundaries of the daylit regions. Note that the 80-100° zone of latitude is not covered in the north in 1982 or 1983, when the UVS pointed along the orbit plane. Note also the absence of clouds at sub-polar latitudes, a point which contradicts the visual reports by Soviet cosmonauts of NLC at subtropical latitudes [*Vasilyvev et al.*, 1987; *Avaste*, 1993].

In Figure 3, SME data at 265 nm are combined within the 1983 North and 1982 South seasons to form statistical distributions $f(A_l)$ of 265 nm cloud limb albedo, A_l. The data apply to the 83 km altitude bin, where the scattering ratios are largest. This is similar to previous figures [*Olivero and Thomas*, 1986; *Thomas and Olivero*, 1989] but is derived from the new data base. The number of clouds within every bin of A_l-values has been divided by the total number of observing opportunities in the entire season. A PMC season is roughly defined as the domain from 50° to 80°, for both the ascending and descending parts of the orbit, and from 30 days before summer solstice to 90 days after summer solstice. Actually in all the distributions presented here, I have masked out for every season all latitudes and days which are not common to each data set. This was necessary to make meaningful interannual comparisons. In addition I want to compare these results with the Nimbus-7 data which was limited to latitudes less than 80°, and so none of the distributions refer to latitudes greater than 75° (less than 105° on the AM side). Figure 4 shows the cumulative directional albedo distributions for all seasons of SME data, defined by

$$g(A_l) = \int_{A_l}^{\infty} dA_l' f(A_l'). \tag{1}$$

The upper limit for A_l is taken to be infinite for mathematical convenience. The actual upper limit is ≈ 0.2, but at these values $f(A_l)$ makes a negligible contribution to the integral.

Turning to the Nimbus-7 data base, these refer to values of the nadir albedo $A_n(\lambda)$ versus latitude and day number, derived from the 12 spectral channels of earth albedo of the SBUV instrument [*Thomas et al.*, 1991]. The excess spectral albedo for nadir viewing is defined as I_λ^n/F_λ where I_λ^n is the nadir intensity attributable to PMC scattering. I will refer to this quantity evaluated at 273.5 nm as the *nadir albedo*, A_n. The effective SBUV field of view is 200 x 200 km, about 5-6 times larger than that of the UVS. The Nimbus-7 data are available for about 16 orbits per day, or about 32 passes over the continuously-sunlit polar cap. This contrasts with

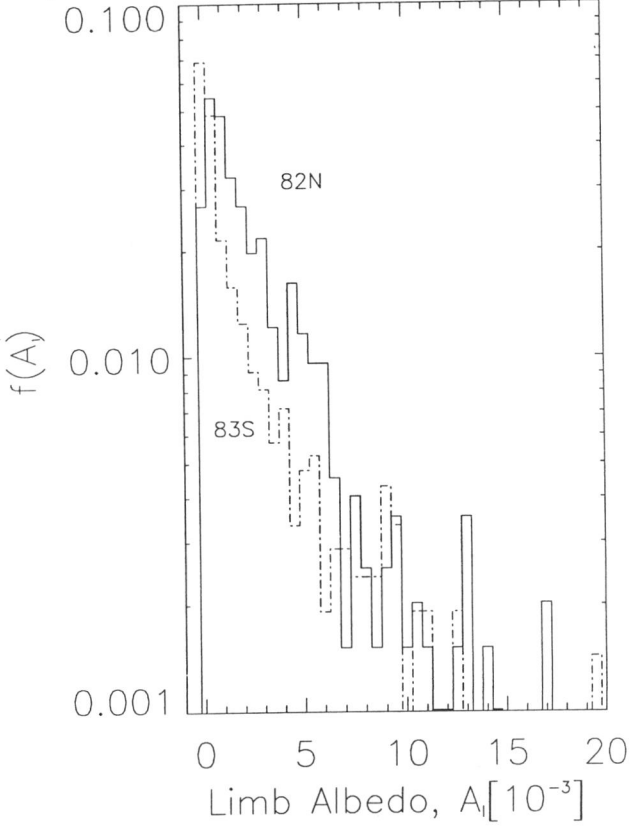

Fig. 3. Brightness distributions of PMC limb albedo derived from the SME data base described in *Thomas and Clancy* [1993]. Limb albedos are grouped within bins of width 5×10^{-4}. Shown are the distributions for the 1982 north (solid curve) and 1983 south (dashed curve) PMC seasons, for latitudinal distances of $10°$ or more from the pole.

the smaller spatial coverage of SME, about 3 to 4 tape-recorded orbits per day, or 6 to 8 total passes over the polar cap. (See *Thomas and Olivero* [1989] for a complete discussion of the longitudinal coverage of SME.) The slant viewing at the limb by the UVS enables clouds to be effectively 'magnified' in brightness by a factor of about 100 [*Thomas and McKay*, 1985]. Combined with the differing backgrounds and counting statistics of the two instruments, this means that the UVS data contain clouds about 30 times dimmer than the weakest cloud discernable in the nadir by the SBUV instrument. The result is that about 10 times more clouds are observed by SME in a given orbital pass. However, the greater number of tape-recorded orbits of Nimbus 7 means that the total number of clouds detected within a season is roughly the same for both data bases. The two data sets are complementary in that SME data provide better statistics for most clouds, whereas the Nimbus 7 data provide better statistics for the brightest clouds. This will be important in comparing the cloud brightness distributions between the two data sets.

The emphasis in our earlier paper [*Thomas et al.*, 1991] was a comparison of seasonally-averaged Nimbus-7 data over the 8-year period 1978-1986. These data showed a pronounced increase of bright cloud occurrence frequency with decreasing solar activity, particularly in the south. Figure 5 shows the frequency distribution of nadir albedo A_n for 8 successive southern seasons, and for three northern seasons. Best-fit straight lines are shown in this semi-log plot to indicate that each distribution follows closely an exponential fall-off. The values of A_n are quite different than those of A_l measured by SME for several reasons: (1) the SBUV instrument observed PMC in the nadir, which causes the clouds to be about a hundred times dimmer, as mentioned earlier; (2) the factor 4π is missing from the definition of SBUV albedo; (3) the scattering angles are different. The scattering angles for SBUV (about $70°$) are intermediate between the northern SME value (about $135°$) and the southern SME value (about $45°$), and this brings in an additional factor between 2 and 4 depending upon the cloud particle scattering phase function; and (4) the filling factors for each field of view are different. The slight difference of wavelengths is unimportant.

Is there an overall number that characterizes PMC activity within a given year? An obvious one is the total number of clouds viewed in a given season, or better still, the overall frequency of occurrence of clouds of all brightness values $g(0)$. However $g(0)$ is dominated by the number of dim clouds, a population that is not well determined since it depends upon the details of the cloud selection algorithm. In particular it depends upon how one determines the cutoff between dim clouds and random fluctuations in background albedo [*Thomas and Clancy*, 1994], and this changes from year to year. A more satisfactory index would involve the *shape* of the albedo distribution. For example, the steeper the slope of the distribution, the fewer bright clouds there are relative to dim clouds. For the Nimbus-7 data (Figure 5) the distributions are approximately exponential, in contrast to the SME data. As we will see later, this is a consequence of the fact that SBUV applies to brighter clouds. The climatological measures that proved to be most useful, both in comparing the two data sets and also in studying solar cycle variations, are the cumulative distributions $g(A_l)$ and $g(A_n)$ at fixed values of albedo.

I now consider the relationship of the Nimbus-7 and SME distributions. Since the two spacecraft observed both polar caps during the same 1982-1986 period, albeit at different local times (see Figure 1), we might expect a nice proportionality between the two data sets. Suppose

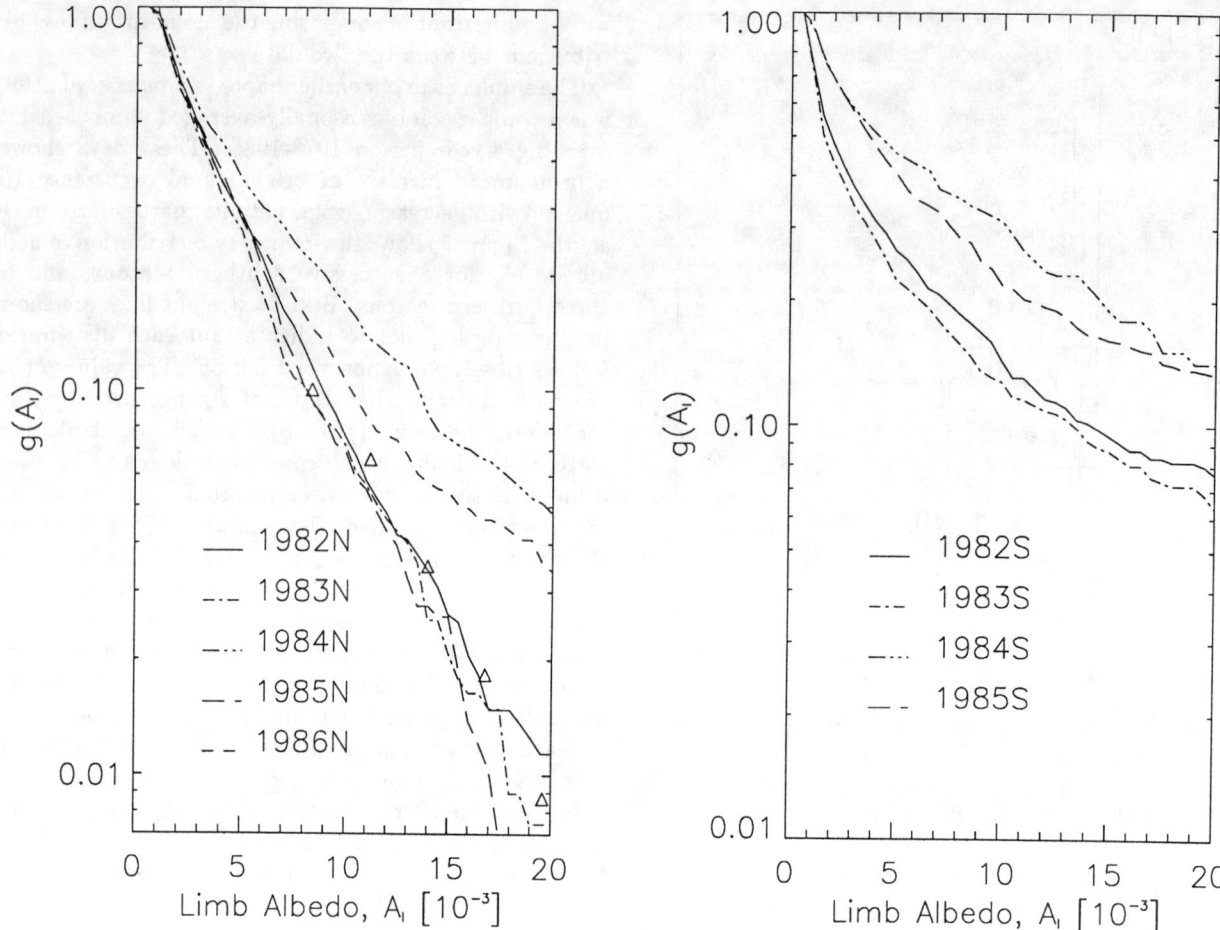

Fig. 4. Cumulative distributions of PMC brightness (defined by equation 1 in text) for all SME seasons versus limb albedo. The points (triangles) are the scaled values of the cumulative distribution $g_{sc}(A_l)$ derived from the SBUV cumulative distribution $g(A_n)$ in 1983 north (see text for the scaling procedure).

a single cloud were to be observed by both the UVS and the SBUV instruments. Making the somewhat dubious assumption that clouds are always spatially uniform over both fields of view, the ratio $R = A_l/A_n$ may be used to obtain the scaled SBUV distribution,

$$f_{sc}(A_l) = f_{SBUV}[A_n(A_l)]\left|\frac{\partial A_n}{\partial A_l}\right| = R^{-1} f_{SBUV}(A_l/R)$$

Adopting a (normalized) exponential approximation, $f_{SBUV} = \beta_n e^{-\beta_n A_n}$, the scaled distribution is written

$$f_{sc}(A_l) = \frac{\beta_N}{R} e^{-(\beta_n/R) A_l}$$

From equation 1, the scaled cumulative distribution is simply

$$g_{sc}(A_l) = e^{-(\beta_n/R) A_l}$$

$g_{sc}(A_l)$ therefore has the slope β_n/R, and can now be overplotted on the corresponding SME distribution for that year, and compared for consistency. Since the SME observing geometries differed between north and south poles, R depends upon which pole is being considered. Unfortunately R is not well determined because truly coincident measurements were not available. By trial and error I found that a value of $R=1400$ gave satisfactory agreeement of the *slopes* of the north SBUV distribution with the north SME distribution. It is also consistent with that expected theoretically for the albedo ratio, if the SBUV filling factor is between 0.3 and 0.6. An example of a scaled SBUV distribution is shown in Figure 4. Note that the faintest detectable SBUV cloud ($A_n = 5 \times 10^{-6}$) is a rather bright cloud as seen by the UVS ($A_l = 7 \times 10^{-3}$). Comparing cumulative distributions shows that the SBUV data are sensitive to only about 10% to 25% of the population of clouds visible to

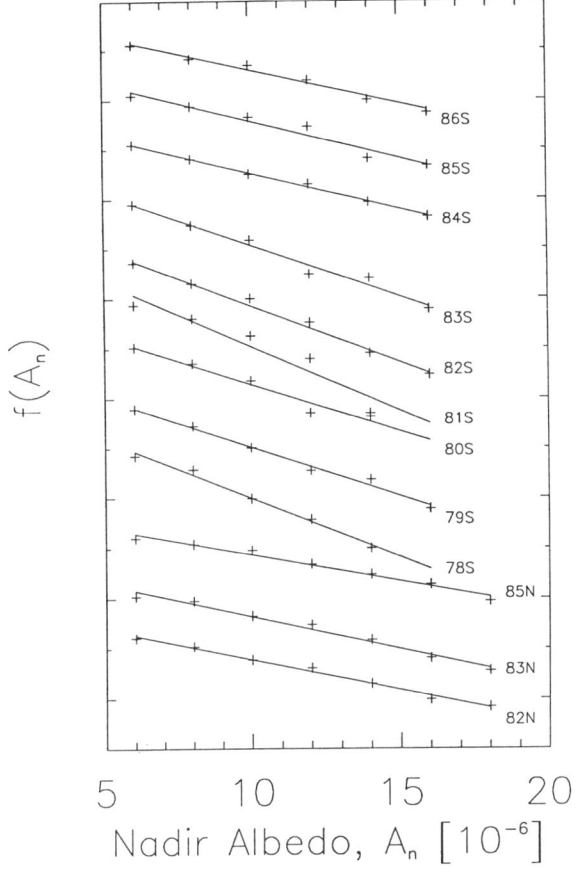

Fig. 5. Brightness distributions of SBUV nadir albedo data $f(A_n)$ versus A_n. The top nine curves are for successive south seasons from 1978 (78S) to 1986 (86S). The bottom three curves are for three north seasons, 1982, 1983 and 1985. Straight lines have been least-square fitted through the points in this semi-log plot. Each successive curve, starting from the bottom, has been multiplied by 10, for convenience of presentation.

the UVS (see Fig. 4). I cannot define a single value of β for the SME distributions because of their non-exponential behaviors. The above exercise has shown that the two distributions are consistent, differing only in their very different sensitivities to albedo.

Why is the UVS apparently not suited to bright cloud detection, when it can detect clouds ten times dimmer than the SBUV? In fact it is not a question of sensitivity, since very bright clouds were occasionally present in the SME data (up to $A_l = 0.2$). However their rare occurrences meant that the statistics are poor at high values of A_l (as shown in Figure 2). Since the SBUV has a greater chance of seeing bright clouds (because of its larger sampling rate) it provides the more suitable data set for the bright (but rare) clouds. Another issue is the the *curvature* of the distributions. While the SBUV data can be fitted to simple exponentials, the SME data exhibit a more curved behavior in semi-log plots. An explanation for this difference will emerge from a comparison with the theoretical distributions. Before considering this question, I will consider how the SME and Nimbus-7 distributions are correlated with solar activity.

CORRELATION WITH SOLAR LYMAN ALPHA FLUX

The 1978-1986 period covered the maximum solar activity of Solar Cycle 21, occurring in 1980-81. It continued to solar minimum activity in 1986-87. Except for 1978 to 1981 there are available accurate measurements of the solar ultraviolet (UV) flux throughout this period, obviating the need to use proxy indicators of UV flux, such as the radio 10.7 cm flux or sunspot number. One of my objectives will be to relate PMC climatological indices to the solar UV flux. Physical conditions near the mesopause are affected primarily by the solar Lyman-alpha (Ly-α) flux at 121.6 nm [*Jensen*, 1989; *Garcia*, 1989]. For this reason I will compare PMC activity only with Ly-α flux, measured directly by the solar instrument on board the SME spacecraft [*Barth et al.*, 1990] from late 1981 to 1987. Prior to 1982, this quantity is determined by the core-to-wing ratio of the MgII line (near 280 nm) measured by SBUV. *Thomas et al.* [1991] found that for PMC-seasonally averaged quantities, this proxy is an excellent indicator of Ly-α flux. I will define the seasonally-averaged flux $F(\text{Ly-}\alpha)$ as the averaged daily values from the SME solar data base for the period from 30 days prior to summer solstice to 60 days afterward. These values are listed for each season in Table 1.

In order to isolate solar cycle changes in each data set, it is convenient to divide the value of each cumulative distribution $g(A_l)$ by the value of $g(A_l)$ (both for fixed A_l) for a solar minimum year for which the solar Ly-α flux was at or near 2.6×10^{12} photons cm^{-2} s^{-1} nm^{-1}. This procedure allows the determination of the *ratios* of the cumulative cloud number for any level of solar activity and for both data sets. All four data sets (north SME, south SME, north SBUV and south SBUV) are ratioed to their particular solar minimum season in Figure 6. The frequency of bright clouds decreased significantly from solar minimum years to solar maximum years. This is particularly true for the SBUV data. I will discuss the smooth curves in Figure 6 after describing the statistical theory of cloud brightness distribution.

THEORY OF THE BRIGHTNESS DISTRIBUTION OF POLAR MESOSPHERIC CLOUDS

The currently-accepted theory of mesospheric clouds can be traced back to *Wegener* [1912]. However only

Table 1. Statistical indices for each PMC season

Year	$A_o(\times 10^{-3})$	$\beta_n(\times 10^5)$	F(Ly-α)	$T_o(K)$	w_o(ppm)	$\delta T(K)$	δw(ppm)
South							
1978		5.4	3.8				
1979		4.5	4.0				
1980		4.4	3.9				
1981		5.9	3.9				
1982	1.9	5.2	3.6	130.5	4.4	3.5	2.1
1983	1.59	4.8	3.1	130.7	4.2	1.5	2.0
1984	4.2	3.3	2.7	128.8	5.4	4.3	3.4
1985	3.7	3.5	2.6	129.1	5.2	3.9	2.0
1986		3.1	2.6				
North							
1979			3.7				
1980			3.8				
1981			3.7				
1982	2.8	2.8	3.6	126.3	7.2	5.4	4.1
1983	2.6	3.0	3.4	126.5	6.9	4.9	3.9
1984	3.3		3.0	125.4	7.9	7.2	5.1
1985	2.9	2.4	2.7	126.3	7.3	5.2	3.6
1986	2.7		2.5	126.4	7.1	6.4	4.8

Statistical indices for each PMC season (defined in text) and seasonally-averaged solar Ly-α flux $\times 10^{12}$ photons cm^{-2}s^{-1}nm^{-1}. The values of A_o are the values of the 'half-power' point albedos for each SME cumulative albedo distribution $g(A_l)$. The values of β_n are the slopes fitted through the SBUV albedo distributions $f(A_n)$ (see Figure 5). The values of δT and δw are taken from the best-fits of the theoretical (error-function) with the SME cumulative albedo distributions $g(A_l)$ and equation 8. T_o and w_o are derived from the relations $T_o(A_o)$ and $w_o(A_o)$ from Figures 7 and 8. Missing values in tables are a result of that quantity not being derivable from the data (for example A_o cannot be derived from SBUV data), or because those data were not available for that season.

with the advent of the computer has it become possible to conduct a realistic simulation of ice particle formation including nearly all relevant physical processes. This effort was pioneered by *Turco et al.* [1982], and was continued and refined by our group [*Jensen et al.*, 1989; *Jensen*, 1989]. The theory includes the processes of dust coagulation and nucleation, deposition, sedimentation and sublimation. The model simulates the life history of three populations – dust particles, ice particles with dust cores, and water vapor concentration. The ice-particle size distribution and the scattered brightness of a cloud due to solar illumination are calculated as a function of time, beginning with a slow inception of super-saturation conditions. The maximum limb albedo at 265 nm occurs about 1 to 2 days in the simulation, depending upon atmospheric conditions [*Jensen*, 1989].

The principal atmospheric 'forcings' of PMC brightness are water vapor concentration and the mesopause temperature. The former is obvious since water is the basic 'fuel' for ice particles – the more water the larger the particles, and the larger the scattering cross-section. The mesopause temperature is the region of highest super-saturation, and in the model is the height at which ice particles are nucleated. However, both the observations and the theory show that the actual cloud height is 4 to 5 km *lower* than the mesopause height [*Thomas and Clancy*, 1994]. The cloud height lies immediately above the height of saturation where the temperature is about 140K. Thus the controlling factor in the model is the mesopause temperature, not the cloud temperature. I will ignore quantities of lesser importance which include the upward wind speed, the cloud particle shape (which affect the terminal velocity), the initial meteoric 'smoke' distribution and the 'contact angle' affecting the efficiency of nucleation.

The model dependences of cloud albedo on mesopause temperature T and water vapor at 70 km are shown in Figures 7 and 8 for the average scattering angles appropriate to the north and south pointing directions of the UVS. Note that the range of predicted limb albedos approximates that observed in the SME data. The available measurements of mesopause temperature and water vapor were reviewed in [*Thomas*, 1991]. *Lübken and Von Zahn* [1991] reviewed high-latitude rocketsonde measure-

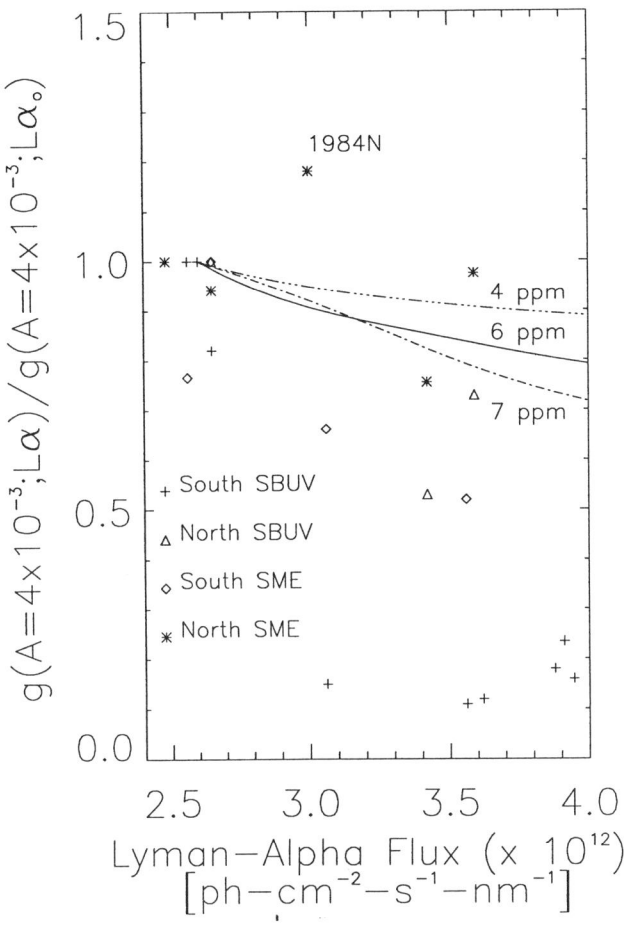

Fig. 6. Cumulative brightness distributions for four different groups of data: south SBUV, north SBUV, south SME, and north SME versus the seasonally-averaged solar Ly-α flux for that season. Each separate group has been divided through by its value $g(A_l = 4 \times 10^{-3};\text{Ly-}\alpha)$ for the solar minimum year for which the solar Ly-α flux was closest to Ly-$\alpha_o = 2.6 \times 10^{12}$ photons cm^{-2}s^{-1}nm^{-1}. (These points are easily recognized by their value of unity.) The three smooth curves are derived from equation 8 using three different model assumptions for the water vapor mixing ratio at 70 km of 4, 6 and 7 ppm. These calculations are for the scattering angle (135°) appropriate to the SME north observing geometry.

ments of mesospheric temperature. The ranges of water vapor and temperature shown in Figures 7 and 8 are representative of the summertime mesopause region [Garcia, 1989]. It is important to note that the water vapor mixing ratio $w(z)$ decreases rapidly with height above z = 70 km [Girard et al., 1988; Gunson et al., 1990] as a result of rapidly increasing solar photodissociation with height. For a nominal solar Ly-α flux of 3.5 ×10^{12} photons cm^{-2}s^{-1}nm^{-1}, Jensen [1989] calculated the cloud-free water vapor distribution with height. The quantity $w(z)$ drops from 6 parts per million by volume (ppm)

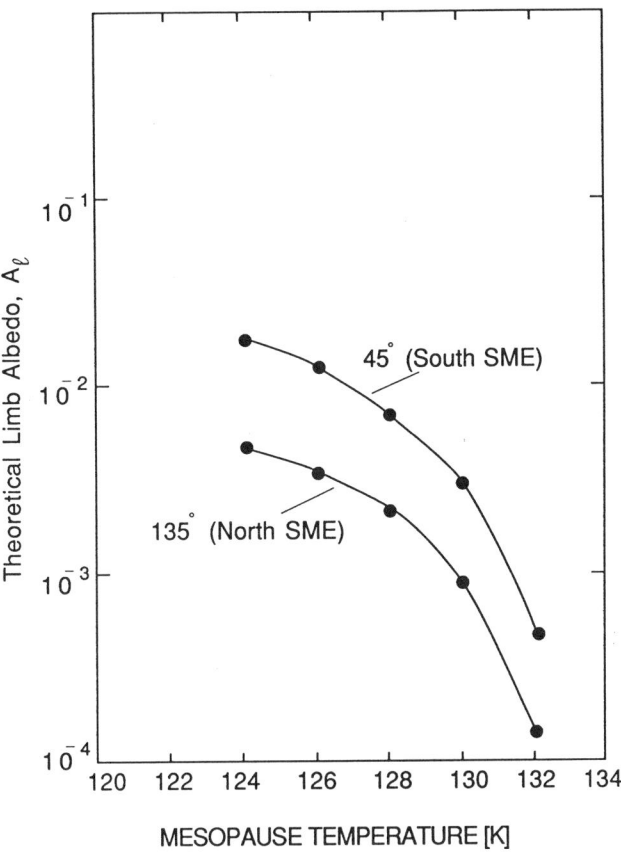

Fig. 7. Theoretical limb albedos calculated for the two scattering geometries appropriate to north and south polar viewing of the SME ultraviolet spectrometer experiment. The parameter varied in the model is the mesopause temperature, holding all other parameters (such as water vapor mixing ratio at 70 km which was assumed to be 6 ppm) fixed in value. These calculations are taken from Jensen [1989], in which all the various assumptions are described.

to 1.40 ppm at 83 km, the mean cloud height. Jensen [1989] used $w(70\ km)$ as a parameter which varied between 3 and 9 ppm to account for the variability of water in the lower mesosphere due to non-solar effects. Since the interest in this paper is how PMC change with solar cycle, one might think that the value at 83 km would be more appropriate. However, if $w(70\ km)$ is simply scaled up and down with no change in the water vapor profile, this in effect changes the 83 km value by the same factor. Thus $w(70\ km)$ (denoted hereafter as simply w) serves 'double-duty' - it describes variations of water vapor of either dynamical or photochemical origin.

To account for the variability of PMC albedo with time and latitude, we could invoke a statistical distribution in which both quantities T and w are variables. Assuming the nominal values apply to the average for an entire

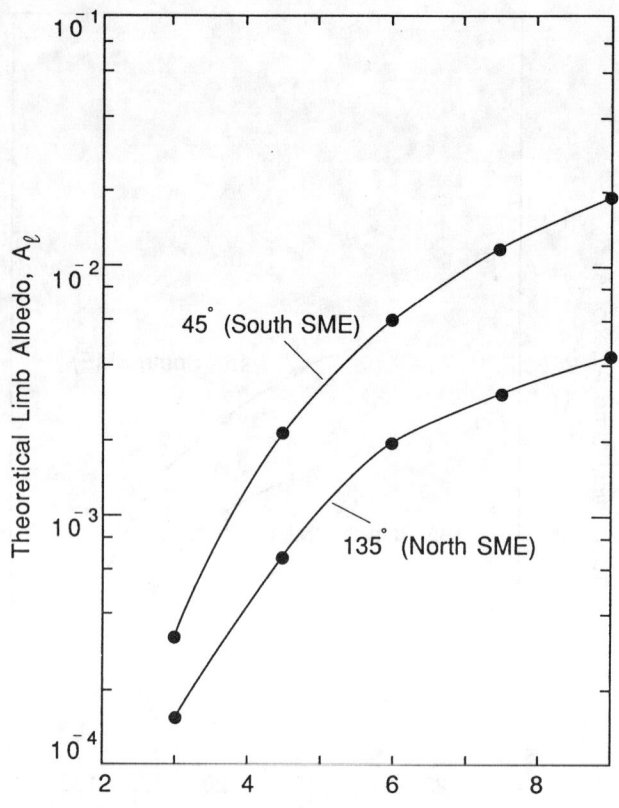

Fig. 8. Same as Figure 7 except that the parameter which is varied is water vapor at 70 km, and mesopause temperature is held constant at 128K (taken from *Jensen* [1989]).

season, we might invoke a separate normal distribution for each variable, so that the joint distribution is given by

$$f(w,T) = f_o e^{-(\frac{T-T_o}{\delta T})^2} e^{-(\frac{w-w_o}{\delta w})^2}.$$

f_o is a normalizing factor, and δT and δw are the associated standard deviations. There are several difficulties with this approach. First, the use of the product of two distributions assumes that T and w are independent, which may not be the case. Second, the number of possible combinations of four variables (T_o, δT, w_o, and δw) is too large to be useful, since the data do not provide that many costraints. Third, it is necessary to have a model prediction for cloud albedo where *both* temperature and water vapor vary through their respective ranges, and these calculations are not currently available.

In this first attempt, I take a simpler approach which illustrates the basic idea, and which provides a fair approximation to the actual distributions. I will assume only *one* stochastic variable is operative, with the other variable held at its nominal value. The two scenarios are:

$$f(T) = f_o T e^{-(\frac{T-T_o}{\delta T})^2} \quad \text{(variable temperature)} \quad (2)$$

$$f(w) = f_{ow} e^{-(\frac{w-w_o}{\delta w})^2} \quad \text{(variable water vapor)} \quad (3)$$

Using the temperature ("$\delta - T$") scenario for now, the equation to transform from the variable T to the variable A_l is

$$f(A_l) = f(T)\left|\frac{\partial T(A_l)}{\partial A_l}\right| = e^{-\left[\frac{T(A_l)-T_o(A_o)}{\delta T}\right]^2}\left|\frac{\partial T(A_l)}{\partial A_l}\right| \quad (4)$$

where the *Jensen* [1989] model results of Figure 7 are used to relate T to A_l, and A_o is the (observed) value of the limb albedo at the nominal value of temperature. The cumulative distribution follows from equation 1,

$$g(A_l) = erf\left[\frac{T(A_l) - T(A_o)}{\delta T}\right] \quad (5)$$

where erf is the error function.

To relate these theoretical distributions to the actual ones, I assume that T_o and δT are constants within a given season, but vary from season to season. It follows from $g(A_l = A_o) = erf(0) = 0.5$ that A_o is the value of the albedo for which half the clouds are brighter and half are dimmer, which I call the *half-power point*. A_o and T_o are listed in Table 1 for all SME seasons. It is interesting that the mean temperatures T_o are higher in the south than in the north. This result is consistent with our earlier results [*Olivero and Thomas*, 1986], derived in an entirely different manner, that the southern PMC are inherently less bright than those in the north. According to the δT scenario, this is due to warmer conditions in the south.

The standard deviations δT may be derived from the cumulative distributions by finding those values which give the best fit of a modified form of equation 5 with the SME data. This modified form is needed to approximate the dependence of $T(A_l)$ over the entire range of observed values of A_l, not just that predicted by theory (see Figures 7 and 8). Assuming the dependence of T on A is exponential over the entire range of A_l, so that $A = A_o e^{-\alpha(T-T_o)}$, it is easily seen that the albedo distribution becomes

$$f(A_l) = \frac{1}{\sqrt{\pi}\alpha\delta T A_l} e^{-\ln^2(A_l/A_o)/(\alpha\delta T)^2} \quad (6)$$

and the cumulative distribution is

$$g(A_l) = erf[\ln(A_l/A_o)/\alpha\delta T]. \quad (7)$$

The distribution $f(A_l)$ in equation 6 is recognized as the *log-normal distribution*, familiar in aerosol theory in a different context. I assume that equation 7 applies to the observed cumulative distribution, and then find the least-squares fit of the quantity $erf[x \ln A_l/A_o]$ to the cumulative distribution $g(A_l)$ where x is an adjustable constant, equal to $(\alpha \delta T)^{-1}$. Equation 7 fits the observed distributions remarkably well over the entire observed range. Two examples are given in Figure 9. The value of δT for each season can then be determined through the relationship

$$\delta T = \frac{T(A_l) - T(A_o)}{x \ln(A_l/A_o)} \qquad (8)$$

where a value of A_l must be selected. I arbitrarily selected $A_l = 4 \times 10^{-3}$ as a representative intermediate value, high enough to be in the middle of the observed range, and yet still within the domain of applicability of the theory. Using different values of A_l affect somewhat the values of δT, but not the relative values, in particular relative values between north and south. The values of δT are also given in Table 1.

It is interesting that the inferred temperature variability is significantly smaller in the south than in the north. This is despite the fact that the southern distributions contain many more brighter clouds than the north. Since southern clouds are viewed in a more favorable scattering geometry, they only *appear* to be brighter. With regard to SBUV data, it is not possible to perform the same fitting of theory with data, since the value of A_o cannot be determined from the data. However it should be noted that the erf function is nearly exponential for $A_l \gg A_o$ (the 1983 North data in Figure 9 is a good example of this). Thus it is reasonable that the SBUV cumulative distribution follows closely an exponential curve (Figure 5). The 'curved' behavior of the SME distributions in Figure 9 is thus a consequence of the variation of the erf function for arguments of the order of unity.

The alternative scenario in which water vapor variability is responsible for cloud brightness variations provides similar expressions for the albedo distributions. The results are

$$f(A_l) = f(w)\left|\frac{\partial w(A_l)}{\partial A_l}\right| = e^{-\left[\frac{w(A_l)-w_o(A_o)}{\delta w}\right]^2}\left|\frac{\partial w(A_l)}{\partial A_l}\right| \qquad (9)$$

$$g(A_l) = erf\left[\frac{w(A_l) - w_o(A_o)}{\delta w}\right] \qquad (10)$$

where the relationship $A_l(w)$ is given by the model (see Figure 8), and the quantities w_o, and δw are analogous to quantities in the δT scenario.

The mean water vapor mixing ratios, w_o derived from the data and the relationship $A_l(w)$, are tabulated in

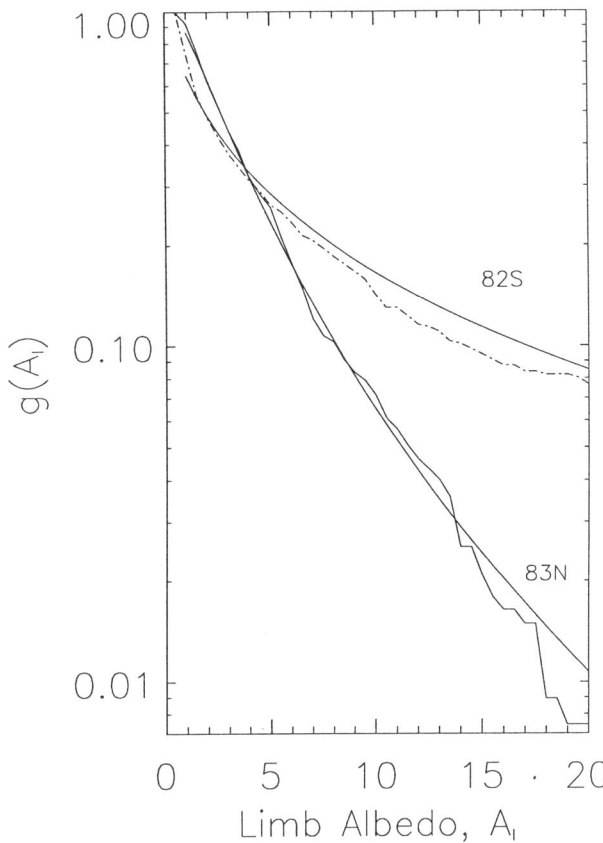

Fig. 9. (Irregular curves) Two cumulative brightness distributions from the SME seasons 1982 south (1982S) and 1983 north (1983N) versus limb albedo A_l (in units of 1×10^{-3}). The smooth curves are the best-fit theoretical cumulative distributions, from either equation 5 (from which mean values of the standard deviation of temperature δT are derived for each season); or from equation 7 (from which values of the standard deviation of water vapor δw are derived for each season).

Table 1, along with the values of δw derived from the same fitting process described in the δT scenario. For the same reasons described earlier, the southern **regions** contain less water vapor than the north, first suggested by *Jensen and Thomas* [1988]. In addition the water vapor standard deviations are smaller in the south than in the north.

Inspection of Table 1 reveals no statistical relationships between solar flux and inferred seasonal temperature (or water), or their associated variances. How is this compatible with the fact that the cumulative distributions for high albedo values *do* show a solar cycle effect? Could the frequency of low-brightness clouds (in contrast to bright clouds) be insensitive to solar cycle? To examine this question quantitatively, I will use the model prediction of how cloud brightness varies with solar Ly-α flux [*Jensen*, 1989; see also *Thomas et al.*, 1991]. *Jensen* found that the modeled limb brightness decreased by a factor of ten when the Ly-α flux varied from 2.5 to 4.5×10^{12} photons cm^{-2} s^{-1} nm^{-1}. This followed from the rapid rate of photodissociation of water vapor near the summertime mesopause, which is not compensated by the relatively slow rate of vertical transport in the model. The relationship A_l(Ly-α) is shown in Figure 9 of *Thomas et al.* [1991] for three assumed values of w_o. This can be used in either the δT or δw scenarios to predict the cumulative distribution variation with solar Ly-α flux. Adopting the δT scenario, the desired expression is

$$g(A_l; Ly\alpha) = erf\left[\frac{T(A_l) - T[A_l(Ly\alpha)]}{\delta T}\right] \quad (11)$$

To compare with the data of Figure 6, I divided equation 11 by its value at a Ly-α flux of 2.6×10^{12} photons cm^{-2} s^{-1} nm^{-1}. The result is shown as the three smooth curves in Figure 6. The curves apply to $A_l = 4 \times 10^{-3}$ and to the northern regions assuming three different water vapor amounts. The results for the southern regions (not shown) are very similar. The reason for the choice of this particular albedo value is simply that this is the largest value predicted by the model. Despite the significant effects of Ly-α on albedo predicted by the model, the theoretical variations are not as pronounced as those shown by even the SME data. Given the large interannual variability (apparently unrelated to the solar cycle) it is then not surprising that the solar cycle effect is relatively weak in the SME data. The fact that the solar cycle is much more pronounced in the SBUV data must be related to the brightness ranges observed by each data set. Unfortunately, the comparison of theory and the SBUV data is not possible (at least for the present) since I do not have theoretical modeling results for the very bright clouds seen by the SBUV instrument. It is plausible however that the same type of curve shown in Figure 6, but applicable to SBUV albedos, would show the much stronger effect evident in the data. (Incidentally, the choice of $A_n = 1.2\times10^{-5}$ for the SBUV data in Figure 6 was arbitrary – any value over the entire range of A_n would have shown similar results.)

DISCUSSION AND CONCLUSIONS

The overall seasonal behavior of Polar Mesospheric Cloud activity is described by several climatological measures derived from the frequency distributions of albedo. In the case of the SME data, the nominal state of the atmosphere is described by the 'half-power point', A_o, the albedo dividing equally the bright clouds and weak clouds. A convenient indicator of the SBUV distribution is the exponent β_n, the slope of the distribution on a semi-log scale. A measure common to both data sets is the cumulative distribution at some fixed value of the limb albedo. In the latter case, both SME and Nimbus-7 data show that more bright clouds occur during seasons of low solar Lyα flux than during seasons of high Lyα flux. This trend is more significant for the SBUV data which apply to the brightest clouds seen in the nadir. I then related the cumulative distribution to atmospheric forcing caused by statistical variations in either temperature or water vapor. A microphysical model related the cloud limb albedo to either mesopause temperature or water vapor amount. The conclusions are:

(1) The distributions of PMC albedo during a single season are successfully modeled by statistical variations of either temperature δT or water vapor δw. Neither δT nor δw are correlated with solar Lyα flux, although PMC cumulative brightness distributions are strongly anti-correlated with solar Lyα flux. In the SME data there are significant differences in the variances between the northern seasons, where δT averaged 5.8K, and in the southern seasons where δT averaged 3.6K. In the δw scenario, the distributions are best fit (on average) with the formulation in which the values of δw are 4.3 ppm in the north and 2.0 ppm in the south.

Support for the notion that temperature changes are mainly responsible for brightness variations within a season comes from an analysis by *Lübken and Von Zahn* [1991] who showed that the rms variation of summertime mesopause temperature at 69° N was 4-5K. This is slightly lower than my value of 5.8K for the north. However 5.8K applies to the entire polar cap regions and over the entire season, and so I consider this good agreement. On the other hand, could water vapor variability be responsible for the brightness variations? Unfortunately no similar data set has yet been published for high latitudes. If water vapor *variability* were constant with latitude, then the results of *Bevilacqua et al.* [1987] and *Tsou et al.*, [1988] are applicable. From microwave emission measurements their retrievals for the mid-latitude northern water vapor mixing ratio yielded $w(70$ km$) = 4.0 \pm 0.5$ ppm for the six-month period from December, 1984 through May, 1985. Although I am not particularly interested in the latitudinal gradient of w in the present work, it is important to note that models, such as *Garcia and Solomon* [1985], are consistent with 4 ppm at 40° latitude and 6 ppm at 70° latitude during summer solstice. The main point is the 0.5 ppm standard deviation, which if scaled linearly to that applicable to 6 ppm, yields a standard deviation of somewhat less than 1 ppm. This value is considerably less than the 4.3 ppm value inferred from the present results. On the basis of this (admittedly weak) objection, I would therefore argue that water vapor variations are less important on a day-to-day basis than temperature variations. On the other hand, variations over the solar cycle may be primarily due to water vapor (see conclusion 3 below).

(2) From the 'half-power point' (A_o) of the SME cumulative brightness distributions, the average temperature T_o (or average water vapor amount w_o) was inferred for each season. Very small interannual variations in these quantities are inferred–the model predicts that T_o in the north varied from a low value of 125.4K in 1984 to a high value of 126.5K in 1983; or alternatively that the northern water vapor at 70 km w_o varied from 6.9 ppmv in 1983 to 7.9 ppmv in 1982. In the south these quantities are significantly different – T_o varied from 128.8K in 1984 to 130.7K in 1983. Thus the north-south differences are 3-4K. As discussed by *Balsley et al.* [1993], the satellite data of *Barnett and Corney* [1985] also confirm this north-south asymmetry. Their data show that at 75° and 80° latitude the southern summer solstice mesosphere at 80 km was 3.5-5K warmer than the corresponding northern mesosphere.

The mesopause temperatures inferred from the northern PMC region (125.5-126.5K) are slightly lower than the values measured by *Lübken and Von Zahn* [1991] in June (128.3K) and July (130.4K) at 69° N. However, these may be consistent since the values derived from the present analysis refer only to the colder cloud regions. According to the current theory of mesopause thermal structure [*McIntyre*, 1989], higher southern mesopause temperatures (129-131K) are consistent with smaller atmospheric gravity wave amplitudes (smaller δT), as pointed out by *Vincent* [1993]. This is indeed the case for a six-year data set (1984-1990) measured by MF radar [*Vincent*, 1993] at Mawson, Antarctica (67° S). Gravity wave amplitudes are significantly lower at Mawson than amplitudes measured by *Wang and Fritts* [1990] at Poker Flat, Alaska (65° N). A still more indirect indication of warmer conditions in the southern summertime mesosphere is the recent result of *Balsley et al.* [1993] who obtained MST radar data throughout the summer of 1992-93, and found a total absence of Polar Mesospheric Summer Echoes (PMSE). Although the mechanism of PMSE formation is still not understood [*Thomas*, 1991; *Goldberg et al.*, 1993], there appears to be a loose association with PMC [*Jensen and Thomas*, 1988; *Inhester et al.*, 1993]. *Balsley* speculated that higher temperatures in the antarctic mesosphere are responsible for the absence of PMSE and for inherently less-bright PMC.

(3) The cumulative distributions from the SME data reveal a modest effect of the solar cycle, whereas the SBUV data, which apply to the brightest clouds ($A_l > 1 \times 10^{-2}$), show a very significant correlation. These rare clouds are more likely (by a factor of nearly ten) to appear during solar minimum conditions than solar maximum conditions. This poses a possible problem since the theory predicts that clouds of all brightness should be affected by changes in the solar Lyα-induced photodissocation of water vapor at cloud level (83 km). I believe that the solution to this problem is that for most clouds the effects of temperature variability (most likely caused by atmospheric gravity waves) dwarf the smaller variability caused by solar-induced changes of water vapor. For only the brightest clouds do the effects of solar variability become apparent, i.e. they rise "above the noise".

As already discussed by *Thomas et al.* [1991] the SBUV data indicated that the southern PMC show more sensitivity to solar variability than the northern PMC. This was not a firm conclusion, since only three complete seasons were available. However this result is now confirmed by the SME data, with five southern seasons showing this same effect (see Figure 6). *Thomas et al.* [1991] explained this difference in terms of the possible differences of water vapor amounts at the two poles. A more plausible explanation is that higher gravity-wave variability in the north tends to mask the solar variability, even for the brightest clouds.

The result that T_o varied by less than 2K over the period (1982-86) when the solar Ly-α flux varied significantly (2.5 to 3.6 $\times 10^{12}$ photons-cm^{-2}-s^{-1}-nm^{-1}) is consistent with the report by *Lübken and Von Zahn* [1991] that the available rocketsonde data indicate only a few K change of T with solar cycle. This fact, coupled with the rather simple physics, that solar Lyman-α flux destroys water vapor less rapidly during solar minimum than solar maximum, is persuasive evidence that long-term water vapor variations are largely responsible for the solar cycle effect on PMC.

An improved description of PMC albedo distributions should involve the vertical motion field. Since vertical motions control both the temperature and water vapor fields, through adiabatic cooling and advective transport, respectively, it may be possible to order brightness variations in terms of a single variable, for example the vertical velocity w at 80 km. In turn, w is directly related to the meriodional velocity v through mass continuity. The current theory of the large-scale circulation of the mesosphere assigns the ultimate responsibility for the unusually cold summertime mesopause to momentum transfer from breaking gravity waves to the mean winds [*Holton*, 1983]. The westerly acceleration of the zonal wind induces an extra Coriolis torque in the equatorward direction, causing a southerly flow v. Through mass continuity this meridional wind induces an upward wind, thus cooling the upper mesosphere. Fortunately, v has been routinely measured by radars for a number of years; in particular there is an excellent data base available from the 65° N Poker Flat VHF 50 MHz radar, at least during the SME era (e.g. see *Hall et al.* [1992] for 1981 and 1982 data). It will be interesting to determine whether v is correlated with PMC albedo, and whether this ground-based measurement might even be a 'predictive' variable for PMC occurrence.

As shown in Figures 4 and 6, the northern 1984 season contained a larger number of bright clouds than either 1985 or 1986, despite the fact that the solar flux was higher in 1984. A possible explanation for this anamoly is that a mesospheric injection of water occurred in April, 1982 from the (17° N) El Chichon volcanic eruption. The mesospheric water vapor was enhanced two years later, due to the long transport time scale. We have argued that violent volcanic eruptions (so-called Plinian type) inject water vapor and/or aerosols into the stratosphere, and eventually make their way to the HLSM with about a two-year time lag [*Thomas et al.*, 1989]. This explains the outbreak of very bright NLC in 1885, two years following the Plinian eruption of Krakatoa in 1883. Unfortunately, the unusual nature of the 1984 season could not be confirmed in the SBUV data because of incomplete coverage. I plan to further examine the available (albeit incomplete) SBUV data to confirm this. (However the ground-based NLC data indicate that 1984 was no different than 1983 or 1985; see *Gadsden* [1990])

A good strategy to observe solar cycle and longer-term effects from orbit is to acquire data on PMC scattering from a nadir-viewing UV imager in sun-synchronous orbit. The data need to be taken every orbit to maximize the chances of observing the comparitively rare very bright clouds. Such clouds are easy to detect even against the bright nadir sky background. Even better would be a limb-viewing imager with a wide field of view.

How do these results relate to the question of possible long-term changes of mesospheric clouds due to the buildup of methane-derived water vapor? A doubling of methane over the past century would have caused water vapor concentrations to have increased significantly (by possibly 50%). Has there been a change in the number of very bright NLC? Since observers have for many years recorded the brightness of NLC on a relative (1-5 scale), it would be interesting to determine from the NLC archives whether 'class-5' clouds show more variability (on 11-year and longer time-scales) than the dimmer clouds. *Gadsden* [1990] showed that the number of nights NLC were visible per season varied with both the solar cycle, and increased in a secular fashion over a still longer time scale (approaching 25 years). Perhaps only the brightest clouds are responsible for this secular increase. Indeed the ground-based reports of the number of NLC per season are probably strongly biased toward bright clouds. It would also be a simple matter to investigate whether class-5 clouds are more prevalent 1-2 years following volcanic eruptions with substantial stratospheric impact.

Additional data for very bright clouds will be contained in the SBUV/2 data for the time period 1984 to present. These data should provide information for PMC variations over solar cycle 22. Fortunately NASA and NOAA plan to continue with their ozone monitoring with SBUV-type instruments for the indefinite future. In addition, the new NASA TIMED satellite mission, planned for the late 1990's, will have three instruments which will detect PMC. Thus present and future data sets will provide a nearly unbroken time series of PMC continuing into the next century.

Acknowledgments. I thank J. Olivero and R. T. Clancy for helpful discussions and suggestions, and Chris Smith for his skillful and diligent programming assistance. I also thank an anonymous refereee for making several helpful suggestions for improving the text. This research was sponsored by a grant from the Aeronomy Program of the NSF, and by grant N00014-90-J-1277 from the SDIO Innovative Science and Technology Program, which was managed by the Office of Naval Research.

REFERENCES

Avaste, O., Noctilucent clouds, *J. Atmos. Terr. Phys., 55*, 133-143, 1993.

Balsley, B.B., R.F. Woodman, M. Sarango, J. Urbina, R. Rodriguez, E. Ragaini, and J. Carey, Southern-Hemisphere PMSE: Where are they?, *Geophys. Res. Lett., 20*, 1983-1986, 1993.

Barnett, J.J. and M. Corney, Middle atmosphere reference model derived from satellite data, *MAP Handbook, 16*, 47-85, 1985.

Barth, C.A., W. K. Tobiska, G. J. Rottman, and O. R. White et al., Comparison of 10.7 cm radio flux with SME solar Lyman Alpha flux, *Geophys. Res. Lett., 17*, 571- 574, 1990.

Bevilacqua, R. M., D. F. Strobel, M. E. Summers, J. J. Olivero, and M. Allen, The seasonal variation of water vapor and ozone in the upper mesosphere: Implications for vertical transport and ozone photochemistry, *J. Geophys. Res., 95*, 883-894, 1990.

Bevilacqua, R. M., W. J. Wilson, and P. R. Schwartz, Measurements of mesospheric water vapor in 1984 and 1985: Results and implications for middle atmospheric transport, *J. Geophys. Res., 92*, 6679-6690,

1987.

Clancy, R. T. and D. W. Rusch, Climatology and trends of mesospheric (58-90 km) temperatures based upon 1982-1986 SME limb scattering profiles, *J. Geophys. Res., 94*, 3377-3393, 1989.

Donahue, T. M., B. Guenther, and J. E. Blamont, Noctilucent clouds in daytime: Circumpolar particulate layers near the summer mesopause, *J. Atmos. Sci., 29*, 1205-1209, 1972.

Gadsden, M. A secular change in noctilucent cloud occurrence, *J. Atmos. Terr. Phys., 52*, 247-251, 1990.

Gadsden, M. and W. Schroder, Noctilucent Clouds, 165 pp., Springer-Verlag, Berlin, 1989.

Garcia, R. R., and S. Solomon, The effect of breaking gravity waves on the dynamics and chemi-cal composition of the mesosphere and lower thermosphere, *J. Geophys. Res., 90*, 3850-3868, 1985.

Garcia, R. R., Dynamics, radiation, and photochemistry in the mesosphere: Implications for the formation of noctilucent clouds, *J. Geophys. Res., 94*, 14,605-14,615, 1989.

Girard, A., J. Besson, D. Brard, J. Laurent, M. P. Lemaitre, C. Lippens, C. Muller, J. Vercheval, and M. Ackerman, Global results of the grille spectrometer experiment on board Spacelab 1, *Planet. Space Sci, 36*, 291-300, 1988.

Goldberg, R. A., E. Kopp, G. Witt, and W. E. Swartz, An overview of NLC-91: A Rocket/Radar Study of the Polar Summer Mesosphere, *Geophys. Res. Lett.* (submitted), 1993.

Gunson, M. R., C. B. Farmer, R. H. Norton, R. Zander, C. P. Rinsland, J. H. Shaw, and B.-C. Gao, Measurements of CH_4, N_2O, CO, H_2O, and O_3 in the middle atmosphere by the Atmospheric Trace Molecule Spectroscopy Experiment on Spacelab 3, *J. Geophys. Res, 95*, 13,867-13,882, 1990.

Hall, T. M., J. Y. N. Cho, and M. C. Kelley, A re-evaluation of the Stokes drift in the polar summer mesosphere, *J. Geophys. Res, 97*, 887-897, 1992.

Heath, D. F., A. J. Krueger, H. A. Roeder, and B. D. Henderson, The solar backscatter ultravio-let and total ozone mapping spectrometer (SBUV/TOMS) for Nimbus G, *Opt. Eng., 14*, 323-331, 1975.

Holton, J. R., The influence of gravity wave breaking on the general circulation of the middle atmosphere, *J. Atmos. Sci., 40*, 2497-2507, 1983.

Inhester, B., Klostermeyer, F. J. Lübken, and U. Von Zahn, Evidence for ice clouds causing mesospheric summer echoes, *J. Geophys. Res.*, submitted, 1993.

Jensen, E. J. and G. E. Thomas, A growth-sedimentation model of polar mesospheric clouds: Comparisons with SME measurements, *J. Geophys. Res., 93*, 2461-2473, 1988.

Jensen, E. J., A Numerical Model of Polar Mesospheric Cloud Formation and Evolution, PhD. Thesis, University of Colorado, 176 pp, 1989.

Jensen, E. J., G. E, Thomas, and O. B. Toon, On the diurnal variation of noctilucent clouds, *J. Geophys. Res., 94*, 14,693-14,702, 1989.

Lübken, F. J. and U. Von Zahn, Thermal structure of the mesopause region at polar latitudes, *J. Geophys. Res.*, 20,841-20,857, 1991.

McIntyre, M. E., On dynamics and transport near the polar mesopause in summer, *J. Geophys. Res., 94*, 14,617-14,628, 1989.

Olivero, J. J. and G. E. Thomas, Climatology of polar mesospheric clouds, *J. Atmos. Sci., 43*, 1263-1274, 1986.

Stanford, J. L. and J. S. Davis, A century of stratospheric cloud reports: 1870-1972, *Bull. Am. Met. Soc., 55*, 213-219, 1974.

Stanford, J. L., Possible long-term variations in stratospheric water-vapour content, *Weather, 29*, 107-112, 1974.

Thomas, G.E., Solar Mesosphere Explorer Measurements of Polar Mesospheric Clouds (Noctilucent Clouds), *J. Atmos. Terr. Phys., 46*, 819-824, 1984.

Thomas, G. E. and C. P. McKay, On the mean particle size and water content of polar meso-spheric clouds, *Planet. Space Sci., 33*, 1209-1224, 1985.

Thomas, G. E. and J. J. Olivero, Climatology of polar mesospheric clouds 2. Further analysis of Solar Mesosphere Explorer data, *J. Geophys. Res., 94*, 14,673-14,681, 1989.

Thomas, G. E. and J. J. Olivero, The heights of polar mesospheric clouds, *Geophys. Res. Lett., 13*, 1403-1406, 1989.

Thomas, G. E. and R. T. Clancy, Height and brightness distributions of Polar Mesospheric Clouds, to be submitted, *J. Geophys. Res.*, 1994.

Thomas, G. E., J. J. Olivero, E. J. Jensen, W. Schroder, and O. B. Toon, Relation between increasing methane and the presence of ice clouds at the mesopause, *Nature, 338*, 490-492, 1989.

Thomas, G. E., Mesospheric clouds and the physics of the mesopause region, *Rev. Geophys., 29*, 553-575, 1991.

Thomas, G. E., R. D. McPeters, and E. J. Jensen, Satellite observations of polar mesospheric clouds by the SBUV Radiometer: Evidence of a solar-cycle dependence, *J. Geophys. Res., 96*, 927-939, 1991.

Thomas, G. E., Recent developments in the study of mesospheric clouds, *Adv. Space Res.*, 1993, to be published.

Thomas, G. E., Solar Mesosphere Explorer measurements of polar mesospheric clouds (noctilucent clouds), *J. Atmos. Terr. Phys., 46*, 819-824, 1984.

Tsou, J.-J., J. J. Olivero, and C. L. Croskey, Study of variability of mesospheric H_2O during spring 1984 by ground-based microwave radiometric observations, *J. Geophys. Res., 93*, 5255-5266, 1988.

Turco, R. P., O. B. Toon, R. C. Whitten, R. G. Keesee, and D. Hollenbach, Noctilucent clouds: Simulation studies of their genesis, properties and global influences, *Planet. Space Sci, 30*, 1147-1181, 1982.

Vasil'yev, O. B., C. I. Willmann, N. M. Gavrilov, V. V. Kovalenok, A. I. Lazarev, and N. P. Fast, Studies of Noctilucent Clouds from Space, Gidrometeoizdat, Leningrad, 1987 (Eng. Transl. DIA transl. num. LN038-88, Defense Intelligence Agency, Washington, D. C. 20301, 1987.)

Vincent, R. A., Gravity-wave motions in the mesosphere and lower thermosphere observed at Mawson, Antarctica, *J. Atmos. Terr. Phys.* (submitted), 1993.

Wang, D.-Y. and Fritts, D. C., Mesospheric momentum fluxes observed by the MST radar at Poker Flat, Alaska, *J. Atmos. Sci., 47*, 1512-1521, 1990.

Wegener, A., Die temperatur der obersten atmospharenschichten, *Gerlands Beitr. Geophys, 11*, 102, 1912.

G.E. Thomas, LASP, University of Colorado, Boulder, CO, 80309-0392

Charge Balance at the Summer Polar Mesopause: Ice Particles, Electrons, and PMSE

George C. Reid

Aeronomy Laboratory, NOAA, and Cooperative Institute for Environmental Sciences, University of Colorado, Boulder, Colorado

Narrow sheets of depleted electron density, known as electron 'biteouts', are a common feature of the high-latitude mesopause region in summer. Scavenging of electrons by small particles has long been suspected as the basic cause of these biteouts, and this paper reviews recent work that has attempted to treat that idea quantitatively. The conclusions suggest that the particles consist mainly of ice, with dimensions of the order of 10 nanometers, and that they must be present with concentrations up to a few thousand per cubic centimeter. Their small size implies that they would be subvisible, and not directly related to the larger ice particles responsible for noctilucent and polar mesospheric clouds, which have much smaller concentrations. They may, however, be directly involved in producing the strong polar mesospheric summer echoes (PMSE) that are frequently observed by radars operating at high latitudes during the summer. The subvisible particles may form as a result of the passage of a long-period gravity wave through the region, with its associated vertical motions and adiabatic cooling. The cooling may be sufficient to drop the temperature below the frost point temporarily, forming a transient population of ice particles that sublimate again in the warm phase of the wave. Some of the implications of this mechanism are discussed.

1. INTRODUCTION

The high-latitude mesopause is a uniquely interesting region of the atmosphere, especially during the summer months, when it records the lowest temperatures that occur anywhere in the earth's atmosphere. These conditions of extreme cold are directly responsible for noctilucent and polar mesospheric clouds, and are probably also responsible in less direct and still poorly understood ways for the phenomena of electron 'biteouts' and the strong polar mesospheric summer echoes (PMSE) that are observed by coherent backscatter radars. This paper will summarize the observations and current theories of the formation of electron biteouts, and will briefly discuss their possible relationship to noctilucent clouds and PMSE.

2. OBSERVATIONS

The first well documented observation of a biteout was described by *Pedersen et al.* [1970], who measured electron and positive ion concentrations in the mesosphere over Andøya, Norway, in June 1966. A deep narrow minimum in the electron density was centered at about 87 km, with a concentration about a factor of 10 lower than the value that would have been expected by interpolating between the values on either side. A much weaker depletion in positive-ion concentration was observed at the tip of the rocket, with a more pronounced ion depletion in the wake. *Pedersen et al.* [1970] pointed out that no electron biteouts of this kind had ever been seen during winter and equinox flights from the same location, and suggested that the biteout was related to the special conditions near the summer mesopause that gave rise to noctilucent clouds. No NLC were visible at the time, however, and the altitude of the biteout was several kilometers higher than the usual altitude at which NLC are seen. They suggested that small subvisible particles might be responsible for removing the electrons in the biteout layer.

Since then, several rocket flights through the high-latitude summer mesopause have reported similar biteouts [e.g., *Johannessen and Krankowsky*, 1972; *Ulwick et al.*, 1988]. A particularly striking example is shown in Figure 1 [*Ulwick et al.*, 1988], obtained during the STATE campaign

The Upper Mesosphere and Lower Thermosphere:
A Review of Experiment and Theory
Geophysical Monograph 87
This paper is not subject to U.S. copyright.
Published in 1995 by the American Geophysical Union

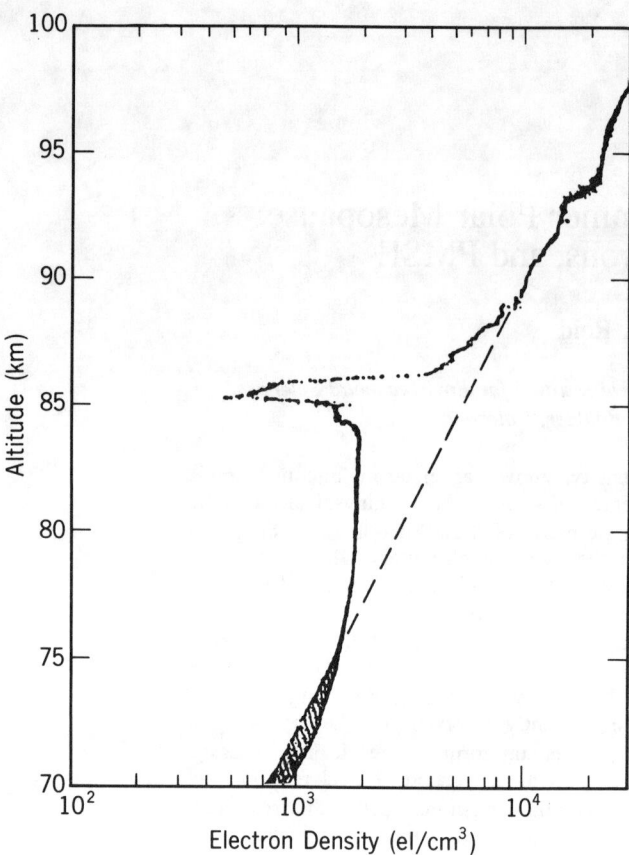

Fig. 1. Electron density profile measured by a rocket probe over Poker Flat, Alaska, on June 17, 1983 (from *Ulwick et al.* [1988]).

at Poker Flat, Alaska, in June 1983. Again the biteout is centered near 86 km, and the peak depletion in electron density is about an order of magnitude below the interpolated profile shown by the broken line. The biteout has a fairly abrupt upper boundary, but a weaker depletion region extends several kilometers below the sharp lower edge. Electron-density gradients at the upper and lower edges of the main biteout are extremely steep.

Individual rocket flights, however, provide only a snapshot of conditions, and give us no information on the spatial structure or time duration of the biteouts. In principle, radars can provide us with such information, if the existence of a biteout can be detected. Powerful incoherent scatter radars operating in high latitudes, such as the EISCAT radar, should be able to detect biteouts, provided that an incoherent scatter echo can be obtained from the upper mesosphere. The nature of the signal would depend on the characteristics of the radar and on the depth and altitude range of the biteout, but would be expected to show a region of weak or non-existent echo situated between regions of stronger echoes above and below. Unfortunately the strong PMSE echoes often mask the incoherent scatter signal during the summer months, and even in their absence the incoherent scatter signal from heights below 90 km is too weak to be measured unless the electron density is enhanced by energetic particle precipitation [e.g., *Röttger and La Hoz*, 1990]. *Röttger et al.* [1990], however, have reported a series of measurements of the mesospheric electron density profile between about 72 and 95 km in July 1988 with the EISCAT UHF radar, showing what appears to be a biteout near 85 km. Figure 2, replotted from *Röttger et al.* [1990], shows a sequence of 4 profiles of echo signal strength at intervals of 15-20 minutes. The drop in signal strength near 85 km reaches its maximum development in the second profile, and has almost disappeared in the last profile.

The overall large-scale shape of the profile itself changed significantly during the time period, probably due to changes in the flux and spectrum of the precipitating particles, but it is highly unlikely that the smaller-scale changes in the biteout region, occupying only one or two kilometers in height, could have been produced in this way. Localized changes in atmospheric conditions are a more reasonable explanation.

The time scale for growth and decay of the electron-density anomaly at 85 km was about an hour, which is a reasonable value for the period of a vertically propagating gravity wave at these altitudes. With only one such example, the link to a gravity wave is obviously speculative, but it is consistent with what is known about the existence and properties of gravity waves in this region, based on the visible structure of noctilucent clouds as well as on other direct and indirect evidence. It would clearly be desirable to obtain many more measurements of this kind, with the addition of horizontal spatial information on the extent of the biteout regions if possible.

3. ELECTRON AND ION SCAVENGING BY AEROSOLS

The suggestion by *Pedersen et al.* [1970] that the biteout they described might be caused by scavenging of electrons by small particles still seems to be the most likely explanation. Since the biteouts appear during the extremely cold conditions of the high-latitude summer mesopause, when temperatures are known to drop below the local frost point, the particles responsible are likely to be either ice-coated or wholly ice, and are thus closely related to the noctilucent cloud particles. As we shall show, however, NLC particles are too big to account for the biteouts, and the ice particles responsible must almost certainly represent a more transient subvisible population.

The theory of the attachment of electrons and ions to small particles was treated rigorously by *Natanson* [1960], and was later modified by *Parthasarathy and Rai* [1966] and

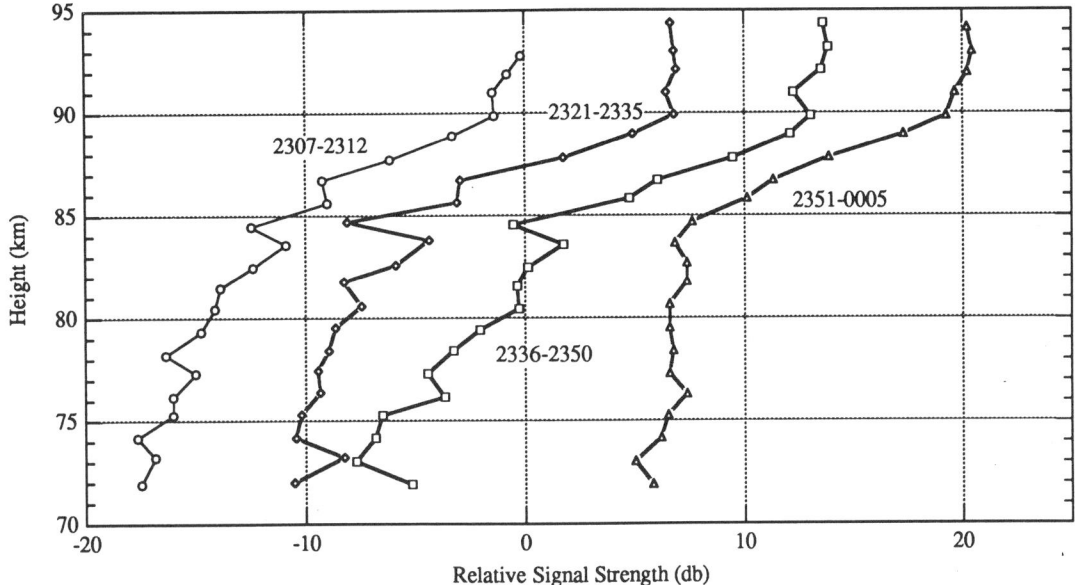

Fig. 2. Incoherent scatter power measured at 933 MHz by the EISCAT radar on July 1, 1988 (replotted from data in *Röttger et al.* [1990]).

Hodges [1969]. The case of attachment to a charged aerosol particle was difficult to treat, since the coulomb field due to the static charge becomes superimposed on the field due to the charge induced by the approaching ion or electron. An empirical correction to the earlier results was applied by *Parthasarathy* [1976], who derived relatively simple expressions for the rate of capture of ions and electrons by neutral and charged aerosol particles. His results were applied by *Reid* [1990] to the case of the electron biteouts, assuming a single size for the aerosol particles, and this treatment was in turn extended by *Jensen and Thomas* [1991], who incorporated the results into a version of their growth-sedimentation model of noctilucent clouds [*Jensen and Thomas*, 1988]. While a range of particle sizes undoubtedly exists, we shall discuss only the single-size (monodisperse) case in the outline given here, since the physics is easier to see, and since there may not be a very direct relationship between the subvisible biteout particles and the larger visible NLC particles.

The derived capture rates for a wide range of values of the aerosol particle radius are shown in Figure 3, expressed in units of $cm^3 s^{-1}$. In physical terms, the quantities shown are the number of captures per second per aerosol particle in a plasma of density 1 ion or electron per cm^3. The capture rates are dimensionally equivalent to a recombination rate, and a typical dissociative recombination rate of 10^{-6} $cm^3 s^{-1}$ for simple cluster positive ions is shown for comparison. Where the curves lie above the recombination line, aerosol particles are more efficient than recombination at removing either ions or electrons, assuming that the concentrations of aerosols and ions (and electrons) are the same.

For the range of particle sizes considered here, a negatively charged particle is only likely to carry a maximum of two unit charges, and a positively charged particle only one. This fact implies that the concentration of particles needed to produce a biteout of the depth shown in Figure 1 must be comparable with the undisturbed electron density. Under these conditions, Figure 3 shows that the attachment of electrons to either neutral or positively charged particles is much more efficient than recombination for all sizes. For particles with radius more than about 15 nanometers, even attachment to a negatively charged particle is faster than recombination, leading to a significant population of doubly negatively charged particles. The heavier and slower positive ions become attached less readily than electrons, but attachment still exceeds recombination for radii greater than 10 nm. For particle sizes smaller than this, however, the disappearance of electrons by attachment actually leads to an increase in ion concentration since it inhibits recombination. Positively charged aerosols could be produced by photoemission from neutral particles, but this process has been ignored here since the photoelectric work function for ice is too high to make a significant contribution. Photoemission of electrons from negatively charged aerosols could be a significant contributor to the overall charge balance; in the absence of any quantitative information, however, it has also been omitted.

Steady-state equations can be set up to describe the charge balance when aerosol particles are present. When the total concentration of aerosol particles is specified, and charge

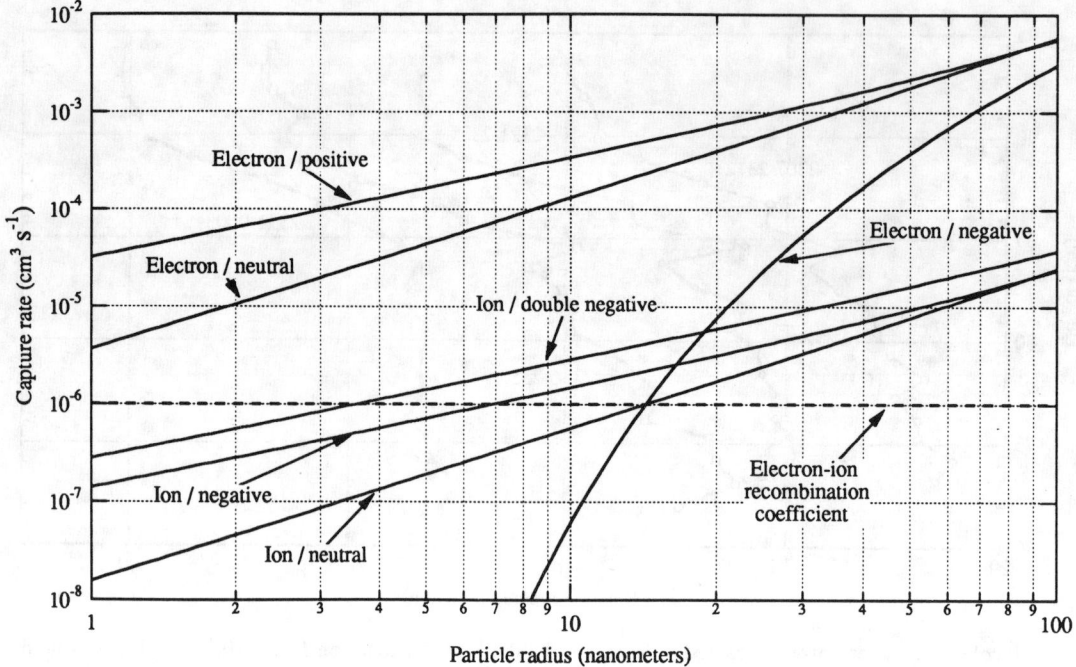

Fig. 3. Calculated rates of capture of electrons and positive ions by neutral and charged aerosols per unit plasma density. A typical electron-ion recombination coefficient for cluster ions is shown for comparison (from data in *Reid* [1990]).

neutrality of the medium as a whole is assumed the equations can readily be solved; details are given by *Reid* [1990]. The six unknowns are the concentrations of electrons, positive ions, and aerosol particles with zero, one positive, one negative, and two negative charges respectively. A typical set of solutions is shown in Figure 4 for total aerosol concentrations up to 4000 cm^{-3} with an assumed radius of 10 nm, and for an ion pair production rate of 10 cm^{-3} s^{-1}, corresponding to a weak particle precipitation event.

The broken line at the top of the figure shows the undisturbed electron and ion concentration in the absence of aerosol particles. As the aerosol concentration increases, the electrons quickly start to disappear, and the concentration of negatively charged aerosols increases proportionately. Only for aerosol concentrations of several thousand per cm^3 does the positively charged component become appreciable, while aerosols with double negative charge never reach significant concentrations. An electron depletion by a factor of 10 below the undisturbed level is reached for an aerosol concentration of about 3000 cm^{-3}, close to the undisturbed electron density as would be expected.

It was pointed out in *Reid* [1990] that the size of the (monodisperse) aerosol particles is constrained by the available water vapor, assuming that the particles are made of ice. Observations and model calculations have shown that the water-vapor mixing ratio at the summer mesopause is

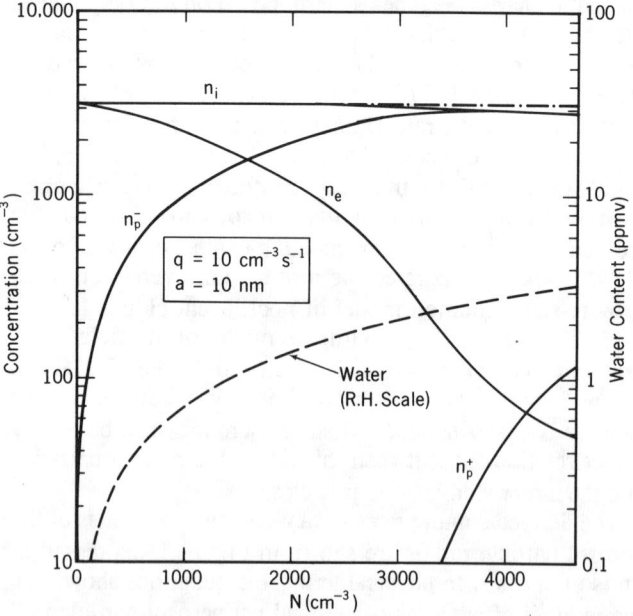

Fig. 4. Steady-state concentrations of electrons, ions, and charged aerosols for an ionization rate of 10 ion pairs cm^{-3} s^{-1} and an assumed aerosol radius of 10 nm. The dash-dot line shows the plasma density in the absence of aerosols, and the broken line and the right-hand scale show the volume mixing ratio of water vapor contained in the aerosols, assuming them to be spherical with the density of normal ice (from *Reid* [1990]).

not likely to be more than a few parts per million [*Arnold and Krankowsky*, 1977; *Garcia and Solomon*, 1985]. Since the concentration of ice particles must be comparable with the undisturbed electron density in order to produce a sizeable biteout, the particles cannot be much larger in radius than about 10 nm, assuming them to be spherical in shape, with the density of normal ice. While neither of these conditions is necessarily true, there is no basis for making any other choice, and they probably represent at least reasonable approximations to the actual shape and composition of the particles. Figure 4 shows the water content as a function of aerosol concentration under these assumptions. While smaller particles would certainly avoid the water-vapor constraint, they would not be able to produce a large electron depletion unless their concentration was extremely high, while larger particles would readily cause a large electron depletion, but would contain unacceptable quantities of water.

4. THE ROLE OF GRAVITY WAVES

The breaking of gravity waves at altitudes somewhat above the mesopause leads to a deposition of momentum at these levels, which in turn forces a global-scale meridional circulation with an upward branch in the summer polar mesosphere, and a downward branch in the winter polar mesosphere. The adiabatic cooling and warming associated with these branches is responsible for the anomalous seasonal temperature distribution in the upper mesosphere, with a cold summer mesopause and a warm winter mesopause at high latitudes.

The high frequency of occurrence of polar mesospheric clouds over the summer poles [*Olivero and Thomas*, 1986] implies that temperatures there are below the local frost point for most of the summer. At the low-latitude edges of the polar caps, however, noctilucent clouds are more sporadic, implying that local saturation conditions occur only occasionally. Much of the time the temperature is probably only slightly above the frost point, and under these conditions the passage of a large-amplitude gravity wave through the region can produce rapid cooling and the formation of a transient ice cloud. Optical observations of airglow emissions from the upper mesosphere over Puerto Rico during the AIDA campaign showed the presence of gravity waves with periods of an hour or two, and with peak-to-peak temperature variations of 10 K or more [*Wiens et al.*, 1993]. Waves with similar periods and amplitudes probably exist at high latitudes, and a drop in temperature of 10 K occurring in a time of the order of an hour could easily produce supersaturated conditions over a region with a horizontal extent of several hundreds of kilometers.

It can be shown [*Hesstvedt*, 1961; *Reid*, 1975] that the rate of growth of the radius of a spherical ice particle under the kinetic theory approximation, which is applicable in the upper mesosphere, is given by

$$\frac{dr}{dt} = \frac{\alpha}{\rho} \left(\frac{m_w}{2\pi k T}\right)^{1/2} (p - p_i) \quad (1)$$

where p and p_i are the ambient and saturation water-vapor pressures, m_w is the mass of a water molecule, ρ is the density of ice, k is Boltzmann's constant, and α is the sticking probability for a water molecule. For supersaturated conditions, $p \gg p_i$, and the radius grows linearly with time.

Assuming a reasonable value of 0.5 for α, a temperature of 140 K, and a water-vapor mixing ratio of 3 ppmv (a pressure of about 1.2×10^{-6} Pa) the rate of growth is about 0.06 nanometers per minute. The radius would reach 5 nm in about 80 minutes, and 10 nm in about 2.5 hours, leading to a substantial electron biteout in either case if the gravity-wave period was comparable with these times.

Sedimentation of the particles out of the cooling region is not likely to be a significant factor. The terminal fall speed of a spherical particle in the upper mesosphere can be shown to be

$$w = \frac{\rho g r}{2n} \left(\frac{\pi}{2 m_a k T}\right)^{1/2} \quad (2)$$

where n and m_a are the concentration and mass of air molecules respectively [*Reid*, 1975]. An updraft of this velocity would thus be sufficient to prevent a particle of radius r from falling. Taking r as 10 nm, we find that a velocity of about 3 cm s^{-1} is all that is needed. Since the temperature variations that occur in a gravity wave are a result of adiabatic cooling and warming, the cold phase in which the ice particles grow must necessarily be accompanied by an updraft. Radar observations in the polar summer mesopause region have in fact measured vertical velocities much larger than those needed to support particles of the sizes considered here [e.g., *Fritts et al.*, 1990].

In their application of the growth-sedimentation model, *Jensen and Thomas* [1991] concluded that it would be difficult to grow a particle to the size needed to cause a substantial electron biteout by gravity-wave cooling, since the particle would fall out of the region before reaching the required size. The updraft in their model had a nominal value of 2 cm s^{-1}, however, [*Jensen and Thomas*, 1988] and did not take account of the fact that the decreasing temperature phase of a gravity wave, which is the region in which the particles grow, must also be accompanied by upward velocities that are significantly larger than the ambient values. The downward phase propagation of gravity waves would also help to keep the particles within the cold region as they fall.

5. RELATIONSHIP TO NOCTILUCENT CLOUDS AND PMSE

Long-period gravity waves are therefore a likely cause of electron biteouts. The small ice particles responsible for the biteouts are transient, evaporating in the warm phase of the wave as quickly as they appear in the cold phase. The larger particles responsible for visible NLC and PMC require much longer times to grow, and presumably can only grow if conditions remain supersaturated for long periods, and vertical velocities are sufficient to keep them from sedimenting. The relationship between the small subvisible particles and the large visible ones is thus not a direct one. Although the subvisible particles may exist in the presence of the larger visible particles, the conditions for their existence are less severe, and they may exist, together with the accompanying biteouts, without the presence of visible clouds.

Polar mesospheric summer echoes (PMSE) have been the subject of several studies since their discovery with the Poker Flat VHF radar in Alaska [*Ecklund and Balsley*, 1981], but their great strength and the wide range of frequencies over which they have been observed are still poorly understood. Possible mechanisms that have been suggested fall into three main categories: coherent scatter from small-scale turbulent irregularities in electron density, incoherent Thomson scatter, and quasi-specular or Fresnel reflection from steep electron-density gradients. The first of these runs up against the formidable problem that the required irregularities, with scale sizes half the radar wavelength, could only exist if electron diffusion were somehow inhibited. *Kelley et al.* [1987] and *Cho et al.* [1992a] have proposed that the necessary restraint could be supplied by the existence of heavy hydrated cluster ions or charged aerosols respectively, but this mechanism does not seem to be capable of explaining the very small irregularities needed to explain the echoes seen at UHF frequencies as high as 1.29 GHz [*Cho et al.*, 1992b].

Mechanisms involving incoherent scatter must involve some major enhancement of the normal incoherent scatter return from the upper mesosphere, which is extremely weak. Such enhancement mechanisms have been suggested by *Havnes et al.* [1990] and *Cho et al.* [1992b], both of whom invoked aerosols to provide the enhancement. The aerosols suggested by *Havnes et al.* [1990] are quite different from those discussed in this paper, consisting of particles with a metallic coating that carry multiple positive charges resulting from photoemission. Those suggested by *Cho et al.* [1992a,b], however, are similar to the particles discussed here, but their suggestion that the predominantly downward motions inferred from Doppler measurements of PMSE represent the fall speed of the particles would imply somewhat larger sizes. A fairly wide range of sizes is, of course, possible, and has been explored in the model of *Jensen and Thomas* [1991].

The difficulties inherent in attempting to explain PMSE by scattering theory have led to the third category of explanation, i.e., that some kind of partial reflection from regions of steep electron-density gradients is involved [e.g., *Röttger and La Hoz*, 1990]. Electron-density gradients are indeed extremely steep at the edges of electron biteouts, but whether or not they can be steep enough to explain the UHF radar measurements is unclear. Observations of strong off-vertical echoes at 50 MHz [*Kelley et al.*, 1990] argue against actual specular reflection as a mechanism, but the existence of tilted structures within an extended sheet of steep gradients could possibly account for the observations.

In summary, the observations have shown a clear relationship between electron biteouts and PMSE, but the detailed mechanism underlying this relationship remains uncertain. Neither PMSE nor electron biteouts, however, have been shown to have a direct connection with the visible noctilucent or polar mesospheric clouds, lending at least circumstantial support to the suggestion that different populations of ice particles are involved.

6. NUCLEATION OF THE BITEOUT PARTICLES

The problem of nucleation of the 'biteout' particles has not been addressed here. In principle, nucleation could take place on either dust particles or ions, both of which are likely to be present in the mesopause region. The analysis presented here has implicitly assumed that the condensation nuclei are dust particles (or meteoric 'smoke' particles [*Hunten et al.*, 1980]) that do not themselves contribute to the charge balance. If the nuclei were heavy cluster ions, each electron that became attached would also remove an ion, and electron biteouts would be accompanied by ion biteouts of comparable size. While there is experimental evidence for positive ion depletions in the same altitude region as electron biteouts [e.g., *Johannessen and Krankowsky*, 1972], they appear to be much less pronounced. More observations are needed, however, before the nucleation issue can be settled.

If the nucleation centers are small (~ 1 nm) dust or smoke particles, they will themselves be predominantly negatively charged, so that a weak electron biteout is likely to be a semi-permanent feature of the mesopause region even in the absence of saturation and ice formation. Since negative ions are not expected to exist in significant numbers at mesopause altitudes, detection of such a permanent imbalance between electron and positive-ion concentrations would be a powerful argument in support of the analysis presented here.

7. SUMMARY

The sharply bounded layers of strongly depleted

electron density, known as electron biteouts, that have been observed in the vicinity of the high-latitude summer mesopause, are likely to be a consequence of electron scavenging by small subvisible ice particles, with radii of the order of 10 nm. These particles could be a transient population resulting from a temporary reduction of the temperature below the frost point by passage of a long-period gravity wave, and are not necessarily directly linked to the larger particles that form visible noctilucent and polar mesospheric clouds. The strong polar mesospheric summer echoes (PMSE) that are recorded by VHF and UHF radars during the local summer may be at least partially related to the extremely steep electron-density gradients that are observed at the upper and lower edges of the biteouts, in agreement with earlier suggestions. Testing of these mainly theoretical arguments is undoubtedly going to require more extensive study of the polar summer mesosphere by well targeted rocket and ground-based experimental campaigns.

REFERENCES

Arnold, F., and D. Krankowsky, Water vapour concentrations at the mesopause, *Nature, 268*, 218-219, 1977.

Cho, J.Y.N., T.M. Hall, and M.C. Kelley, On the role of charged aerosols in polar mesosphere summer echoes, *J. Geophys. Res., 97*, 875-886, 1992a.

Cho, J.Y.N., M.C. Kelley, and C.J. Heinselman, Enhancement of Thomson scatter by charged aerosols in the polar mesosphere: measurements with a 1.29 GHz radar, *Geophys. Res. Lett., 19*, 1097-1100, 1992b.

Ecklund, W.L., and B.B. Balsley, Long-term observations of the arctic mesosphere with the MST radar at Poker Flat, Alaska, *J. Geophys. Res., 86*, 7775-7780, 1981.

Fritts, D.C., U.-P. Hoppe, and B. Inhester, A study of the vertical motion field near the high-latitude summer mesopause during MAC/SINE, *J. Atmos. Terrest. Phys., 52*, 927-938, 1990.

Garcia, R.R., and S. Solomon, The effect of breaking gravity waves on the dynamics and chemical composition of the mesosphere and lower thermosphere, *J. Geophys. Res., 90*, 3850-3868, 1985.

Havnes, O., U. de Angelis, R. Bingham, C.K. Goertz, G.E. Morfill, and V. Tsytovich, On the role of dust in the summer mesopause, *J. Atmos. Terrest. Phys., 52*, 637-643, 1990.

Hesstvedt, E., Note on the nature of noctilucent clouds, *J. Geophys. Res., 66*, 1985-1987, 1961.

Hodges, R.R., Ion pair annihilation by aerosols in the lower ionosphere, *J. Geophys. Res., 74*, 2223-2228, 1969.

Hunten, D.M., R.P. Turco, and O.B. Toon, Smoke and dust particles of meteoric origin in the mesosphere and stratosphere, *J. Atmos. Sci., 37*, 1342-1357, 1980.

Jensen, E.J., and G.E. Thomas, A growth-sedimentation model of polar mesospheric clouds: comparison with SME measurements, *J. Geophys. Res., 93*, 2461-2473, 1988.

Jensen, E.J., and G.E. Thomas, Charging of mesospheric particles: implications for electron density and particle coagulation, *J. Geophys. Res., 96*, 18,603-18,616, 1991.

Johannessen, A., and D. Krankowsky, Positive-ion composition measurement in the upper mesosphere and lower thermosphere at a high latitude during summer, *J. Geophys. Res., 77*, 2888-2901, 1972.

Kelley, M.C., D.T. Farley, and J. Röttger, The effect of cluster ions on anomalous VHF backscatter from the summer polar mesosphere, *Geophys. Res. Lett., 14*, 1031-1034, 1987.

Natanson, G.L., On the theory of the charging of amicroscopic aerosol particles as a result of capture of gas ions, *Sov. Phys. Tech. Phys., 5*, 538-551, 1960.

Olivero, J.J., and G.E. Thomas, Climatology of polar mesospheric clouds, *J. Atmos. Sci., 43*, 1263-1274, 1986.

Parthasarathy, R., Mesopause dust as a sink for ionization, *J. Geophys. Res., 81*, 2392-2396, 1976.

Parthasarathy, R., and D.B. Rai, Effect of meteoric dust on the effective recombination coefficient in the lower ionosphere, *Radio Sci., 1*, 1401-1407, 1966.

Pedersen, A., J. Tröim, and J.A. Kane, Rocket measurements showing removal of electrons above the mesopause in summer at high latitude, *Planet. Space Sci., 18*, 945-947, 1970.

Reid, G.C., Ice clouds at the summer polar mesopause, *J. Atmos. Sci., 32*, 523-535, 1975.

Reid, G.C., Ice particles and electron 'biteouts' at the summer polar mesopause, *J. Geophys. Res., 95*, 13,891-13,896, 1990.

Röttger, J., and C. La Hoz, Characteristics of polar mesosphere summer echoes (PMSE) observed with the EISCAT 224-MHz radar and possible explanations of their origin, *J. Atmos. Terrest. Phys., 52*, 893-906, 1990.

Röttger, J., M.T. Rietveld, C. La Hoz, T. Hall, M.C. Kelley, and W.E. Swartz, Polar mesosphere summer echoes observed with the EISCAT 933-MHz radar and the CUPRI 46.9-MHz radar, their similarity to 224-MHz radar echoes and their relation to turbulence and electron density profiles, *Radio Sci., 25*, 671-687, 1990.

Ulwick, J.C., K.D. Baker, M.C. Kelley, B.B. Balsley, and W.L. Ecklund, Comparison of simultaneous MST radar and electron density probe measurements during STATE, *J. Geophys. Res., 93*, 6989-7000, 1988.

Wiens, R.H., S.P. Zhang, R.N. Peterson, G.G. Shepherd, C.A. Tepley, L. Kieffaber, R. Niciejewski, and J.H. Hecht, Simultaneous optical observations of long-period gravity waves during AIDA '89, *J. Atmos. Terrest. Phys., 55*, 325-340, 1993.

G.C. Reid, Aeronomy Laboratory, NOAA, 325 Broadway, Boulder, CO 80303

Ionic Nucleation of Ice Particles in Noctilucent Clouds

Takuya Sugiyama

Department of Physics, Faculty of Science, Kyoto University

Formation of ice particles in the polar summer mesosphere is simulated with heavy proton hydrates as the origin. In time dependent simulations, stable oscillations of the cloud formation are found in periods of 3 to 4 days, a cloud being bright after a delay of 1.3 days from an active nucleation. Oscillation is found to be a new criterion to determine whether embryonic nuclei of the clouds are in situ origin or external origin.

1. INTRODUCTION

Meteor debris is accepted as the major embryonic nuclei for ice particles in noctilucent clouds (NLCs). Observations of heavy proton hydrates, $H^+(H_2O)_n$, (PHs) in the neighborhood of a NLC [*Björn and Arnold*, 1981], however, have suggested a possible contribution of PHs to the origin of the clouds. *Arnold* [1980] proposed a scheme in which electron recombination converts heavy PHs to neutral embryonic nuclei for ice particles. With a simple estimation of hydration time given by *Arnold and Joos* [1979], *Turco et al.* [1982] found an onset of probable oscillatory formation of clouds in their time dependent simulations. Recently, *Sugiyama* [1994] developed the ion-recombination nucleation scheme of *Arnold* [1980] and has attempted to simulate NLC formation as a whole through PH origin. In this paper, with a review of the *Sugiyama's* [1994] work, a cloud formation through PHs is compared with that of an external origin.

2. MODEL OF PROTON HYDRATIONS

Evaluations of proton hydration in the mesosphere require all coefficients of the master equation of clustering because the spectrum deviates largely from equilibrium because the recombination shortcircuits the hydration:

$$H^+(H_2O)_{n-1} \underset{b_n}{\overset{f_{n-1}}{\rightleftharpoons}} H^+(H_2O)_n \underset{b_{n+1}}{\overset{f_n}{\rightleftharpoons}} H^+(H_2O)_{n+1}$$

$$\downarrow \alpha_{n-1} N_e \qquad \downarrow \alpha_n N_e \qquad \downarrow \alpha_{n+1} N_e$$

$$\text{neutral products} \qquad (1)$$

where N_e is the electron concentration, and α_n is the recombination coefficient of the nth PH with electrons. The forward and backward hydration frequency f_n and b_n are related to the reaction coefficients $k_{f,n}$ and $k_{b,n}$ by

$$f_n = k_{f,n}[H_2O][M], \qquad b_n = k_{b,n}[M] \qquad (2)$$

where brackets denote concentration of species and M represents the third body. For the heavy PH production to be effective, low temperature and high humidity are required to fulfill $f_n > b_n$ and $f_n > \alpha_n N_e$. The reaction coefficients of PHs up to the hydration order of $n = 6$ have been given in laboratory experiments by *Lau et al.* [1982], among which f_5 and b_6 are shown in Figure 1. Coefficients up to $n = 1000$ are modeled with the nucleation theory and with a help of an approximation, in the unimolecular reaction theory, which assumes that the dissociation rate is determined by the excess energy in reaction. Figure 1 illustrates the model reaction frequencies for $n \leq 12$. In the reaction model, for a cluster ($n > 8$) at low temperatures ($T < 130\,\text{K}$) with $[M] = 10^{14}\,\text{cm}^{-3}$, $k_{f,n}[M]$ approaches $k_L = 10^{-9}\,\text{cm}^{-3}\,\text{s}^{-1}$ (Langevin rate in the ion-molecular reaction). This is because a heavy cluster at low temperatures captures every colliding water molecule by imparting the released binding energy to its large number of degrees of

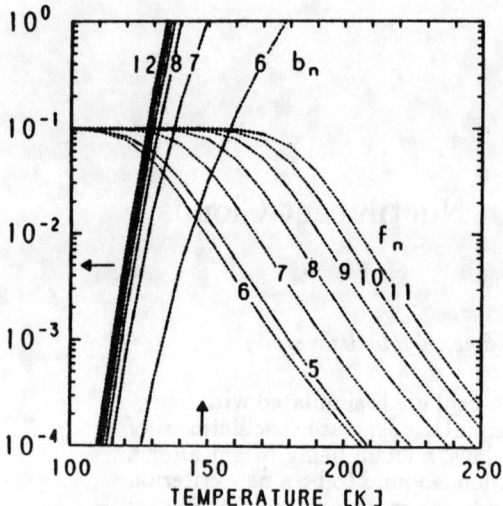

Fig. 1. Model proton hydration frequencies in the case of $[M]=10^{14}\,cm^{-3}$ and $[H_2O]=1$ ppmv. Here f_5 and b_6 are laboratory data by *Lau et al.* [1982]. The horizontal arrow indicates a typical recombination frequency in the mesosphere. The vertical arrow indicates the equilibrium temperature for bulk ice.

freedom, whereas a small cluster captures a water molecule only by transferring the excess energy to the third body. Recombination coefficients of PHs with $n \leq 6$ have been reported by *Leu et al.* [1973], and $\alpha_n = 5 \times 10^{-6}\sqrt{300/T}\,cm^3\,s^{-1}$ is assumed for $n \geq 7$.

An example of model PH spectra in the mesopause is shown in Figure 2. In the abscissa, n^* is a critical size of a neutral cluster in the background neutral system, in other words, the instability of $(H_2O)_{n^*}$ caused by its surface energy thermodynamically balances the supersaturated water vapor. In the spectrum, a critical size for ion-recombination nucleation n_R^* is defined by $n_R^* = n^* + n_v$, where n_v is the number of molecules evaporated from a PH at recombination with electron. Provided that 13.6eV is released at recombination, $n_v \sim 30$ is obtained because the binding energy is ~ 0.4 eV per water molecule in bulk ice. In a scheme of ion-recombination nucleation, electron recombination converts PHs with $n > n_R^*$ into neutral clusters $(H_2O)_n$ with $n \geq n^*$ which get runaway growth in the neutral system. Then the nucleation rate J for neutral embryonic nuclei is defined by the sum of recombination rates of PHs above $n \geq n_R^*$. In a system with a steady ionization rate Q, we can define a nucleation efficiency η by J/Q. As a result, an ionic system with recom-

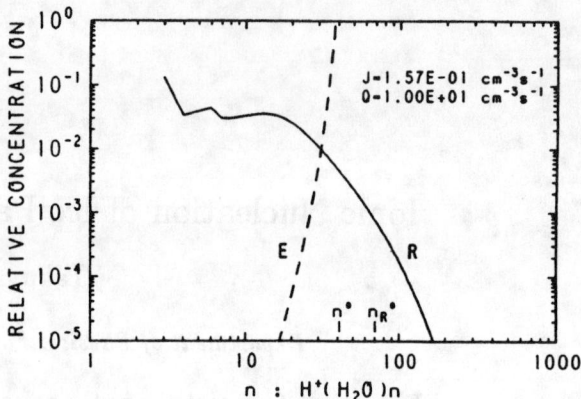

Fig. 2. PH size spectrum, relative to total, obtained by the hydration model for $Q = 10\,cm^{-3}\,s^{-1}$ at $T = 125$ K, $[M] = 10^{14}\,cm^{-3}$ and $[H_2O] = 2.5$ ppmv. The dashed line, labeled E, is an equilibrium spectrum ($N_e = 0$) and the solid line, labeled R, is a spectrum in the ion-recombination case ($N_e = 1.2 \times 10^3\,cm^{-3}$). A resultant nucleation rate is $J = 1.57 \times 10^{-1}\,cm^{-3}\,s^{-1}$.

bination develops into a steady engine to produce neutral embryos for condensation.

3. OPTIMUM IONIZATION RATE FOR IONIC NUCLEATION

In the scheme of ion-recombination nucleation, there is an optimum Q_m which maximizes J, approximately evaluated by

$$\frac{1}{2}\frac{n_R^*}{k_{f,n_R^*}[M][H_2O]} \approx \frac{1}{\sqrt{Q_m \alpha_{n_R^*}}}, \quad (3)$$

where the left-hand side, except for a factor $1/2$, is a hydration time to produce n_R^*th ion cluster, and the right-hand side is a recombination time of n_R^*th PH with electrons. Using $k_{f,n_R^*}[M] = k_L$ and $\alpha_{n_R^*} = 5 \times 10^{-6}\,cm^3\,s^{-1}$, $Q_m \sim 8\,cm^{-3}\,s^{-1}$ is obtained for $n_R^* = 100$ and $[H_2O] = 3 \times 10^8\,cm^{-3}$.

Assuming that the ion-recombination nucleation occurs within a thin layer (nucleation layer) located above a NLC layer, a rough estimate of required J for NLC particles $r = 70$ nm in radius with concentration $N_{P,r} = 100\,cm^{-3}$ is as follows. In an uniform atmosphere with $[M]=10^{14}\,cm^{-3}$, a particle 70 nm in radius descends with a velocity of $v_r = 40\,cm\,s^{-1}$. Then the particle flux at the cloud region is $N_{P,r}v_r = 4 \times 10^3\,cm^{-2}\,s^{-1}$, which equals a flux of embryonic nuclei F_E produced at a nucleation layer because of the conservation of flux. If the depth of a nucleation layer l is 400 m, then $J = 0.1\,\text{cm}^{-3}\,\text{s}^{-1}$ is required from $F_E = Jl$. At the bottom of the nucleation layer, neutral clusters $r^* \sim 1\,\text{nm}$ are accumulated with $N_{P,r^*} \sim 7 \times 10^3\,\text{cm}^{-3}$, since their descent speed is $v_{r^*} \sim 0.6\,\text{cm s}^{-1}$.

Figure 3 shows a variation of J against Q at temperatures around $T \sim 125\,\text{K}$ with $[H_2O] = 2\,\text{ppmv}$. The ionization rate of $Q \sim 10\,\text{cm}^{-3}\,\text{s}^{-1}$ is optimum for nucleation of $J \sim 0.1\,\text{cm}^{-3}\,\text{s}^{-1}$. Thus Figure 3 suggests that the quiettime mesopause may produce a rather stable flux of embryonic nuclei against variations of Q. A question has arisen, however, of how the supersaturation condition needed to maximize the nucleation rate is maintained in spite of the large variability in the mesosphere.

4. EMBRYO FLUX FROM A NUCLEATION LAYER

A model nucleation layer is set to have $\Delta T = 3.5$ K in temperature dip and 600 m in thickness of Gaussian e^{-1} full width at the altitude of 88.5 km, where the July mean temperature shows a minimum of $T = 128.5\,\text{K}$ [Lübken and von Zahn, 1989]. Produced embryo fluxes F_E from the model nucleation layer are calculated with models of hydration, growth-sedimentation and charging of ice particles. As shown in Figure 4, F_E is a strong function of $[H_2O]$ since each hydration is in proportion to $[H_2O]$. Further, F_E increases rapidly above a in Figure 4, because neutral embryonic nuclei capture electrons even if they are as small as 1 nm in radius, and heavy PHs are "acceleratedly" formed when electrons are depressed. Production of neutral embryonic nuclei through heavy PHs and depression of electrons are in a positive feedback. A depression of electrons, however, decreases the conversion of heavy ions into neutral embryos, so that the scheme of ion-recombination nucleation is also responsible for the saturation of F_E occurring above b in Figure 4. In the general scheme of ionic nucleation, F_E should approach the utmost limit of $J = Q$, but it may not be the case in the mesosphere (see section 5). Resultant concentrations of ice particles at the bottom of the nucleation layer (circles in Figure 4) are also strongly dependent on $[H_2O]$. How the mesopause maintains a supersaturation condition to bring about embryos for a "stable" cloud is a crucial question, which will be resolved in the time dependent simulations.

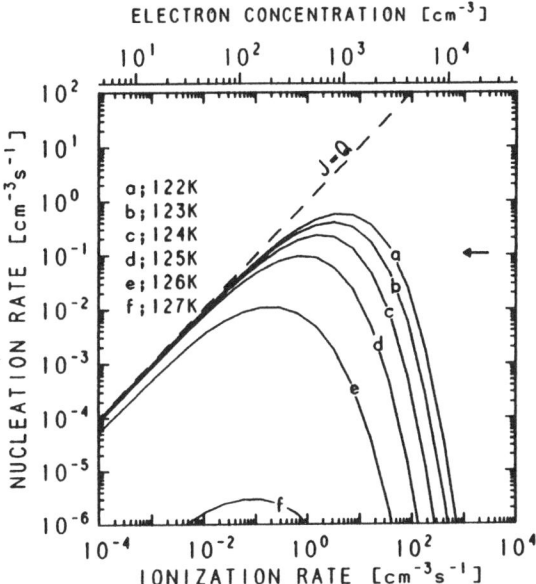

Fig. 3. Nucleation rate versus ionization rate in the system with $[M] = 10^{14}\,\text{cm}^{-3}$ and $[H_2O] = 2\,\text{ppmv}$ at around $T = 125\,\text{K}$. The broken line shows $J = Q$ which is the upper limit of the ionic nucleation. The horizontal arrow is a rough estimate for J in a NLC (see text). Concentrations of electrons are shown at the top of the figure.

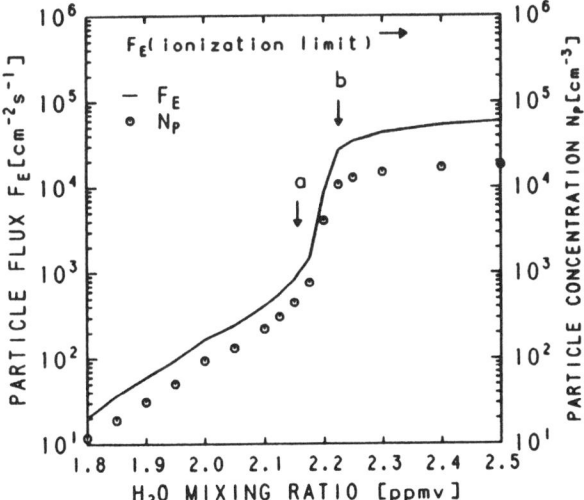

Fig. 4. Model results of embryo fluxes at the bottom of a nucleation layer as a function of H_2O mixing ratio. $Q = 10\,\text{cm}^{-3}\,\text{s}^{-1}$ is applied. The horizontal arrow is the utmost flux in the ionization limit given by Ql with $l = 600\,\text{m}$. Circles are concentrations of particles at the bottom of the nucleation layer, with their axis in the right-hand side.

5. OSCILLATIONS OF CLOUD FORMATION THROUGH PHS

In the mesosphere temperatures are higher at lower altitudes, and the growth of embryos is limited by the descent time down to the cloud bottom where supersaturation ratio equals to 1. *Reid* [1975] showed that it is difficult for a spherical ice particle to grow to $r \sim 100$ nm in the mesosphere, and he suggested the shape could be non-spherical. *Turco et al.* [1982] showed the onset of probable oscillatory formation of a cloud through PHs in their time dependent simulations. *Jensen and Thomas* [1988] showed a "flash" supply of embryos severely changes the water vapor profile.

The substantial consumption of H_2O owing to growth of ice particles solves the question of how the mesosphere materializes rather "stable" clouds against the strong dependence of nucleation rate on $[H_2O]$. This is shown in simulations of cloud formation as a whole, with models of the embryo production, growth-sedimentation, and evaporation and relaxation of H_2O by one-dimen-sional diffusion. The following are assumed in the models; a boundary $[H_2O]$ at the altitude of 70 km is fixed to be 4 ppmv; diffusion coefficient is $D_{zz} = 20\,\text{m}^2\,\text{s}^{-1}$; updraft wind velocity is $2\,\text{cm}\,\text{s}^{-1}$ below 86 km and linearly damped to zero at 88.5 km; and photodissociation rate of H_2O molecule is $10^{-6}\,\text{s}^{-1}$. Supersaturation ratio is about 200 at the altitude of 88.5 km where $T_{min}=125$ K, and it equals to 1 at about 82 km. The embryo flux as shown in Figure 4 is assumed to appear at 88.5 km. A result is shown in Figure 5a where a cloud formation oscillates with a period of 4 days, which is the sum of 1.3 days for the sedimentation and 2.7 days for the relaxation of H_2O by diffusion. In our calculation, steady formation of clouds occurs when $[H_2O]$ at 70 km is lower than 3.5 ppmv; otherwise, the clouds oscillate stably. The oscillation is caused both by the strong sensitivity of nucleation rate to $[H_2O]$ and by the finite time of diffusive relaxation of H_2O

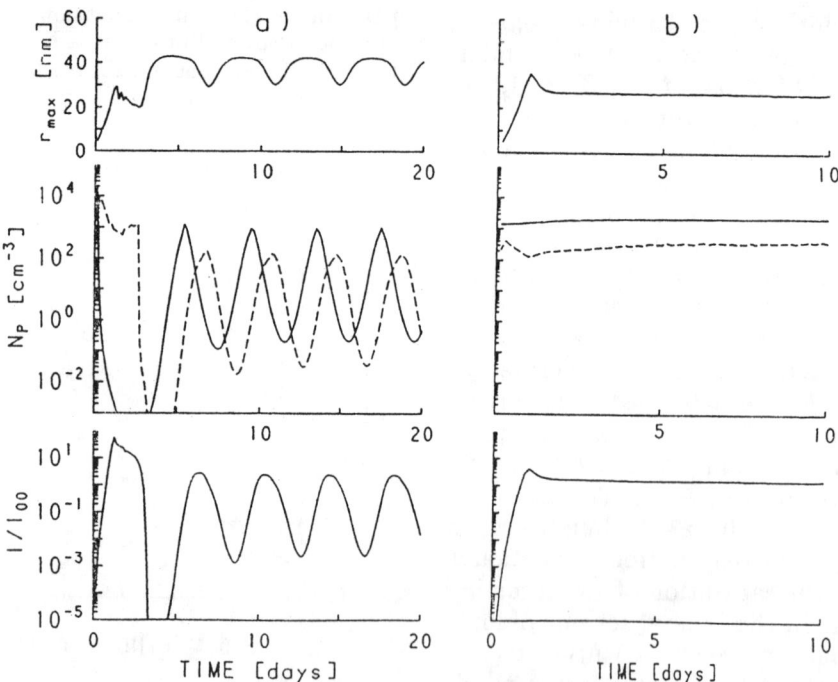

Fig. 5. a) Results of time dependent simulations of a cloud formation where nucleation begins with sudden decrease of temperatures. Top: Maximum radius of mature ice particles. Middle: Concentration of particles at the bottom of the nucleation layer (solid line), and those at the maximum radius (dashed line). Bottom: Scattering intensity obtained by $\int_0^\infty \int_0^\infty N_{P,r}(z)r^6 dr dz$ divided by an intensity of a reference cloud 1 km in thickness with a monochromatic particle size of $r = 70$ nm and $N_{P,r} = 1\,\text{cm}^{-3}$. b) Same as Figure 5a except but embryo flux at 88.5 km is fixed to be $F_E = 2 \times 10^3\,\text{cm}^{-3}\,\text{s}^{-1}$.

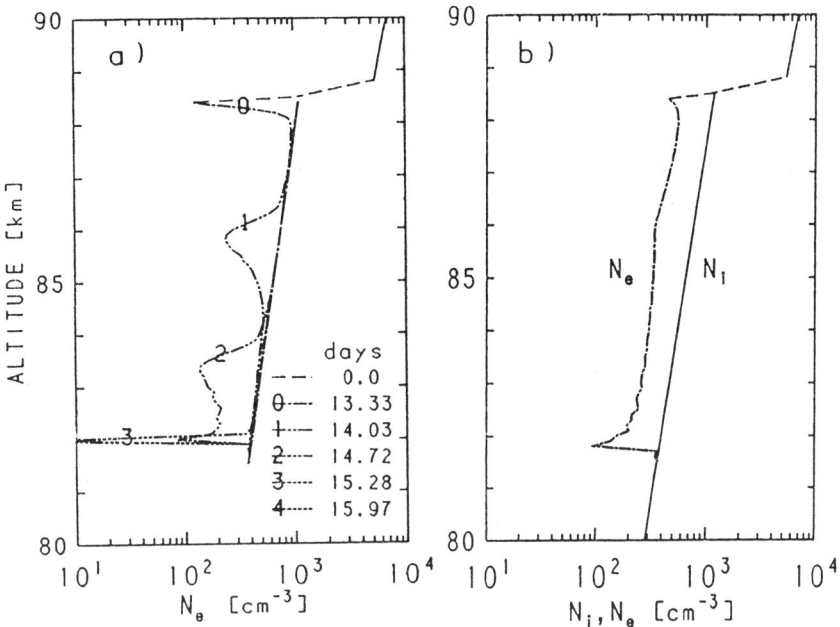

Fig. 6. a) Electron profiles in a similar simulation with Figure 5a except but $D_{zz} = 40\,\text{m}^2\,\text{s}^{-1}$. Oscillation period of cloud formation is 3.3 days. Times from the onset of nucleation for labels 0-4 are denoted in the figure. Solid line above 89 km is the region of major primary ions. b) Electron and positive ion profiles in the same simulation in Figure 5b. Times from the onset of nucleation is 9 days.

vapor. An "excess" production of nuclei is easily occurs which itself results in a pause of the nucleation activity through the effect of "freeze-drying".

For comparison, we made a simulation of a cloud formation with externally supplied embryonic nuclei. Figure 5b shows a result of a simulation with a steady flux of $F_E = 2 \times 10^3\,\text{cm}^{-3}\,\text{s}^{-1}$ occurring at the altitude of 88.5 km. A transient variation of a cloud begins at the onset of embryo supply, but they decay in ∼1 day. The cloud becomes steady due to an upward H_2O flow to compensate the deposition of water vapor into ice particles. Figure 5 stresses that the oscillatory formation is a characteristic feature of the in situ origin of NLC particles.

6. VARIATION OF ELECTRON CONCENTRATION

Figure 6a shows time variations of electron profile in the similar simulation in Figure 5a except for $D_{zz} = 40\,\text{m}^2\,\text{s}^{-1}$. Owing to the large diffusion coefficient, the period of cloud formation is shortened to 3.3 days and a large supply of water vapor to the nucleation layer brings about an "accelerated" nucleation. And electron bite-out in our simulation occurs in the region of active nucleation due to the charging of embryos, and it develops to depressions of electrons in the course of particle descent. The electron bite-out ceases in the bright phase of a cloud. An external supply of a steady embryo flux, on the other hand, brings about extended electron depressions which endure as long as a large supply of embryos continues (Figure 6b). The two figures in Figure 6 suggest that a major contribution of PHs to the cloud particles is not plausible if electron bite-outs or depressions endure more than 2 days. We anticipate strong radar returns from these regions of steep profiles of electrons where electron irregularities, caused by the vertical motions of neutral turbulence, are largely enhanced.

7. CONCLUSIONS

Our simulations stress that there is a neat connection between the electron bite-out and the oscillatory formation of NLCs because heavy PHs are acceleratedly formed in the deficiency of electrons. The hypothesis of PH origin of NLCs will become more plausible if following observational evidences are obtained.

1) Radar echoes from a fixed altitude, electron

bite-out, and brightness of NLCs and polar mesospheric clouds oscillate with a period of 3 to 4 days.

2) Echoes from heavy PH regions precede a bright period of a cloud by ~1 day, and the echoes cease when a cloud is bright.

Future developments are required both in microscopic and macroscopic modeling of the cloud formation. An improved clustering model or laboratory data for PH formation even up to $n = 20$ will greatly improve accuracy of the whole discussions. We anticipate spatial structures of the cloud formation to appear in a two-dimensional diffusion model.

Acknowledgments. The author thanks H. Hasegawa and T. Kozasa for tutorial discussions on the nucleation theory. He also thanks Y. Muraoka for useful discussions. This research was supported by the grant-in-aid for "Computer Experiments and Data Analysis" by Radio Atmospheric Science Center, Kyoto University.

REFERENCES

Arnold, F., Ion-induced nucleation of atmospheric water vapor at the mesosphere, *Planet. Space Sci.*, *28*, 1003-1009, 1980.

Arnold, F., and W. Joos, Rapid growth of atmospheric cluster ions at the cold mesopause, *Geophys. Res. Lett.*, *6*, 763-766, 1979.

Björn, L. G., and F. Arnold, Mass spectrometric detection of precondensation nuclei at the arctic summer mesopause, *Geophys. Res. Lett.*, *8*, 1167-1170, 1981.

Jensen, E. J. and G. E. Thomas, A growth-sedimentation model of polar mesospheric clouds: comparison with SME measurements, *J. Geophys. Res.* *93*, 2461-2473, 1988.

Lau, Y. K., S. Ikuta, and P. Kebarle, Thermodynamics and kinetics of the gas-phase reactions, $H_3O^+(H_2O)_{n-1} + H_2O = H_3O^+(H_2O)_n$, *J. Amer. Chem. Soc.*, *104*, 1462-1469, 1982.

Leu, M.T., M. A. Biondi, and R. Johnsen, Measurements of the recombination of electrons with $H_3O^+ \cdot (H_2O)_n$-Series Ions, *Phys. Rev. A, Gen. Phys.*, *7*, 292-298, 1973.

Lübken, F. J., and U. von Zahn, Thermal structure of the mesopause region at polar latitudes, *J. Geophys. Res.*, *96*, 20841-20857, 1991.

Reid G. C., Ice cloud at the summer polar mesopause, *J. Atoms. Sci.*, *32*, 523-535, 1975.

Sugiyama, T., Ion-recombination nucleation and growth of ice particles in noctilucent clouds, *J. Geophys. Res.*, *99*, 3915-3929, 1994.

Turco, R. P., O. B. Toon, R. C. Whitten, R. G. Keesee, and D. Hollenbach, Noctilucent clouds: simulation studies of their genesis, properties and global influences, *Planet. Space Sci.*, *30*, 1147-1181, 1982.

T.Sugiyama, Department of Physics, Faculty of Science, Kyoto University, Kyoto, 606-01, Japan

Changes in the Concentration of Mesospheric O_3 and OH During a Highly Relativistic Electron Precipitation Event

R. A. Goldberg[1], C. H. Jackman[2], D. N. Baker[1], and F. A. Herrero[1]

NASA/Goddard Space Flight Center, Greenbelt, Maryland

Highly relativistic electron precipitation events (HREs) can provide a major source of energy affecting ionization levels and minor constituents in the mesosphere. Based on satellite data, these events are most pronounced during the minimum of the solar sunspot cycle, increasing in intensity, spectral hardness and frequency of occurrence as solar activity declines. Furthermore, although the precipitating flux is modulated diurnally in local time, the noontime maximum is very broad, exceeding several hours. Since such events can be sustained up to several days, their integrated effect in the mesosphere can dominate over those of other external sources such as relativistic electron precipitation events (REPs) and auroral precipitation. In this work, the effects of HRE relativistic electrons on the neutral minor constituents OH and O_3 are modeled during a modest HRE, to estimate their anticipated impact on mesospheric heating and dynamics. The data to be discussed and analyzed were obtained by rocket at Poker Flat, Alaska on May 13, 1990 during an HRE observed at midday near the peak of the sunspot cycle. Solid state detectors were used to measure the electron fluxes and their energy spectra. An x-ray scintillator was included to measure bremsstrahlung x-rays produced by energetic electrons impacting the upper atmosphere; however, these were found to make a negligible contribution to the energy deposition during this particular HRE event. Hence, the energy deposition produced by the highly relativistic electrons dominated within the mesosphere and was used exclusively to infer changes in the middle atmospheric minor constituent abundances. By employing a two-dimensional photochemical model developed for this region at Goddard Space Fight Center, it has been found that for this event, peak modifications in the neutral minor species occurred near 80 km. A maximum enhancement for OH was calculated to be over 40% at the latitude of the launch site, which in turn induced a maximum depletion of O_3 in excess of 30%. Since this particular HRE occurred near solar maximum, it was of modest intensity and spectral hardness, parameters which could grow significantly as solar minimum is approached. Estimates of mesospheric OH enhancement and O_3 depletion have also been made for more intense HRE events, as might be expected during the declining phase of the solar cycle. The findings imply that the energy deposition from highly relativistic electrons during more intense HREs could modulate the concentration of important minor species within the mesosphere to much higher levels than estimated for the observed HRE. By causing O_3 destruction, the electron precipitation can also modify the penetration depth of solar UV radiation, which may affect thermal properties of the mesosphere to depths approaching 60 km.

1) Laboratory for Extraterrestrial Physics
2) Laboratory for Atmospheres

The Upper Mesosphere and Lower Thermosphere:
A Review of Experiment and Theory
Geophysical Monograph 87
This paper is not subject to U.S. copyright.
Published in 1995 by the American Geophysical Union

INTRODUCTION

Highly relativistic electron precipitation events (HREs) may provide the major source of ionization energy from precipitating electrons affecting mesospheric constituents. Based on satellite data, these events are most pronounced during the declining phase of the solar cycle, increasing in intensity, spectral hardness, and frequency of occurrence as the solar cycle reaches minimum [e.g., *Baker et al.*, 1987, 1993]. Since such events can be sustained up to several days, their integrated effect in the mesosphere can potentially dominate other sources of energy input such as relativistic electron precipitation events (REPs) and auroral precipitation events (APEs). Here, the effects of an HRE on mesospheric neutral minor constituents such as ozone (O_3) during a measured HRE are modeled, to demonstrate its impact on these constituents.

It has been known for some time that solar protons impact the middle atmosphere and affect its neutral and electrodynamic structure. Rocket measurements by Weeks et al. [1972] during the large solar proton event (SPE) of November, 1969 were used by Swider and Keneshea [1973] in a model calculation to demonstrate that solar proton events (SPEs) could modify mesospheric HO_X, O_3, and other minor constituents. For the great SPE in August, 1972, Heath et al. [1977] showed a remarkable depletion in stratospheric O_3 by measuring solar backscattered ultraviolet radiation with the SBUV instrument aboard Nimbus IV. This O_3 depletion required several weeks for recovery following the event [*Jackman and McPeters*, 1987]. Thomas et al., [1983], Solomon et al. [1983], McPeters and Jackman [1985], and Jackman and McPeters [1985] have also investigated and/or modeled significant middle atmospheric O_3 depletions caused by several other SPEs of varying magnitudes. More recently, Johnson and Luhmann [1993] have shown evidence for SPE modulation of mesospheric dynamical structure, comparing Poker Flat MST Radar data with SPE occurrences during the early 1980s. Finally from the electrodynamic viewpoint, ionization enhancements from SPEs are sufficiently intense and well defined to have caused a renaming of SPEs as polar cap absorption events (PCA), in response to the major ionization effects caused by SPEs at high magnetic latitudes.

REPs and APEs are relatively short-lived compared to SPEs. Since REPs and APEs are much less intense and do not usually penetrate as deeply into the atmosphere, they are more likely to have their maximum effect in the mesosphere; i.e., maximum electron energies between 0.5 and 1.0 MeV usually restrict energy deposition by REPs and APEs to atmospheric altitudes above 60 km [*Goldberg et al.*, 1984; *Jackman*, 1991]. Thorne [1980] has argued that REP events might have as significant an impact on the middle atmosphere as SPEs, because of their much higher frequency of occurrence. However, his definition of REPs may have been expanded to include all relativistic electron precipitations, including those associated with auroral geomagnetic disturbances and possibly HREs, which are usually characterized by a harder energy spectrum.

HREs represent the most intense and spectrally hard electron events observed to date. Details of these events as observed at geosynchronous satellite altitudes can be found in Baker et al. [1979, 1986]. From comparisons with lower altitude satellites [*Imhof et al.*, 1991], and from the rocket study to be evaluated here, it is apparent that a significant fraction of the outer zone (high altitude) electrons associated with an HRE reach the middle atmosphere [*Herrero et al.*, 1991; *Baker et al.*, 1993] and strongly influence the electrodynamics of that region [*Goldberg et al.*, 1994]. Since HREs can sustain their activity for several days (albeit with a diurnal variability having a wide local noontime maximum of several hours) and recur over several solar rotations, and since they may cover a broader region in longitude and latitude than the auroral zone, their impact on the middle atmosphere should be large.

This paper considers an HRE that occurred in May 1990 as reported earlier [*Herrero et al.*, 1991; *Baker et al.*, 1993; *Goldberg et al.*, 1994]. It is concerned with the influence that the HRE electrons had on the minor constituents OH and O_3 within the mesosphere, the region where most of the HRE electron energy from this event was absorbed. It is based on results obtained from rocket payloads launched during the event from Poker Flat Research Range, Alaska. The event studied was found to be relatively modest, which is to be expected immediately following the maximum of the solar cycle. Nonetheless, the results from a GSFC two-dimensional photochemical model imply that the modest fluxes observed in this event were capable of modifying the OH and O_3 concentrations in the mesosphere by a measurable amount.

EXPERIMENT DESCRIPTION

General Considerations

In May of 1990 multiple payloads were launched in a rocket experiment to measure the relativistic electron fluxes reaching the middle atmosphere during an HRE, and to investigate the influence of HREs on the electrodynamical and neutral properties of the middle atmosphere. Since the relativistic electron precipitation events are thought to be increasingly frequent and more intense following the solar maximum period, it was the goal of the May 1990

Table 1. Characteristics of Taurus-Orion flight trajectories.

ROCKET		LAUNCH		APOGEE		
Number	Type	Date	Time (UT)	Altitude (km)	Time (TSL, sec)	Horizontal Range (km)
33.059	Taurus Orion	13 May 90	2129:02	130.4	184	44.0
33.060	Taurus Orion	14 May 90	2026:00	127.9	182	45.5

experiment to first measure the intensity and energy spectrum of these relativistic electrons in the middle atmosphere for a modest event near solar maximum. Poker Flat, Alaska was selected because it lies at the equatorward edge of the auroral zone, where representative components of precipitating electron showers are normally expected to occur, and where there is the opportunity to coordinate measurements with those from a ground station and from a geostationary spacecraft on approximately the same magnetic field line.

The linear prediction filter of Baker et al. [1990] was used to anticipate when relativistic electrons should be at their maximum levels. This technique utilizes ground-based geomagnetic indices as input time series and gives a reasonable level of confidence for predicting enhancements months in advance of their occurrence. Real-time data from operational spacecraft in synchronous orbit were also available to help judge the specific time when the rocket launches should occur. These data were accessed via the NASA/SPAN computer network and were resident on computers at the Los Alamos National Laboratory and the NOAA Space Environment Laboratory.

Two Taurus-Orion rockets and one Orion rocket were launched into an HRE event of modest intensity near local noon on May 13-14, 1990, when outer zone trapped electrons were locally at peak diurnal intensity. Apogee for each Taurus-Orion rocket trajectory was about 130 km, providing full coverage of the mesosphere and part of the lower thermosphere. Each Taurus-Orion payload contained a mix of solid state and x-ray detectors to measure the energetic radiation at and below about 130 km altitude, and an assortment of plasma probes and an electric field symmetric boom array to measure middle atmospheric electrodynamic response to the impinging radiation. In addition, Taurus-Orion 33.059, launched on May 13, 1990, was immediately followed by an Orion payload (30.035), containing assorted instrumentation to measure the electrodynamic properties of the middle atmosphere in more detail than on the Taurus-Orion. This work makes use of the flux measurements obtained on 33.059 and 33.060 to evaluate the expected perturbations of the measured HRE on OH and O_3. The electrodynamic properties of the atmosphere obtained from this study can be found in Goldberg et al. [1994] and will not be discussed further in this paper. Table 1 lists the Taurus-Orion rocket flights with their relevant specifications.

Critical Instrument Descriptions

Most instruments flown on the Taurus-Orion and Orion payloads have been described in elsewhere [*Herrero et al.*, 1991; *Goldberg et al.*, 1994]. The electron spectrometers and the x-ray scintillator are briefly described here, because of their prime use in measuring the electron energy flux. Each Taurus-Orion payload included two silicon (Li-drifted) solid state electron spectrometers to detect low energy electrons (0.09-1.0 MeV, 12 differential channels) and high energy electrons (0.4 to 3.8 MeV, 8 differential channels), respectively. In addition, each payload carried an x-ray NaI scintillator detector to measure bremsstrahlung x-rays and electrons above 120 keV. Each scintillator was fitted with a 125 micron Be window and optically mounted to a ruggedized photo multiplier [cf. *Goldberg et al.*, 1984 for a detailed description). Electronic sampling of the pulse amplitudes permitted sampling in the integral spectral ranges >5, >10, >20, >40, and >80 keV. These detectors were also sensitive to charged particles that can penetrate the Be window, which has a threshold for electrons near 120 keV.

During each Taurus-Orion flight, the payload was despun to approximately 1.3 Hz and then oriented at 45° to the Earth's magnetic field with a magnetic attitude control system. This permitted the solid state and x-ray detectors to sweep in look angle from 0° to 90° magnetic pitch angle during each revolution of the payload, since they were

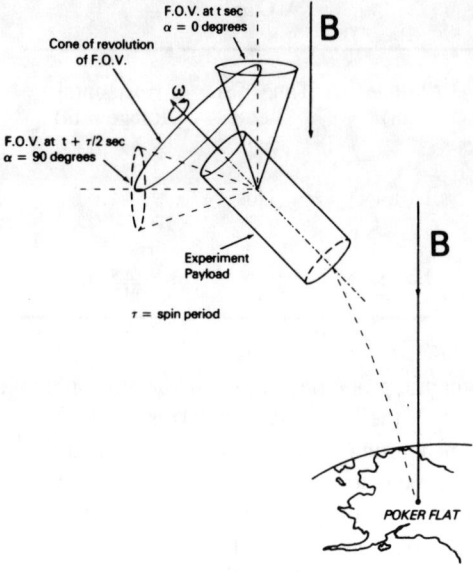

Fig. 1. Schematic view of payload geometry demonstrating how a 0-90° pitch angle sweep for spectrometers and scintillator detectors is induced by payload spin.

A detailed description of the energy deposition analysis has been provided in Goldberg et al. [1984] and specifically for this flight in Goldberg et al. [1994]. Briefly, the count rates from each channel are combined to produce an exponential integral spectrum of the form:

$$J(>E) = J_0 \int_E^\infty e^{-E/E_0} dE \quad (1)$$

where J_0 is the flux at energy $E = 0$, and E_0 is folding energy. Figure 3 provides a sample of these spectra obtained during period "3" between 230 and 231 seconds into the flight. The vertical bars on each point represent the one-sigma range for the statistical sampling rate. The upper and lower panels depict spectra obtained by the low and high energy spectrometers, respectively. The curves labeled A, B, C, and D represent least squares fits to the measured points to the left and right of each breakpoint, which appear at about 350 keV (upper) and 1700 keV (lower). The values for J_0 and E_0 for each linear segment are displayed in the upper right hand corner of each panel. From this plot we observe that the total spectrum can be described with three linear components having breakpoints near 0.35 and 1.7 MeV, since curves B and C are overlapping regions and roughly equivalent. Additional plots for events "1" and "2"

mounted at 45° to the spin axis. A schematic diagram depicting this function is illustrated in Figure 1. Although the fields of view for the x-ray and low energy electron detectors were large (~30° half angle), the instruments still provided useful information concerning the pitch angle distribution of impacting electrons. For the high energy electron spectrometer, the field of view was only 13.5° half angle, thereby providing higher instrumental pitch angle resolution.

DATA ANALYSIS AND RESULTS

Figure 2 depicts the count rates measured by three representative channels on each electron spectrometer during the flight of 33.059. The upper panel (a) refers to the low energy spectrometer, the lower panel (b) to the high energy spectrometer. Also shown are the trajectory of the flight and magnetic pitch attitude (Fig. 2a). The latter shows that the payload stabilized near 115 sec (~105 km) into the flight. Of note are the wide burst region encompassing "1", the narrow burst region "2", and the more nominal steady region "3". From the energy spectra measured during each of these periods, we have been able to calculate the energy deposition within the middle atmosphere produced by the measured flux.

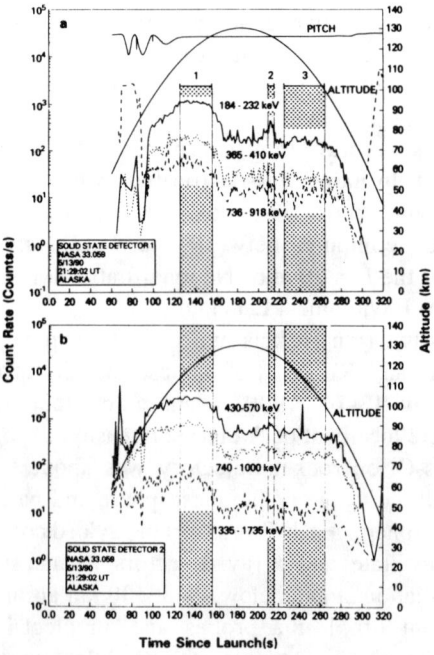

Fig. 2. Compressed view of counting rates (33.059) for three select channels on the low energy (top panel) and high energy (bottom panel) spectrometers. Payload pitch attitude (arbitrary units) and trajectory are also provided.

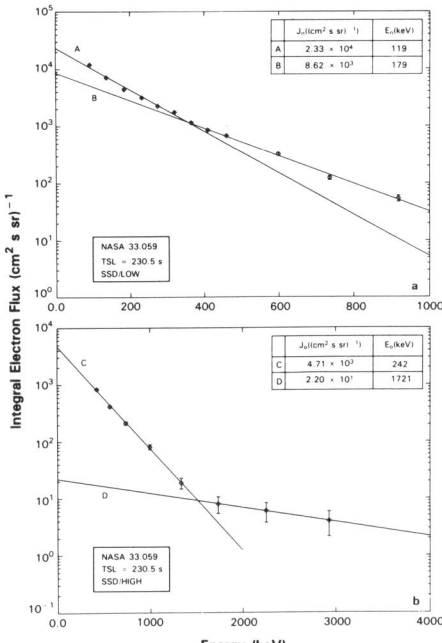

Fig. 3. Sample spectra for non-burst period "3" for 230-231 sec TSL on 33.059. Spectra for the low (top) and high (bottom) energy spectrometers are provided. The tables (upper right) give values of E_0 and J_0 for segments A and B (low energy) and C and D (high energy).

during the 33.059 flight show similar features but with increased fluxes, particularly at the lower energies. Corresponding plots were made for 33.060, but in this case from energy spectrum values averaged over the entire flight above approximately 105 km (110-260 seconds into the flight), since the electron flux at all measured energies was nearly uniform and without bursts during this period. Finally, it has been shown in Goldberg et al. [1994] that the pitch angle distribution in most cases was nearly constant; hence the flux was treated as isotropic when calculating the energy deposition.

Figure 4 shows the energy deposition profiles for the three periods "1", "2", and "3" during the 33.059 flight on May 13, 1990, and for the period in which the 33.060 payload was above 105 km on May 14, 1990, as determined from the energy deposition model of Jackman described in the Appendix of Goldberg et al. [1984]. These profiles were determined exclusively from the electron spectrometers since the contribution from bremsstrahlung x-rays was found to be negligible [Goldberg et al., 1994]. This finding is unique, since x-rays have always been observed in significant quantity during auroral precipitation and REPs [e.g., Goldberg et al., 1984; Goldberg et al., 1990]. Also shown

is the anticipated cosmic ray contribution adjusted for site location and phase of the sunspot cycle [Nicolet, 1975]. The peak near 80 km is probably artificially produced by cutting off the electron spectrum near 90 keV, which is the threshold for each of the low energy electron spectrometers. For example, if the sample spectrum shown in Figure 3 was extrapolated to lower energies assuming the slope remained unchanged, one would expect to see the ion-pair production rates continue to grow above 80 km as noted in Goldberg et al. [1984]. Alternatively, the near absence of low energy (5-80 keV) x-rays may signify an electron energy spectrum which decays rapidly below the 90 keV threshold of the solid state detector, thereby providing validation for that portion of the curves as shown in Figure 4 above 80 km.

From Figure 4, it is evident that the non-burst periods for 33.059 and 33.060 are similar in shape, with 33.059 exhibiting a more intense and somewhat spectrally harder electron flux. The latter can be seen through the greater depth of penetration for the 33.059 period "3" profile. Burst "2" during 33.059 shows an enhancement of flux above the non-burst periods, but mainly at the lower energies (higher altitudes). Finally, burst period "1" shows an enhancement above all other periods at all energies, but with the strongest enhancement above 60 km (below about 1 MeV). Hence, the event appears to be more intense and variable during the

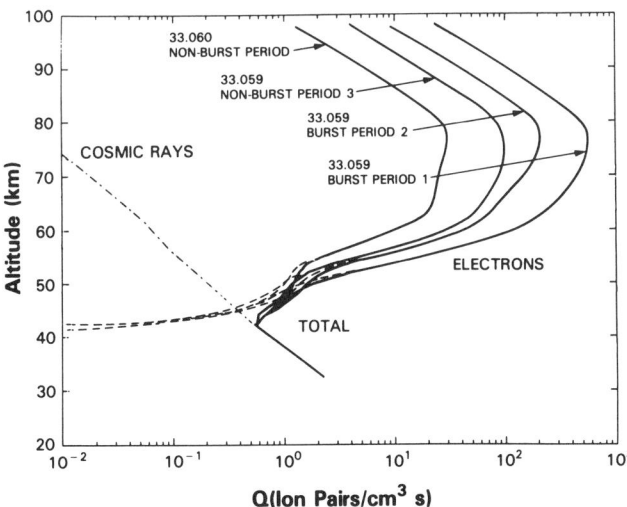

Fig. 4. Energy deposition profiles in terms of ion pair production rate caused by measured electrons. Shown are periods "1", "2", and "3" during the flight of 33.059 and the value during 33.060. Also included is the calculated effect from cosmic rays corrected for latitude and time of solar cycle. Maxima near 80 km are probably caused by limiting the electron spectrum to values above 90 keV, which is the low energy spectrometer threshold value.

flight of 33.059 on May 13, 1990, with the sample measured on May 14 representing a period during the decay of the HRE. In spite of the modest character of this HRE, these energy deposition profiles, particularly burst "1", match or exceed values measured by us during all previous aurorally active or REP events. The energy deposition profiles shown here have been used to calculate the OH and O_3 modulations discussed in the next section.

MODELING RESULTS

A NASA 2-D photochemical model has been used to calculate the mesospheric OH enhancement and O_3 depletion anticipated for this event from two of the energy deposition curves presented in Figure 4. This model is described in detail by Douglass et al. [1989] and Jackman et al. [1990]. Briefly, it employs a latitudinal range from 85°N to 85°S in 10° increments, and an altitude range from 1000 to 0.0024 mbar (0 to about 90 km) with approximately a two km resolution. Around 50 species are considered including O_x, NO_x, HO_y, Cl_y, and Br_y. The 130 chemical reactions included are as specified in JPL 90-1 [De More et al., 1990]. Well known chemical-family approximations are used to reduce the number of transported species in the model.

Twenty two (22) species or families (O_x, NO_y, Cl_y, and Br_y) are transported in the model simulations used here. These include O_x (O_3, $O(^1D)$, $O(^3P)$), NO_y (N, NO, NO_2, NO_3, HO_2NO_2, N_2O_5, $ClONO_2$, but not including HNO_3), Cl_y (Cl, ClO, HOCl, HCl, $ClONO_2$), Br_y (Br, BrO, HBr, $BrONO_2$), HNO_3, N_2O, CH_4, H_2, CO, CH_3OOH, $CFCl_3$, CF_2Cl_2, CH_3Cl, CCl_4, CH_3CCl_3, CH_3Br, $CHClF_2$, $C_2Cl_3F_3$, $C_2Cl_2F_4$, C_2ClF_5, $CBrClF_2$, and $CBrF_3$. The HO_x (H, OH, HO_2) species, H_2O_2, and the hydrocarbons CH_3, CH_3O, CH_3O_2, CH_2O, and CHO are calculated using photochemical equilibrium assumptions. The CO_2 mixing ratio is set at 330 ppmv and the H_2O distribution is fixed using LIMS measurements and other model calculations [cf. Jackman et al., 1987]. Ground boundary conditions for the trace gases are taken from WMO [1992, Table 8-6b], for the 1990 steady state. The residual circulation and diffusion specification is the Dynamics A formulation described in Jackman et al. [1991].

HO_x is the primary species family affecting the loss of ozone during particle precipitation events within the mesosphere. The HO_x constituents are produced by complicated ion chemistry resulting from the production of ion pairs. Each ion pair results in the formation of two HO_x species up to an altitude of approximately 70 km. Above 70 km, the HO_x species produced per ion pair depend quite strongly on the ionization rate and the duration of the particle precipitation event [Solomon et al., 1981]. The 2-D model employed here does not develop the ion chemistry internally. Instead, it uses the computations of HO_x per ion pair as provided in Solomon et al. [1981, Figure 2]. HO_x enhancements can lead to O_3 depletion through several catalytic processes; these are dominated by the following catalytic process above 60 km altitude [Jackman and McPeters, 1987]:

$$H + O_3 \longrightarrow OH + O_2 \quad (2)$$
$$OH + O \longrightarrow H + O_2. \quad (3)$$

The H atom, destroyed in reaction (2), is regenerated in reaction (3) and can continue cycling through reactions (2) and (3) until either H or OH reacts with another constituent in the mesosphere. The net of reactions (2) and (3) causes O_3 and O to transform into $2O_2$.

The NASA 2-dimensional model was originally developed for use in multi-year simulations. As a result, many of the parameters are restricted to temporal variations with a minimum resolution of one day. Since HREs have a time scale of several days, but with a diurnal variability, it was necessary to use average daily values for various parameters such as the ion-pair production rate (energy deposition, Figure 4) instead of tracking it or modeling it on a smaller time-step basis. In order to use this model, it is therefore assumed that cumulative effects are considered to reach an equilibrium value within one day, which seems reasonable based on the time constants of the various processes occurring in the region of interest. Furthermore, many of the

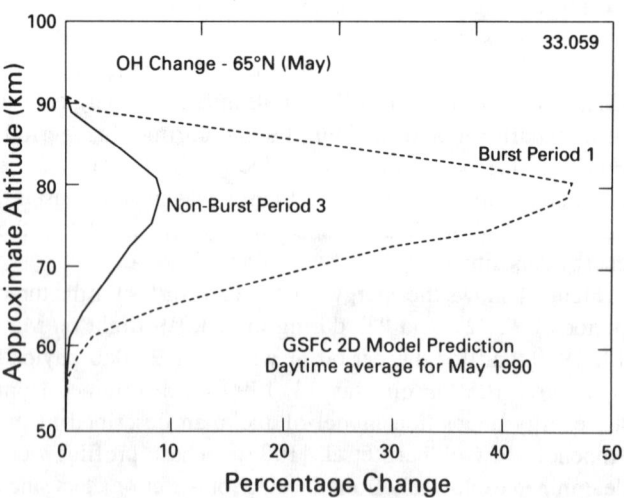

Fig. 5. The GSFC 2-D model prediction for the percent change of the OH mesospheric vertical profile at Poker Flat, Alaska (65°N) in May following bombardment by relativistic electrons with an intensity and spectral distribution equivalent to burst period "1" and non-burst period "3" in the measured HRE.

parameters are highly latitude dependent because of the rapidly changing length of day with latitude except at equinox; hence this work models the effects of the measured radiation specifically at the latitude of Poker Flat Research Range (65.1°N, 147.5°W, magnetic L = 5.5) on the date of the launch.

The four curves presented in Figure 4 all have a similar shape, with the energy deposition exhibiting a monotonic decay at all altitudes between the 33.059 burst "1" and the profile obtained from 33.060. The results obtained from the 2-D photochemical model discussed in this section concentrate on two of the four periods, viz. periods "1" and "3". Figure 5 displays the vertical profile for percentage change in OH at the Poker Flat site, assuming daytime average conditions on the date of the 33.059 launch. Figure 6 shows the resulting profile for O_3. These parameters have been calculated for the two periods assuming a one day steady source. This appears reasonable since the model approaches an equilibrium value within a few hours, which is the daily half width of the event maximum about noon. Hence, the model predicts that for a daytime average for May 1990 under burst "1" conditions, the OH would have increased by over 40% near 80 km, leading to a depletion of O_3 in excess of 30% within the same region.

The HRE and more common natural sources of HO_x compete to drive the atmospheric abundance of HO_x constituents at different times. The model background H_2O is crucial to determination of an HRE influence on the atmosphere. We have employed the H_2O climatology of Remsberg et al. [1989] for the spring-Northern Hemisphere at mid-latitudes in our 2-D model simulations. Table 2 provides the H_2O concentration profile used in our 2-D

Table 2. H_2O mixing ratios used in our simulations with the NASA/GSFC 2D model.

Altitude (mbar)	Altitude (km)	H_2O mixing ratio (ppmv)
0.23	59	6.0
0.17	61	6.0
0.13	63	5.2
0.098	65	5.0
0.074	67	4.3
0.056	69	3.8
0.042	71	3.2
0.031	73	2.7
0.024	75	2.1
0.018	77	1.9
0.013	79	1.7
0.010	81	1.5
0.0076	83	1.3
0.0057	85	1.1
0.0043	87	0.92
0.0032	89	0.78
0.0024	91	0.64

model above 0.23 mb (59 km). Since the most common natural sources of HO_x are dependent on sunlight, either from photolysis of H_2O or the reaction of $O(^1D)$ with H_2O, and since for our simulation, the HRE source of HO_x is assumed to be independent of solar zenith angle, it follows that an HRE event should have its maximum effect at the maximum solar zenith angles within its range of occurrence [e.g. see *Solomon et al.*, 1983; *Jackman and McPeters*, 1985].

DISCUSSION AND CONCLUSIONS

For the HRE under consideration here, Herrero et al. [1991] demonstrated that a significant portion of the relativistic electron flux observed at geosynchronous altitude reaches the middle atmosphere. Baker et al. [1993] expanded that concept and discussed the HRE effects on the stratosphere. More recently, Goldberg et al. [1994] have demonstrated the electrodynamic effects of the relativistic electron flux from this HRE on the middle atmosphere through comparisons with simultaneous measurements of electrical conductivity and related parameters. This study has considered the HRE as an energy source, and evaluated its impact on trace constituents OH and O_3 using the GSFC 2-D photochemical model. It is found that even a modest HRE of the magnitude observed and discussed here can enhance OH several percent leading to an associated depletion in O_3. For more intense HREs, as are anticipated further into the solar cycle decline, these effects could be much higher.

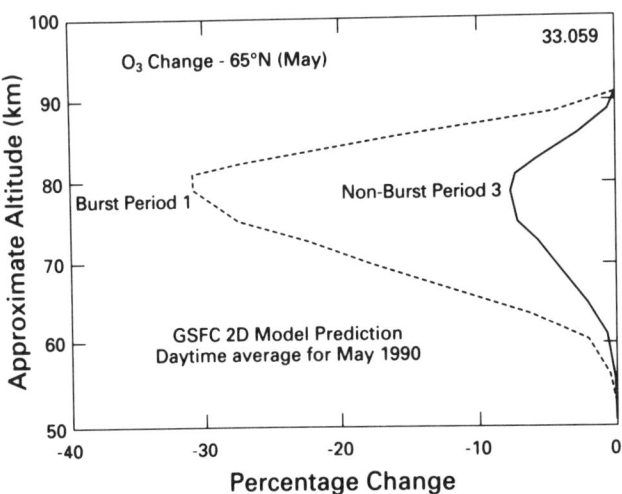

Fig. 6. The vertical profile for the percent change of O_3 under the same conditions as Figure 5.

Since HREs are more intense than REPs and APEs and can continue for several days, their integrated effect on mesospheric ionization and minor constituents is probably much greater than that of REPs and APEs. For the relatively weak event measured during this experiment, the calculated decrease of O_3 was about 30% for daytime average conditions, but could reach much higher values near twilight. For more intense events, larger depletions of O_3 would be expected. Such depletions in O_3 would reduce the absorption of UV within the mesosphere, thereby allowing the UV radiation to penetrate to lower altitudes, leading to possible modifications of the temperature, chemistry and dynamics of the region.

Acknowledgments. We thank Neil Armour, Roy Hagemeyer, Bill Davis, Benjamin Harris, Floyd Hunsaker, and Paul Rozmarynowski for their efforts in payload instrument design and construction, and Patti Twigg for her excellent assistance in various aspects of the data analysis. We also appreciate the highly professional efforts of the NASA/Wallops Flight Facility payload team including Charles Manion, Jack Gum, Herb Morgan, Wayne Raemer, and Carl Wipprecht, which led to the success of this program.

REFERENCES

Baker, D. N., P. R. Higbie, R. D. Belian, and E. W. Hones, Jr., Do Jovian electrons influence the terrestrial outer radiation zone?, *Geophys. Res. Lett., 6,* 531, 1979.

Baker, D. N., J. B. Blake, R. W. Klebesadel, and P. R. Higbie, Highly relativistic electrons in the Earth's outer magnetosphere. 1: Lifetimes and temporal history, 1979-1984, *J. Geophys. Res., 91,* 4265, 1986.

Baker, D. N., J. B. Blake, D. J. Gorney, P. R. Higbie, R. W. Klebesadel, and J. H. King, Highly relativistic electrons: A role in coupling to the middle atmosphere, *Geophys. Res. Lett., 14,* 1027, 1987.

Baker, D. N., R. L. McPherron, R. W. Klebesadel, and T. E. Cayton, Linear prediction filter analysis of relativistic electron properties at L = 6.6, *J. Geophys. Res., 95,* 15133, 1990.

Baker, D. N., R. A. Goldberg, F. A. Herrero, J. B. Blake, and L. B. Callis, Satellite and rocket studies of relativistic electrons and their influence on the middle atmosphere, *J. Atmos. Terr. Phys., 54,* 1619, 1993.

De More, W. B., S. P. Sander, D. M. Golden, M.J. Molina, R. F. Hampson, M. J. Kurylo, C. J. Howard, and A. R. Ravishankara, Chemical kinetics and photochemical data for use in stratospheric modeling, *JPL Publication 90-1,* 217 pages, 1990.

Douglass, A. R., C. H. Jackman, and R. S. Stolarski, Comparison of model results transporting the odd nitrogen family with results transporting separate odd nitrogen species, *J. Geophys. Res., 94,* 9862, 1989.

Goldberg, R. A., C. H. Jackman, J. R. Barcus, and F. Sørass, Nighttime auroral energy deposition in the middle atmosphere, *J. Geophys. Res., 89,* 5581, 1984.

Goldberg, R. A., C. L. Croskey, L. C. Hale, J. D. Mitchell, and J. R. Barcus, Electrodynamic response of the middle atmosphere to auroral pulsations, *J. Atmos. Terr. Phys., 52,* 1067, 1990.

Goldberg, R. A., D. N. Baker, F. A. Herrero, S. P. McCarthy, P. A. Twigg, C. L. Croskey, and L. C. Hale, Energy Deposition and Middle Atmosphere Electrodynamic Response to a Highly Relativistic Electron Precipitation Event, *J. Geophys. Res., 98,* In Press, 1994.

Heath, D. F., A. J. Krueger, and P. J. Crutzen, Solar proton event: Influence on stratospheric ozone, *Science, 197,* 886, 1977.

Herrero, F. A., D. N. Baker, and R. A. Goldberg, Rocket measurements of relativistic electrons: New features in fluxes, spectra, and pitch angle distributions, *Geophys. Res. Lett., 18,* 1481, 1991.

Imhof, W. L., H. D. Voss, J. Mobilia, D. W. Datlowe, J. P. McGlennon, and D. N. Baker, Relativistic electron enhancement events: Simultaneous measurements from synchronous and low altitude satellites, *Geophys. Res. Lett., 18,* 397, 1991.

Jackman, C. H., and R. D. McPeters, The response of ozone to solar proton events during solar cycle 21: A theoretical interpretation, *J. Geophys. Res., 90,* 7955, 1985.

Jackman, C. H., and R. D. McPeters, Solar proton events as tests for the fidelity of middle atmospheric models, *Physica Scripta, T18,* 309, 1987.

Jackman, C. H., A. R. Douglass, R. B. Rood, R. D. McPeters, and P. E. Meade, Effect of solar proton events on the middle atmosphere during the past two solar cycles as computed using a two-dimensional model, *J. Geophys. Res., 95,* 7417, 1990.

Jackman, C. H., Effects of energetic particles on minor constituents of the middle atmosphere, *J. Geomag. Geoelectr., 43,* 637, 1991.

Jackman, C. H., A. R. Douglass, K. F. Brueske, and S. A. Klein, The influence of dynamics on two-dimensional model results: Simulations of ^{14}C and stratospheric aircraft NO_X injections, *J. Geophys. Res., 96,* 22559, 1991.

Johnson, R. M., and J. G. Luhmann, Poker Flat MST radar observations of high latitude neutral winds at the mesospause during and after solar proton events, *J. Atmos. Terr. Phys., 55,* 1203, 1993.

McPeters, R. D., and C. H. Jackman, The response of ozone to solar proton events during solar cycle 21: The observations, *J. Geophys. Res., 90,* 7945, 1985.

Nicolet, M., On the production of nitric oxide by cosmic rays in the mesosphere and stratosphere, *Planet. Space Sci., 23,* 637, 1975.

Remsberg, E. E., J. M. Russell, III, and C. Y. Wu, An interim reference model for the variability of the middle atmosphere H_2O vapor distribution, *Handbook for the Middle Atmosphere Program, 31,* 50, 1989.

Solomon, S. D. W. Rusch, J. -C. Gerard, G. C. Reid, and P. J. Crutzen, The effect of particle precipitation events on the neutral and ion chemistry of the middle atmosphere. II. Odd hydrogen, *Planet. Space Sci., 29,* 885, 1981.

Solomon, S., G. C. Reid, D. W. Rusch, and R. J. Thomas, Mesospheric ozone depletion during the solar proton event of July 13, 1982; Part II. Comparison between theory and measurements, *Geophys. Res. Lett., 10*, 257, 1983.

Swider, W., and T. J. Keneshea, Decrease of ozone and atomic oxygen in the lower mesosphere during a PCA event, *Planet. Space Sci., 21*, 1969, 1973.

Thomas, R. J., C. A. Barth, G. J. Rottman, D. W. Rusch, G. H. Mount, G. M. Lawrence, R. W. Sanders, G. E. Thomas, and L. E. Clemens, Mesospheric ozone depletion during the solar proton event of July 13, 1982, Part 1, Measurement, *Geophys. Res. Lett., 10*, 253, 1983.

Thorne, R. M., The importance of energetic particle precipitation on the chemical composition of the middle atmosphere, *Pure Appl. Geophys., 118*, 128, 1980.

Weeks, L. H., R. S. CuiKay, and J. R. Corbin, Ozone measurements in the mesosphere during the solar proton event of November 2, 1969, *J. Atmos. Sci., 29*, 1138, 1972, 1972.

WMO, Scientific assessment of ozone depletion: 1991, World Meteorological Organization, Global Ozone Research and Monitoring Project, *Report No. 25*, Ed. by D. L. Albritton, R. T. Watson, S. Solomon, R. F. Hampson, and F. Ormand, 1992.

R. A. Goldberg, F. A. Herrero, Laboratory for Extraterrestrial Physics, Code 690, NASA/Goddard Space Flight Center, Greenbelt, MD 20771

C. H. Jackman, Laboratory for Atmospheres, Code 916, NASA/Goddard Space Flight Center, Greenbelt, MD 20771

D. N. Baker, now at University of Colorado, Dwayne Physics Bldg., Campus Box 392, Boulder, Co. 80309

Nitric Oxide in the Lower Thermosphere

Charles A. Barth

*Laboratory for Atmospheric and Space Physics and
Department of Astrophysical, Planetary, and Atmospheric Sciences
University of Colorado, Boulder, Colorado*

Nitric oxide is produced in the lower thermosphere by solar soft x-rays and extreme ultraviolet radiation. In the auroral region, electron bombardment and Joule heating cause an increase in the density of nitric oxide. Nitric oxide influences the temperature structure of the thermosphere and the composition of the ionosphere. Ultraviolet spectrometer measurements of nitric oxide may be used to determine the density distribution. A technique using observations of an optically-thick and an optically-thin band yield density determinations that are relatively free of uncertainties in the knowledge of the sensitivity of the instrument. Analysis of the rotational structure of the nitric oxide gamma bands may be used to determine the temperature of the lower thermosphere. Model calculations of the high latitude thermosphere show that there is a non-linear dependence of the nitric oxide density with the flux of auroral electrons. Model calculations of the low latitude thermosphere give a linear relationship between the solar soft x-ray flux and the nitric oxide density. A sensitivity test of the model shows that the calculated nitric oxide density is extremely sensitive to the value that is used for the branching ratio for the production of excited nitrogen atoms in the dissociative recombination of ionized nitric oxide. Satellite observations of the distribution of nitric oxide in the lower thermosphere indicate that there is always a larger amount in the auroral regions and that at high and low latitudes there is great variability.

1. INTRODUCTION

Nitric oxide is an important chemical constituent of the lower thermosphere. Its presence is an indicator of the injection of energy into the lower thermosphere. Its density is highly variable and is controlled by variations in the energy output of the sun. At low latitudes, the density of nitric oxide is determined by solar radiation in the soft x-ray (1.8-5.0 nm) and extreme ultraviolet (17.0-102.6 nm) portion of the spectrum. At polar latitudes, the nitric oxide density is controlled by the flux of auroral electrons and the Joule heating that occurs during auroral activity. The sun is controlling the nitric oxide in the auroral regions through the interaction of the solar wind on the earth's magnetosphere (*Barth*, 1992).

Nitric oxide plays an important role in the ionosphere at all latitudes, mainly because the energy necessary to ionize this molecule is less than the energy needed to ionize the major constituents of the thermosphere, molecular oxygen, atomic oxygen, and molecular nitrogen. Nitric oxide is a source of ionization in the D-region of the ionosphere; in fact, the need of an ionizable constituent in this region of the atmosphere was the first clue as to the presence of nitric oxide in the earth's atmosphere. Solar Lyman alpha radiation which is able to penetrate through the molecular oxygen to the D-region ionizes nitric oxide. Variations in the density of nitric oxide lead to variations in the electron density in the D-region. At higher altitudes, in the E and F_1 regions of the ionosphere, nitric oxide is an important participant in the ion-molecule and charge exchange reactions. Although neutral nitric oxide is a minor constituent in this part of the atmosphere, ionized nitric oxide is a major constituent because of its low ionization potential.

Nitric oxide plays an important role in the energy balance of the thermosphere. Since it is a heteronuclear molecule, it is able to radiate in the infrared portion of the

The Upper Mesosphere and Lower Thermosphere:
A Review of Experiment and Theory
Geophysical Monograph 87
Copyright 1995 by the American Geophysical Union

spectrum, while molecular nitrogen and molecular oxygen, being homonuclear molecules, are not. Particularly, at times of high solar and geomagnetic activity, the nitric oxide thermal emission at 5.3 μm contributes to the cooling of the lower thermosphere. Nitric oxide may be used as a tracer of atmospheric motions in the lower thermosphere. Since the photochemical lifetime of nitric oxide is about one day, nitric oxide density variations may be used to detect gravity waves. Also, the rotational structure of the nitric oxide fluorescence spectrum may be used to determine atmospheric temperatures in the lower thermosphere (*Barth and Eparvier*, 1993).

Nitric oxide chemically reacts with ozone to form nitrogen dioxide which in turn reacts with atomic oxygen to reform nitric oxide. This is a catalytic cycle which destroys ozone while leaving the odd-nitrogen intact. Any nitric oxide that is transported downward from the lower thermosphere into the mesosphere and stratosphere may participate in the catalytic destruction of ozone. An opportune time for downward transport to take place is during polar night when photodissociation of nitric oxide is not occurring (*Solomon et al.*, 1982).

2. MEASUREMENT TECHNIQUES

The most frequently used method of measuring nitric oxide in the thermosphere is the measurement of the fluorescent scattering of solar radiation by the nitric oxide molecule. This technique is well-suited for remote sensing observations from rockets and from satellites. The most intense of the nitric oxide emissions is the (1,0) gamma band at 215.0 nm. The oscillator strength of this band is moderately large and other airglow emissions in this wavelength range are relatively weak. Ultraviolet spectrometers and photometers have the capability to measure densities of nitric oxide as low as about 10^6 molecules/cm^3. This was the technique used in the rocket experiment that led to the discovery of nitric oxide in the thermosphere (*Barth*, 1964) and in the first satellite measurements from OGO-4 (*Rusch and Barth*, 1975). Global observations of nitric oxide have been made from the Atmosphere Explorer satellites (*Barth et al.*, 1973; *Stewart and Cravens*, 1978; *Cravens and Stewart*, 1978) and from the Solar Mesosphere Explorer (SME) satellite (*Barth et al.*, 1988; *Barth*, 1990). In the use of the remote-sensing technique, what is actually measured is the column emission rate along the path of the field-of-view of the instrument. The column emission rate, $4\pi I$, is related to the column density, N, of nitric oxide molecules by the emission rate factor, g, (*Chamberlain*, 1961).

$$4\pi I = g N$$

$$g = F \frac{\pi e^2}{mc^2} \lambda^2 f \frac{A_{v'v''}}{\Sigma A_{v'v''}}$$

Knowledge of the nitric oxide density is dependent upon the knowledge of the solar flux, F, which is measured from a rocket or satellite, and the oscillator strength, f, of the absorption transition and the branching ratio, $A_{v'v''}/\Sigma A_{v'v''}$, of the emission transition, both of which are measured in the laboratory. The rest of the terms in this equation are physical constants that are well-determined. Most important, the measurement of the column emission rate is dependent upon knowledge of the sensitivity of the instrument which is susceptible to systematic errors in the calibration procedure and to changes in the sensitivity during the launch and operation of the instrument.

An ultraviolet remote sensing technique has been developed which is not sensitive to the calibration of the airglow instrument nor to the calibration of the solar instrument which measures the incident solar radiation (*Eparvier and Barth*, 1992). This technique uses the simultaneous measurement of an optically-thick band and an optically-thin band of the nitric oxide emission spectrum. Using this method, the determination of the column density of nitric oxide is primarily dependent upon the oscillator strength, f, of the optically-thick emission band which is measured in the laboratory to a high degree of precision (*Farmer et al.*, 1972).

$$f = 8.09 \times 10^{-4} \pm 5\%$$

A demonstration of this technique is shown in the results from a rocket experiment conducted at the Poker Flat Research Range in March 1989 (*Eparvier and Barth*, 1992). The observed ultraviolet airglow spectrum which shows eight nitric oxide gamma bands between 210.0 and 250.0 nm is plotted in Figure 1. The ultraviolet spectrometer viewed the airglow horizontally using a telescope to obtain an enhanced emission rate from the long path length. The (1,0) band at 215.0 nm which is the most intense was chosen as the optically-thick band and the (0,1) which is relatively free from overlapping spectral features was chosen as the optically-thin band. The results of the rocket observation are shown in Figure 2 where the apparent slant column density is plotted from the (1,0) and the (0,1) bands. The slant column density from the (0,1) band is interpreted as the true column density since this emission is optically-thin. The slant column density from the (1,0) band is interpreted as an apparent slant column density since this band is optically-thick. The apparent slant column

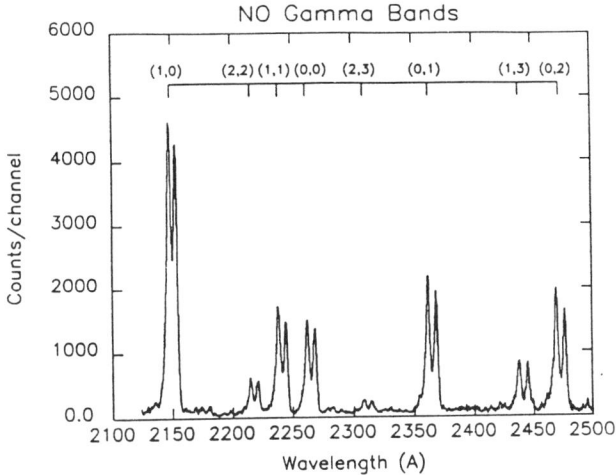

Fig. 1. Spectrum of the earth's dayglow between 2125 and 2500 Å from the sum of the spectral scans made between 100 and 110 km. The eight gamma bands are : (1,0) 214.9 nm, (2,2) 221.6 nm, (1,1) 223.9 nm, (0,0) 226.2 nm, (2,3) 230.9 nm, (0,1) 236.3 nm, (1,3) 243.9 nm, (0,2) 247.1 nm.

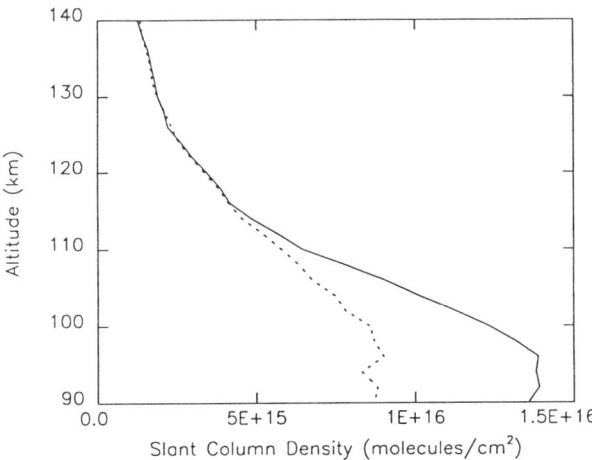

Fig. 2. Nitric oxide slant column densities from rocket observations of the (1,0) band (dotted line) and (0,1) band (solid line).

density determined from the (1,0) band falls below the true slant column density deduced from the (0,1) band below about 115 km. This difference is produced by the attenuation of the (1,0) band emission as it traverses the optically-thick nitric oxide. The essential point of this technique is that the difference between these two measurements is dependent only upon the transmission function which is primarily determined by the oscillator strength, f. Tables of the transmission functions of the nitric oxide (1,0) gamma band as a function of density and temperature have been calculated by Eparvier (1991). Figure 3 shows a plot of the transmission functions of the (1,0) and (0,1) bands as a function of the true column density for the conditions appropriate to the time of the rocket flight. The transmission functions are dependent on the temperature of the nitric oxide because of the rotational structure of the absorption band. The atmospheric temperature may be determined from a model atmosphere or from an analysis of the rotational structure of the nitric oxide gamma bands. Once the sensitivity of the instrument has been verified by the optically-thick-optically-thin band technique, the nitric oxide density may be determined over an extended altitude range using the measured column emission rate and the emission rate factor, g. The emission rate factor, g, which has a weak dependence on temperature, has been calculated by Eparvier (1991) taking into account the rotational structure of the nitric oxide molecule and the detailed structure of the solar spectrum. The value obtained for 300°K is:

$$g = 6.37 \times 10^{-6} \text{ photon/sec molecule}$$

The volume density of nitric oxide as determined from the column density in Figure 2 is shown in Figure 4. Nitric oxide has a maximum density at about 100 km. At higher altitudes, the density decreases with increasing altitude because, in part, of the decreasing density of the atmosphere. In Figure 5, the ratio of the nitric oxide density to the density of all of the constituents of the thermosphere is shown as a mixing ratio. The steep gradient below 100 km is the result of the absence of a source of nitric oxide.

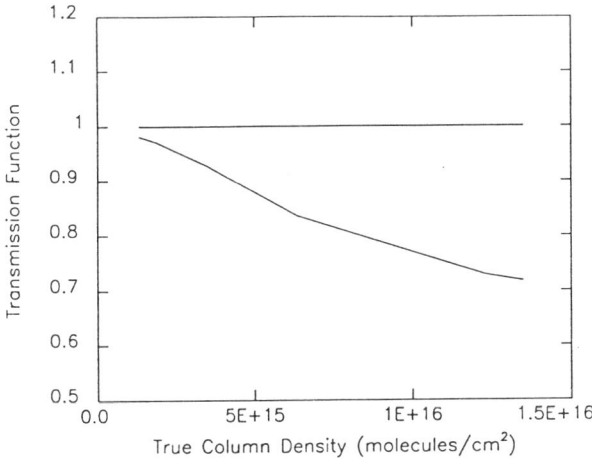

Fig. 3. Transmission functions for the (1,0) band (lower line) and the (0,1) line (upper line).

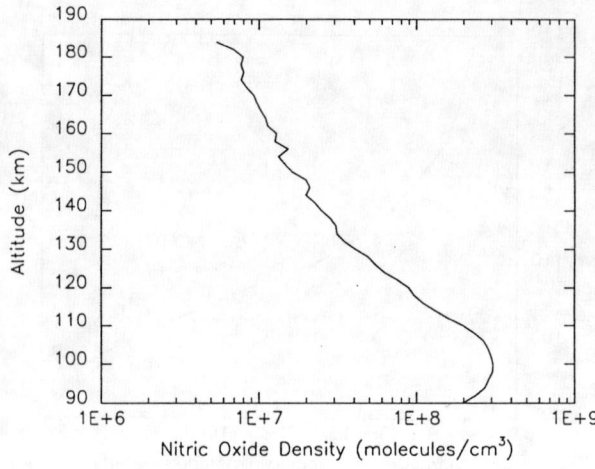

Fig. 4. Nitric oxide volume density measured at Poker Flat Research Range, March 7, 1989.

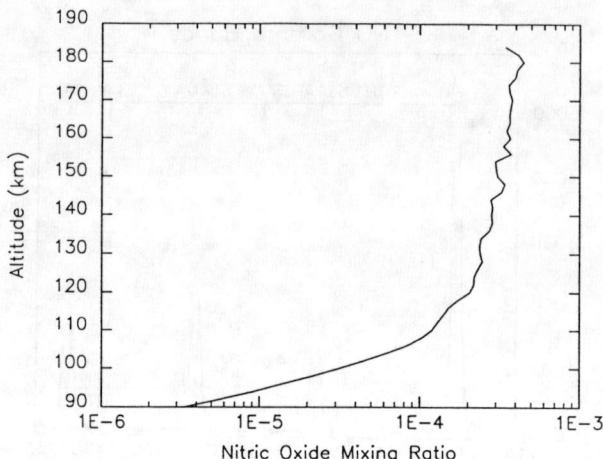

Fig. 5. Nitric oxide mixing ratio from Poker Flat rocket observations.

3. TEMPERATURE MEASUREMENT

The airglow spectrum of the nitric oxide gamma system has a well-developed rotational structure. Since the excited nitric oxide molecules are in rotational equilibrium with the atmospheric molecules, an analysis of the rotational structure of the emission band may be used to determine the temperature of the atmosphere (*Barth and Eparvier*, 1993). The gamma bands are the result of transitions between the $A^2\Sigma^+$ and $X^2\Pi$ states of the nitric oxide molecule. Each band consists of hundreds of rotational lines arranged in twelve branches. Synthetic spectra of the rotational structure may be calculated using knowledge of the line strengths of the rotational transitions, the populations of the rotational levels, and detailed structure of the solar spectrum (*Eparvier and Barth*, 1992). Figure 6 shows the rotational structure of the (1,1) band at a temperature of 280°K and 700°K. Also shown in Figure 6 are the observed spectra of the (1,1) gamma band measured in the rocket experiment at altitudes of 110 and 140 km. The synthetic spectra have been adjusted to the resolution of the rocket spectrometer and fit to the observed spectra. The agreement is excellent, and these temperatures are in agreement with the temperatures of the MSIS-86 model for the conditions of the rocket flight (*Hedin*, 1987).

4. PHOTOCHEMICAL THEORY

There have been a number of discussions of the sources of thermospheric nitric oxide since the introduction of ideas involving excited nitrogen atoms (*Norton and Barth*, 1970) and photoelectrons (*Strobel et al.*, 1970). The following discussion is taken from a recent paper (*Barth*, 1992).

The principal source of nitric oxide in the lower thermosphere is the chemical reaction between excited atomic nitrogen and molecular oxygen.

$$N(^2D) + O_2 \rightarrow NO + O \qquad R1$$

The reaction of ground state atomic nitrogen with molecular oxygen is significant only at higher altitudes in the thermosphere.

$$N(^4S) + O_2 \rightarrow NO + O \qquad R2$$

The principal loss mechanism for nitric oxide is its reaction with ground state atomic nitrogen.

$$NO + N(^4S) \rightarrow N_2 + O \qquad R3$$

Other loss mechanisms are the reaction of nitric oxide with ionized molecular oxygen and the photodissociation of nitric oxide by solar ultraviolet radiation near 190.8 nm.

$$NO + O_2^+ \rightarrow NO^+ + O_2 \qquad R4$$

$$NO + h\nu \rightarrow N(^4S) + O \qquad R5$$

The main sources of excited atomic nitrogen atoms in the lower thermosphere are two ionospheric reactions: the dissociative recombination of ionized nitric oxide and the ion-molecule reaction between ionized molecular nitrogen and atomic oxygen.

$$NO^+ + e \rightarrow N(^2D) + O \qquad R6$$

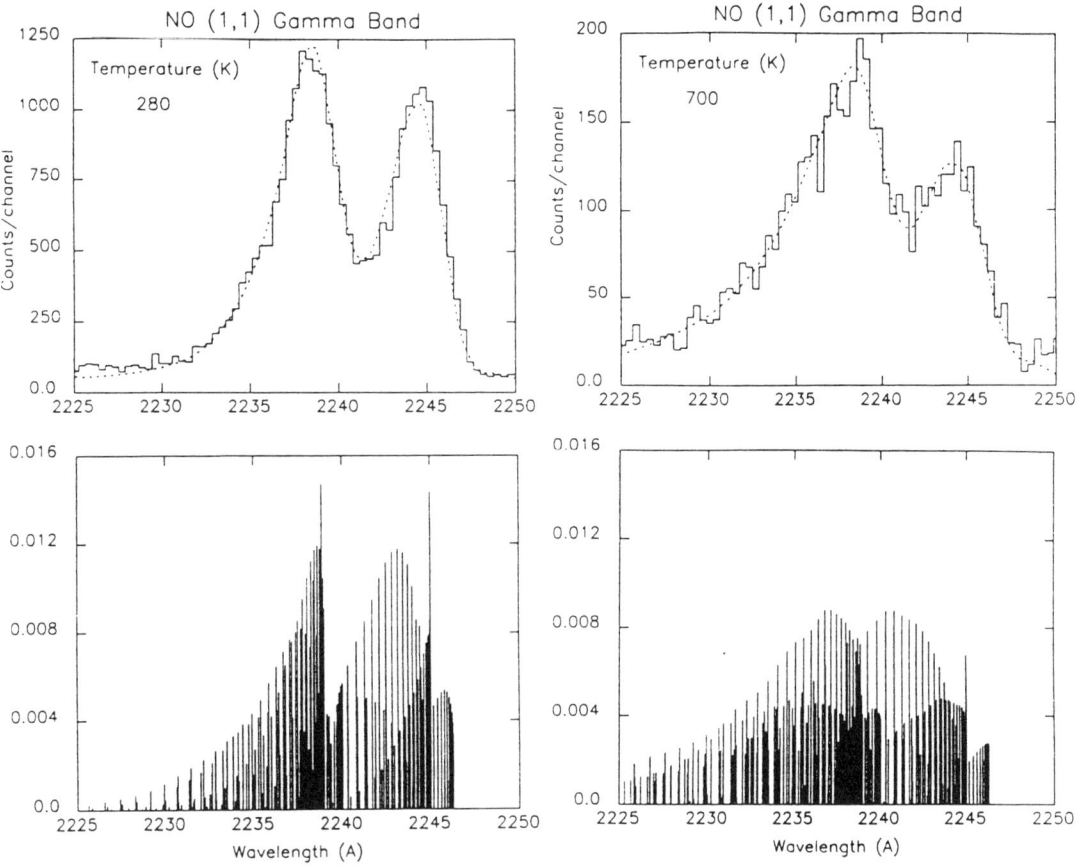

Fig. 6. Temperature determination from spectra of the (1,1) gamma band at 110 km (280 °K) and 140 km (700 °K). Synthetic spectra of the (1,1) band at 280 °K and 700 °K.

$$N_2^+ + O \rightarrow N(^2D) + NO^+ \qquad R7$$

The next most important source is the photoelectron impact dissociation of molecular nitrogen.

$$N_2 + pe \rightarrow N(^2D) + N \qquad R8$$

where pe indicates the flux of photoelectrons.

During auroral bombardment, auroral secondary electrons are an important source of excited atomic nitrogen.

$$N_2 + ae \rightarrow N(^2D) + N \qquad R9$$

where ae indicates the flux of auroral secondary electrons.

An important loss mechanism for excited atomic nitrogen is the deactivation reaction with atomic oxygen.

$$N(^2D) + O \rightarrow N(^4S) + O \qquad R10$$

In the lower thermosphere, there are many sources of ground state atomic nitrogen. Reactions R6, R8, and R9 all produce ground state atoms simultaneously with the production of excited state atoms. The branching ratio into these two states is extremely crucial for the production of nitric oxide. Reactions R5 and R10 are sources of ground state atomic nitrogen. The most important loss mechanism is reaction R3, the reaction of ground state atomic nitrogen with nitric oxide destroying both of them.

5. MODEL

A number of one-dimensional, time-dependent models for nitric oxide that include transport and ionospheric processes have been developed. The one that is used here was originally written by Cleary (1986) and modified by Siskind et al. (1990). This model which includes 35 chemical and ionospheric reactions is described in Barth (1992) where the reaction rates are tabulated. The input parameters are time of day, latitude, longitude, solar radio flux ($F_{10.7}$), geomagnetic activity index (Ap), solar soft x-ray flux, and the auroral electron flux and energy.

To help understand the processes that determine the

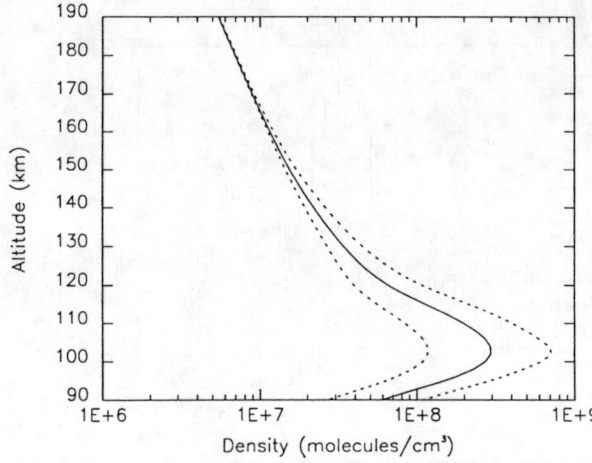

Fig. 7. Densities of nitric oxide for various magnitudes of the auroral electron flux. The solid line represents a flux of 1.5 ergs/sec cm^2. The dashed lines are for fluxes of 0.5 and 3.5 ergs/sec cm^2.

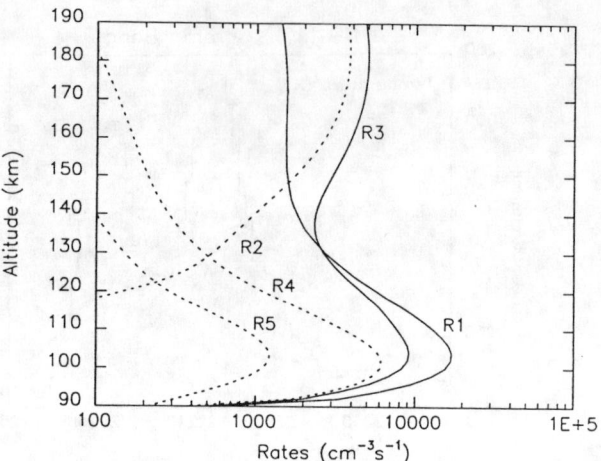

Fig. 8. Production and loss rates for nitric oxide for an auroral electron flux of 1.5 ergs/sec cm^2. The production processes R1 and R2 and the loss processes R3, R4, and R5 are identified by the equations in the text.

density of nitric oxide in the lower thermosphere, the model has been applied to two different physical situations: the first corresponds to a high-latitude location during a time of auroral activity (for example, Poker Flat Research Range) and the second corresponds to a low-latitude location during a time of high soft x-ray flux from the sun (for example, White Sands Missile Range). The input parameters for these two model calculations are given in Table 1.

The results of the model calculation for high-latitude conditions are shown in Figures 7 and 8. In Figure 7, the calculated nitric oxide density is given for three values of the auroral electron flux, 0.5, 1.5, and 3.5 ergs/sec cm^2. The nitric oxide density produced by 1.5 ergs/sec cm^2 of electron flux corresponds to the amount of nitric oxide observed in the Poker Flat rocket experiment (see Figure 4). The maximum in nitric oxide density occurs near 103 km and decreases at higher altitudes with the decreasing density of the atmosphere. This figure demonstrates the variability of nitric oxide with auroral activity. At 103 km, the nitric oxide density increases with increasing auroral flux; however, the relationship is non-linear. An increase in auroral electron flux from 0.5 to 1.5 ergs/sec cm^2 (a factor of 3.0) produces an increase in nitric oxide density of about a factor of 2.4, while a further increase from 1.5 to 3.5 ergs/sec cm^2 (a factor of 2.3) also produces a nitric oxide increase of 2.4.

Figure 8 shows that the principal production mechanism for nitric oxide at low altitudes is reaction R1, the reaction of $N(^2D)$ atoms with molecular oxygen, and the principal loss mechanism is R3, the reaction of $N(^4S)$ atoms with nitric oxide. At low altitudes, there are the additional loss mechanisms, reaction R4, the charge transfer reaction of nitric oxide with ionized molecular oxygen, and reaction R5, the photodissociation of nitric oxide by solar ultraviolet radiation. The first of these loss reactions leads to a recycling of the odd-nitrogen through ionized nitric oxide to excited atomic nitrogen which in turn produces nitric oxide. The second of the loss reactions leads to a loss of nitric oxide since the ground state atomic nitrogen that is formed reacts with nitric oxide destroying both forms of odd-nitrogen.

The principal sources of excited nitrogen atoms are the ionospheric reactions R6 and R7; however, another important source is reaction R9, the auroral electron impact dissociation of molecular nitrogen. The auroral electrons also produce the ions that participate in reactions R6 and R7.

The results of the model calculations for low-latitude conditions are shown in Figures 9 and 10. Figure 9 gives the nitric oxide density distribution for three values of the solar soft x-ray flux, 0.5, 1.0 and 2.0 ergs/sec cm^2. The nitric oxide density peaks at 102 km and decreases more rapidly than the atmospheric density immediately above that

TABLE 1. Model Calculations

parameter	high latitude	low latitude
latitude	65°N	32°N
longitude	147°W	105°W
local time	12 noon	12 noon
F$_{10.7}$	200	200
Ap	30	30
x-rays	1 ergs/sec cm^2	1 ergs/sec cm^2
auroral flux	1.5 ergs/sec cm^2	

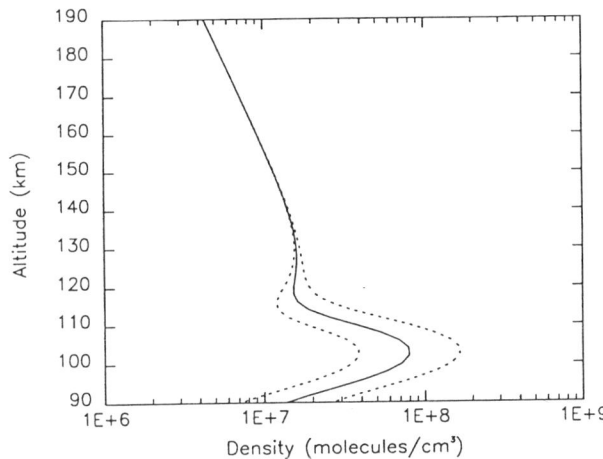

Fig. 9. Densities of nitric oxide for various magnitudes of the solar soft x-ray flux. The solid line represents a flux of 1.0 ergs/sec cm^2. The dashed lines are for fluxes of 0.5 and 2.0 ergs/sec cm^2.

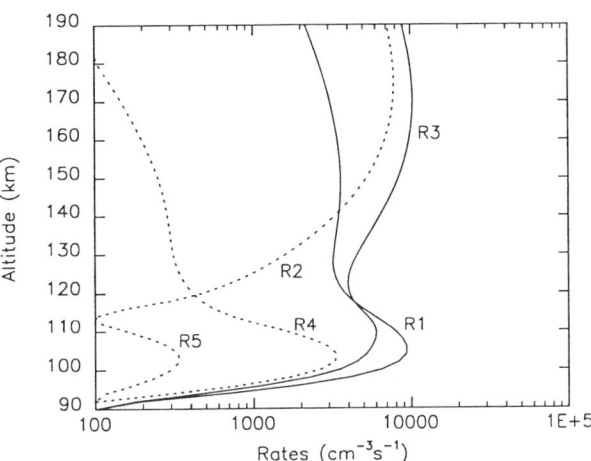

Fig. 10. Production and loss rates for nitric oxide for a solar soft x-ray flux of 1.0 ergs/sec cm^2. The production processes R1 and R2 and the loss processes R3, R4, and R5 are identified by the equations in the text.

altitude. At higher altitudes, the nitric oxide density follows the atmospheric density. At 102 km, the nitric oxide density responds directly to changes in the solar soft x-ray flux. The relationship is approximately linear with a doubling of the soft x-ray flux producing a doubling of the nitric oxide density.

The production and loss rates for the low-latitude case are shown in Figure 10. At 102 km, reaction R1, the reaction of excited atomic nitrogen with molecular oxygen, is the principal source of nitric oxide and reaction R3, the reaction of ground state atomic nitrogen with nitric oxide, is the main loss mechanisms. The principal role of the soft x-rays is to produce photoelectrons. These photoelectrons in turn create an increased number of molecular nitrogen ions and excited nitrogen atoms through reaction R7. The dissociative recombination of ionized nitric oxide, reaction R6, creates an increased number of excited nitrogen atoms. In addition, the photoelectrons produced from the soft x-rays create excited nitrogen atoms by the electron impact dissociation of molecular nitrogen, reaction R8 (*Siskind et al.*, 1990).

6. SENSITIVITY

A study was made of the sensitivity of the model to uncertainties in the reaction rates and branching ratios of the chemical and ionospheric reactions used in the model (*Barth*, 1992). Of the ten reactions described in the theory section, it was found that the nitric oxide density calculated with the model was insensitive to uncertainties in the rates of four of them, reactions R2, R3, R4, and R6. The model calculation of nitric oxide is very sensitive to the uncertainty of the main production mechanism, R1. A change of twenty percent in the reaction rate leads to a change of twenty percent in the calculated nitric oxide density. The model is less sensitive to uncertainties in the reaction rates of reactions R5, R10, and R7. A change of twenty percent in their reaction rates gives a change in the nitric oxide density of ten percent for the first two and of seven percent for the third. Most important, the model is extremely sensitive to the branching ratios used in reactions R6, R8, and R9. Changing the branching ratio of reaction R6, the dissociative recombination of ionized nitric oxide, from 0.75 to 0.90 changes the calculated amount of nitric oxide by a factor of 2. This effect is clearly demonstrated in Figure 11. At 102 km, the amount of nitric oxide calculated by the model for a branching ratio of 0.90 is twice that calculated for a branching ratio of 0.75. Similarly, changing the branching ratio of reactions R8 and R9 from 0.50 to 0.60 produces a change of eighty percent in the calculated nitric oxide density. The reason for this great sensitivity is that a branching ratio of 0.75 for reaction R6 means that seventy-five percent of the reactions produce excited nitrogen atoms,

$$NO^+ + e \rightarrow N(^2D) + O \qquad R6$$

and twenty-five percent produce ground state nitrogen atoms.

$$NO^+ + e \rightarrow N(^4S) + O \qquad R6$$

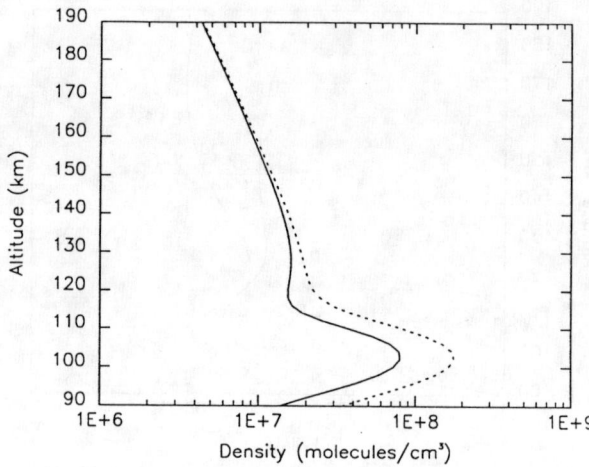

Fig. 11. Densities of nitric oxide for different values of the branching ratio for the dissociative recombination of ionized nitric oxide. The solid line represents a branching ratio of 0.75 and the dashed line a ratio of 0.90.

The excited nitrogen atoms react with molecular oxygen in reaction R1 to produce nitric oxide and the ground state nitrogen atoms react with nitric oxide in reaction R3 destroying it. Any small change in the value used in the branching ratio leads to large changes in the calculated nitric oxide density (Norton and Barth, 1970). The precise measurement of these branching ratios in the laboratory is extremely difficult and uncertainties remain in the knowledge of their precise value (Kley et al., 1977).

7. GLOBAL DISTRIBUTION

The global distribution of nitric oxide has been measured from the Solar Mesosphere Explorer over a period of four and a half years (Barth, 1990; Barth, 1992). An important result of these observations is the recognition that the density of nitric oxide in the lower thermosphere is highly variable. In the polar regions, the variability of nitric oxide is associated with auroral activity. There is an approximate correlation between the geomagnetic index Ap and the density of nitric oxide. There is always more nitric oxide at auroral latitudes (65° geomagnetic latitude) than at low latitudes. In equatorial regions, the nitric oxide density has a 27-day periodicity that is correlated with solar activity indices such as the solar 10.7 cm radio flux (Barth et al., 1988). The equatorial nitric oxide also varies with the 11-year solar cycle. Even at low latitudes, there is a contribution to the nitric oxide density from geomagnetic activity.

The physical processes producing the temporal and latitudinal variation are listed in Figure 12 which shows a

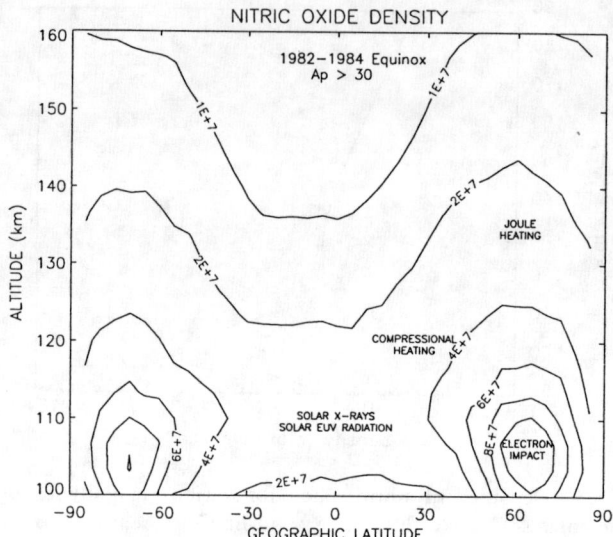

Fig. 12. Nitric oxide densities for the 1982—1984 equinox periods as a function of geomagnetic latitude and altitude for times of high geomagnetic activity. The locations of the major sources of nitric oxide are indicated on this contour plot. between 100 and 110 km.

latitude-altitude plot of nitric oxide density for a time of high solar activity and high auroral activity. The greatest density of nitric oxide occurs at an altitude of about 105 km and geomagnetic latitude of 65°. This nitric oxide is produced by the electron impact of auroral electrons precipitating into the atmosphere at 65° geomagnetic latitude. The altitude is determined by the energy of the primary auroral electrons. The altitude distribution of nitric oxide during auroral electron bombardment is shown in Figure 7. At high altitudes in the auroral region, the atmosphere is heated by Joule heating associated with the presence of electric fields during auroral activity. Because reaction R2, the reaction of ground state atomic nitrogen with molecular oxygen, is very sensitive to temperature, the nitric oxide density increases. The Joule heating also causes the atmosphere to expand producing compressional heating equatorward of the auroral region (Roble et al., 1987). The behavior of nitric oxide during an auroral storm has been modeled (Siskind et al., 1989a; Siskind et al., 1989b). At low latitudes, the nitric oxide is produced by the action of solar soft x-rays and solar extreme ultraviolet radiation on the atmosphere. The altitude distribution of nitric oxide at low latitudes from these sources is plotted in Figure 9. The observations in Figure 12 show that low-latitude nitric oxide is affected by auroral activity through the process of compressional heating. This means that the variation of nitric oxide density at low latitudes should have a weak dependence on geomagnetic indices such as Ap in addition to a strong dependence on solar indices such as $F_{10.7}$.

8. SUMMARY

Rocket and satellite observations of nitric oxide in the lower thermosphere show that its density is highly variable. This variation is caused by variations in the output of the sun. Solar soft x-rays and extreme ultraviolet radiation produce the variations in the nitric oxide density at low latitudes. Geomagnetic activity which is induced by changes in the solar wind causes large changes in the nitric oxide density at polar latitudes. The changing nitric oxide density influences the temperature structure of the thermosphere and the composition of the ionosphere. Ultraviolet spectrometer observations of the spectrum of the nitric oxide gamma bands may be used to determine the temperature of the thermosphere and the absolute density of the nitric oxide.

REFERENCES

Barth, C.A., Rocket measurement of the nitric oxide dayglow, *J. Geophys. Res., 69*, 3301, 1964.

Barth, C.A., Reference models for thermospheric NO, *Adv. Space Res. 10*, 103, 1990.

Barth, C.A., Nitric Oxide in the Lower Thermosphere, *Planet. Space Sci., 40*, 315, 1992.

Barth, C.A., and F.G. Eparvier, A Method of Measuring the Temperature of the Lower Thermosphere, *J. Geophys. Res., 98*, 9437, 1993.

Barth, C.A., W.K. Tobiska, D.E. Siskind, and D.D. Cleary, Solar-terrestrial coupling: Low-latitude thermospheric nitric oxide, *Geophys. Res. Letters, 15*, 92, 1988.

Barth, C.A., D.W. Rusch, and A.I. Stewart, The UV nitric-oxide experiment for Atmosphere Explorer, *Radio Science, 8*, 379, 1973.

Cleary, D.E., Daytime high-latitude rocket observations of the NO, γ, δ, and ε bands, *J. Geophys. Res., 91*, 11337, 1986.

Chamberlain, J.W., *Physics of the Aurora and Airglow*, Academic Press, New York, 1961.

Cravens, T.E., and A.I. Stewart, Global morphology of nitric oxide in the lower E region, *J. Geophys. Res., 83*, 2446, 1978.

Eparvier, F.G., Rocket measurements of self-absorption of the nitric oxide (1,0) gamma band in the daytime thermosphere, PhD thesis, University of Colorado, 1991.

Eparvier, F.G., and C.A. Barth, Self-absorption theory applied to rocket measurements of the nitric oxide (1,0) gamma band in the daytime thermosphere, *J. Geophys. Res., 97*, 13,723, 1992.

Farmer, A.J.D., V. Hasson, and R.W. Nicholls, Absolute oscillator strength measurements of the ($v''=0$, $v'=0$-3) bands of the ($A^2\Sigma$-$X^2\Pi$) γ-system of nitric oxide, *J. Quant. Spectrosc. Radiat. Transfer, 12*, 627, 1972.

Hedin, A.E., MSIS-86 thermospheric model, *J.Geophys. Res., 92*, 4649, 1987.

Kley, D., G.M. Lawrence, and E.T. Stone, The yield of N(^{2}D) atoms in the dissociative recombination of NO$^+$, *J. Chem. Phys., 66*, 4157, 1977.

Norton, R.B., and C.A. Barth, Theory of nitric oxide in the Earth's atmosphere, *J. Geophys. Res., 75*, 13903, 1970.

Rusch, D.W., and C.A. Barth, Satellite measurements of nitric oxide in the polar region, *J. Geophys. Res., 80*, 3719, 1975.

Roble, R.G., J.M. Forbes, and F.A. Marcos, Thermospheric dynamics during the March 22, 1979, magnetic storm: 1. Model simulations, *J. Geophys. Res., 92*, 6045, 1987.

Siskind, D.E., C.A. Barth, and R.G. Roble, The response of thermospheric nitric oxide to an auroral storm. 1. Low and middle latitudes, *J. Geophys. Res., 94*, 16,885, 1989a.

Siskind, D.E., C.A. Barth, D.S. Evans, and R.G. Roble, The response of thermospheric nitric oxide to an auroral storm. 2. Auroral latitudes., *J. Geophys. Res., 94*, 16,899, 1989b.

Siskind, D.E., C.A. Barth, and D.D. Cleary, The effect of solar soft x-rays on thermospheric nitric oxide, *J. Geophys. Res., 95*, 4311, 1990.

Solomon, S., P.J. Crutzen, and R.G. Roble, Photochemical coupling between the thermosphere and the lower atmosphere, 1, Odd nitrogen from 50 to 120 km, *J. Geophys. Res., 87*, 7206, 1982.

Stewart, A.I., and T.E. Cravens, Diurnal and seasonal effects in E region low-latitude nitric oxide, *J. Geophys. Res., 83*, 2453, 1978.

Strobel, D.F., D.N. Hunten, and M.B. McElroy, Production and diffusion of nitric oxide, *J. Geophys. Res., 75*, 14307, 1970.

Charles A. Barth, LASP, University of Colorado, Boulder, CO 80309-0590

The Role of Fast N(^{4}S) Atoms and Energetic Photoelectrons on the Distribution of NO in the Thermosphere

J.-C. Gérard

LPAP, Institut d'Astrophysique, Université de Liège, Belgium

V. I. Shematovich and D. V. Bisikalo

Institute of Astronomy of the Academy of Sciences, Moscow, 109017, Russia

Ground state N(^{4}S) nitrogen atoms are produced with excess kinetic energy by direct N$_2$ dissociation as well as exothermic chemical reactions. A stochastic model is used to calculate their steady state energy distribution function in the thermosphere. Inelastic collisions with O$_2$ produce nitric oxide with a much higher efficiency than thermal nitrogen atoms. It is shown that the consideration of this additional source of NO is significant. Numerical simulations for the low latitude thermosphere at solar maximum activity conditions show that N$_2$ dissociation and ionization by fast photoelectrons produced by solar soft X rays are a significant source of N(^{2}D) and fast N(^{4}S) atoms which are precursors of NO in the lower thermosphere. A twofold increase of the solar soft X rays irradiance is shown to double the NO peak density as observed with a 27-day period during high solar activity phases.

1. INTRODUCTION

Large variations of the nitric oxide peak density were observed with the Solar Mesosphere Explorer (SME) satellite from 1982 to 1985 during the declining phase of the previous solar cycle [Barth et al., 1988; Fesen et al., 1990; Barth, 1989, 1994]. These resonance scattering measurements showed a decrease by a factor of 7.5 as solar activity dropped during the 4-year period of the observations. Using SME data collected only during days of low geomagnetic activity (Ap ≤ 15), Gérard et al. [1990] and Fesen et al. [1990] derived a peak density decrease by a factor of 2.5 to 4, depending on season and latitude from the same data base. In addition, a variation by nearly a factor of 2 was also observed with a 27-day period during the high solar activity phase in 1982. It was interpreted [Barth et al., 1988; Fuller-Rowell, 1993] as a consequence of a solar rotation induced change of the XUV and soft X ray radiation interacting with the E region and the lower thermosphere.

The SME data set is complemented by a set of rocket measurements by the University of Tokyo using the same observational technique. These observations were made from Japon at a fixed latitude of 31° N at sunset between 1981 and 1987 [Kuze and Ogawa, 1988]. They cover $F_{10.7}$ solar indices ranging from 71 to 259. They also show a solar activity dependence of the maximum NO density near 110 km. However, the increase from low to very high activity conditions is only a factor of 3.5. In addition, their absolute NO densities are larger than the SME values, possibly a consequence of an uncorrected optical thickness contribution in the SME limb viewing geometry (Eparvier and Barth, 1992).

The large solar cycle and solar rotation dependence of the NO peak prompted Barth et al. [1988] and Siskind et al. [1989] to emphasize the role of the solar soft X rays ($\lambda < 200$ Å) in the solar control of nitric oxide in the lower thermosphere. This part of the solar

TABLE 1 : list of important odd nitrogen reactions used in the model.

Number	Reaction	Rate coefficient [a]
R1.	$N_2 + e \to N + N^{(+)} + e + (e)$	$f(N(^2D)) = 0.54$
R2.	$N_2 + h\nu \to N(^4S) + N(^2D)$	$f(N(^2D)) = 0.5$
R3.	$NO^+ + e \to N(^4S,^2D) + O$	$4.2(-7) \, (300/T_e)^{0.85}$; $f(N(^2D)) = 0.75$
R4.	$N_2^+ + O \to N(^2D) + NO^+$	$1.4(-10) \, (300/T_i)^{0.44}$; $f(N(^2D)) = 1$
R5.	$N_2^+ + e \to N(^4S) + N(^2D)$	$4.2(-7) \, (300/T_e)^{0.85}$; $f(N(^2D)) = 0.5$
R6.	$N(^2D) + O_2 \to NO + O(^3P,^1D)$	$6(-12)$ [b]
R7.	$N(^4S) + O_2 \to NO + O$	$4.4(-12) \, \exp(-3220/T)$
R8.	$N_f(^4S) + O_2 \to NO + O$	see text
R9.	$N(^4S) + NO \to N_2 + O$	$3.4(-11)$
R10.	$O_2^+ + NO \to NO^+ + O_2$	$4.4(-10)$
R11.	$N(^2D) + NO \to N_2 + O$	$6.7(-11)$
R12.	$N(^2D) + O \to N(^4S) + O$	$6.7(-13)$
R13.	$N(^2D) + e \to N(^4S) + e$	$6.0(-10) \, (T_e/300)^{0.5}$
R14.	$N(^2D) \to N(^4S) + h\nu$	$1.07(-5) \, s^{-1}$
R15.	$O^+ + N_2 \to N(^4S) + NO^+$	$1.57(-12)$
R16.	$N^+ + O_2 \to N(^4S) + O_2^+$	$4.0(-10)$
R17.	$NO + h\nu \to N(^4S) + O$	

- [a] In $cm^3 \, s^{-1}$ unless otherwise specified
- [b] $6(-12)$ should read 6×10^{-12}

spectrum varies strongly with solar activity (and rotation) and plays an important role in the production of the precursors of nitric oxide. Indeed, photoionization of N_2, O_2 and O in the lower thermosphere produces photoelectrons with energies of hundreds of electronvolts. These energetic electrons can dissociate and ionize N_2 and provide an additional source of N_2^+ ions and N atoms. Both species, in turn, contribute to the NO production in the lower thermosphere. Two sources of thermospheric NO have been indentified :

$$\begin{aligned} N(^4S) + O_2 &\to NO + O \\ N(^2D) + O_2 &\to NO + O \end{aligned} \quad (1)$$

The first reaction is strongly temperature dependent and has a significant activation energy but the second one is faster and has no activation energy. Metastable $N(^2D)$ atoms are formed by various processes involving the break of the N_2 molecule bond either directly (photon or fast electron collisions) or indirectly (chemical reactions). A list of the main processes involving odd nitrogen formation is given in Table 1 . Only one process leads to the destruction of odd nitrogen in the lower thermosphere and involves the mutual destruction of N and NO.

$$NO + N(^4S) \to N_2 + O \quad (2)$$

Consequently, the NO peak density depends critically on the relative production of ground state $N(^4S)$ atoms (which are mostly a sink of NO) and $N(^2D)$ metastable atoms (which are a source of NO). On the basis of the physical properties of the N_2 molecule, Zipf and McLaughlin [1978] concluded that photodissociation of N_2 yields $N(^4S)$ and $N(^2D)$ atoms with the same efficiency. The electron impact dissociation of N_2 slightly favors the production of $N(^2D)$ as does the NO^+ dissociative recombination and the $N_2^+ + O$ charge transfer (reactions R3 and R4 in Table 1). Solomon [1983] suggested that the non-thermal $N_f(^4S)$ atoms produced by dissociation of N_2 react with O_2 at a rate much faster than thermal atoms :

$$N_f(^4S) + O_2 \to NO + O \quad (3)$$

The steady state energy distribution of these "hot" nitrogen atoms was calculated by Shematovich et al. [1991, 1992] by solving the Boltzmann equation including elastic and inelastic collisions with ambient atmospheric constituents. The effect on the odd nitrogen photochemistry and the NO distribution was investigated using a one-dimensional photochemical-transport model [Gérard et al., 1992; 1993; Shematovich et al., 1992]. They included all major sources of fast N atoms, namely,

$$N_2 + \{^{h\nu}_e\} \rightarrow N(^4S) + N(^4S, ^2D, ^2P)(+e)$$
$$NO^+ + e \rightarrow N(^4S) + O + 2.75\,eV$$
$$O^+ + N_2 \rightarrow NO^+ + N(^4S) + 1.09\,eV \quad (4)$$
$$N^+ + O_2 \rightarrow O_2^+ + N(^4S) + 2.48\,eV$$
$$N(^2D) + O \rightarrow O + N(^4S) + 2.38\,eV$$

The initial kinetic energy of the N atoms is a continuum with a maximum value of 1.65 eV for the N_2 predissociation by solar photons between 800 and 1000 Å. In fact, a signifiant amount of fast atoms is only produced below 1.26 eV. The main contribution arises from the CIII solar line at 977 Å. The $N(^4S)$ atoms produced by the four chemical reactions listed before carry an initial excess energy of 1.47, 0.74, 1.72 and 1.27 eV respectively. The main processes leading to the production of fast $N(^4S)$ atoms and odd nitrogen are sketched in Figure 1. Previous studies of the effect on the hot N atoms on the NO distribution indicated that a signifiant amount of fast $N(^4S)$ react with O_2 in the region of the maximum NO concentration. This question was examined independently by Lie-Svendsen et al. [1991] for the case of auroral precipitation using a different approach. The general conclusion [Shematovich et al., 1992; Gérard et al., 1993] is that the shape and magnitude of the NO peak in the E-region is significantly affected by the consideration of the fast N contribution. An increase of the peak density by about 45 % to over 60 % is predicted, dependent on solar activity and detailed parameters used in the calculation. The introduction of this contribution relaxes the requirements on the yield of $N(^2D)$ atoms near 110 km and reconciles the laboratory and theoretical studies of the $N(^2D)$ yield with the observed NO distribution in the lower thermosphere. In this study, we illustrate the role of both the fast photoelectrons produced by solar soft X rays and the additional NO source provided by the $N_f(^4S) + O_2$ reaction for high solar activity conditions.

2. THE MODEL

The one-dimensional model used to calculate the steady state energy function distribution of non thermal atoms was described in details by Shematovich et al. [1992]. It involves the solution of the non linear Boltzmann equations for fast $N(^4S)$ atoms using a Monte-Carlo computer simulation :

$$\frac{\partial}{\partial t} F_f(t,c) = Q_f(t,c) + \sum_{M=O_2,N_2,O} \mathcal{I}_{el}(F_f, F_M) + \mathcal{I}_{reac}(F_f, F_{O_2}) \quad (5)$$

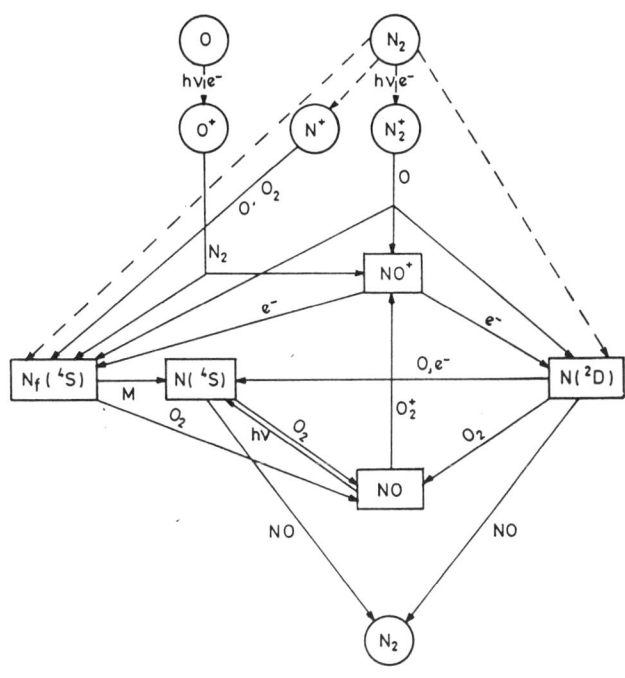

Fig. 1. Schematic diagram of photochemical processes leading to the production of non-thermal $N_f(^4S)$ atoms and controlling nitric oxide in the thermosphere. The dotted lines indicate ionization and dissociation of N_2 by photons and photoelectrons. M denotes elastic collisions.

where $F_f = n[N(^4S)]\, f_{N(^4S)}(t,c)$ is the energy distribution function for the hot nitrogen atoms and $f_{N(^4S)}(t,c)$ is the distribution function for the hot nitrogen atoms describing the distribution of atom velocities c. The function Q_f denotes the fast N production rate in processes (4) and the non linear collision term $\mathcal{I}_{el}$ and $\mathcal{I}_{reac}$ describe the thermalization and chemical processes respectively. The photoelectron spectrum is also calculated using the same Monte-Carlo method.

All the calculations are made for the equatorial lower thermosphere, high solar activity conditions ($F_{10.7}$ = 245) and quiet geomagnetic conditions. The major constituent and temperature vertical distribution are provided by the MSIS-86 model [Hedin, 1987]. The extreme ultraviolet and soft X rays solar irradiances are taken from Tobiska [1991].

The calculations of XEUV absorption by the main atmospheric constituents were made using a compilation of recent laboratory measurements of photoabsorption and photoionization cross sections of O, O_2 and N_2 [Conway, 1989]. The calculation of electron impact on the main constituents is made using the analytical approximations of excitation, dissociation and ionization cross-section given by Jackman et al. [1977] and Peterson et al. [1973]. For electron impact dissociation of N_2, the partial and total cross-sections presented by Zipf and McLaughlin [1978] are adopted.

We use the same rate coefficients as adopted by Roble et al. [1987] (Table 1). The quenching coefficient of $N(^2D)$ by O (R12) is taken as 6.7×10^{-13} cm^3 s^{-1} as measured recently by Fell et al. [1990]. Another important parameter, the branching ratio for the production of $N(^2D)$ atoms by photoelectron impact dissociation of N_2 (R1) was set to 0.54 [Shematovich et al., 1992]. The metastable $N(^2P)$ atoms are assumed to be efficiently quenched by atomic oxygen into $N(^2D)$ atoms. Consequently, it is reasonable to assume that most nitrogen atoms formed into the $N(^2P)$ level radiationally or collisionally cascade into the $N(^2D)$ level.

3. RESULTS

The key quantity controling the efficiency of reaction 3 as an additional source of NO is the velocity distribution of the fast $N(^4S)$ atoms. As mentioned before, this contribution was calculated using different methods by Shematovich et al. [1991] and Lie-Svendsen [1991]. In our calculations, we use the energy-dependent cross section for reaction 3 and derived by Polak et al. (1984) for energies above 0.85 eV. For the energy range between from the threshold E_o to 0.85 eV, the cross section is written in the form suggested by LeRoy (1969)

$$\sigma(E) = A(E - E_o)^n \exp[-m(E - E_o)] \quad (6)$$

where A, n, m and E_o are parameters determined by fitting the theoretical thermal reaction rate

$$k(T) = (\pi\mu)^{-0.5} (2/k_B T)^{1.5} \int_0^\infty \sigma(E) \, E \, \exp(-E/kT) \, dE \quad (7)$$

coefficient k of reaction 3. In (7), k_B denotes the Boltzmann constant and μ the reduced mass of the $N(^4S)$ - O_2 system. The best fit was obtained with the values $A = 5.3 \times 10^{-16}$ cm^2, $n = 1.7$, $m = 0.5$ eV^{-1} and E_o = 0.25 eV. The total energy dependent cross section is illustrated in figure 2. It is based on expression (6) for $E < 0.85$ eV and on Polak et al's value for higher energies. For comparison, a plot of the cross section obtained by Lie-Svendsen et al. (1991) by a fit to the available experimental values of k(T) over a range of temperatures is also presented. The results are very close for all energies.

Lie-Svendsen et al (1991) reached the conclusion that, in the case of the nightside aurora excited by ener-

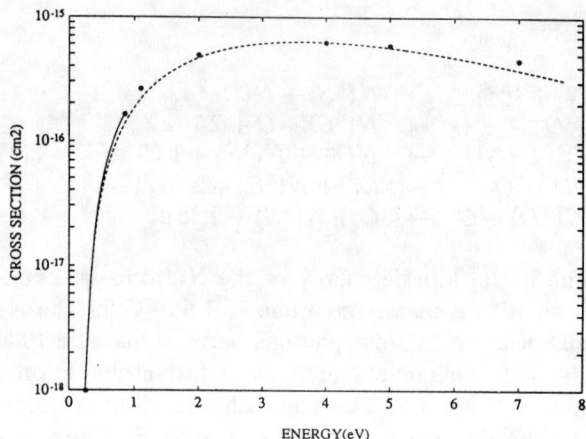

Fig. 2. Cross section of the $N_f(^4S) + O_2$ (reaction 3) process used in this model. The circles indicate values derived from Polak et al. (1984) and the solid dashed line is obtained from (6) (see text). For comparison, the cross section adopted by Lie-Svendsen et al. (1991) in their model is also shown (dashed line).

getic electrons, the contribution remains low. In contrast, Shematovich et al. [1991, 1992] and Gérard et al. [1993] found significant NO increases in the lower thermosphere. Since the two groups modeled different situations (aurora and daytime low latitudes) using different elastic collision cross sections and neutral atmosphere, it was found difficult to identify the sources (if any) of discrepancies. A more straightforward way to test the two methods is to compare the fraction of $N(^4S)$ hot atoms reacting with O_2 to form NO as a function of the initial atom velocity but using the same atmospheric density, composition and collision cross sections. This comparison is shown in Figure 3 which also illustrates the importance of second and higher order collisions. For this test, an elastic collision cross section of 1×10^{-15} cm^2 and the inelastic collision cross section energy dependence given by Lie-Svendsen et al. were adopted for the calculations.

The calculated reacting fraction increases from 0 near 2×10^5 cm s^{-1} to about 0.25 for $v = 8 \times 10^5$ cm s^{-1}. The agreement between this model and Lie-Svendsen's calculations is excellent, indicating that both methods provide powerful tools to study the role of non-thermal atoms in the thermosphere. The relative importance of the N_2 ionization by EUV and soft X ray photons and by photoelectrons is illustrated in Figure 4. The two contributions are nearly equal at 120 km but the electron impact contribution dominates below this altitude and peaks near 110 km where is exceeds the photoionization contribution by about a factor of 7. Con-

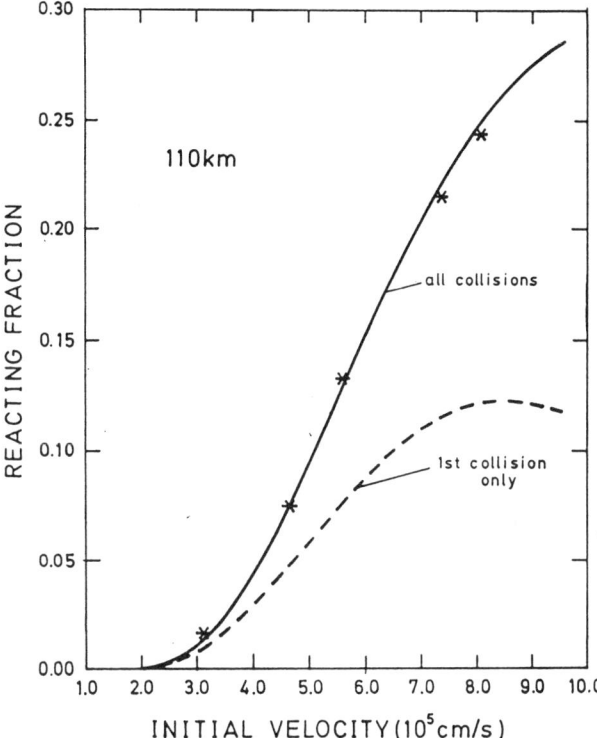

Fig. 3. Fraction of fast N(^{4}S) atoms reacting with O$_2$ at 110 km. The solid line was calculated by Lie-Svendsen and Rees [private communication, 1991] including all collisions whereas the dashed line take first collisions only into account. The stars were calculated with the method described in this paper

The steady state nitric oxide density distribution shows a peak near 110 km, in agreement with rocket and satellite observations (Figure 6). The importance of the hot N(^{4}S) contribution is shown in the comparison of two cases. In the first one, the full model (including hot N) is used and the NO peak reaches 6.2×10^7 cm^{-3} for the high solar activity conditions (F10.7 = 243) adopted in this run. A second case was run for identical conditions except that the effect of reaction (3) was removed. In this case, the NO maximum drops from 6.3 to 3.9×10^7 cm^{-3} as a consequence of the removal of the hot N source of NO.

The importance of the fast photoelectrons produced by soft X ray ionization of N$_2$, O$_2$ and O is illustrated by doubling the solar irradiance below 200 Å to simulate the modulation by a solar rotation at solar maximum activity. Consequently, the integrated irradiance in the 18-200 Å range is increased from 1.61 to 3.22 erg cm^{-2} s^{-1}. As a result, the N$_2$ photoionization rate (Figure 4) shows a moderate increase at high altitude but increases by a factor of 2 at 110 km where soft X rays are most efficiently absorbed by N$_2$. In contrast, the ionization by photoelectrons is nearly doubled at all altitudes since soft X rays below 200 Å are the main contributions to the fast photoelectron production, independent of altitude. Similary, the effect of the soft X ray doubling on the N(^{2}D) production from N$_2$ is shown in Figure 5. The photodissociation component is unaffected since

sequently, most of the N$_2^+$ and N$^+$ ion production at the NO peak (and consequently the ionospheric sources of N(^{2}D) and N(^{4}S) as well) is controled by the impact of secondary electrons. A detailed analysis shows that the main contribution to the photoelectron impact ionization results from the high energy tail of the photoelectron spectrum due to the soft X ray part of the solar spectrum as suggested by Barth et al. [1988] and Siskind et al. [1990].

The vertical distribution of the N(^{2}D) atom production by R1 and R2 is shown in Figure 5. The photoelectron contribution dominates between 120 km and 180 km but both are nearly equal at 110 km. The N$_2$ predissociation in the 800-1000 Å region produces equal amounts of N(^{4}S) and N(^{2}D) atoms and reactions R9 and R11 are the main sinks of nitric oxide. Consequently, only the fraction of hot N(^{4}S) atoms reacting with O$_2$ to form NO is a net source of NO due to R2. Since we use a N(^{2}D) quantum yield of 0.54 for R1, both the N(^{2}D) and the hot N(^{4}S) channels are net sources of NO in this case.

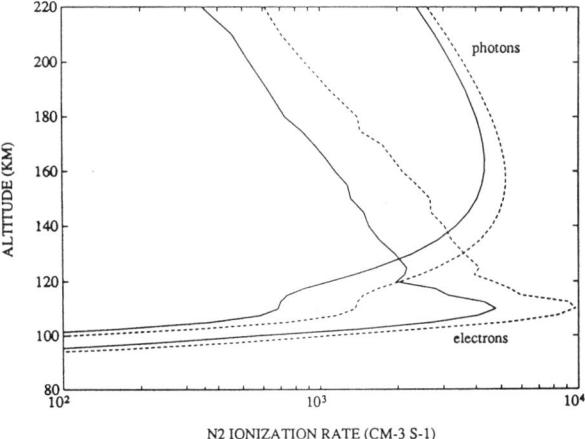

Fig. 4. Log$_{10}$ of the N$_2$ ionization rate calculated for 0° latitude and high solar activity level (F10.7 = 245). The curves show the direct photoionization component and the photoelectron impact contribution. The calculations were made using both the nominal [Tobiska, 1992] solar irradiance (solid lines) and a twofold increase in soft X ray ($\lambda < 200$ Å)(dashed curves).

only photons in the 800-1000 Å range contribute to the N$_2$ predissociation. The electron impact dissociation rate (R1) increases by a factor of 2 at 110 km and by 35 % at 220 km where the unmodified higher wavelength EUV radiation also makes an important contribution to the production of the photoelectrons able to dissociate N$_2$ and produce N(^{2}D) atoms. The total effect on the nitric oxide distribution is shown in Figure 6, with and without the hot N(^{4}S) atom contribution. In both cases, the NO peak concentration is nearly doubled when the soft X ray are multiplied by two. It reaches 1.2×10^8 cm^{-3} for the 2 times X ray case with the hot N(^{4}S) chemistry included.

4. CONCLUSIONS

Results described before and in previous studies show that the consideration of the N(^{4}S) non thermal atoms is important in the description of processes affecting the nitric oxide production in the E region. As previously suggested in the literature, our calculations show that fast photoelectrons produced by soft X ray ionization of the lower thermospheric constituents is an important source of N$_2$ dissociation and ionization. Consequently, additional N(^{2}D) and fast N(^{4}S) atoms are produced by this variable solar flux component and the calculated NO distribution predicted using measured reaction coefficients and cross sections now shows reasonable agreement with the observations. The NO peak density is shown to scale almost linearly with the soft

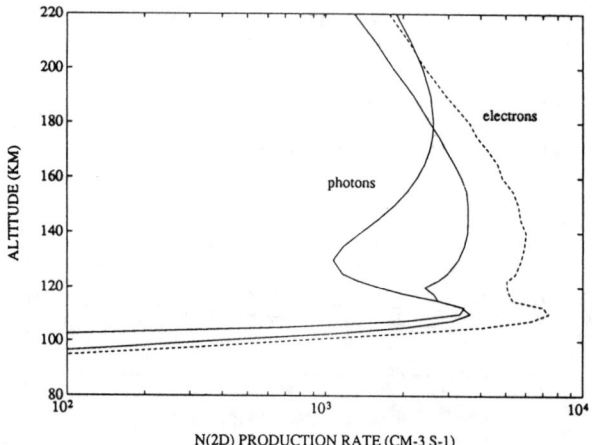

Fig. 5. Log$_{10}$ of the N(^{2}D) production rate by direct N$_2$ dissociation for the same conditions as Figure 4. The line labeled photons shows the solar EUV (800-1000 Å) predissociation (R2) and the line labeled electrons the photoelectron contributions (R1). The dashed curve is for a twofold increase in soft X ray intensity.

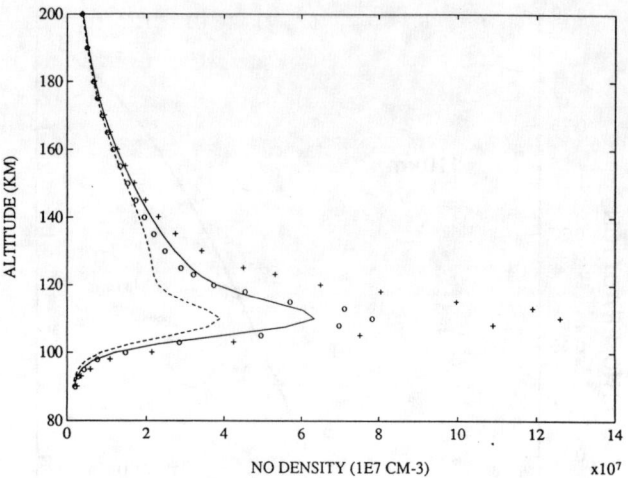

Fig. 6. Vertical distribution of the NO density for the same conditions as Figure 4 with or without the hot N(^{4}S) source included. The solid line shows the profile calculated with the non thermal N(^{4}S) contribution and the dashed line without it. The circle and plus curves show the distribution for a doubled soft X ray intensity with (plus) and without (circles) the fast atom chemistry.

X ray flux. In particular, the twofold variation of the NO at 110 km during solar rotation is well reproduced by doubling the soft X ray flux below 200 Å.

Acknowledgments. J.C. Gérard is supported by the Belgian Foundation for Scientific Research (FNRS) and V.I. Shematovich and D.V. Bisikalo are funded by the Russian Foundation for Fundamental Research (grant 93-02-2847). This work was supported by FRFC grants 2.4505.87 and 2.4539.93

REFERENCES

Barth, C. A., Reference models for thermospheric NO, *Adv. Space Res.*, 10, 103, 1989.

Barth, C.A., Reference models for thermospheric nitric oxide, *Adv. Space Res.*, 1994, in press.

Barth, C. A., W. K. Tobiska, D. E. Siskind, and D. D. Cleary, Solar-terrestrial coupling: low-latitude thermospheric nitric oxide, *Geophys. Res. Lett.*, 15, 92, 1988.

Conway, R.A., Photoabsorption and photoionization cross section of O, O$_2$ and N$_2$ for photoelectron production calculations : A compilation of recent laboratory measurements, *Report 6155, Naval Research Laboratory*, Washington, 1989.

Eparvier, F. J., and C. A. Barth, Self absorption theory applied to rocket measurements of the nitric oxide (1, 0) γ band in the daytime thermosphere, *J. Geophys. Res.*, 97, 13723, 1992.

Fell, C. , J.I. Steinfeld and S. Miller, Quenching of N(^{2}D) by O(^{3}P), *J. Chem. Phys.*, 92, 4768, 1990.

Fesen, C.G., D.W. Rusch and J.C. Gérard, The latitudinal gradient of the NO peak density, *J. Geophys. Res.*, 95, 19053, 1990.

Fuller Rowell, T.J., Modeling the solar cycle change in nitric oxide in the thermosphere and upper mesosphere, *J. Geophys. Res.*, 98, 1559-1570, 1993.

Gérard, J.C., Thermospheric odd nitrogen, *Planet. Space Sci.*, 40, 337, 1992.

Gérard, J.C., V.I. Shematovich and D.V. Bisikalo, Non thermal nitrogen atoms in the Earth's thermosphere 2. A source of nitric oxide, *Geophys. Res. Lett.*, 18, 1695, 1991.

Gérard, J.C., V.I. Shematovich and D.V. Bisikalo, Effect of hot N(^{4}S) atoms on the NO solar cycle variation in the lower thermosphere, *J. Geophys. Res.*, 98, 1993, in press.

Hedin, A., MSIS-86 thermospheric model, *J. Geophys. Res.*, 92, 4649, 1987.

Jackman, C.H., R.H. Garvey and A.E.S. Green, Electron impact on atmospheric gases I. Updated cross section, *J. Geophys. Res.*, 82, 5081, 1977.

Kuze, A. and Ogawa, T, Solar cycle variation of thermospheric NO: a model sensitivity study, *J. Geomagn. Geoelectr.*, 40, 1053, 1988.

LeRoy, R.L., Relationship between Arrhenius activation energy and excitation functions, *J. Phys. Chem.*, 73, 4338, 1969.

Lie–Svendsen, O, M.H. Rees, K. Stamnes and E.C. Whipple, The kinetics of "hot" nitrogen atoms in upper atmosphere neutral chemistry, *Planet. Space Sci.*, 39, 929, 1991.

Peterson, L.R., T. Sawada, and J.N. Bass and A.E.S. Green, Electron energy deposition in a gaseous mixture, *Comp. Phys. Comm.*, 5, 239, 1973.

Roble, R.G., E.C. Ridley and E.C. Dickinson, On the global mean structure of the thermosphere, *J. Geophys. Res.*, 92, 8745, 1987.

Shematovich, V.I., D.V. Bisikalo and J.C. Gérard, Non thermal nitrogen atoms in the Earth's thermosphere 1. Kinetics of hot N(^{4}S), *Geophys. Res. Lett.*, 18, 1691, 1991.

Shematovich, V.I., D.V. Bisikalo and J.C. Gérard, The thermospheric odd nitrogen photochemistry: role of non thermal N(^{4}S) atoms, *Ann. Geophys.*, 10, 792, 1992.

Siskind, D. E., C. A. Barth, and R. G. Roble, The response of thermospheric nitric oxide to an auroral storm, 1., Low and mid latitudes, *J. Geophys. Res.*, 94, 16,885, 1989.

Siskind, D.E., C.A. Barth and D.D. Cleary, The possible effect of solar soft X Rays on thermospheric nitric oxide, *J. Geophys. Res.*, 95, 4311, 1990.

Solomon, S., The possible effects of translationally excited nitrogen atoms on lower thermospheric odd nitrogen, *Planet. Space Sci.*, 33, 135, 1983.

Tobiska, W.K., Revised solar extreme ultraviolet flux model, *J. Atmos. Terr. Phys.*, 53, 1005, 1991.

Zipf, E.C. and R.W. McLaughlin, On the dissociation of nitrogen by electron impact and by the EUV photoabsorption, *Planet. Space Sci.*, 26, 449, 1978.

J.-C. Gérard, LPAP, Institut d'Astrophysique, Université de Liège, 5, avenue de Cointe, B-4000 Liège, Belgium.

V. I. Shematovich and D. V. Bisikalo, Institute of Astronomy of the Academy of Sciences, 48, Pjatnitskaja, Moscow, 109017, Russia.

The State of O$_2$ in the Mesopause Region

Donal P. Murtagh

Meteorologiska Institutionen, Stockholms Universitet, Stockholm, Sweden.

Excited states of O$_2$ are formed in the mesopause region through the recombination of atomic oxygen. The relative amounts of each state present and the mechanisms through which they come into being have, for a long time, been the subject of controversy. Recent years have seen some improvement in the state of our knowledge and this paper attempts to illustrate this. One thing that is clear from ground based studies of the airglow is that the relative populations of the states are far from constant but depend greatly on the amount of atomic oxygen present and on where it is. The study identifies the O$_2$(A') state as the major precursor of the O$_2$ Atmospheric band emission in the nightglow while both O$_2$(c) and O$_2$(A') must act as precursors to the greenline.

1. INTRODUCTION

In the region of the mesopause the collision frequency is such that molecular oxygen produced in excited states through the association of oxygen atoms can live long enough to release energy through emission of radiation. This emission of radiation occurs despite the fact that all of the excited states are, in fact, metastable and have no allowed electrical dipole transitions. Emissions from the three high lying states constitute the UV nightglow [*Chamberlain*, 1961]. The molecular states involved and the names of the various band systems are given in Figure 1.

The excited states can be produced either directly as a product of the three body association of oxygen atoms or indirectly through energy transfer processes involving states that are produced directly. The direct production of the individual states is determined mainly by their relative statistical weights as well as by the form of the potential energy curves at large inter nuclear distances. Wraight [1982] and Smith [1984] have carried out calculations and obtained values for the production fractions. Bates [1988] has revised these values slightly based upon new calculations of the shape of the very loosely bound $^5\Pi_g$ state. These values are given in Table 1. Bates [1988] has however argued that the $^5\Pi_g$ state will thermally dissociate again and therefore is unlikely to take any further part in the photochemistry.

2. WHAT DO THE MEASUREMENTS TELL US?

The study of long-lived metastable states continues to present a considerable challenge to laboratory scientists and much of the information available on the excitation processes has and continues to come from the atmospheric laboratory. This natural laboratory has to be probed either in-situ by rocket-borne instruments or through passive remote sensing of the region using the emissions from the species under investigation.

2.1. Rocket Measurements

By flying narrow band filter photometers through the emitting layers it is possible to deduce the volume emission rate of a particular emission [see e.g. *Murtagh et al.* 1982]. However in the UV-region it is difficult to isolate emission from a particular state which often implies that a number of more or less serious assumptions has to be made in order to extract the feature of interest. The Herzberg I bands can be measured with some ease if reasonable assumptions about the vibrational distribution can be made [*Murtagh et al.* 1986a] and even the (5-2) band of the Chamberlain system can be satisfactorily isolated [*Murtagh et al.* 1986b]. However these assumptions can seriously limit the extent to which the data can be used to exactly determine excitation mechanisms. There are currently no rocket measurements of the volume emission rate for the Herzberg II system as individual bands cannot be isolated without the use of a spectrograph and synthesis of spectra.

In the atmosphere the oxygen emission features may be excited directly through association of oxygen atoms

$$O + O + M \xrightarrow{\alpha k_2} O_2^* + M \qquad (1)$$

and the volume emission rate for O_2^*, $V(O_2^*)$ at a particular height in the atmosphere is given by

$$V(O_2^*) = A_1^* \cdot \frac{\alpha k_2 [O]^2 [M]}{\left(A_2^* + k_{N_2}^*[N_2] + k_{O_2}^*[O_2] + k_o^*[O]\right)} \qquad (2)$$

where A_1^* is the Einstein A coefficient for the transition and A_2^* is the inverse lifetime of the emitting state and $k_{N_2}^*$, $k_{O_2}^*$ and k_o^* are the rates for quenching of O_2^* by N_2, O_2 and O respectively. α is the production efficiency for the unspecified excited state O_2^*. When a specific state is intended we will use another greek letter such as γ or δ. Alternatively they may be excited via an indirect mechanism such as

$$\left. \begin{array}{l} O + O + M \xrightarrow{\alpha k_2} O_2^* + M \\ O_2^* + X \xrightarrow{k_T} O_2^\uparrow + X^* \end{array} \right\} \qquad (3)$$

where $O_2^\uparrow$ is another state of O_2 and X^* may represent an excited transfer agent. In this case the volume emission rate, V for either $O_2^\uparrow$ or X^* is given by

$$V = A_1 \cdot \frac{\alpha k_2 [O]^2 [M]}{\left(A_2^* + k_{N_2}^*[N_2] + k_{O_2}^*[O_2] + k_o^*[O]\right)} \cdot \frac{k_T[X]}{\left(A_2 + k_{N_2}[N_2] + k_{O_2}[O_2] + k_o[O]\right)} \qquad (4)$$

where the first term represents the photochemical equilibrium concentration of the precursor state O_2^* and the second term represents the transfer step and the loss of the emitting state. A_1 and A_2 refer to the emitting state as do the unsuperscripted quenching rates. In the case that X is an oxygen atom then equation (3) describes the Barth mechanism for the excitation of the greenline. The coefficient k_2 is the rate of association of oxygen atoms in the presence of a third body M and α is the efficiency for producing the state in question. The value for k_2 used in this

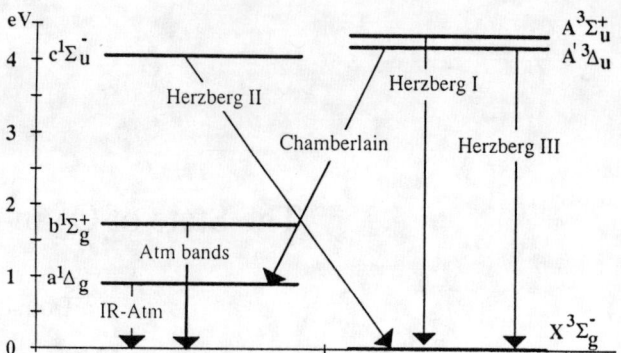

Fig. 1. The electronic states and main band systems for O_2.

model is that due to Campbell and Gray [1973]; $k_2 = 4.7 \times 10^{-33}(300/T)^2$ cm^{6}s^{-1}. Bates [1988] has argued that the actual rate coefficient may be twice this value since the $O_2(^5\Pi g)$ formed under the laboratory conditions would have immediately redissociated. We have chosen to use the former value for consistency with previous papers but note that a revision of this value only implies a scaling of the coefficients in the empirical model. Since k_T and α are often unknown quantities these are combined into the empirical quenching coefficients for the precursor as follows

$$V = A_1 \cdot \frac{k_2 [O]^2 [M]}{\left(Q_R^* + Q_{O_2}^*[O_2] + Q_O^*[O]\right)} \cdot \frac{[X]}{\left(A_2 + k_{N_2}[N_2] + k_{O_2}[O_2] + k_o[O]\right)} \qquad (5)$$

where

$$Q_R^* = \frac{A_2^*}{\alpha k_T}, \quad Q_{O_2}^* = \frac{k_{O_2}^*}{\alpha k_T} \qquad (6)$$

and similarly for Q_O^*. The corresponding term $Q_{N_2}^*$ has been dropped here since it is impossible to separate quenching by N_2 and O_2 from rocket measurements, i.e. we assume that O_2 is responsible for all molecular quenching. This has been further discussed by McDade et al. [1986].

In the case of states populated by energy transfer reactions and where there are laboratory measurements for the quenching rates of the emitting state then a rocket measurement of the volume emission rate and the atomic oxygen density can be used to obtain Q_O^* and $Q_{O_2}^*$. The Q_R^* term is often more difficult to obtain as it should dominate at higher altitudes where the signal to noise in the

TABLE 1. The statistical weights and theoretical production fractions for the various states of O_2.

State	$X^3\Sigma_g^-$	$a^1\Delta_g$	$b^1\Sigma_g^+$	$c^1\Sigma_u^-$	$A'^3\Delta_u$	$A^3\Sigma_u^+$	$^5\Pi_g$
Statistical Weight	3	2	1	1	6	3	10
Fraction	0.12	0.08	0.04	0.04	0.23	0.12	0.38
Bates 88	0.12	0.07	0.03	0.04	0.18	0.06	0.50

measurements is low and O quenching is often still dominant and therefore masks the radiative quenching. This is particularly true for the A' and c states that have quite long lifetimes. This approach was introduced by Thomas [1981] and was applied by McDade et al.[1986] to obtain quenching coefficients for the precursors for the OI greenline (X = O, and X* emitting) and the O_2 atmospheric bands (X = O_2). Recently we have used a similar approach [Stegman and Murtagh, 1991; Murtagh et al., 1991] to study the quenching of the $O_2(A')$ state from the only two available rocket measurements of the (5-2) band made on Stockholm University payloads in the S35/Oxygen [Witt et al. 1984] and ETON [Greer et al., 1986] campaigns. A direct excitation mechanism was assumed in analogy to the production of the A state and all molecular quenching assigned to O_2. Since the radiative loss is small it is not possible to separate this term from the collisional quenching and it is therefore impossible to obtain a value for the production efficiency for the $O_2(A')$ state which we designate γ from these measurements. The relative quenching rates obtained are given in Table 2.

Spectrographic measurements of the UV-nightglow have been made from rockets on only a very few occasions. The first photographic spectrum was obtained by Hennes [1966] and a photoelectric spectrum by Sharp and Siskind [1989]. The Hennes spectrum was subsequently analysed by Degen [1969] to yield a vibrational population distribution for the $O_2(A)$ state. However when the significant presence of the other UV-band systems was fully realised Slanger and Huestis [1981] analysed the photometric atlas of Broadfoot and Kendall [1968] (hereafter B&K) to obtain vibrational population distributions for all three upper states. Two attempts [Murtagh et al., 1986a; Kita et al., 1988] to study possible variations of the vibrational distributions with altitude have been made using filter photometers that were weighed by the selected wavelength intervals toward emissions predominantly from high and low vibrational levels. Both studies concluded that there was very little variation. Siskind and Sharp [1990] have analysed their rocket spectra and found indications that the vibrational distribution is more relaxed at lower altitudes but because of the difficulties separating the contributions from the various band systems and the low signal to noise ratio this analysis cannot be considered conclusive. A measurement with higher resolution would be required to settle this question. An interesting feature is however the dip in the vibrational populational distribution at v=5 for the low altitude spectrum, a feature also noted by Stegman and Murtagh [1991].

2.2. Satellite Measurements

Very recently two satellite measurements of the Herzberg band systems have become available. The first, by Eastes et al. [1992] is a nadir measurement with a resolution of 0.59 nm. Again the low resolution hinders the separation of the various emission systems but they do extract a population distribution in general agreement with previous results. It is difficult to assess, from the figures presented, just how good or bad the fits actually are but a general impression is that the valleys between the main peaks are much deeper in the synthetic spectrum than in the measurements. The derived ratio of the HI: HII system intensities was 4.5:1 but no estimate of the absolute intensities is given. The second set of spectra by Owens et al. [1993] are a limb measurment from the ATLAS 1 mission on the space shuttle. These spectra with a resolution of 0.23 nm have the potential to answer some of the questions raised above but the synthetic spectral fits so far applied are inadequate for the task as they fail to represent the rotational structure clearly present in the measured spectra. This limitation has been recognised by the authors.

2.3. Ground-based Measurements

Because of the difficulties involved in making spectrally resolved rocket measurements of the UV-nightglow high resolution ground-based observations combined with monitoring of the standard airglow emission such as the oxygen greenline provide a necessary complement to the

TABLE 2. The ratio of the quenching coefficients by O and O_2 for various states of O_2

State	k_O (cm³s⁻¹)	k_{O_2} (cm³s⁻¹)	A (s⁻¹)	Ratio $k_O:k_{O_2}$	Measurement Technique
Greenline precursor				15	Rocket[a]
A-band precursor				2.9	Rocket[a]
$O_2(A^3\Sigma_u^+)$	$< 9 \times 10^{-12}$	1.6×10^{-11}	11	<1	Rocket[b,d,e]
$O_2(A'^3\Delta_u)$	$2.4 \times 10^{-10}\gamma$	$9.8 \times 10^{-11}\gamma$	1.4	~2.5	Rocket, Ground-based[c]
$O_2(c^1\Sigma_u^-)$	$3.3 \times 10^{-9}\delta$	$6.6 \times 10^{-12}\delta$	0.66	500	Ground-based[c]

γ and δ are the production efficiencies for the states $O_2(A'^3\Delta_u)$ and $O_2(c^1\Sigma_u^-)$ respectively
[a] McDade et al. [1986]
[b] Murtagh et al. [1986a]
[c] Stegman and Murtagh [1991]
[d] k_{O_2} corrected to A = 11 s⁻¹ after Bates [1989]
[e] k_O from Kenner and Ogryzlo [1980] and consistent with being an upper limit in the analysis of the rocket measurements.

rocket measurements. The significance of the analysis of the B&K atlas has already been discussed above. The Stockholm University ground-based optical facility boasts a high throughput 1m Ebert-Fastie scanning spectrometer giving a resolution of 0.13 nm at 340 nm [*Stegman*, 1988] which facilitates the separation of the indivdual band contributions. Results from these measurements that have been made during the period 1985-1989 have been reported by Stegman and Murtagh [1988,1991]. The main results indicated a surprising stability in the vibrational population distributions, with a distinct dip at v=5, under a wide variety of atmospheric conditions ranging from very high (> 400R) greenline intensities to very low ones (< 20R). The range of greenline intensities is indicative of variation in the oxygen atom concentrations by a factor of 3 or more. In addition they confirmed the basic premises of the excitation models for the $O(^1S)$, $O_2(A)$ and $O_2(A')$ states derived from the rocket measurements and allowed the derivation of quenching parameters, albeit with considerable uncertainty, for the $O_2(c)$ state. These quenching parameters are summarised in Table 2. The ratio $k_O:k_{O_2}$ represents the relative rate of quenching of the state by oxygen atoms compared to a molecular quencher (O_2, N_2) expressed in terms of molecular oxygen quenching. It should be emphasised that any conclusions reached later in this paper are not dependent on this assumption and can be easily rephrased, should future laboratory measurements make it possible to partition the molecular quenching into an O_2 and an N_2 part. A recent modern day equivalent of the B&K spectrum has recently been published by Johnston and Broadfoot [1993]. Using spectrographs with ICCD detectors a similar atlas with a resolution ranging from 0.4 to 1 nm has been obtained in only a few hours compared to nearly 54 hours. The authors have undertaken an analysis of the band systems present in the spectral region between 300 and 3800 nm and obtain the total system intensity ratios HI:Ch:HII = 2.8:1:<1 at a greenline intensity of 320 R and an Atmospheric (0-1) band intensity of 350 R. A vibrational population consistent with the results of Slanger and Huestis [1981] but lacking the dip at v=5 mentioned above was derived. The values for the intensity ratios are not inconsistent with the results of Stegman and Murtagh [1991] but inspection of the comparison between the measurements and the syntetic spectra presented in the paper indicates that considerable problems in the treatment of the background signal arising from the zodiacal and integrated starlight and the transmission of the atmosphere remain and make the results rather uncertain. A detailed discussion of the question of the vibrational development of the $O_2(A)$ state must await improved data sets and a consistent treatment of the spectral analysis.

3. DISCUSSION

The results obtained from the various measurements provide data for a more general model of the nightglow such as that described in Murtagh [1989] and Stegman and Murtagh [1991]. Using this model the changes in the relative populations of the excited states of O_2 can be studied in more detail. Figure 2 illustrates this by presenting the populations of the three *ungerade* states of

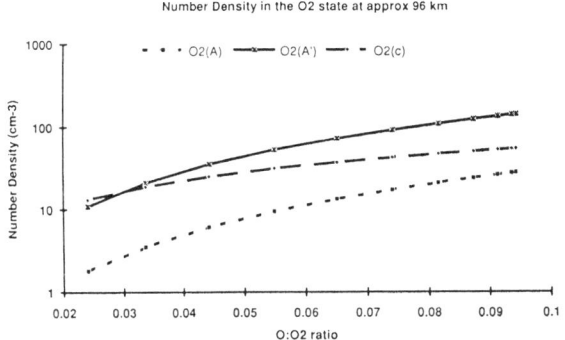

Fig. 2. The variation of the populations of the three upper excited states of O_2 with the $O:O_2$ ratio.

O_2 as a function of changing $O:O_2$ ratio as, for instance, would occur with changing the altitude of the O-layer or background atmospheric conditions. An interesting observation is the clear dominance of the A' state at values of $O:O_2$ typical of the airglow layer maximum (0.07).

When the excitation mechanisms for the greenline and the O_2 atmospheric bands are discussed in terms of energy transfer mechanisms then the identity of the precursor state is a subject of considerable discussion. Witt et al. [1979] would like to identify the precursor in both cases as the $O_2(c)$ state while Krasnopolsky [1981, 1986] and Wraight [1982] consider the $O_2(^5\Pi)$ state as the omnipotent precursor. Bates [1992] argues for a combination of the $O_2(c)$ and $O_2(A)$ states as the precursor for the A-band and $O_2(c)$ for the greenline. In a recent laboratory study Wildt et al. [1991] have been able to show that when laser excitation is used to selectively populate a given level of any of the three ungerade states emission occurs in the atmospheric bands. The efficiency of the energy transfer process was very high and it appeared that N_2 was even more effective than O_2, at least for transfer from the $O_2(A)$ state. To assess the consequences of these laboratory experiments on our understanding of the atmospheric production process we modelled the effect of assuming that all molecular quenching of the $O_2(A, A'$ and c) states leads to the production of $O_2(b)$. A second assumption in this exercise is that the A' and c states are formed in the recombination reaction with the efficiencies given in Table 1. This assumption is necessary because the rate of quenching by O_2 for these states is only known with respect to the production rate (see Table 2). For the A state the value of 3% derived from rocket measurements was used. However this was derived assuming that the quenching of the A state by atomic oxygen is negligible, if the upper limit value in Table 2 is employed then the production efficiency must be increased to 6% which is consistent with Table 1 but without causing any change in the results. The result is shown in Figure 3 for the A-band measurement made on P229H during the ETON campaign [*Greer et al.*, 1986]. The agreement between the modelled and observed profiles is better than deserved considering the assumptions involved but the main conclusion would seem to be that the A' state is the main precursor under atmospheric conditions with a 20% contribution from the

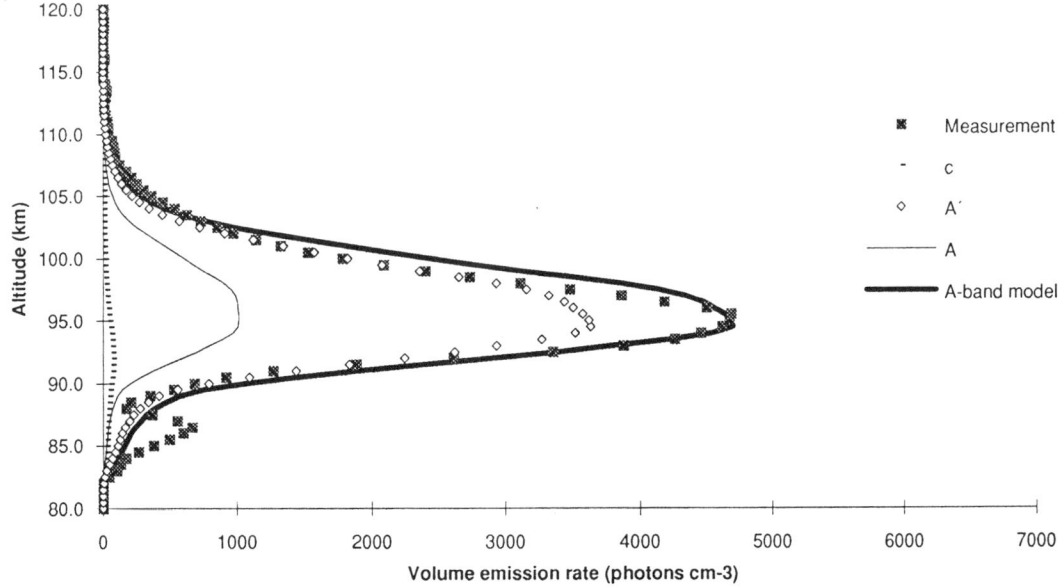

Fig. 3. The relative contributions to the production of the O_2 (b) state by the three possible precursor states assuming unit transfer efficiency.

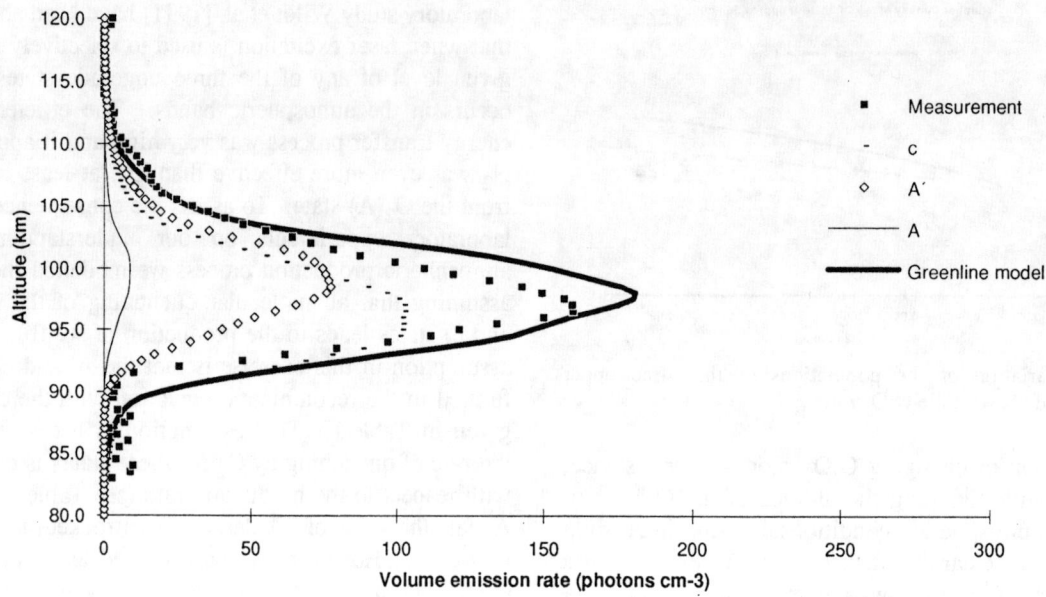

Fig. 4. The relative contributions to the production of $O(^1S)$ by the three possible precursor states assuming a transfer efficiency of 4% in all cases.

A state of O_2. This is also in agreement with the observation that the A-band precursor and the $O_2(A')$ state have the same ratio of quenching by atomic to molecular oxygen. The agreement was equally good for the other measurements of the A-band made during the ETON campaign when the appropriate oxygen profiles are used. Allowing $O_2(b)$ to be directly populated with the efficiency given in Table 1 would increase the modelled intensity by 10-20% allowing something of a relaxation on the requirements that all molecular quenching of the ungerade states produces the b state.

No similar experiments have been reported regarding the excitation of the oxygen greenline but as a purely hypothetical exercise, we again allowed all three states to produce $O(^1S)$ when quenched by atomic oxygen with the same efficiency. In order to reproduce the observed intensities this efficiency need only be 4%. Figure 4 presents the results. The c and A' states appear to be the major contributors with only a minor contribution from the A state. Further adjustment of the parameters would be meaningless computer juggling and would not change the main conclusion that at least these two states must contribute, an argument that is strengthen by the need to reconcile the $k_O:k_{O_2}$ required for the greenline precursor with the same ratios for the states that may be identified as such.

4. CONCLUSIONS

The combination of rocket and ground-based studies has contributed substantially to our knowledge about the excited states of O_2. Although several questions, particularly regarding the variation of rate constants with vibrational quantum number, still remain open a consistent picture of the excitation mechanisms and quenching rates for the various states is appearing. The long radiative lifetime and the relatively slow quenching of $O_2(A')$ state makes it dominant in the airglow region in terms of abundance and it appears to be the main precursor state for the production of the mesospheric atmospheric band emissions. Further the results obtained concerning the relative quenching of the three ungerade states by atomic oxygen and molecular species rule out the identification of a single such state as the greenline precursor but rather would tend to indicate that both the c and A' state are involved.

Acknowledgements. This work was supported by the Swedish Natural Science Research Council and the Swedish National Space Board. We would like to thank J. Stegman and G. Witt for helpful discussions and suggestions.

REFERENCES

Bates, D. R., Excitation and quenching of the oxygen bands in the nightglow, *Planet. Space Sci.*, 36, 875-881, 1988.

Bates, D. R., Oxygen band system transition arrays *Planet. Space Sci.*, 37, 881-887, 1989.

Bates, D. R., Nightglow emissions from oxygen in the lower thermosphere. *Planet. Space Sci.*, 40, 211-221, 1992.

Broadfoot, A. L. and K. R. Kendall, The airglow spectrum 3 100 - 10 000 Å. *J. Geophys. Res.*, 73, 426-428, 1968.

Campbell, I. M. and C. N. Gray, Rate constants of $O(^3P)$ recombination and association with $N(^4S)$. *Chem. Phys. Letts.* 18, 607-609, 1973.

Chamberlain, J. W., *Physics of the Aurora and Airglow*, Academic Press Inc., 1961.

Degen, V, Vibrational populations of $O_2(A^3\Sigma_u^+)$ and synthetic spectra of the Herzberg bands in the night airglow. *J. Geophys. Res.*, 74, 5145-5154, 1969.

Eastes, R. W., R. E. Hauffman and F. J. Leblanc, NO and O_2 Ultraviolet nightglow and spacecraft glow from the S3-4 satellite. *Planet. Space Sci.*, 40, 481-493, 1992.

Greer, R. G. H., D. P. Murtagh, I. C. McDade P. H. G., Dickinson, L Thomas., D. B. Jenkins, J. Stegman E. J., Llewellyn, G. Witt, D. J. Mackinnon, and E. R. Williams, ETON 1: A database pertinent to the study of energy transfer in the oxygen nightglow. *Planet. Space Sci.*, 34, 771-788, 1986.

Hennes, J. P., Measurement of the ultraviolet nightglow spectrum. *J. Geophys. Res.*, 71, 763-770, 1966.

Johnston, J. E. and A. L. Broadfoot, Midlatitude observations of the night airglow: Implications to quenching near the mesopause. *J. Geophys. Res.*, 98, 21593-21603, 1993.

Kenner, R. D. and E. A. Ogryzlo, Deactivation of $O_2(A^3\Sigma_u^+)$ by O_2, O and Ar. *Int. J. Chem, Kinetics* 12, 501, 1980.

Kita, K., N. Iwagami, T. Ogawa, A. Miyashita and H. Tanabe, Height distributions of the night airglow emissions in the O_2 Herzberg I system and oxygen green line from a simultaneous rocket observation. *J. Geomag. Geoelectr.*, 40, 1067-1084, 1988.

Krasnopolsky, V. A., Excitation of oxygen emission in the terrestrial planets. *Planet. Space Sci.*, 34, 511, 1981.

Krasnopolsky, V. A., Oxygen emissions in the airglow of Earth, Venus and Mars. *Planet. Space Sci.*, 34, 511, 1986.

McDade, I. C., D. P. Murtagh, R. G. H. Greer, P. H. G. Dickinson, G. Witt, J. Stegman, L. Thomas and D. B. Jenkins, ETON 2: Quenching parameters for the proposed precursors of $O_2(b^1\Sigma_g^+)$ and $O(^1S)$ in the terrestrial nightglow. *Planet. Space Sci.*, 34, 78, 19869.

Murtagh, D. P., R. G. H. Greer, I. C. McDade, E. J. Llewellyn and M. Bantle, Representative volume emission profiles from rocket photometer data. *Ann. Geophys.* 2, 4, 467-474, 1984.

Murtagh, D. P., I. C. McDade, R. G. H. Greer, J. Stegman, G. Witt and E. J. Llewellyn, ETON 4: An experimental investigation of the altitude dependence of the $O_2(A^3\Sigma_u^+)$ vibrational populations in the nightglow. *Planet. Space Sci.*, 34, 811, 1986a.

Murtagh, D. P., G. Witt and J. Stegman, Triplet emissions in the nightglow. *Can. J. Phys.* 64, 1587-1593, 1986b.

Murtagh, D. P., A self-consistent model of the most common nightglow emissions. *Proc 9th ESA symposium on European rocket and balloon programmes and related research, Lahnstein FRG 3-7 April 1989*, ESA SP-291, 167-171, 1989.

Murtagh, D. P., J. Stegman and G. Witt, The excitation of the Chamberlain and Herzberg II bands in the terrestrial nightglow. *Proc 10th ESA Symposium on European Rocket and Balloon Programmes and Related Research, Mandelieu-Cannes, France, 27-31 May*, ESA SP-317, 127-130, 1991.

Owens, J. K., D. G. Torr, M. R. Torr, T. Chang, J. A. Fennelly, P. G. Richards, M. F. Morgan, T. W. Baldridge, C. W. Fellows, H. Dougani, W. Swift, A. Tejada, T Orme., G. A. Germany and S. Yung, Mesospheric nightglow spectral survey taken by the ISO spectral imager on ATLAS 1. *Geophys res. Letts.*, 20, 515-518, 1993.

Siskind, D. E. and W. E. Sharp, A vibrational analysis of the $O_2(A^3\Sigma_u^+)$ Herzberg I system using rocket data. *Planet. Space Sci.*, 38, 11, 1399-1408, 1991.

Sharp, W. E. and D. E. Siskind, Atomic emission in the ultraviolet nightglow. *Geophys res. Letts.*, 16, 1453-1456, 1989.

Slanger, T. G. and D. H. Huestis, $O_2(c^1\Sigma_u^- - X^3\Sigma_g^-)$ emission in the terrestrial nightglow. *J. Geophys. Res.*, 86, A5, 3551, 1981.

Smith, I. W. M., The role of electronically excited states in recombination reactions. *Int. J. chem. Kinetics* 16, 423, 1984.

Stegman, J., Ground-based facility for spectrophotometric studies of the airglow and aurora. *Report AP-29*, Dept of Meteorology, Stockholm University, 1988.

Stegman, J. and D. P. Murtagh, High resolution spectroscopy of the oxygen UV airglow. *Planet. Space Sci.*, 36, 9, 927-934, 1988.

Stegman, J. and D. P. Murtagh, The molecular oxygen band systems in the UV nightglow: measured and modelled. *Planet. Space Sci.*, 39, 4, 595-609, 1991.

Thomas, R. J., Analysis of atomic oxygen, the greenline and Herzberg bands in the lower thermosphere. *J. Geophys. Res.*, 86, 206, 1981.

Wildt, J., G.Bednarek, E. H. Fink and R. P. Wayne, Laser excitation of the $A^3\Sigma_u^+$, $A'^3\Delta_u$ and $c^1\Sigma_u^-$ states of molecular oxygen. *Chem. Phys.* 156, 497-508, 1991.

Witt, G., J. Stegman, D. P. Murtagh, I. C. McDade, R. G. H. Greer, P. H. G. Dickinson and D. B. Jenkins, Collisional energy transfer and the excitation of $O_2(b^1\Sigma_g^+)$ in the atmosphere. *J. Photochem.* 25, 365-378, 1984.

Witt, G., J. Stegman, B. H. Solheim and E. J. Llewellyn, A measurement of the $O_2(b^1\Sigma_g^+ - X^3\Sigma_g^-)$ Atmospheric band and the $OI(^1S)$ green line in the nightglow. *Planet. Space Sci.*, 27, 341-350, 1979.

Wraigt, P. C., Association of atomic oxygen and airglow excitation mechanisms. *Planet. Space Sci.*, 30, 251, 1982.

D. P. Murtagh, Meteorologiska institutionen, Stockholms Universitet, S-106 91 Stockholm, Sweden.

Measurements of Electron Density Profiles in the Ionosphere Using Artificial Periodic Inhomogeneities

V. V. Belikovich, E. A. Benediktov, A. V. Tolmacheva

Radiophysical Research Institute, Nizhny Novgorod, Russia

We describe an electron concentration profile measurement method using artificial periodic inhomogeneities (API) of ionospheric plasma. API are created by the action of high-power, high-frequency radio waves on the ionosphere.

Back scatter of a probe pulse signal from API is observed from a height where the wave lengths of the disturbance (λ_1) and probe (λ_2) signals are equal: $\lambda_1 = \lambda_2$ or $f_1 n_1 = f_2 n_2$ (here f_1 and f_2 — frequencies, n_1 and n_2 — refractive indices of the disturbing and probing wave respectively). If the polarizations of the two waves are different the resonance condition for the given frequencies f_1, f_2 is fulfilled only for a certain electron density. The corresponding height is obtained from the time delay of the pulsed probe signal. This method is capable of measuring electron densities below the F region maximum including the $E - F$ interlayer valley and upper D region with high accuracy. The profiles $N(h)$ measured at Nizhny Novgorod are presented. The base factors defining the errors of measurements are considered.

Introduction

Artificial periodic inhomogeneities (API) in ionospheric plasma density are created by standing waves formed due to the interference between incident and reflecting waves and exist in all height intervals up to the reflection point. This well-proven technique is used by HF modification facilities.

Reception of back scattered pulse signals from API from different height levels makes possible the measurement of electron concentration in the ionosphere. The method of measuring $N(h)$ profiles in the ionosphere using API (or the resonance scattering method) was proposed by *Belikovich et al.*, [1978]. Examples of measured profiles were shown in *Belikovich et al.*, [1986] for January 1984 and in the recent work by *Benediktov et al.*, [1993] for November 1990.

General Description

Bragg back scatter occurs if the spatial period of the irregularities is equal to half the wavelength of the probe signal. The spatial period of the API is equal to half the wavelength of the disturbance signal. So the resonance back scatter is observed from the height where the equality of disturbing and probe wavelengths takes place, $\lambda_1 = \lambda_2$, or

$$f_1 n_1 = f_2 n_2 \qquad (1)$$

Here f_1 and f_2 are the frequencies, n_1 and n_2 are the refractive indices of the disturbing and probe waves in the ionospheric plasma. If f_1 is not equal to f_2, the synchronism condition (1) is fulfilled only for waves with different polarizations. The refractive index is equal to

$$n^2 = 1 - \frac{X}{1 - Y_T^2/2(1 - X) \pm \sqrt{Y_T^4/4(1 - X)^2 + Y_L^2}} \qquad (2)$$

Here $X = f_e/f$, f_e is the Langmuir frequency, $Y_T = Y \sin\theta$, $Y_L = Y \cos\theta$, $Y = f_H/f$, f_H is the gyrofrequency of electrons, θ is the angle between wave vector and magnetic field and f is one of the used frequencies (f_1 or f_2), the signs plus or minus depend on the wave polarization. The equation (2) is given from a well-known formula by Appleton [*Davies*, 1969]. The only assumption used was that we did not take into account the collisions ($\nu_e^2 \ll (\omega_{1,2} - \omega_L)^2$).

The refractive indices depend on the electron concentration in the ionospheric plasma and diminish when $N(h)$ increases. So the wavelength in the medium varies. The resonance condition gives information about the electron density N depending on the disturbance and probe frequencies, and the gyrofrequency. In particular, for a quasi longitudinal approach the condition (1) is seen as:

$$f_e f_L = |f_2 - f_1|(f_2 - f_L)(f_1 + f_L) \qquad (3)$$

where $f_L = f_H \cos\theta$. We know these frequencies so we can obtain N at the height where synchronism of wavelengths takes place. The corresponding height is defined by the time

The Upper Mesosphere and Lower Thermosphere:
A Review of Experiment and Theory
Geophysical Monograph 87
Copyright 1995 by the American Geophysical Union

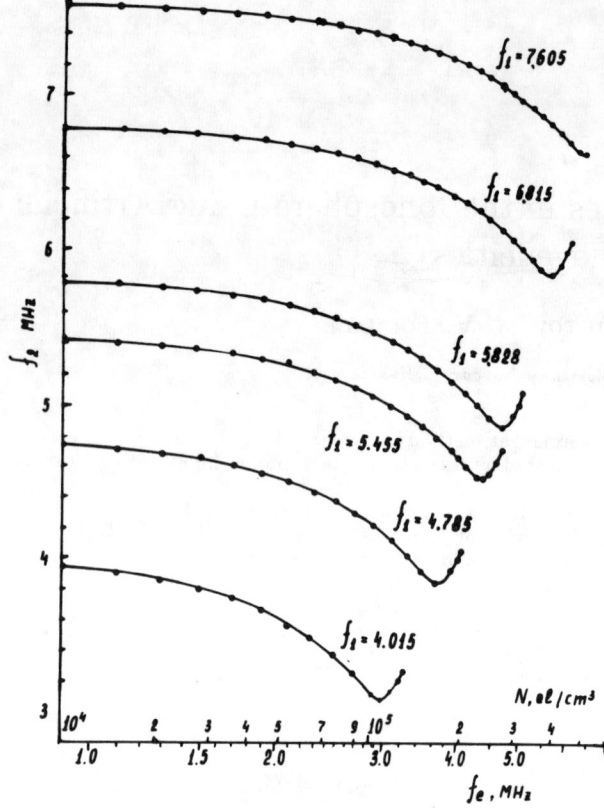

Fig. 1 f_e and N calculations for given f_1 (X-mode) and f_2 (O-mode); $f_1 = 4.015, 4.785, 5.455, 5.828, 6.815, 7.605$ MHz.

delay of the pulsed probe signal. It is not difficult to see that by varying f_1 and keeping f_2 at a constant value (or vice versa) one can obtain a sort of ionogram since the virtual height h_v from which Bragg scatter is observed changes depending on $\Delta f = |f_1 - f_2|$.

Examples of f_e or N calculations for given f_1 and f_2 are shown in Figure 1. Here $f_1 = 4.015, 4.785, 5.455, 5.828, 6.815, 7.605$ MHz (X - mode); the f_2 values (O - mode) are plotted on the vertical axis, f_e and corresponding concentrations N in the logarithmic scale are on the horizontal axis. The calculations were based on equation (1) using formula (2) for the refractive indices of anisotropic plasma. It is valid for heights $h \geq 90$ km. Also the Airy function effects in the vicinity of the reflection point of the disturbance wave have not been taken into account. From Figure 1 we can see that concentration measurements for $f_1 = 5.828$ MHz are possible from 10^4 up to 10^5 cm^{-3} if f_2 varies from 5.75 up to 5.44 MHz (the range of scanning is less than 0.5 MHz).

At first glance the method of electron concentration measurements using API seems very similar to the widely applied ionosonde method. But there are some advantages to our method worth noting here. For example it can be used for the electron concentration measurements both in the upper part of the D region and in the $E - F$ interlayer valley. The common shortcoming of both methods is the measurement of virtual heights. The calculation of actual height from virtual ones is the most difficult in the region close to the reflecting level. However in the E layer and interlayer valley the virtual heights from which the scatter echoes come, the Langmuir frequencies are essentially less than the operating frequencies and the difference between virtual and actual heights could be taken into account with more accuracy than for the ionosonde method.

In general the height interval in which the $N(h)$ profile can be obtained is limited. So the maximum altitude is limited by the disturbance wave height of reflection and it cannot surpass the height of the F layer maximum. The lowest altitude boundary is defined by the minimum possible difference $\Delta f_{min} = |f_1 - f_2|$, which depends on the probe pulse frequency spectrum width.

Examples of Measurements

API were created by the HF modification facility radiating 20 MW ERP into the X-mode at $f = 5.828$ MHz. The diagnostics of API were made using the partial reflection (PR) technique [*Belrose and Burke*, 1964]. The PR installation is located close to the HF modification facility. The frequencies of probe radio waves varied from 5.45 MHz up to 5.75 MHz in 0.03 MHz steps. We received O - mode back scatter pulse signals. As a rule the amplitude of back scattered signal was well above the noise level so no averaging was needed. The frequency stepping was performed manually, so it took several minutes to measure the $N(h)$ profile.

The examples of the measured profiles are shown below. Figure 2 presents profiles obtained on 13.11.1990 using

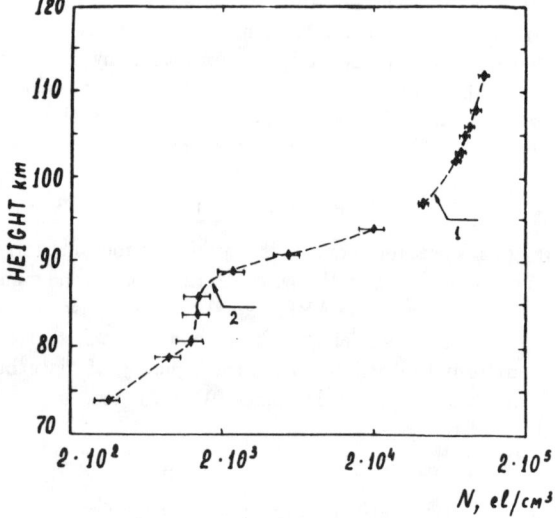

Fig. 2 The $N(h)$ profiles obtained on 13.11.1990 using API (curve 1) and by the partial reflection technique (curve 2).

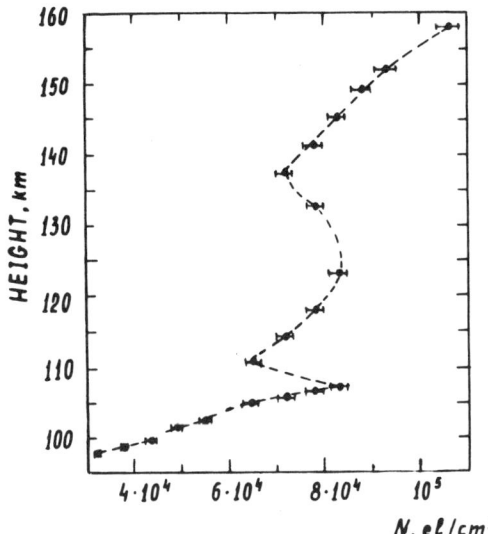

Fig. 3 The $N(h)$ profile for 19.11.1991 (08.50 LT).

API (curve 1), and by the partial reflection method (curve 2). In our opinion there is reasonable agreement between these curves and the reliability of measurements using API is confirmed. Figure 3 shows an $N(h)$ profile obtained on 19.11.1991 (08.50 LT). The E layer maximum ($h = 120 - 123$ km), the valley in the height region 125 - 145 km, and the maximum of the E_s layer ($h = 106$ km) are clearly seen on the profile. The error bars are shown in Figure 2 and Figure 3.

Figure 4 illustrates the dynamics of the $N(h)$ profiles for 17.12.1991 from 09.00 up to 12.30 LT. At 09.00 the smooth increase in $N(h)$ was observed, at 09.35 the valley in region 120 - 140 km appeared, then a thinner structure inside the valley arises and the picture stabilizes a little but the E layer near the maximum (110 - 115 km) becomes thinner. Probably the alternation of concentration near the maximum E layer is connected with the appearance of an Es near the same height ($f_b E_s$ varies from 2.2 up to 3 MHz).

ACCURACY OF MEASUREMENTS

The question about the accuracy of measured $N(h)$ profiles using API was analyzed. The base factor determining the accuracy of the obtained profile is the count of the height level from which the probe signal was back scattered. The accuracy of height count is determined by the length of the sounding pulse $L_1 = c\tau/2$ (for $\tau = 25\,\mu s$, $L_1 = 3.75$ km). However the synchronism region L_2 may be longer or shorter than L_1 and it broadens the back scatter signal. If there is an altitude gradient of electron concentration, the theoretical expression for L_2 obtained by *Belikovich and Mareev* [1987] is as follows

$$L_2 = [\frac{c^2 k_0 l}{4\pi(f_1^2 - f_2^2)}]^{1/2} \quad (4)$$

where $l = [(1/N_0)\frac{dN}{dh}]^{-1}$, k_0 is the wave number in the synchronism point, and $k_0 = k_1 = k_2$. If the height alternation of electron concentration is smooth ($l \sim 40 - 50$ km near valley), the value of $L_2 \sim 4 - 5$ km and it is more than L_1. For large gradients ($l \sim 10 - 20$ km for E_s) L_2 diminishes up to 2 - 3 km. The relative error of measuring is equal to

$$\Delta N/N \approx \Delta h/l, \quad (5)$$

where Δh - error of height count. The formula for calculating the electron concentration includes f_1, f_2, and f_H. The work frequencies are given by highly stable frequency generators. The error of f_H depends on the accuracy of measurements or the model of the geomagnetic field and is not more than 2 - 3 per cent.

The technical characteristics of the facility make it possible to determine the height with accuracy $\Delta h \approx 0.5$ km for measuring N at the lower region (E_s, E layers) where $l \sim 10 - 20$ km and $L_2 \sim 2 - 3$ km and $\Delta N/N \leq 3 - 5$ per cent. When the alternation of N is more smooth (above E layer maximum, in the valley) the back scatter signal is broadened on height and Δh may approach to 1 - 2 km. However the value of error $\Delta N/N$ is equal 2 - 5 per cent also (it is seen from (4)). Hence the errors of electron concentration measurement is not more than 5 per cent at all heights and the average error is equal to about 3 per cent.

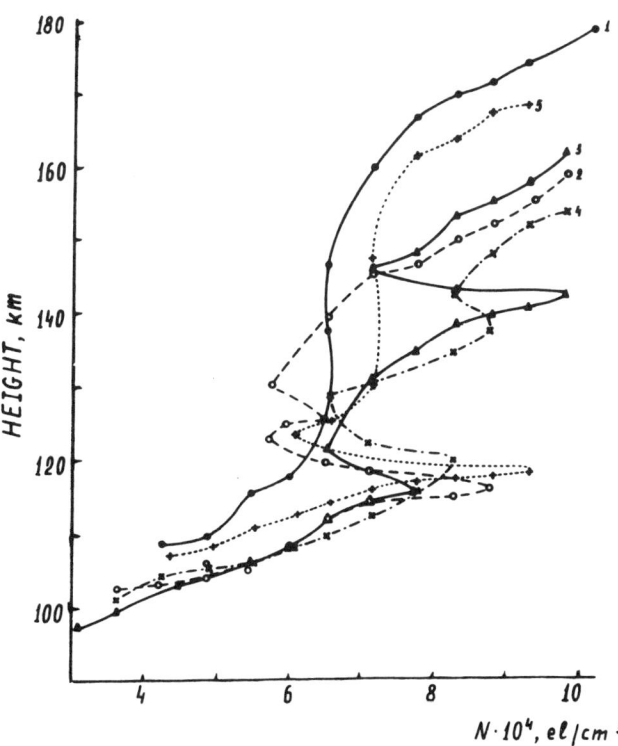

Fig. 4 The temporal variations of the $N(h)$ profiles for 17.12.1991: curve 1 — 09.00 LT, curve 2 — 09.35 LT, curve 3 — 10.00 LT, curve 4 — 10.45 LT and curve 5 — 12.35 LT.

Conclusion

The results of the measurements of $N(h)$ profiles using API and the analysis of accuracy and the possibilities of application allow us to note several important advantages of the method.

It is possible to measure $N(h)$ profiles in large height intervals from 90 km up to the F layer maximum, including the interlayer valley. The simultaneous application of the partial reflection technique and API allow us to broaden the height range of $N(h)$ profile measurement to include the D region (down to ~ 60 km).

The method makes it possible to study the temporal variations of N and in particular to investigate sporadic layer dynamics and thin structures in the valley.

The method has a high accuracy and good height resolution comparable to rocket methods, but using API allows us to carry out regular measurements of $N(h)$ and to study dynamic phenomena in the ionosphere.

Acknowledgements. This work is supposed by Russian Fund of Fundamental Investigation under grant 93-05-9661.

References

Belikovich, V. V., E. A. Benediktov, G. G. Getmantsev, M. A. Itkina, G. I. Terina, and A. V. Tolmacheva, Ionospheric Electron Density Measurement Using Radio Wave Scattering from Artificial Plasma Inhomogeneities, *Radiophysics and Quantum Electronics*, 21, 853, 1978.

Belikovich, V. V., E. A. Benediktov, G. I. Terina, Diagnostics of the Lower Ionosphere by the Method of Resonance Scattering of Radio Waves. *J. Atmos. Terr. Phys.*, 48, 1247, 1986.

Belikovich, V. V., E. A. Mareev, Radio Wave Scattering by Artificial Quasi periodic Irregularities in the Ionosphere, *Radiophysics and Quantum Electronics*, 30, 631, 1987.

Belrose, J. S., M. J. Burke, A Study of the Lower Ionosphere Using Partial Reflections. 1.Experimental Technique and Method of Analysis. *J. Geophys. Res.*, 69, 2799, 1964.

Benediktov, E. A., V. V. Belikovich, N. P. Goncharov, A. V. Tolmacheva, Studies of the Lower Ionosphere Performed at Nizhny Novgorod Using the Decametre Radiowave Back Scatter, *J. geophys. Research*, 1993 (to appear).

Davies, K., *Ionospheric Radio Waves*, Space Disturbance Laboratory and University of Colorado, Waltham, Massachusetts — Toronto — London, 1969.

Aspects of Mesospheric Simulation in a Comprehensive General Circulation Model

Kevin Hamilton

*Geophysical Fluid Dynamics Laboratory/N.O.A.A.
Princeton University*

The simulation of mesospheric circulation in the 40 level GFDL "SKYHI" troposphere-stratosphere-mesosphere general circulation model is examined. The model is shown to produce a "reversed" equator-pole temperature gradient near the upper summer mesosphere with consequent closing off of the summertime easterly jet. This results from the mean flow driving associated with explicitly resolved internal gravity waves. The mean easterlies in the summer upper mesosphere become weaker as the horizontal resolution in the model is improved, presumably reflecting a more complete representation of the gravity wave spectrum. The rate of loss of the eddy kinetic energy in the model due to subgrid scale dissipation is computed. The model results in the mesosphere compare reasonably well with the limited data available concerning the turbulent dissipation rates obtained from *in situ* rocket experiments.

Also discussed here for the first time are simulations with a diurnally-varying version of the SKYHI general circulation model. The diurnal tidal signal appears to be of realistic amplitude in the model, but it is striking that the other eddy components (gravity waves, Rossby normal modes, quasi-stationary planetary waves) are strong enough to dominate the tide in any instantaneous map of the global mesospheric wind field. The mean flow driving (Eliassen-Palm flux divergence) associated with the diurnal tide is also computed. The present results for this quantity differ considerably in detail from earlier calculations which were based on more idealized tidal theories. The tidal driving of the mean flow in the SKYHI model is an important, though not dominant, effect in the summer mesosphere region.

1. INTRODUCTION

Comprehensive, nonlinear, three-dimensional, time-dependent general circulation models (GCMs) have proved to be valuable research tools in tropospheric and stratospheric meteorology, but relatively little effort has been devoted to mesospheric general circulation modelling. The reasons for this neglect no doubt include the complicated physical and chemical processes that may be important for determining the mesospheric circulation. Such complications as the deviations from local thermodynamic equilibrium (LTE) for the calculation of radiative heating rates, and the large diurnal cycle in the concentration of radiatively-active trace constituents are not normally considered by meteorological modellers. To date those GCMs that have been used to model the atmosphere from the ground through the mesosphere have all been fairly straightforward adaptations of more conventional meteorological models (Fels et al., 1980, Hunt, 1991).

Among extant meteorological models the GFDL

The Upper Mesosphere and Lower Thermosphere:
A Review of Experiment and Theory
Geophysical Monograph 87
This paper is not subject to U.S. copyright.
Published in 1995 by the American Geophysical Union

"SKYHI" GCM (Fels et al., 1980; Mahlman and Umscheid, 1987) has perhaps been the most valuable for mesospheric research. While not a complete chemical/dynamical model of the mesosphere, the SKYH GCM has produced some interesting results concerning aspects of the atmospheric flow that traditionally were the concern of aeronomers. In particular Miyahara et al. (1986), Hamilton and Mahlman (1988) and Hayashi et al. (1989) have demonstrated the crucial role of explicitly resolved gravity waves in the global scale momentum budget of the SKYHI mesospheric simulation. The present paper is an informal survey of some recent research concerning the mesospheric circulation in the SKYHI model. The emphasis will be on issues of interest to aeronomers, and among the results presented will be the first obtained from a diurnally-varying version of the SKYHI model.

The discussion below begins in Section 2 with a brief description of the SKYHI model. Section 3 then reviews the basic simulation of zonal-mean circulation produced by the model. Section 4 considers aspects of the high frequency eddies simulated in SKYHI. The model analysis to be discussed in this section is inspired by recent developments in mesospheric observational techniques, some of which will no doubt be the focus of other papers in this conference proceedings. In particular, global synoptic maps of the mesospheric horizontal flow field in models will soon have observational counterparts from UARS Doppler-radiometer measurements of wind. Similarly, the parameterized subgrid-scale energy dissipation in a model can now be compared with *in situ* observations of turbulent dissipation rates. The brief presentation of model results in Section 4 is meant as a first step towards a detailed analysis of the nature of mesospheric high frequency variability using models and observations. The paper continues with a brief description of the solar tides in the SKYHI model and a preliminary consideration of tidal effects on the mesospheric mean flow (Section 5). Conclusions are summarized in Section 6.

2. THE SKYHI MODEL

The GFDL SKYHI GCM is designed to simulate the global atmosphere from the ground to near the mesopause. In SKYHI the primitive equations are discretized on a horizontal grid. In this paper three different versions with $3°\times3.6°$, $2°\times2.4°$ and $1°\times1.2°$ latitude-longitude resolution will be considered (referred to as N30, N45 and N90 respectively). In each case the vertical coordinate is terrain-following near the ground and converges to a pure pressure coordinate above 353 mb. 40 levels in the vertical are employed, with the altitude spacing between levels gradually increasing with height. The mesospheric levels are at 0.0096 mb, 0.0308 mb, 0.0697 mb, 0.131 mb, 0.222 mb, and 0.349 mb, corresponding roughly to heights of 80, 73, 68, 63, 59, and 55 km. At the top of the domain the usual "rigid lid" boundary condition (i.e. the pressure velocity is zero at zero pressure) is used along with a linear damping applied to eddy motions at the highest full model level (see Fels et al., 1980). The model is integrated with explicit leapfrog time differencing using time steps of 60 s, 180 s and 225 s for N90, N45 and N30, respectively. A realistic topography and land-sea distribution are employed. The model includes a sophisticated computation of the longwave and shortwave heating rates. Mean ozone amounts are prescribed, but locally the mixing ratio above 35 km is allowed to vary linearly with temperature in order to account for the photochemical acceleration of radiative eddy damping (see Fels et al., 1980). The cloud field used in the radiation code is also prescribed. The radiation code includes absorption of solar radiation by O_3, O_2 and CO_2 and cooling through IR emission by CO_2 and O_3. The treatment of the radiative transfer is essentially based on local thermodynamic equilibrium and ignores effects that may be significant above ~70 km, notably the airglow and chemiluminescence of the excited products of solar absorption (e.g., Mlynczak and Solomon, 1991).

All the model simulations reported here are seasonally-varying, but the treatment of the diurnal cycle varies among the different experiments. In some integrations diurnally-averaged solar heating rates are employed, while in other runs the zenith angle used in the solar heating undergoes a realistic diurnal cycle (updated every hour).

The SKYHI model includes a representation of the hydrological cycle, with parameterizations of evaporation, moist convection and stable precipitation. A dry convective adjustment and a Richardson number dependent vertical diffusion are also employed. A deformation-dependent nonlinear horizontal subgrid-scale diffusion is included (Andrews et al., 1983). No attempt is made to parameterize the effects of subgrid-scale gravity waves. In the experiments described here a realistic climatological annual cycle of the sea surface temperature is prescribed. Surface temperatures and soil wetness over land are calculated each time step using a simple prognostic soil model.

3. BASIC RESULTS FOR THE CONTROL SIMULATION

The analysis of the diurnally-averaged control runs discussed here will use 10 consecutive years of data from the middle of an N30 model integration, 2 years from an N45 integration, and 1 year from an N90 integration. In each case it is reasonable to assume that the results are essentially independent of the initialization employed (see Hamilton, 1993, for a discussion of the initialization of each run). Figure 1 shows the December-February (DJF) mean of the zonally-averaged zonal wind for each of these simulations along with the gradient winds in balance with the observed climatological pressure field reported by Fleming et al. (1988). To first order the simulations all look realistic, with subtropical jets in the troposphere of each hemisphere, a polar night westerly jet in the winter stratosphere and mesosphere and an easterly jet in the summer middle atmosphere.

When examined closely, however, there are evident deficiencies in the simulated zonal wind structure. In the winter upper stratosphere and mesosphere the westerly vortex in the model has roughly the right strength, but is unrealistically confined to high latitudes. There is a great deal of interannual variability in the Northern Hemisphere (NH) winter middle atmospheric circulation in both the real world and the SKYHI simulations (due largely to the intermittent occurrence of sudden warmings). The 10 years of N30 simulation used in Fig. 1 are enough to give a fairly good estimate of the long term mean statistics, but since only much shorter N45 and N90 integrations are available, the NH differences seen in Fig. 1 among the three different model resolutions may not be statistically significant.

The easterly jet in the SKYHI simulated summer hemisphere mesosphere is unrealistically intense and does not completely close off in any of the simulations. This problem becomes progressively less severe with increasing horizontal resolution, however. There is little interannual variability in the seasonal mean summertime circulation, so these differences among the models are almost certainly

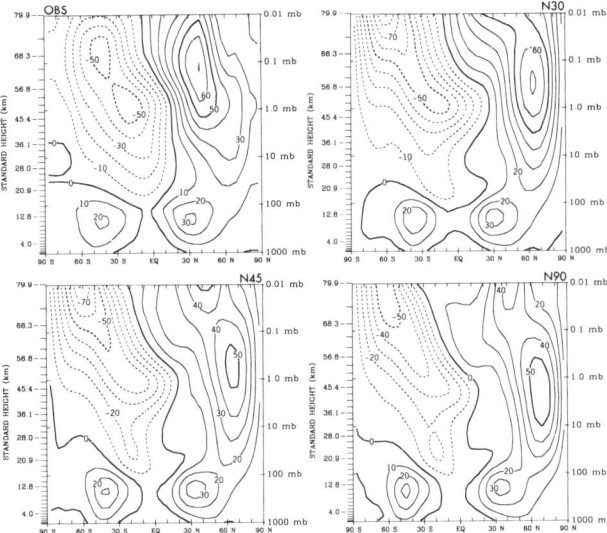

Fig. 1. December-February climatology of the zonal-mean wind from observations (Fleming et al., 1988) and from simulations with the N30, N45 and N90 versions of the SKYHI model. Contour labels are in m-s^{-1} and positive values represent westerly (eastward) winds.

real. The only eddies that are thought to be significant in the extratropical mesosphere are inertia-gravity waves (e.g., Andrews et al., 1987). Thus the strength of the mesospheric easterly jet is determined by a balance between radiative processes that try to establish a warm summer pole (and consequent easterly shear in the summer hemisphere) and the opposing gravity wave drag. Miyahara et al. (1986) and Hayashi et al. (1989) showed that increasing horizontal resolution leads to an increase in the upward gravity wave momentum fluxes. The direct consequence of this can be seen in Figure 2 which shows the force per unit mass acting on the zonal-mean zonal flow due to Eliassen-Palm flux divergence from all eddies. The top panel is for the N30 simulation and displays the strong westerly driving of the mean flow in the summer mesosphere of as much as 20 m-s^{-1}day^{-1}. This is due almost entirely to transient waves (i.e. deviations from the monthly mean). The middle panel shows the change in this quantity in the N45 integration and the bottom panel shows this difference for the N90 integration. As resolution is increased the westerly drag in the summer hemisphere mesosphere rises significantly. This is consistent with the reduced strength of the easterly jet seen in the high resolution simulations in Fig. 1.

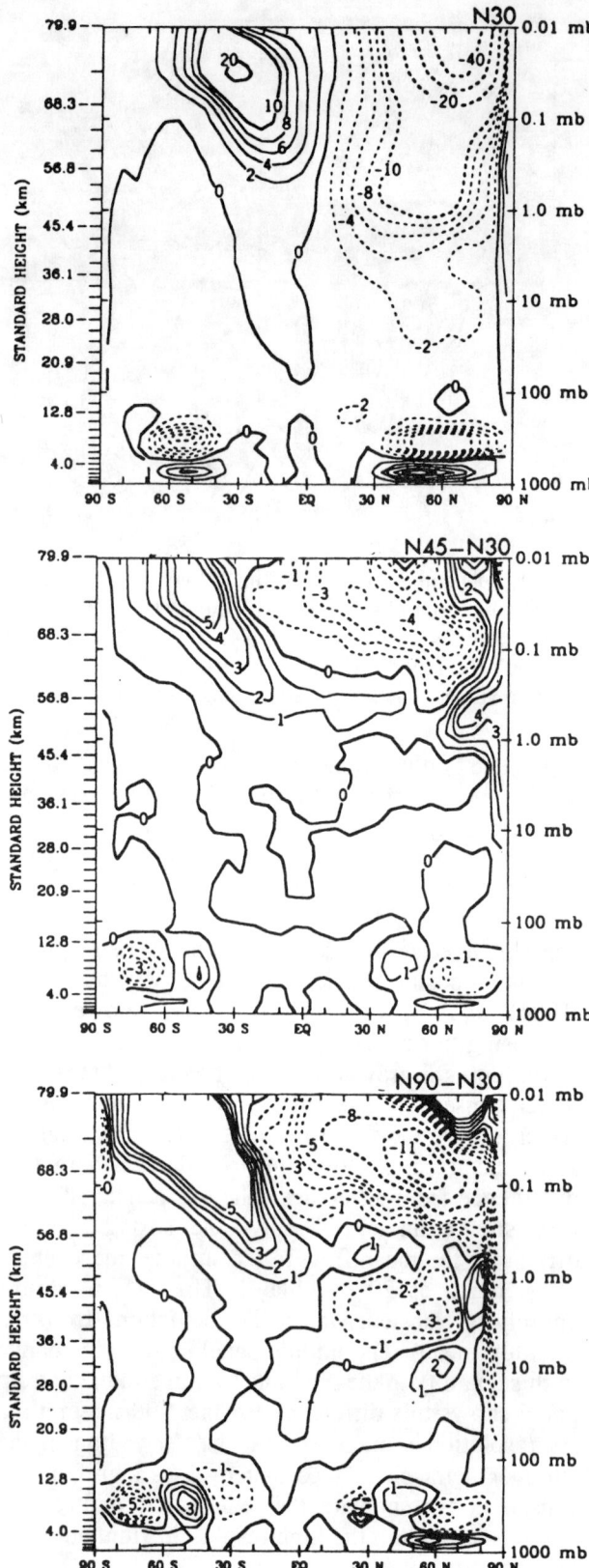

The Eliassen-Palm flux divergence in the extratropical NH (i.e. winter) mesosphere in each panel of Fig. 2 represents contributions from both quasi-stationary planetary waves and transient gravity waves. The net effect of the wave fluxes in winter is to act as a drag on the westerly mean jet. Given the large interannual variability, no claim is made that the differences in the winter hemisphere among the three resolutions seen in Fig. 2 are statistically significant.

The discussion has thus far been limited to the boreal winter season. The results in the NH summer are similar to those in SH summer, but the SH winter simulation in SKYHI is less satisfactory than in NH winter. In particular, the simulated SH polar night jet is unrealistically strong, even at N90 resolution (see Hamilton, 1993). This deficiency is shared by other stratospheric GCMs, and this problem is a matter of active investigation.

4. HIGH FREQUENCY VARIABILITY AND TURBULENT DISSIPATION

Figure 3 shows instantaneous snapshots of the simulated horizontal wind for two mesospheric levels (0.0308 and 0.222 mb) at 12 GMT January 1. This figure comes from a diurnally-varying simulation carried out with the N30 version of the model. The strong mean westerlies in the winter hemisphere and easterlies in the summer hemisphere are apparent. Superposed on these features are strong eddy motions of various scales. In the high latitude NH there is a pronounced zonal wavenumber one pattern (particularly strong at 0.222 mb). This reflects the effects of topographically forced planetary waves in perturbing the westerly polar vortex. This aspect of the synoptic wind field generally changes fairly slowly (timescales of weeks).

Fig. 2. The December-February climatology of the zonal force per unit mass exerted on the mean flow by the Eliassen-Palm flux divergence associated with all eddies. The top panel gives the result for the N30 simulation. The middle panel shows the difference in this quantity between the N45 simulation and the N30 simulation, while the bottom panel shows the difference between the N90 simulation and the N30 simulation. Positive values represent westerly (eastward) forcing of the mean flow. Contour labels are in m-s^{-1}d^{-1}. The contour values in the top panel are 0, 2, 4, 6, 8, 10, 20, 30, ... In the bottom two panels the contour values are 1, 2, 3, 4, 5, 8, 11, 14...

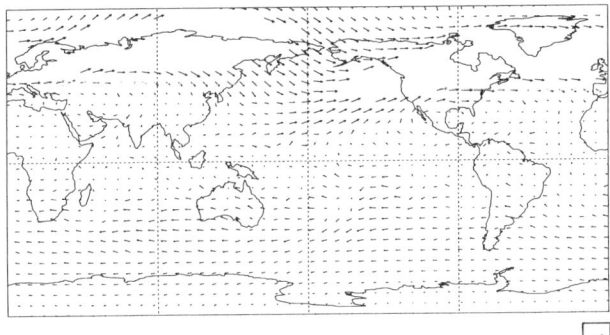

Fig. 3. Instantaneous horizontal wind field at 12 GMT on January 1 during one year of the integration of the N30 SKYHI model including a diurnal cycle. Results from the second highest model level (0.0308 mb, top) and the fifth highest model level (0.222 mb, bottom). The vector length shown in the box in the bottom right hand corner represents 120 m-s^{-1}. No vectors are plotted for those points where the wind speed exceeds 120 m-s^{-1}.

Away from the polar latitudes there is a more complex mix of scales evident, and the pattern does change significantly even during a single day. Part of this variability is due to the solar tide (as will be shown below). However, the striking aspect of this figure is the extent of variability that is not tidally related. The flow field contains variability on all the resolved scales, but eddies with scales of many thousands of km (but less than global) are prominent. This overall picture appears to be in agreement with reported UARS Doppler-radiometer measurements of the winds near mesopause levels. These observations (Morton et al., 1993) also show very significant large scale eddies that cannot be identified as tidal components. Further analysis of the detailed information available from the present model simulation should allow a determination of the nature of the large scale mesospheric eddies (i.e. the relative contributions from normal mode Rossby waves, inertia-gravity waves, equatorial planetary waves etc.).

The upward propagation of linear planetary and gravity waves in a compressible atmosphere is accompanied by a growth in wave amplitudes (as measured by the perturbation wind and temperature). This growth eventually leads to some kind of nonlinear breaking and the consequent generation of turbulence. Within the turbulent regime there is presumably a cascade to smaller scale motions that are ultimately dissipated by molecular viscosity and heat diffusion. The turbulent dissipation of kinetic energy in the mesosphere and lower thermosphere has been estimated using various observational techniques. Hocking (1985) reviewed several such estimates for the 80-120 km range. For winter high latitudes his review suggests a mean value of the order of 100 mW-kg^{-1} for the turbulent dissipation at 80 km (although with a great deal of variability among individual profiles). More recently Lubken et al. (1993) have used a new technique to estimate energy dissipation in the 60-100 km height range. Analysis of data taken during several rocket flights in mid/ late winter at Andoya (69°N) suggested rather lower values for the dissipation rate than the earlier observations. In particular Lubken et al. find dissipation rates of the order of 10 mW-kg^{-1} at 80 km and suggest that this value falls off rapidly to ~1 mW-kg^{-1} in the lower mesosphere.

Given the limited spatial resolution in a GCM, an explicit simulation of the isotropic turbulence regime is obviously not possible. However, the subgrid scale mixing parameterizations are included in an attempt to account for the effects of unresolved scales of motion on the resolved flow. The energy lost by the resolved flow to the subgrid scale parameterizations is assumed to cascade to ever smaller scales and end up as isotropic turbulence (which is ultimately dissipated by molecular viscosity). Thus it is reasonable to compare the model energy dissipation due to parameterized subgrid scale processes with the observed estimates of small scale turbulent dissipation. There are caveats to be noted, however. One concerns time scales. If the cascade from grid scale (which may be ~100's of km) to the dissipation range in the real atmosphere takes sufficiently long then the actual dissipation may be significantly nonlocal. The other complication is that the subgrid scale parameterizations remove not only kinetic energy, but also potential energy (this can occur particularly in the application of dry convective adjustment). This energy loss from the resolved scales might also ultimately end up as kinetic energy of isotropic turbulence. In the present paper only the direct dissipation of resolved scale kinetic energy by the subgrid scale mixing parameterizations will be computed.

Figure 4 shows this dissipation rate averaged (from daily instantaneous samples) over the month of February and averaged around a latitude circle near 70°N for each of the N30, N45 and N90 integrations. The results are plotted only up to the second highest model level (0.0308 mb) since the artificial eddy damping imposed at the top level distorts the results there. For all three models the dissipation rate rises rapidly with height in the mesosphere, in agreement with the observations of Lubken et al. The dissipation rates also increase with model resolution, presumably reflecting the more complete spectrum of gravity waves in the higher resolution model. When extrapolated in a straightforward manner to 0.01 mb (~80 km) the N90 results would imply dissipation rates ~20 mW-kg^{-1}, a figure which lies between the observational estimates of Lubken et al. (1993) and Hocking (1985) for that level. This suggests that the overall strength of the gravity wave energy fluxes in the extratropical mesosphere is captured in fairly realistic manner by the GCM.

5. SOLAR TIDES AND THEIR EFFECT ON THE MEAN FLOW

Solar tides are often regarded by meteorologists as little more than a curiosity, but tides are a very prominent component of the circulation in the upper atmosphere. The inclusion of the diurnal cycle of radiation in the SKYHI model now allows an examination of the simulation of tides within a troposphere-stratosphere-mesosphere GCM. A 7 year diurnally varying integration with the N30 SKYHI model has now been completed. The results below are all from the final 5 years of this run.

The effects of the diurnal cycle are evident in Figure 5, which shows a 15 day timeseries of the surface pressure at a model gridpoint near the equator in the SKYHI model simulation. The behavior seen here is quite typical of observed barograph traces at low latitudes, i.e the pressure peaks twice each day around 1000 and 2200 local time. The semidiurnal pressure range seen in Fig. 5 is ~2.5 mb, in agreement with low latitude observations (e.g., Chapman and Lindzen, 1970). Preliminary analysis

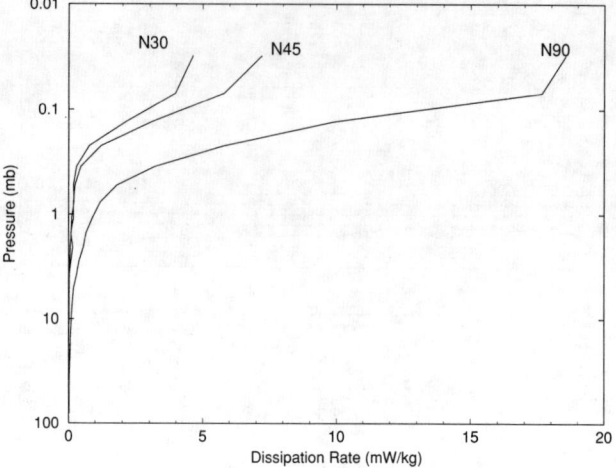

Fig. 4. Dissipation of kinetic energy by the subgrid scale parameterizations in the SKYHI model. Results are for February means and averages around a latitude circle near 70°N (70.5°N for N30, 69°N for N45, and 69.5°N for N90). Values are given for each of N30, N45 and N90 resolutions and are plotted up to the second highest model level (0.0308 mb).

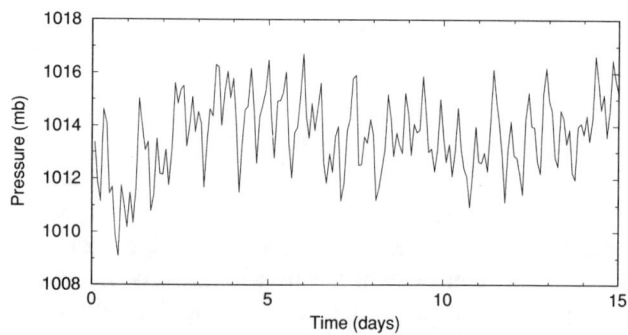

Fig. 5. A 15 day times series of 2 hourly values of the surface pressure at a near-equatorial gridpoint in the diurnally-varying N30 SKYHI model simulation.

suggests that the geographical distributions of both the diurnal and semidiurnal surface pressure oscillations in the model are rather realistic.

The top panel in Figure 6 shows the horizontal wind at the second highest model level at 1200 GMT minus that at 0000 GMT. The results represent a composite over the month of January in 5 consecutive years. The bottom panel shows the same quantity for the fifth level from the top (0.222 mb). At each level the dominant feature is a zonal wave

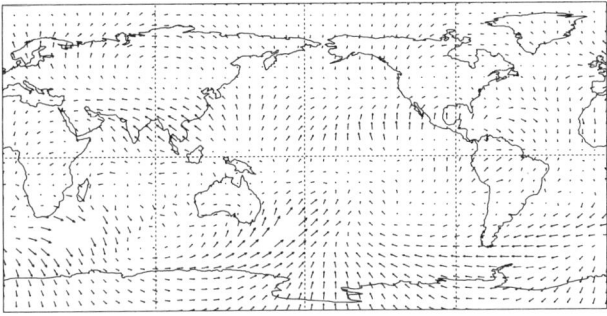

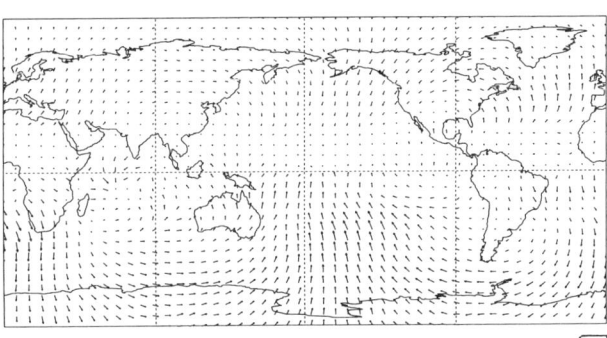

Fig. 6. The horizontal wind at 12 GMT minus that at 00 GMT averaged over five Januaries in the diurnally-varying N30 SKYHI integration. Results from the second highest model level (0.0308 mb, top) and the fifth highest model level (0.222 mb, bottom). The vector in the box shown at lower right represents 25 m-s^{-1}.

one pattern which is most intense in the summer hemisphere. This is consistent with theoretical expectations that the diurnal tide at these high altitudes should be almost entirely sun-synchronous. The peak-to-peak amplitude of the diurnal wind oscillation indicated in the summer hemisphere midlatitudes would appear from Fig. 6 to be almost 20 m-s^{-1}. This is generally consistent with available observations at mesospheric heights (Chapman and Lindzen, 1970). This amplitude is rather less than the magnitude of the total eddy wind field seen in a typical snapshot of the model fields (note the difference in the scaling of the wind vectors as plotted in Fig. 3 and Fig. 6). If this accurately reflects the behavior of the real atmosphere, the difficulty observers have faced in obtaining stable tidal parameters from short data records is readily understandable. The vertical structure of the diurnal tide is examined in more detail in Figures 7 and 8, which show the amplitude and phase of the sun-synchronous diurnal harmonic of the zonal wind at

selected latitudes based on the 5 year July composite. The overall increase of amplitude with height seen at each latitude in Fig. 7 is consistent with that of earlier observations and theoretical models. As in the January results noted above, in July the diurnal wind oscillation is strongest in the summer hemisphere. There is little vertical phase propagation evident at 45°N, suggesting a dominance by vertically trapped modes in the diurnal tidal response at this latitude. At the equator and at 45°S, some phase variation in the vertical is evident, but there is not the simple monotonic phase progression expected from purely upward propagating tides. The complicated phase pattern suggests that the propagating components of the tide may be partially reflected from the upper boundary (this problem is even more apparent in meteorological simulation models with lower lids, e.g., Zwiers and Hamilton, 1986). The adequacy of the present formulation clearly needs to be checked more carefully (e.g., with linear tidal theory calculations, or through a direct comparison with a GCM that explicitly includes the lower thermosphere).

During the last 25 years there have been a number of papers that have considered the question of the effects of the heat and momentum transports by the solar tides on the zonal-mean circulation. In the tropical lower thermosphere, the diurnal tide is expected to break nonlinearly, resulting in a strong easterly mean flow forcing in this region (Hines, 1972; Miyahara, 1978a,b, 1984). In the mesosphere one expects less dramatic effects, but a number of papers have raised the possibility of significant tidal mean flow driving.

Meyer (1970) speculated that the semiannual oscillation of the mean wind at the stratopause may be driven by tidal eddy fluxes, while Groves (1983) suggested that the tidal fluxes may contribute to the semiannual oscillation observed near the mesopause. Fels and Lindzen (1974) determined the mean flow driving at the equator from the diurnal tide, as computed using somewhat idealized thermal excitation in classical tidal theory (i.e a linear calculation assuming a motionless basic state). They found that the mean flow driving had an oscillatory behavior with height (with wavelength ~25 km). The magnitude of mean flow forcing rose to a peak of about 0.7 m-s^{-1}d^{-1} in the mesosphere. Fels and Lindzen found that the computed mean flow driving was very sensitive to the number of Hough modes retained in the tidal theory calculation. This might be

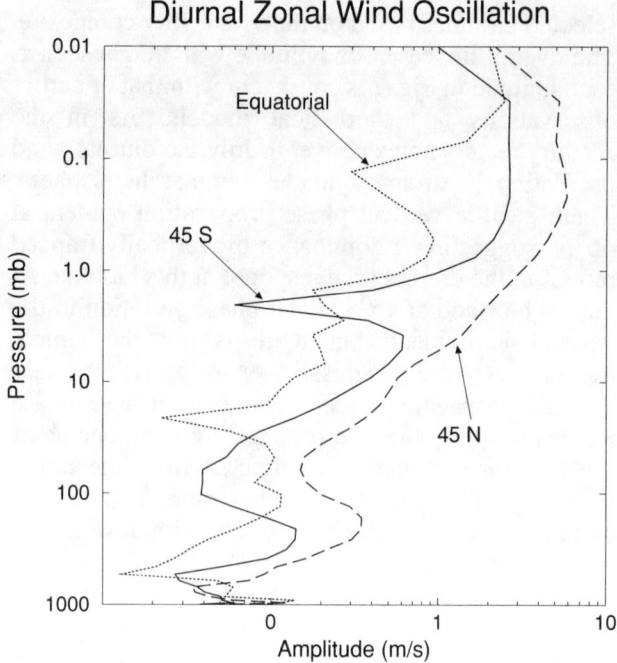

Fig. 7. The amplitude of the sun-synchronous diurnal oscillation of the zonal wind at each of 40 model levels and at 3 latitudes. Results based on data during the month of July in five consecutive years of N30 SKYHI integration.

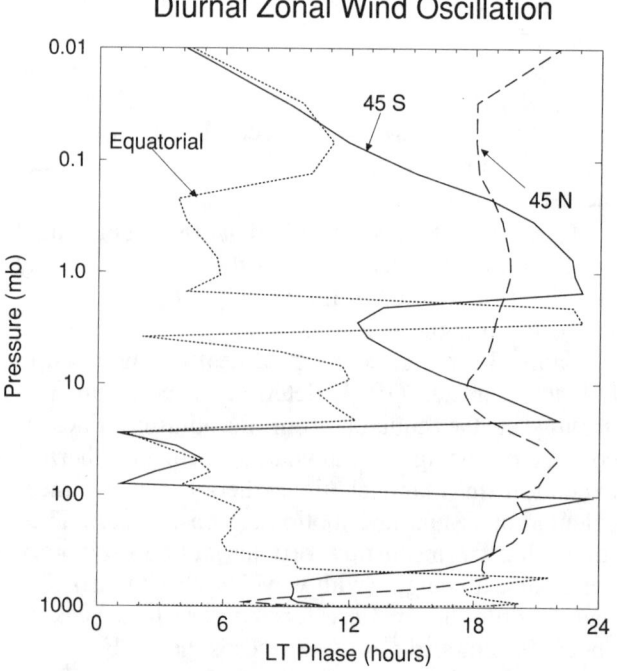

Fig. 8. As in Fig. 7, but for the local time phase of the zonal wind oscillation.

a hint that the tidal Eliassen-Palm flux divergence may be strongly dependent on other aspects of the tidal calculation (mean flow effects, detailed spatial distribution of heating, presence of dissipation etc.). Miyahara (1978a,b) and Hamilton (1981) extended the Fels and Lindzen analysis to the global domain. Hamilton (1981) found that at equinox the eddy forcing from the diurnal tide has two off-equatorial peaks (at ~15° latitude). In all these studies a detailed calculation of the tidal eddy forcing (Eliassen-Palm flux divergence) was attempted, but rather *ad hoc* models were used to estimate the actual mean flow changes that can be attributed to the tides. A GCM has the potential to resolve this problem, since both the tides and all the other circulation features affecting the mean flow are computed self-consistently. The study of tidal effects in SKYHI is still a project in progress, but a preliminary look at tidal influence on the mean flow will be presented here.

The present analysis began by using the 5 years of SKYHI data (sampled every 2 hours) to form a composite in local time within each calendar month. Then the diurnal harmonic was computed. These monthly mean diurnal harmonics (and the zonal-mean fields for each month) were then used to compute the Eliassen-Palm flux divergence. The finite difference approximations employed in this calculation were carefully formulated to be consistent with the GCM numerics (see Andrews et al., 1983). Figure 9 shows the result (expressed as a force per unit mass) averaged over the equinoctial months of September and October. The equatorial values of the eddy driving on the equator are smaller than those found by Fels and Lindzen (1974) or Hamilton (1981). The tendency for this quantity to peak off the equator is even more pronounced than in Hamilton's (1981) calculations. These results suggest that the tidal mean flow driving is strongly dependent on the mean flow conditions and that the "classical" tidal theory is not adequate to accurately compute this quantity. The peak magnitude of the eddy force per unit mass seen in Fig. 9 is just over 1 $m \cdot s^{-1} d^{-1}$. Figure 10 shows the same quantity averaged over the solstitial months of June and July. Here the tendency for the forcing to peak off the equator is again evident, although near solstice there is a very large interhemispheric asymmetry apparent. The largest accelerations are nearly 3 $m \cdot s^{-1} d^{-1}$ and occur between 20° and 30° in the summer hemisphere. These numbers can be compared to the

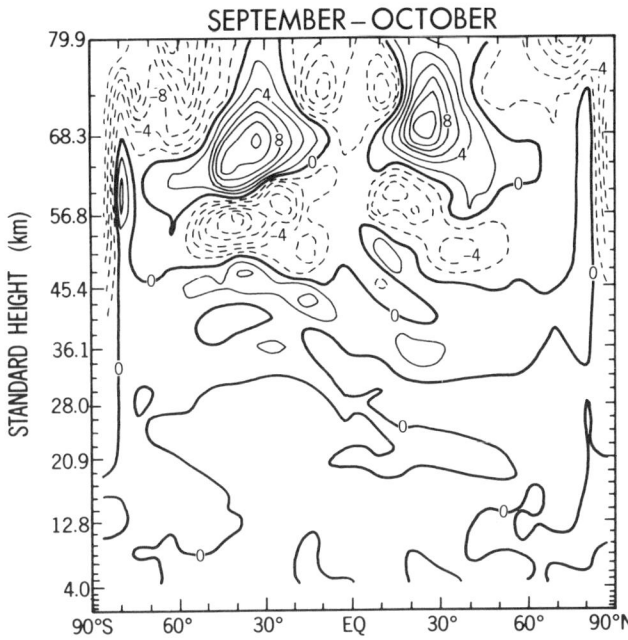

Fig. 9. The zonal forced per unit mass exerted on the zonal-mean zonal wind by the divergence of the Eliassen-Palm flux associated with the solar diurnal tide in the N30 SKYHI run. Results represent a mean over 5 September-October periods. The contour interval is 0.2 m-s^{-1}d^{-1} and positive values (solid contours) denote a net westerly (eastward) forcing of the mean flow.

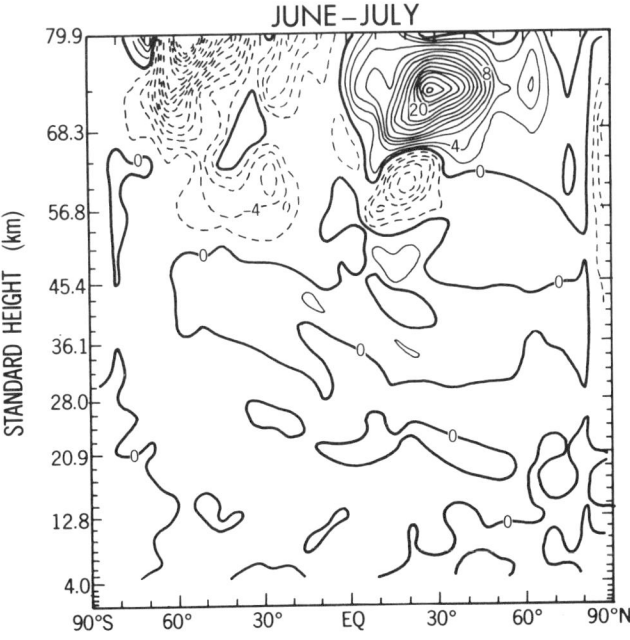

Fig. 10. As for Fig. 9, but for a mean over 5 June-July periods.

total force per unit mass exerted on the mean flow by all the eddies in the nondiurnal N30 simulation (top panel of Fig. 2). In the 20°-30° latitude range in the upper mesosphere of the summer hemisphere the values in Fig. 2 are between 10 and 20 m-s^{-1}d^{-1}. This comparison suggests that tidal forcing of the mean flow may be a significant, though far from dominant, effect in the summer mesosphere. Research is now underway to compare the actual seasonal evolution of the mean flow in the diurnal and nondiurnal SKYHI models.

6. CONCLUSION

This paper has briefly reviewed some aspects of mesospheric circulation as simulated by a comprehensive meteorological model. The dependence of the simulated zonal-mean fields on model resolution, the typical appearance of large-scale transient eddy features in the mesospheric horizontal wind field, the dissipation of kinetic energy by subgrid scale turbulence, the solar tides and their influence on the mean flow were examined. Investigation is continuing on all these topics and more detailed publications are planned. The use of a comprehensive model like SKYHI enables the investigation of mean flow, tidal and planetary wave behavior in an atmosphere strongly influenced by a stochastic field of vertically-propagating gravity waves (which are represented in an entirely self-consistent manner). This allows one to avoid many of the *ad hoc* approximations that are inevitable in simpler theoretical models of the middle atmosphere.

Despite its sophistication, the SKYHI model as it stands still ignores some aspects of the atmospheric physics that are likely important for mesospheric dynamics. The lack of a diurnal cycle in the model O_3 field and the neglect of non-LTE effects in the treatment of radiative transfer are two obvious examples. The imposition of a model lid near 80 km is also a matter of concern for mesospheric simulation. Efforts are now underway to develop a version of the SKYHI code with improved treatment of non-LTE effects that will extend into the lower thermosphere.

Acknowledgments. The author thanks R.J. Wilson for his assistance with the model integrations. He also thanks J.D. Mahlman, L.J. Donner and two referees for their comments on the manuscript of this paper.

REFERENCES

Andrews, D. G., J. D. Mahlman, and R. W. Sinclair, Eliassen-Palm diagnostics of wave-mean flow interaction in the GFDL SKYHI general circulation model, *J. Atmos. Sci., 40*, 2768-2784, 1983.

Andrews, D.G., J.R. Holton and C.B. Leovy, *Middle Atmosphere Dynamics*, Academic Press, New York, 489 pp., 1987.

Chapman, S., and R.S. Lindzen, *Atmospheric Tides*, D. Riedel, Hingham, Massachusetts, 200 pp., 1970.

Fels, S.B., and R.S. Lindzen, The interaction of thermally excited gravity waves with mean flows, *Geophys. Fluid Dyn., 6*, 149-191, 1974.

Fels, S. B., J. D. Mahlman, M. D. Schwarzkopf, and R. W. Sinclair, Stratospheric sensitivity to perturbations in ozone and carbon dioxide: radiative and dynamical response, *J. Atmos. Sci., 37*, 2265-2297, 1980.

Fleming, E.L., S. Chandra, M.R. Schoeberl and J.J. Barnett, Monthly mean global climatology of temperature, wind, geopotential height, and pressure for 0-120 km, *NASA Technical Memorandum 100697*, 85 pp., 1988.

Fritts, D. C., Gravity wave saturation in the middle atmosphere: A review of theory and observation, *Rev. Geophys. Space Phys., 22*, 275-308, 1984.

Groves, G.V., Energy fluxes of the (1,1,1) atmospheric oscillation. *Planet. Space Sci., 31*, 67-71, 1983.

Hamilton, K., *Numerical studies of wave-mean flow interaction in the stratosphere, mesosphere and lower thermosphere.* Ph.D. Thesis, Princeton University, Princeton, New Jersey, 386 pp., 1981.

Hamilton, K., What we can learn from general circulation models about the spectrum of middle atmospheric motions, *Coupling Processes in the Lower and Middle Atmosphere* (E. Thrane et al., eds.), Kluwer Academic Publishers, Dordrect, pp. 161-174, 1993.

Hamilton, K., and J. D. Mahlman, General circulation model simulation of the semiannual oscillation of the tropical middle atmosphere, *J. Atmos. Sci., 45*, 3212-3235, 1988.

Hayashi, Y., D. G. Golder, J. D. Mahlman, and S. Miyahara, The effect of horizontal resolution on gravity waves simulated by the GFDL SKYHI general circulation model, *Pure Appl. Geophys., 130*, 421-443, 1989.

Hines, C. O., Momentum deposition by atmospheric waves and its effects on thermospheric circulation, *Space Research XII*, 1157-1161, 1972.

Hocking, W.D., Turbulence in the altitude region 80-120 km, *Middle Atmosphere Program Handbook, 16*, 290-304, 1985

Hunt, B.G., A simulation of the gravity wave characteristics and interactions in a diurnally varying model atmosphere, *J. Meteor. Soc. Japan, 68*, 145-161, 1991

Lubken, F.J., W. Hillert, G. Lehmacher, U. von Zahn, T. Blix, E. Thrane, H.-U. Widdel, G.A. Kokin, and A.K. Knyazev, Morphology and sources of turbulence in the mesosphere during DYANA, *J. Atmos. Terr. Phys.*, in press, 1993.

Mahlman, J.D. and L.J. Umscheid, Comprehensive modeling of the middle atmosphere: the influence of horizontal resolution, *Transport Processes in the Middle Atmosphere* (G. Visconti ed.),D. Riedel, Hingham, Massachusetts, pp. 251-266, 1987.

Meyer, W.D., A diagnostic numerical study of the semiannual variation of the zonal wind in the tropical stratosphere and mesosphere, *J. Atmos. Sci., 27*, 820-830, 1970.

Miyahara, S., Zonal mean winds induced by vertically propagating tidal waves in the lower thermosphere, Part I, *J. Meteor. Soc. Japan, 54*, 91-98, 1978a.

Miyahara, S., Zonal mean winds induced by vertically propagating tidal waves in the lower thermosphere, Part II, *J. Meteor. Soc. Japan, 54*, 548-558, 1978b.

Miyahara, S., Zonal mean winds induced by solar diurnal tides in the lower thermosphere for a solstice condition, *Dynamics of the Middle Atmosphere* (J.R. Holton and T. Matsuno, eds.), D. Reidel, Dordrecht, 181-197, 1984.

Miyahara, S., Y. Hayashi, and J. D. Mahlman, Interaction between gravity waves and planetary-scale flow simulated by the GFDL SKYHI general circulation model, *J. Atmos. Sci., 43*, 1844-1861, 1986.

Mlynczak, M.G., and S. Solomon, On the efficiency of solar heating in the middle atmosphere, *Geophys. Res. Lett.*, 18, 1201-1204, 1991.

Morton, Y.T., R.S. Liebermann, P.B. Hays, D.A. Ortland, A.R. Marshall, D. Wu, W.R. Skinner, M.D. Burrage, D.A. Gell and J.-H. Yee, Global mesospheric winds observed by the high resolution doppler imager on board the upper atmosphere research satellite, *Geophys. Res. Lett., 20*, 1263-1266, 1993.

Zwiers, F., and K. Hamilton, Simulation of solar tides in the Canadian Climate Centre general circulation model, *J. Geophys. Res., 91*, 11877-11896, 1986.

Kevin Hamilton, Geophysical Fluid Dynamics Laboratory/National Oceanic and Atmospheric Administration, Princeton University, P.O. Box 308, Princeton, NJ 08542.

A Numerical Spectral Model for the Mean Zonal Circulation and the Tides in the Middle and Upper Atmosphere

K. L. Chan*, H. G. Mayr**, J. G. Mengel*, and I. Harris**

Applied Research Corporation, Landover, MD

**Goddard Space Flight Center, Greebelt, MD*

A three-dimensional, time-dependent, spectral model has been developed that solves the fully non-linear Navier Stokes equations without any of the commonly adopted hydrostatic or anelastic approximations. The marching in time is implicit, except for the terms that are non-linear and involve the Coriolis force. Horizontal variations of the dependent variables are expressed in terms of spherical harmonics and, through spectral transforms, the non-linearities are evaluated in physical space. The equations are formulated such that the total mass and total angular momentum are conserved to round-off error. Simplified, linear versions of the model have been derived to describe: (i) the stationary, axisymmetric mean circulation with zonal wave number m = 0, and (ii) stationary thermal tides by expressing the time derivatives as multiples of $i\omega$, where ω is the frequency with the period of a day. In both cases the same eddy diffusion and Rayleigh friction parameterizations are used for internal consistency. These solutions are obtained implicitly in one step by inverting the coupling matrices for a sufficiently large number of spherical harmonics that assure convergence. The models simulate the essential features of the observed mean circulation and tides in the mesosphere and thermosphere under solstice and equinox conditions. When the model is used in the time marching mode, the results obtained reproduce the stationary solutions, albeit with relatively long integration times. Non-linear processes are then also accounted for and some of the effects are discussed here.

1. INTRODUCTION

In modeling the atmosphere, our objective is to develop an understanding of the important processes involved and to provide a self consistent physical description of the observations. We shall concentrate on the region of the middle atmosphere (stratosphere and mesosphere) and upper atmosphere (or thermosphere). This region can be considered as a system that is, to some extent, closed but is strongly coupled internally and extremely complicated. To make progress, one usually must concentrate on a limited aspect of the dynamical regime. Our own strategy is to work with a family of models that complement each other. For this purpose, we exploit the analytical value of the spectral approach which naturally decomposes the variable fields into separate zonal components. The modeling concept we adopt is rooted in a spectral code recently developed (Chan et al., 1994a) which is 3D, time dependent, non-linear, and has few limitations, so that it can describe many of the phenomena observed. (The model is also non-hydrostatic, which is not relevant for the present application.) In a separate paper we discuss in depth some results from this model that describe the mean zonal circulation and tides (Chan et al., 1994b). A linear axisymmetric version of the code (with implicit spherical harmonic coupling) was applied to the convecting atmosphere of the Sun and reproduced the observed internal angular velocity distribution (Chan and Mayr, 1990); and a non-linear version of the code was used to successfully simulate the spin-up of the Venusian atmosphere (Chan and Mayr, 1990).

In Section II, we present a brief review of the field. In Section III, the salient features of the numerical code and some of its model derivatives are discussed. In Section IV, we present results for the zonal mean and for the solar diurnal as well as semidiurnal tides that illustrate the capability of the model.

2. BACKGROUND

To provide a background for our modeling effort, we review here briefly the phenomenology and processes that characterize the combined region of the middle atmosphere (stratosphere and mesosphere) and upper atmosphere (or thermosphere). The region of interest is one in which many of the important processes are known to undergo fundamental change. What causes these changes is not well understood nor is it known precisely where they do occur. For example, in the middle atmosphere turbulent mixing or eddy diffusion dominates such that the relative concentrations of atmospheric species, controlled by transport processes, tend to be independent of altitude. Above what is referred to as the turbopause near 100 km, molecular diffusion begins to dominate so that the composition changes with altitude, the heavier species concentrated lower down and the lighter ones at higher altitudes. The process that induces turbulent mixing in the mesosphere is believed to be related to gravity wave breaking which is not well understood, and it is not known where and how the transition to molecular diffusion occurs, how it changes globally, annually or on other time scales. Associated with eddy and molecular diffusion are heat and momentum transfer processes which impact the energetics and dynamics of the region. Thus, radiative energy transfer tends to dominate over eddy heat conduction in the middle atmosphere. Except for the 70 - 80 km altitude region, the principal energy loss is CO_2 cooling by infrared radiation. Thermal emission from the fundamental band of NO becomes the major radiative cooling agent above 120 km, while molecular heat conduction becomes increasingly more important at higher altitudes and eventually dominates in the upper thermosphere. Dynamically, the consequence is that in the middle atmosphere, depending on the time scale of the phenomenon, the winds tend to be geostrophic due to the Coriolis force from the Earth's rotation which causes the motions to be directed across pressure gradients; while in the upper atmosphere or thermosphere the winds become increasingly more ageostrophic and are directed along pressure gradients, due to the importance of molecular viscosity and ion drag. The energy sources from the Sun that drive these processes also change throughout the region. In the middle atmosphere, the absorbed radiation is in the UV which varies with solar activity, and the absorber is primarily ozone, strongly influenced by chemical processes. In the lower thermosphere, O_2 absorption in the Schumann-Runge continuum is the main heat source; while in the tenuous thermosphere at higher altitudes, the solar input derives primarily from EUV radiation which is highly variable, and from magnetospheric sources at high latitudes which vary sporadically. These sources do not carry much energy but the resulting specific heat input per gas molecule is large and so are its variations with solar and magnetic activity, thus accounting for the large temperature and its variability that characterize the thermosphere. The energetic photons in the EUV and auroral particles also ionize the atmosphere to produce the ionospheric plasma. Although the density of the ionosphere is comparatively small, its effects on the neutral atmosphere are large due to the actions of the geomagnetic field and the electric fields from the solar wind and the dynamo process which become important in the lower thermosphere.

In addition to these energy and momentum sources that drive the system from the outside, non-linear processes are of fundamental importance and produce internal energy and momentum sources or sinks that further contribute to the rich phenomenology. In order of decreasing spatial and temporal scales (which are usually positively correlated), the major components of the atmospheric motions are categorized as: global circulation (zonally averaged; annual and semi-annual components), planetary waves (azimuthal wave number less than 6; periods of one day to one month), thermal tides (trapped or propagating; diurnal, semi-diurnal, and ter-diurnal variations), acoustic/gravity waves (horizontal wave lengths from a few to a few thousand km; periods from a few minutes to a few hours), and three dimensional turbulence (with spatial scales less than a few km and temporal scales less than a few minutes). This dynamical system is extremely complex, as illustrated graphically in Figure 1.

The meridional component of the mean (zonally averaged) global circulation generally takes on the form of cells with the fluid rising in the region with excess differential heating. The pattern of this circulation changes with annual and semi-annual cycles, and is affected by internal energy and momentum sources due to tides and waves. In the middle atmosphere, the geostrophic zonal winds vary seasonally and change direction from summer to winter, with peak velocities of about 60 m/s near 45° latitude at 60 km. In the dissipative thermosphere, the circulation pattern changes significantly or can even reverse direction during large geomagnetic storms.

The global circulation significantly influences the energy

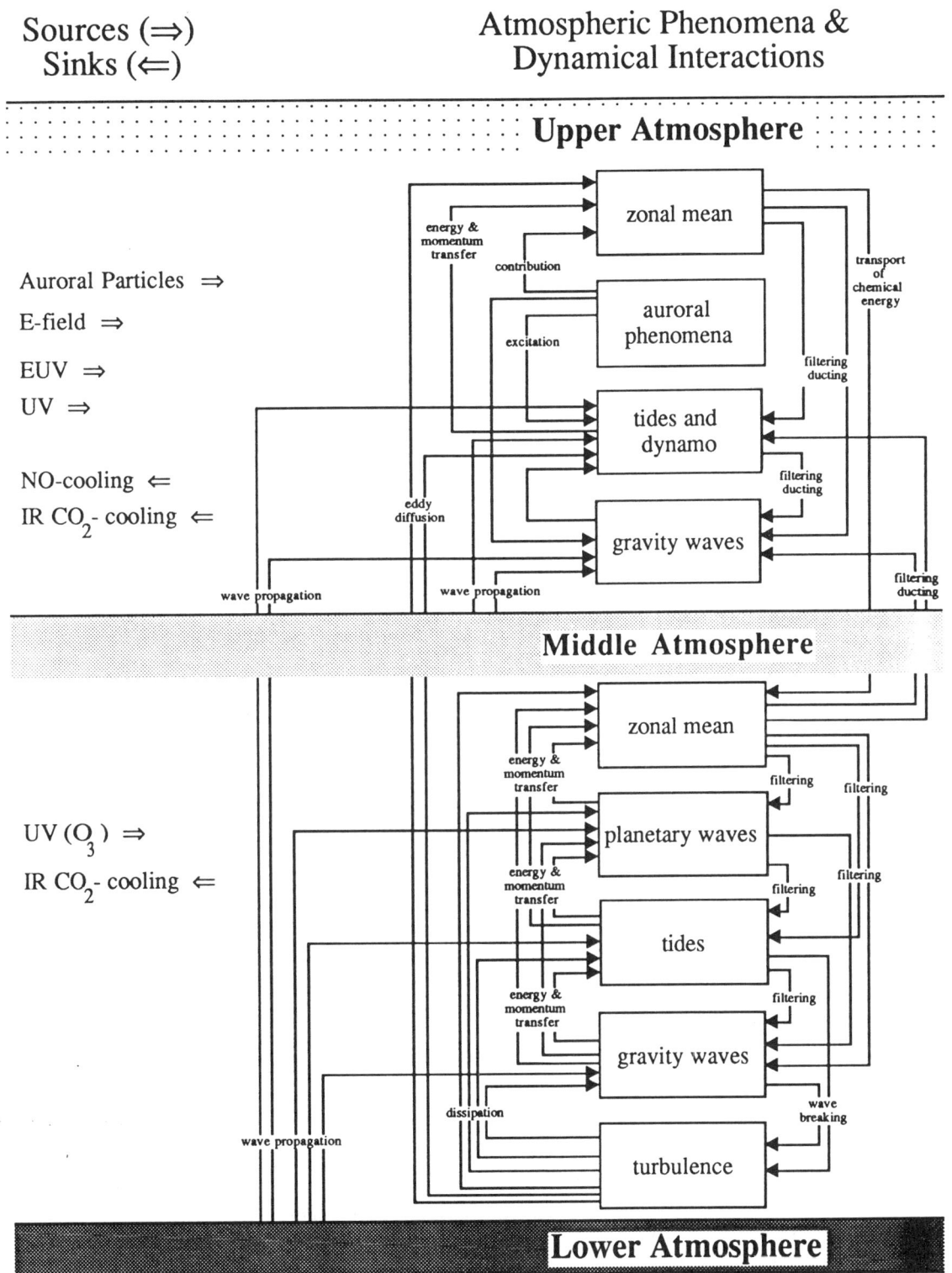

Fig. 1: Schematic illustrating the important energy sources and sinks, and the dynamical interactions in the middle and upper atmosphere.

balance and temperature distribution of the atmosphere. An important example is the 'mesospheric temperature anomaly' near 90 km, where the polar temperature in summer is as much as 80 K lower than in winter, contrary to expectation from radiative considerations. The energy budget for this temperature anomaly is controlled by a diabatic circulation (Murgatroyd and Goody, 1958, Murgatroyd and Singleton, 1961; Dunkerton, 1978) which is believed to be maintained by breaking gravity waves emanating from the lower atmosphere (Holton, 1982; Garcia and Solomon, 1985). On the other hand, the mean zonal circulation also significantly affects the propagation of the gravity waves. Such non-linear coupling is a common feature of the dynamics characterizing the region.

The mean meridional circulation is also very effective in changing the chemical composition of the atmosphere. An outstanding example of this effect is seen in the thermospheric distribution of He which increases by more than a factor of 20 from the summer to the winter pole. A smaller but similar kind of effect is also observed in the distribution of atomic oxygen at altitudes below 300 km, which in turn affects the thermospheric pressure field that drives the winds and significantly modifies the temperature distribution. Atomic oxygen transported from the summer into the winter hemisphere recombines there, and the latent energy released contributes to the above described mesospheric temperature anomaly as suggested by Kellogg (1961). Effects like these are also induced by the meridional circulation driven by auroral heating. Strong coupling between energetics, dynamics and chemistry is important due to transport of O_3 in the middle atmosphere and NO in the lower thermo-sphere.

In reviewing the physical properties of atmospheric tides, we refer to the books of Chapman and Lindzen (1970), Kato (1980), and Volland (1988). Thermal tides are sub-harmonic oscillations of a solar day and are primarily excited by the diurnal variations of the insolation absorption. In the middle atmosphere, the principal sources driving the diurnal and semi-diurnal tides are due to water vapor absorption of the solar near-IR radiation below 15 km and by the ozone absorption of UV radiation between 30 - 60 km altitude. The much less studied terdiurnal tide (third harmonic) may have magnitudes comparable to that of the diurnal component at mid latitudes (Cevolani and Bonelli, 1985). Aside from direct solar driving, this component is in part excited by the non-linear interaction between the diurnal and semi-diurnal tides (Tetenbaum et al., 1989). Tides can be decomposed into trapped and propagating components identified as Hough modes (Chapman and Lindzen, 1970; Kato, 1980). When the propagating modes travel upwards, their amplitudes experience exponential growth with height due to the decrease in the ambient density, and they become the most prominent dynamical feature of the upper mesosphere and lower thermosphere. The momentum and energy carried by the tides are then deposited into the mean flow, and the resulting circulation is comparable in magnitude to that generated by direct solar heating (Groves and Forbes, 1984).

Mean flows and horizontal temperature gradients significantly influence the propagation of tides (Lindzen and Hong, 1974), and this interaction produces cross-coupling among the Hough modes (Walterscheid and Venkateswaran, 1979). Propagating modes with growing amplitudes can also be generated by trapped modes through the interaction with the mean flow (Vial, 1986). An important feature of the tides in the transition region between the mesosphere and lower thermosphere is their day-to-day and geographic variability (Vincent and Ball, 1977; Groves, 1980). Differences between tides observed at Poker Flat (Tetenbaum et al., 1986) and at Mawson (McLeod and Vincent, 1985) indicate significant hemispheric asymmetries.

As discussed by Volland (1988), the tides generated above 150 km are primarily driven by EUV absorption and tend to be evanescent due to the importance of molecular viscosity and ion drag. Due to collisional momentum transfer between the neutrals and ions, atmospheric momentum is redistributed to produce higher order tidal components, and the same process is believed to be responsible also for the anomalous variations in the temperature and wind fields related to the equatorial ionization anomaly (Raghavarao et al., 1991). In the thermosphere at high latitudes, the solar wind induced electric fields as modeled by Volland (1973) and the ions they accelerate drive a two cell vortex circulation which has a statistically significant regular diurnal component with large wind velocities of several hundred m/s.

The diurnally varying wind system generated in the thermosphere also affects the composition, but the feedback on the temperature structure is less important than in the case of the annual and magnetic storm variations. Correlations between the horizontal wind field of the diurnal tides and the diurnal variations in ion density (drag) give rise to an internal momentum source which accelerates the zonal circulation in the thermosphere. This source can produce the small superrotation observed at low latitudes.

3. THEORETICAL MODELING

In modeling the atmosphere, we are facing a complex system which is driven by a variety of variable external sources and contains many interacting components. The observations and their coverage of the space and time domains being limited, this amounts to a complicated puzzle in which some of the important pieces are still missing. An essential objective of theory is to fill in these pieces, to develop a coherent picture in which the phenomenology can be explained. To that end, we must rely on numerical modeling and simulation and numerical experimentation which are all complementary. The ability to simulate numerically an observed phenomenon means that a number of requirements are more or less satisfied. First, the basic processes that produce the phenomenon are known and accounted for in the theory. Second, the physical and numerical treatments of the mechanisms involved are correct. Third, the parameterizations of external or unresolved processes are appropriate. Fouth, the measurements are sufficiently accurate, and the data can be interpreted in a way consistent with the theory. Simulation is therefore important for linking phenomena to fundamental physical principles and for developing a coherent interpretation of the observations. In performing a simulation, however, some processes cannot be explicitly calculated due to limited scope, domain, or resolution of the model. These processes are then parameterized in the model. To obtain realistic parameterization for a complicated dynamical process, it is useful to perform numerical experiments which explore the parametric behavior and to develop scaling relationships. In studying the dynamics of the mesosphere and thermosphere, we need to perform global-scale simulations that require parameterizations of small-scale wave breaking processes, which in turn need to be studied by high resolution numerical experiments.

Existing thermospheric circulation models (Dickinson et al., 1975, 1981, 1984; Fuller-Rowell and Rees 1980, 1981, Roble et al., 1982, 1988) do not go below 90 km. On the other hand, models of the stratosphere and mesosphere usually do not go above 120 km (e. g. Matsuno, 1982; Garcia and Solomon, 1985; Akmaev et al., 1992). The thermosphere and the atmosphere below are generally not coupled in these models. Yet from all we know about the interactions between these regions, it is essential now to treat it as a coupled system in order to make progress.

In our numerical model, the region of interest is the middle and upper atmosphere, a region for which a great deal of work has been done to treat radiative and photochemical processes (Dickinson, 1973, 1984; Lacis and Hansen, 1974; Strobel, 1978; Kockarts, 1980; Hinteregger et al., 1977; Fels and Schwarzkopf, 1981; Richards et al., 1982; Zhu, 1990; Allen et al. 1984; Torr, 1985; Roble and Ridley, 1987), ion-neutral momentum coupling (Volland and Mayr, 1973; Volland, 1973; Forbes and Garrett, 1976), and to parameterize wave interactions (Lindzen, 1981; Miyahara et. al., 1986; Rind et al., 1988; Garcia, 1991).

Distinct from the upper thermosphere which is characterized by strong dissipation, the mesosphere and lower thermosphere represent a much more formidable fluid dynamical problem. This region of the atmosphere is not weakly non-linear like the thermosphere but is strongly non-linear. Thus, a time dependent, 3D, fully non-linear, transformed spectral model has been developed which is formulated in terms of spherical harmonics. Details concerning the formulation of this approach and test results for this code can be found in Chan et al (1994a). This model does not invoke the hydrostatic approximation (not relevant in the present context) nor the anelastic one, and all the waves (e.g. Rossby-, gravity-, acoustic- waves) associated with the linearized hydrodynamic equations are described rigorously. The computation can be performed efficiently because an implicit method is used to avoid severe time step restrictions imposed by the fast traveling waves. However, in this model one approximation is made. To assure certain important conservation properties, the horizontal variation of density is neglected in the nonlinear advection terms (but retained in the linear terms). This limits the non-linearity of the advection to quadratic order so that alias-free transformations can be performed, and the prognostic equations can be put in simple closed forms similar to those derived for standard spectral GCMs. De-aliasing then keeps the total mass and total angular momentum conserved, both formally and numerically. This model has the capability to account for the essential processes involved and is able to describe the important interacting phenomena.

The adopted spectral approach has the well known advantage of being more efficient in low and moderate resolutions and, unlike the finite difference method, does not break down in the polar regions. As pointed out earlier, the spectral approach naturally decomposes the variable fields into separate zonal components that can be studied individually and interactively. The equations of motion for each of these components are explicitly formulated to build the model; thus it is straightforward to generate simplified versions of the code that describe an individual component. This property has been exploited to develop a family of sub-models that are both very efficient and versatile to use and provide insight. Since they have a common root, these models can be readily reassembled into a GCM, allowing us to develop step by step an increasingly more complete picture. Leaving synthesis to a later stage in our analysis, we shall study here the mean circulation and tides.

To understand the model and numerical procedures adopted, it is helpful to review briefly some of the properties of spherical harmonics which are used in our spectral formulation. Spherical harmonics make up a complete and orthogonal set of basis functions. Under certain simplifying assumptions which have some validity in describing the atmosphere dynamics, they also are the eigenfunctions of the linear system. This means that a source in the form of a particular spherical harmonic then will excite only an atmospheric response corresponding to that one harmonic. Perturbations in the other spherical harmonics will not be excited, which simplifies the analysis to such an extent that it effectively reduces the dimension of the problem. In addition to linearity, the principal condition for this simplification to be valid is that the Coriolis force can be neglected. Other conditions, less severe, require that the viscosity varies only with altitude. These simplifying assumptions generally apply to acoustic gravity waves which have periods much less than a day so that the Coriolis force is not important. They also apply to some extent to the dynamical condition encountered in the thermosphere where the time constant for dissipation is short compared to a day.

Making use of the above described property of spherical harmonics (P_ℓ^m), the equations of motion are expressed in compact form

$$\frac{\partial}{\partial t}\vec{S}(P_\ell^m) - \mathbf{L}\vec{S}(P_\ell^m) = \mathbf{L}_c \vec{S}(P_{\ell-1}^m, P_\ell^m, P_{\ell+1}^m) +$$
$$+ \mathbf{N}\vec{S}(P_m^m, ... P_\ell^m, ... P_{\ell+n}^m) + \vec{Q} \quad (1)$$

where $\vec{S}$ represents the solution vector, $\mathbf{L}$ is the linear operator associated with acoustic gravity waves and dissipation, $\mathbf{L}_c$ is the linear operator associated with the Coriolis force, $\mathbf{N}$ is an operator that accounts for all the remaining processes associated with non-linearities and horizontal as well as temporal variations in viscosity, diffusivity and ion drag, which are evaluated in physical space; and $\vec{Q}$ is the external driver. Specifically, the time discretized approximation of Eq. 1 can be written as

$$\left\{\frac{1}{\Delta t} - \beta \mathbf{L}\right\}\left(\vec{S}_{t+1}(P_\ell^m) - \vec{S}_t(P_\ell^m)\right) =$$
$$= \mathbf{L}\vec{S}_t(P_\ell^m) + \mathbf{L}_c \vec{S}_t(P_{\ell-1}^m, P_\ell^m, P_{\ell+1}^m) +$$
$$+ \mathbf{N}\vec{S}_t(P_m^m, .. P_\ell^m, .. P_{\ell+n}^m) + \vec{Q}_t \quad (2)$$

where the t index denotes the time level, Δt the time step, and β determines the degree to which the scheme is implicit. In each step of the time marching procedure, the solutions for individual spherical harmonics are obtained sequentially by inverting the corresponding matrix on the left hand side of Eq. 1. These spherical harmonic solutions are then used to update the right hand side for the next time step. For $\beta > 0.5$, the numerical procedure is stable even when the time steps are larger than those imposed by the Courant conditions associated with acoustic gravity waves. As discussed earlier, the matrix on the left hand side of Eq. 2 does not couple different spherical harmonics; it only couples the variables at neighboring altitude levels. Thus, this matrix has a tri-diagonal structure with small blocks and can be inverted in a straightforward and manner. The large time steps in this scheme and the fast inversion of the matrix make the solution of Eq. (2) very efficient. Most of the computer time, however, is used to evaluate the non-linear terms, $\mathbf{N}$, which require forward and backward transformations between the spectral and physical domains.

The matrix elements for the Coriolis terms, $\mathbf{L}_c$, and the Jacobian for the non-linear terms, $\mathbf{N}$, are not included in the implicit treatment of Eq. 2, because they would couple the neighboring spherical harmonics and thus increase the block size of the tri-diagonal matrix by a large factor equal to the order of spherical harmonics (say n). This would increase the computer time for inversion by a factor n^3, far exceeding the time needed to evaluate the non-linearities, which is usually prohibitive.

An implicit treatment of the Coriolis terms, however, becomes very practical when a stationary solution of the form $e^{im\omega t}$ is sought by solving the linearized equations

$$im\omega\vec{S}(P_\ell^m) - \mathbf{L}\vec{S}(P_\ell^m) - \mathbf{L}_c \vec{S}(P_{\ell-1}^m, P_\ell^m, P_{\ell+1}^m) = \vec{Q} \quad (3)$$

The large matrix on the left hand side, which couples all the spherical harmonics, then only needs to be inverted once. This approach provides a solution that is in steady state explicitly, unlike the time marching solution which must converge to that state. In the linear regime, this approach also proves to be more efficient.

In our analysis, we have used both the time marching and the steady state ansatz to describe the mean zonal circulation (m=0) and the diurnal as well as semidiurnal tides (m=1, 2, $\omega=2\pi/day$), as illustrated in Figure 2. Although the two solution procedures are applied in computer codes that have a common root they are sufficiently different, so that a comparison of the results is a meaningful test of the model. Comparing a solution obtained marching in time with one derived for steady state also provides a measure of convergence. In practical applications, the fast steady state or $i\omega$ solution is used as an initial condition for the time marching procedure in which the non-linearities are introduced.

4. NUMERICAL RESULTS

4.1. Mean Circulation

Much work has been done describing the two dimensional circulation of the stratosphere and mesosphere (e.g., Garcia and Solomon, 1983, 1985). Energetically important minor species such as O_3 and O are redistributed by the meridional circulation, thus tying together dynamics, chemistry and energetics. The dynamics involves wave breaking and wave

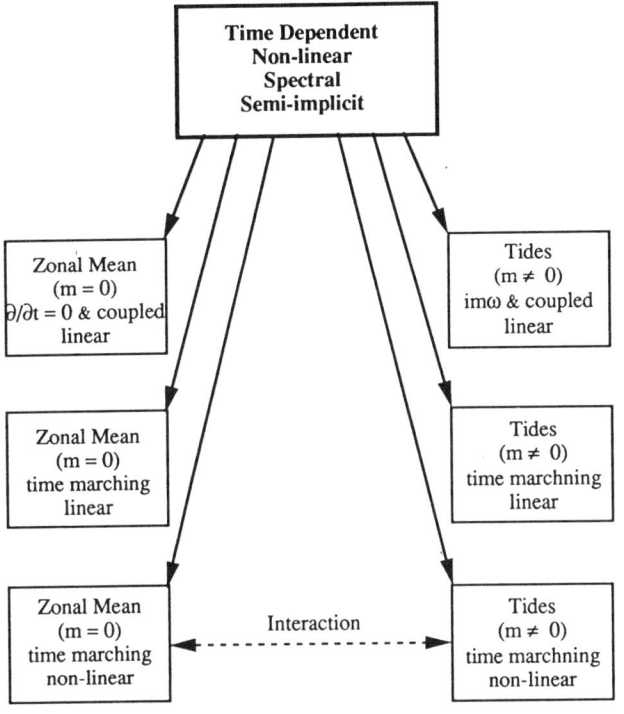

Figure 2: Schematic illustrating the organization of the modeling.

mean flow interactions (Lindzen, 1981; Holton, 1982; Garcia and Solomon, 1985), parameterized in terms of eddy diffusion and Rayleigh friction.

Our model is simplified by adopting the globally averaged temperature and density variations taken as functions of height. Emphasis is given to the stratosphere and mesosphere. But the height integration is carried out from the surface up to 400 km, allowing for sponge layers that minimize the artificial constraint from boundary conditions. At the surface the vertical and horizontal winds are set to zero; at the top the vertical gradients of temperature and horizontal winds are set to zero, accounting for the effect of molecular diffusion (heat conduction and viscosity).

A two dimensional spectral model is first derived by restricting the solution to a zonal wave number, $m = 0$. The assumption is made that a time independent solution for perpetual summer is appropriate, which may not be entirely valid at lower altitudes where the important time constants are long. Accurate solutions are obtained in a few minutes of VAX computer time by solving simultaneously for a total of 10 spherical harmonics, which is equivalent to a resolution of about 12°(assuming 3 grid points per wavelength), compared to that of 7.2° by Garcia and Solomon (1985). With 12 harmonics, the results are not perceptibly different. This fast solution permits us to perform sensitivity studies for a number of external parameters.

The heat source driving the circulation is obtained from Strobel (1978) and contains components associated with the O_3 and O_2 concentrations of the atmosphere which absorb the UV radiation from the Sun. For the source in the stratosphere, the heating efficiency is set to one, assuming that the dissociation energy is locally recovered through recombination. For the absorption of the UV bands above 70 km, however, only the net heating is accounted for, assuming that the dissociation energy is effectively removed through transport processes. For these heat input calculations, the ozone distribution given by Strobel is slightly modified by removing the small density peak near 80 km. The computed heating rates are shown in the upper panel of Figure 3 for solstice condition. Radiative cooling is taken from Dickinson (1973) and Wehrbein and Leovy (1982) and is shown in the lower half of Figure 3.

Following the earlier work of Garcia and Solomon (1983), we first assumed that the Rayleigh friction and eddy diffusion coefficients vary monotonically with altitude inversely proportional to the square root of the density, with $K = 6 \times 10^5$ cm^2/sec at 80 km and limited between 10^4 and 3×10^6. The Prandtl number is taken to be 3. The computed temperature variations, as well as zonal, meridional and vertical velocities, were found to be in good agreement with those obtained by Garcia and Solomon (1983) and showed that the winter minus summer temperature difference in the 80-90 km range is only about 30K, less than half of the observed value. In a later paper, Garcia and Solomon (1985) introduced more realistic parameterizations for gravity wave breaking and produced results in substantial agreement with observations. We were also able to do that by fine-tuning the eddy diffusion and Rayleigh friction parameters, and the adopted height distributions are shown in the lower portion of Figure 3.

As mentioned earlier and illustrated on the left hand side of Figure 2, we can also use the time marching procedure to obtain a solution. After an integration time of 2 months, the linear steady state and time marching solutions differ by less than 5%. When the non-linearities are included, the integration time does not increase significantly but the amount of computer time required increases by more than a factor of 10. The computed temperature variations and zonal winds from a non-linear calculation are presented in Figure 4. The anomalous mesospheric temperature minimum in the summer hemisphere is reproduced, and the zonal wind velocities

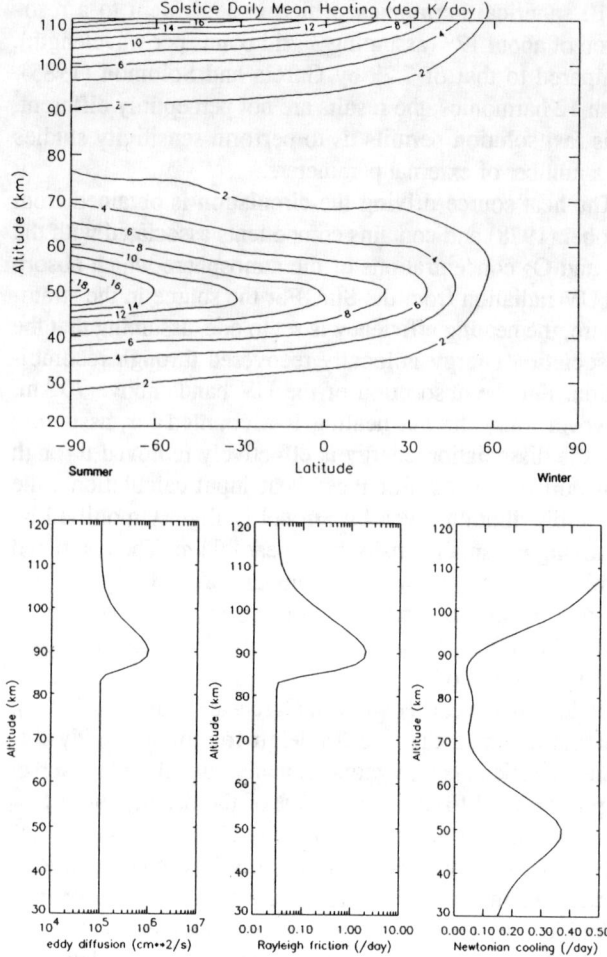

Figure 3: Heating rates for the zonal mean during solstice condition and adopted rates for eddy diffusion, Rayleigh friction and Newtonian cooling.

are comparable to those observed. In Figure 5 we compare the linear and non-linear solutions for the vertical temperature gradients at the summer and winter poles. Below 60 km, the effect is negligible, due to the importance of radiative cooling. At higher altitudes where transport processes are more important, however, the non-linearities have the effect of reducing the temperature variations.

4.2. Tides

Based on classical tidal theory, it is known that Hough functions are the normal modes of oscillation and that propagating as well as trapped components can be excited (Lindzen, 1970). This analytical representation in terms of normal modes is only valid under simplifying assumptions, which require that the atmosphere is inviscid and without background winds and horizontal temperature variations. Under more realistic conditions, numerical integration is required. For this purpose, finite difference models have been employed by Forbes and Lindzen (1976a,b), Forbes and Hagan (1979), Fesen et al. (1986) and others. Spectral models have been employed in different forms, using spherical harmonics (e.g., Volland and Mayr, 1973; Mayr and Harris, 1977; Walterscheid and Venkateswaran 1979) or using a combination of Hough modes and spherical harmonics (Mayr et al., 1990). In a model similar to that presented here, the spectral approach has been applied in a time marching procedure by Aso (1993) to describe the tides of the middle atmosphere.

To illustrate the capability of our model, we show results for the fundamental diurnal tide with zonal wave number $m = 1$ and a period of 24 hours. A steady state ($im\omega$) solution is obtained by solving simultaneously for a sufficiently large number of spherical harmonics (of order $l = 10$; with $l = 8$, the differences are small). The tide is excited thermally in the water vapor layer near the surface, in the ozone layer at 50 km and in the thermosphere (Forbes and Garrett, 1976). Dissipation due to molecular and eddy diffusion as well as Rayleigh friction and ion drag is accounted for. Figure 6 shows the computed amplitudes and phases of the zonal winds and relative temperature variations at latitudes of 3°, 16°, 41°, and 60° using the eddy diffusion (K) and Rayleigh friction parameterizations adopted for the mean zonal circulation shown in Fig. 3. The results are in reasonable agreement with those obtained by Forbes (1982a).

An important test of the model is to demonstrate that, under certain simplifying assumptions, it reproduces the results from classical tidal theory. Neglecting dissipation associated with eddy diffusion and Rayleigh friction as well as the vertical component of the Coriolis force, results are shown in Figure 7 for a propagating tide excited with an S_1^1 source in the water vapor layer near the surface. Up to 100 km, the amplitudes grow, their ratios for different latitudes remain constant, and the phases coincide, revealing a constant vertical wavelength of about 35 km. This demonstrates that only the S_1^1 mode is excited, as expected from theory. At higher altitudes, however, molecular diffusion and ion drag become important, thus exciting other modes as seen from the changing amplitude ratios and phase progression. At altitudes above 200 km dissipation becomes so important that the tide becomes evanescent as discussed by Volland (1988).

As in the case of the mean circulation, we also solved the tidal equations by marching in time. In agreement with the work of Fesen et al. (1986), Kähler (1989), and Miyahara and Wu (1989), the solution converged within two model months, when the differences relative to the steady state

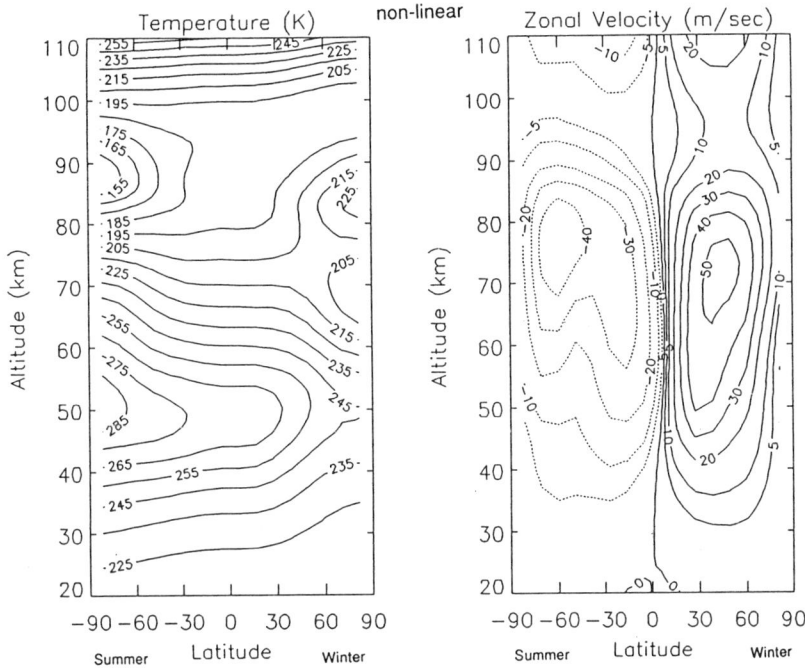

Figure 4: Computed zonal mean temperature distribution and wind field for solstice condition.

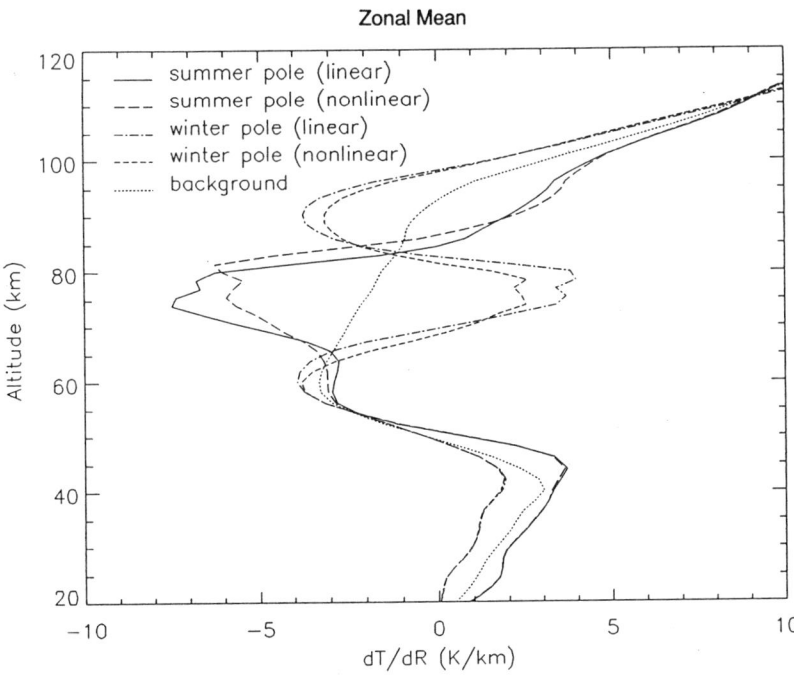

Figure 5: Comparison between vertical temperature variations for the linear and non-linear results.

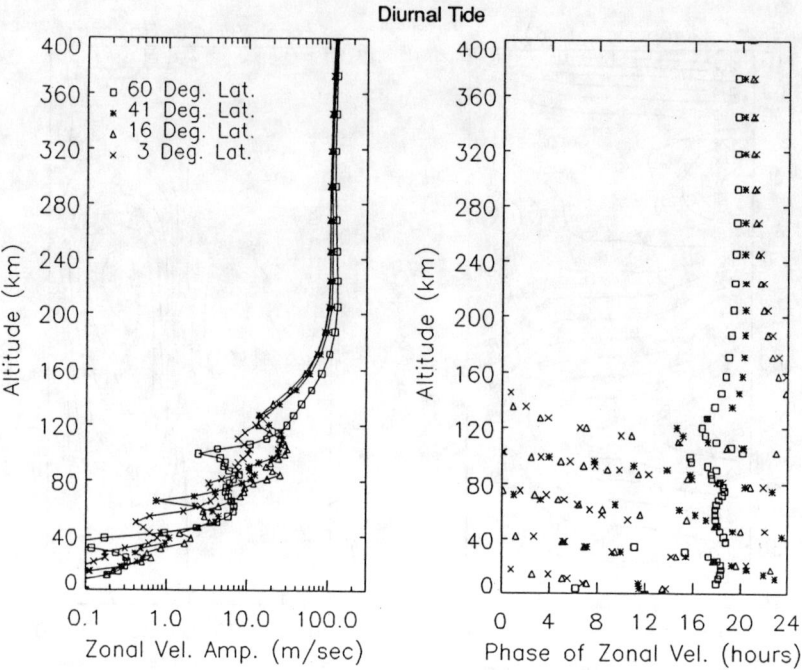

Figure 6: Computed amplitude and phase at different latitudes plotted versus altitude for the fundamental diurnal tide (m = 1, period of 24 hours).

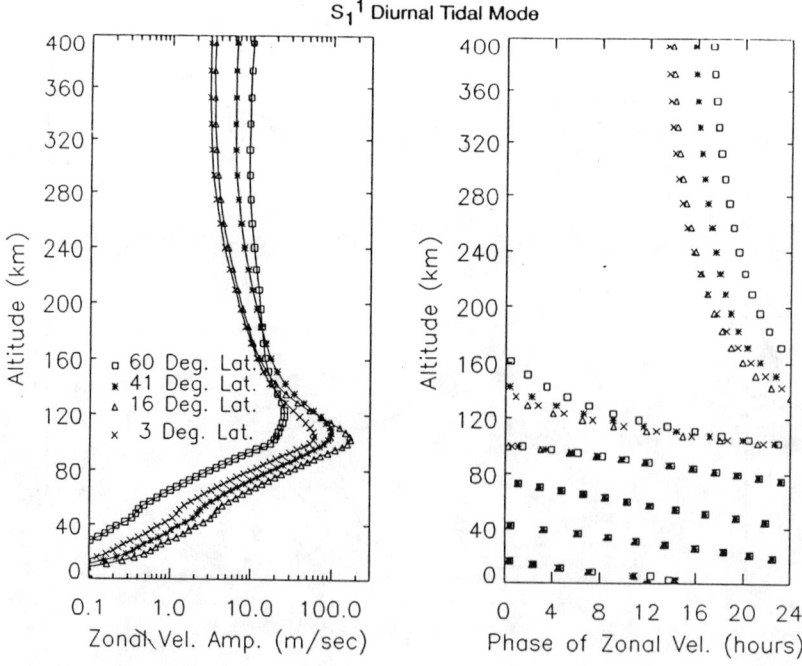

Figure 7: Simulation of the fundamental Hough mode S_1^1 excited in the water vapor layer near the ground, suppressing Rayleigh friction and eddy diffusion.

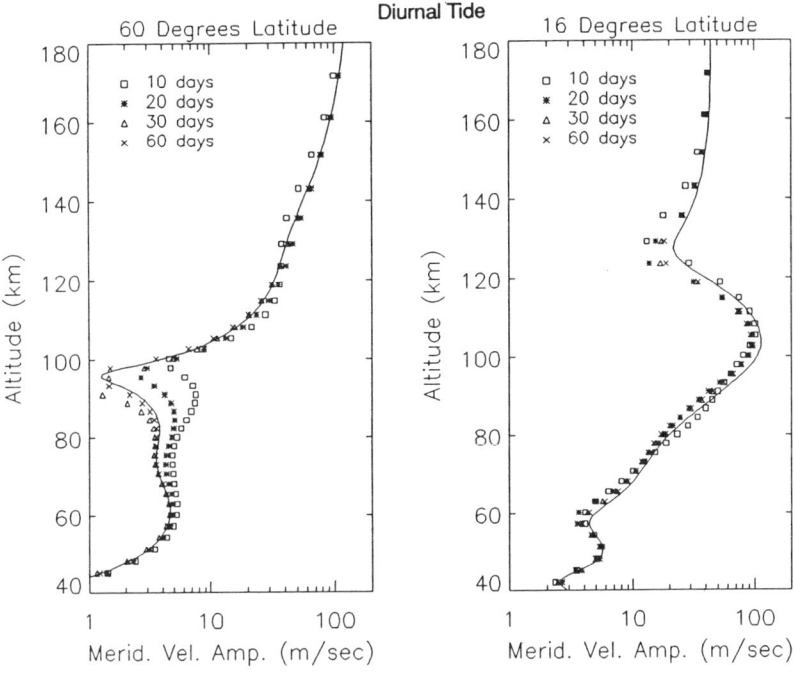

Figure 8: Comparison between the zonal velocity amplitudes derived from a stationary iω solution (solid lines) and after different integration times. Note that the solution tends to converge more rapidly at low latitudes where the propagating modes dominate.

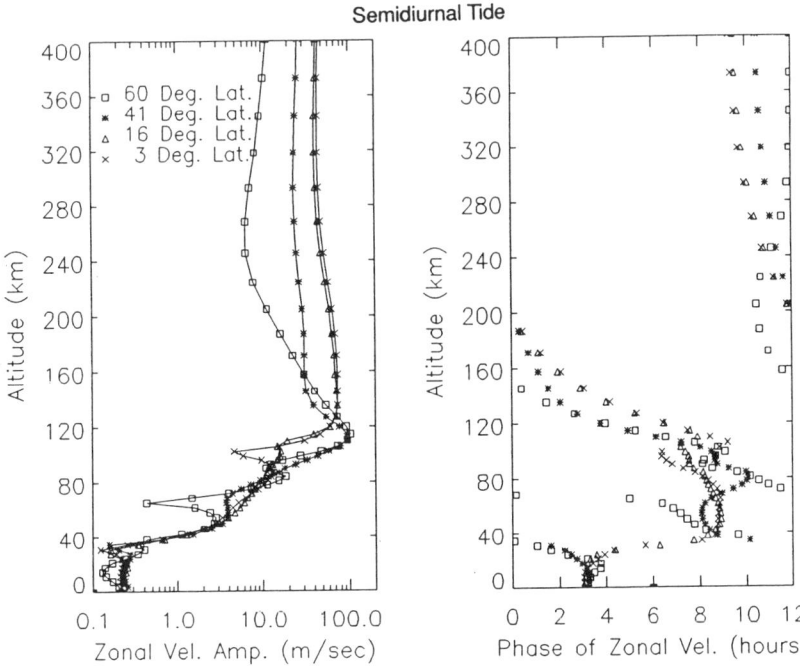

Figure 9: Computed amplitude and phase at different latitudes plotted versus altitude for the semidiurnal tide (m = 2, period of 12 hours).

solution are generally less than 10%. The propagating modes which are excited preferentially near the equator, however, tend to converge more rapidly than the trapped modes at higher latitudes. At both latitudes, though, convergence is fast in the thermosphere above 100 km. This is illustrated in Figure 8 which shows at low and high latitudes the computed amplitudes for the meridional winds in steady state and for different integration times.

For the semidiurnal tide (m = 2, and period of 12 hours), we show in Figure 9 the computed amplitudes and phases of the zonal winds at different latitudes, again using the eddy diffusion (K) and Rayleigh friction parameterizations from Figure 3. Below 120 km, the results are in excellent agreement with those obtained by Forbes (1982b). The differences at higher altitudes are attributed to a limitations of our model, which does not yet account for the ion drag momentum coupling with the fundamental tide which is known to be important.

The limited scope of this paper does not allow us to present results that account for the coupling with the mean zonal circulation. This subject was discussed in a separate paper (Mayr et al., 1993). It was shown there that above a certain threshold in the background temperature variations, close to the realistic condition, a baroclinic type instability develops in the summer hemisphere near the mesopause which generates planetary waves.

Acknowledgements: The authors are indebted to two anonymous referees for valuable comments.

REFERENCES

Akmaev, R. A., V. I. Fomichev, N. M. Gavrilov, and G. M. Shved, Simulation of the zonally mean climatology of the middle atmosphere with a three-dimensional spectral model for solstice and equinox conditions, *J. Atm. Terr. Phys.*, 54, 119, 1992

Allen, M., J. I. Lunine, and Y. L. Yung, The vertical distribution of ozone in the mesosphere and lower thermosphere, *J. Geophys. Res.*, 89, 4841, 1984

Aso, T., Time dependent numerical modelling of tides in the middle atmosphere, *J. Geom.. Geoel.*, 45, 41, 1993

Cevolani, G., and P. Bonelli, Tidal activity in the middle atmosphere, *Nuovo Cimento*, 8C, 461, 1985

Chan, K. L., H. G. Mayr, J. G. Mengel, and I. Harris, A 'stratified' spectral model' for stable and convective atmospheres, *J. Comp. Phys.*, in press, 1994a

Chan, K. L., H. G. Mayr, J. G. Mengel, and I. Harris, A spectral approach for studying middle and upper atmospheric phenomena, *J. Atmos. Terr. Phys.*, in press, 1994b

Chan, K. L., and H. G. Mayr, Differential rotation as an axisymmetric resonant mode of convection, in *The Sun and Cool Stars: Activities, Magnetism, Dynamos*, ed. I. Tuominen, D. Moss, and G. Rüdiger, p 178, Berlin, Springer Verlag, 1990

Chan, K. L., and H. G. Mayr, Modeling the circulation in a Venus like atmosphere, *EOS*, 71, 549, 1990

Chapman, S., and R. S. Lindzen, *Atmospheric Tides*, D. Reidel, Dordrecht, 1970

Dickinson, R. E., Infrared radiative cooling in the mesosphere and lower thermosphere, *J. Atm. Terr. Phys.*, 46, 995, 1984

Dickinson, R. E., E. C. Ridley, and R. G. Roble, Thermospheric general circulation with coupled dynamics and composition, *J. Atm. Sci.*, 41, 205, 1984

Dickinson, R., E., E. C. Ridley, and R. G. Roble, A three dimensional general circulation model of the thermosphere, *J. Geophys. Res.*, 86, 1499, 1981

Dickinson, R. E., E. C. Ridley, and R. G. Roble, Meridional circulation in the thermosphere I. Equinox conditions, *J. Atm. Sci.*, 32, 1737, 1975

Dickinson, R. E., Method of parameterization for infrared cooling between altitudes of 30 and 70 kilometers, *J. Geophys. Res.*, 78, 4451, 1973

Dunkerton, T., On the mean meridional mass motion of the stratosphere and mesosphere, *J. Atm. Sci.*, 35, 2325, 1978

Fels, S. B., M. D. Schwarzkopf, An efficient, accurate algorithm for calculating CO_2 15 µm band cooling rates, *J. Geophys. Res., 86*, 1205, 1981

Fesen, C. G., R. E. Dickinson, and R. G. Roble, Simulation of thermospheric tides at equinox with the National Center for Atmospheric Research General Circulation Model, *J. Geophys. Res.*, 91, 4471, 1986

Forbes, J. M., Atmospheric tides: 1. Model description and results for the diurnal component, *J. Geophys. Res.*, 87, 5222, 1982a

Forbes, J., M., Atmospheric tides: 2. The solar and lunar semi-diurnal component, *J. Geophys. Res.*, 87, 5241, 1982b

Forbes, J. M., and M. E. Hagan, Tides in the joint presence of friction and rotation: An f plane approximation, *J. Geophys. Res.*, 84, 803, 1979

Forbes, J. M., and H. B. Garrett, Solar tides in the thermosphere, *J. Atm. Sci.*, 33, 2226, 1976

Forbes, J. M., and R. S. Lindzen, Atmospheric solar tides and their electrodynamic effects. I. The global S_q current system, *J. Atm. Terr. Phys.*, 38, 897, 1976a

Forbes, J. M., and R. S. Lindzen, Atmospheric solar tides and their electrodynamic effects. II. The equatorial electrojet, *J. Atm. Terr. Phys.*, 38, 911, 1976b

Fuller-Rowell. T. J., and D. Rees, A three dimensional, time dependent simulation of the global dynamical response of the thermosphere to a geomagnetic storm, *J. Atm. Terr. Phys., 43*, 701, 1981

Fuller-Rowell, T. J., and D. Rees, A three-dimensional time-dependent global model of the thermosphere, *J. Atm. Sci., 37*, 2545, 1980

Garcia, R. R., Parameterization of planetary wave breaking in the middle atmosphere, *J. Atm. Sci., 48*, 1405, 1991

Garcia, R., R., and S. Solomon, The effect of breaking gravity waves on the dynamical and chemical composition of the mesosphere and thermosphere, *J. Geophys. Res., 90*, 3850, 1985

Garcia, R. R., and S. Solomon, A numerical model of the zonally averaged dynamical and chemical structure of the middle atmosphere, *J. Geophys. Res., 88*, 1379, 1983

Groves, G. V., and J. M. Forbes, Equinox tidal heating of the upper atmosphere, *Planet. Space. Sci., 32*, 447, 1984

Groves, G. V., Seasonal and diurnal variations of middle atmosphere winds, *Phil. Trans. Roy. Soc., London, A296*, 19, 1980

Hinteregger, H. E., D. E. Bedo, J. E. Manson, and D. R. Skillman, EUV flux variation with solar rotation observed during 1974-1976 from the AE-C satellite, Space Res. XVII, 533, 1977

Holton, J. R., The role of gravity wave induced drag and diffusion in the momentum budget of the mesosphere, *J. Atm. Sci., 39*, 791, 1982

Kähler, M. , The effects of non-linearity on thermal tides in a 3-D numerical model, *J. Atm. Terr. Phys., 51*, 101, 1989

Kato, S., *Dynamics of the Upper Atmosphere*, D. Reidel, Dordrecht, 1980

Kellog, W. W., Chemical heating above the polar mesopause in winter, *J. Meteorol., 18*, 373, 1961

Kockarts, G., Nitric oxide cooling in the terrestrial thermosphere, *Geophys. Res. Lett., 7*, 137, 1980

Lacis, A. A., and Hansen, J. E., A parameterization for the absorption of solar radiation in the Earth's atmosphere, *J. Atm. Sci., 31*, 118, 1974

Lindzen, R. S., Turbulence and stress owing to gravity wave and tidal breakdown, *J. Geophys. Res., 86*, 9707, 1981

Lindzen, R. S., and S. Hong, Effects of mean winds and horizontal temperature gradients on solar and lunar semidiurnal tides in the atmosphere, *J. Atm. Sci., 31*, 1421, 1974

Lindzen, R. S., Internal gravity waves in atmospheres with realistic dissipation and temperature Part I. Mathematical development and propagation of waves into the thermosphere, *Geophys. Fluid Dyn., 1*, 303, 1970

McLeod, R., and R. A. Vincent, Observations of winds in the Antarctic summer mesosphere using the spaced antenna technique, *J. Atm. Terr. Phys., 47*, 567, 1985

Matsuno, T., A quasi on-dimensional model of the middle atmosphere circulation with internal gravity waves, *J. Meteor. Soc. Jpn., 60*, 215, 1982

Mayr, H. G., K. L. Chan, and J. G. Mengel, Generation of 5-day planetary waves in the summer mesopause, *EOS, 75*, 461, 1993

Mayr, H. G., I. Harris, and F. A. Herrero, The dynamo of the diurnal tide and its effect on the thermospheric circulation, *Planet. Space Sci., 38*, 301, 1990

Mayr, H. G., and I. Harris, Diurnal variations in the thermosphere. 2. Temperature, composition, and winds, *J. Geophys. Res., 82*, 2628, 1977

Miyahara, S., and D. Wu, Effects of solar tides on the mean circulation in the lower thermosphere: soltice condition. *J. Atm. Terr. Phys., 51*, 635, 1989

Miyahara, S., Y. Hayashi, and J. D. Mahlman, Interactions between gravity waves and planetary-scale flow simulated by the GFDL "SKYHI" General Circulation Model, *J. Atm. Sci., 43*, 1844, 1986

Murgatroyd, R. J., and F. Singleton, Possible meridional circulation in the stratosphere and mesosphere, *Quart. J. Roy. Met. Soc., 87*, 125, 1961

Murgatroyd, R. J., and R. M. Goody, Sources and sinks of radiative energy from 30 to 90 km , *Quart. J. Roy. Meteor. Soc., 84*, 225, 1958

Raghavarao, R., L. E. Wharton, N. W. Spencer, H. G. Mayr, and L. H. Brace, An Equatorial Temperature and Wind Anomaly (ETWA), *Geophys. Res. Lett., 18*, 1193, 1991

Richards, P. G., M. R. Torr, and D. G. Torr, Nitric oxide cooling in the thermosphere, *Planet. Space Sci., 30*, 515, 1982

Rind, D., R. Suozzo, N. K. Balachandran, A. Lacis, and G. Russell, The GISS Global Climate Middle Atmosphere Model. Part I: Model Structure and Climatology, *J. Atm. Sci., 45*, 329, 1988

Roble, R. G., E. C. Ridley, A. D. Richmond, and R. E. Dickinson, A coupled thermosphere/ionosphere general circulation model, *Geophys. Res. Lett., 15*, 1325, 1988

Roble, R. G., and E. C. Ridley, An auroral model for the NCAR thermospheric general circulation model (TGCM), *Ann. Geophys., 5A*, 369, 1987

Roble, R. G., R. E. Dickinson, and E. C. Ridley, Global circulation and temperature structure of the thermosphere with high latitude plasma convection, *J. Geophys. Res., 87*, 1599, 1982

Strobel, D. F., Parameterization of atmospheric heating rate from 15 to 120 km due to O_2 and O_3 absorption of solar radiation, *J. Geophys. Res., 83*, 6225, 1978

Tetenbaum, D., F. Vial, A. H. Manson, R. Giralsez, and M. Massebeuf, Non-linear interaction between the diurnal and semidiurnal tides; terdiurnal and diurnal secondary waves, *J. Atm. Terr. Phys., 51*, 627, 1989

Tetenbaum, D., S. K. Avery, and A. C. Riddle, Observations of mean winds and tides in the upper mesosphere during 1980-1984, using the Poker Flat, Alaska, MST radar as a meteor radar, *J. Geophys. Res., 91*, 14539, 1986

Torr, D. G., *Photochemistry of Atmospheres*, p 165, Academic Press, New York, 1985

Vial, F., Numerical simulation of atmospheric tides for solstice condition, *J. Geophys. Res., 91*, 8955, 1986

Vincent, R. A., and S. Ball, Tides and gravity waves in the mesosphere at mid- and low- latitudes, *J. Atm. Terr. Phys., 39*, 965, 1977

Volland, H., *Atmospheric Tidal and Planetary Waves*, Kluwer Academic Publishers, Dordrecht, Boston, London, 1988

Volland, H., A semiempirical model of large-scale magnetospheric electric fields, *J. Geophys. Res., 78*, 171, 1973

Volland, H., and H. G. Mayr, Numerical study of the three-dimensional diurnal variations in the thermosphere, *Ann. Geophys., 29*, 61, 1973

Walterscheid, R. L., and S. V. Venkateswaran, Influence of mean zonal motion and meridional temperature gradients on the solar semidiurnal atmospheric tide: A spectral study, Part I: Theory, *J. Atm. Sci., 36*, 1623, 1979

Wehrbein, W. M., and C. B. Leovy, An accurate radiative heating and cooling algorithm for use in a dynamical model of the middle atmosphere, *J. Atm. Sci., 39*, 1532, 1982

Zhu, X., Carbon dioxide 15 μm band cooling in the upper middle atmosphere calculated by Curtis matrix interpolation, *J. Atm. Sci., 47*, 755, 1990

Energy Spectrum of the NCAR-TIGCM Lower Thermosphere

R. G. Raskin[1], A. G. Burns, T. L. Killeen

Space Physics Research Laboratory, University of Michigan, Ann Arbor, Michigan

R. G. Roble

High Altitude Observatory, Boulder, Colorado

The spatial energy spectrum of the 110-150 km region is evaluated using a base case run of the NCAR-TIGCM. Available potential energy, divergent kinetic energy, and rotational kinetic energy are decomposed into spherical harmonic modes and the relative sizes of the reservoirs are compared to those observed in the lower atmosphere. The lower thermospheric spectrum is dominated by the signals of diurnal and semidiurnal tides, but dissipation occurs within the study region. The total potential energy exceeds the available potential energy by three orders of magnitude, but unlike the tropospheric case, more energy resides as kinetic energy than as available potential energy, a result that is typical of propagating tides. A modified definition of available potential energy apprpriate for thermospheric heights is presented.

INTRODUCTION

Two fundamental properties of the atmosphere are its abilities to transform energy from one form to another and to redistribute energy spatially. These processes are part of an atmospheric energy cycle, in which energy from the sun is ultimately converted into motion and heat. A summary of the extensive research on spectral energetics of the lower atmosphere is found in [*Wiin-Nielsen and Chen*, 1993]. By contrast, less effort has gone into analogous studies of the upper atmosphere, due in part to the lack of observational data sets that are global in extent. The most comprehensive analysis to date is [*Roble et al.*, 1987], who used the NCAR Thermospheric-Ionospheric General Circulation Model (TIGCM) to evaluate the global average energetics of the thermosphere.

The purpose of the present study is to evaluate the spatial energy spectrum of the 110-150 km region of the NCAR-TIGCM using a base case run. The sizes of the thermospheric energy reservoirs are compared with each other and with values observed in the troposphere to reveal the similarities and differences between the corresponding energy cycles. The thermospheric energy decomposition to be considered is an incomplete one, as it focuses upon the forms of energy included in lower atmospheric analyses: available potential energy (APE), total potential energy (TPE) (consisting of internal plus gravitational energy), divergent kinetic energy (DKE), and rotational kinetic energy (RKE). In this first attempt to understand the spectral energetics of the region, chemical energy and electrical energy are not included. The energy fields are decomposed in terms of the two-dimensional spherical harmonics, to represent a decomposition by spatial scale. The source, sink, and exchange terms for the spectral energy reservoirs are evaluated in [*Raskin*, 1992].

---

1. Currently at: Institute of Marine Sciences, University of California, Santa Cruz, CA 95064.

The Upper Mesosphere and Lower Thermosphere:
A Review of Experiment and Theory
Geophysical Monograph 87
Copyright 1995 by the American Geophysical Union

The NCAR-TIGCM is described in [*Dickinson et al.*, 1984] and [*Roble et al.*, 1988]. The model's vertical domain extends from 97 km to approximately 500 km using log-pressure coordinates. Horizontal resolution is 5 degrees in latitude and longitude, and finite differences are used to approximate the derivative terms. Semidiurnal tides are simulated through empirically derived forcing functions at the lower boundary using Hough functions H(2,2) through H(2,6) [*Fesen et al.*, 1985]. The TIGCM run adopted for this study corresponds to December solstice, solar minimum (F10.7 solar flux = 67, 30-day average F10.7 flux = 72) and low geomagnetic activity (cross-cap potential= 30 kV). A sensitivity analysis to changes in geophysical conditions is deferred to a later study.

At present, direct comparisons with global observational data are limited due to the scarcity of observation sites. Comparison of TIGCM tidal fields with radar data obtained from three sites during the first Lower Thermosphere Coupling Study (LTCS) shows fair agreement [*Fesen and Roble*, 1991]. As the Upper Atmospheric Research Satellite (UARS) and Thermosphere, Ionosphere, Mesosphere: Energetics and Dynamics (TIMED) missions are able to provide data with global coverage, comparisons with model output can be carried out.

AVAILABLE POTENTIAL ENERGY IN THE THERMOSPHERE

Available potential energy is defined as the maximum amount of total potential energy (gravitational plus internal) that can be converted into kinetic energy through adiabatic rearrangement of the mass. In particular, APE should be interpreted as the upper limit on the conversion though the mechanisms of adiabatic ascent and pressure gradient forcing. In the lower atmosphere, this conversion represents the only significant source of kinetic energy, but in the thermosphere, ion drag represents a kinetic energy source that bypasses the APE reservoir entirely. During periods of strong geomagnetic forcing and where highly diabatic conditions are present, the relevance of the APE concept is not known, and is not addressed in this paper.

While it is well known that conversion of APE into kinetic energy is induced by temperature gradients, in the upper atmosphere potential energy can also be released through gradients of mean molecular weight. This effect can be verified from the hydrostatic relation in log-pressure coordinates:

$$\frac{\partial \phi}{\partial Z} = RT \qquad (1)$$

where ϕ is geopotential, R is the species-dependent gas constant, T is temperature, and Z is the vertical coordinate:

$$Z = -\ln(P/P_0) \qquad (2)$$

where P is pressure and P_0 is a reference pressure level. Equation (1) implies that local variations of RT on a pressure surface induce a slope in the ϕ-Z plane, and hence, a horizontal pressure gradient force. Therefore, a modified definition of available potential energy appropriate for regions of variable mean molecular weight will be used for this study:

$$A = \frac{1}{2\overline{S}}\left[(RT)'\right]^2 \qquad (3)$$

where S is the modified static stability parameter:

$$S = \frac{\partial(RT)}{\partial Z} + \kappa RT \qquad (4)$$

and κ is the ratio of specific heat at constant pressure to that at constant volume. An overbar denotes a global average along the constant pressure surface and a prime indicates a deviation from this average. Equation (3) reduces to its conventional form involving variance of temperature alone [*Lorenz*, 1967] if mean molecular weight is constant. Use of (3) rather than the usual form tends to increase the thermospheric APE because temperature and mean molecular weight tend to be negatively correlated [e.g. *Roble*, 1992]; hence, R and T are positively correlated. Available potential energy is always defined relative to a particular (and arbitrary) reference state; in this case, the reference pressure surface is one of equally distributed mass, that is, with RT everywhere constant, and no horizontal motion. Use of another reference state would generally produce a different value for APE.

Equation (3) is derived below based upon the parcel method, following [*Wiin-Nielsen and Chen*, 1993]. Available potential energy is defined as the work performed by the atmosphere in bringing a parcel into the specified reference state adiabatically. The restoring force on a displaced parcel is:

$$F = -\frac{1}{\rho}\frac{\partial P}{\partial z} - g \approx \frac{\overline{\rho}g}{\rho} - g = \frac{RT - \left(\overline{RT}\right)}{\overline{H}} \qquad (5)$$

where ρ represents density, the lower case z represents geometric height, H represents scale height, and g is the gravitational force. The quantities RT and $\overline{RT}$ in (5) represent parcel and environmental values, respectively, at the *displaced* level. These values are related to their common value $(RT)_{ref}$ at the reference level through:

$$RT \approx (RT)_{ref} - R\frac{g}{c_p}\delta z$$

$$\approx (RT)_{ref} - \kappa \overline{RT}\delta Z \qquad (6)$$

and

$$\overline{RT} \approx (RT)_{ref} + \frac{\partial(\overline{RT})}{\partial Z}\delta Z. \quad (7)$$

In (6), the adiabatic lapse rate has been assumed for the parcel displacement, and a conversion to log-pressure coordinates has been made through

$$\delta z \approx \overline{H}\delta Z. \quad (8)$$

From (4), (6), and (7),

$$(RT)' = RT - \overline{RT} \approx -\overline{S}\delta Z. \quad (9)$$

Combining (5) and (9), the energy released in returning the parcel to the reference level is:

$$A = -\int_0^{z'} F dz = \int_0^{z'} \overline{S}\delta Z \frac{dz}{\overline{H}} = \int_0^{z'} \overline{S}\delta Z \, dZ. \quad (10)$$

Assuming a slowly changing mean static stability and using (9) again, produces the desired form:

$$A \approx \frac{\overline{S}}{2} Z'^2 = \frac{(RT)'^2}{2\overline{S}}. \quad (11)$$

SPECTRAL ENERGY EQUATIONS

Available Potential Energy

Any scalar variable can be expressed as linear combinations of spherical harmonics [*Hobson*, 1955]:

$$q(\theta,\lambda) = \sum_{n=0}^{\infty} \sum_{m=-n}^{n} q_{m,n} Y_n^m(\theta,\lambda) \quad (12)$$

where θ is geographic latitude and λ is longitude. The functions $Y_n^m(\theta,\lambda)$ are the spherical harmonics indexed by zonal wave number m, representing spatial scale in the zonal direction alone, and total wave number n, representing total spatial scale independent of any preferred direction. It is worth noting that the total wave number is also independent of the placement of the poles; hence, the power spectrum as a function of n alone, is identical in geographic and geomagnetic coordinates. The spherical harmonic com-ponents are obtained by projection:

$$q_{m,n} = \frac{1}{2\pi} \int_0^{2\pi} \int_{-\pi/2}^{\pi/2} q(\theta,\lambda) Y_n^{*m} d\lambda \cos\theta d\theta, \quad (13)$$

where * is the complex conjugate. By Parseval's relation,

$$\frac{1}{4\pi} \int_0^{2\pi} \int_{-\pi/2}^{\pi/2} q^2 d\lambda \cos\theta d\theta = \sum_{n=0}^{\infty} \sum_{m=0}^{n} (1-\frac{\delta_{m0}}{2})|q_{m,n}|^2 \quad (14)$$

where δ is the Kronecker delta function.

The available potential energy residing within an atmospheric layer (bounded by two pressure surfaces) can be decomposed using (14) as:

$$\frac{1}{4\pi}\frac{P}{g} \int_0^{2\pi} \int_{-\pi/2}^{\pi/2} A \, d\lambda \cos\theta d\theta = \frac{P}{g}\sum_{n=1}^{\infty} \sum_{m=0}^{n} (1-\frac{\delta_{m0}}{2}) A_{m,n} \quad (15)$$

where

$$A_{m,n} = \frac{1}{2\overline{S}} \left|\{RT\}_{m,n}\right|^2 \qquad n \neq 0. \quad (16)$$

The factor P/g represents the conversion from per unit mass to per unit area, as can be seen from the identity:

$$P/g = \rho H \quad (17)$$

The per area representation produces commensurable quantities at differing vertical levels.

For comparison purposes, the total potential energy (from which APE derives) is expressible as

$$\frac{1}{4\pi}\frac{P}{g} \int_0^{2\pi} \int_{-\pi/2}^{\pi/2} c_p T \, d\lambda \cos\theta d\theta \quad (18)$$

where c_p is the specific heat at constant pressure.

Kinetic Energy

Any *vector* function **u**:

$$\mathbf{u}(\theta,\lambda) = [u \; v] \quad (19)$$

can be represented in terms of *vector* spherical harmonics:

$$\mathbf{u}(\theta,\lambda) = \sum_{n=0}^{\infty} \sum_{m=-n}^{n} \mathbf{u}_{m,n} \mathbf{U}_n^m. \quad (20)$$

The coefficients are themselves vectors:

$$\mathbf{u}_{m,n} = \begin{bmatrix} u_{m,n}^D & u_{m,n}^R \end{bmatrix} \quad (21)$$

consisting of irrotational (hereafter referred to as divergent) and non-divergent (hereafter referred to as rotational) components, represented by superscripts D and R, respectively. The basis functions are the Hermitian matrices:

$$\mathbf{U}_n^m = \frac{1}{\sqrt{n(n+1)}} \begin{bmatrix} \frac{imY_n^m}{\cos\theta} & \frac{dY_n^m}{d\theta} \\ -\frac{dY_n^m}{d\theta} & \frac{imY_n^m}{\cos\theta} \end{bmatrix} ; \quad (22)$$

therefore, the multiplication in (20) is of a vector with a matrix, producing a vector, as required. (In [*Swarztrauber*, 1993], the rows of matrix $\mathbf{U}_n^m$ are written as vectors **B** and **C**, and the elements of $\mathbf{u}_{m,n}$ are denoted b and c.) The coefficients are obtained by projection:

$$\mathbf{u}_{m,n} = \int_0^{2\pi}\int_{-\pi/2}^{\pi/2} \mathbf{u}(\theta,\lambda)\mathbf{U}_n^{*m} \cos\theta \, d\theta \, d\lambda \quad (23)$$

where the asterisk represents the complex conjugate of the transpose of the matrix.

Using the relation

$$K = \frac{1}{2}\|\mathbf{u}\|^2 \\ = \frac{1}{2}\left(\left|u^D\right|^2 + \left|u^R\right|^2\right) \quad (24)$$

the globally integrated kinetic energy residing within an atmospheric layer is given by

$$\frac{1}{4\pi}\frac{P}{g}\int_0^{2\pi}\int_{-\pi/2}^{\pi/2} K \, d\lambda \, \cos\theta \, d\theta = \frac{P}{g}\sum_{n=0}^{\infty}\sum_{m=0}^{n}(1-\frac{\delta_{m0}}{2})K_{m,n} \quad (25)$$

where

$$K_{m,n} = \frac{1}{2}\|\mathbf{u}_{m,n}\|^2 \\ = \frac{1}{2}\left(\left|u_{m,n}^D\right|^2 + \left|u_{m,n}^R\right|^2\right) \\ = K_{m,n}^D + K_{m,n}^R \quad (26)$$

Thus, spectral kinetic energy decomposes into divergent and rotational components. The former is directly forced by the pressure gradient force (including tidal effects); the latter is most effectively impacted by geomagnetic effects at high latitudes [*Thayer*, 1990], as well as by the Coriolis force (which involves an exchange of divergent and rotational kinetic energy). Further discussion of the divergent-rotational energy decomposition is provided in [*Wiin-Nielsen and Chen*, 1993].

RESULTS

Four vertical levels in the NCAR-TIGCM are evaluated at unit scale height increments ranging from Z= -5 (approx. 110 km) to Z= -2 (approx. 150 km). Only UT=0 conditions are used for this steady-state run, as a sensitivity analysis to UT showed very little change in large-scale energetics throughout the day [*Raskin*, 1992]. Thus, to a first approximation, longitudinal variations are equivalent to time variations, with zonal wave number 1 spectra representing the diurnal tide and zonal wave number 2 spectra representing the semidiurnal tide. This approximation breaks down in the presence of geomagnetic activity, where spatial and temporal variations are not interchangable.

The importance of the semidiurnal tides is highlighted in Figure 1a, showing the spectrum of available potential energy density at the four vertical levels. Plotted for each (m,n) are the quantities:

$$\frac{P}{g}(1-\frac{\delta_{m0}}{2})A_{m,n} \quad \frac{P}{g}(1-\frac{\delta_{m0}}{2})K_{m,n}^D \quad \frac{P}{g}(1-\frac{\delta_{m0}}{2})K_{m,n}^R$$

using equations (16) and (26). Progressing upward from the Z= -5 to Z= -2 surfaces, the semidiurnal tide loses its dominance. This effect is clearly visible in the power spectrum of DKE density (Figure 1b), which exhibits strong dissipation between z= -5 and z= -3, particularly for higher values of n. In contrast, the RKE (Figure 1c) spectrum shows relatively little evidence of the tide. The broader RKE spectrum is a consequence of the strong rotational wind at the smaller scales of the high latitude region.

The same data grouped by zonal wave number m are presented in Figure 2. Individual bars within the stacked bar graph are additive; hence, the total APE at Z= -5 is about 100 mJ m^{-2}. Virtually all the thermospheric energy resides at zonal wave numbers m= 0, 1, and 2 as apparently molecular processes are damping out the small-scale variations. APE and DKE show a sharp reduction of energy density with height that is related to the decreasing mass of the atmosphere. However, a non-dissipative upward flux of tidal energy at wave 2 would imply a *constant* energy density

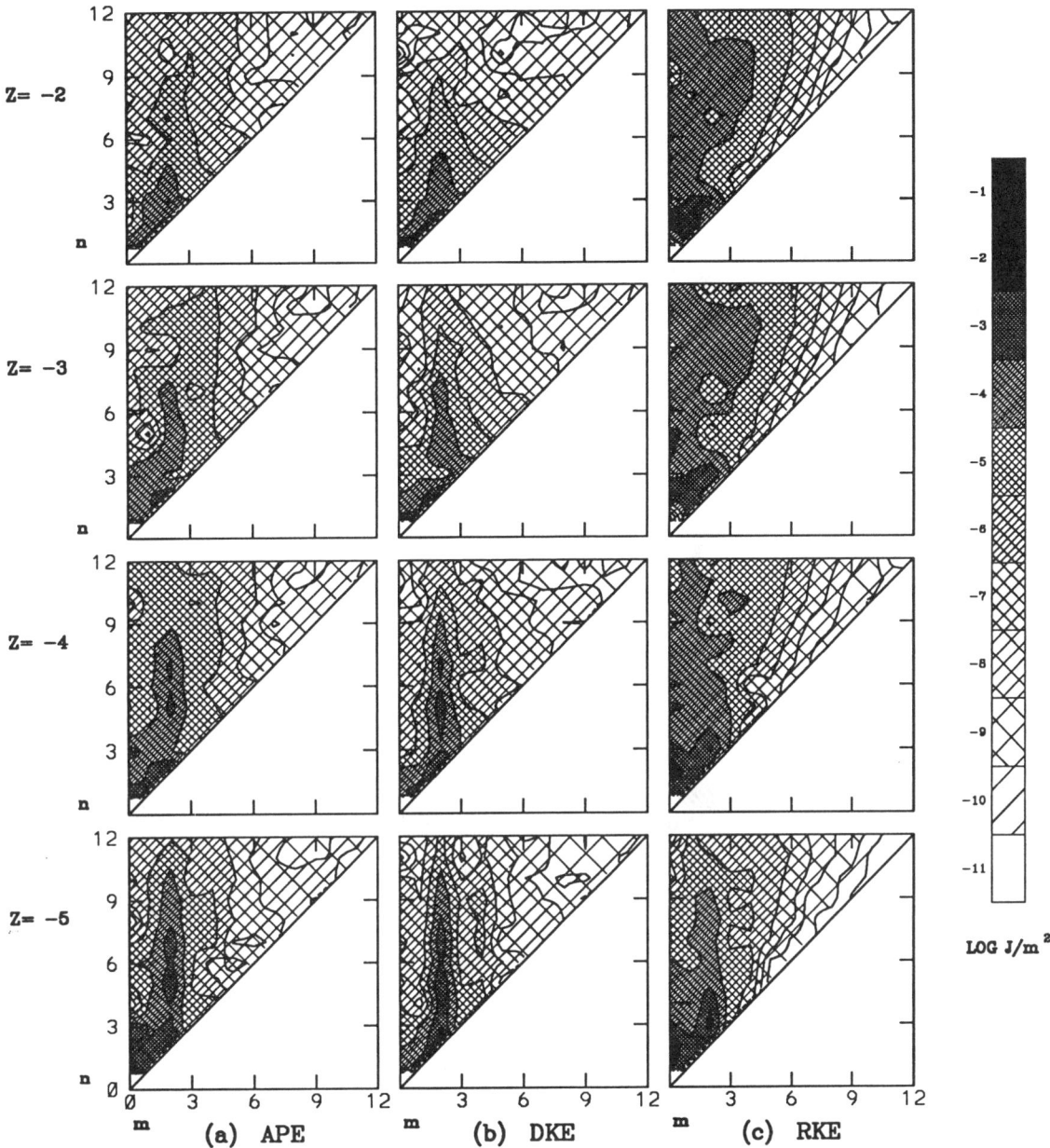

Figure 1. Energy density spectrum of available potential energy as a function of m and n at vertical levels: z= -5, z= -4, z= -3, and z=-2, for a) available potential energy; b) divergent component of kinetic energy; and c) rotational component of kinetic energy.

with height; hence, dissipation is occurring. The rotational kinetic energy density exhibits a much smaller decrease with height because the high latitude winds are increasing with altitude. A diurnal tide at the upper pressure surface is also visible.

At each vertical level, the total kinetic energy exceeds the available potential energy, a situation that is opposite that observed in the troposphere. The total kinetic energy is seen to exceed the available potential energy by factors ranging from about 2.2 at Z= -5 to 7 at Z= -2. In contrast, the observed tropospheric available potential energy exceeds the kinetic energy by factors of about 3 or 4 [Lorenz, 1967]. This reversal occurs (in part) because vertically propagating tides carry more of their energy as kinetic energy than poten-

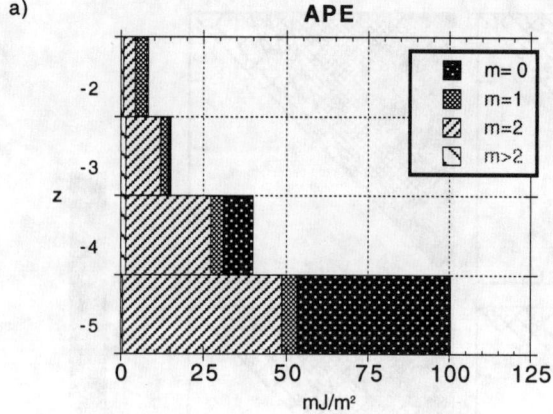
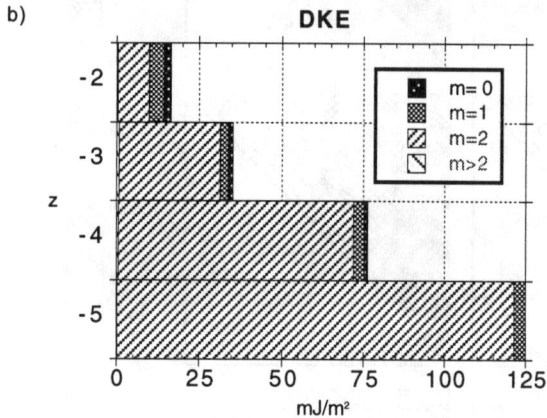
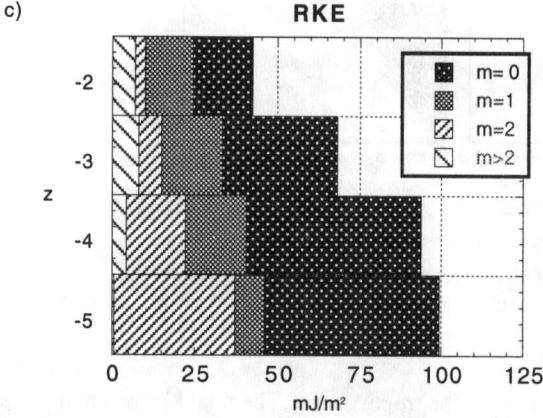

Figure 2. Energy density spectrum as a function of vertical level, grouped by zonal wave number m for : a) available potential energy; b) divergent kinetic energy; c) rotational kinetic energy.

tial energy [*Richmond*, 1975], with a typical APE:(DKE+RKE) ratio of between 2 and 3. Observations of the lower *stratosphere* also appear to show more kinetic energy than available potential energy due to the mechanical forcing of upwardly propagating planetary waves [*Dopplick*, 1971].

Comparison of Figures 2b and 2c reveal an RKE:DKE ratio varying from unity at the lowest level to about 3 at the highest pressure surface. These ratios are very different from those observed in the lower atmosphere, where non-divergent flow is usually an excellent approximation [*Wiin-Nielsen and Chen*, 1993].

Figure 3 shows the corresponding data of Figure 2 in terms of energy intensity, or energy per unit mass (that is, without the P/g factor). Wave 2 amplitudes increase with height through z= -3 despite the corresponding decrease in kinetic energy density. Hence, the semidiurnal wind is increasing with height, but at a slower rate than that consistent with constant energy density and non-dissipation.

Finally, a comparison is made between total potential energy (TPE) and the above energy density quantities. Figure 4 shows that the total energy exceeds the available portion (cf. Figure 2a) by factors of between 2000 to 3000; whereas typically observed ratios in the troposphere are 500 to 1000 [*Lorenz*, 1967]. Similarly, comparison of Figure 4 with Figures 2bc shows that the total potential energy exceeds the kinetic energy by factors of between 500 and 1000, whereas, the observed tropospheric ratio is approximately 2000 [*Lorenz*, 1967]. Thus, relative to the troposphere, kinetic energy represents a larger fraction of the TPE (and in fact, exceeds the APE); whereas, APE represents a smaller proportion of TPE.

CONCLUSION

For this study, the energy spectrum of the 110-150 km region has been analyzed using output from the NCAR-TIGCM for a base case run. The zonal wave number 2 spectrum (representing semidiurnal tides) experiences dissipation within the region, particularly in the higher order modes. In contrast to the lower atmosphere, the kinetic energy exceeds the available potential energy, as upwardly propagating tides carry most of their energy in the kinetic energy component. The total potential energy is three orders of magnitude larger.

It should be remembered that these results are for the TIGCM and not necessarily the real thermosphere, for which global-scale data are limited. A much larger observational database is required to verify that the energy cycles of the thermosphere are properly represented in the model.

In a separate study [*Raskin*, 1992], the source, sink, and

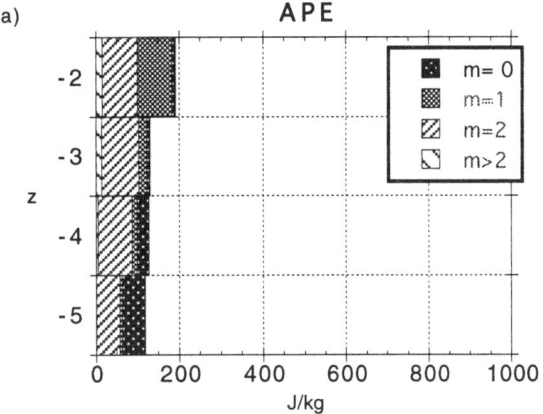

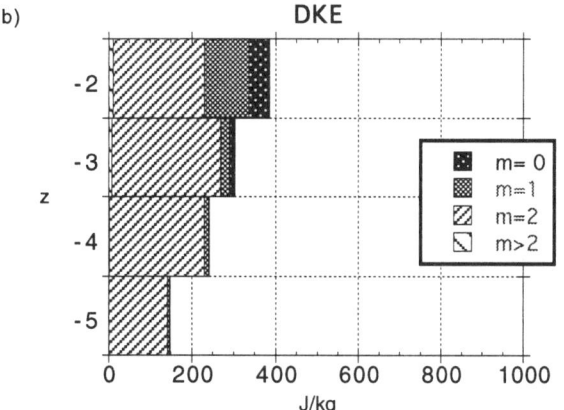

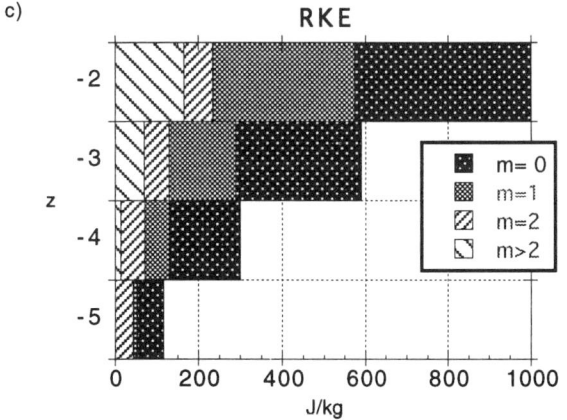

Figure 3. Same as Figure 2 but for energy intensity (energy per unit mass) spectrum.

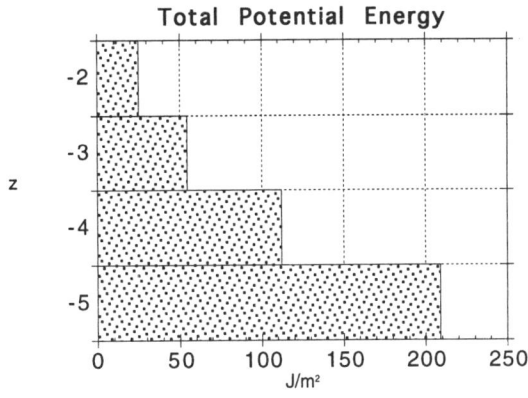

Figure 4. Total potential energy as a function of height.

exchange terms for the spectral energy reservoirs are computed, to complete the analysis of the energy cycle. The present work can also be extended to include analyses of the chemical and electric energy reservoirs. The magnitude of the chemical energy reservoir is much larger than the available potential and kinetic energy reservoirs considered here; however, only a small portion of the oxygen recombination energy is actually converted into available potential energy. Therefore, it is desirable to develop a relevant measure of available chemical energy. Electrical energy is effectively stored in the magnetosphere, rather than in the thermosphere, and thus a complete treatment of energetics must consider the coupled thermosphere-magnetosphere system.

Acknowledgments. This work was supported through Geophysical Laboratory Grant # F19628-89K-0026 and National Science Foundation Grant # ATM-8918476. Two anonymous reviewers provided many helpful comments to improve the scientific quality and clarity of this paper.

REFERENCES

Dickinson, R. E., E. C. Ridley, and R. G. Roble, Thermospheric general circulation with coupled dynamics and composition, *J. Atm. Sci.*, 41, 205-219, 1984.

Dopplick, T. G., The energetics of the lower stratosphere including radiative effects, *Quart. J.. Roy. Met. Soc.*, 97, 209, 1971.

Fesen, C. S. and R. G. Roble, Simulations of the September 1987 lower thermospheric tides with the National Center for Atmospheric Research Thermospheric-Ionospheric General Circulation Model, *J. Geophys. Res.*, 96, 1173-1180, 1991.

Fesen, C. S., R. E. Dickinson, and R. G. Roble, Simulation of thermospheric tides at equinox with the NCAR thermospheric general circulation model, *J. Geophys. Res.*, 91, 4471- 4489, 1986.

Hobson A., *The theory of spheroidal and ellipsoidal harmonics*, Chelsea Publ., New York, 500 pp., 1955.

Lorenz, E. N. *The nature and theory of the general circulation of the atmosphere*, World Meteorological Organization, #218, Geneva, Switzerland, 1967.

Raskin, R. G., Spectral energetics of the lower thermosphere, Ph.D. Dissertation, Department of Atmospheric, Oceanic, and Space Sciences, University of Michigan, Ann Arbor, 1992.

Richmond, A. D., Energy relations of atmospheric tides and their significance to approximate methods of solution for tides with dissipative forces, *J. Atmos. Sci.*, 32, 980-987, 1975.

Roble, R. G., The polar lower thermosphere, *Planet. Space Sci.*, 40, 271-297, 1992.

Roble, R. G., E. C. Ridley, A. D. Richmond, and R. E. Dickinson, On the global mean structure of the thermosphere, *J. Geophys. Res*, 92, 8745-8758, 1987.

Roble, R. G., E. C. Ridley, A. D. Richmond, and R. E. Dickinson, A coupled thermosphere-ionosphere general circulation model, *Geophys., Res. Let.*, 15, 1325-1328, 1988.

Swarztrauber P. N., The vector harmonic transform method for solving partial differential equations in spherical geometry, *Mon. Weath. Rev.*, 121, 3415-3437, 1993.

Thayer, J., Neutral wind vortices in the high-latitude thermosphere, Ph.D. Dissertation, Department of Atmospheric, Oceanic, and Space Sciences, University of Michigan, Ann Arbor, 1990.

Wiin-Nielsen A. and T-C. Chen, *Fundamentals of atmospheric energetics*, Oxford University Press, New York, 352 pp., 1993.

R. G. Raskin, Institute of Marine Sciences, University of California, Santa Cruz, CA 95064.

SHARC, a Model for Calculating Atmospheric Infrared Radiation Under Non-Equilibrium Conditions

R. L. Sundberg, J. W. Duff, J. H. Gruninger, L. S. Bernstein,
M. W. Matthew, S. M. Adler-Golden, and D. C. Robertson

Spectral Sciences, Inc., Burlington, Massachusetts

R. D. Sharma and J. H. Brown

Phillips Laboratory, Geophysics Directorate/OS, Hanscom AFB, Massachusetts

R. J. Healey

Yap Analytics, Inc., Lexington, Massachusetts

A new computer model, SHARC, has been developed by the U. S. Air Force for calculating high-altitude atmospheric IR radiance and transmittance spectra with a resolution of better than 1 cm^{-1}. Comprehensive coverage of the 2 to 40 μm (250 to 5,000 cm^{-1}) wavelength region is provided for arbitrary lines of sight in the 50-300 km altitude regime. SHARC accounts for the deviation from local thermodynamic equilibrium (LTE) in state populations by explicitly modeling the detailed production, loss, and energy transfer processes among the contributing molecular vibrational states. The calculated vibrational populations are found to be similar to those obtained from other non-LTE codes. The radiation transport algorithm is based on a single-line equivalent width approximation along with a statistical correction for line overlap. This approach calculates LOS radiance values which are accurate to ±10% and is roughly two orders of magnitude faster than the traditional LBL methods which explicitly integrate over individual line shapes. In addition to quiescent atmospheric processes, this model calculates the auroral production and excitation of CO_2, NO, and NO^+ in localized regions of the atmosphere. Illustrative comparisons of SHARC predictions to other models and to data from the CIRRIS, SPIRE and FWI field experiments are presented.

1. INTRODUCTION

The calculation of infrared (IR) radiance and transmittance spectra is important in many areas of atmospheric science, including modeling the energy budget, analyzing data from remote sounding experiments, and understanding molecular excitation and production processes. The US Air Force has developed a number of computer codes that can address these applications, including LOWTRAN [Kneizys et al., 1988], MODTRAN2 [Anderson et al., 1993], and FASCODE [Clough et al., 1986]. In the upper altitude regime, more sophisticated models are required to describe deviations from local thermodynamic equilibrium (LTE) as radiative and collisional processes become comparable. An initial non-LTE (NLTE) computer model, HAIRM [Degges and D'Agati, 1985], resulted from analyses of the 1977

The Upper Mesosphere and Lower Thermosphere:
A Review of Experiment and Theory
Geophysical Monograph 87
Copyright 1995 by the American Geophysical Union

SPIRE rocket measurements [Stair et al., 1985; Sharma and Healey, 1991]. In addition, high resolution radiative transport codes, NLTE [Sharma et al., 1983; Wintersteiner et al., 1992] for the quiescent atmosphere and AARC [Winick et al., 1987] for auroral conditions, were developed. However, a rapid and unified code spanning a wide range of altitudes and conditions has been lacking.

This paper describes a new NLTE code, SHARC, [Sharma et al., 1991] which calculates upper atmospheric IR radiation and transmittance. Its combination of speed and spectral resolution (0.5 cm^{-1}) should make it useful for many applications. The calculational model includes seven IR-active species in the 2-40 μm wavelength region for arbitrary line-of-sight (LOS) paths between 50 and 300 km. It incorporates the bands of NO, CO_2, O_3, H_2O, OH, CO, and CH_4 found in the quiescent atmosphere, including isotopic bands of CO_2 and H_2O. It also accounts for auroral production and excitation of CO_2, NO, and NO^+ caused by the flux of energetic solar electrons.

In accounting for NLTE effects, molecular vibrational state populations are calculated by explicitly solving the chemical equations for the excitation and relaxation of each vibrational state. As in other NLTE atmospheric models (e.g., [Wintersteiner et al., 1992; Lopez-Puertas et al., 1986a; Lopez-Puertas et al., 1986b]), steady-state kinetics are assumed for the quiescent atmospheric processes of collisional excitation, de-excitation, energy transfer, radiative decay, chemical production, and illumination by the sun, earth, and atmosphere. The additional production and excitation mechanisms resulting from electron deposition during auroral storms are included in SHARC using a time-dependent kinetic model.

A profile of vibrational state populations is input to the LOS spectral radiance module, which performs line-by-line (LBL) radiation transport calculations using an equivalent width formalism. This approach uses an algebraic approximation for the total absorption by a single isolated line, resulting in a considerable enhancement in speed compared to the standard LBL grid method [Clough et al., 1986], by which we mean explicit integration over individual line shapes. Corrections are made for line overlap. As will be discussed below, typical radiance differences with this approach are less than $\pm 10\%$.

SHARC is available for use by the scientific community and may be obtained from the Air Force Phillips Laboratory/GPOS, 29 Randolph Rd., Hanscom AFB, MA 01731.

2. CODE DESCRIPTION

The diagram in Figure 1 illustrates SHARC's modular structure and overall calculational sequence. The input module is a menu-driven user interface. Atmospheric temperature and species density profiles are specified via an external file. Profiles are required for the IR-active species (NO, CO_2, H_2O, O_3, CO, OH, and CH_4), the major atmospheric species to which they are collisionally coupled (N_2, O_2, O), and atomic hydrogen, whose reaction with O_3 provides the main source of OH(v). Other input parameters include LOS specifications and, if desired, the coordinates of a localized auroral region through which the LOS may pass.

Altitude profiles of excited vibrational state populations are calculated in the chemical kinetics and radiative transfer modules and saved in a data file for later use. Under auroral conditions, a time-dependent chemical model then calculates the additional production of CO_2, NO, and NO^+ arising from interactions of auroral electrons. To generate the desired LOS spectrum, the vibrational state populations are fed to the spectral radiance module, which outputs radiance and transmittance spectra and in-band intensities.

Condensed descriptions of the calculations are given below. Additional information may be found in the SHARC Users Manual [Sharma et al., 1991]. Code upgrades in progress or planned include: a generalized generator for atmospheric profiles of IR-active species based on scaling of MSIS [Hedin, 1991] profiles; extending the spectral region into the near IR and visible regions; and provisions for other environments which can be described as multiple, distinct atmospheric regions, for use in modeling the solar terminator, tidal and gravity waves, or other atmospheric inhomogeneities.

2.1. Quiescent Chemical Kinetics

The quiescent chemical kinetics are handled primarily through separate reaction sets for each IR-active molecule; these sets are solved individually in each layer of the atmosphere to obtain the excited vibrational state population profiles. Atmospheric layers are assumed to be horizontally uniform and typically 2 to 10 km in height. The reaction database files contain the list of vibrational states for each molecule and the equations for the chemical reactions and rate constants written in symbolic form. These equations include chemical formation of excited vibrational states, collisional deactivation and excitation (satisfying detailed balance), spontaneous emission, and the radiative excitation processes associated with absorption of radiation from the sun and from the atmosphere. The size of the database is indicated in Table 1.

The SHARC CHEMKIN module, which is based on the Sandia CHEMKIN general-purpose chemical kinetics code

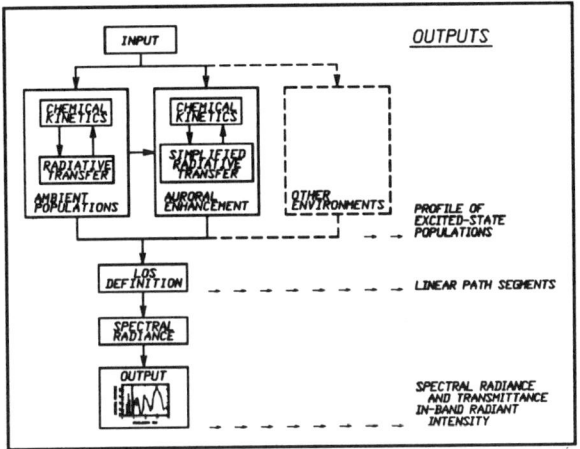

Fig. 1. SHARC module structure and calculational sequence.

[Kee et al., 1980], reads the reaction files and sets up the time-dependent differential rate equations. The quiescent vibrational state number density [M*] in each atmospheric layer is obtained from the solution to the steady-state equations. The species $CO_2(\nu_3)$, $H_2O(\nu_2)$, and possibly OH(v) are indirectly coupled to each other via resonant energy transfer processes involving $N_2(v=1)$ [Kumer and James, 1974; Kumer, 1977a; Kumer, 1978; Lopez-Puertas et al., 1986a,b]. The steady-state equations for these species are linearized by equating the ground vibrational state number densities to the total number densities and then solving for the $N_2(v=1)$ population.

The collisional rate constants were obtained from the recent literature and from Taylor's review [1974] of measurements prior to 1974; details are given elsewhere [Robertson et al., 1991]. The solar excitation rates are derived using the LOS radiation transport model discussed below. For solar zenith angles greater than 90°, MODTRAN calculations [Anderson et al., 1993] are used to account for the attenuation of sunlight by the atmosphere below 50 km. The earthshine excitation rates are expressed in terms of an effective blackbody temperature corresponding to the altitude where the vibrational band becomes optically thick in a nadir view.

The calculation of the radiative excitation rate, r_a, of molecular vibrational states due to emission originating from atmospheric layers in the 50-300 km altitude range is performed in a subroutine dubbed NEMESIS and follows the treatment of Kumer and co-workers [Kumer and James, 1974; Kumer, 1977a]. For a given vibrational band, the population enhancements of the vibrational states are expressed via a set of linear equations,

$$[M^*_i] - [M^*_i]_o = \alpha_i \omega_i \sum_j P_{ji} \omega_j [M^*_j] \quad , \quad (1)$$

where α_i is the probability for absorption of a photon entering layer i, P_{ji} is the probability that a photon emitted from layer j will be absorbed in layer i, and ω_i is the branching ratio for re-emission. Both the P matrix elements and the α_i depend on the populations of the lower vibrational states. Therefore, the calculation proceeds in stages to solve for successively higher-energy states. The first CHEMKIN run (with all $r_a = 0$) defines the ground state populations $[M_i]$ and the initial excited state populations $[M^*_i]_o$ for transitions to the ground state. The results are used in NEMESIS to evaluate the P matrix and α_i and to compute the radiative excitation rates r_a for those transitions. CHEMKIN is then rerun including the r_a to generate the corresponding populations for the next set of vibrational transitions. The CHEMKIN/NEMESIS sequence is repeated until solutions for the highest energy states are obtained.

The major calculational effort in NEMESIS is computing the P matrix elements, which involves a multidimensional integral over the location, direction, vibration-rotation line, frequency location within the line, and propagation distance of the emitted photons within each layer. The integral is evaluated with the aid of Monte Carlo sampling using trial "photons". The calculation assumes semi-infinite plane-parallel geometry and uses the Voigt line shape.

The CHEMKIN/NEMESIS calculations were verified by comparing vibrational temperatures for NLTE vibrational states of CO_2, which are both solar and earthshine pumped, with calculations by other workers. These profiles of vibrational temperatures are consistent with the results of Wintersteiner et al. [1992] and Lopez-Puertas et al. [1992], who used different computer algorithms but similar chemical kinetic models and atmospheric profiles. Typical results from SHARC are shown in Figure 2. The vibrational states are labeled using the HITRAN [Rothman et al., 1992] convention. The fine structures in the profiles are numerical noise from the Monte Carlo calculations.

TABLE 1. Quiescent Radiating Species in SHARC

Molecule	Isotopes	States	Reactions	Bands
CO_2	3	84	664	113
H_2O	4	32	128	36
O_3	1	30	176	45
NO	1	3	5	3
CO	1	3	9	3
OH	1	10	32	24
CH_4	1	12	52	13

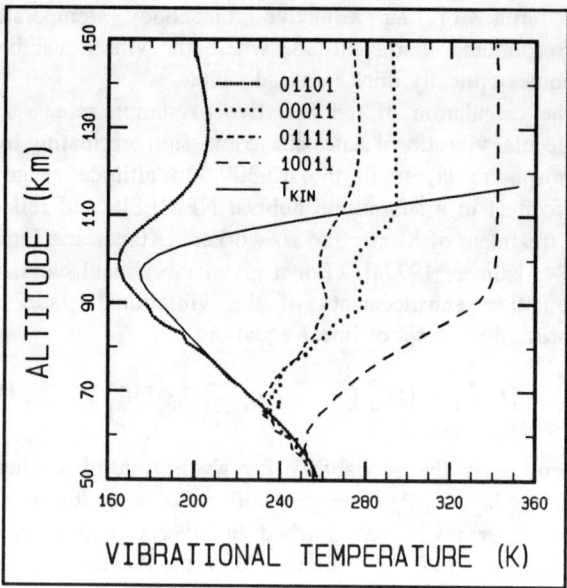

Fig. 2. Calculated vibrational temperatures for CO_2 states for a solar zenith angle of 82°.

2.2. Auroral Kinetics and Radiation Transport

The auroral kinetics model calculates the enhancements of CO_2, NO, and NO^+ radiation resulting from energy deposited in the upper atmosphere by auroral electrons. Approximately 600 time- and energy-dependent rate equations are used to calculate the secondary electron distribution (in fourteen energy bins) and the subsequent reactive and energy-transfer processes. The energy deposition model for the primary electrons is based on work by Grün, Rees, and Strickland [Grün, 1957; Rees, 1964; Strickland et al., 1983] as implemented in the Air Force Phillips Laboratory code AARC [Winick et al., 1987]. The chemical reactions and energy transfer processes used in SHARC are described in Robertson et al. [1991] and Kumer [1977b].

The integration of the time-dependent differential equations is accomplished using the Gear algorithm supplied with the Sandia CHEMKIN code [Kee et al., 1980]. The use of time-dependent kinetics results in an improvement over the steady-state treatment used, for example, in AARC [Winick et al., 1987]. In particular, SHARC predicts an increase in the NO(v=1) production efficiency per ion pair with time during the aurora. This results from collisional excitation of the increasing concentration of ground state NO during the aurora. In AARC the efficiency remains constant.

The emissions from NO and NO^+ are both optically thin and prompt. (Prompt emission is characterized by excited states that radiatively decay before they are collisionally quenched.) Therefore, coupling of the radiative transfer and collisional processes is not required, and the solutions to the excitation mechanisms for each layer suffice. However, CO_2 (ν_3) emission below 100 km is both optically thick and delayed, so that coupling the CO_2 time-dependent chemical kinetics to a radiative transfer scheme is required. With the present kinetic scheme, a full treatment would require either solving approximately 4000 coupled differential equations or repeating the radiative excitation calculations (CHEMKIN/NEMESIS) at each time increment. Since these are impractical, a modified escape function approximation [Kumer, 1977b], which assumes that photons emitted from a given layer either escape the atmosphere or are re-absorbed within the layer, has been adopted. To insure the correct limit for negligible auroral excitation, SHARC includes a correction which forces a match to the exact NEMESIS solution for ambient conditions. This correction consists of a time-independent source or sink term added to the differential equation for excitation of CO_2 (ν_3).

2.3. Line-of-Sight Spectral Radiance Model

In order to determine the LOS spectral radiance, the LOS properties are represented in terms of a sequence of homogeneous segments. The spectral radiance contribution, $I_{ij}(\nu)$, of a single emission line, j, to a single segment, i, is modeled by

$$I_{ij}(\nu) = \frac{1}{\Delta \nu} R_{ij} [\xi_i W_{ij} - \xi_{i-1} W_{(i-1)j}] \quad , \quad (2)$$

where ν is the central frequency of the calculational spectral interval of width $\Delta \nu$, R_{ij} is the emission source function, W_{ij} is the cumulative equivalent width for the LOS path from the observer through the i'th segment, and ξ is a correction factor which accounts for line overlap. The total path spectral radiance is determined by direct summation of I_{ij} over all segments and lines.

The emission source term is derived directly from the kinetically determined excited-state vibration-rotation populations and is given by

$$R_{ij} = \frac{C_1 \nu^3}{\pi} \frac{\gamma_{ij}}{1-\gamma_{ij}} \quad , \quad (3)$$

where C_1 is the first radiation constant and γ_{ij} is the ratio of the upper-to-lower state vibration-rotation populations. This expression for R_{ij} reduces to the Planck blackbody function in the limit in which all the vibration-rotation state populations are specified by the local gas temperature, i.e., LTE conditions.

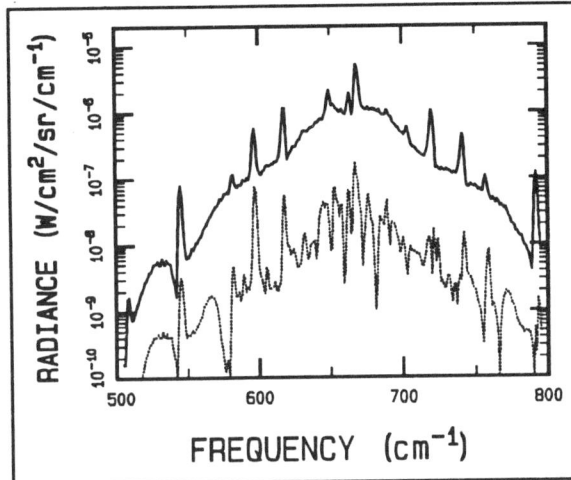

Fig. 3a. SHARC (solid) and SHARC minus LBL difference (dotted) spectra for CO_2 radiance for a 50 km limb view.

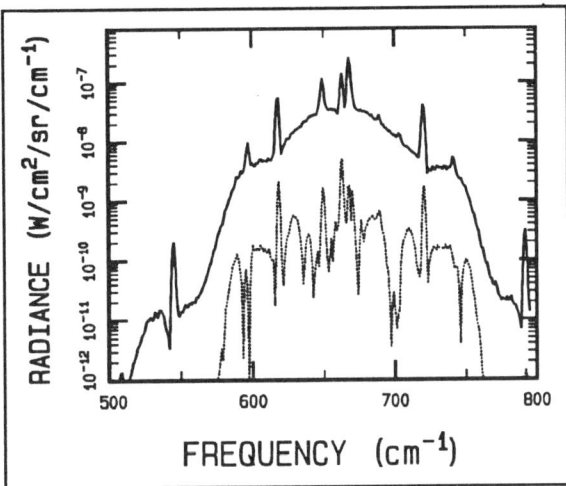

Fig. 3b. SHARC (solid) and SHARC minus LBL difference (dotted) spectra for CO_2 radiance for a 75 km limb view.

In the limit of no line overlap ($\xi=1$) Equation (2) is exact provided that the equivalent widths are exactly determined, such as by explicit frequency integration over the line shape function. However, this is computationally time consuming. By using the approximations described below, a rapid evaluation can be achieved with only a modest sacrifice of accuracy (about $\pm 10\%$).

The equivalent widths W_{ij} for single lines are calculated using the Rodgers-Williams [1974] approximations for Voigt line shapes,

$$W_{ij}^2 = W_v^2 = \alpha_D \frac{2}{\ln 2} \left[W_L^2 + W_D^2 - \left(\frac{W_L W_D}{W_W}\right)^2 \right] \quad (4)$$

where the subscripts V, D, L, and W refer to the Voigt, Doppler, Lorentz, and weak-line limits, and $\alpha_D(cm^{-1})$ is the Doppler linewidth. W_D, W_L, and W_W are calculated from the approximations given by Ludwig et al.. [1973], which are accurate to within $\pm 8\%$ or better, using Curtis-Godson [Curtis, 1952; Godson, 1953] path-averaged line parameters. The line strengths and air-broadened half-widths are tabulated in a file generated from the HITRAN atlas [Rothman et al., 1992] that has been supplemented with lines for NO^+ and higher vibrational states of NO and O_3.

The line overlap correction is based on the statistical overlap approximation in which it is assumed that there is no positional correlation among the lines in the calculational spectral interval. This approximation accounts for line wing overlap within the spectral interval but does not include the contribution of line wings from lines centered outside the interval. For ℓ lines in the interval $\Delta\nu$, the correction factor is given by

$$\xi_i = \Delta\nu \left[1 - \prod_j^\ell \left(1 - \frac{W_{ij}}{\Delta\nu}\right) \right] / \sum_j W_{ij}. \quad (5)$$

In applying this spectral radiance model, it is implicitly assumed that the calculational spectral interval width, $\Delta\nu$, is larger than the largest single-line equivalent width for the entire LOS. This limits the calculational spectral resolution to about 0.5 cm^{-1} for tangent paths near the lower atmosphere boundary of 50 km. However, higher spectral resolution can be used for non-limb paths or limb paths with higher-altitude tangent points.

The accuracy of this LOS radiance algorithm has been explored through comparisons to standard high-resolution LBL grid calculations for typical atmospheric conditions. A number of different regimes were investigated, including vertical and horizontal viewing geometries, optically thin and thick lines, multiple overlapping lines and bands, and LTE and NLTE conditions. Figures 3a and 3b show SHARC limb radiance calculations for just CO_2 in the ν_2 spectral region at tangent altitudes of 50 and 75 km, respectively. Each figure also includes the absolute value of the difference spectrum which was obtained by subtracting a LBL grid method calculation from the SHARC calculations. A major source of error in the SHARC calculations is the statistical line overlap approximation. This approximation under-corrects for line overlap when the emission lines in a given spectral interval are nearly degenerate. This can be seen in the 50 km limb calculation shown in Figure 3a for the Q branch near 597 cm^{-1} and leads to a 20% over prediction for the radiance over a 2 cm^{-1} spectral interval. The line overlap approximation

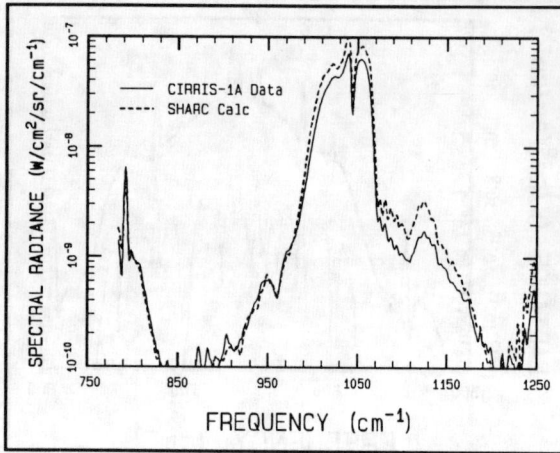

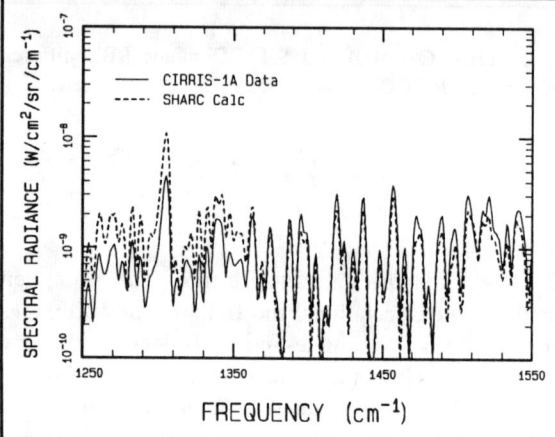

Fig. 4. Quiescent nighttime limb spectrum from the CIRRIS-1A experiment [Ahmadjian et al., 1990] near 64 km tangent height and SHARC simulation.

over-corrects for line overlap when the emission lines in a spectral interval are nearly equally spaced. This occurs in the calculation shown in Figure 3a in the Q branch near 667 cm^{-1} and leads to an error of ±10%. For LOS's and bands which have less line overlap, such as the 75 km limb calculation in Figure 3b, peak errors remain below ±10%. The SHARC calculation is roughly two orders of magnitude faster than a LBL calculation using FASCODE.

3. DATA COMPARISONS

While comparisons with other computer codes can verify the numerical algorithms, upper atmospheric IR data are needed to assess the reasonableness of the various model assumptions, including kinetic mechanisms, rate constants, and properties of the atmosphere. To illustrate typical levels of agreement between field data and model calculations in

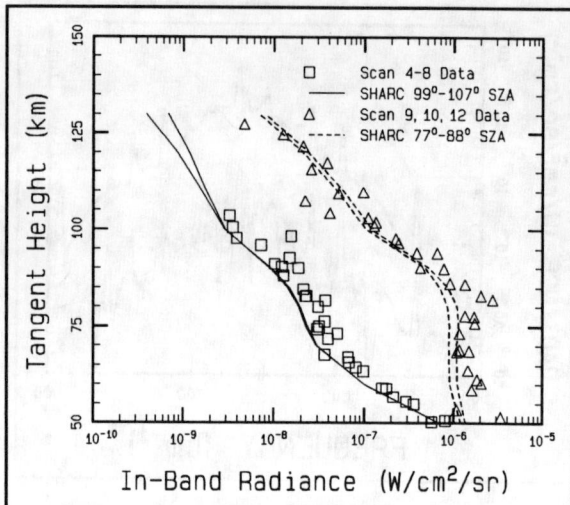

Fig. 5. CO_2 4.3 μm band limb radiance measured in the SPIRE rocket experiment [Stair et al., 1985] and predicted by SHARC for different solar zenith angles near the dawn terminator.

diverse applications, several comparisons of SHARC predictions and data from US Air Force-sponsored field experiments are presented.

The CIRRIS-1A experiment [Ahmadjian et al., 1990] conducted on the Space Shuttle STS-39 during April, 1991, collected the first extensive set of limb IR radiance spectra over a wide altitude range using a sensitive cryogenic Michelson interferometer with a spectral resolution of 1 cm^{-1}. A typical quiescent nighttime spectrum in the 780-1550 cm^{-1} spectral region is shown in Figures 4a and 4b; the tangent height at the center of the detector FOV is about 64 km. For clarity in plotting, the resolution of the original spectrum has been degraded to 3 cm^{-1} FWHM (full width, half maximum). The major features are CO_2 hot bands near 791 cm^{-1} and 961 cm^{-1}, the ν_3 (1042 cm^{-1}) and ν_1 (1103 cm^{-1}) bands of O_3, the ν_4 band of CH_4 near 1311 cm^{-1}, and resolved lines of the ν_2 band of H_2O, which dominate the spectrum beyond around 1370 cm^{-1}. The overlaid SHARC calculation, performed for similar viewing conditions, reproduces the observed spectral structure throughout this region. The differences in the absolute intensities of some bands are up to a factor of two and can be ascribed mainly to differences in species concentrations between the actual atmosphere and the model atmosphere, which is derived from the NRL Trace Gas Climatology database [Summers et al., 1992]. Differences of up to ±30% can also result from the instrument pointing uncertainty of up to ±2 km.

Data for upper atmospheric CO_2 (ν_3) emission and its dependence on solar angle is provided in Figure 5, which shows the limb radiance measured in an earlier rocket experiment, SPIRE [Stair et al., 1985], launched from Poker Flat Research Range, Alaska near the dawn terminator. Different phenomena dominate the excitation depending on the altitude regimes. During the day, the primary isotope is solar-excited above 110 km, while at lower altitudes the radiance is enhanced by pumping of the 2.7 μm band and by emissions from hot bands and minor isotopes [Sharma and Wintersteiner, 1985; Kumer, 1983]. As the density increases, the widening line shape makes the single-quantum excitation of the primary isotope again important with contributions from multiple quantum bands at 1.6, 2.0, and 2.7 μm [Sharma and Wintersteiner, 1985]. The SPIRE data and the SHARC calculations agree to within the data scatter, which averages around a factor of two.

An example of SHARC's capability to calculate auroral radiance enhancements is shown in Figure 6. A nitric oxide spectrum from the Field-Widened Interferometer rocket experiment [Espy et al., 1988], which observed an IBC Class II aurora, is compared with a SHARC model calculation at a resolution of 1 cm^{-1}. The auroral calculation uses energy deposition parameters described by Picard et al. [1987]. The reaction of metastable nitrogen atom, N(^{2}D), with molecular oxygen has been assumed to be the only source of NO chemiluminescence in the SHARC auroral model. The overall band shape, which includes strong aurorally induced hot band contributions, is reproduced well by the calculation, except near the edges of the band where NLTE rotational populations, not currently modeled, yield enhanced high rotational lines. The calculated absolute radiance is a factor of four lower than the data. Solomon [1983] and Gèrard et al. [1991] have suggested that the reaction of translationally hot N(^{4}S) with O$_2$ may be an important contributor to NO formed in the thermosphere. Sharma et al. [1993] have presented calculations that show that the reaction of N(^{4}S) atoms with O$_2$ accounts for recent observations of highly rotationally excited NO vibrational emissions in the dayglow [Armstrong et al., 1993; Smith and Ahmadjian, 1993]. Finally, the reactions of N(^{4}S) and N(^{2}D) with O$_2$ may lead to a quantitative explanation of NO formed in recent artificial auroral experiments [Lipson et al., 1993]. Given the uncertainties associated with characterization of the aurora and as well as the role of N(^{4}S) in the formation of NO, the agreement between the model and data is encouraging.

Numerous other data and model comparisons have been and are currently being performed for all the major atmospheric IR emission bands during daytime, nighttime, and auroral conditions. The results will be reported in future papers.

4. CONCLUSION

A new computer model, SHARC, has been developed by the Air Force for rapid LBL calculation of NLTE upper atmospheric IR radiance and transmittance spectra with a resolution of 1.0 cm^{-1} or better. SHARC treats the important molecular vibrational bands from 2 to 40 μm (250 to 5,000 cm^{-1}) for arbitrary lines of sight in the 50-300 km altitude range, accounting for the detailed production, loss, and energy transfer processes among the important vibrational states. Calculated vibrational temperatures agree with results from other NLTE codes [Wintersteiner et al., 1992; Winick et al., 1987; Lopez-Puertas et al., 1986b], and the equivalent-width spectral algorithm used in LOS radiation transport calculations results in a considerable time savings over grid LBL methods that explicitly integrate over the full line shape.

Comparisons of SHARC radiance predictions with field measurements, especially those from CIRRIS 1A, are ongoing. Comparisons performed to date indicate agreement to within typically a factor of two or better for most emission bands, including the CO$_2$ 4.3 μm feature, which poses a good test for different aspects of the code. Planned upgrades include a solar terminator module, extension of the spectral region down to the near IR, and a spectral-spatial clutter option [Sundberg et al., 1994]. Data simulations using SHARC have applications to remote sensing of the upper atmosphere, such as deriving density profiles for species such as NO, O$_3$, and H$_2$O, which have significant atmospheric variability.

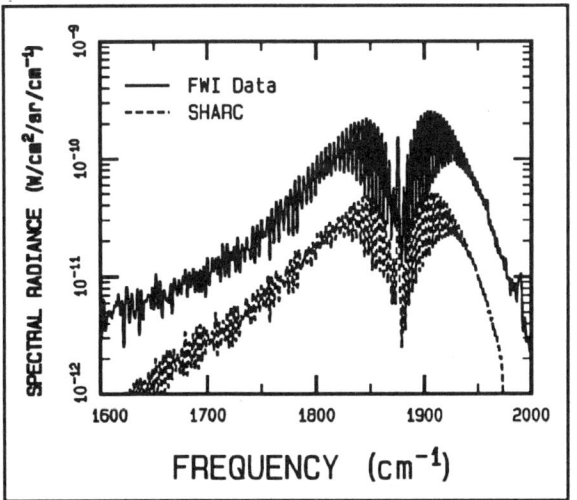

Fig. 6. Calculated and observed [Espy et al., 1988] NO spectrum for zenith viewing through a class II aurora from 90 km altitude.

Acknowledgements. The authors wish to thank Dr. A.J. Ratkowski and Dr. W.A.M. Blumberg (Phillips Laboratory/GPOS) for their support during the development of SHARC and Mr. R.M. Nadile of the Phillips Laboratory/GPOB for providing the emission spectrum from the CIRRIS 1A measurements. The authors also wish to acknowledge Dr. V.I. Lang of the Aerospace Corp. for reviewing the chemical kinetic database. This work was funded by the Ballistic Missile Defense Organization (formerly SDIO) under PMA 1105.

REFERENCES

Ahmadjian, M., R. M. Nadile, J. O. Wise, and B. Bartschi, "CIRRIS-1A Space Shuttle Experiment," *J. Spacecraft*, 27, 669 (1990).

Anderson, G. P., J. H. Chetwynd, J.-M. Theriault, P. Acharya, A. Berk, D. C. Robertson, F. X. Kneizys, M. L. Hoke, L. W. Abreu, and E. P. Shettle, "MODTRAN2: Suitability for Remote Sensing," *SPIE Proc.*, 1968, SPIE, Box 10, Bellingham, WA 98227 (1993).

Armstrong, P. S., S. J. Lipson, J. R. Lowell, W. A. M. Blumberg, D. R. Smith, R. M. Nadile, and J. A. Dodd, "Analysis of comprehensive CIRRIS 1A observations of nitric oxide in the thermosphere," *Eos Trans AGU*, 74, 225 (1993).

Clough, S. A., F. X. Kneizys, E. P. Shettle, and G. P. Anderson, "Atmospheric Radiance and Transmittance: FASCOD2", *Proc. of the Sixth Conference on Atmospheric Radiation*, pp. 141-144, Am. Met. Soc., Boston, MA (1986).

Curtis, A. R., "A Statistical Model for Water Vapour Absorption," *Q. J. R. Meteorol. Soc.*, 78, 638-640 (1952).

Degges, T. C. and A. P. D'Agati, "A User's Guide to the AFGL/Visidyne High Altitude Infrared Radiance Model," AFGL-TR-85-0015 (NTIS No. ADA 161432) (1985).

Espy, P. J., C. R. Harris, A.J. Steed, J. C. Ulwick, R. H. Haycock, and R. A. Straka, "Rocketborne Interferometer Measurement of Infrared Auroral Spectra," *Planet. Space Sci.*, 36, 543 (1988).

Gérard, J.-C., V. I. Shematovich, D. V. Bisikalo, "Non thermal nitrogen atoms in the Earth's thermosphere 2. a source of nitric oxide," *Geophys. Res. Lett.*, 18, 1695-1698 (1991).

Godson, W. L., "The Evaluation of Infrared-Radiative Fluxes due to Atmospheric Water Vapour," *Q. J. R. Meteorol. Soc.*, 79, 367-379 (1953).

Grün, A. E., "Luminescenz-photometrische Messungen der Energieabsorption im Strahlungsfeld von Elektronenquellen Eindimensionaler Fall im Luft," *Z. Natürforsch.*, 112a, 89-95 (1957).

Hedin, A. E., "Extension of the MSIS Thermospheric Model into the Middle and Lower Atmosphere," *J. Geophys. Res.*, 96, 1159 (1991).

Kee, R. J., J. A. Miller, and T. H. Jefferson, "CHEMKIN: Problem-Independent, Transportable, Fortran Chemical Kinetics Code Package," Sandia Rpt. No. SAND80-8003, Sandia National Laboratory, Livermore, CA 94550 (1980).

Kneizys, F. X., G. P. Anderson, E. P. Shettle, W. O. Gallery, L. W. Abreu, J. E. A. Selby, J. H. Chetwynd, and S. A. Clough, "Users Guide to LOWTRAN 7," AFGL-TR-88-0177, (NTIS No. ADA 206773) (1988).

Kumer, J. B., "Atmospheric CO_2 and N_2 Vibrational Temperatures at 40- to 140-km Altitude," *J. Geophys. Res.*, 82, 16 (1977a).

Kumer, J. B., "Theory of the CO_2 4.3 μm Aurora and Related Phenomena," *J. Geophys. Res.*, 82, 2203 (1977b).

Kumer, J. B., and T. C. James, "CO_2 (001) and N_2 Vibrational Temperatures in the $50 \leq Z \leq 130$ km Altitude Range," *J. Geophys. Res.*, 79, 638 (1974).

Kumer, J. B., R. M. Nadile, and W. Grieder, "Detailed Analysis of 4.3 μm Earthlimb Data," *SPIE Proceedings*, 430, 244, SPIE, Box 10, Bellingham, WA 98227 (1983).

Kumer, J. B., A. T. Stair, Jr., N. Wheeler, K. D. Baker, and D. J. Baker, "Evidence for an $OH^{\dagger\nu} \rightarrow N_2^{\dagger\nu} \rightarrow CO_2(\nu_3) + h\nu(4.3 \mu m)$ Mechanism for 4.3-μm Airglow," *J. Geophys. Res.*, 83, 4743-4747 (1978).

Lipson, S. J., P. S. Armstrong, J. R. Lowell, W. A. M. Blumberg, D. E. Paulsen, M. J. Fraser, W. T. Rawlins, D. B. Green, R. E. Murphy, and J. A. Dodd, "Mission-wide EXCEDE III nitric oxide spectroscopic analysis," *Eos Trans AGU*, 74, 225 (1993).

Lopez-Puertas, M., M. A. Lopez-Valverde, C. P. Rinsland, and M. R. Gunson, "Analysis of the Upper Atmosphere $CO_2(\nu_2)$ Vibrational Temperatures Retrieved From ATMOS/Spacelab 3 Observations," *J. Geophys. Res.*, 97, 20469 (1992).

Lopez-Puertas, M., R. Rodrigo, A. Molina, and F. W. Taylor, "A Non-LTE Radiative Transfer Model for Infrared Bands in the Middle Atmosphere, I. Theoretical Basis and Application to CO_2 15 μm Bands," *J. Atmos. Terr. Phys.*, 48, 729 (1986a).

Lopez-Puertas, M., R. Rodrigo, J. J. Lopez-Moreno, and F. W. Taylor, "A Non-LTE Radiative Transfer Model for Infrared Bands in the Middle Atmosphere, II, CO_2 (2.7 and 4.3 μm) and Water Vapor (6.3 μm) Bands and $N_2(1)$ and $O_2(1)$ Vibrational Levels," *J. Atmos. Terr. Phys.*, 48, 749 (1986b).

Ludwig, C. B., W. Malkmus, J. E. Reardon, and J. A. Thomson, *Handbook of Infrared Radiation From Combustion Gases*, SP-3080, Scientific and Technical Information Office, NASA, Washington DC (1973).

Picard, R. H., J. R. Winick, R. D. Sharma, A. S. Zachor, P. J. Espy, and C. R. Harris, "Interpretation of infrared measurements of the high-latitude thermosphere from a rocketborne interferometer," *Adv. Space Res.*, 7, 23-30 (1987).

Rees, M. H., "Auroral Ionization and Excitation by Incident Energetic Electrons," *Planet. Space Sci.*, 111, 1209-18 (1964).

Robertson, D. C., P. K. Acharya, S. M. Adler-Golden, L. S. Bernstein, F. Bien, J. W. Duff, J. H. Gruninger, R. L. Sundberg, R. J. Healey, J. M. Sindoni, P. M. Bakshi, A. Dalgarno, and B. Zygelman, "Investigations into Atmospheric Radiative Processes in the 50-300 km Regime," PL-TR-91-2137, (NTIS No. ADA 251588) (1991).

Rodgers, C. D., and A. P. Williams, "Integrated Absorption of a Spectral Line with the Voigt Profile," *J. Quant. Spectrosc. Radiat. Transfer*, 14, 319 (1974).

Rothman, L. S., R. R. Gamache, A. Goldman, L. R. Brown, R. A. Toth, H. M. Pickett, R. L. Poynter, J. M. Flaud, C. Camy-Peyret, A. Barbe, N. Husson, C. P. Rinsland, and M. A. H. Smith, "The HITRAN Molecular Database: Editions of 1991 and 1992," *J. Quant. Spectrosc. Radiat. Transfer, 48*, 469 (1992).

Sharma, R. D., J. W. Duff, R. L. Sundberg, L. S. Bernstein, J. H. Gruninger, D. C. Robertson, and R. J. Healey, "Description of SHARC-2, The Strategic High-Altitude Radiance Code, PL-TR-91-2071, (NTIS No., ADA 239008) (1991).

Sharma, R. D. and R. J. Healey, "Earthlimb Emission Analysis of Spectral Infrared Rocket Experiment (SPIRE) Data at 2.7 μm - A 10 Year Update," *SPIE Proc., 1540*, SPIE, Box 10, Bellingham, WA 98227 (1991).

Sharma, R. D., R. D. Siani, M. K. Bullitt, and P. P. Wintersteiner, "A Computer Code to Calculate Emission and Transmission of Infrared Radiation Through Non-Equilibrium Atmospheres," AFGL-TR-83-0168 (NTIS No. ADA 137162) (1983).

Sharma, R. D., Y. Sun, and A. Dalgarno, "Highly rotationally excited nitric oxide in the terrestrial thermosphere," *Geophys. Res. Lett., 20*, 2043-2045 (1993).

Sharma, R. D. and P. P. Wintersteiner, "CO_2 Component of Daytime Earth Limb Emission at 2.7 μm," *J. Geophys. Res., 90*, 9789 (1985).

Smith, D. R., and M. Ahmadjian, "Observation of Nitric Oxide Rovibrational Band Head Emissions in the Quiescent Airglow During the CIRRIS-1A Space Shuttle Experiment," *Geophys. Res. Lett., 20*, 2679 (1993).

Solomon, S., "The possible effects of translationally excited nitrogen atoms on lower thermospheric odd nitrogen," *Planet. Space Sci., 31*, 135-139 (1983).

Stair, A. T. Jr., R. D. Sharma, R. M. Nadile, D. J. Baker, and W. F. Grieder, "Observations of Limb Radiance with Cryogenic Spectral Infrared Rocket Experiment," *J. Geophys. Res., 90*, 9763 (1985).

Strickland D. J., J. R. Jasperse, and J. A. Whalen, "Dependence of Auroral FUV Emissions on the Incident Electron Spectrum and Neutral Atmosphere," *J. Geophys. Res., 88*, 8051-62 (1983).

Summers, M. E., "Zonally Averaged Trace Constituent Climatology," Report No. NRL-MR-7641-93-7416, Naval Research Laboratory, Washington, DC (1993).

Sundberg, R. L., J. Gruninger, P. De, and J. Brown, "Infrared Radiance Fluctuations in the Upper Atmosphere," *SPIE Proc., 2223*, SPIE, Box 10, Bellingham, WA 98227 (1994).

Taylor, R. L., "Energy Transfer Processes in the Stratosphere," *Can. J. Chem., 52*, 1436 (1974).

Winick, J. R., R. H. Picard, R. D. Sharma, R. A. Joseph, and P. P. Wintersteiner "Radiative Transfer Effects on Aurora Enhanced 4.3 Micron Emission," *Adv. Space Res., 7*, 17-21 (1987).

Wintersteiner, P. P., R. H. Picard, R. D. Sharma, J. R. Winick, and R.A. Joseph, "Line-by-Line Radiative Excitation Model for the Non-Equilibrium Atmosphere: Application to CO_2 15-μm Emission," *J. Geophys. Res., 97*, 18083 (1992).

R. L. Sundberg, J. W. Duff, J. H. Gruninger, L. S. Bernstein, M. W. Matthew, S. M. Adler-Golden, D. C. Robertson, Spectral Sciences, Inc., 99 South Bedford Street, Burlington, MA 01803-5169.

R. D. Sharma and J. H. Brown, Phillips Laboratory, Geophysics Directorate/OS, 29 Randolph Road, Hanscom AFB, MA 01731-3010.

R. J. Healey, Yap Analytics, Inc., 594 Marrett Road, Lexington, MA 02173.

WINDII on UARS — Status Report and Preliminary Results

G.G. Shepherd

Institute for Space and Terrestrial Science, York University, Toronto, Canada

G. Thuillier

Service D'Aéronomie du CNRS, Verrières-le-Buisson, France

B.H. Solheim, Y.J. Rochon, J. Criswick, W.A. Gault, R.N. Peterson, R.H. Wiens and S.-P. Zhang

Institute for Space and Terrestrial Science, York University, Toronto, Canada

WINDII, the WIND Imaging Interferometer, was launched on the Upper Atmosphere Research Satellite on September 12, 1991. This joint project, sponsored by the Canadian Space Agency and the French Centre National d'Etudes Spatiales, in collaboration with NASA, has the responsibility of measuring the global wind pattern at the top of the altitude range (80 - 110 km) covered by UARS. In fact, WINDII measures wind, temperature and emission rate over the altitude range 80 to 300 km from a selection of visible region airglow emissions. We present a brief review of the Doppler imaging concept, its implementation in WINDII and the current status of validation and data processing. Preliminary results on emission rates and winds from the atomic oxygen 557.7 nm emission are presented in the context of validation overpasses of an airglow ground station.

1. INTRODUCTION

WINDII, the WIND Imaging Interferometer, takes advantage of airglow emission lines photochemically produced from species in the upper mesosphere (OH), the lower thermosphere (O_2, $O(^1S)$) and the middle thermosphere ($O(^1S)$, $O(^1D)$ and $O^+(^2P)$) for the measurement of winds and temperatures in the altitude range 80 to 300 km. The instrument is based on the concept of a field-widened Michelson interferometer first described by Boucheraine and Connes (1963) and implemented in a configuration later conceived by Hilliard and Shepherd (1966). The development of an imaging instrument intended for flight on Spacelab is given by Shepherd et al.

(1985) and the WINDII instrument is based on an extension of that described by Thuillier and Shepherd (1985). A detailed description of WINDII has been presented by Shepherd et al. (1993a) so only a brief summary is given here.

For a single Gaussian line the output of a Michelson interferometer is a damped cosinusoid, where the cosinusoid frequency corresponds to the centre frequency of the line, and the Gaussian envelope is the Fourier Transform of the spectral line shape. To obtain such an "interferogram" by direct measurement would require collecting about one million data points, requiring a very complex instrument that would not be practical for our application. However, a very simple configuration is capable of producing the information required. If the path difference is set to some reasonably large value and then is scanned over only one fringe the Doppler velocity is manifested as a phase shift of this fringe. The modulation depth, which Michelson defined as fringe "visibility", namely the ratio of amplitude to mean value, is

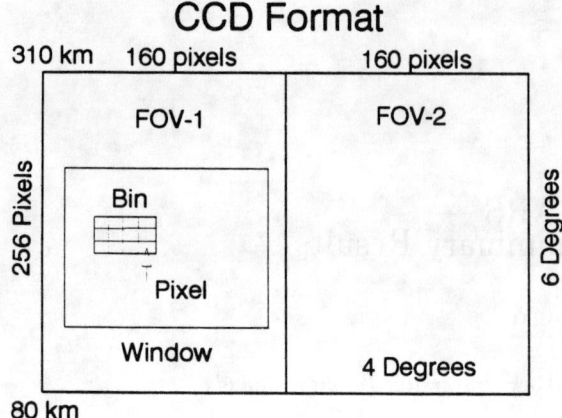

Figure 1. Format of the CCD array used in WINDII.

directly related to the temperature, which can thus be determined.

This measured velocity is the component along the line of sight. In order to obtain two components of the horizontal wind, WINDII employs two view directions, at 45° and 135° from the velocity vector on the anti-sun side of the spacecraft. Images taken simultaneously in both directions are recorded side-by-side on a single CCD in a format shown in Figure 1, and are referred to as Fov (Field of view) 1 and 2. Pairs taken roughly 7 min apart are combined to give orthogonal wind vectors from the same volume of atmosphere, on the assumption that the wind does not change in that time. WINDII thus views a track off to one side of the spacecraft, and since it is mounted on the anti-sun side it views southward when UARS is flying forward, and northward when WINDII is flying in the reverse "backward" orientation. The orientation is changed approximately once a month, which is the time it takes for the sun to move from one side of the spacecraft to the other. Thus WINDII covers a range from 42° latitude in one hemisphere, to 72° in the other hemisphere for any given month.

In this paper we review the status of WINDII performance and data analysis, using as a focus a single day from a more extended validation experiment conducted at Bear Lake, Utah. The combination of spacecraft and ground-based results provides a powerful approach to understanding the global atmospheric behavior, as well as providing validation of the first results from WINDII.

2. BRIEF INSTRUMENT DESCRIPTION

The interferometer, shown in Figure 2. is similar in design to the one made for the earlier WAMDII instrument (Shepherd, *et al.* 1985) but there are some important differences. The beamsplitter consists of two cemented half-hexagons with a low-polarizing, semi-reflecting dielectric multilayer on one of the diagonal faces. It is made of BK7 glass and the entrance and exit faces are 7.6 cm square. Both arms are solid glass (LF5 and LaFN21) except for a small gap at the end of the LF5 arm. The gap is necessary to allow the small mirror motion which is required to scan one or two fringes, and also provides an extra degree of freedom in the design of the Michelson.

The moving mirror is cemented to three piezoelectric pillars which in turn are cemented to the end of the arm. The position and tilt of the moving mirror are sensed by three small capacitors consisting of electrodes deposited on the end of the LF5 arm and on the ends of glass pillars cemented to the mirror. These capacitors form part of a bridge which provides an error signal to the circuit that controls the voltage on the piezoelectrics. The moving mirror assembly and control circuits were provided by Queensgate Instruments, U.K. The beamsplitter and glass arms were manufactured by Interoptics, Ottawa.

The Michelson interferometer is designed to have a large field of view and to be thermally compensated with respect to phase throughout the required spectral region of 552 to 763 nm. The thermal compensation is accomplished by balancing the thermal coefficient of the scanning mirror mounting against that of the rest of the interferometer, in the way described by Thuillier and Shepherd (1985), and implemented for the first time in WINDII. The other change from WAMDII was to incorporate a wedge in the air gap, balanced by the glass in the other arm, reducing the contrast of the secondary

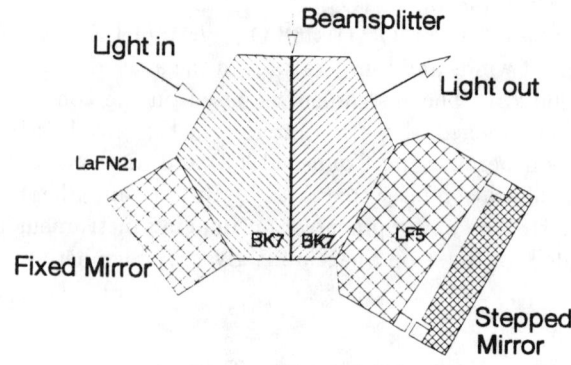

Figure 2. The configuration of the Michelson interferometer used in WINDII.

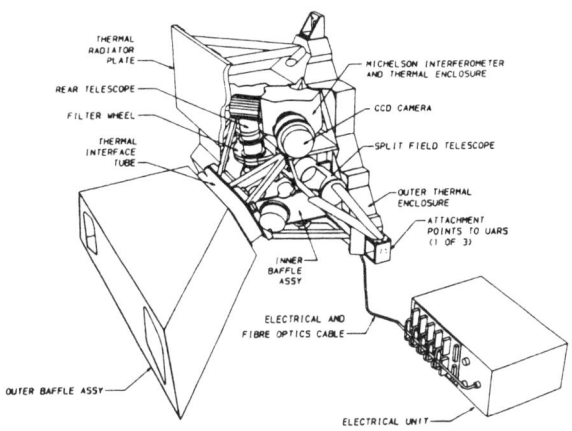

Figure 3. Drawing of the assembled WINDII instrument.

fringes caused by reflections at the air/glass surface to allow much improved Doppler temperature results.

The principal elements of WINDII's optical system are a baffle, a telescope with two objectives and a field combiner, a filter wheel, another telescope, the Michelson interferometer and the CCD camera. The two orthogonal fields of view, each $4° \times 6°$ in object space, are imaged side-by-side at the detector. Inside the baffle, two mirrors are deployed during calibration to block the incoming light and allow the instrument to view the calibration sources. The optical train has been folded into a compact shape by means of several plane mirrors. The configuration of the assembled instrument is shown in Figure 3.

The main function of the baffle is to limit the amount of scattered light reaching the optics when the earth is sunlit so as to permit daytime measurements of the airglow. To achieve maximum baffle length within the allowed instrument envelope, the two fields of view are criss-crossed inside the body of the baffle. The last vane in each section is located at the aperture of the first telescope and behind the apertures are the fold mirrors that direct the light from the two fields of view towards the telescope's objective lenses. During daytime, the entrance apertures are stopped down to a narrow slot shape parallel to the horizon and to the bottom edge of the first baffle vane. The first vane is so placed that, for the average spacecraft altitude, the scattered sunlight from clouds at the top of the troposphere is prevented from entering the aperture. The loss of aperture area during daytime is compensated for by the increased brightness of the emissions.

The WINDII detector consists of a CCD camera with an $f/1$ lens and a thinned, back-illuminated CCD, RCA part number SID501 EX. The pixel elements are about 30 μm square and the device consists of 512x320 pixels. In the WINDII application, one half of the array is used as storage and is masked off to avoid illumination. The image area then consists of 256 rows and 320 columns. Because of the two fields-of-view in WINDII, the image area is further divided in two so that 256 rows and 160 columns are used for each Fov. Thus the two Fovs are imaged side-by-side on the CCD image area, as was illustrated in Figure 1.

3. SCOPE OF WINDII MEASUREMENTS

WINDII observes five different emissions in order to cover the full desired range of wind measurement. Seven filters are used, including one which measures white light background for the 557.7 nm emission. Another measures background for the O^+ emission in the upper part of the image, but is also used to observe OH emission in the lower part of the window. Two additional hydroxyl filters are used, using different rotational lines. For the O_2 Atm band the different rotational lines are spread out across the image. Thus rotational temperatures may be measured for both molecular emissions. The scope of WINDII measurements is summarized in Table 1.

At the present time, the validation of results from $O(^1S)$ airglow is nearly complete. Although less attention has been paid to $O(^1D)$, its optical characteristics are very similar, so that much of the same validation ap-

Table 1: SCOPE OF WINDII MEASUREMENTS

Species	Wind Velocity	Number Density	Doppler Temp.	Rotational Temp.	Rayleigh Temp.	Other
$O(^1S)$	X	X	X			Aurora
$O(^1D)$	X	X	X			Aurora
$O_2\,(^1\Sigma)$	X	X		X		
$OH(^2\Pi)$	X	X	X	X		N_2 1PG
$O^+(^2P)$	X	X				
Background		X			X	PMC

plies to both. The OH emission is also similar, since we have employed interference filters that isolate a single line; although it was necessary to consider the relative phases of the lambda-doubled components. They differ by about 80°, but still yield good contrast. An examination of limited amounts of data indicate the the OH winds will be good, and the Doppler temperatures appear useful as well. Work on rotational temperatures is in progress. For O_2 a much narrower interference filter is required, about 0.1 nm. As a consequence the rotational lines are separated into ring patterns on the CCD (see Shepherd et al., (1993a) for actual images. The O_2 PP(7) and PQ(7) blended lines were put closely in phase by a careful selection of the interferometer optical path difference. This allows four wind values to be obtained from a single image, one from each rotational line. The reduction algorithm is considerably more complex than for the atomic emissions, but is producing good results in off-line software, and soon will be in production. Excellent results are being obtained on Rayleigh scattering temperatures and noctilucent clouds, and preliminary results have been obtained on aurora. Finally, the O^+ emission requires a separate analysis algorithm, and its development is still in progress.

4. WINDII VALIDATION

Although considerable effort was made to characterize the WINDII instrument on the ground, so that it would provide accurate data in orbit, a program has also been established to validate the values obtained in orbit, under flight conditions. The overall approach makes use of three methods of comparison measurement: 1) those using a similar type of Michelson interferometer on the ground; 2) those using the same method, namely optical Doppler airglow measurement, but with a Fabry-Perot spectrometer rather than a Michelson interferometer and 3) those using a completely different method such as a radar.

Under method 1) an extensive validation program is being conducted with the French MICADO instrument at the Observatoire d'Haute Provence (OHP), an instrument that has already established a successful program of wind and temperature measurements (Thuillier and Hersé, 1991). A new Canadian instrument called GWAMDI has been assembled at York University around the original WAMDII Michelson interferometer and is being operated at Bear Lake, Utah. This site was chosen because at 42° latitude it is at the turning point of WINDII's fields of view, as the satellite turns around at 57°. The latitude of OHP is also close to 42°, ensuring at least one overpass per night. The problem

in comparing optical measurements made in space with those made on the ground is that the integrating paths through the emission layer are different. Since WINDII data are inverted to vertical profiles it is possible to use these data to simulate the results of integrating along the path seen from the ground; it is these simulations that will be compared with the ground-based observations. Where the emitting atmosphere is horizontally homogeneous the results are expected to agree; but in the presence of gradients, or wave structures, they may not. In order to assess this, ground-based imagers for the emissions observed by WINDII have been deployed at the validation sites.

5. WIND COMPARISON

In Figure 4. we show wind, temperature and emission rate profiles obtained from a nighttime pass over Bear Lake at 0710 UT on January 27, 1993. The inverted volume emission rate for the 557.7 OI emission is shown in the upper left; it peaks at 95 km with with a value of 140 photons cm^{-3} s^{-1}. The Doppler temperature is shown

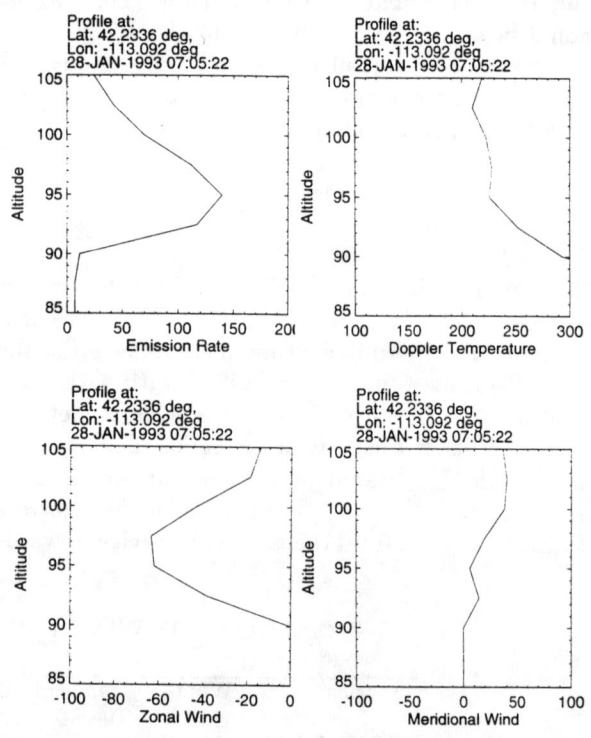

Figure 4. Emission rate (upper left), temperature (upper right), zonal wind (lower left) and meridional wind (lower right) profiles obtained from WINDII, over Bear Lake, during a nighttime pass at 0705 UT on January 27, 1993 (not Jan. 28 as indicated).

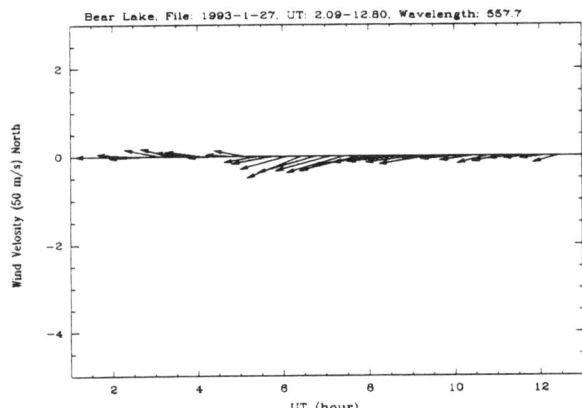

Figure 5. GWAMDI ground-based wind measurements for the night of January 27, 1993 at Bear Lake.

in the upper right, with a value of 225 K at the volume emission rate peak. The high temperature at 90 km should be ignored as the emission rate is very low here. In the lower frames we show the meridional and zonal wind profiles taken close to Bear Lake. For comparison purposes, we show the corresponding GWAMDI data for the full night, in Figure 5. These data show a strong zonal westward wind during the time of the overpass, of about 100 m/s and a weak southward flow. WINDII detected a somewhat weaker westward zonal flow of 62 m/s at the emission rate peak, with higher values above and below this peak, as shown in Figure 4. The meridional wind measured by WINDII, also shown in Figure 4 were very light, at 6 m/s, but northward rather than southward as seen by GWAMDI.

To directly compare GWAMDI and WINDII winds we need to integrate the WINDII altitude profiles along the GWAMDI line of sight; i.e. to compute an average wind through the 557.7 nm emission layer, weighted by the volume emission rate. The results of this are shown in Figure 6, for a sequence of WINDII measurements made as the FOVs passed over Bear Lake. Here we do not use zonal and meridional wind, but rather compare the wind components measured separately with Fov 1 and 2. W1 indicates the weighted WINDII wind for Fov-1 while W2 indicates the same for Fov-2. G1 indicates the component of the GWAMDI wind along the WINDII Fov-1 direction, so that W1 and G1 may be directly compared. We have chosen this approach to validation measurement because the zero-wind offset may be different for the two fields of view. Similarly, G2 may be compared to W2. BL indicates the location of Bear Lake. The results are reasonable, but not identical. The zero-wind phase values for WINDII were obtained before launch, using OI 557.7 nm emission line lamps obtained from Resonance Ltd; it is satisfying that these values do give reasonable agreement in orbit more than two years after the ground measurements. Small adjustments can be made if the evidence from external evidence so indicates, but to date there has been no consistent evidence indicating a change. The WINDII data show that the wind pattern to the east of Bear Lake is different from that to the west, so precise agreement is not expected.

While the ground-based optical comparisons are encouraging, we have found that with the restrictions of good weather, and darkness, the number of satisfactory comparisons has been limited. Furthermore, even when conditions seem ideal, inconsistent results may be obtained for individual cases, probably due to gradients, or gravity wave structure. We are therefore accumulating a larger database of both optical and radar data for comparison.

6. TEMPERATURE AND EMISSION RATE COMPARISON

At the same time, temperatures have been measured with a Mesopause Oxygen Rotational Temperature Imager (MORTI), using the O_2 (0,1) band as described by Wiens et al. (1991). The temperatures and emission rates for the night of January 27, 1993 are shown in Figure 7. The time of the WINDII overpass is shown; the WINDII temperature at the emission rate peak was 225 K, comparing very closely with the MORTI value of 222 K. It is interesting that at this time the emission rate is rising rapidly, while the temperature has already reached a maximum; normally the emission rate and

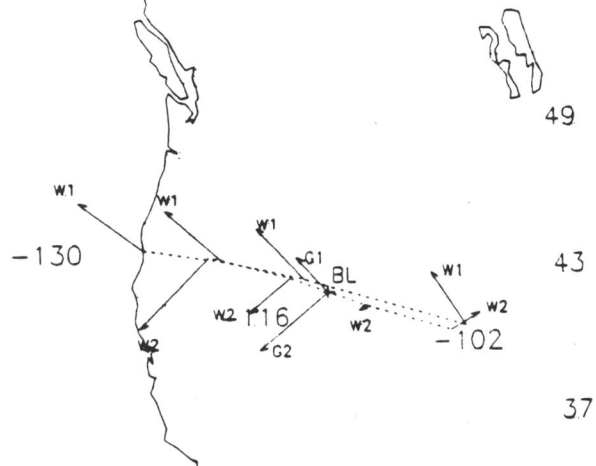

Figure 6. Comparison of WINDII line-of-sight winds W1 and W2 with the corresponding GWAMDI wind components G1 and G2. BL indicates Bear Lake, and the data are for the 0705 UT overpass on January 27, 1993.

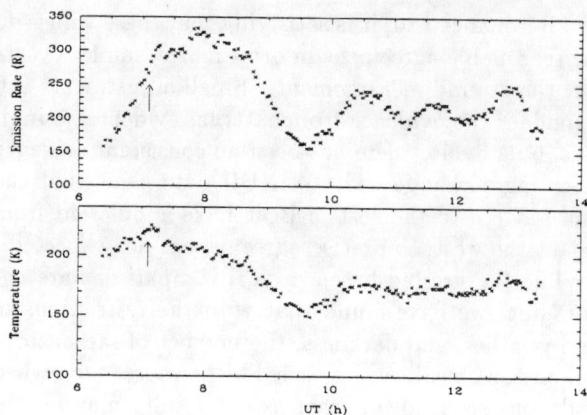

Figure 7. O$_2$ Atm band rotational temperature and emission rates at Bear Lake for the night of January 27, 1993.

temperature are closely in phase, as they are later in the night. The integrating path for WINDII is very different from the one for the ground based instruments, so for the two to agree there must be local homogeneity. To assess this, a CCD imager called YASAC (York All Sky Airglow Camera) was operated at Bear Lake. A YASAC pair of images for the time of the overpass is shown in Figure 8; with the OI 555.5 nm image on the left indicating a normal Van Rhijn effect with no wave structure, while the 677.7 nm broadband image on the right (used to monitor the presence of aurora) shows only stars and the Milky Way. Thus wave activity was minimal at 0710 UT, even though the emission rate was rising as seen in Figure 7.

7. MEASUREMENTS ALONG THE SPACECRAFT TRACK

A convenient way to show latitude variability is to generate contours of volume emission rate from a sequence of vertical emission profiles acquired along the track. This gives the highest data resolution, with good continuity between profiles, although one must remember that not all profiles will have been obtained at the same longitude, or local time. The results of this for the 557.7 nm emission at night, for the orbit that passed over Bear Lake at 0705 UT on January 23, 1993 are illustrated in Figure 9. The latitudes covered are -30 to 42, with Bear Lake at the extreme righthand side; the altitude range shown is 85 to 105 km. Two features are of interest. First, while the classical "layer" of atomic oxygen airglow is considered to be a uniform layer some 10 km thick, centred at 97 km, we see here large scale variability in the emission rate at 95 km, from less than 25 photons cm^{-3} s^{-1} at the equator to 175 photons cm^{-3} s^{-1} at 35°. In general the "layer" has a relatively flat topside (less so in this particular case), but a highly variable bottomside. Bear Lake is somewhat poleward of a bright "cell" of emission located close by at 35°. The second remarkable feature is a "gap" in the emission at the equator, where the volume emission rate drops to about 20 photons cm^{-3} s^{-1}, about one tenth of that at mid latitudes. These two features are characteristic of all of the limited amount of data that has been examined to date. Since the 557.7 nm emission is a direct result of the recombination of atomic oxygen, with one photon indicating the loss of two O atoms, this pattern closely reflects the atomic oxygen density. ISIS-II satellite results had demonstrated the existence of a maximum of 557.7 nm emission at mid-latitudes, at times near solstice, with a miminum at the equator (Cogger, et al., 1981); the WINDII results are consistent with this, but show that the smooth pattern observed by ISIS results from these more localized discrete features. The origin of the equatorial gap appears to be due to the action of the strong diurnal tide, whose equatorial effects have been modelled by Forbes et al. (1993). These tidal signatures have have been observed with HRDI (Morton et al., 1993), and may be seen in Figure 10, which shows the meridional wind measured during the same overpass as for the other data shown here. At 100 km one can see a northward wind cell of about 50 m/s (positive wind) south of the equator and a southward wind of about 100 m/s north of the equator. The zero wind line is located somewhat north of the equator, at about 10°. These signatures are dramatically evident in the daytime, when the vertical range of emission is greater, but are clearly observable here, in this thin nighttime layer.

8. PLANETARY SCALE DISTURBANCES

A useful way to study the longitudinal variations of emission rate is to select data for a single latitude band

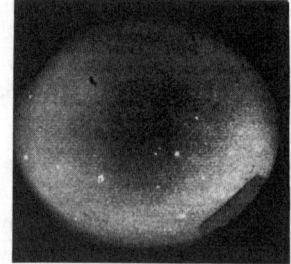 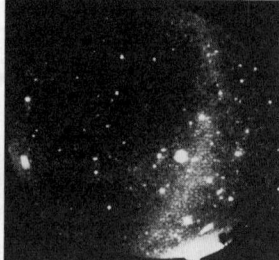

Figure 8. A YASAC all-sky image at the time of the WINDII overpass, showing 557.7 nm emission on the left, and broadband 677.4 nm emission on the right. The latter shows only stars.

for a single day. Since the orbit precession is about 20 min per day, all data points in such a band have essentially the same local time. For the 557.7 nm emission at night, the results of this for volume emission rate for all of the data for January 27, 1993 are shown in Figure 11 for the latitude band 40-42°N. The vertical axis extends from 85 to 105 km and the contours are for 50, 100 and 150 cm^{-3} s^{-1}. Volume emission rates of 150 photons cm^{-3} s^{-1} are evident at 70°E and 80°W, with a minimum of less than 50 photons cm^{-3} s^{-1} between 10°E to 40°W, a peak to valley ratio of 4:1. Examination of the wind results show that both the meridional and zonal wind correlate with this structure. Comparing results for successive days show that these structures normally move relatively little from one day to another. This is consistent with the fact that ground-based observers rarely see a factor of four variation in emission rate on a single night. Nightly variations seem to be due to gravity waves (Wiens et al., 1993) with a variation more like 30%. Large changes in emission rate are attributed to changes in the position or structure of these planetary scale disturbances by Shepherd et al. (1993a), who describe their characteristics in more detail. The volume emission rates over Beak Lake peak at 100 cm^{-3} s^{-1} and show rapid variation with longitude; as confirmed by an examination of a sequence of profiles. This structure may be related to the rapid rise in emission rate seen by MORTI and shown in Figure 7.

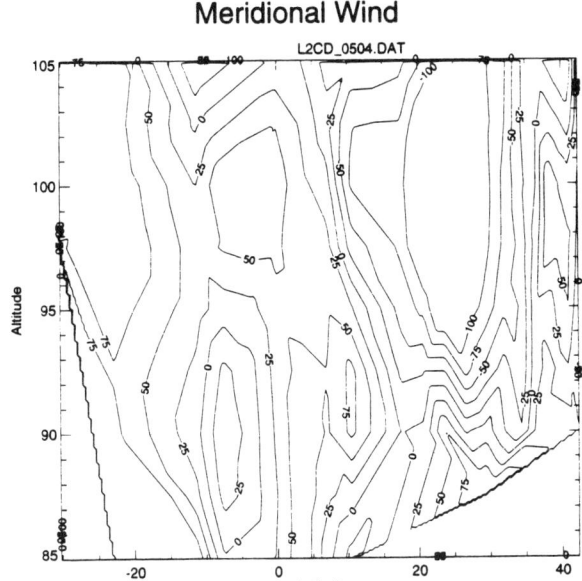

Figure 10. Meridional wind along the spacecraft track from the nighttime 557.7 nm emission for the 0705 UT Bear Lake overpass on January 27, 1993.

9. ERRORS

An analysis and discussion of errors for WINDII is a large topic, beyond the scope of this paper. However a few comments are made here to assist the reader in interpreting the results presented. The random errors have been presented and discussed by Shepherd et al. (1993b); for the nighttime green line these are 4 m/s and 11 K. Systematic errors for the wind arise primarily from the determination of zero wind phase; for the results presented here this quantity is currently validated to about 10 m/s, but this limit can be improved through further validation. The systematic error for temperature is dominated by the accuracy of background subtraction; for the nighttime data shown here this error is small compared to the random error.

10. DISCUSSION AND CONCLUSIONS

The WINDII Science Team placed requirements on WINDII that would allow it to measure winds and temperatures from about 80 to 300 km, using the OH, O$_2$ Atm, O(^{1}S), O(^{1}D), and O$^+$(^{2}P) emissions. When this was done, in 1985, one could not be sure that these requirements would be met. However, with careful attention to the incorporation of excellent surface flatness and thermal stability of the Michelson interferometer, high system transmission and low CCD noise level, the requirements have all been met or exceeded. The single-

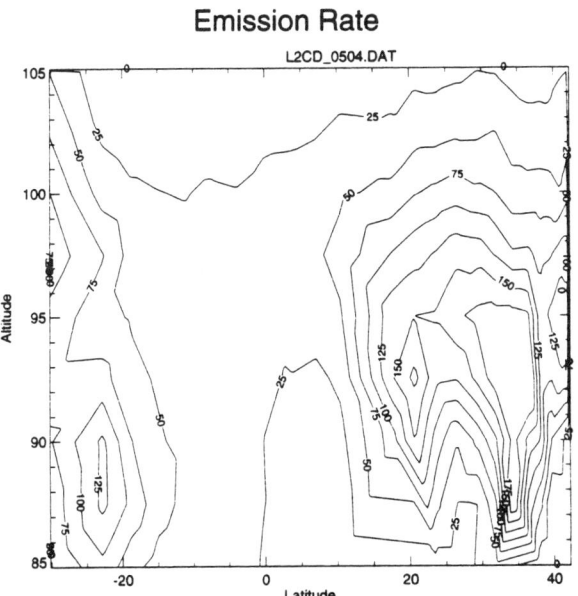

Figure 9. Latitudinal variation of atomic oxygen 557.7 nm emission for the 0705 UT Bear Lake overpass on January 27, 1993.

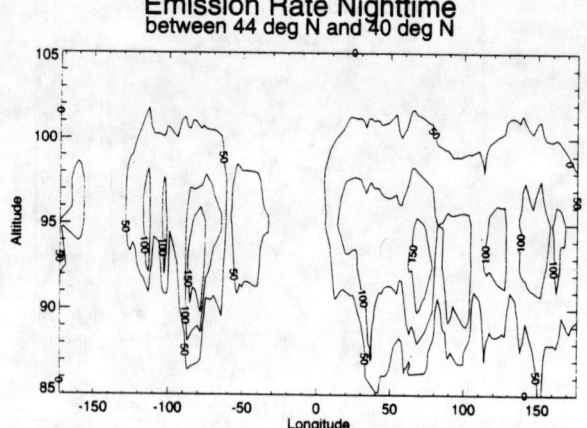

Figure 11. Volume emission rate for the 557.7 nm emission at night, for a latitude band 38-43°N, plotted versus longitude for January 27, 1993.

stage baffle, with its knife edge and the technique of inserting a daytime aperture worked extremely well, and since the launch on Sept. 12, 1991 has permitted excellent measurements of the atomic oxygen 557.7 nm emission in the daytime, as well as at night.

Both latitudinal and longitudinal variations of the emission rate of the OI 557.7 nm emission are much more dramatic than expected. Large scale planetary scale disturbances show a peak to valley ratio of 4:1, and are correlated with structure in the meridional and zonal wind. The latitudinal structure is characterized by a cellular type structure at mid-latitudes, and a "gap" in emission at the equator; the latter seems to be associated with the diurnal tide, which is strongly manifested near the equator.

WINDII had accumulated over twelve million images by October, 1993, providing us with a comprehensive view of the upper atmosphere that will keep aeronomers busy for years to come.

Acknowledgments. The WINDII project was jointly sponsored by the Canadian Space Agency and the Centre National d'Etudes Spatiales, in collaboration with NASA. The authors are grateful for their dedicated commitment and loyal support. They are indebted to the prime contractor, AIT Corporation, and the principal subcontractor responsible for the design and fabrication of the flight instrument, CAL Corporation.

REFERENCES

Bouchareine, P. and P. Connes, Interferometer with compensated field for Fourier transform spectroscopy, *J. Phy.*, 24, 2, 134, 1963.

Cogger, L.L., R.D. Elphinstone and J.S. Murphree, Temporal and latitudinal 5577 airglow variations, *Can. J. Phys.* 59, 1296, 1981.

Forbes, J.M., R.G. Roble and C.G. Fesen, Acceleration, heating, and compositional mixing of the thermosphere due to upward propagating tides, *J. Geophys. Res.* 98, 311, 1993.

Hilliard, R.L. and G.G. Shepherd, Wide-angle Michelson interferometer for measuring Doppler line widths, *J. Opt., Soc. Am.* 56, 362, 1966.

Morton, Y.T., R.S. Lieberman, P.B. Hays, D.A. Ortland, A.R. Marshall, D. Wu, W.R. Skinner, M.D. Burrage, D.A. Gell and J.-H. Yee, Global mesospheric tidal winds observed by the high resolution Doppler imager on board the upper atmosphere research satellite, *Geophys. Res. Lett.* 20, 1263, 1993

Shepherd, G.G., W.A. Gault, D.W. Miller, Z. Pasturczyk, S.F. Johnston, P.R. Kosteniuk, J.W. Haslett, D.J.W. Kendall, and J.R. Wimperis, WAMDII: Wide-angle Michelson Doppler imaging interferometer for Spacelab, *Appl. Opt.*, 24, 1571, 1985.

Shepherd, G.G. *et al.*, WINDII - the WIND Imaging Interferometer on the Upper Atmosphere Research Satellite, *J. Geophys. Res.*, 98, 10,725, 1993a.

Shepherd, G.G., G. Thuillier, B.H. Solheim, S. Chandra, L.L. Cogger, M.-L. Duboin, W.F.J. Evans, R.L. Gattinger, W.A. Gault, M. Hersé, A. Hauchecorne, C. Lathuillere, E.J. Llewellyn, R.P. Lowe, H. Teitelbaum and F. Vial, Longitudinal structure in atomic oxygen concentrations observed with WINDII on UARS, *Geophys. Res. Lett.*, 20, 1303, 1993b.

Thuillier, G. and G.G. Shepherd, Fully compensated Michelson interferometer of fixed path difference, *Appl, Opt.*, 24, 1599, 1985.

Thuillier, G. and M. Hersé, Thermally stable field compensated Michelson interferometer for measurement of temperature and wind of the planetary atmospheres, *Appl. Opt.*, 30, 1210, 1991.

Wiens, R.H., S.P. Zhang, R.N. Peterson and G.G. Shepherd, MORTI: a mesopause oxygen rotational temperature imager, *Planet. Space Sci.*, 39, 1363, 1991.

Wiens, R.H., S.-P. Zhang, R.N. Peterson, G.G. Shepherd, C.A. Tepley, L. Kieffaber, R. Niciejewski and J.H. Hecht, Simultaneous optical observations of long-period gravity waves during AIDA '89, *J. Atmos. Terr. Phys.*, 55, 325, 1993.

G.G. Shepherd, B.H. Solheim, Y.J. Rochon, J. Criswick, W.A. Gault, R.N. Peterson, R.H. Wiens and S.-P. Zhang, Institute for Space and Terrestrial Science, 4700 Keele Street, York University, North York, Ontario, Canada M3J 1P3.

G. Thuillier, Service D'Aeronomie du CNRS, B.P. 3, 91371 Verrières-le-Buisson, France.

Preliminary Results from the Imaging Spectrometric Observatory Flown on ATLAS 1

D. G. Torr[1], M. R. Torr[2], M. F. Morgan[1], T. Chang[1], J. K. Owens[2],
J. A. Fennelly[1] P. G. Richards[3] and T. W. Baldridge[2]

[1]*The University of Alabama In Huntsville, Optical Aeronomy Laboratory, Department of Physics and the Center for Space Plasma and Aeronomic Research, Huntsville, Alabama 35899*

[2]*NASA/Marshall Space Flight Center, Huntsville, Alabama 35812*

[3]*The University of Alabama In Huntsville, Optical Aeronomy Laboratory, Department of Computer Science, Huntsville, Alabama 35899*

The Imaging Spectrometric Observatory (ISO) which comprises an array of five spatial-spectral imaging spectrometers was flown on the ATLAS 1 mission during March 24 to April 2, 1992. The instrument acquired data on the ionosphere-thermosphere-mesosphere (ITM) altitude region between 60 - 300 km and over the wavelength range 30 to 832 nm. Observational sequences were designed to provide spectroscopic data needed to address several specific outstanding problems. In particular measurements were made of: 1) the OH near-ultraviolet $A^2\Sigma - X^2\Pi$ (0,0) band in the dayglow and these data were used to retrieve the altitude profile of the concentration of OH between 70 and 80 km; 2) the O_2 Atmospheric, Herzberg and Chamberlain bands, which are being used to determine the vibrational distributions of their parent states, particularly of the O_2 $A^3\Sigma_u^+$, $A'^3\Delta_u$ and $c^1\Sigma_u^-$, states as a function of altitude; the measurements of the O_2 bands are also used to determine the temperature (Tn) profile in the mesosphere; 3) the $O^+(^2P)$ 732 nm emission which was used to retrieve the concentrations of O, O_2, N_2 and Tn as a function of altitude; 4) the brightnesses of thermospheric and mesospheric dayglow emissions, which have been compared with results from our comprehensive Field Line Interhemispheric Plasma (FLIP) model that was run in a predictive mode prior to the mission. A comparison of the model results with 12 key measured airglow emissions yielded good agreement.

1. INTRODUCTION

The Imaging Spectrometric Observatory (ISO) launched on the ATLAS 1 mission on March 24, 1992 comprises an array of five half-meter modified Czerny-Turner grating spectrometers with two-dimensional intensified CCD array detectors to record simultaneously spectral and spatial information. Altitude is imaged in the dimension parallel to the slit and spectral information normal to the slit, when placed perpendicular to the horizon. The field of view of the instrument is 0.65° along the slit which images approximately 20 km at mesospheric altitudes resolved to ~ 2 km. In addition, the instrument has a front mirror which can be used to scan the limb to obtain altitude profiles of emissions with the slit in the horizontal position. The instrument is programmable via its onboard microprocessors which offers a large selection of operating modes. The wavelength range 30 to 832 nm is covered at ~ 0.15 to 1.0 nm resolution in a programmable series of grating steps.

The sensitivity, from the pre-flight calibration, varies from ~ 1 count/R-s in the near-ultraviolet (NUV) to ~ 0.02 counts/R-s in the extreme-ultraviolet (EUV) and near-IR. The wavelength range above 200 nm was calibrated in the laboratory using a NIST standard calibrated tungsten halogen lamp. For the wavelength region between 100 and 200 nm a calibrated deuterium lamp was used. The EUV wavelength range was calibrated using a Samson flow lamp with various gases to generate various emission lines which were spectrally separated with an EUV/FUV monochromator. Due to time pressure to meet a launch deadline the EUV and FUV calibration was not comprehensive and we will carry out a post flight calibration, if resources are available.

At each grating step a segment of the spectrum is spectrally and spatially imaged. Approximately 100 nm were

The Upper Mesosphere and Lower Thermosphere:
A Review of Experiment and Theory
Geophysical Monograph 87
Copyright 1995 by the American Geophysical Union

Table 1. Summary of ISO Instrument Parameters

Parameters	Spec.1	Spec.2	Spec.3	Spec.4	Spec.5
Wavelength range (nm)	307.0-311.4	382.1-832.3	213.8-428.2	100.3-238.9	30.0-139.3
Spectral window per grating step (nm)	4.36	48.2	23.8	17.7	18.1
Spectral resolution (nm)	0.05	0.66	0.23	0.17	0.54
Field of view (degrees)	0.65 x 0.007	0.65 x 0.007	0.65 x 0.007	0.65 x 0.007	0.65 x 0.007
Typical sensitivity (slit) (counts/R-s)	1.0	0.4	1.2	0.15	0.04

simultaneously imaged for the gratings selected for the ATLAS 1 mission. Table 1 provides a brief summary of the ISO key parameters. Figure 1 shows an example of a spectrum obtained by the ISO (during a shuttle roll) at a tangent height of ~ 190 km. A summary of the conditions under which the observations were taken is given in Section 3.4.1 in Table 2.

For the ATLAS 1 mission the decision was made to acquire a database that could be used to address several outstanding scientific problems in the mesosphere, thermosphere, ionosphere region. The purpose of this paper is to present examples of the data taken during some of the special observational sequences designed to address several goals, and results of theoretical modeling of these observations.

2.1 The Global Ionosphere-Thermosphere

The primary objectives of the investigation are to provide altitude profiles of the concentrations of O, O_2, N_2, O^+, O_2^+, N_2^+, and the neutral temperature, Tn. To achieve these objectives, a coordinated study was carried out with the ground-based community who operated the global network of ionosonde and incoherent scatter radar stations and several airglow observatories. A large amount of data was gathered by the participating investigative groups. While there is no formal cooperative project to utilize the data, the UAH group will (within current resource constraints) provide data to researchers for ISO data analysis.

The ISO was turned on during overflights of the groundbased stations. The operational mode selected for this study was designed to obtain altitude profiles of a limited number of key emissions needed for the retrieval of neutral and ion composition and Tn with ~ 5 to 10° latitudinal resolution. This was achieved by using the front mirrors on the instrument to scan in tangent ray height (TRH). The emissions selected were OII 83.4, 732; OI 130.4, 135.6, 557.7/297.2, 630 nm line features, and the N_2^+ First Negative (0,0) band at 391.4 nm. An example of the basic approach to be used for the retrieval is given below, where we present the first results on the retrieval of neutral composition and temperature from remote sensing of thermospheric emissions.

We use the 732 nm emission of $O^+(^2P)$ for the retrieval due to its simple photochemistry and its sensitivity to O and N_2.

The dominant production source of the $O^+(^2P)$ ion is photoionization of O by solar EUV radiation at wavelengths of 66.5 nm and less. A second source is photoelectron impact ionization of O. The $O^+(^2P)$ ion is lost by radiative decay and quenching by O, N_2, and electrons. Quenching by N_2 and O becomes competitive with radiative decay below 300 km. It is this increase of quenching due to N_2 which allows for the retrieval of N_2 profiles from the ISO 732 nm measurements.

Except for changes in cross sections and reaction rates given below, the local photochemical model of the $O^+(^2P)$ ion and the inversion algorithm are the same as was used in the twilight ground-based retrieval technique by *Fennelly et al.* [1991, 1993a,b]. *Chang et al.* [1993] have derived new quenching rates for O and N_2 from Atmosphere Explorer data with values of 4.0 x 10^{-10} cm^3 s^{-1} and 3.4 x 10^{-10} cm^3 s^{-1}, respectively. The O quenching rate is almost eight times larger, while the N_2 quenching rate is nearly the same as the previously determined rates of *Rusch et al.* [1977]. The errors on these reaction rates are ~ ± 50%. In future work we will assess the effect on the retrieved quantities of errors related to these quenching rates. Photoionization and photo-absorption cross sections for O, N_2 and O_2 have recently been compiled by *Fennelly and D. G. Torr* [1992] which include more detailed structure and autoionization. These were the only changes made to the model, except for the adaptation to orbital geometry.

Because no solar EUV measurements were made during this mission, the solar EUV flux needed for calculating the photoionization rate of O is derived from the $F_{10.7}$ cm flux as described by *Fennelly et al.* [1993a,b]. *Richards et al.* [1994] discuss uncertainties associated with EUV flux proxy models.

The neutral thermospheric densities and temperature in the retrieval algorithm are calculated using the Bates-Walker model [*Walker*, 1965]. This simple model requires input values for [O], [N_2], and temperature at 120 km, the exospheric temperature, and a shape factor (related to the temperature derivative at 120 km). This procedure reduces the number of unknowns to a manageable number (five). The inversion technique solves for a linear scaling constant for each of the input parameters' spatial variations. This results in retrieved Bates-Walker input parameters which can then be used to construct thermospheric vertical profiles for O, N_2 and temperature anywhere in the area viewed by ISO at a particular

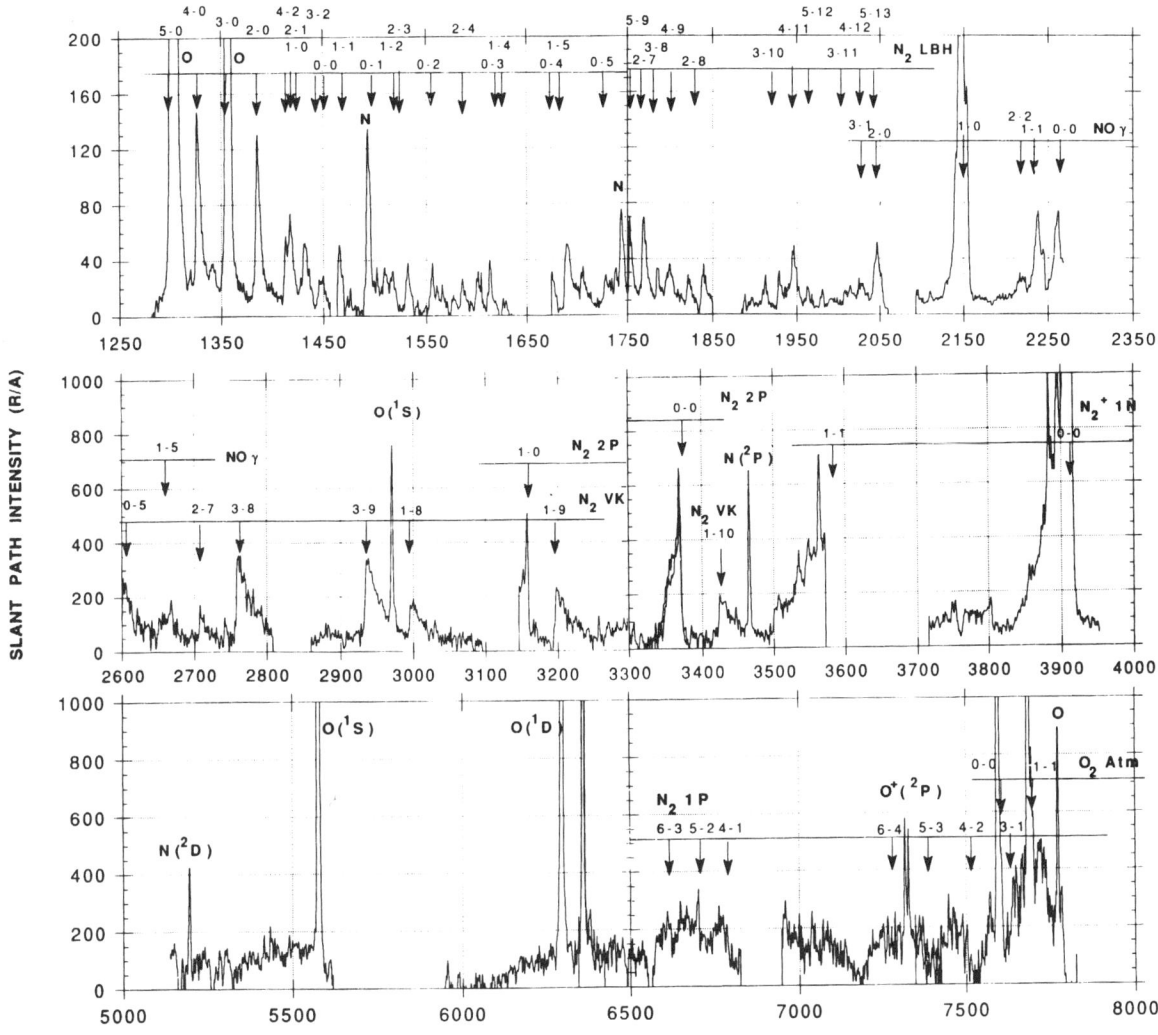

Figure 1: Example of an ISO spectrum obtained during a shuttle roll at a tangent ray height of approximately 190 km.

Table 2. Summary of Conditions for Roll Observation Sequence

Parameter	Value
Date	March 27, 1992
GMT	07:18-07:43
Local time	5:22-10:53 h
Solar zenith angle	99.6-55.5°
F10.7	181
F10.7 90-day average	172
Ap	11
Latitude	10.5S-57N
Longitude	27.9W-60.8E

time. These input parameters have spatial structure over the geographical extent of the airglow measurements, which are assumed fixed over the time the measurements are made - less than 6.5 minutes in this case. We have assumed the relative spatial variation given by MSIS-86 [*Hedin*, 1987] for these input parameters.

Examples of the 732 nm spectra measured by the ISO at tangent heights between 219 to 255 km at mid-latitudes are shown in Figure 2. The shuttle altitude was ~ 300 km. The data are absolutely calibrated line-of-sight intensities in R/A. Measured 732 nm intensity profiles as a function of tangent height are shown in Figure 3a along with the emissions modeled using the MSIS-86 neutral thermospheric values and those derived from the retrieved atmosphere. Figures 3b, c and d show sample vertical profiles of temperature, and O and N_2 densities, respectively, retrieved from this data set. The

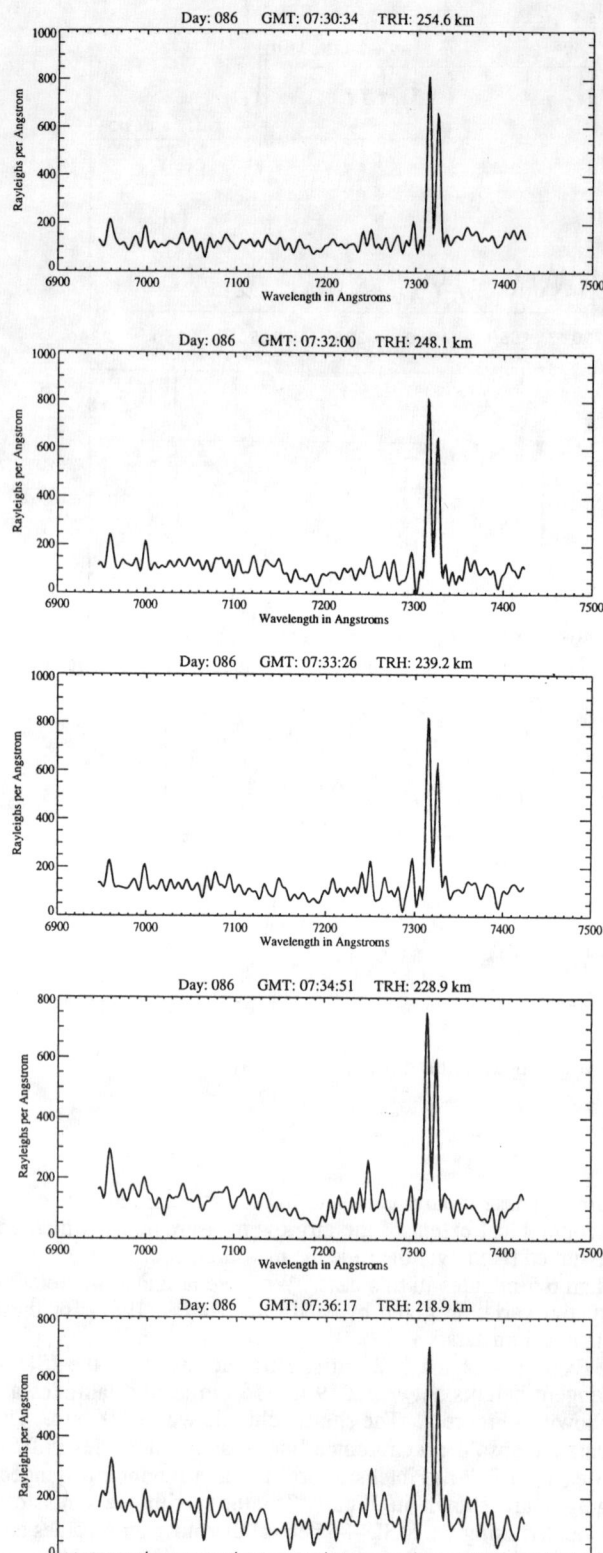

preliminary retrieved values are in reasonable agreement with the MSIS-86 values. The difference above 400 km is about a factor of 2 which is not unreasonable because the variability at these heights can be large. An earlier comparison by *Fennelly et al.* [1993a] used a preliminary value for the F10.7 cm flux which yielded larger values for T_∞ from the MSIS model.

Even though we have no way of guaranteeing the uniqueness of the recovered results, we have exhaustively tested the algorithm on realistic simulated data. Results of these tests have been reported by *Fennelly et al.* [1993a,b].

The values used for the solar EUV flux in this study have also been used to accurately model the following emissions as a function of height (120 to 300 km): OI 135.6 and 777.4 nm; several N_2 bands (Lyman-Birge-Hopfield 160 to 180 nm; Vegard Kaplan at 320 nm; N_2 Second Positive at 337.1 nm); $O(^1S)$ 297.2 and 557.7 nm; the N_2^+ First Negative bands (358.4, 391.4 and 427.8 nm); $N(^2D)$ 520 nm, and $O(^1D)$ 630 nm. These results are presented in section 3.4. The good agreement obtained with the ISO data strongly supports the photochemistry that evolved during the Atmosphere Explorer program and our F10.7 cm proxy model for the solar EUV flux.

These preliminary results on retrieved thermospheric O, N_2, and temperature show the method to be a promising tool for remote sensing of the thermosphere, although much remains to be done, for example, the utilization of other emissions to broaden the retrieved database. One limitation of the ATLAS 1 data is the lack of *in situ* measurements that could be used for validation of the retrieval. We are planning to use the Atmosphere Explorer (AE) data to validate the technique and to estimate the errors on the retrieved thermospheric densities and temperature. The AE data has both 732 nm airglow emission measurements and *in situ* measurements of the quantities retrieved by our technique.

2.2 Measurements of the Hydroxyl Radical and its Importance in Mesospheric Oxygen-Hydrogen Chemistry

The photochemistry of the oxygen-hydrogen family that controls ozone in the mesosphere is thought to be conceptually well understood (see *Allen et al*, 1984). However, it has never been quantitatively tested because key odd hydrogen constituents, such as atomic hydrogen (H) and the hydroxyl radical (OH) have never been simultaneously measured with their source and sink molecules. There have been problems in reproducing the measured O_3 values from the Solar Mesosphere Explorer with current photochemical models, and without measurements of the active radicals it is difficult to identify the source of the problem.

Ozone in the mesosphere is produced by the three-body recombination of O and O_2. It is destroyed by photolysis and the well known HOx catalytic cycles of which the main one in the mesosphere is

Figure 2. ISO spectral limb scans of dayglow 732 nm emission. The line-of-sight distance to the tangent ray height at 218.9 km is 1172 km.

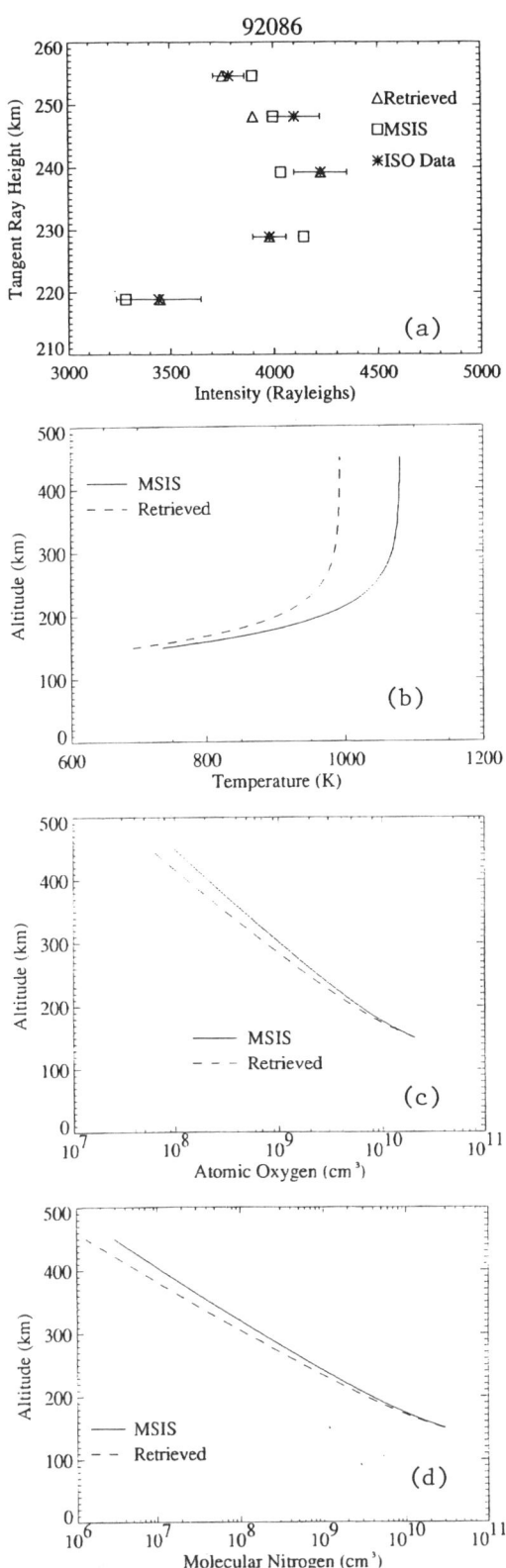

$$O_3 + H \rightarrow OH + O_2$$
$$\underline{OH + O \rightarrow H + O_2}$$
$$O + O_3 \rightarrow O_2 + O_2$$

To verify this photochemistry requires measurements of O, O_3, H and OH. In section 3.3 we discuss how we plan to obtain daytime O densities from the ISO data. Several instruments in the ATLAS payload provided measurements of ozone. Hydrogen was measured by the Atmospheric Lyman Alpha Emissions Experiment (ALAE) [*Bertaux et al.*, 1993] using deuterium measurements as a proxy. The ISO measured OH emissions that arise from resonance scattering of sunlight in the transition $A^2\Sigma - X^2\Pi$ around 307 nm. Thus, the ATLAS 1 database includes the necessary measurements needed to test the basic photochemistry of mesospheric ozone, odd hydrogen (H, OH, HO_2) and water vapor. Water vapor was measured by three of the ATLAS 1 instruments.

The hydroxyl radical is a trace constituent of prime importance in photochemistry throughout the lower and middle atmosphere. It is central to oxidation chemistry in the troposphere. In the lower stratosphere, it contributes to the catalytic destruction of ozone. Perhaps most importantly in the stratosphere, OH mediates the partitioning of chlorine and nitrogen compounds (which also catalyze ozone destruction) between active radicals and inactive reservoirs. Measurements of OH densities needed to support modeling efforts are sparse, especially in the mesosphere. Two sounding rocket observations [*Anderson*, 1971a,b] are the only previous direct daytime observations of hydroxyl above 50 km. These instruments observed the OH $A^2\Sigma - X^2\Pi$ (0,0) band in resonance scattering. Several techniques for measurement of OH profiles in the stratosphere have been developed. Among these are balloon-borne LIDAR [*Heaps and McGee*, 1985], far infrared spectroscopy [*Carli et al.*, 1989], a series of balloon-borne, *in situ* measurements [*Stimpfle at al.*, 1990 and references therein; *Anderson*, 1976], and high resolution UV spectroscopic observations of resonance fluorescence [*D. G. Torr et al.*, 1987]. Ground-based determinations of the OH column abundance using interferometric solar UV absorption measurements [*Burnett and Burnett*, 1989, and references therein] have been conducted since 1977 at a variety of sites. These measurements provide a unique and valuable data base on variability of hydroxyl abundance, but do not give any direct indication of its altitude distribution.

Improved data on the distribution of mesospheric OH would be valuable. With only two previous observations, the range of diurnal and geographic variations has been poorly sampled. The lack of more extensive mesospheric data leaves an important void in the testing of models of mesospheric chemistry [*Allen et al.*, 1984; *Rodrigo et al.*, 1986; *Bjarnason et al.*, 1987]. In addition, observations of mesospheric OH may lead to improved understanding of stratospheric chemistry.

Figure 3. (a) The 732 nm emission: measured, modeled using MSIS-86, and modeled from retrieved atmosphere. (b) Retrieved vertical temperature profile compared to MSIS-86. (c) Retrieved vertical O profile compared to MSIS-86. (d) Retrieved vertical N_2 profile compared to MSIS-86.

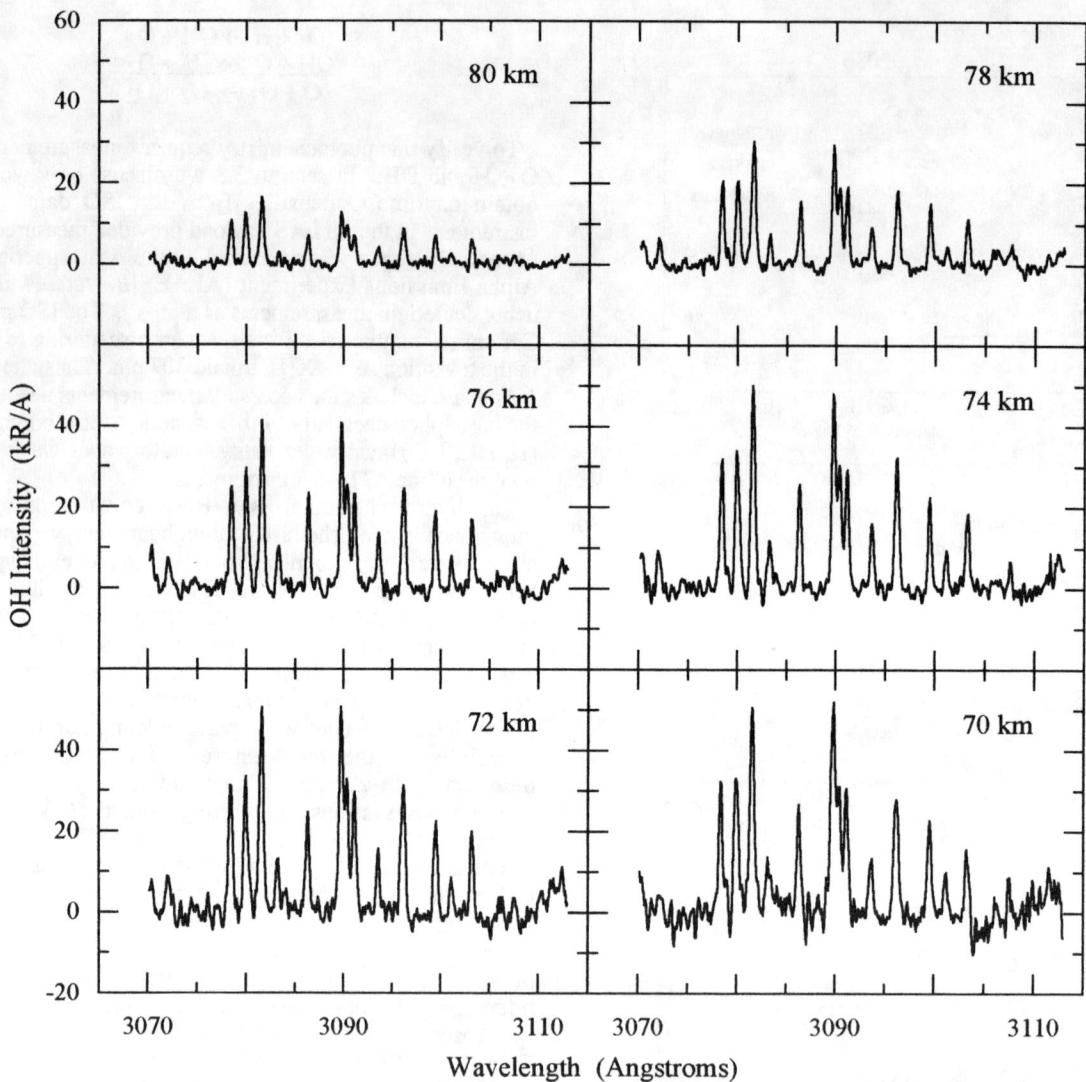

Figure 4. OH $A^2\Sigma - X^2\Pi$ (0,0) band spectra extracted from limb imaging observations. Tangent height given in upper right corner of each frame.

In the mesosphere the oxygen-hydrogen chemistry is isolated in a clean atmosphere. The relative simplicity of mesospheric chemistry will make it much easier to test models of oxygen-hydrogen chemistry, given the required mesospheric data. The opportunity to address the need for data on the distribution of mesospheric OH on a global scale and over a range of local conditions was the motivation for the measurements of mesospheric hydroxyl conducted during the ATLAS 1 mission.

Atmospheric OH $A^2\Sigma - X^2\Pi$ resonance fluorescence occurs when ground state OH is resonantly excited by incident ultraviolet sunlight. Subsequent radiative decay produces the characteristic $A^2\Sigma - X^2\Pi$ spectrum. The volume emission rate of the scattering is proportional to the OH $X^2\Pi$ density. Thus, determination of OH resonance scattering intensities can provide information on ground state OH densities. One of the ISO spectrometers (spectrometer 1) is dedicated to measurement of OH $A^2\Sigma - X^2\Pi$ (0,0) spectra around 307 nm. It is an f/3.5 diffraction grating imaging spectrometer with spectral resolution of 0.5 nm over range 306.9 nm-311.3 nm.

The measurements reported here were taken with shuttle Atlantis pitched nose downward to point the center of the ISO field of view on a slant path with a fixed tangent ray height near 75 km, with the entrance slit perpendicular to the horizon. The 12 rows of spectral data then provide a simultaneous altitude scan. The rows are separated by 0.055°, which equates to 1.65 km in tangent height at an orbiter altitude of 300 km. The first and last rows are only partially illuminated, leaving 10 spectra covering an altitude range of 16 km in each 4-second image frame. Here, 28 successive frames are averaged to yield the spectra shown in Figure 4. The data were taken on March 30,

1992, at GMT 09:53. The tangent point position was 39°N, 82°E; the solar zenith angle was 57°.

The spectrum observed on each spatial segment consisted of Rayleigh scattering plus OH resonance fluorescence, integrated along the line of sight viewed by the segment. To determine the OH intensity, the OH spectrum was separated from the bright and highly structured Rayleigh scattering spectrum [see *Morgan et al.*, 1993].

Figure 5 shows the density profile derived from the spectra shown in Figure 4. The error bars indicate the 3-sigma relative uncertainty in the densities due to detection noise and extraction residuals. An additional absolute uncertainty of about 50%, due to the g-factor and the absolute calibration of the spectrometer, is not included in the error bars shown. Systematic errors due to the assumption that no OH is present above the highest tangent height observed are small, since the observations extend to altitudes 4 km above the highest significant OH emissions. Simulated retrievals show that these errors should be less than 5%.

The uncertainty in the altitude is less than ± 2 km. The relative accuracy is better than 0.2 km.

The preliminary OH density profile shown in Figure 5 is the first measurement of mesospheric ground state OH densities since Anderson's rocket measurements in 1969 and 1971. It is the first direct measurement of atmospheric hydroxyl densities from orbit.

Figure 5 also shows model results for OH in the mesosphere at noon [*Allen*, 1984; *Rodrigo, et al.*, 1986]. The measurement appears to match the results of Allen reasonably well, but the model results are for substantially different conditions such as time of day, season, and solar activity.

The result reported here is only a sample of a large set of mesospheric OH observations, which will fuel many further studies. A picture of the distribution of OH in the mesosphere throughout the northern hemisphere should begin to emerge as more data are analyzed. When these results are eventually combined with measurements of other mesospheric constituents, notably ozone, atomic hydrogen, and water vapor, made by other groups during ATLAS 1, comprehensive testing of models of mesospheric chemistry will be possible.

2.3 Photochemistry of the Metastable States of Oxygen

The three-body recombination of atomic oxygen results in the formation of intermediate states of O_2, namely the $A^3\Sigma_u^+$, $A'^3\Delta_u$ and $c^1\Sigma_u^-$ which radiate in the Herzberg I, Chamberlain and Herzberg II bands respectively. These intermediate states are believed to be the precursors which lead to the formation of the $O_2(b^1\Sigma_g^+, a^1\Delta_g)$ and $O(^1S)$ metastable states which are the sources of the three brightest mesospheric emissions: the O_2 Atmospheric (0,0) band at 762 nm, the O_2 Infrared I system at 1270 nm and the "green line" at 557.7 nm [see recent papers by *Lopez-Gonzalez et al.*, 1992 and *Bates*, 1992 and the references cited therein]. Considerable uncertainty exists regarding the chemical reaction rates and branching ratios of this photochemical scheme. Accurate profiles of the column abundances of the emitting states are needed to quantify the photochemistry.

The ISO was configured to obtain the first near-simultaneous complete set of measurements of key emissions as a function of height from the various band systems and lines that arise from excited states of O_2 and $O(^1S)$. The principal goals of this investigation are to utilize the nighttime observations, when many of the weak bands are clearly observable, to quantify the photochemistry. With this knowledge in hand, we anticipate that it should be possible to use daytime observations of the $O(^1S)$ and $O_2(b^1\Sigma_g^+)$ emissions at 557.7 and 762.0 nm, which are the only visible emissions observable above the bright underlying Rayleigh scattered continuum to retrieve height profiles of O and O_2. Naturally this will require a knowledge the cross-sections, rate coefficients *etc.* needed to determine the production and loss rates accurately. *Leko et al.* [1993] have presented arguments that suggest the information content of the overall nighttime database acquired with the ISO is sufficient, when coupled with recent laboratory measurements of rate coefficients, to adequately constrain uncertainties in the photochemistry. Figure 6a shows a set of spatially resolved simultaneously acquired images of the $O(^1S)$ emission at 557.7 nm and Figure 6b shows two spectra from a daytime limb scan of the 557.7 nm feature. The varying spectral resolution is a measure of the image quality across the focal plane. Currently, no technique exists for the retrieval of atomic oxygen in the daytime mesosphere mainly because of the lack of measurements of $O(^1S)$ emission. Atomic oxygen is fundamental to the photochemistry of almost all processes that occur in the mesosphere. The ISO data suggest that the daytime measurements should be adequate to test an algorithm for the retrieval of [O] from the $O(^1S)$ emission.

To illustrate the full mesospheric spectrum, Figure 7 shows one nighttime orbit where the gratings were stepped through the wavelength range 260 to 830 nm, viewing a tangent ray height of 78.8 km. The accuracy of the altitude determination is better than ± 2 km for all the results reported in this section. The

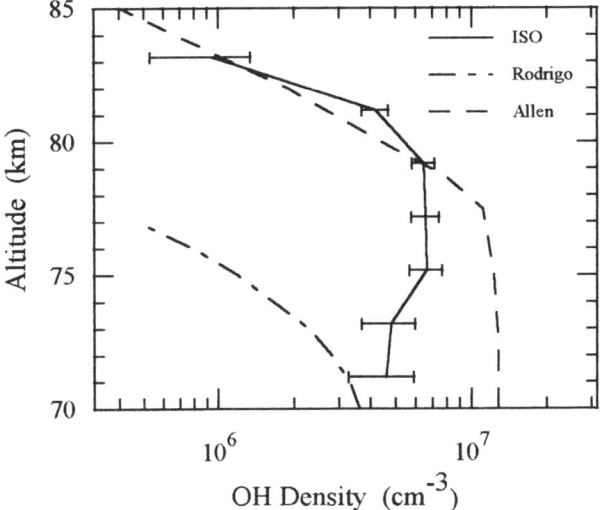

Figure 5. OH $X^2\Pi$ density profile derived from $A^2\Sigma - X^2\Pi$ (0,0) band intensity measurements (solid line); error bars are 1 sigma. OH profiles at noon from model results: *Allen* [1984], dotted line; *Rodrigo et al.* [1986], dash dot line.

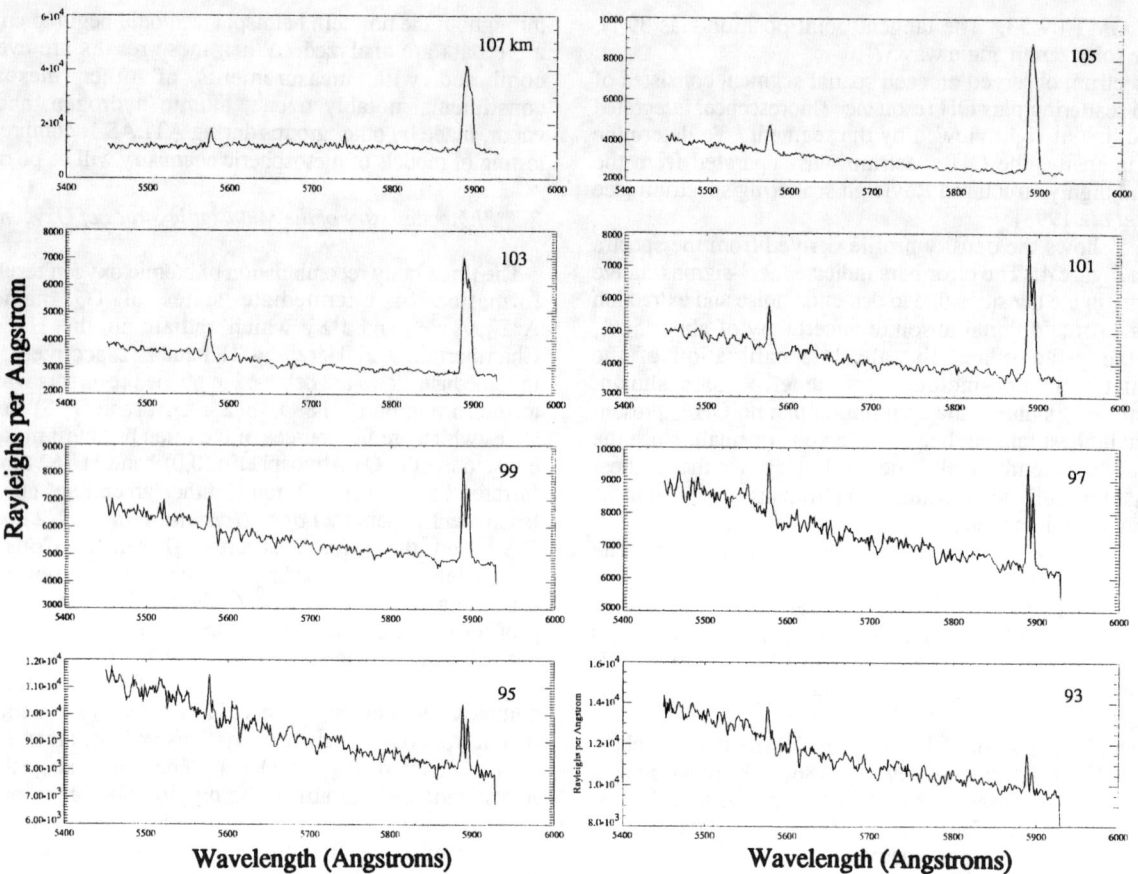

Figure 6. (a) An example of a partial spectral-spatial image of the O(^{1}S) 557.7 nm feature in the daytime mesosphere superimposed on the Rayleigh continuum. The full spatial image covers 20 km resolved to 2 km. Also seen is the sodium doublet at 589 nm. Final orbit attitude information is not yet available.

spectrum shown in the two upper panels in Figure 7 provides a detailed record of the Herzberg I, II and Chamberlain bands. The bottom panel shows the O$_2$ Atmospheric (0,0) band at 762 nm, the O(^{1}S) feature at 557.7 nm and the OH Meinel bands originating from v′ = 4 to 9. In what follows we focus on the segment covering the wavelength range 275 to 300 nm.

Nine measurements were made on day 90 of 1992 of the 275 to 300 nm segment at tangent heights covering the range 90 to 113 km. Figure 8 shows measurements at four representative heights with absolute intensities that are similar to those reported by *Sharp and Siskind* [1989]. The integration time was 8 seconds per altitude step, with less than 5° latitudinal smearing for the sequence shown. No spectral smoothing was applied to the data. Synthetic spectral fits for each tangent height are also shown. The Hz I/Hz II ratio is > 6 in the spectral range from 275 to 300 nm. For this preliminary study this ratio was determined by incrementally increasing the Hz II band intensity (after obtaining a best fit for the Hz I bands alone) until visually detectable departures from the ISO data became observable. The spectral code used for this purpose does not reproduce the detailed rotational structure which is caused by spin splitting. A collaborative effort with T. Slanger is currently underway to model the spin splitting, but this lies beyond the scope of this paper. Within these constraints it was found that below 85 km altitude the spectra could be fit by a constant temperature of 200 ± 20 K. At 92 and 98 km the data could be fit with temperatures of 230 ± 20 and 250 ± 30 K respectively. The vibrational distributions were determined independently at each tangent height and were found to lie within ± 10% of the mean distribution shown in Figure 9 for the level v′ = 3 to 8. The values shown for v′ = 0, 1, 2, 9 and 10 are consistent with the observations but, because of weak intensities, have no quantitative significance. It can be readily shown that the height independence of the relative column populations implies height independence in the relative vibrational volume number densities.

Figure 10 shows the O$_2$(A$^3\Sigma_u^+$) column abundance (N) derived as a function of altitude for all of the levels v′ = 3 to 8 using the expression

$$N = \frac{I_{v'v''}}{A_{v'v''} n_{v'}}$$

where v′ and v″ designate the upper and lower vibrational

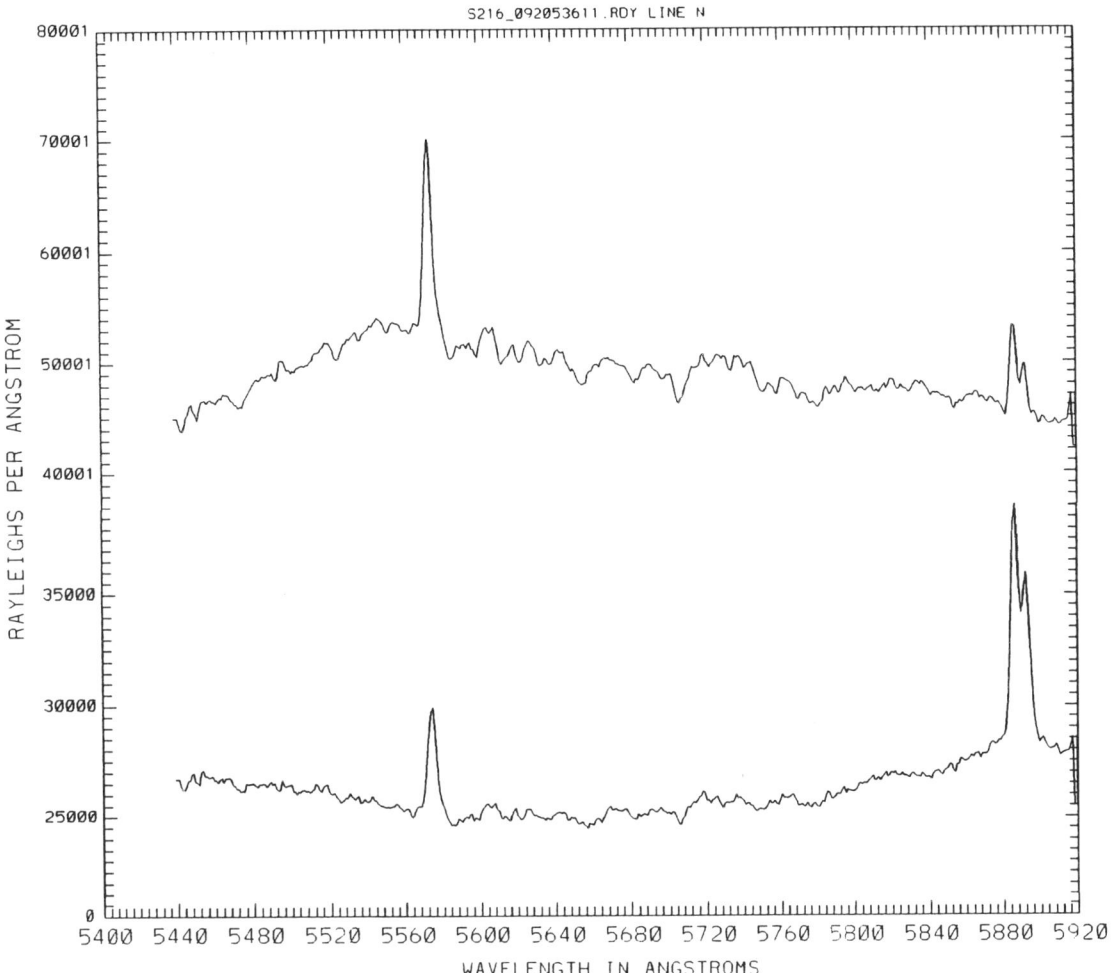

Figure 6. (b) Examples of dayglow limb observations with horizontal slit of the O(^{1}S) 557.7 nm and sodium emissions. The difference in tangent heights is 10.3 km. Final orbit attitude information is not yet available to determine the absolute tangent heights.

levels of the transition respectively, I is the corresponding measured band intensity, $n_{v'}$ the relative (fractional) population of the v' level determined by synthetic spectral fitting the data, and $A_{v'v''}$ is the Einstein coefficient for the v' - v" transition. In practice N is the average value computed using the brightest spectrally pure emissions in the observational spectral window. Values for the Einstein coefficients were taken from *Bates* [1992]. The results presented in Figure 10 correct an error in the values given by *Owens et al.* [1993]. The altitudes have also been revised using final post flight orbit attitude data.

The absence of any significant dependence of the relative population on altitude for the height range shown is in agreement with the theoretical calculations of *Lopez-Gonzalez et al.* [1992], the photometric rocket results of *Murtagh et al.* [1986] *and Kita et al.* [1988]. On the other hand, *Siskind and Sharp* [1990], detected measurable differences for v' = 5 and 7 to 9 above 95 km. At these heights the relative populations at v' = 7 were 0.85 and 0.55 respectively and at v' = 8 were 0.25 and 0.05 which is not necessarily inconsistent with model calculations which show a height dependence in the distributions above about 100 km.

In current models the A, A' and c states are continuously formed by three-body atomic recombination into some vibrational level higher than v' = 10, followed by collisional deactivation into lower levels. The populations of the lower levels are then controlled by single or multiple quantum vibrational quenching, electronic quenching, and spontaneous emission. Calculations by *McDade et al.* [1982], for example, used rate coefficients independent of vibrational level, and their model predicted large variations in the relative vibrational distribution with altitude. *Murtagh et al.* [1986] were the first to explain the altitude invariance of the Hz I distribution, and

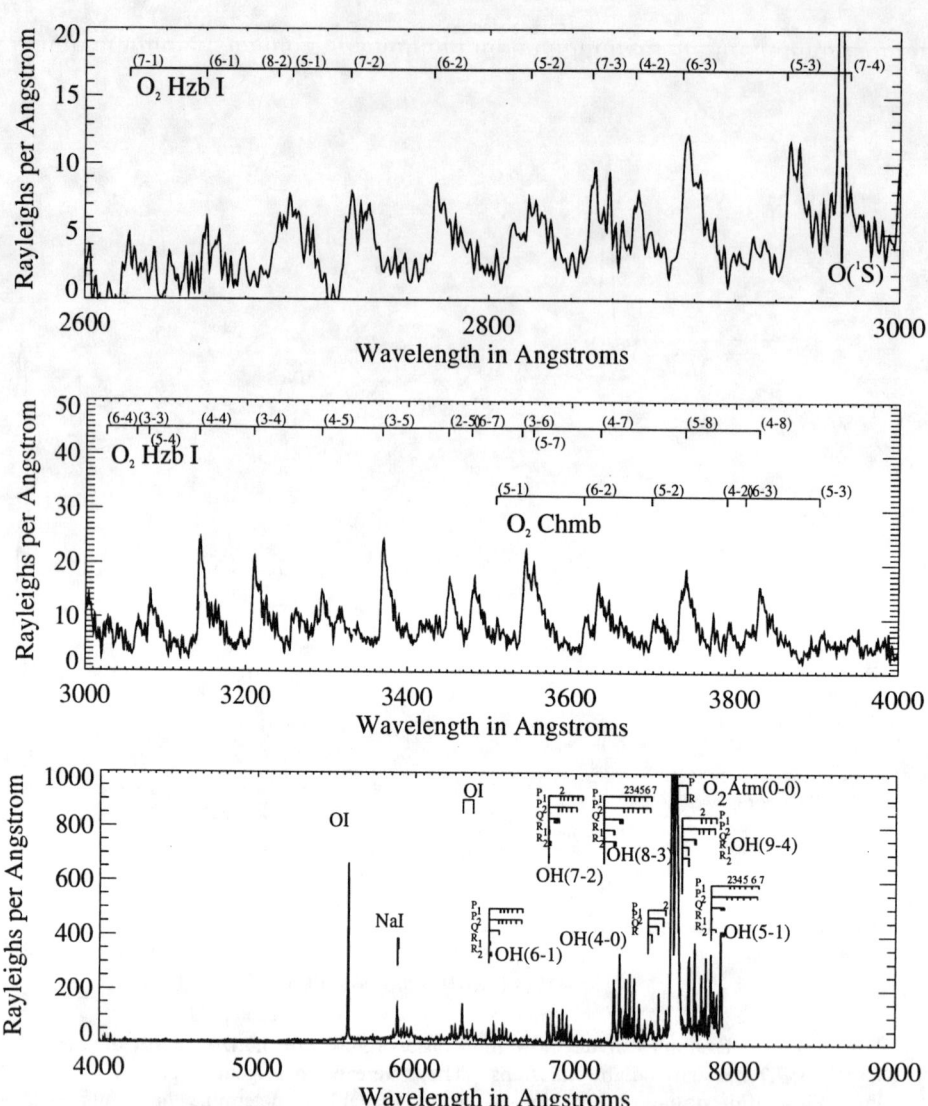

Figure 7: Spectral images of mesospheric nightglow over New Guinea measured by the ISO on ATLAS 1 at ~ GMT 18:05 on day 88, 1992 at an altitude of ~ 80 km. (Note: This Figure replaces Figure 1 of Owens et al, [1993].)

pointed out that the variation with altitude would essentially vanish if the electronic quenching coefficients were faster than the laboratory measurements had suggested.

An *ad hoc* scheme used by *Lopez Gonzalez et al.* [1992] not only yields an approximate altitude independent population distribution, but also reproduces a deficit in the population at v' = 4 and 5 observed by *Stegman and Murtagh* [1991]. The ISO observations (see Figure 9) suggest a slight depletion in v' = 6.

The rocket observational database used in the analysis discussed above was acquired mainly by photometric methods with large filter bandpasses. The only available altitude dependent spectral observations [*Siskind and Sharp*, 1990] exhibited low signal to noise ratios which precluded obtaining detailed information on the altitude profiles of the Herzberg I bands. The ISO data provide spatially resolved spectra which should allow very much tighter constraints to be placed on the model parameters. The examples presented here represent only a sample of a larger diverse set of mesospheric observations, which will stimulate many other studies, providing powerful tests of the atomic oxygen recombination chemistry.

2.4 Thermospheric Photochemistry

2.4.1 *Modeling Key Thermospheric Emissions*. During the ATLAS 1 mission the ISO obtained the first detailed spectral atlas of the dayglow over a broad wavelength range (60

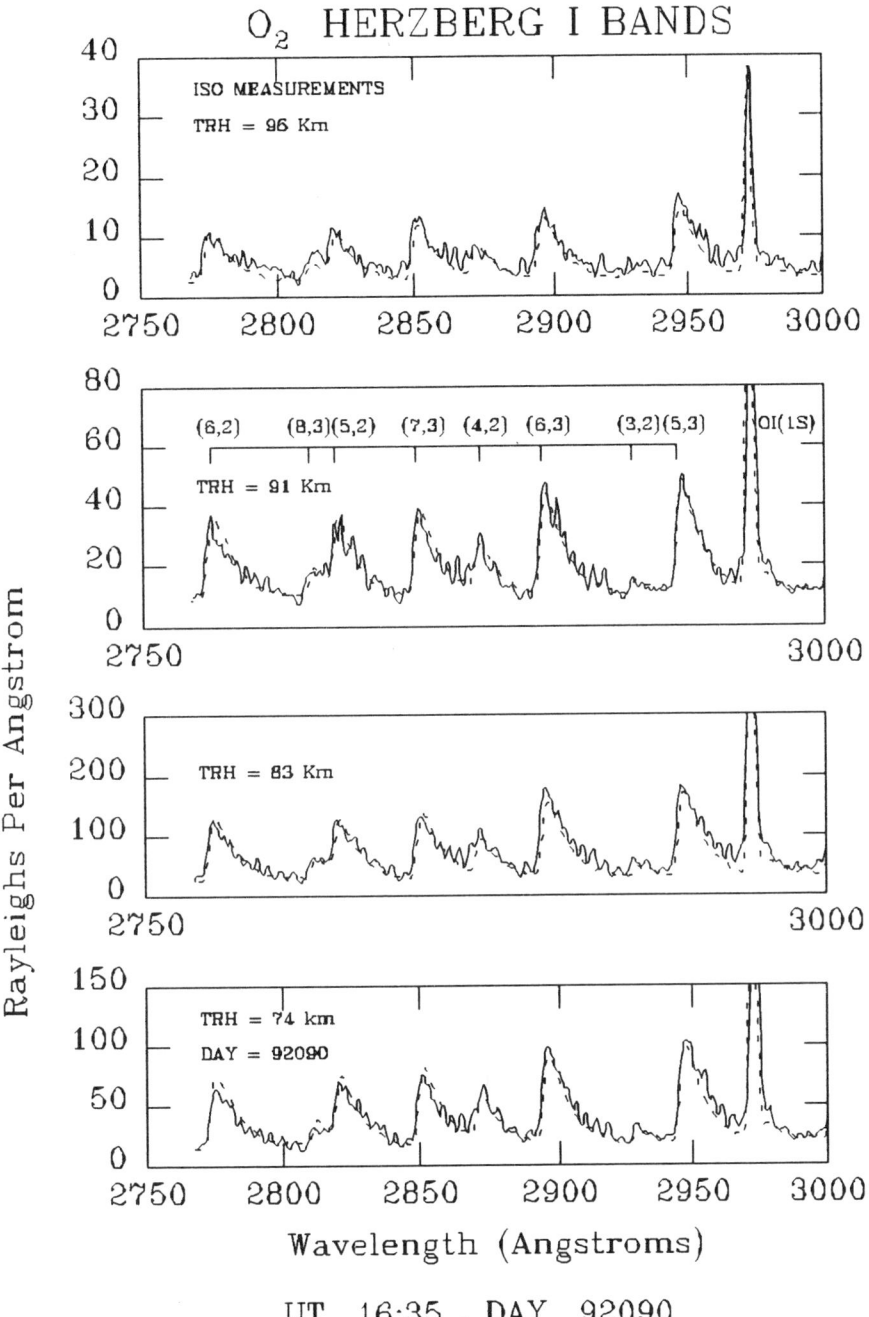

Figure 8: Synthetic spectral fits (dashed) to measurements (full line) of the Herzberg I bands at four heights, observed on day 90, 1992 at ~ GMT 16:35 over South Central Australia. (Note: The tangent ray heights shown correct the values given by Owens *et al*, [1993].)

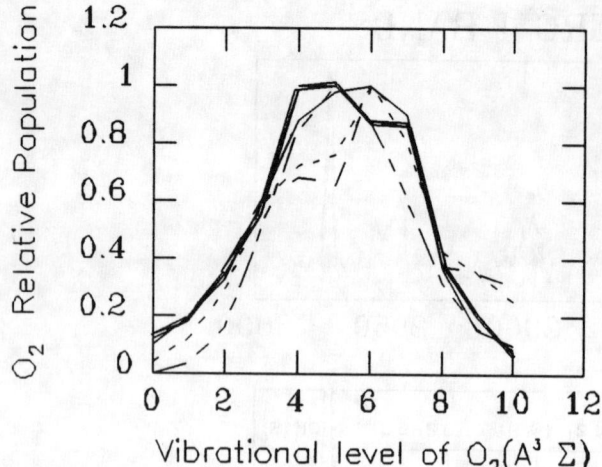

Figure 9: Vibrational populations. Thick solid curve: ISO distribution; Long dashed lines: *Stegman and Murtagh* [1991]; Thin solid line: *Slanger and Heustis* [1981]; Long dash, three short dashes: *Lopez-Gonzales et al.* [1992]: Model with electronic deactivation vibration independent rate coefficients; Dotted line: *Lopez-Gonzales et al.* [1992]: Model with electronic deactivation vibration dependent.

nm to 840 nm). Spectral emissions over this wavelength range were obtained as a function of altitude, allowing comparisons to be made with current thermospheric photochemistry models. Much of the present understanding of thermospheric photochemistry is based on the work done using the multi-instrument complement onboard the Atmosphere Explorer satellites flown in the 1970's. It is therefore of considerable interest to answer the question of how well that basic photochemistry predicts a large number of different airglow emissions measured almost 20 years later. In this section we report the results of running our comprehensive thermospheric FLIP model [see *M. R. Torr et al.*, 1990; *Richards et al.*, 1986] for conditions appropriate to the time of the ATLAS 1 mission [*D. G. Torr et al.*, 1993] and a comparison of the slant path intensities computed (as a function of altitude, latitude, longitude, and local time) with a dozen major emissions measured in the course of a particular observing sequence. Bearing in mind that the model has been run in a predictive mode, and that no attempt has yet been made to iterate the fit to the data, it is found that the agreement is good, indicating that the major processes controlling the thermospheric airglow are relatively well understood. By modeling a large number of emissions simultaneously, a diverse set of production and loss mechanisms are tested, providing valuable multiple constraints on the theory.

For the study, we selected a particular observation sequence in which the instrument line-of-sight was scanned down through the atmosphere by executing a special orbiter roll maneuver about the nose tail axis of the vehicle. The vehicle was moving to the east and north. The line of sight of the instrument was pointed in the wake direction with the slit parallel to the horizon. The solar and geomagnetic indices are given in Table 2.

Figure 1 shows the coverage at a selected height in the spectral domain. However, at each wavelength the data were also acquired in the altitude domain. An example is shown in Figure 11. From data such as these, we can determine slant path intensity profiles for any of the emission features of interest.

The FLIP model is well suited for studies of the airglow as it calculates all the excited states (including metastable and vibrational) of the atoms and molecules of the region, ionized or neutral. The concentrations of all the minor gases are computed from 80 km in one hemisphere, along the magnetic field line to 80 km in the conjugate hemisphere. Thus no artificial upper boundary is assumed for photoelectron and heat flows. Details of the photochemistry and the model are given elsewhere [*M. Torr et al.*, 1990]. The major neutral atmosphere is provided from MSIS-86 [*Hedin*, 1987] for the appropriate solar and magnetic conditions, as well as season, geographic location, and local time. We project a line of sight through the four-dimensional model output and compute line-of-sight intensities for any selected emission at any point along the orbit.

In the comparisons shown below, the model was run for the Ap and F10.7 conditions prevailing on the day in question. No attempt has been made to refine the fit to the data. Our objective was to see how well the model would reproduce the measurements without iterating parameters. The results of the comparison of the measured and modeled slant path intensities are given in Figures 12 and 13. While there is some interesting detail in the 630 nm and 520 nm results near 250 km, the fit is generally rather good for all 12 emissions.

2.4.2. *The N_2^+ First Negative Bands.* In an analysis of N_2^+ First Negative emissions observed on Spacelab 1, *M. Torr et al.* [1992] reported intensities about a factor of 5 to 10 larger

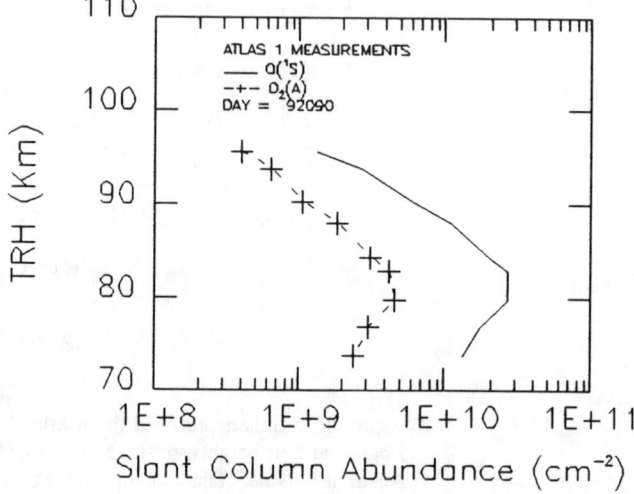

Figure 10: The $O_2(A)$ slant column abundance derived as a function of tangent ray height using the ISO vibrational distribution shown in Figure 3. Also shown is the simultaneously measured $O(^1S)$ abundance. (Note: This Figure updates Figure 4 of Owens *et al*, [1993].)

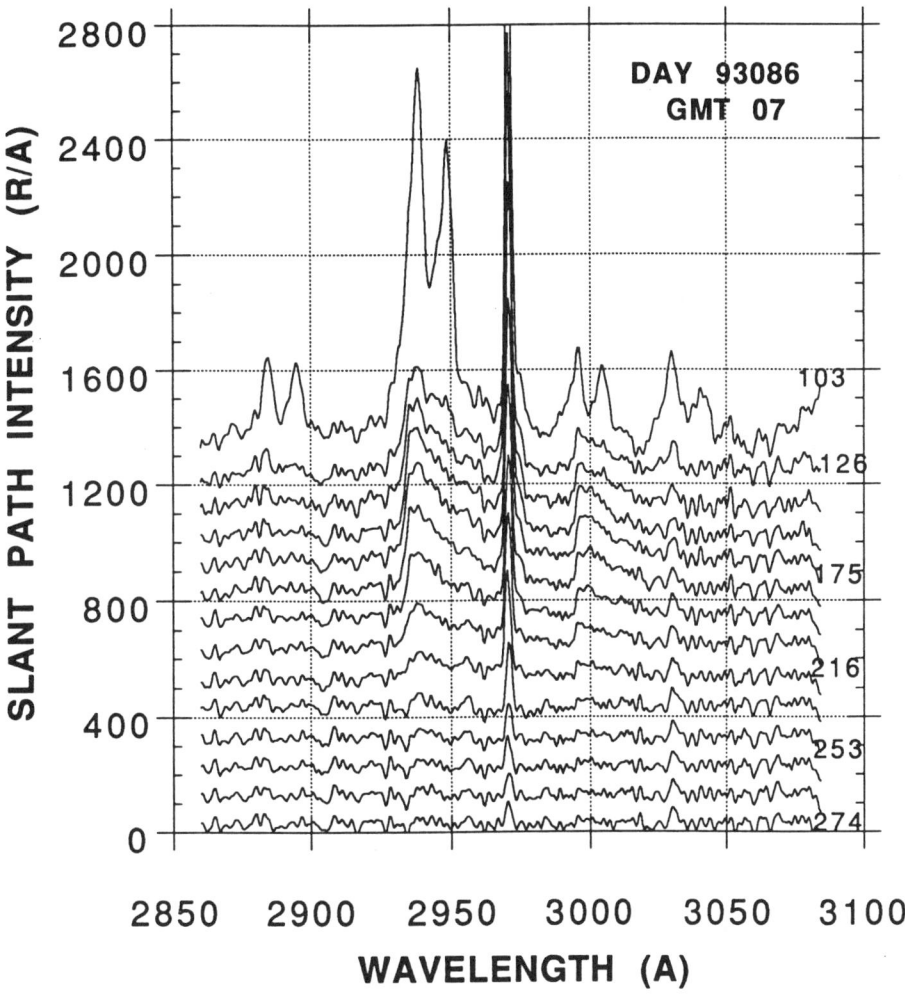

Figure 11: Example of the spectra obtained at different tangent ray heights in the course of the roll maneuver.

than those predicted by the FLIP model, suggesting serious plasma contamination of the shuttle environment. *M. Torr et al.* [1993a] have repeated the study using ISO ATLAS 1 measurements of the 391.4 nm emission. Figure 14 shows examples of the measured N_2^+ First Negative spectra and Figure 12 shows a comparison of experimental and theoretical slant path intensities. In this case good agreement is obtained between measured and modeled intensities. The observations are consistent with the Atmosphere Explorer photochemistry with the N_2^+ dissociative recombination rate coefficient a constant at ~ 2.7×10^{-7} cm^{3}s^{-1} [*Abdou et al.*, 1984; *D. Torr*, 1985], in agreement with recent laboratory measurements by *Canosa et al.* [1991]. *M. Torr et al.* [1993a] also published results on the $N_2^+(B^2\Sigma_u^+)$ vibrational distribution. All efforts to model these distributions, including a variety of ways of distributing the products of the reaction $O^+(^2D) + N_2 \rightarrow N_2^+ (X)_{v > 0} + O$ have failed to reproduce the observations. Work is continuing on this problem.

2.4.3. *N(^{2}P) Dayglow*. *M. Torr et al.* [1993b] have reported an analysis of the first measurements of the N(^{2}P) feature at 346.6 nm in the dayglow. Photodissociation of N_2 is identified as the major source with radiative decay to N(^{2}D) and quenching by O as the main sinks. The rate coefficient for quenching by O was determined to be ~ 3×10^{-11} cm^{3}s^{-1}.

3. CONCLUSION

The Imaging Spectrometric Observatory (ISO) was used to acquire a comprehensive database of emissions from the ionosphere, thermosphere and mesosphere. In particular, the results obtained have demonstrated the ability to make measurements in the mesosphere and lower thermosphere during the daytime. Measurements of $O^+(^2P)$ 732 nm emission were used to retrieve thermospheric composition and temperature between 120 and 400 km. Measurements of the $A^2\Sigma - X^2\Pi$ (0,0) OH band were used to obtain the

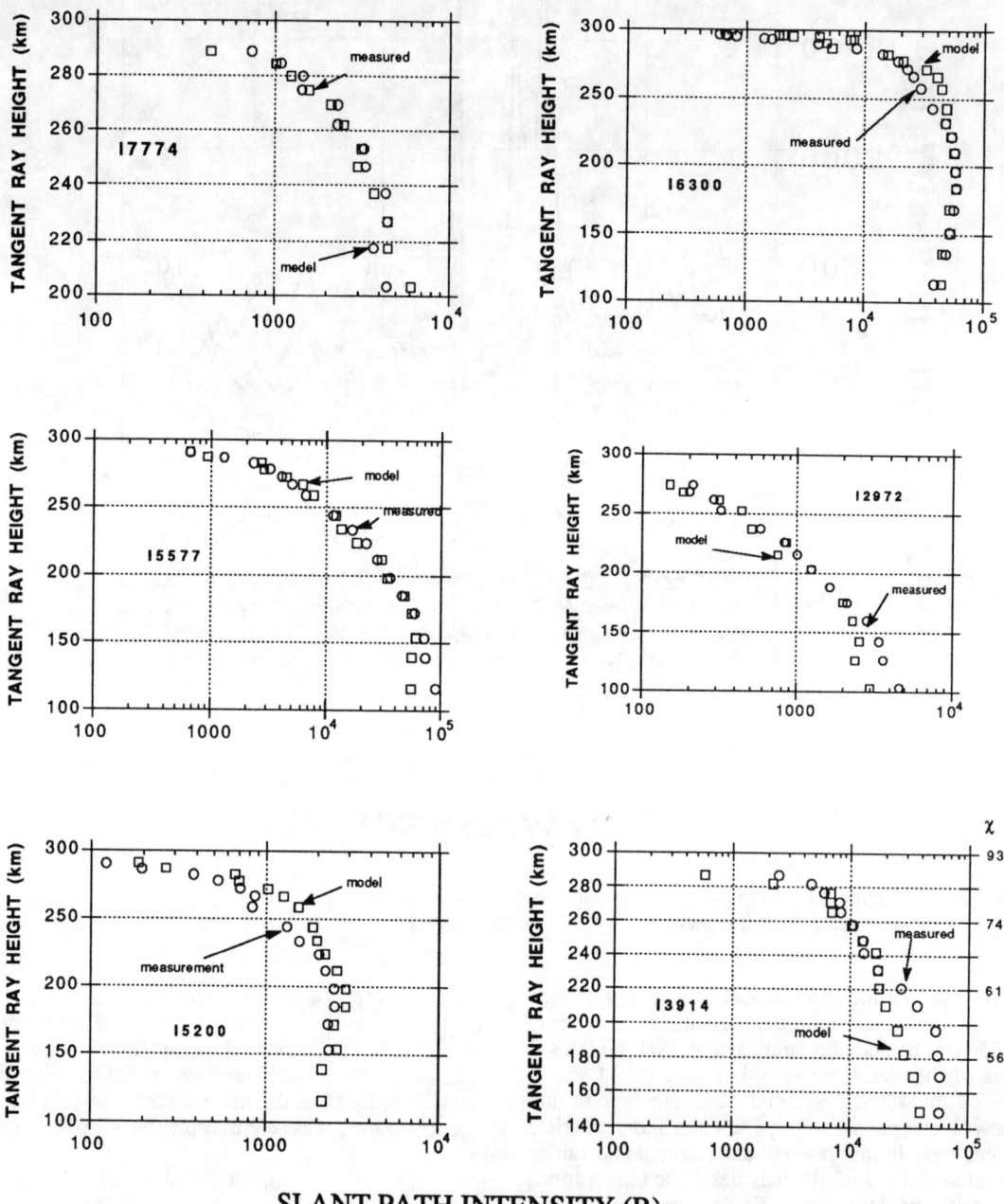

Figure 12: Comparison of measured and modeled slant path intensity for: OI 7774A, OI 6300A, OI 5577A, OI 2972A, NI 5200A and N_2^+ 1N 3914A. The variation of solar zenith angle is shown on the right hand axis of the 3914A panel.

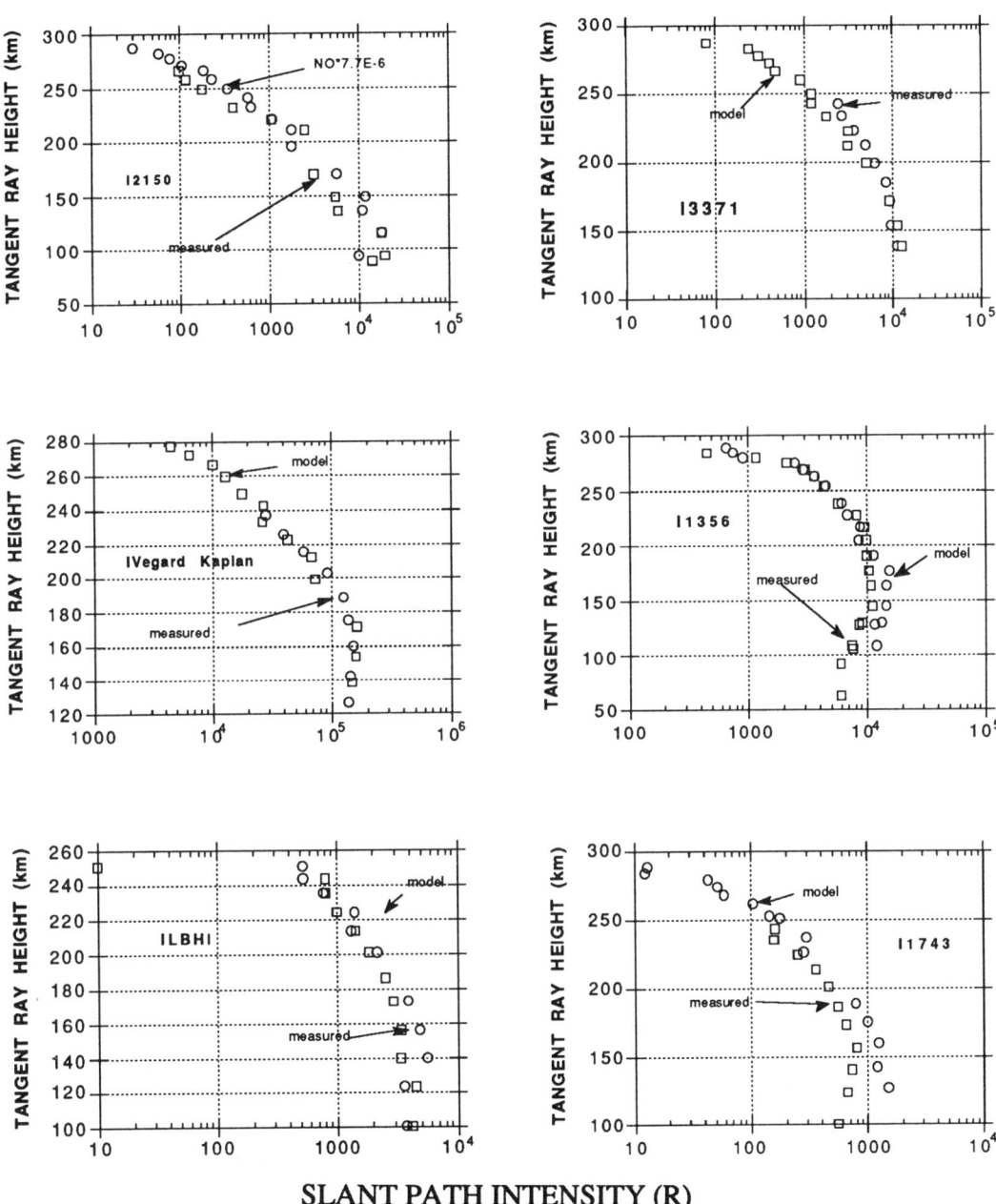

Figure 13: Comparison of measured and modeled slant path intensity for: NO γ 2150A, N_2 2P 3371A, N_2 V-K (0-6, 0-7), OI 1356A, N_2 LBH 1670-1850 A, NI 1743A.

The thermosphere measurements were made as part of a coordinated semi-global study with the groundbased ionospheric community, with ATLAS 1 providing composition and temperature. The OH measurements, when combined with measurements of other mesospheric constituents made by other ATLAS 1 instruments, (notably ozone, atomic hydrogen, and water vapor) should provide a comprehensive database for testing models of mesospheric chemistry.

Spectral-spatial images were obtained of the Herzberg 1 (and II) and Chamberlain bands of O_2, and the $O(^1S)$ 557.7 or 297.2 nm emissions. The band emissions were used to determine the vibrational populations of the associated electronic states. The results confirmed earlier rocket photometric measurements of the altitude invariance of the concentration of OH in the mesosphere as a function of altitude. distribution between ~ 70 and ~ 100 km. The quality of the data is good enough to separate the Hz I and II bands in the future, once synthetic spectral fitting codes have been upgraded to include the fine structure observed, which is attributed to spin splitting. The nature of the altitude dependence of the $c^1\Sigma_u^-$ (Herzberg II) vibrational populations should shed considerable light on the processes which quench the c^1 state. We are in the process of developing an algorithm for retrieving atomic oxygen from measurements of $O(^1S)$ emissions.

Several studies have been conducted of thermospheric minor constituents. Measurements of the N_2^+ first negative bands revealed higher vibrational temperatures than those predicted by resonance fluorescence scattering, which suggests that the reaction of $O^+(^2D)$ with N_2 may be a significant source of N_2^+ vibrational excitation. While model results are consistent with the absolute intensities, it has proved difficult to account for the observed vibrational distributions, even taking the reaction of N_2 with $O^+(^2D)$ into account. Measurements of the $N(^2P)$ emission at 346.6 nm were compared with model results, and the major sources and sinks evaluated.

The studies reported in this paper cover only a small subset of those planned for the ISO database. It is hoped that in the future there will be greater community utilization of this database.

Acknowledgments. The space-based component of work was funded by NASA Contract NAS8-37106 to the University of Alabama in Huntsville and the groundbased coordination effort by NSF Grant ATM-9018165. We thank the Atlantis crew, the ISO team, the MSFC POCC team and Dr. H. Rassoul and team of the Florida Institute of Technology for the support provided during the mission. Dr. Richards was supported under grant NAGW-996 to the University of Alabama in Huntsville.

REFERENCES

Abdou, W. A., D. G. Torr, R. G. Richards, M. R. Torr and E. L. Breig, *J. Geophys. Res.*, 89, 9069-9079, 1984.

Allen, Mark, Jonathan I. Levine and Yuk L. Yung, The vertical distribution of Ozone in the mesosphere and lower thermosphere, *J. Geophys, Res.*, 89, 4841-4872, 1984.

Anderson, James G., Rocket-borne ultraviolet spectrometer measurement of OH resonance fluorescence with a diffusive

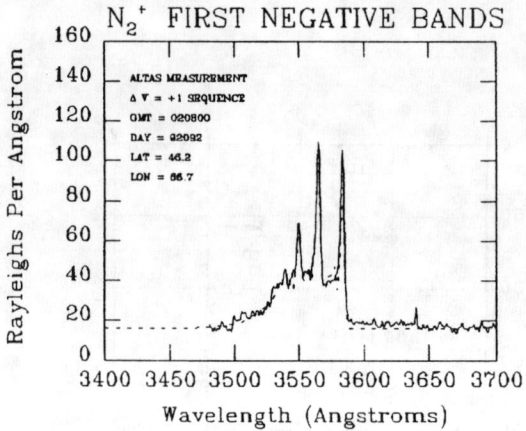

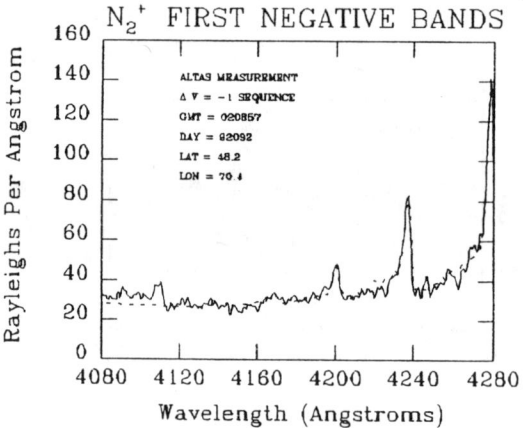

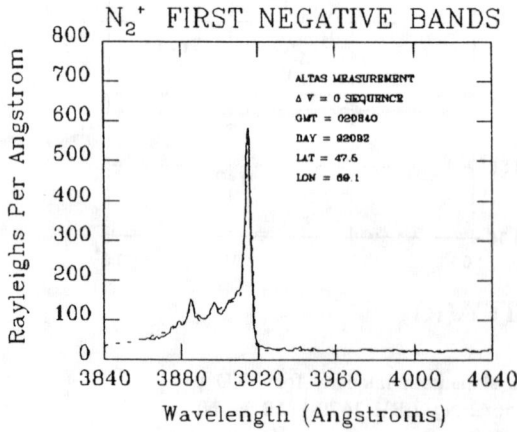

Figure 14: Example of the First Negative bands of N_2^+ observed by the ISO on ATLAS 1 at 300 km viewing into the wake.

transport model for mesospheric photochemistry, *J. Geophys. Res.*, 76, 4634-4652, 1971a.

Anderson, James G., Rocket measurement of OH in the mesophere, *J. Geophys. Res.*, 76, 7820-7824, 1971b.

Anderson, James G., The absolute concentration of OH($X^2\Pi$) in the earth's atmosphere, *Geophys. Res. Lett.*, 3, 165-168, 1976.

Bates, D. R., Nightglow emissions from oxygen in the lower thermosphere, *Planet. Space Sci.*, 40, 211-221, 1992.

Bertaux, Jean-Loup, Eric Quemerais, Florence Goutail, Observations of atomic deuterium in the mesosphere from ATLAS 1 with ALAE instrument, *Geophys. Res. Lett.*, 20, 507-510, 1993.

Bjarnason, Gudmundur G., Susan Solomon and Rowlando R. Garcia, Tidal influences on vertical diffusion and diurnal variability of ozone in the mesosphere, *J. Geophys. Res.*, 92, 5609-5620, 1987.

Burnett, Elizabeth Beaver, Clyde R. Burnett and Kenneth R. Minschwaner, Periodic Behaviors in the observed vertical column abundances of atmospheric hydroxyl, *Geophys. Res. Lett.*, 16, 1285-1288, 1989.

Canosa, A., J. C. Gomet, B. R. Towe and J. L. Queffelec, Flowing afterglow Langmuir probe measurement of the $N_2^+(v=0)$ dissociative recombination rate coefficient, *J. Chem. Phys.*, 94, 7159, 1991.

Carli, Bruno, Massimo Carlotti, Bianca M. Dinelli, Francesco Mencaraglia and Jae H. Park, The Mixing ratio of the stratospheric hydroxyl radical from far infrared emission measurements, *J. Geophys. Res.*, 94, 11049-11058, 1989.

Chang, T., D. G. Torr, P. G. Richards and S. C. Solomon, Re-evaluation of the $O^+(^2P)$ reaction rate coefficients derived from Atmosphere Explorer-C observations, *J. Geophys. Res.*, in press, 1993.

Fennelly, J. A. and D. G. Torr, Photoionization and photoabsorption cross sections of O, N_2, O_2 and N for aeronomic calculations, *Atomic Data and Nuclear Data Tables*, 51, 321, 1992.

Fennelly, J. A., D. G. Torr, P. G. Richards, M. R. Torr and W. E. Sharp, A method for retrieval of atomic oxygen density and temperature profiles from ground-based measurements of the $O^+(^2D-^2P)$ 7320-Å twilight airglow, *J. Geophys. Res.*, 96, 1263, 1991.

Fennelly, J. A., D. G. Torr, M. R. Torr, P. G. Richards and S. Yung, Retrieval of Thermospheric Atomic Oxygen, Nitrogen and Temperature From the 732 nm Emission Measured by the ISO on ATLAS 1, *Geophys. Res. Lett.*, 20, 527-530, 1993a.

Fennelly, J. A., D. G. Torr, P. G. Richards and M. R. Torr, Simultaneous retrieval of the solar EUV flux and neutral thermospheric O, O_2, N_2 and Temperature from twilight airglow, submitted to *J. Geophys. Res.*, April 1993b.

Heaps, William and Thomas J. McGee, Progress in stratospheric hydroxyl measurement by balloon-borne LIDAR, *J. Geophys. Res.*, 90, 7913-7921, 1985.

Hedin, A. E., MSIS-86 thermospheric model, *J. Geophys. Res.*, 92, 4649, 1987.

Kita, K., N. Iwagami, T. Ogawa, A. Miyashuta and H. Tanabe, Height distributions of the night airglow emissions in the O_2 Herzberg I system and oxygen green line from simultaneous rocket observation, *J. Geomagn. Geolect.* 40, 1067, 1988.

Leko, J., D. G. Torr, T. Chang, M. R. Torr, M. F. Morgan, P. G. Richards, T. Baldridge and H. Rassoul, Implications of large variability observed on ATLAS 1 in mesospheric oxygen airglow for atomic oxygen, Presented at the Fall Meeting of the American Geophysical Union in San Francisco, California, December 1993, Abstract Published in *EOS*, 74, No. 43, 472, October 26, 1993.

Lopez-Gonzalez, M. J., J. J. Lopez-Moreno and R. Rodrigo, Altitude and vibrational distribution of the O_2 ultraviolet nightglow emissions, *Planet. Space Sci.*, 40, 913-928, 1992.

McDade, I. C., E. J. Llewellyn, R. G. H. Greer and D. P. Murtagh, The altitude dependence of the $O_2(A^3\Sigma_u^+)$ vibrational distribution in the terrestrial nightglow, *Planet. Space Sci.*, 30, 1133, 1982.

Morgan, F., D. G. Torr and M. R. Torr, Preliminary Measurements of Mesospheric OH $X^2\Pi$ by ISO on ATLAS 1, *Geophys. Res. Lett.*, 20, 511-514, 1993.

Murtagh, D. P., I. C. McDade, R. G. H. Greer, J. Stegman, G. Witt and E. J. Llewellyn, ETON 4: An experimental investigation of the altitude dependence of the $O_2(A^3\Sigma_u^+)$ vibrational population in the nightglow, *Planet. Space Sci.*, 34, 811, 1986.

Richards, P. G. and D. G. Torr, An investigation of the consistency of the ionospheric measurements of the photoelectron flux and solar EUV flux, *J. Geophys. Res.*, 89, 5625, 1984.

Richards, P. G. and D. G. Torr, Thermal coupling of conjugate ionospheres and the tilt of the earth's magnetic field, *J. Geophys. Res.*, 91, 9017-9021, 1986.

Richards, P. G., J. A. Fennelly and D. G. Torr, EUVAC: A solar EUV flux model for aeronomic calculations, *J. Geophys. Res.*, in press, 1994.

Rodrigo, R., J. J. Lopez-Moreno, M. Lopez-Puetras, F. Moreno and A. Molina, Neutral atmospheric composition between 60 and 220 km: a theoretical model for mid-latitudes, *Planet. Space Sci.*, 34, 723-743, 1986.

Rusch, D. W., D. G. Torr and P. B. Hays, The OII (7319-7330-Å) dayglow, *J. Geophys. Res.*, 82, 719, 1977.

Sharp W. E. and D. E. Siskind, Atomic emission in the ultraviolet nightglow, *Geophys. Res. Lett.*, 16, 1453-1456, 1989.

Siskind, D. E. and W. E. Sharp, A vibrational analysis of the $O_2(A^3\Sigma_u^+)$ Herzberg I system using rocket data, *Planet. Space Sci.*, 38, 1399-1408, 1990.

Slanger and Heustis, $O_2(c^1\Sigma_u^- \rightarrow X^3\Sigma_g^-)$ emission in the terrestrial nightglow, *J. Geophys. Res.*, 86, 3551-3554, 1981.

Stegman, J. and D. P. Murtagh, The molecular oxygen band systems in the u.v. nightglow: measured and modelled, *Planet. Space Sci.*, 39, 595-609, 1991.

Stimpfle, R. M., O. O. Wennberg, L. B. Lapson and J. G. Anderson, Simultaneous, *in situ* measurements of OH and HO_2 in the stratosphere, *Geophys. Res. Lett.*, 17, 1905-1908, 1990.

Torr D. G., The Photochemistry of the upper atmosphere, *The Photochemistry of Atmospheres*, 165-278, Ed. by J. S. Levine, Academic Press, 1985.

Torr, D. G., M. R. Torr, W. Swift, J. Fennelly and G. Liu, Measurements of OH($X^2\Pi$) in the stratosphere by high resolution UV spectroscopy, *Geophys. Res. Lett.*, 14, 937-940, 1987.

Torr, M. R., Scientific objectives of the ATLAS 1 shuttle mission,

Geophys. Res. Lett., 20, 487-490, 1993.

Torr, M. R. and D. G. Torr, Ionization frequencies for solar cycle 21: Revised, *J. Geophys. Res., 90,* 6675, 1985.

Torr, M. R., D. G. Torr, P. G. Richards and S. Yung, A mid- and low-latitude model of thermospheric emissions: 1. $O^+(^2P)$ 7320 Å and $N_2(2P)$ 3371 Å, *J. Geophys. Res., 95,* 21147-21168, 1990.

Torr, M. R., D. G. Torr and P. G. Richards, The N_2^+ first negative system in the dayglow from the spacelab 1, *J. Geophys. Res., 97,* 17075-17095, 1992.

Torr, M. R., D. G. Torr, T. Chang, P. G. Richards, T. W. Baldridge, J. K. Owens, H. Dougani, C. W. Fellows, W. Swift, S. Yung and J. Hladky, The First Negative Bands of N_2^+ in the Dayglow from the ATLAS 1 Shuttle Mission, *Geophys. Res. Lett., 20,* No. 6, 523-526, March 1993a.

Torr, M. R., D. G. Torr and P. G. Richards, $N(^2P)$ in the Dayglow: Measurement and Theory, *Geophys. Res. Lett., 20,* No. 6, 531-534, March 1993b.

Walker, J. C. G., Analytic representation of upper atmosphere densities based on Jacchia's static diffusion models, *J. Atmos. Sci., 22,* 462, 1965.

---

[1]The University of Alabama In Huntsville, Optical Aeronomy Laboratory, Department of Physics and the Center for Space Plasma and Aeronomic Research, Huntsville, Alabama 35899

[2]NASA/Marshall Space Flight Center, Huntsville, Alabama 35812

[3]The University of Alabama In Huntsville, Optical Aeronomy Laboratory, Department of Computer Science, Huntsville, Alabama 35899

Spatial Variability in O(^{1}S) and O$_2$($b^1\Sigma_g^+$) Emissions as Observed with the Wind Imaging Interferometer (WINDII) on UARS

W. E. Ward[1], E. J. Llewellyn[2], Y. Rochon[1], C. C. Tai[1], W. S. C. Brooks[2], B. H. Solheim[1] and G. G. Shepherd[1]

[1] *Institute for Space and Terrestrial Science, York University, 4700 Keele St., North York, Canada, M3J 1P3*

[2] *Institute of Space and Atmospheric Studies, University of Saskatchewan, Saskatoon, Saskatchewan, Canada, S7N 0W0*

On October 30, 1992, the Wind Imaging Interferometer on the Upper Atmosphere Research Satellite (UARS), observed the oxygen green line emission and several lines in the O$_2$ atmospheric band on alternate measurements. The resulting data permits a comparison of the spatial variations of these two emissions. In this short paper results from the dayside and nightside segment of one orbit during this day are presented and briefly analyzed. Particular features which stand out are the close spatial correlation between both emissions during the night, the considerable spatial variability of the O$_2$($b^1\Sigma_g^+$) emission during the day, and the increased O(^{1}S) emission during the day relative to the night.

1. INTRODUCTION

The Wind Imaging Interferometer on UARS is capable of observing a number of important airglow emissons in the vicinity of the mesopause [Shepherd et al. 1993a, see also Shepherd et al., 1994, this volume]. While its prime objective is to measure winds using Doppler shifts in these emissions, it also provides the emission rates which in themselves provide information on constituent concentrations and ultimately on transport effects. Since the analysis of the WINDII data set is still in its early stages, only the spatial morphology of the emission rate of two emissions, the O(^{1}S) and the ^{P}Q(7) line of the (0-0) band of the O$_2$($b^1\Sigma_g^+$) transition are presented in this paper. The data are from October 30, 1992. During this day WINDII was viewing each of these emission alternately, thus providing the opportunity to compare their spatial distributions.

The close linkage between the excitation mechanisms of these emissions in the nightglow, has been outined in several papers [McDade et al., 1986, Murtagh et al., 1990]. The current consensus is that both emissions are the result of a two step process involving the recombination of atomic oxygen (termed the Barth mechanism in the case of the O(^{1}S) emission) and have peaks near the mesopause, the green line peak, on average, being the higher of the two. Given this relationship between the emissions, their spatial variations should be similar.

For the dayglow, this close connection does not exist. In the vicinity of the mesopause, the oxygen green line emission is thought to be due to the same mechanism as during the night. In the lower thermosphere, dissociative recombination of O$_2^+$ and photoelectron impact make a contribution to the emission and dominate near 150 km. The O_2 atmospheric band system is the result of resonance scattering of sunlight, energy transfer from $O(^1D)$ and a small contribution from the night-time Barth type mechanism [Bucholtz et al., 1986], with the former two mechanisms dominating. In the vicinity of the mesopause, the green line emission is expected to be closely tied to the atomic oxygen concentration and the O_2 atmospheric emissions to the O_2 concentration.

2. MEASUREMENT DESCRIPTION

WINDII is an imaging wide angle Michelson interferometer, which, for each measurement, takes a sequence of images, each at a different path difference. From these images, profiles of wind, temperature and volume emission rate are obtained, for the volume of air sampled during the measurement. For each measurement, the atmosphere is sampled at two locations on the earth's limb: 45° and 135° relative to the spacecraft track. Details of the instrument and measurement process may be found in papers by Shepherd et al. [1993a] and Shepherd et al. [1994, this volume].

The UARS is in a 57° inclination orbit with an orbital period of about 100 minutes. For the emissions being discussed here, each measurement takes approximately 30 seconds, so that the spatial resolution for each emission is approximately 5° of latitude. In order to provide volume emission rate profiles for the green line, three profiles are taken: a thermospheric profile at 4 km resolution, followed by a mesospheric profile at 2 km resolution, and then a second thermospheric profile. All three profiles are used in the inversion of the data. For O_2 atmospheric band emissions, a narrow band filter is used to spatially separate light from the ^PQ and ^PP, 5 and 7 lines (0-0 band) on the detector so that all four lines are viewed simultaneously. All images in each measurement are used to determine the line of sight quantities corresponding to each emission. The line of sight intensities are then inverted to obtain volume emission rate profiles. The vertical resolution is 2 km.

In this paper, results from the night-time ascending portion of one orbit (UT from 10:29 to 11:03, LT from 21:43 to 4:18) and the subsequent day-time descending portion of the same orbit (UT from 11:05 to 11:48, LT from 5:03 to 13:39) for the forward field of view are presented. Figures 1 and 2 show the night-time results and Figures 3 and 4 the day-time results for the $O(^1S)$ and the $^PQ(7)$ line of the 0-0 band of the O_2 atmospheric band emission respectively. There are three frames in each figure. The largest frame shows height versus latitude plots of the emission rate, the upper right frame shows an individual height profile, and the lower right frame shows the observation points along the satellite track used for the contour plot. The contouring was accomplished using IDL with linear interpolation. Similar results to those presented here were obtained with the other field of view and, in the case of O_2, with the other observed lines in the band. The volume emission rate are in general precise to about ±1 % for the mean observed emission rate, and are accurate to about ± 10%.

3. DISCUSSION

3.1. Measurements during Night

The results presented in Figures 1 and 2 for nightglow data show the spatial distribution of both atomic and molecular emission to have analogous form with maxima in midlatitudes near ±40°. This structure resembles that reported by Shepherd et al. [1993b, 1994, this volume] and although the observations presented here are for equinox conditions rather than solstice conditions the conclusions presented there also hold in this case. Because of the similarity in excitation mechanism and dependence on atomic oxygen concentration for both emissions, the resemblance between the two distributions is expected. It is interesting to note that the O_2 emission increases more rapidly in the early morning (40° N) than the green line.

An additional feature of these figures which supports previous analysis [Llewellyn and McDade, 1984] is that the emission layer peak for the green line is in general the higher of the two. This can be seen both in the contour plots and in the individual emission profiles provided in the figures.

The height dependence of the emission is also of interest. Apart from regions of enhanced emission the heights of the emission peak are close to the accepted values of 97 km for $O(^1S)$ and 94 km for O_2 [Murtagh et al., 1990, Tarasick and Evans, 1993]. For regions of enhanced emission the heights of the layer peak appear several kilometers below these accepted heights. The tendency of the emission at the peak to be greater when the altitude of the peak is lower was noted previously by Ward et al. [1994].

3.2. Measurements during Day

As noted in the introduction, the daytime excitation mechanisms for these two emissions are sufficiently different that one would expect differences in their spatial

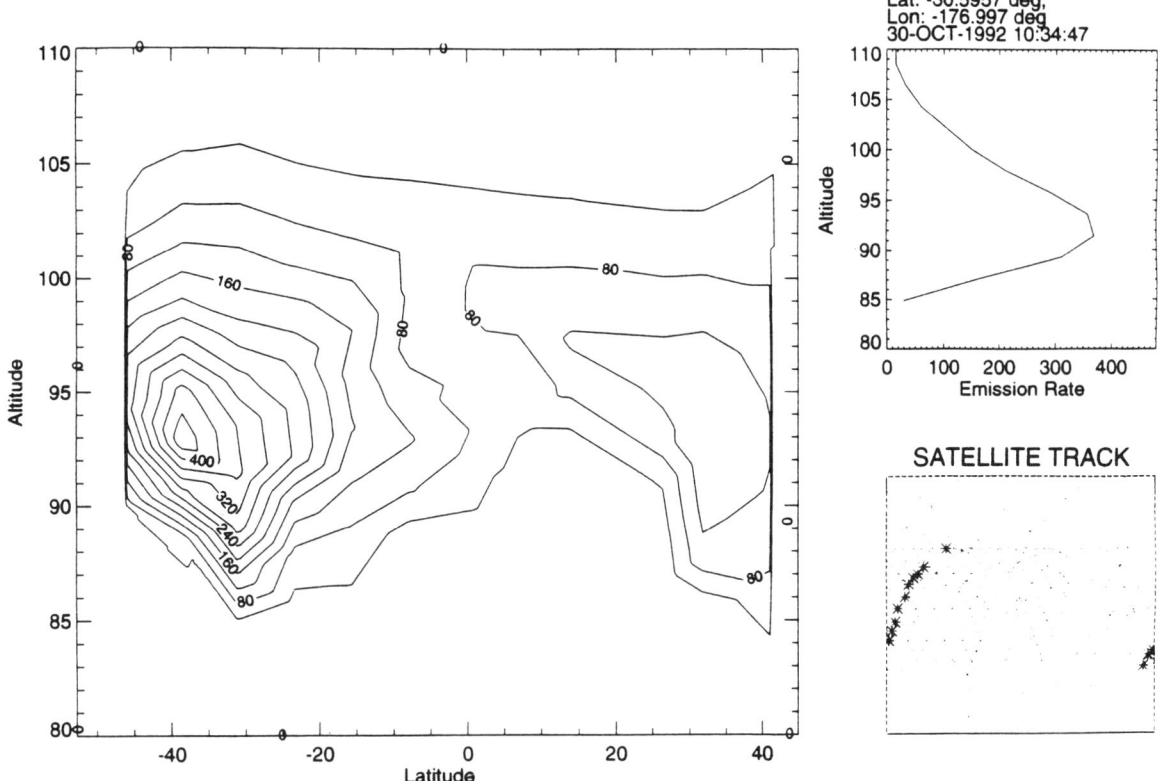

Fig. 1. Observations of the O(^{1}S) emission in the nightglow for a portion of an orbit on October 30, 1992. The large panel shows the emission rate (photons/cm^3/s) along the satellite track as measured in the forward field of view. The upper right panel shows an individual emission profile at -31° latitude, -177° longitude and 10:34:47 UT. The derived volume emission rates at successive altitudes are joined by straight lines. The panel on the lower right shows the satellite track and the location of the profiles used in generating the contour plot.

morphology. Figures 3 and 4 show this to be the case.

For the green line emission (Figure 3), two maxima appear, one in the lower thermosphere and the other near the mesopause at about 100 km. These correspond to maxima in the photo-electron impact/dissociative recombination and oxygen recombination mechanisms respectively. The former peak has been extensively discussed in the literature [see Bates, 1990 for references]. There seems to have been little discussion of the latter peak apart from the early analysis of Wallace and McElroy [1966] where the emission layer is described as being centered at about 95 km with a half width of 25 km.

The latitudinal variation in the emission rate in the upper peak is due to the variation in local time along the satellite orbit (and to a lesser extent the variation with latitude of the pathlength for solar radiation) and is expected. The photo-electron impact/dissociative recombination of O_2^+ peaks at local noon and is dependent on the amount of solar radiation present. It is less clear why the mesospheric peak should follow a similar pattern. Furthermore, it is apparent that although the mechanism in the lower peak is the same as that in the night, the emission rate is significantly greater and does not have the same morphology. This result suggests the need for an additional mechanism for this emission that is dependent on the presence of solar radiation.

The contour plot for the atmospheric band emission is considerably more variable than the green line plot and considerably broader than the corresponding nightglow features. Measurements of this emission have been reviewed by Tarasick and Evans [1993]. However, they do not show any volume emission rate profiles so com-

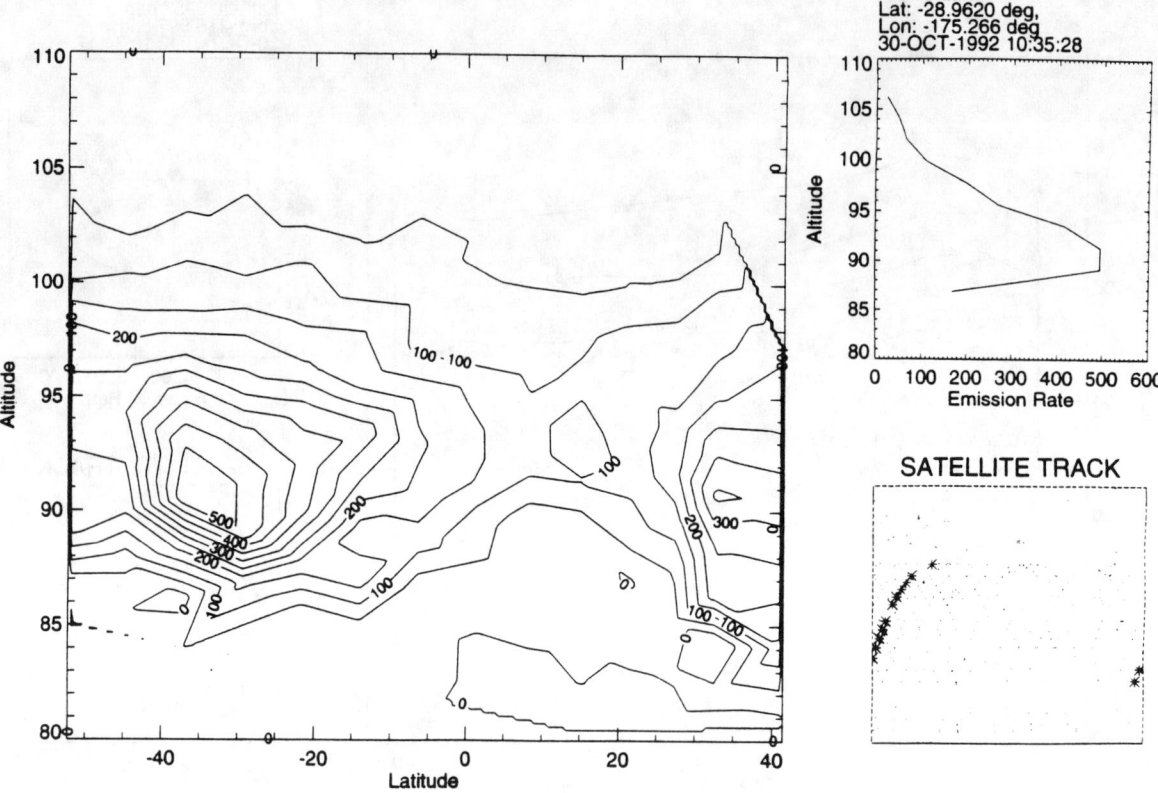

Fig. 2. As in Figure 1 except that the emission being observed here is the $^PQ(7)$ line of the 0-0 band of the O_2 atmospheric band. The profile is at -29° latitude, -175.3° longitude and 10:35:28 UT.

parison to the present results is difficult. There is a sharp cutoff at dawn (40° N) where the main excitation mechanisms cease and the recombination of atomic oxygen becomes the main source. This emission appears bright over a larger range of latitudes than does the mesospheric green line. The height of the emission peak decreases toward dawn (40° N), although whether this is a local time effect or latitudinal effect cannot be distinguished here.

The variability of this emission is of particular interest. An extreme example of this is shown in the individual emission rate profile shown in Figure 4 where a double layer is clearly seen. We believe this to be the first time such a feature has been observed in this emission in the dayglow. This signature is seen in all four O_2 lines observed with WINDII and in both fields of view so it is unlikely to be a temperature effect and is similar in form to double layers observed in $O_2(^1\Delta)$ in nightglow by Evans et al. [1972] and Witt et al. [1984]. Given the dependence of this emission in this region on the concentration of molecular oxygen, it seems that a more likely explanation is variability in O_2. Work, currently in progress [McDade et al., 1992], suggests features such as this may be a signature of gravity waves.

4. SUMMARY

Several of the emission features noted above are typical of this day of observation and worth noting. As expected, the night-time $O(^1S)$ and $O_2(b^1\Sigma_g^+)$ emissions show a high degree of spatial correlation. The $O(^1S)$ emission peak in general lies higher than the $O_2(b^1\Sigma_g^+)$ emission peak. The day-time mesosperic green line emission is generally brighter than the night-time green line emission. Finally, the day-time $O_2(b^1\Sigma_g^+)$ emission peak appears to show a height variation, increasing from dawn (40° N) to mid-day (-40 ° S). This may be a spatial or temporal variation.

The primary conclusion, which is becoming more and more evident as WINDII data is analysed is that there is considerable variability in constituents in the mesopause region. This is illustrated in the results presented here, with the nightglow results suggesting variability in

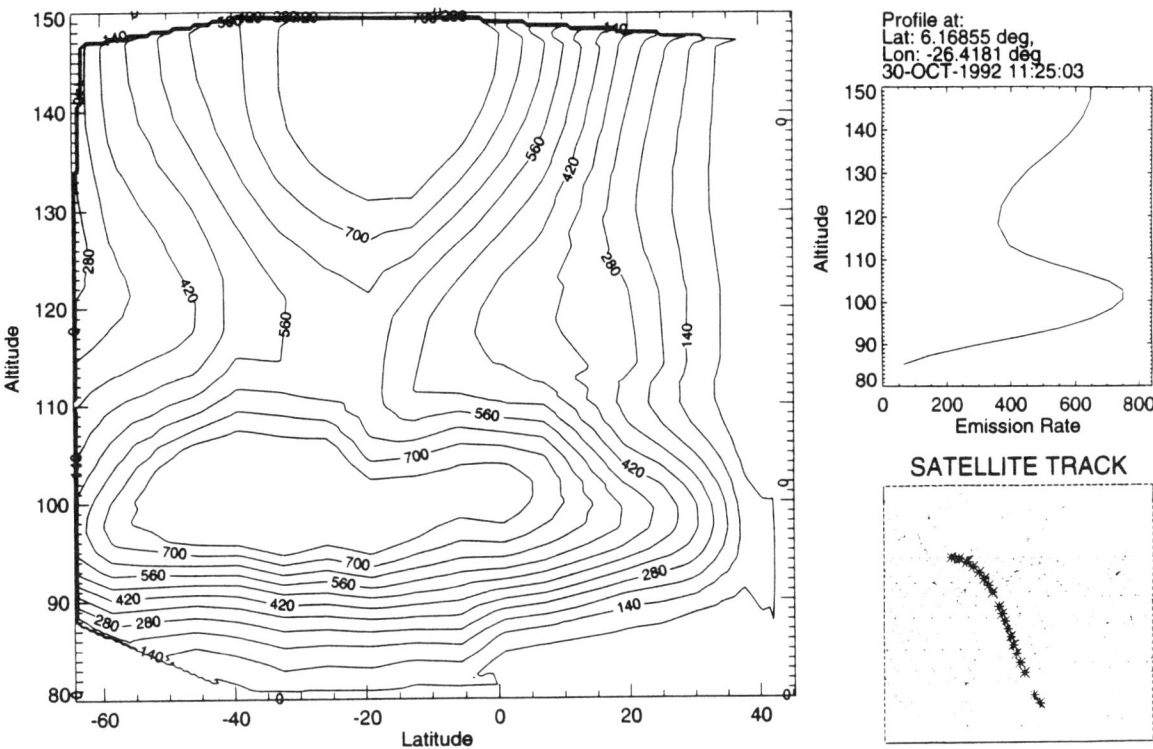

Fig. 3. As in Figure 1 except that the emission being observed here is the $O(^1S)$ emission in the dayglow. The profile is at 6.1° latitude, -26.4° longitude and 11:25:03 UT.

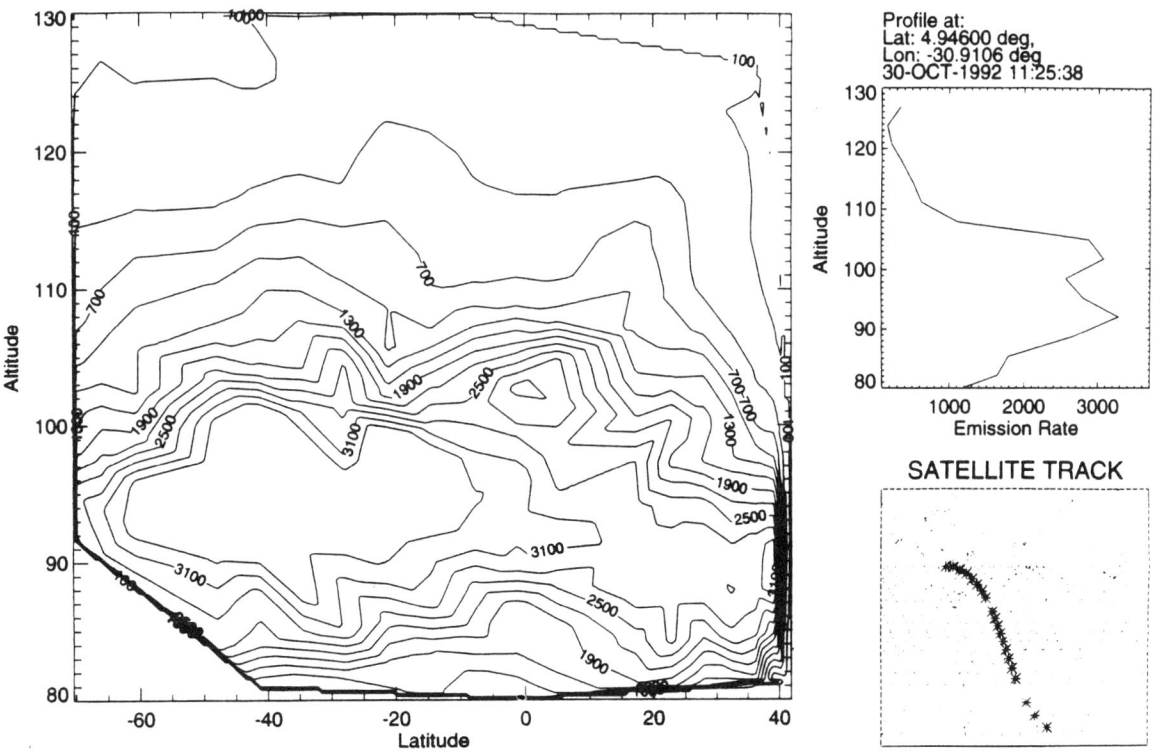

Fig. 4. As in Figure 1 except that the emission being observed here is the $^PQ(7)$ line of the 0-0 band of the O_2 atmospheric band in the dayglow. The profile is at 4.9° latitude, -30.9° longitude and 11:25:38 UT.

atomic oxygen and the dayglow results suggesting variability in molecular oxygen. The reason for this variability remains an outstanding problem. Whether it is due to dynamical effects, diffusive effects, photochemical effects or some combination of these, remains to be determined. As validated winds from WINDII become available, the answers to this and other questions are certain to be clarified.

Acknowledgments. The WINDII project was jointly sponsored by the Canadian Space Agency and the Centre Nationale d'Études Spatiales. The authors are grateful for their dedicated commitment and loyal support. The scientific analysis of the data is supported by the Natural Sciences and Engineering Research Council of Canada. The Institute for Space and Terrestrial Science is a designated Centre of Excellence supported by the Province of Ontario's Technology Fund.

REFERENCES

Bucholtz, A., W. R. Skinner, V. J. Abreu, and P. B. Hays, The dayglow of the O_2 atmospheric band system, *Planet. Space Sci.*, 34, 1031-1035, 1986.

Bates, D. R., Oxygen green and redline emission and O_2^+ dissociative recombination, *Planet Space Sci.*, 28, 889-902, 1990.

Evans, W. J. F., E. J. Llewellyn, and A. J. Vallance-Jones, Altitude distribution of the $O_2(^1\Delta)$ nightglow emission. *J. Geophys. Res.*, 77, 4899-4901, 1972.

Llewellyn, E. J., and I. C. McDade, Singlet molecular oxygen in planetary atmospheres, *J. Photochem.*, 25, 379-388, 1984.

McDade, I. C., W. S. C. Brooks, and E. J. Llewellyn, Gravity wave induced structure in night O_2 atmospheric band emission (abstract), *Eos Trans. AGU, Fall Meeting Suppl.*, 73, 43, 102, 1992.

McDade, I. C., D. P. Murtagh, R. G. H. Greer, P. H. G. Dickinson. G. Witt, J. Stegman, E. J. Llewellyn, L.Thomas, and D. B. Jenkins, ETON 2: quenching parameters for proposed precursors of $O_2(b^1\Sigma_g^+)$ and $O(^1S)$ in the terrestrial nightglow, *Planet. Space Sci.*, 34, 789-800, 1986.

Murtagh, D. P., G. Witt, J. Stegman, I. C. McDade, E. J. Llewellyn, F. Harris, and R. G. H. Greer, An Assessment of proposed $O(^1S)$ and $O_2(b^1\Sigma_g^+)$ nightglow excitation parameters, *Planet. Space Sci.*, 38, 43-53, 1990.

Shepherd, G. G. et al., WINDII the Wind Imaging Interferometer on the Upper Atmosphere Research Satellite, *J. Geophys. Res.*, 98, 10,725-10,750, 1993a.

Shepherd, G. G., et al., Longitudinal structure in atomic oxygen concentrations observed with WINDII on UARS, *Geophys. Res. Lett.*, 20, 1303-1306, 1993b.

Shepherd, G. G., G. Thuillier, B. H. Solheim, Y. J. Rochon, J. Criswick, W. A. Gault, R. N. Peterson, R. H. Wiens, and S.-P. Zhang, WINDII on UARS, Status Report and Preliminary Results, this volume.

Tarasick, D. W., and W. F. J. Evans, A Review of $O_2(a^1\Delta_g)$ and $O_2(b^1\Sigma_g^+)$ Airglow Emissions, *Adv. Space Res.*, 13, 145-148, 1993.

Wallace, L. and M. B. McElroy, The Visual Dayglow, *Planet Space Sci.*, 677-708, 1966.

Ward, W. E., Y. J. Rochon, C. McLandress, D. Y. Wang, J. R. Criswick, B. H. Solheim, and G. G. Shepherd, Correlations between the mesospheric $O(^1S)$ emission peak intensity and height and temperature at 98 km. using WINDII data, it Adv. Space Res, in press, 1994.

Witt, G., J. Stegman, D. P. Murtagh, I. C. McDade, R. G. H. Greer, P. H. G. Dickinson, and D. B. Jenkins, Collisional energy transfer and the excitation of $O_2(b^1\Sigma_g^+)$ in the atmosphere, *J. Photochem.*, 25, 365-378, 1984.

W. E. Ward, Y. Rochon, C. C. Tai, B. H. Solheim and G. G. Shepherd Institute for Space and Terrestrial Science, Centre for Research in Earth and Space Science, York University, North York, Ontario, Canada M3J 1P3.

E. J. Llewellyn, W. S. C. Brooks Institute of Space and Atmospheric Studies, University of Saskatchewan, Saskatoon, Saskatchewan, Canada, S7N 0W0.

A Sequential Estimation Technique for Recovering Atmospheric Data From Orbiting Satellites

D. A. Ortland, P. B. Hays, W. R. Skinner, M. D. Burrage,
A. R. Marshall, and D. A. Gell

Department of Atmospheric, Oceanic, and Space Sciences, University of Michigan, Ann Arbor, MI

An algorithm is described for constructing a sequential estimation of atmospheric state parameters from sets of measurements made at different locations and times such as from an orbiting satellite. This method combines the equations that relate the measurements to the atmospheric state parameters into a single system by constraining the estimates to one another as a means of filtering out measurement noise. The degree to which the set of estimates are related and how this can be adjusted is discussed. Examples of sequential estimation are applied to the recovery of mesospheric winds from measurements taken by the High Resolution Doppler Imager (HRDI). The quality of the estimates will depend on the magnitude of the measurement noise and the variability of the atmospheric state.

1. INTRODUCTION

This paper presents a summary of a general method, known as sequential estimation or Kalman filtering, that may be used to recover atmospheric state parameters from measurements made by remote sounding instruments aboard satellites. The technique described here was developed for the purpose of recovering winds, temperatures and aerosol scattering coefficients from measurements made by the High Resolution Doppler Imager (HRDI), flown on the Upper Atmosphere Research Satellite, but is applicable to many situations where one wishes to recover data from a series of related measurements. It is particularly applicable to inversion problems where the measurement noise is high and good climatological constraints are not available. Sequential estimation is a type of constrained linear inversion or optimal estimation, as described in *Rodgers* [1976]. This paper expands upon ideas that were introduced in section 6d of *Rodgers* [1976] and presents a description of the filtering behavior of a specific and simple form of sequential estimator. A problem encountered by most inversion methods is how to reduce the effect of measurement noise on the retrieval. Sequential estimation provides a way to do this by using a minimum of climatological constraints. The goal of this paper is to show how this is achieved. A more detailed study will be given in *Ortland et al.* [in preparation].

2. OPTIMAL ESTIMATION

We begin by describing the inversion problem within the framework of linear algebra. Further information on inversion theory may be found in the excellent treatments of *Menke* [1989], *Rodgers* [1976, 1990], *Tarantola* [1987], and *Daley* [1991]. Assume that the atmospheric state one wishes to recover may be represented by an M-dimensional vector $\mathbf{m} = (m_1, m_2, \cdots, m_M)$. For example, the components m_i may represent the perturbation of temperature from a basic state or wind components at various altitudes over a specific location on the Earth. A set of measurements at this location may be represented by a D-dimensional vector $\mathbf{d} = (d_1, d_2, \cdots, d_D)$ where the component d_j may represent the perturbation of a radiance measurement from a model value at a particular wavelength or a Doppler shift measured when viewing the atmosphere with a line of sight at one of D tangent heights. The equation that relates the measurements and atmospheric state vector is assumed to be linearized and approximated by a matrix equation $\mathbf{d} = \mathbf{Gm} + \varepsilon$ where $\mathbf{G}$ is a D x M matrix of weights and ε is the vector of errors in measurement **d**. Assume these errors are Gaussian with covariance matrix S_d.

It is useful to view the equation **d** = **Gm** as a linear transformation from the vector space X consisting of all possible state vector profiles to the vector space Y consisting of all possible measurement profiles. The matrix **G** provides a forward model of the measurement system, since for a given atmospheric state profile **m** the transformed vector **d**=**Gm** is the predicted measurement. Conversely, given a measurement profile **d**, one wishes to determine the best estimate **m** of the atmospheric state profile such that **d**=**Gm** within experimental error.

In general, there is no unique solution to the equation **d**=**Gm** because there may be a subspace of X called the null space which contains the set of all $\mathbf{m}_{null}$ such that $\mathbf{Gm}_{null} = 0$. Thus, if **m** is a solution to the equation **d**=**Gm**, then so is $\mathbf{m} + \mathbf{m}_{null}$. Moreover, there may also be vectors in X which are 'nearly null.' If one does a singular value decomposition of **G** [see *Menke*, 1989, section 7.6], then the nearly null vectors are the ones whose components have singular values that are smaller than the measurement error. Measurements provide very little information on these components of the atmospheric state profile, and in the presence of measurement noise, these components are greatly amplified by unconstrained inversion methods.

If an inverse problem is underdetermined by the presence of null or nearly null solutions, then one must add *a priori* information to the problem in order to determine these components. One way to do this is to assume that the atmospheric state profile **m** that one wishes to estimate is a member of a Gaussian statistical ensemble with mean vector $\mathbf{m}_0$ (the *a priori* constraint) and covariance about the mean given by the symmetric matrix S_m. This means that we consider **m** as a random variable from a distribution whose density function is given by

$$P(\mathbf{m}) = \frac{1}{\sqrt{(2\pi)^M \det(S_m)}} \cdot \exp\left[-\frac{1}{2}(\mathbf{m}-\mathbf{m}_0)^t S_m^{-1}(\mathbf{m}-\mathbf{m}_0)\right]. \quad (1)$$

The condition that **d**=**Gm**+ε means that the conditional probability density function $P(\mathbf{d}|\mathbf{m})$ is given by

$$P(\mathbf{d}|\mathbf{m}) = \frac{1}{\sqrt{(2\pi)^M \det(S_d)}} \cdot \exp\left[-\frac{1}{2}(\mathbf{Gm}-\mathbf{d})^t S_d^{-1}(\mathbf{Gm}-\mathbf{d})\right]. \quad (2)$$

The most likely atmospheric state profile is the one that maximizes the distribution $P(\mathbf{m}|\mathbf{d}) = P(\mathbf{m})P(\mathbf{d}|\mathbf{m})/P(\mathbf{d})$, by Bayes theorem. Taking the product of Eqs. (1) and (2), we see that the most likely value for **m** is the one that minimizes the quadratic

$$Q(\mathbf{m}) = (\mathbf{m}-\mathbf{m}_0)^t S_m^{-1}(\mathbf{m}-\mathbf{m}_0) \\ + (\mathbf{d}-\mathbf{Gm})^t S_d^{-1}(\mathbf{d}-\mathbf{Gm}). \quad (3)$$

The minimum is attained by

$$\mathbf{m}_{est} = \left(\mathbf{G}^t S_d^{-1} \mathbf{G} + S_m^{-1}\right)^{-1} \left(\mathbf{G}^t S_d^{-1} \mathbf{d} + S_m^{-1} \mathbf{m}_0\right), \quad (4a)$$

which we shall refer to as the *optimal estimate*. The covariance matrix of this estimate is given by

$$S_{est} = \left(\mathbf{G}^t S_d^{-1} \mathbf{G} + S_m^{-1}\right)^{-1}. \quad (4b)$$

3. SEQUENTIAL ESTIMATION

Suppose that we have a sequence $\{\mathbf{d}_I\}_{I=1,...,S}$ of measurement profiles, with the position of $\mathbf{d}_I$ given by some parameter p_I. For example, p_I might be a position coordinate along a satellite track, or it may be the time that a measurement was taken. If the corresponding atmospheric state profiles $\mathbf{m}_I$, where $\mathbf{d}_I = \mathbf{G}_I \mathbf{m}_I$, are changing smoothly with position p, then one can use the whole set of measurements to help estimate the atmospheric state at a single position. The sequential estimator described below is one way to do this.

The sequential estimator is based on the optimal estimate (4). Instead of using climatology for $\mathbf{m}_0$ and S_m, one uses the estimate previously obtained at an adjacent position in the sequence. There are two ways to do this, one by proceeding 'forward' through the sequence by using $\mathbf{m}_0 = \mathbf{m}_{est,I-1}^F$ as a constraint to obtain $\mathbf{m}_{est,I}^F$ (where the superscript F denotes the forward estimate) and one by proceeding 'backward' through the sequence by using $\mathbf{m}_0 = \mathbf{m}_{est,I+1}^B$ as a constraint to obtain $\mathbf{m}_{est,I}^B$ (where B denotes the backward estimate).

To start each of these sequences one may use $\mathbf{m}_0$ from climatology to get $\mathbf{m}_{est,1}^F$ and $\mathbf{m}_{est,S}^B$. The object here is to use a large enough covariance matrix S_m so that the climatological constraint will only be applied to the null components of the state profile and not influence the determination of the other components. This constraint will be 'forgotten' as we proceed along the sequence. For the sake of simplicity, in the rest of the paper we shall ignore the influence of the climatological constraints at each end of the sequence. This can be done by starting the forward and backward smoothing with an unconstrained inversion if there is no null space or by restricting to a subspace in profile space X that is complimentary to the null space if there is one.

The simple sequential estimator described here is based on the assumption that the difference $\Delta \mathbf{m} = \mathbf{m}_I - \mathbf{m}_{I-1}$ between adjacent atmospheric state profiles at a fixed position I is an unbiased random vector, i.e. that its expectation value $\langle \Delta \mathbf{m} \rangle = 0$. It is possible to construct a sequential estimator that takes into account the possibility that $\langle \Delta \mathbf{m} \rangle \neq 0$, but we shall see that it is not really necessary to do so. Let $\Delta p = p_{I-1} - p_I$ be the difference of the sequence position parameters, and let $\Delta \mathbf{m}_1$ be the difference between profiles separated by a unit distance of the sequence position. If it is assumed that $(d\mathbf{m}/dp)(p)$ is uncorrelated in p (white noise), it follows that the covariance matrix of $\Delta \mathbf{m}$ is [see *Gelb*, 1986, section 3.7]

$$\langle \Delta \mathbf{m}(\Delta \mathbf{m})^t \rangle = \Delta p \mathbf{S}_{\Delta m}, \text{ where } \mathbf{S}_{\Delta m} = \langle \Delta \mathbf{m}_1 (\Delta \mathbf{m}_1)^t \rangle. \quad (5)$$

From Eq. (4b), the estimate $\mathbf{m}^F_{est,I-1}$ has covariance matrix

$$\mathbf{S}^F_{est,I-1} = \left(\mathbf{G}^t_{I-1} \mathbf{S}^{-1}_{d_{I-1}} \mathbf{G}_{I-1} + \mathbf{S}^{-1}_{m_{I-1}} \right)^{-1}. \quad (6)$$

This means $\mathbf{m}_{est,I-1} = \mathbf{m}_{true,I-1} + \varepsilon$, where ε is a random error vector with expectation value $\langle \varepsilon \rangle = 0$ and covariance $\mathbf{S}^F_{est,I-1}$. The difference

$$\mathbf{m}^F_{est,I-1} - \mathbf{m}_{true,I} = \mathbf{m}_{true,I-1} - \mathbf{m}_{true,I} + \varepsilon = \Delta \mathbf{m} + \varepsilon \quad (7)$$

is thus a random vector with zero expectation value and covariance

$$\mathbf{S}_m = \Delta p \mathbf{S}_{\Delta m} + \mathbf{S}^F_{est,I-1}. \quad (8)$$

Thus when $\mathbf{m}_0 = \mathbf{m}^F_{est,I-1}$ is used as the constraint to obtain the estimate $\mathbf{m}^F_{est,I}$, its covariance is given by $\mathbf{S}_m$.

These assumptions are admittedly *ad hoc*, but the result is a method of constructing an *a priori* estimate of $\mathbf{m}_I$ from $\mathbf{m}^F_{est,I-1}$ that depends on their separation Δp in the sequence. One cannot simply use $\mathbf{S}_m = \mathbf{S}^F_{est,I-1}$, however. As explained below [see Eq. (15)], this will result in the equality of all the estimates at each position.

The matrix $\mathbf{S}_{\Delta m}$ should be derived from climatology [see *Daley*, 1991], but can also be approximated by a simple model. The choice completely determines how the sequential estimator will work. The simplest model to use is $\mathbf{S}_{\Delta m} = \sigma^2 \mathbf{I}$, where σ is a parameter that can be employed to adjust the degree of smoothing along the sequence. The parameter σ may be interpreted as the expected standard deviation of the atmospheric parameters for unit change in position p. The model used in HRDI data processing is given by

$$(\mathbf{S}_{\Delta m})_{ij} = \sigma^2 \exp\left[-(z_i - z_j)^2 / \rho^2 \right] \quad (9)$$

where ρ is another parameter that can be used to adjust the degree of smoothness of the estimated atmospheric state profile, and where z_i is a coordinate (e.g. altitude) for the elements of the profile. The parameter ρ may be interpreted as the expected correlation 1/e-width between elements of the atmospheric state profile.

The value of σ essentially controls how similar the individual estimates of the atmospheric profiles will be to each other, and ρ controls how smooth each profile will be. The estimates will be optimal if these values are set based on knowledge of the statistics of the actual atmospheric state. Proper adjustment of these parameters may be attained by carrying out an extensive validation program [see *Burrage et al.*, 1993]. For the HRDI mesospheric wind inversion, the parameters σ and ρ were chosen to minimize the variance between HRDI estimates and radar measurements (*Burrage et al.*, in preparation).

The *sequential estimate* $\mathbf{m}_{est,I}$ is constructed by using both the forward and backward estimates $\mathbf{m}^F_{est,I-1}$ and $\mathbf{m}^B_{est,I+1}$ as *a priori* constraints. In this way, all of the measurements in the sequence are used to obtain the estimate at a given position. Moreover, if our assumption that there is no bias in $\Delta \mathbf{m}$ at any given position is false, then this bias will tend to cancel since it will usually have opposite sign in the forward and backward estimates. This is the main reason for using both the forward and backward estimates. Thus, let

$$\begin{aligned} \mathbf{m}_1 &= \mathbf{m}^F_{est,I-1} \\ \mathbf{m}_2 &= \mathbf{m}^B_{est,I+1} \\ \mathbf{S}_1 &= \mathbf{S}^F_{est,I-1} + (p_I - p_{I-1})^2 \mathbf{S}_{\Delta m} \\ \mathbf{S}_2 &= \mathbf{S}^B_{est,I+1} + (p_{I+1} - p_I)^2 \mathbf{S}_{\Delta m} \end{aligned} \quad (10)$$

The 'most likely' *a priori* estimate is then

$$\begin{aligned} \mathbf{m}_0 &= \left(\mathbf{S}_1^{-1} + \mathbf{S}_2^{-1} \right)^{-1} \left(\mathbf{S}_1^{-1} \mathbf{m}_1 + \mathbf{S}_2^{-1} \mathbf{m}_2 \right) \\ \mathbf{S}_m &= \left(\mathbf{S}_1^{-1} + \mathbf{S}_2^{-1} \right)^{-1}. \end{aligned} \quad (11)$$

The sequential estimate $\mathbf{m}_{est,I}$ is computed by substituting (11) into the optimal estimate formulae (4).

4. SMOOTHING PROPERTIES

All of the measurements in the sequence contribute to the estimate $\mathbf{m}_{est,I}$ at one position. Since the estimation process is linear, this contribution may be expressed as a weighted sum

$$\mathbf{m}_{est,I} = \sum_{J=1}^{S} \mathbf{W}_{I,J} \mathbf{d}_J, \qquad (12)$$

where each $\mathbf{W}_{I,J}$ is a matrix which gives the contribution of the measurement at position J to the estimate at position I. As one moves away from position I, the weighting matrix $\mathbf{W}_{I,J}$ gets smaller at a rate which is completely controlled by the choice of $S_{\Delta m}$.

A way to investigate the filtering properties of the sequential estimator is provided by inserting $\mathbf{d}_J = \mathbf{G}_J \mathbf{m}_{true,J} + \varepsilon_J$ into (12) (where ε_J is random noise). This gives

$$\mathbf{m}_{est,I} = \sum_{J=1}^{S} \mathbf{R}_{I,J} \mathbf{m}_{true,J} + \sum \mathbf{W}_{I,J} \varepsilon_J \qquad (13)$$

where $\mathbf{R}_{I,J} = \mathbf{W}_{I,J} \mathbf{G}_{I,J}$. This is related to the model resolution matrix of *Menke* [1989] and *Rodgers* [1990]. Equation (13) expresses the estimate as a weighted sum of the true atmospheric state parameters plus a weighted sum of the random errors. It can be shown by induction and use of the formulae (4) that $\Sigma_{J=1,S} \mathbf{R}_{I,J} = 1$. Thus, the sequential estimates are horizontally smoothed versions of the true profiles plus some noise. The object one would like to achieve is to make $\Sigma_{J=1,S} \mathbf{R}_{I,J} \mathbf{m}_{true,J}$ as close to $\mathbf{m}_{true,I}$ as possible while at the same time ensuring that $\Sigma_{J=1,S} \mathbf{W}_{I,J} \varepsilon_J$ is as close to zero as possible. For example, in the case where G has an inverse, if one lets σ in (9) go to infinity, then the resolution matrices approach

$$\mathbf{R}_{I,I} = 1, \quad \mathbf{R}_{I,J} = 0 \text{ for } I \neq J. \qquad (14)$$

This gives the best resolution if there is no measurement noise. But if (14) holds, then we must have $\mathbf{W}_{I,I} = \mathbf{G}_I^{-1}$, $\mathbf{W}_{I,J} = 0$, $I \neq J$. Thus $\mathbf{m}_{est,I} = \mathbf{m}_{true,I} + \mathbf{G}_I^{-1} \varepsilon_I$, which has the greatest error magnification as discussed in Section 2. The opposite extreme occurs when one sets σ in (9) equal to zero. The resolution matrices then become

$$\mathbf{R}_{I,J} = \frac{1}{S} \mathbf{1}, \quad \mathbf{W}_{I,J} = \frac{1}{S} \mathbf{G}_J^{-1}, \qquad (15)$$

which gives $\mathbf{m}_{est,I}$ as a simple average of the $\mathbf{m}_{true,J}$. The random error is minimized in this case, but the resolution is poor.

As an illustration, consider the situation where all of the matrices and vectors in the above formulae are scalars. Also, for simplicity, consider only the forward estimates, and assume that all $G_I = 1$, all $S_{d,I}$ are equal, and $p_I - p_{I-1} = 1$ for all I. One then obtains a simple linear combination

$$m_{est,I}^F = \sum_{J=1}^{I} r_{I,J}(\lambda) m_{true,I}, \text{ where } \sum_{J=1}^{I} r_{I,J}(\lambda) = 1, \quad (16)$$

and where the values of $r_{I,J}$ are completely determined by the value of $\lambda = S_{\Delta m} / S_d$. If $\lambda = 0$ (i.e. $S_{\Delta m} = 0$ or $S_d \to \infty$) then one finds that $r_{I,J} = 1/I$, so that $m_{est,I}^F$ is the average of all $m_{true,J}$ up to I, plus noise with variance $S_I^F = S_d / I$. If λ is large ($S_d = 0$ or $S_{\Delta m} >> S_d$), then $r_{I,I} \approx 1$ and $r_{I,J} \approx 0, I \neq J$, so $m_{I,est}^F \cong m_{I,true}$, plus noise with variance $S_I^F \cong S_d$. For intermediate values of λ, plots of $r_{I,J}$ are shown in Figure 1. Thus, the degree of smoothing is increased as $\lambda \to 0$.

Of course, if all the $m_{I,true}$ are equal, then $m_{I,est}^F$ will not depend on the choice of $S_{\Delta m}$, but its variance will. Thus the appropriate choice is $S_{\Delta m} = 0$, in order to minimize the variance. In general, one would choose $S_{\Delta m}$ in a way that

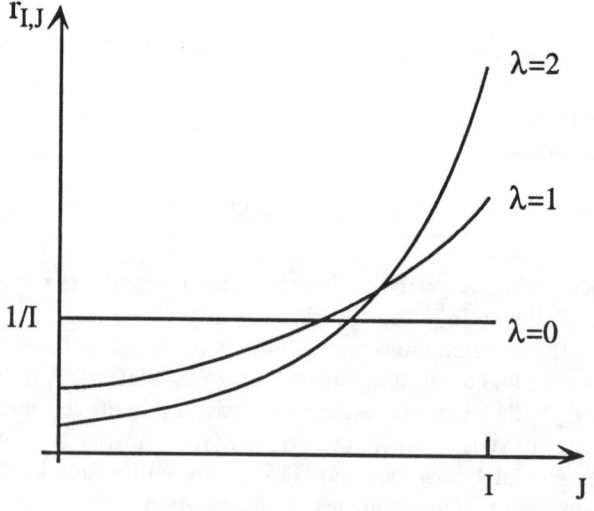

Fig. 1. Contribution weights of the true parameter values along a sequence from 1 to I to the estimated value at position I for various values of a smoothing parameter.

is appropriate for resolving the expected variability in $m_{I,true}$ along the sequence while keeping the variance in the estimate that is due to the measurement noise at a minimum. This is the approach taken in tuning the inversion of HRDI data, where we expect to be able to resolve tidal or planetary wave structures, but not gravity waves.

For the case of vector profiles where the kernel G and the measurement covariance matrices S_d are independent of the position in the sequence, one can find a decomposition of the profiles into components, each of which are smoothed independently as in the scalar example above. This follows from the theorem that two symmetric matrices (in this case $G^t S_d^{-1} G$ and $S_{\Delta m}^{-1}$) may be simultaneously diagonalized. Each component will be smoothed differently, with the components with greater vertical structure smoothed more. The degree of smoothing for each component is controlled by the ratio of the expected variance of that component along the sequence to the variance of the measurement noise, as in the one dimensional example above. A component with a small ratio will have contribution weights $r_{I,J} \approx 1/S$ for all J, and will only appear in the estimate if it persists in the actual atmospheric state profile over several positions in the sequence. Otherwise, it will not be discernible from the measurement noise, and will be filtered out of the estimate. The decomposition of the profiles into individually smoothed components may not be as illuminating as one may wish, however. It serves to help understand how the sequential estimator works, but does not answer quantitative questions like how well particular wave structures may be resolved. These can be answered by simply applying the resolution matrices R to the wave structures.

To qualitatively evaluate the performance of the sequential estimator, it is helpful to substitute the estimated atmospheric state parameters back into the forward model to see what their predicted measurements are. These should be within the error bars of the actual measurements. If they are not, too much smoothing is being applied, while if Gm is too close to the measurements, too much noise is being allowed into the estimates. A more rigorous test that can be made [see *Marks and Rodgers*, 1993] is to check the value of

$$\chi^2 = (d - Gm_{est})^t S_d^{-1} (d - Gm_{est}) \\ + (m_{est} - m_0)^t S_m^{-1} (m_{est} - m_0), \quad (17)$$

where m_0 and S_m are given by (11). This should be close to the dimension of the measurement profile.

5. EXAMPLES

To illustrate the inversion of measurements with the sequential estimator, we will present some examples of observations made by HRDI. This instrument makes high resolution measurements of emission spectra in the mesosphere while viewing the limb of the earth with various line of sight tangent heights. The measured Doppler shift of a spectrum from its rest position may be expressed as a weighted average of the wind velocity component in the viewing direction at positions along the line of sight. The Doppler shifts for a limb scan, d, are expressed in terms of the velocity component profile, m, located at the average tangent point as a matrix equation $d=Gm$. The wind velocity profiles are then sequentially estimated as described above. The sequence consists of all the limb scans taken during the dayside half of the orbit. There are about 75 scans in two complimentary look directions so as to resolve both components of the horizontal wind vectors.

Figure 2 shows the results of the sequential estimation for one of the scans in a half orbit. These figures show how the nature of the estimated profile depends on the parameters σ and ρ in the covariance model (9). Smaller values of σ tighten the constraint between adjacent profiles while larger values of ρ tighten the constraint between adjacent elements of an individual profile. The values used to produce the estimates in the figure are: $\sigma=20$ (m/sec) per degree angle along the orbit track and $\rho=5$ km for 2a; $\sigma=20$ and $\rho=10$ for 2b; $\sigma=5$ and $\rho=5$ for 2c; $\sigma=5$ and $\rho=10$ for 2d. The estimates in 2a and 2c both have high frequency components, but these components have lower amplitude in 2c because of the tighter constraint to adjacent profiles. The profiles in 2b and 2d are smoothed versions of their counterparts in 2a and 2c, respectively, because higher values of ρ were used. Notice that the estimate standard deviations, as denoted by the error bars, grow smaller as the constraints are tightened. This is in accordance with the expected tradeoff between resolution of high frequency components and reduction of random errors.

The solid lines of the plots in Figure 3 give the predicted Doppler shifts Gm for each of the corresponding estimates in Figure 2, while the dots with error bars are the actual Doppler shift measurements. Notice that the high frequency structure evident in the measured Doppler shift profile between 90 and 100 km has been filtered out of all four estimates. This is because the value of ρ is not small enough to allow oscillations with this frequency into the estimates. The closest match between d and Gm is in Figure 3a, because this corresponds to the least constrained estimate, but there is really very little difference between

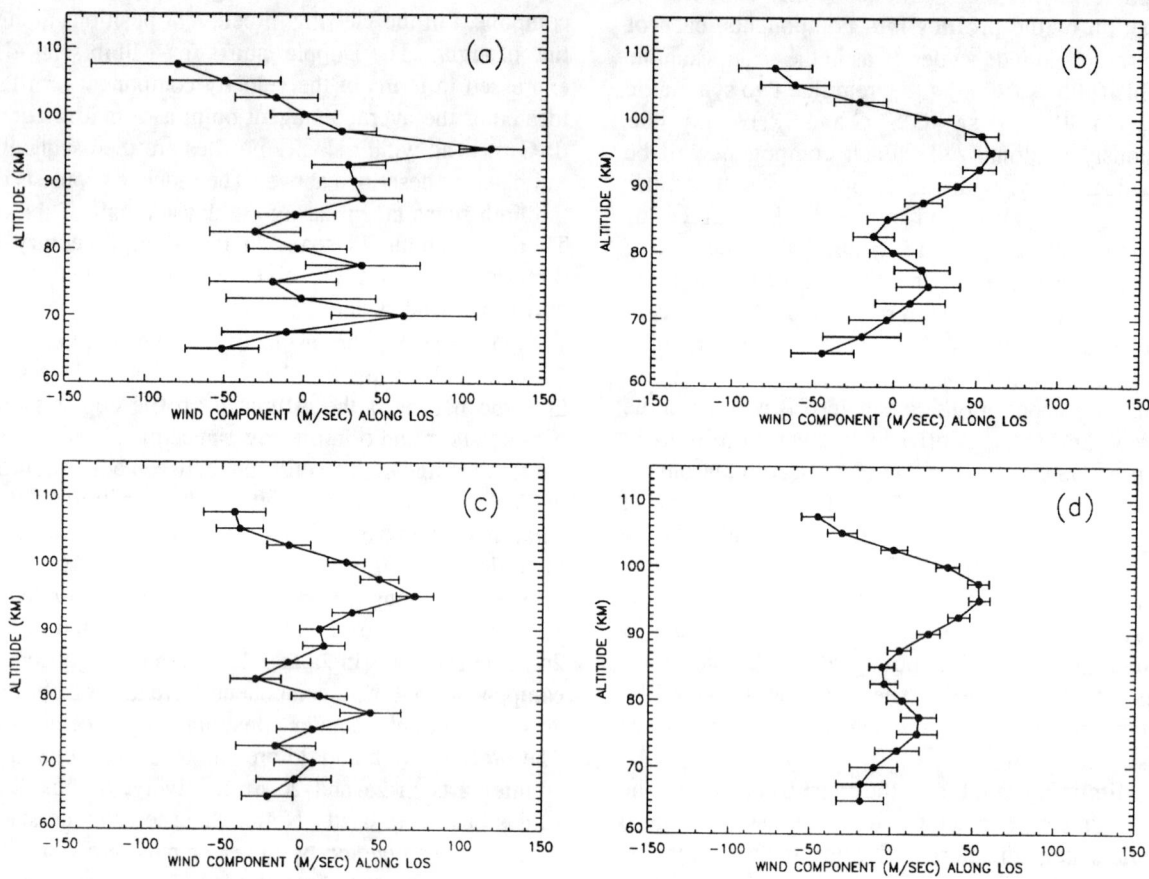

Fig. 2. Estimates of a wind profile using smoothing parameters: a) σ=20 (m/sec) per degree angle along the orbit track and ρ=5 km; b) σ=20 and ρ=10; c) σ=5 and ρ=5; d) σ=5 and ρ=10.

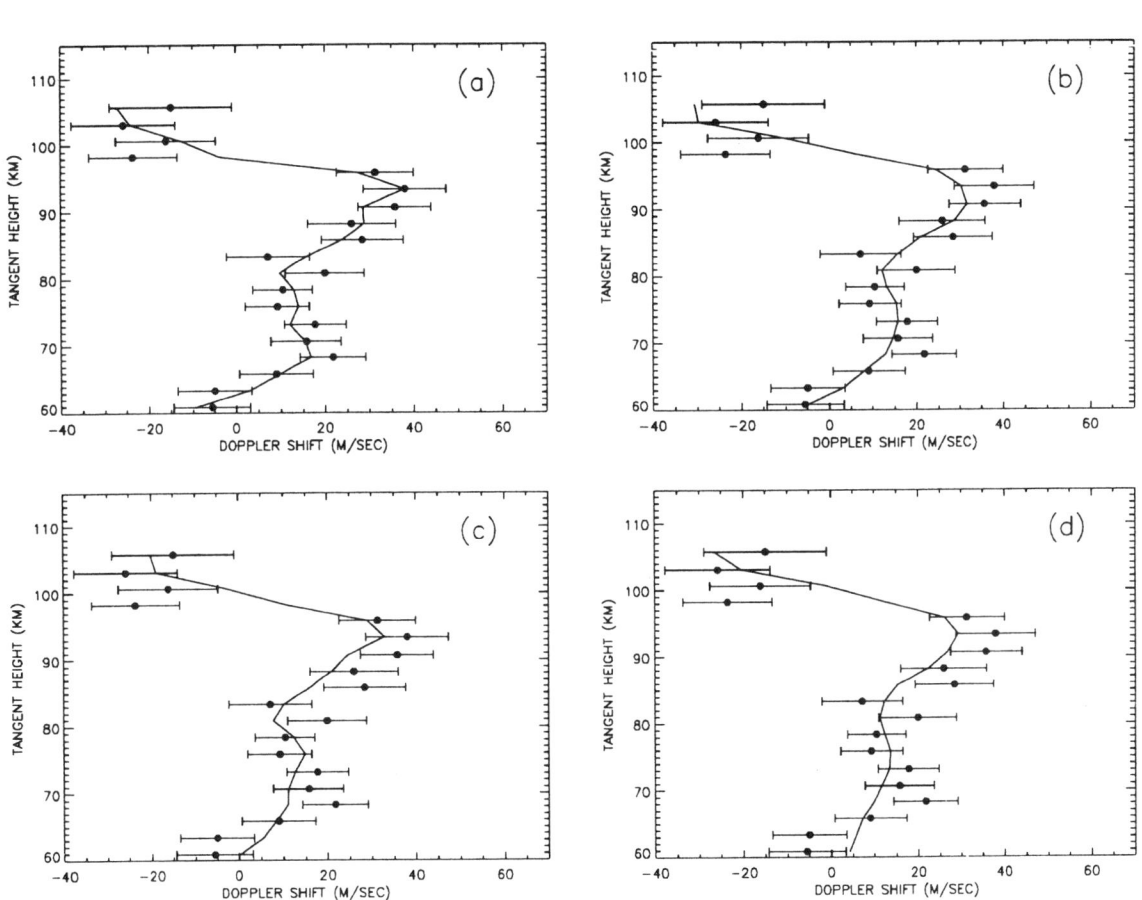

Fig. 3. The predicted Doppler shift measurement profiles Gm (solid lines) for each of the estimates in Figure 2, compared with the actual measurements (dots) and their error bars.

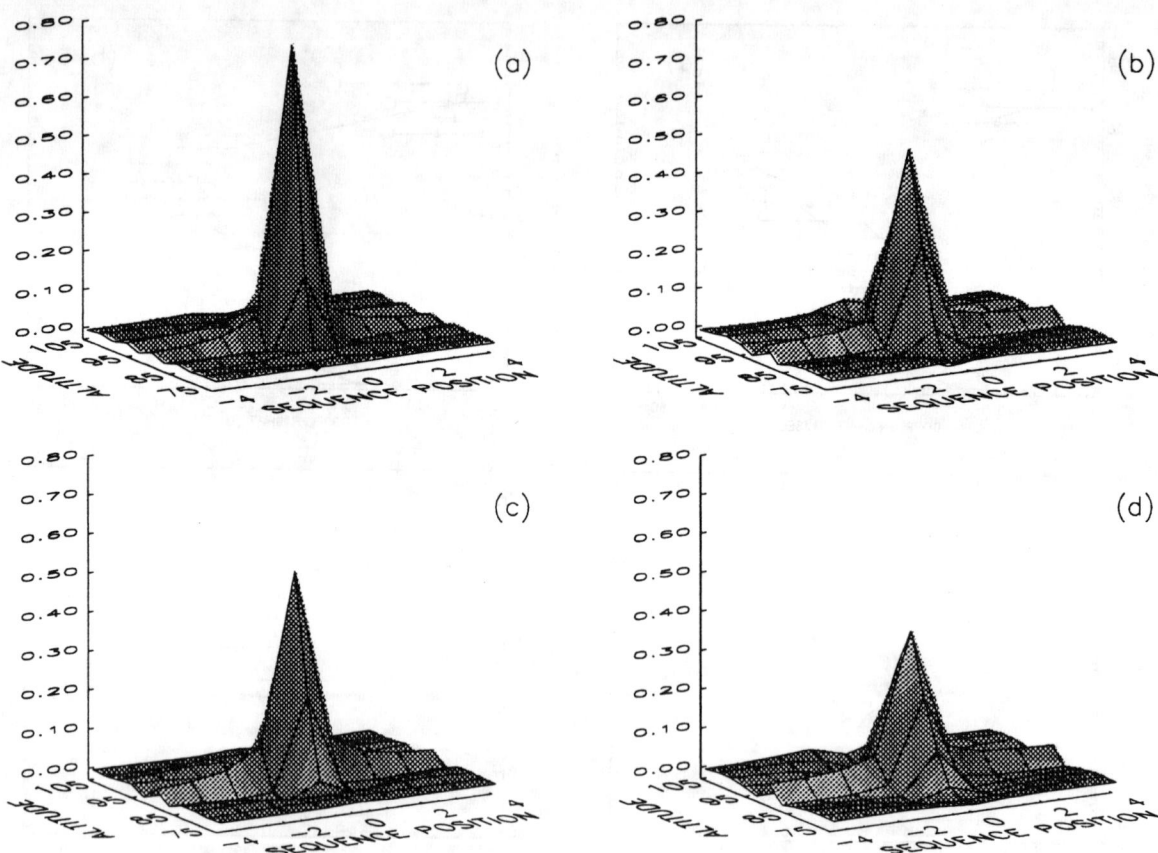

Fig. 4. Contribution weights for each of the estimates at 90 km in Figure 2. These weights show how much the true wind at a given altitude and relative sequence position contributes to the estimated wind at 90 km.

the **G m** for all four estimates. This shows that the components of the profile that make up the difference between the various estimates is poorly determined by the measurements, and their determination must come from *a priori* information.

Some plots of the weighting matrices $\left(\mathbf{R}_{IJ}\right)_{ij}$ are shown in Figure 4. They show how much the true wind at a given altitude and relative sequence position contributes to the estimated wind at 90 km for each of the sequential estimation parameters used to generate the estimates in Figure 2. The indices I and J denote the position in the sequence of the estimate and the contributing true profile, respectively, so that J-I is the relative position. The indices i and j are altitude indices for the estimate and the true profile, respectively. These plots are for fixed I and i, with J-I changing along the x axis and j along the y axis. One can see how much the contribution to the estimate from adjacent profiles increases as the value of σ is decreased, while larger values of ρ allow more contribution from adjacent altitudes in the same individual profile.

Acknowledgments. We would like to thank Deborah Eddy for help in preparing the manuscript, and to thank the reviewers for their valuable comments. This work was supported by NASA Contract No. NAS 5-27751.

REFERENCES

Burrage, M. D., W. R. Skinner, A. R. Marshall, P. B. Hays, R. S. Lieberman, S. J. Franke, D. A. Gell, D. A. Ortland, Y. T. Morton, F. J. Schmidlin, R. A. Vincent, and D. L. Wu, Comparison of HRDI measurements with radar and rocket observations, *Geophys. Res. Lett.*, 20, 1259-1262, 1993.

Daley, R., *Atmospheric Data Analysis*, Cambridge University Press, New York, 1991.

Gelb, A., *Applied Optical Estimation*, MIT Press, Cambridge, 1986.

Marks, C.J. and C.D. Rodgers, A retrieval method for atmospheric composition from limb measurements, *J. Geophys. Res.*, 98, 14939-14953, 1993.

Menke, William, *Geophysical Data Analysis: Discrete Inverse Theory*, Academic Press, New York, 1989.

Rodgers, C.D., Characterization and error analysis of profiles retrieved from remote sounding measurements, *J. Geophys. Res..*, 95, 5587-5595, 1990.

Rodgers, C.D., Retrieval of atmospheric temperature and composition from remote measurements of thermal radiation, *Rev. Geophys. & Space Phys.*, 14, 609-624, 1976.

Tarantola, A., *Inverse Problem Theory*, Elsevier, Amsterdam, 1987.

D.A. Ortland, P.B. Hays, W.R. Skinner, M.D. Burrage, A.R. Marshall, and D.A. Gell, Department of Atmospheric, Oceanic, and Space Sciences, University of Michigan, Ann Arbor, MI 48109-2143

Satellite Observations of Neutral Density Cells in the Lower Thermosphere at High Latitudes

Geoffrey Crowley

The Johns Hopkins University, Applied Physics Laboratory, Laurel, Maryland

Jacqueline Schoendorf and Raymond G. Roble

High Altitude Observatory, NCAR, Boulder, Colorado

Frank A. Marcos

Geophysics Directorate, Phillips Laboratory, Hanscom Air Force Base, Bedford, Massachusetts

Although density variations at high latitudes have been observed for many years, their interpretation has been difficult and haphazard. Recently, *Crowley et al.* [1989a,b] discovered a cellular structure in the high-latitude lower-thermosphere neutral density in simulations using the NCAR Thermospheric General Circulation Model. The structure extends upwards from about 120 km into the upper thermosphere. This paper reports data–model comparisons that confirm the existence of the cellular pattern. The S85-1 satellite measured total atmospheric density in a sun-synchronous, almost circular orbit near 200 km, with equatorial crossings at 2240 and 1040 UT. Steady-state Thermosphere–Ionosphere General Circulation Model (TIGCM) simulations for a range of magnetic activity levels under solar minimum equinox conditions were used to predict the density variations to be encountered by the spacecraft. The TIGCM predicts that large density perturbations should be observed along the satellite track due to the cellular structure. The most obvious manifestations of the cells are density peaks predicted near 70°Λ on the dayside and nightside. The density deviation along the satellite track is expected to reach values of almost 40% for magnetically active conditions. Density data from about 60 passes of the S85-1 satellite were available from September 1984. About 20% of the orbits contained clear evidence of density peaks near 70°Λ, as predicted by the TIGCM. Furthermore, the amplitude of the observed density perturbations increased with magnetic activity, as predicted by the computer simulations. The discovery of this cellular structure is important because it provides for the first time a framework in which to interpret diverse and sometimes apparently conflicting high-latitude density observations. The cells are also important because they define the structure of the virtually unexplored lower thermosphere at high latitudes. This work sets the stage for a number of upcoming satellite missions that will focus on the lower thermosphere.

1. INTRODUCTION

The neutral atmosphere at altitudes between 100 and 220 km remains relatively unexplored [e.g., *Crowley*, 1991]. Data from the lower thermosphere are sparse, since neither balloons nor most rockets reach lower thermospheric altitudes, and the high atmospheric density at these heights imposes short lifetimes on satellites. Neutral densities in this region were first deduced by extrapolating the densities from higher altitudes based on satellite drag and mass spectrometers. Remote sensing from the ground by radar and optical methods has led to significant progress in the study of winds and temperatures in the lower thermosphere; however, the accurate measurement of neutral densities is not yet possible

The Upper Mesosphere and Lower Thermosphere:
A Review of Experiment and Theory
Geophysical Monograph 87
Copyright 1995 by the American Geophysical Union

using these techniques. The incoherent scatter radar technique permits densities to be estimated below about 115 km [*Reese et al.*, 1991] or above 225 km [*Bauer et al.*, 1970], and optical techniques are still in the developmental stage [*Fennelly et al.*, 1991].

Despite the difficulties of making *in situ* density measurements, a number of instruments have been flown in the 120–220 km altitude region during the past quarter century. Pioneering measurements were made using low-altitude Air Force satellites such as LOGACS [*Pearson*, 1973], which had a lifetime of 4 days. The Atmosphere Explorer series of satellites [*Dalgarno et al.*, 1973] permitted measurements routinely down to about 155 km. Long lifetimes in the low perigee phase (up to one year) were achieved with high eccentricity orbits to minimize time in high-drag regions, but the low-altitude data were sparse. Additional measurements have been carried out by accelerometers on Air Force satellites designated S3-1 [*Marcos et al.*, 1977], S3-4 [*Marcos and Champion*, 1982], SETA-1 [*Marcos and Forbes*, 1985], and S85-1 [*Kayser*, 1988; *Crowley et al.*, 1989b]. S3-1 was in a highly elliptical orbit for 9 months. S3-4, SETA-1, and S85-1 were in low eccentricity orbits with lifetimes of 4 months, 3 weeks, and 6 weeks, respectively. In addition, suborbital Space Shuttle accelerometer measurements [*Blanchard et al.*, 1989] have provided several density profiles below 160 km. However, the shuttle descent is shallow, and the measured "vertical" profiles necessarily include some horizontal structure.

Interpretation of the density data from high latitudes has always been difficult because of the extreme variability encountered there. *Crowley et al.* [1989a,b] recently discovered a cellular structure in the lower thermospheric densities at high latitudes using the NCAR Thermospheric General Circulation Model (TGCM). *Crowley et al.* [1993] established the morphology of the structures in detail, showing that they extend upwards from about 120 km into the upper thermosphere. The discovery of this cellular structure is important because it provides for the first time a framework in which to interpret diverse and sometimes apparently conflicting high-latitude density observations. Because of the variability in the high-latitude densities, the cellular structure found in the TGCM simulations has not previously been recognized in satellite data, either in statistical averages or from inspection of individual passes. Furthermore, the cellular structure is not apparent in the MSIS models [e.g., *Hedin*, 1987], probably because of the low order of spherical harmonic terms used to represent the neutral densities. *Crowley et al.* [1993] showed that the number, location, and amplitude of the density cells depend on altitude and magnetic activity.

This paper focuses on the 200-km altitude regime, where under magnetically active conditions, the structure consists of a four-cell pattern of high and low densities. Steady-state Thermosphere–Ionosphere General Circulation Model (TIGCM) simulations are used to predict the density variations to be encountered by the S85-1 satellite near 200-km altitudes. The satellite data confirm the existence of the density cells over a range of magnetic activity levels. This paper also demonstrates how the morphology of the cell structure deduced from computer simulations can be used generically as a framework to interpret high-latitude density data.

2. THE S85-1 SATELLITE

An accelerometer was flown on the U.S. Air Force Space Test Program mission S85-1 during the summer and fall of 1984. The mission was described by *Kayser* [1988]. The accelerometer provided estimates of the total neutral density [*Marcos et al.*, 1977; *Marcos and Forbes*, 1985]. The orbit of S85-1 was sun synchronous with equatorial crossings at about 2230 and 1030 local solar time (LST). Near the end of the mission, the orbit was circularized at 200 km. This interval included the 8-day period from 17 to 24 September 1984 (day 84261 to 84268) designated as the Equinox Transition Study (ETS). The geophysical conditions during the ETS were described in detail by *Carlson and Crowley* [1989] and by *Crowley et al.* [1989a]. The 8-day ETS interval included the most disturbed day of the year (84267) and a day about half as disturbed (84263), separated and preceded by quiet days. The neutral density data from the S85-1 accelerometer were examined for evidence of high-latitude density cells. Data were available from about 60 passes.

3. TIGCM PREDICTIONS

Thermospheric General Circulation Models (TGCMs) have been developed to study the basic global circulation, temperature, and compositional structure of the thermosphere and its response to solar and auroral activity [*Fuller-Rowell and Rees*, 1980; *Dickinson et al.*, 1981]. The development of the NCAR–TGCM has been described in a series of papers by *Dickinson et al.* [1981] and *Roble et al.* [1982, 1987a, 1988]. *Roble et al.* [1987b] developed a global average (1-D) model of the coupled thermosphere–ionosphere system, which formed the basis [*Roble et al.*, 1988] for a new 3-D Thermosphere–Ionosphere General Circulation Model (TIGCM). Recently, this model has been superseded by the thermosphere–ionosphere–electrodynamic GCM, which includes self-consistent electrodynamic interactions between the thermosphere and ionosphere [*Richmond et al.*, 1992].

For this study, steady-state diurnally reproducible simulations were generated for equinox solar minimum conditions appropriate for the September 1984 period. The solar flux was modeled by an average $F_{10.7}$ flux of 72 and a daily flux of 67. Geomagnetically quiet, moderate, and active conditions were represented by cross-cap potentials of 30, 60, and 90 kV, with corresponding hemispheric power inputs of 5, 11, and 33 GW, respectively. *Foster et al.* [1986] investigated the relationship between K_p and cross-cap potential, and showed that cross-cap potentials of 30, 60, and 90 kV correspond to K_p values of 2^-, 4^-, and 6^-. The steady-state TIGCM simulations were used to predict the density variations expected in the S85-1 data due to the high-latitude cellular features.

Figure 1 illustrates the effect of magnetic activity on the cellular structure at 200-km altitude. This polar plot depicts densities in geographic coordinates, with an outer latitude circle of 45°N. Under magnetically quiet conditions (Figure 1a), only two cells are formed. A high-density region (referred to as the noon cell) is most often located in or near the noon sector, although it protrudes into the nighttime sector for most UTs, as shown here. The low-density cell is always located in the dawn sector. The density in the center of the cells deviates from the hemispheric mean density by about 7–12%. Under moderately active conditions (Figure 1b), three cells are formed: in addition to the noon and dawn cells, a high-density cell forms in the midnight sector. In this case, the density in the noon cell deviates from the hemispheric mean density by about 20%. For magnetically active conditions (Figure 1c), four cells are present: there is an additional low-density region in the dusk sector between the noon and midnight high-density cells. In this case, the noon cell density is about 30–40% higher than the hemispheric average density. Once formed, the cells remain approximately fixed with respect to the geomagnetic pole.

By superimposing a satellite trajectory in the 1040–2240 Local Time plane, it is possible to predict the density variation to be observed by the S85-1 satellite. For the 30-kV case (Figure 1a), the satellite first encounters the nightside low-latitude tidal structure. At about 80° latitude, it observes an extension of the noon high-density cell into the nightside. On the dayside, the satellite passes through the center of the noon high-density cell and then reaches lower latitudes, where it sees further tidal structures. For the 60-kV case (Figure 1b), the satellite crosses the edge of the midnight cell before penetrating the center of the noon cell. For the 90-kV case (Figure 1c), the dusk low-density cell is also observed.

The density cells are fixed in magnetic coordinates, and a polar orbiting satellite such as the S85-1 spacecraft samples different parts of the structure at different UTs. Figure 2 indicates the range of magnetic latitudes and local times sampled by the S85-1 satellite during 24 hours. For the sake of clarity, passes are only shown every 4 hours, and only the

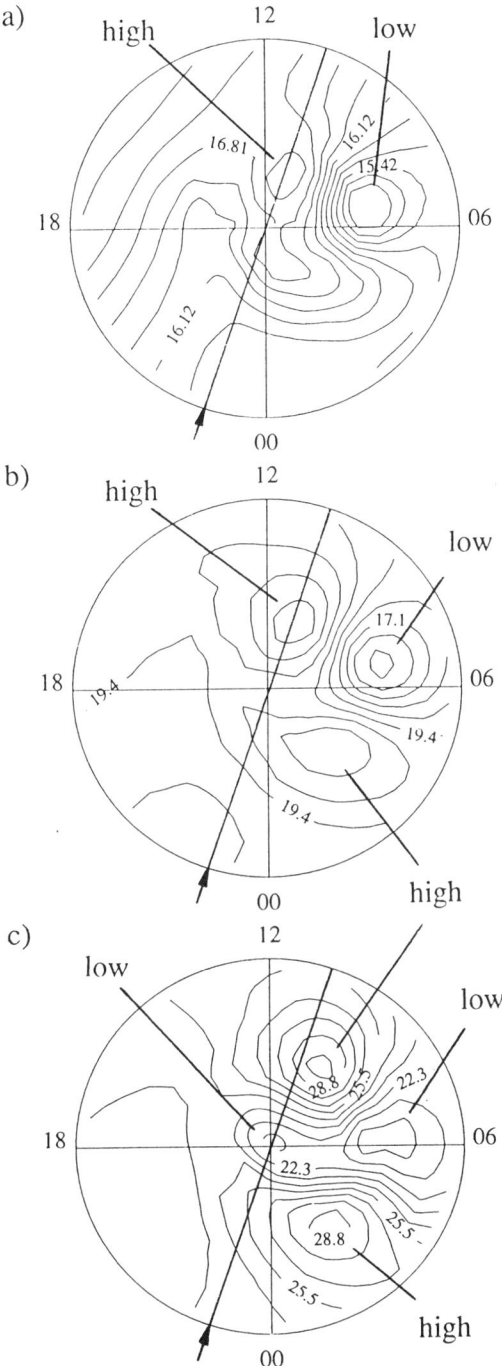

Fig. 1. TIGCM predictions of neutral mass density showing 2-, 3-, and 4-cell patterns in geographic coordinates at 200-km altitude and 12 UT from (a) quiet time (30 kV), (b) moderate (60 kV), and (c) active (90 kV) conditions, respectively. A satellite trajectory in the 2240–1040 Local Time plane is superimposed on each figure. The outer latitude circle corresponds to 45°N. Intermediate latitude circles are omitted for clarity.

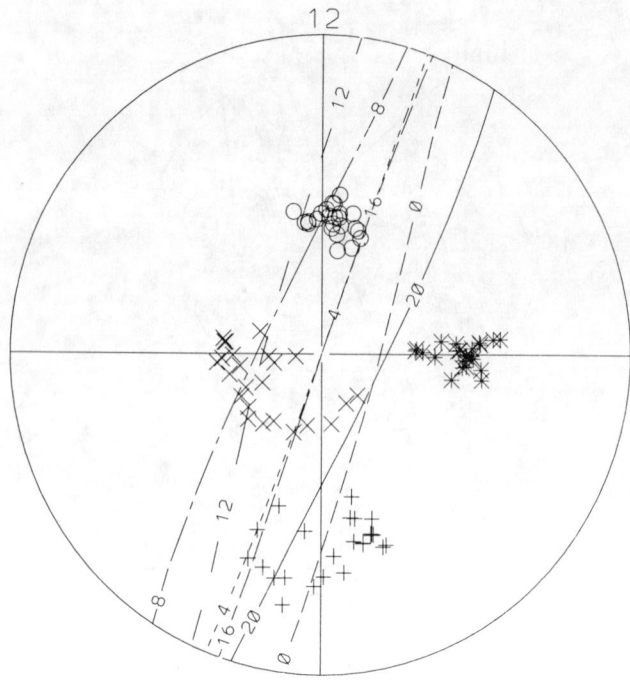

Fig. 2. Location of cell centers in magnetic coordinates for one UT day, in the Northern Hemisphere at 200 km for magnetically active equinox solar minimum conditions [after *Crowley et al.*, 1993]. The symbols O, *, + and × refer to the location of the noon (high), dawn (low), midnight (high), and dusk (low) cell centers, respectively, at 24 UTs. The outer latitude circle is at 55°Λ. Six S85-1 orbits in the 2240–1040 Solar Local Time frame are superimposed at 4-hour intervals, emphasizing the wide range of magnetic local times sampled by the spacecraft. Consequently, the satellite samples much of the high-latitude density structure in 24 hours.

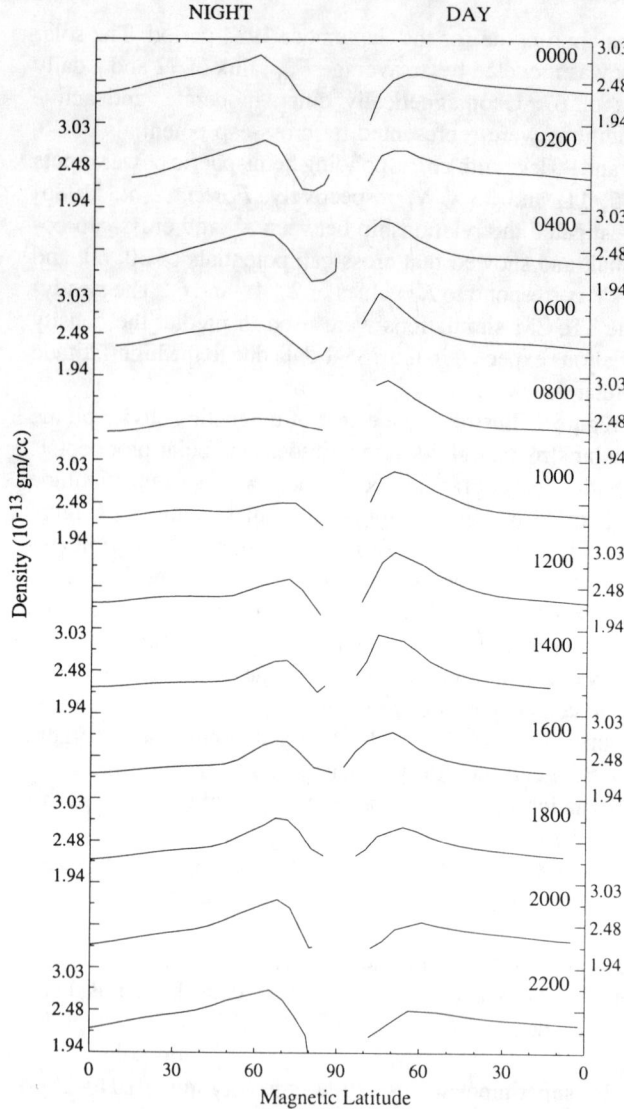

Fig. 3. Predicted density variation as a function of magnetic latitude for alternate UTs for a polar orbiting satellite at 200 km in the 1040–2240 Local Time plane, based on a TIGCM simulation for active conditions (90 kV). The left side of the figure corresponds to the nightside, and the right side corresponds to the dayside. The gaps near the pole arise because the satellite does not generally fly directly over the magnetic pole.

outer latitude circle (55°Λ) is included in the figure. Superimposed on the figure are the locations of the centers of the four cells at 200 km during magnetically active conditions. The figure implies that passes between 06 and 12 UT will sample both the dusk and noon cells. In contrast, passes near 20 and 24 UT will cut through the midnight cell, but will miss the dusk and noon cells.

Figures 3–5 summarize the TIGCM predicted density variations to be observed by the S85-1 satellite (1040–2240 Local Time plane at 200 km) as a function of magnetic latitude for every other UT. The left side of each figure corresponds to the nightside, whereas the right side corresponds to the dayside. Figure 3 indicates the type of density variation expected for magnetically active cases (90 kV). Note that the highest density may be on the dayside (04–16 UT) or the nightside (18–02 UT), depending on UT and whether the satellite intersects the center of the noon or midnight cell, respectively. On the nightside, the satellite should generally observe higher densities as it travels from lower to higher latitudes, until a peak density is observed between 60 and 70° magnetic latitude (Λ) due to the midnight cell. Near the pole, low densities will be sampled at all UTs. The density should then rise to a second maximum on the dayside at 60–70°Λ due to the noon cell, although at some UTs the satellite will only sample the edge and not the center of the noon high-density cell. Figure 3 indicates that

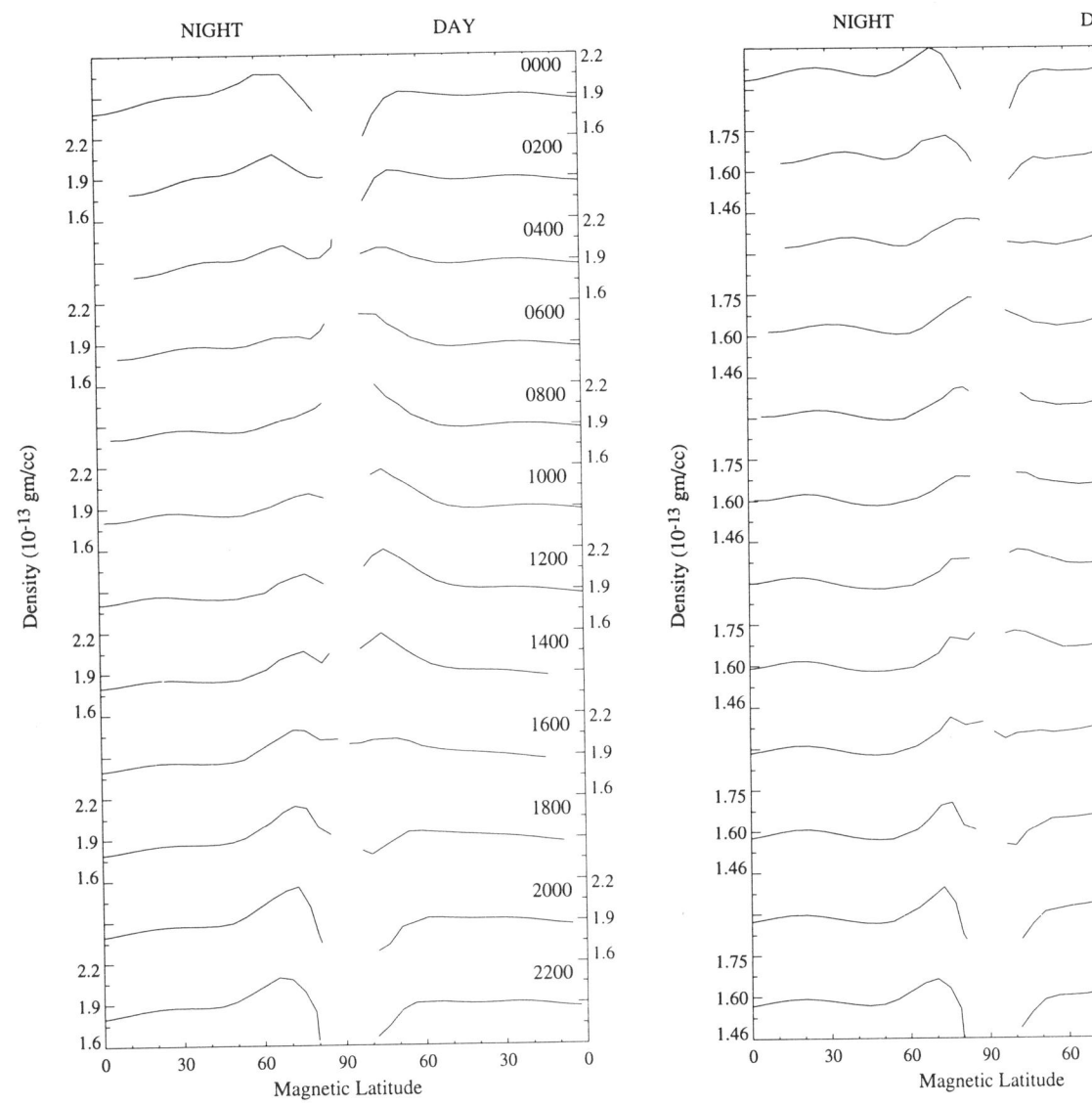

Fig. 4. Same as Figure 3, but for moderate (60 kV) activity.

Fig. 5. Same as Figure 3, but for quiet (30 kV) conditions.

the satellite should encounter density variations on the order of 20–35% during any orbit in active times.

For the 60-kV case (Figure 4), the predicted variation with latitude and UT has a similar form but a smaller amplitude (<17%), and the density peaks tend to be at higher latitudes (70–80°Λ). At 30 kV (Figure 5), the predicted density peaks are less than 8% above the background density, and are more difficult to identify since this is comparable with the tidal effects at 200-km altitude. Despite the small amplitude, there are many similarities between Figures 4 and 5, including the location of the peaks at latitudes of 70–80°Λ.

We note that the TIGCM simulations for quiet conditions (Figure 1a) contain only two cells: the noon high-density and dawn low-density cells. At first glance, one would not therefore expect the satellite to observe a density peak on the nightside. Closer examination of the TIGCM simulations reveals an extension of the high-density cell into the nighttime sector, resulting in a peak in the nightside satellite densities for certain UTs (18–04 UT) in Figure 5.

4. DATA ANALYSIS

4.1. Active Conditions

The S85-1 density data were examined for evidence of the high-latitude structure predicted by the TIGCM. Ten out of 25 active time orbits ($K_p \geq 3^0$) contained strong evidence of cellular structure at high latitudes. For example, Figure 6a shows the satellite track in magnetic coordinates (outer latitude 55°Λ) at 1645 UT on day 84264 ($K_p = 3^+$) superimposed on the four active cell locations predicted by the TIGCM (cf. Figure 2). On this pass, data were collected only on the dayside, beginning near the dawn–dusk meridian. This figure indicates that the satellite should see both the dusk low-density cell and the noon high-density cell. The lower panel of Figure 6 illustrates the density variation measured by the satellite as a function of magnetic latitude, with the nightside on the left and the dayside on the right. Near the pole, the measured density fell to a minimum, as predicted. The figure also contains strong evidence of a high-density cell in the noon sector near 75–80°, where the density rose to about 35×10^{-14} gm/cm^3. This increase of about 35% over the background level is of a comparable magnitude and in a similar location to that predicted by the TIGCM. Unfortunately, there is insufficient data to draw any conclusions about the nightside. In addition to the dayside density peak at 80°Λ, the satellite density data also contain structure at lower latitudes. The large-scale feature covering 15–55°Λ with a maximum density near 40° is ubiquitous throughout the data set. It displays a regular UT variation and is probably caused by upward-propagating semidiurnal tides. Typically, a similar high-density region is observed on the nightside at low latitudes. The tidal effects predicted by the TIGCM were much smaller in magnitude (see Figure 5), indicating that the TIGCM tidal inputs were inadequate. The small density maximum between 55 and 70°Λ on the dayside in Figure 6b is not repeatable, and its cause is not known.

Another detailed example of the high-latitude cellular structure is presented in Figure 7 for the satellite pass at 2203 UT on day 84265 ($K_p = 3^0$). The upper panel indicates that the satellite probably cut through the midnight cell and skirted the edge of the noon cell. Consequently, a significant density variation should have been observed at high latitudes on this pass. Figure 7b shows that the satellite detected high densities both on the nightside and the dayside at about 75°Λ. The density maxima were separated by a region of low densities, as predicted by the TIGCM. The maximum density near 75°Λ is about 28×10^{-14} gm/cm^3 on both the dayside and the nightside, which is about 25% higher than the density over the pole.

Figure 8 summarizes the ten S85-1 passes during active conditions ($K_p \geq 3^0$), which contained strong evidence of the density cells predicted by the TIGCM. Each pass contains a high-density peak near 70–80°Λ on the dayside and/or 60–80°Λ on the nightside. The amplitude of the peaks increases with increasing magnetic activity, as discussed later. Most of the positive evidence came from passes obtained during

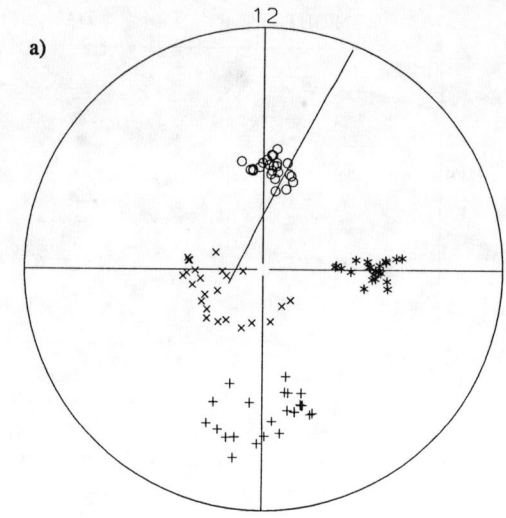

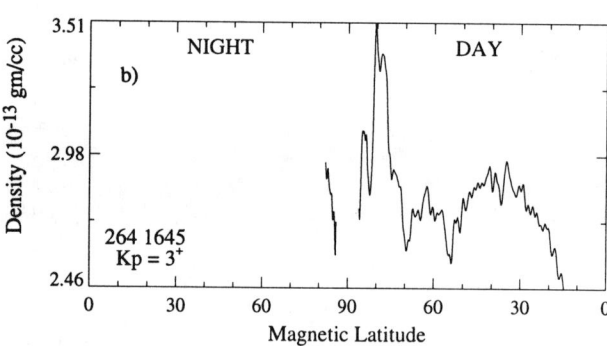

Fig. 6. (a) Satellite orbit for day 84264, 1645 UT ($K_p = 3^+$) superimposed on magnetic coordinates, outer latitude 55°Λ. The symbols ○, *, +, and × refer to the location of the noon (high), dawn (low), midnight (high), and dusk (low) cell centers, respectively, at 24 UTs; (b) corresponding density variation measured by satellite.

moderate magnetic activity. However, many passes contained so much structure that no single peak could be singled out as being due to the density cells. The remaining passes contained too little data to be able to make a statement about the high-latitude structure. There were no active passes in which there were data but no structure.

4.2. Quiet Conditions

Although the TIGCM predictions are climatological in nature, the satellite is subject to "space weather," including smaller scale transient effects. In practice, the small density peaks of 8–17% predicted by the TIGCM for quiet conditions (Figures 4 and 5) are easily masked by other structures in the variable high-latitude environment, including tides

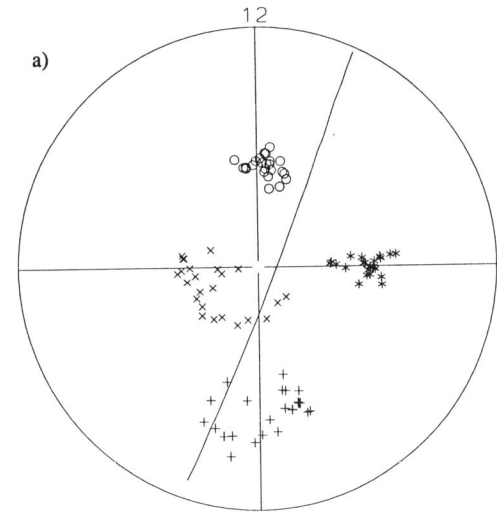

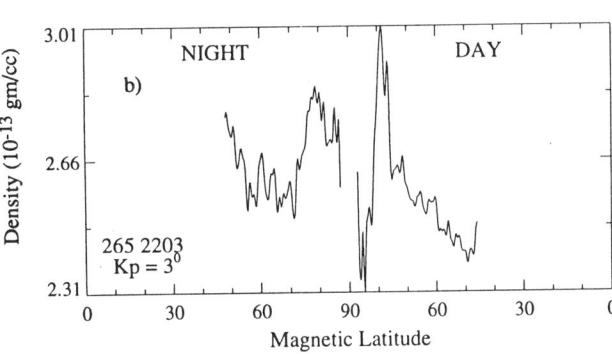

Fig. 7. (a) Satellite orbit for day 84265, 2203 UT ($K_p = 3^0$) superimposed on magnetic coordinates, outer latitude 55°Λ. The symbols ○, *, +, and × refer to the location of the noon (high), dawn (low), midnight (high), and dusk (low) cell centers, respectively, at 24 UTs; (b) corresponding density variation measured by satellite.

and gravity waves. From a total of 40 passes for magnetically quiet conditions ($K_p < 3^0$), many had limited data coverage, and most contained sufficient structure that no single peak could reasonably be identified. From this very limited data set, four passes contained density peaks in the same location and of the same scale and amplitude as those predicted by the TIGCM.

4.3. *Magnetic Activity Dependence of Cell Amplitudes*

It is interesting to examine the increase in density perturbation in the cells measured by the satellite as a function of K_p, and to compare it with that predicted by the TIGCM. Examination of Figures 6–8 reveals that the measured density often falls to local minima both equatorward and pole-

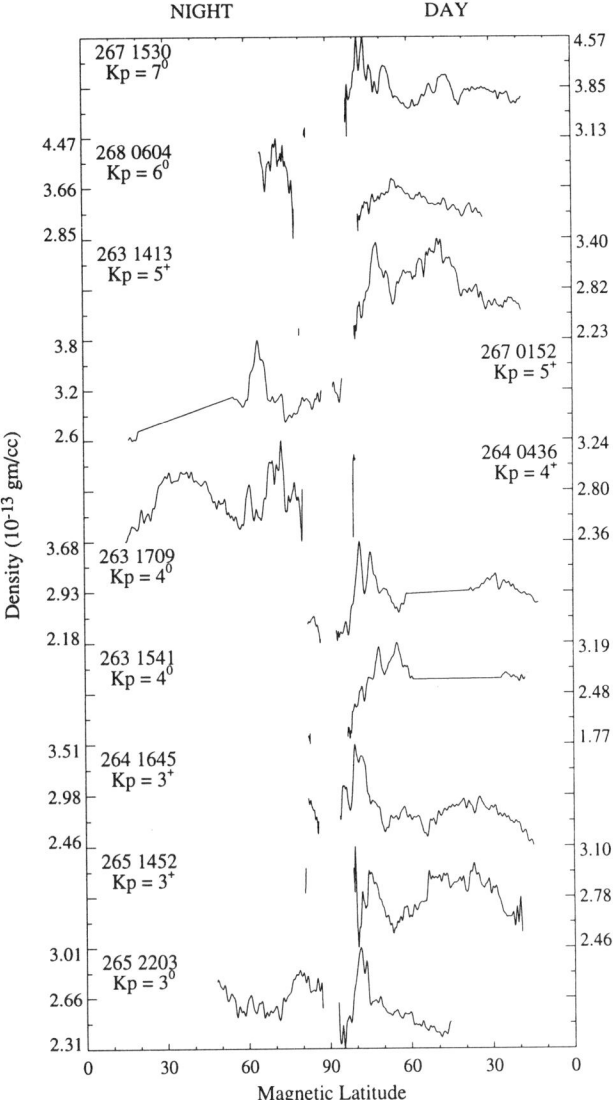

Fig. 8. Summary of satellite data from 10 active passes showing evidence of cell structure.

ward of the density maxima located at 60–80°Λ, but the poleward minimum tends to be deeper than the equatorward minimum. It is therefore difficult to determine a single meaningful value for the magnitude of the density changes predicted for a satellite pass. We have thus considered the noon and midnight cell maxima separately, and have also identified the local minima poleward and equatorward of these maxima. Potentially, four amplitudes could be predicted and measured for each pass, although data gaps usually precluded making all four measurements. Figure 9 displays the amplitude variation as a function of K_p for the 10 active cases and 4 quiet cases where density features could

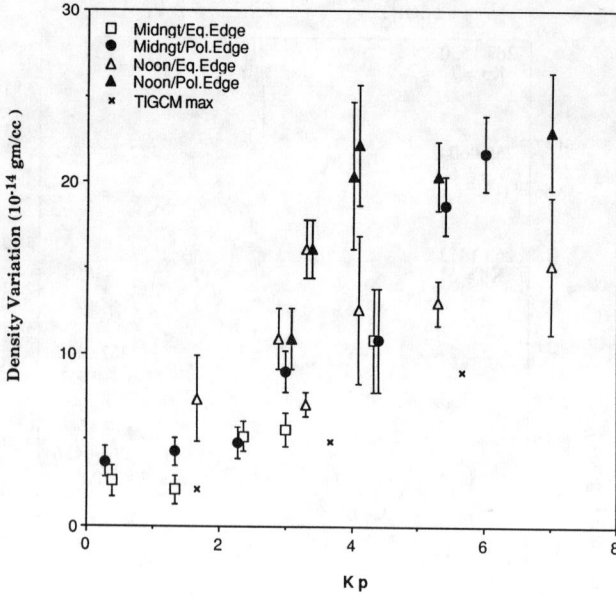

Fig. 9. Magnitude of high-latitude density variation observed by the S85-1 satellite as a function of K_p.

clearly be identified near 60–80°Λ. Different symbols are used to represent the poleward and equatorward edges of the midnight and noon density cells. The error bars provide an estimate of the uncertainty in the amplitude measurement. Part of the uncertainty comes from the presence of small-scale (~1°Λ) structure superposed on the broader (~10°Λ) features of interest here. The data reveal a dramatic increase in the amplitude of the density perturbations observed by the satellite as a function of K_p. The measured perturbations range from 3×10^{-14} g/cm^3 at $K_p = 0^0$ to 23×10^{-14} g/cm^3 at $K_p = 7^0$. In general, the density change is greater on the poleward side of the maxima (solid symbols) than on the equatorward side (open symbols). This is most evident for the higher activity levels, suggesting that the amplitude of the perturbations on the poleward edge of the structures increases with K_p about twice as rapidly as the amplitude on the equatorward edge.

A similar exercise was applied to the TIGCM predictions of Figures 3–5. Although four separate amplitudes could be defined for each predicted pass, they were not included if the density changed slowly with latitude (i.e., the edges of the cells were not sharply defined in the model). This tended to rule out the equatorward edge of the cells; see, for example, 02 UT in Figure 4. Since the amplitude of the predicted density perturbation varies with UT, the UT values corresponding to the satellite passes for $K_p = 2^-$, 4^-, and 6^- were selected from the model. The predicted magnitudes were smaller than those observed, and although the predicted density perturbations increased with magnetic activity, the rate of increase with K_p was smaller than that observed by the satellite.

To test whether this difference was simply due to the UT selection process, the model predictions were examined for all UTs. The maximum amplitudes (regardless of UT or whether the cell was a midnight or noon cell) are depicted in Figure 9 as an "x" for the 30-, 60-, and 90-kV simulations. The points are plotted at corresponding K_p values of 2^-, 4^-, and 6^- [*Foster et al.*, 1986]. Even confining our analysis to the maximum amplitudes predicted by the model, the simulation cannot be brought into agreement with the observations. The result indicates that UT selection is not responsible for the difference between the model and the data.

One possible explanation of this difference is that the forcing applied to the TIGCM is too small to represent the desired simulation conditions, and the choice of K_p corresponding to the TIGCM inputs (Figure 9) is too high. Closer examination reveals that in addition to the density perturbations, the absolute densities in the model are also significantly smaller than those measured by the satellite. As a percentage of the mean density along the satellite track, the model density deviations are much closer to the observed density variations, although the discrepancy cannot be fully accounted for. The amplitude differences require further investigation, and a more detailed discussion is premature at this point. Suffice to say that the data confirm the model prediction that the amplitude of the cellular structure increases with magnetic activity.

5. CONCLUSIONS

The analysis of neutral mass density data from the S85-1 satellite accelerometer confirms the existence of density cells discovered in the high-latitude lower thermosphere by *Crowley et al.* [1989a,b]. The NCAR–TGCM predicts that the structure extends from about 120 km into the upper thermosphere, its form and amplitude varying with height and magnetic activity. This paper focuses on the cell structure at 200-km altitudes. Under magnetically active conditions, the structure consists of a four-cell pattern of high and low densities near altitudes of 200 km.

Steady-state TIGCM simulations for solar minimum equinox conditions were used to predict the high-latitude density variations to be encountered by the S85-1 spacecraft. Simulations were performed for a range of magnetic activity levels and provided a theoretical framework within which to interpret the satellite data. The satellite flew in a circular polar orbit near 200-km, with equatorial crossings at 2230 and 1030 Local Time. The TIGCM predicts that large density perturbations should be observed along this satellite track due to the cellular structure. The most obvious manifestations of the cells are the density peaks predicted near 70°Λ on the dayside and nightside. The density deviation along the satellite track is expected to reach values of almost 40% for magnetically active conditions.

The S85-1 satellite data confirm the existence of the high-latitude neutral density cells predicted by the NCAR–TIGCM. Density data from about 60 passes of the S85-1 satellite were available from September 1984. About 20% of the orbits contained clear evidence of density peaks near 70°Λ, as predicted by the TIGCM. Furthermore, the amplitude of the observed density perturbations increased with magnetic activity, as predicted by the computer simulations. There were no passes in which there were data but no structure. However, many passes contained so much structure that no single peak could be singled out as being due to the density cells. The remaining passes contained too little data to be able to make a statement about the high-latitude structure. The magnitude of the observed perturbations is significantly larger (factor of 2) than that predicted by the TIGCM. The reason for this discrepancy is not understood and requires further research.

The work reported here is of vital importance for studies of the lower thermosphere. In particular, it provides a theoretical framework in which to interpret high-latitude data obtained from previous satellite missions. Interpretation of density data from high latitudes has always been difficult because of the extreme variability encountered there. Owing to this variability, the cellular structure found in the TIGCM simulations has not previously been recognized in satellite data, either in statistical averages or from inspection of individual passes. This study also provides an important framework in which to understand density and composition data from several upcoming missions that will study the lower thermosphere, including TIMED (Thermosphere Ionosphere Mesosphere Energetics and Dynamics), RAIDS (Remote Atmospheric and Ionospheric Detection System), ADS (Atmospheric Density Specification Experiment), and MSX (Midcourse Space Experiment).

Acknowledgments. This work was supported by NSF grant ATM–8901131.

REFERENCES

Bauer, P., P. Waldteufel, and D. Alcayde, Diurnal variations of the atomic oxygen density and temperature determined from incoherent scatter measurements in the ionospheric F-region, *J. Geophys. Res.*, 75, 4825–4832, 1970.

Blanchard, R. C., E. W. Hinson, and J. Y. Nicholson, High resolution accelerometer package experiment results: atmospheric density measurements between 60–160 km, *J. Spacecraft and Rockets*, 28, 173–180, 1989.

Carlson, H. C., and G. Crowley, The Equinox Transition Study: an overview, *J. Geophys. Res.*, 94, 16,861–16,868, 1989.

Crowley, G., B. A. Emery, R. G. Roble, H. C. Carlson, and D. J. Knipp, Thermospheric dynamics during September 18–19, 1984: 1. Model simulations, *J. Geophys. Res.*, 94, 16,925–16,944, 1989a.

Crowley, G., B. A. Emery, R. G. Roble, H. C. Carlson, J. E. Salah, V. B. Wickwar, K. L. Miller, W. L. Oliver, R. G. Burnside, and F. A. Marcos, Thermospheric dynamics during September 18–19, 1984: 2. Validation of the NCAR thermospheric general circulation model, *J. Geophys. Res.*, 94, 16,945–16,959, 1989b.

Crowley, G., Dynamics of the earth's thermosphere, *Rev. Geophys.*, supplement, 1143–1165, 1991.

Crowley G., J. Schoendorf, R. G. Roble, and F. A. Marcos, Neutral density cells in the lower thermosphere at high latitudes: TIGCM simulations, *J. Geophys. Res.*, in press, 1993.

Dalgarno A., W. B. Hanson, N. W. Spencer, and E. R. Schmerling, The Atmosphere Explorer mission, *Radio Sci.*, 8, 263–266, 1973.

Dickinson, R. E., E. C. Ridley, and R. G. Roble, A three-dimensional, time-dependent general circulation model of the thermosphere, *J. Geophys. Res.*, 86, 1499–1512, 1981.

Fennelly, J. A., D. G. Torr, P. G. Richards, M. R. Torr, and W. E. Sharp, A method for the retrieval of atomic oxygen density and temperature profiles from ground-based measurements of the $O^+(2D-2P)$ 7320-Å twilight airglow, *J. Geophys. Res.*, 96, 1263–1273, 1991.

Foster, J. C., J. M. Holt, R. G. Musgrove, and D. S. Evans, Ionospheric convection associated with discrete levels of particle precipitation, *Geophys. Res. Lett.*, 13, 656–659, 1986.

Fuller-Rowell, T. J., and D. Rees, A three-dimensional, time-dependent global model of the thermosphere, *J. Atmos. Sci.*, 37, 2545–2557, 1980.

Hedin A. E., MSIS-86 thermospheric model, *J. Geophys. Res.*, 92, 4649–4662, 1987.

Kayser, D. C., Measurements of thermospheric meridional winds from the S85-1 spacecraft, *J. Geophys. Res.*, 93, 9979–9986, 1988.

Marcos, F. A., C. R. Philbrick, and C. J. Rice, Measurements of atmospheric mass density by accelerometers, mass spectrometers and ionization gauges, *Space Res.*, 17, 329–343, 1977.

Marcos, F. A., and K. W. Champion, Satellite density measurements with the rotatable calibration accelerometer (ROCA), *AFGL-TR-82-0091*, Air Force Geophysics Laboratory, Hanscom AFB, Bedford, Mass., 4 Mar 1982.

Marcos, F. A., and J. M. Forbes, Thermospheric winds from the satellite electrostatic triaxial accelerometer system, *J. Geophys. Res.*, 90, 6543–6552, 1985.

Pearson, J. A., The low-G accelerometer calibration system orbital accelerometer system, *Rep. TR-0074(4260-10)-1*, Aerospace Corp., Los Angeles, Calif., 1973.

Reese, K. W., R. M. Johnson, and T. L. Killeen, Lower thermosphere neutral densities from Sondrestromfjord incoherent scatter radar during LTCS-1, *J. Geophys. Res.*, 96, 1091–1098, 1991.

Richmond, A.D., E. C. Ridley, and R. G. Roble, A thermosphere/

ionosphere general circulation model with coupled electrodynamics, *Geophys. Res. Lett., 19,* 601–604, 1992.

Roble, R. G., R. E. Dickinson, and E. C. Ridley, Global circulation and temperature structure of the thermosphere with high-latitude plasma convection, *J. Geophys. Res., 87,* 1599–1614, 1982.

Roble, R. G., J. M. Forbes, and F. A. Marcos, Thermospheric dynamics during the March 22, 1979 magnetic storm: 1. Model simulations, *J. Geophys. Res., 92,* 6045–6068, 1987a.

Roble, R. G., E. C. Ridley, and R. E. Dickinson, On the global mean structure of the thermosphere, *J. Geophys. Res., 92,* 8745–8758, 1987b.

Roble, R. G., E. C. Ridley, A. D. Richmond, and R. E. Dickinson, A coupled thermosphere/ionosphere general circulation model, *Geophys. Res. Lett., 15,* 1325–1328, 1988.

G. Crowley, The Johns Hopkins University Applied Physics Laboratory, Johns Hopkins Rd, Laurel, MD 20723-6099.

J. Schoendorf and R. G. Roble, High Altitude Observatory, NCAR, P.O. Box 3000, Boulder, CO 80307.

F. A. Marcos, Geophysics Directorate, Phillips Laboratory, Hanscom Air Force Base, Bedford, MA 01731.

Recommended Drag Coefficients for Aeronomic Satellites

Mildred M. Moe and Steven D. Wallace

University of California, Irvine, Irvine, California

and

Kenneth Moe

Space and Missile Systems Center/XRF, Los Angeles Air Force Base,

Los Angeles, California

Densities derived from accelerometer measurements depend on the drag coefficient assigned to the satellite. Although laboratory measurements increase our understanding of gas-surface interactions, they are not adequate to determine the appropriate drag coefficient because it is not known how the surface conditions at any particular altitude relate to the heterogeneous chemisorption and physisorption revealed by measurements in the laboratory. Therefore it is necessary to rely on drag and accommodation coefficients which have been measured in orbit. We use knowledge of these coefficients from our recent review of satellite measurements, and insights gained from laboratory measurements, to construct a table showing how the accommodation coefficient of compact satellites varies wtih altitude and solar activity. We then insert the accommodation coefficients in theoretical calculations to provide recommended drag coefficients for a variety of satellite shapes in low earth orbit. By using all of the information on aerodynamic coefficients measured by previous satellites, we can minimize errors in density measurements made by future satellites. Soon a new aeronomic satellite, STEP-1, will be flown. If its large flat plates sometimes are oriented at several different angles to the airstream, our knowledge regarding the dependence of drag coefficients on altitude and angle of incidence can be improved.

1. INTRODUCTION

In order to derive atmospheric densities from accelerometer measurements, it is necessary to assume a drag coefficient, C_d, because aeronomic satellites do not incorporate a method of measuring it. Consequently, the accuracy of the inferred densities can be no better than the accuracy of the assumed value of C_d. Its numerical value depends on the reference area, which often is chosen to be the cross sectional area of the satellite normal to the airstream. With that choice, the value of C_d which is customarily used for convex satellites of compact shapes is 2.2 [Cook, 1965, 1966]. By "compact shape" we mean that the ratio of the satellite's maximum to minimum diameter is less than 1.5, and the satellite does not have large external structures like solar paddles. A C_d in the neighborhood of 3.0 to 3.5 is used for long cylindrical satellites that fly like an arrow, depending on the length-to-diameter ratio of the satellite and the air temperature [Sentman, 1961; DeVries, 1972; Moe, et al., 1993].

The Upper Mesosphere and Lower Thermosphere:
A Review of Experiment and Theory
Geophysical Monograph 87
Copyright 1995 by the American Geophysical Union

The fundamental processes which determine the drag coefficient are momentum exchange and chemical reactions at the satellite surface, so we discuss laboratory studies of these processes in the following section. Then we summarize what has been learned about momentum transfer from orbital measurements, which we recently reviewed [Moe, et al., 1993]. From the orbital measurements of surface-particle interaction, we infer the appropriate parameters to use in calculating C_d at various altitudes for satellites in low-earth orbit. By low-earth orbit we mean an orbit with a perigee altitude below 1000 km and an eccentricity value less than 0.3. Because satellite measurements of the appropriate parameters have not been made above 325 km, we mention some theoretical studies which may be useful in extrapolating the parameters to higher altitudes. We use the measured parameters to calculate C_d for satellites of several simple shapes. Finally, we suggest some theoretical and experimental efforts that could improve our knowledge of satellite drag coefficients.

2. LABORATORY MEASUREMENTS

The exchange of momentum and energy when gases strike solid surfaces has been studied in the laboratory for eight decades. This research has shown that the exchange depends on many factors: the mass of the substrate atoms; the particular crystal face; the mass, fractional coverage, and binding energy of adsorbed molecules; and the mass, energy, and angle of incidence of the incoming molecules [Saltsburg, et al., 1967; Thomas, 1980; Boffi and Cercignani, 1986]. It would be almost impossible to measure all of these factors in orbit, then reproduce them in laboratory experiments. This is the reason we believe that in-orbit measurements of gas-surface interactions are essential.

On the other hand, laboratory measurements have yielded much information on the interaction of gases with surfaces, and have provided models which can be fitted to the satellite measurements of gas-surface interactions. An example is the laboratory data from O'Keefe and French [1969], which is copied in Figure 1a. It shows that when Argon (mass 40) with a kinetic energy of 1.35 eV strikes a surface covered with adsorbed Hydrogen (mass 2), much of the Argon is reemitted in the quasi-specular direction; but when the surface is covered with adsorbed gases which are heavier or bind more strongly to the surface, the quasi-specular peak is reduced, indicating a stronger interaction with the surface. Figure 1b shows the data from Gregory and Peters' [1987] experiment on the Space Shuttle Flight STS-8. Here atomic oxygen is reflecting from a surface which probably is coated with atomic oxygen chemisorbed

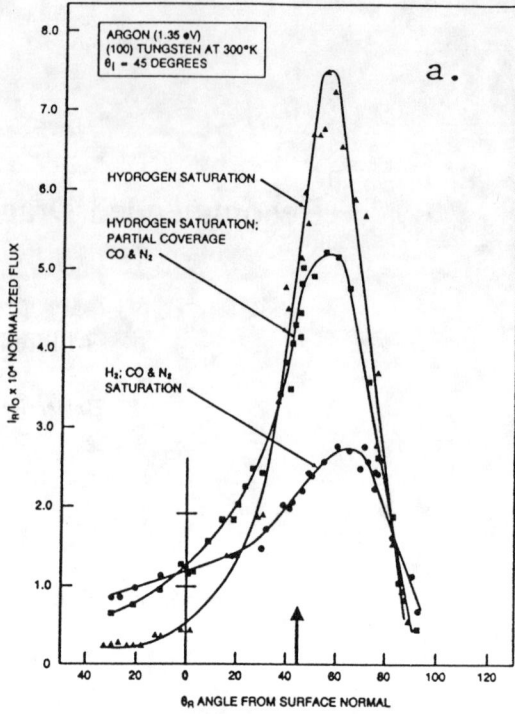

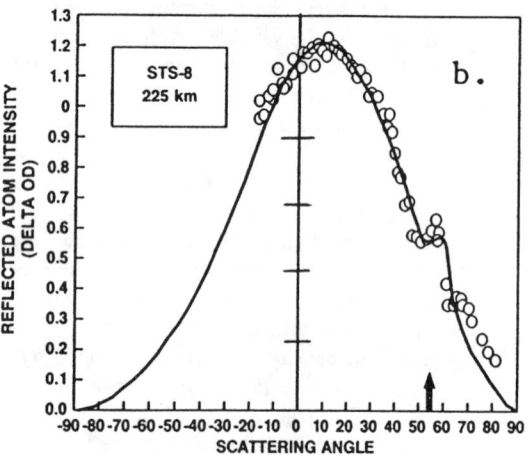

Fig. 1. This composite figure shows how the relative mass, binding energy, and fractional coverage of adsorbed molecules affect the angular distribution of reflected molecules: (a) The laboratory measurement of O'Keefe and French shows argon with a kinetic energy of 1.35 e.v. reflecting from the 100 crystal face of tungsten which is partly coated with weakly bound, lighter molecules. As the fractional coverage and mass of the physisorbed molecules increase, the high quasispecular peak is gradually reduced. In contrast, (b), from the satellite experiment of Gregory and Peters, shows that when atomic oxygen strikes a carbon surface coated with chemisorbed oxygen, which has a binding energy comparable with the incident kinetic energy of 5 e.v., the quasispecular component has been reduced to 2%. (see text).

with a binding energy comparable with the incident kinetic energy. In this case, the quasi-specular component is only 2% of the reflected beam. (The vertical arrows indicate the specular angle of reflection, i.e. the angle at which all the molecules would be reemitted if the reflection were 100% specular.) This figure illustrates the controlling influence of adsorption on the interaction of gases with surfaces.

Our understanding of adsorption has increased greatly since World War II because of the application of new analytic techniques, such as low-energy electron diffraction (LEED), Auger electron spectroscopy (AES), field ion microscopy, flash desorption, and many others. Knowledge obtained by these techniques can be applied to satellite problems, even though the techniques are too complicated to be employed in space. By using some of these techniques, Gomer [1967] showed that carbon monoxide adsorbs on tungsten in several different states. Farnsworth [1970] investigated the complicated chemical reactions of carbon monoxide and molecular oxygen on a nickel surface. Czanderna [1970] showed that molecular oxygen adsorbs on silver in several states, some atomic and some molecular. At the present time, an enormous body of knowledge has accumulated regarding the chemisorption and the physisorption of oxygen on the transition metals [Brundle and Broughton, 1990]. This experimental knowledge of surface chemical reactions can be combined with theoretical models of adsorption [Honig, 1967; Steele, 1967; Newns, 1969; Sparnaay, 1970; Ross, 1967; Roberts and McKee, 1978; Davenport and Estrup, 1990] to organize and rationalize the data obtained on gas-surface interactions from satellite experiments. Such studies could improve greatly the interpretation of future measurements of density and composition in the thermosphere.

3. ORBITAL MEASUREMENTS

The analysis of both satellite drag and mass spectrometric measurements requires understanding the surface reactions of atomic oxygen and carbon. The universal contaminant, carbon, interacts with the principal thermospheric constituent, atomic oxygen, to produce new species, some of which stick to the surface and modify the energy relationships when the airstream strikes a satellite surface. Evidence of these reactions was unmistakable in the Ogo 6 satellite measurements [Hedin, et al., 1973], in the comparison of rocket measurements by cooled and uncooled mass spectrometers [Offermann and Grossmann, 1972], and in the Long Duration Exposure Facility (LDEF) experiments [Levine, 1991]. A laboratory calibration by Offermann and Trinks [1971] also showed that carbon monoxide and carbon dioxide were produced within the mass spectrometer when molecular oxygen flowed in. Laboratory

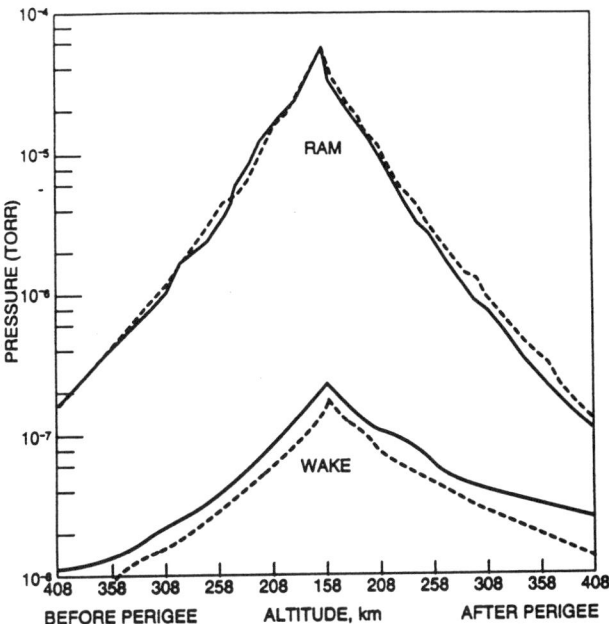

Fig. 2. History of the pressure in a gage aboard the Spades Satellite [after Carter, et al, 1969]. The loci of the maximum (ram) and minimum (wake) pressures during each spin cycle are shown for orbital revolutions 25 (solid lines) and 109 (dashed lines). These curves show that net adsorption occurs whenever the satellite descends toward perigee. Because adsorbed molecules increase energy accommodation, this figure suggests that C_d also changes as the satellite descends and ascends.

simulations in which atomic oxygen flowed into a mass spectrometer resulted in the formation of molecular oxygen, carbon monoxide, and carbon dioxide [Lake and Nier, 1973; Lake and Mauersberger, 1974].

Efforts to model adsorption at satellite surfaces [Moe and Moe, 1967, 1969; Hedin, et al., 1973] have utilized the Langmuir isotherm [Brunauer, et al., 1967] because of its simplicity and its ability to fit the data. Heterogeneous adsorption has been represented by a Langmuir "patch model" [Moe, et al., 1972]. Despite these modest successes, it should be realized that the commonly used isotherms (including Langmuir's) are models of physical adsorption. A more penetrating analysis of the combination of chemical and physical adsorption which occurs at satellite surfaces might require the application of the Hamiltonian representation of the energy relationships [Goldstein, 1980] which can be solved in the Hartree-Fock approximation [Newns, 1969; Einstein, et al., 1980]. Nevertheless, it is clear from the satellite data and the analyses which have been performed that the surface coverage of adsorbed molecules varies continuously as the satellite moves up and down in its orbit. This can be seen in Figure 2, which shows the

TABLE 1. Accommodation Coefficients (A diffuse angular distribution is assumed.)

$F_{10.7}$	75	75	75	150	150	150	225	225	225
Altitude (km)	α	M	Temp (K)	α	M	Temp (K)	α	M	Temp (K)
150	1.00	22	526	1.00	23	618	1.00	24	710
175	1.00	21	629	1.00	22	760	1.00	23	900
200	0.99	20	693	0.99	22	842	0.99	23	1027
225	0.98	19	728	0.98	21	887	0.99	22	1105
250	0.96	18	745	0.97	20	920	0.98	22	1156
275	0.94	17	757	0.95	19	938	0.96	21	1190
300	0.92	16	763	0.93	18	949	0.94	20	1213
325	0.89	15	766	0.92	17	955	0.93	19	1230
∞			779			973			1270

maximum (RAM) and minimum (WAKE) pressure in the spinning pressure gage on the Spades satellite [Carter, et al., 1969]. The minimum pressure is supplied by desorption from the walls of the gage, so we deduce from this and similar data from other satellite instruments that the surface coverage varies continuously. Because energy exchange at surfaces depends strongly on surface coverage, we infer that drag coefficients also vary as the satellite moves up and down. We also deduce from Figure 2 that the surface coverage varies much more slowly with altitude than the ambient atmospheric density.

4. RECOMMENDED PARAMETERS

Until a detailed Hamiltonian analysis can be performed for some representative satellite surfaces and compared with the satellite measurements, we suggest the use of a simplified empirical description utilizing the energy accommodation coefficient α [Wachman, 1962] defined by

$$\alpha = (E_i - E_r) + (E_i - E_W) \quad (1)$$

where E_i is the kinetic energy carried to a unit area of the surface by the incident molecules, E_r the kinetic energy carried away from the unit area by the reflected molecules, and E_W the kinetic energy the reflected molecules would carry away from the surface if they were reemitted at the temperature of the surface (or wall). This coefficient has been measured in orbit by three paddlewheel satellites [Moe et al., 1993], and by the ratio of lift to drag on the S3-1 satellite [Ching, et al., 1977]. The angular distribution of reemitted molecules has been measured only on the STS-8 at 225 km, suggesting that only 2% of the oxygen atoms were reflected quasi-specularly at that altitude [Gregory and Peters, 1987]. Because we lack sufficient information on angular distributions, and because high accommodation coefficients are associated with nearly diffuse angular distributions in laboratory experiments [Saltsburg, et al., 1967], we have used values of α which were derived from satellite data by assuming a diffuse distribution. Using data from the three satellites (S3-1, Proton 2, Ariel 2) in low-earth orbit, we show in Table 1 as a function of altitude our estimate of the accommodation coefficient, α, mean molecular mass, M, and atmospheric temperature for three levels of solar activity, as represented by the 10.7 solar radio emissions, $F_{10.7}$. Since none of the three satellites carried an accelerometer, the drag forces and torques represented an integration over the region around perigee. Therefore we assigned the measured values of α to an altitude 2/3 of a scale height above the perigee altitude [King-Hele, 1966]. The values of α for $F_{10.7}$ = 75 are based on the data in Figure 3 of Moe et al. [1993]; the values for higher solar activity are based on the assumption that the accommodation coefficient is determined by the amount of atomic oxygen adsorbed on the surface of the satellite. The relationship is highly nonlinear: Large changes in surface coverage have only a small effect on α when it approaches 1.00. For this reason, this coefficient changes more slowly with altitude when high solar activity has caused the atmosphere to expand upward. The values we give in the tables are for use in calculating drag coefficients for satellites of compact shapes in low-earth orbit. There is insufficient evidence in the open literature to make such an estimate for the sides of a long cylindrical satellite, or for a satellite in a highly eccentric orbit [Moe, et al., 1993].

We have terminated Table 1 at 325 km because we do not have measurements of α above that altitude. We expect that at higher altitudes, where α is likely to be lower, a quasi-specular component of reemitted molecules should be included in calculations of drag coefficients. Methods of

TABLE 2. Drag Coefficient of a Satellite with L/D = 1

The cumulative effect of the terms in equation (2) is shown for a cylindrical satellite capped by a flat plate which faces the airstream. The parameters are: α = 0.95, M = 19, satellite velocity, V_i = 7600 m/s, local atmospheric temperature T = 938 K, and satellite wall temperature T_w = 300 K. A completely diffuse angular distribution of reemission is assumed.

TERM 1	TERMS 1 & 2	TERMS 1 & 2 & 3
2.000	2.317	2.627

incorporating a quasi-specular component have been developed by Schamberg [1959], Nocilla [1963], Boring and Humphris [1970], Fredo and Kaplan [1981], and Herrero [1983].

5. DRAG COEFFICIENTS

The measurement of atmospheric drag on a satellite leads to a determination of atmospheric density through the well-known relation

$$F_d = \frac{1}{2} \rho C_d A V^2$$

which also defines the drag coefficient. Here F_d is the drag force, ρ is the atmospheric density, A is a suitable reference area of the satellite, and V is the velocity of the airstream relative to the satellite. A is usually taken to be the projected area of the satellite normal to the velocity vector. Under most circumstances, the values of A and V are precisely known, so the measurement of F_d by an accelerometer determines the product ρC_d. Any uncertainty in C_d produces an uncertainty in the density ρ. Suppose that the air molecules had no thermal motion and all of the incident molecules stuck to the surface. (The thermal velocity is a small fraction of the satellite velocity at altitudes below 300 km.) Then, according to Newton's second law, the drag force would be $\rho A V^2$. Therefore C_d = 2. If now we compute the effect of the molecules reemitted from the satellite surface, but still ignore any random thermal velocities, we obtain a correction term K which depends on the accommodation coefficient, the angular distribution of reemitted molecules, and the satellite shape. Adding the contribution of the reemitted molecules to that of the incident molecules, we have C_d = 2 + K. If, in addition, we take account of the random thermal motions of the molecules, we obtain a different correction term G which depends, in addition to the other parameters, on another parameter, the speed ratio, S, which is the ratio of the satellite orbital speed to the most probable speed of air molecules. Now we can write C_d = 2 + G. For satellites of compact shapes, thermal motions typically increase C_d by 5% to 15%, but thermal motions can increase the C_d of long cylinders by 30% to 50% [Sentman, 1961; Mazzella, et al., 1983; Moe, et al., 1993].

We rewrite the expression for C_d = 2 + G as

$$C_d = 2 + K + H \qquad (2)$$

where $H = G - K$. This is done to show why drag measurements of density are capable of higher accuracy than other thermospheric measurements: The expression for C_d begins with the constant, 2. For compact satellites in the lower thermosphere, this constant term is an order of magnitude larger than the other terms. An example of the size of the correction terms in Equation (2) is given in Table 2. It shows C_d for a short cylindrical satellite capped by a flat plate which faces the airstream. Its length-to-diameter ratio (L/D) is 1.00. The parameters, which correspond to conditions at an altitude of 275 km, are given in Table 2. In the second column one sees the effect of the molecules reemitted from the front plate of the satellite when random motions are ignored (hyperthermal approximation). This quantity was calculated using Schamberg's [1959] model. The third column includes the effect of random thermal motions of the air [Sentman, 1961]. (This table is based on the assumption that molecules are reemitted diffusely.) By contrast with drag measurements of density, other thermospheric measurements of density (mass spectrometric, extinction) require calibration factors, knowledge of chemisorption, and photoabsorption cross sections which are not dominated by a large constant term which is known a priori.

Drag coefficient calculations which assume a completely diffuse angular distribution of reemitted molecules are computationally simple but not entirely satisfying, because of the laboratory measurements discussed earlier, which are illustrated by Figure 1. Imbro et al. [1975] explored several combinations of diffuse and quasi-specular distributions, but could not reach a definite

TABLE 3. Drag Coefficients for the Case $\alpha = 1.00$

Temp.	M	Flat Plate	Cylinder + Flat Plates (L/D =1)	Flat Plates	Cone $\theta = 30°$	Cone $\theta = 60°$	Sphere
500 K	18	2.131	2.332	2.187	2.069	2.114	2.094
500 K	22	2.117	2.300	2.169	2.062	2.103	2.084
1000 K	18	2.139	2.424	2.232	2.077	2.122	2.106
1000 K	22	2.124	2.382	2.208	2.069	2.109	2.094
1500 K	18	2.147	2.496	2.269	2.085	2.130	2.118
1500 K	22	2.131	2.447	2.241	2.075	2.116	2.103

TABLE 4. Drag Coefficients for the Case $\alpha = 0.95$

Temp.	M	Flat Plate	Cylinder + Flat Plates (L/D =1)	Flat Plates	Cone $\theta = 30°$	Cone $\theta = 60°$	Sphere
500 K	18	2.353	2.555	2.362	2.180	2.307	2.242
500 K	22	2.347	2.530	2.349	2.177	2.302	2.237
1000 K	18	2.361	2.646	2.407	2.188	2.315	2.254
1000 K	22	2.354	2.612	2.388	2.184	2.308	2.247
1500 K	18	2.369	2.718	2.444	2.196	2.323	2.266
1500 K	22	2.361	2.677	2.421	2.190	2.315	2.257

TABLE 5. Drag Coefficients for the Case $\alpha = 0.90$

Temp.	M	Flat Plate	Cylinder + Flat Plates (L/D =1)	Flat Plates	Cone $\theta = 30°$	Cone $\theta = 60°$	Sphere
500 K	18	2.480	2.682	2.462	2.244	2.417	2.327
500 K	22	2.476	2.658	2.450	2.241	2.413	2.323
1000 K	18	2.488	2.773	2.506	2.252	2.425	2.338
1000 K	22	2.482	2.740	2.489	2.248	2.420	2.332
1500 K	18	2.496	2.845	2.544	2.260	2.433	2.351
1500 K	22	2.489	2.805	2.522	2.254	2.426	2.342

conclusion because of the limitations of previous satellite data. We have explored likely errors caused by these uncertainties for the satellite shape and parameters of Table 2. If the value of α were 0.97 instead of 0.95, and the angular distribution remained diffuse, the value of C_d would be 2.560. If 90% of the molecules were diffusely reemitted with $\alpha = 1.00$ and 10% were quasi-specularly reemitted with $\alpha = 0.50$, then C_d would be 2.513. If the molecules striking the front plate were all diffusely reemitted with $\alpha = 0.95$, while 10% of the molecules which struck the cylindrical sides were quasi-specularly reemitted, then C_d would be 2.608. These exploratory calculations suggest that near 275 km, uncertainties in the accommodation coefficient and angular distribution could cause the calculated drag coefficient of compact satellites to be uncertain by about 4%. The uncertainties will increase rapidly with altitude.

On the other hand, at altitudes near 200 km, similar calculations suggest an uncertainty of about 1% in the drag coefficient.

As an example to assist experimenters who want to use more realistic drag coefficients in reducing density measurements, we have constructed Tables 3, 4, and 5 for a range of parameters, using Sentman's [1961] equations, which assume a completely diffuse angular distribution for the reemitted molecules. Despite the uncertainties discussed above, the tables are given to 3 decimal places to prevent rounding errors when they are interpolated. C_d is given for several simple shapes, using accommodation coefficients of 1.00, 0.95, and 0.90, respectively for three ambient temperatures and two values of the mean molecular mass. An orbital velocity, V_i, of 7,600 m/s and a satellite surface (wall) temperature, T_w, of 300 K are assumed. (The airstream, indicated by the arrow, is perpendicular to the surface of the flat plate.) The accommodation coefficient can be chosen from Table 1. An experimenter can use Table 1 and the exact shape of his satellite to calculate C_d at a range of altitudes, using Sentman's [1961] equations, as we have done.

6. DISCUSSION

We have presented tables which illustrate a method for estimating C_d for compact satellites in low-earth orbit. The tables are based on satellite measurements of accommodation coefficient and the theoretical model of Sentman [1961]. From laboratory measurements, we expect that the assumption of a completely diffuse angular distribution will cause errors of only a few percent at altitudes below 300 km, where the accommodation coefficient is above 0.90. Many of the remaining uncertainties could be resolved by the STEP-1 aeronomic satellite which will provide (in 1994) the rare opportunity to measure C_d in orbit. Its flat plates (solar panels) facilitate the measurement of lift and drag by its three-axis accelerometer. If the STEP-1 is flown with its flat plates in a number of different orientations during geomagnetically quiet times, it will provide a large body of information to resolve uncertainties regarding accommodation coefficients and angular distributions of reemitted molecules.

It may be that Hamiltonian methods will be used in the future to analyze the gas-surface interactions encountered in past and future laboratory and satellite experiments. Combining a realistic theory of chemisorption with the experimental data could lead to better measurements by accelerometers and mass spectrometers.

7. CONCLUSIONS

By combining satellite measurements of accommodation coefficient with Sentman's [1961] equations, we have constructed tables of recommended drag coefficients for several simple compact satellite shapes in low-earth orbit. By using this method with the exact shape to calculate C_d, experimenters can improve the accuracy of density measurements by accelerometers. The method of calculation itself can be improved by careful measurements with future aeronomic satellites, and by future developments in which the theory of chemisorption is combined with experimental data.

REFERENCES

Boffi, V., and C. Cercignani (Editors), *Rarefied Gas Dynamics*, Proc. 15th Intl. Symp., B.G. Teubner, Stuttgart, passim, 1986.

Boring, J.W., and R.R. Humphris, Drag coefficients for free molecule flow in the velocity range 7-37 km/s, *AIAA Journal*, 8, 1658-1662, 1970.

Brunauer, S., L.E. Copeland, and D.L. Kantro, The Langmuir and BET theories, in E.A. Flood (editor), *The Solid-Gas Interface*, Vol. 1, Marcel Dekker, Inc., N.Y., p. 73, 1967.

Brundle, C.R., and J.Q. Broughton, The initial interaction of oxygen with well-defined transition metal surfaces, in D.A. King and P.D. Woodruff (Editors), *The Chemical Physics of Solid Surfaces and Heterogeneous Catalysis*, Elsevier, Amsterdam, vol. 3, Pt. A, pp. 131-388, 1990.

Carter, V.L., B.K. Ching, and D.D. Elliott, Atmospheric density above 158 kilometers ..., *J. Geophys. Res.*, 74, 5083, 1969.

Ching, B. K., D. R. Hickman, and J. M. Straus, Effects of atmospheric winds and aerodynamic lift on the inclination of the orbit of the S3-1 satellite, *J. Geophys. Res.*, 82, pp. 1474-1480, 1977.

Cook, G. E., Satellite drag coefficients, *Planet. Space Sci.*, 13, p. 929, 1965.

Cook, G. E., Drag coefficients of spherical satellites, *Ann. Geophys.*, 22, p. 53, 1966.

Czanderna, A.W., Chemisorption studies on clean silver powders ..., in G. Goldfinger (Editor), *Clean Surfaces*, Marcel Dekker, N.Y., pp. 133-152, 1970.

Davenport, J.W., and P.J. Estrup, Hydrogen on Metals, in D.A. King and P.D. Woodruff (Editors), *The Chemical Physics of Solid Surfaces and Heterogeneous Catalysis*, Elsevier, Amsterdam, vol. 3, Pt. A, pp. 1-37, 1990.

DeVries, L.L., Analysis and interpretation of density data from the Low-G accelerometer calibration system (LOGACS), in *Space Research 12*, North-Holland, Amsterdam, pp. 777-780, 1972.

Einstein, T.L., J.A. Hertz, and J.R. Schrieffer in J.E. Smith (Editor), Theory of Chemisorption, *Topics in Current Physics*, vol. 19, Springer Verlag, Berlin, 1980.

Farnsworth, H.E., Characterization and preparation of atomically clean surfaces ..., in G. Goldfinger (Editor), *Clean Surfaces*, Marcel Dekker, N.Y., pp. 77-95, 1970.

Fredo, R.M., and M.H. Kaplan, Procedure for obtaining

aerodynamic properties of spacecraft, *J. of Spacecraft and Rockets*, *18*, 367-373, 1981.

Goldstein, H., *Classical Mechanics*, Addison Wesley, 1980.

Gomer, R., Chemisorption, in H. Saltsburg, J.N. Smith, Jr., and M. Rogers (Editors), *Fundamentals of Gas-Surface Interactions*, Academic Press, N.Y. and London, pp. 182-215, 1967.

Gregory, J. C., and P. N. Peters, A measurement of the angular distribution of 5 eV atomic oxygen scattered off a solid surface in earth orbit, in *Rarefied Gas Dynamics*, vol. 15, Proc. Intl. Symposium, 1987.

Hedin, A.E., B.B. Hinton, and G.A. Schmitt, Role of gas-surface interactions ..., *J. Geophys. Res.*, *78*, 4651-4668, 1973.

Herrero, F.A., The drag coefficient of cylindrical spacecraft in orbit at altitudes greater than 150 km, NASA TM 85043, Goddard Space Flight Center, 1983.

Honig, J.M., Adsorption theory from the viewpoint of order-disorder theory, in E.A. Flood (Editor), *The Solid-Gas Interface*, Marcel Dekker, N.Y., pp. 371-412, 1967.

Imbro, D. R., M. M. Moe, and K. Moe, On fundamental problems in the deduction of atmospheric densities from satellite drag, *J. Geophys. Res.*, *80*, 3077-3086, 1975.

King-Hele, D.G., Methods for determining air density from satellite orbits, *Annls. Geophys.*, *22*, 40, 1966.

Lake, L.R., and A.O. Nier, Loss of atomic oxygen in mass spectrometer ion sources, *J. Geophys. Res.*, *78*, 1645-1653, 1973.

Lake, L.R., and K. Mauersberger, Investigation of atomic oxygen in mass spectrometer ion sources, *Intl. J. Mass Spectrom. and Ion Phys.*, *13*, 425-436, 1974.

Levine, A.S. (Editor), LDEF - 69 Months in Space, NASA CP-3134, passim, 1991.

Mazzella, A.J., Jr., R.J. Leong, R.W. Fioretti, and W.A. Smith, Analysis for satellite triaxial accelerometers, AFGL-TR-83-0009, Phillips Laboratory, Hanscom AFB, MA 01731, 1983.

Moe, K. and M.M. Moe, The effect of adsorption on densities measured by orbiting pressure gauges, *Planet. Space Sci.*, *15*, 1329-1332, 1967.

Moe, M.M., and K. Moe, The roles of kinetic theory and gas-surface interactions ..., *Planet. Space Sci.*, *17*, 917-922, 1969.

Moe, K., M.M. Moe, and N.W. Yelaca, Effect of surface heterogeneity on the adsorptive behavior ..., *J. Geophys. Res.*, *77*, 4242-4247, 1972.

Moe, M.M., S.D. Wallace, and K. Moe, Refinements in determining satellite drag coefficients ..., *J. of Guidance, Control, and Dynamics*, *16*, 441-445, 1993.

Newns, D.M., Self-consistent model of hydrogen chemisorption, *Phys. Rev.*, *178*, 1123-1135, 1969.

Nocilla, S., The surface re-emission law in free molecular flow, in *Rarefied Gas Dynamics*, Third Symposium, vol. 1, 327-346, 1963.

Offermann, D., and U. Grossmann, Neutral composition measurements with a helium-cooled ion source, in *Space Research 12*, North-Holland, Amsterdam, 1972, pp. 665-668.

Offermann, D., and H. Trinks, A rocket-borne mass spectrometer with helium-cooled ion source, *Rev. Sci. Instruments*, *42*, 1836-1843, 1971.

O'Keefe, D.R., and J.B. French, High energy scattering of inert gases ..., in *Rarefied Gas Dynamics*, Proc. 6th Symposium, 1969.

Roberts, M.W., and C.S. McKee, *Chemistry of the Metal-Gas Interface*, Oxford, passim, 1978.

Ross, S., The homotattic surface and adsorption potentials, in E.A. Flood (Editor), *The Solid-Gas Interface*, Marcel Dekker, N.Y., vol. 1, pp. 491-501, 1967.

Saltsburg, H., J.N. Smith, Jr., and M. Rogers, *Fundamentals of Gas-Surface Interactions*, Academic, N.Y., passim, 1967.

Schamberg, R., Analytic representation of surface interaction ..., in *Aerodynamics of the Upper Atmosphere*, The Rand Corporation, Santa Monica, CA, R-339, 1959.

Sentman, L. H., Free molecule flow theory ..., *Technical Report LMSC-448514*, Lockheed Aircraft Corporation, Sunnyvale, California, 1961.

Sparnaay, M.J., Adsorption properties of heterogeneous surfaces, in G. Goldfinger (Editor), *Clean Surfaces*, Marcel Dekker, N.Y., pp. 153-165, 1970.

Steele, W.A., Adsorbate equation of state, in *The Solid-Gas Interface*, Edited by E.A. Flood, Marcel Dekker, N.Y., 1967.

Thomas, L.B., Accommodation of molecules on controlled surfaces ..., in *Rarefied Gas Dynamics*, Proc. 12th Symposium, 1980.

Wachman, H.Y., The thermal accommodation coefficient: A critical survey, *ARS Journal*, *32*, pp. 2-12, January 1962.

M.M. Moe, Department of Physics, University of California, Irvine, CA 92717-4575.

S.D. Wallace, Department of Electrical and Computer Engineering, University of California, Irvine, CA 92717.

K. Moe, Space and Missile Systems Center/XRF, Los Angeles AFB, Los Angeles, CA 90009.